Comprehensive Natural Products Chemistry

Comprehensive Natural Products Chemistry

Editors-in-Chief

Sir Derek Barton†
Texas A&M University, USA

Koji Nakanishi
Columbia University, USA

Executive Editor

Otto Meth-Cohn
University of Sunderland, UK

Volume 4

AMINO ACIDS, PEPTIDES, PORPHYRINS, AND ALKALOIDS

Volume Editor

Jeffery W. Kelly
The Scripps Research Institute, USA

1999

ELSEVIER

AMSTERDAM – LAUSANNE – NEW YORK – OXFORD – SHANNON – SINGAPORE – TOKYO

Elsevier Science Ltd., The Boulevard, Langford Lane, Kidlington, Oxford, OX5 1GB, UK

First edition 1999

Library of Congress Cataloging-in-Publication Data
Comprehensive natural products chemistry / editors-in-chief, Sir Derek Barton, Koji Nakanishi ; executive editor, Otto Meth-Cohn. -- 1st ed.
    p. cm.
  Includes index.
  Contents: v. 4. Amino acids, peptides, porphyrins, and alkaloids / volume editor Jeffery W. Kelly
  1. Natural products. I. Barton, Derek, Sir, 1918-1998. II. Nakanishi, Koji, 1925- . III. Meth-Cohn, Otto.
QD415.C63 1999
547.7--dc21   98-15249

British Library Cataloguing in Publication Data
Comprehensive natural products chemistry
  1. Organic compounds
  I. Barton, Sir Derek, 1918-1998 II. Nakanishi Koji III. Meth-Cohn Otto
572.5

ISBN 0-08-042709-X (set : alk. paper)
ISBN 0-08-043156-9 (Volume 4 : alk. paper)

⊚™ The paper used in this publication meets the minimum requirements of the American National Standard for Information Sciences—Permanence of Paper for Printed Library Materials, ANSI Z39.48–1984.

Typeset by BPC Digital Data Ltd., Glasgow, UK.
Printed and bound in Great Britain by BPC Wheatons Ltd., Exeter, UK.

Contents

The chapter *Cyclosporin: The Biosynthetic Path to a Lipopeptide* by H. Von Döhren and H. Kleinkauf was originally planned to be included in this volume. However, due to unforseen circumstances during the preparation of this volume it now appears as Chapter 20 in Volume 1.

Introduction

For many decades, Natural Products Chemistry has been the principal driving force for progress in Organic Chemistry.

In the past, the determination of structure was arduous and difficult. As soon as computing became easy, the application of X-ray crystallography to structural determination quickly surpassed all other methods. Supplemented by the equally remarkable progress made more recently by Nuclear Magnetic Resonance techniques, determination of structure has become a routine exercise. This is even true for enzymes and other molecules of a similar size. Not to be forgotten remains the progress in mass spectrometry which permits another approach to structure and, in particular, to the precise determination of molecular weight.

There have not been such revolutionary changes in the partial or total synthesis of Natural Products. This still requires effort, imagination and time. But remarkable syntheses have been accomplished and great progress has been made in stereoselective synthesis. However, the one hundred percent yield problem is only solved in certain steps in certain industrial processes. Thus there remains a great divide between the reactions carried out in living organisms and those that synthetic chemists attain in the laboratory. Of course Nature edits the accuracy of DNA, RNA, and protein synthesis in a way that does not apply to a multi-step Organic Synthesis.

Organic Synthesis has already a significant component that uses enzymes to carry out specific reactions. This applies particularly to lipases and to oxidation enzymes. We have therefore, given serious attention to enzymatic reactions.

No longer standing in the wings, but already on-stage, are the wonderful tools of Molecular Biology. It is now clear that multi-step syntheses can be carried out in one vessel using multiple cloned enzymes. Thus, Molecular Biology and Organic Synthesis will come together to make economically important Natural Products.

From these preliminary comments it is clear that Natural Products Chemistry continues to evolve in different directions interacting with physical methods, Biochemistry, and Molecular Biology all at the same time.

This new Comprehensive Series has been conceived with the common theme of "How does Nature make all these molecules of life?" The principal idea was to organize the multitude of facts in terms of Biosynthesis rather than structure. The work is not intended to be a comprehensive listing of natural products, nor is it intended that there should be any detail about biological activity. These kinds of information can be found elsewhere.

The work has been planned for eight volumes with one more volume for Indexes. As far as we are aware, a broad treatment of the whole of Natural Products Chemistry has never been attempted before. We trust that our efforts will be useful and informative to all scientific disciplines where Natural Products play a role.

D. H. R. Barton†   K. Nakanishi   O. Meth-Cohn

Preface

It is surprising indeed that this work is the first attempt to produce a "comprehensive" overview of Natural Products beyond the student text level. However, the awe-inspiring breadth of the topic, which in many respects is still only developing, is such as to make the job daunting to anyone in the field. Fools rush in where angels fear to tread and the particular fool in this case was myself, a lifelong enthusiast and reader of the subject but with no research base whatever in the field!

Having been involved in several of the *Comprehensive* works produced by Pergamon Press, this omission intrigued me and over a period of gestation I put together a rough outline of how such a work could be written and presented it to Pergamon. To my delight they agreed that the project was worthwhile and in short measure Derek Barton was approached and took on the challenge of fleshing out this framework with alacrity. He also brought his long-standing friend and outstanding contributor to the field, Koji Nakanishi, into the team. With Derek's knowledge of the whole field, the subject was broken down into eight volumes and an outstanding team of internationally recognised Volume Editors was appointed.

We used Derek's 80th birthday as a target for finalising the work. Sadly he died just a few months before reaching this milestone. This work therefore is dedicated to the memory of Sir Derek Barton, Natural Products being the area which he loved best of all.

OTTO METH-COHN
Executive Editor

SIR DEREK BARTON

Sir Derek Barton, who was Distinguished Professor of Chemistry at Texas A&M University and holder of the Dow Chair of Chemical Invention died on March 16, 1998 in College Station, Texas of heart failure. He was 79 years old and had been Chairman of the Executive Board of Editors for Tetrahedron Publications since 1979.

Barton was considered to be one of the greatest organic chemists of the twentieth century whose work continues to have a major influence on contemporary science and will continue to do so for future generations of chemists.

Derek Harold Richard Barton was born on September 8, 1918 in Gravesend, Kent, UK and graduated from Imperial College, London with the degrees of B.Sc. (1940) and Ph.D. (1942). He carried out work on military intelligence during World War II and after a brief period in industry, joined the faculty at Imperial College. It was an early indication of the breadth and depth of his chemical knowledge that his lectureship was in physical chemistry. This research led him into the mechanism of elimination reactions and to the concept of molecular rotation difference to correlate the configurations of steroid isomers. During a sabbatical leave at Harvard in 1949–1950 he published a paper on the "Conformation of the Steroid Nucleus" (*Experientia*, 1950, **6**, 316) which was to bring him the Nobel Prize in Chemistry in 1969, shared with the Norwegian chemist, Odd Hassel. This key paper (only four pages long) altered the way in which chemists thought about the shape and reactivity of molecules, since it showed how the reactivity of functional groups in steroids depends on their axial or equatorial positions in a given conformation. Returning to the UK he held Chairs of Chemistry at Birkbeck College and Glasgow University before returning to Imperial College in 1957, where he developed a remarkable synthesis of the steroid hormone, aldosterone, by a photochemical reaction known as the Barton Reaction (nitrite photolysis). In 1978 he retired from Imperial College and became Director of the Natural Products Institute (CNRS) at Gif-sur-Yvette in France where he studied the invention of new chemical reactions, especially the chemistry of radicals, which opened up a whole new area of organic synthesis involving Gif chemistry. In 1986 he moved to a third career at Texas A&M University as Distinguished Professor of Chemistry and continued to work on novel reactions involving radical chemistry and the oxidation of hydrocarbons, which has become of great industrial importance. In a research career spanning more than five decades, Barton's contributions to organic chemistry included major discoveries which have profoundly altered our way of thinking about chemical structure and reactivity. His chemistry has provided models for the biochemical synthesis of natural products including alkaloids, antibiotics, carbohydrates, and DNA. Most recently his discoveries led to models for enzymes which oxidize hydrocarbons, including methane monooxygenase.

The following are selected highlights from his published work:

The 1950 paper which launched Conformational Analysis was recognized by the Nobel Prize Committee as the key contribution whereby the third dimension was added to chemistry. This work alone transformed our thinking about the connection between stereochemistry and reactivity, and was later adapted from small molecules to macromolecules e.g., DNA, and to inorganic complexes.

Barton's breadth and influence is illustrated in "Biogenetic Aspects of Phenol Oxidation" (*Festschr. Arthur Stoll*, 1957, 117). This theoretical work led to many later experiments on alkaloid biosynthesis and to a set of rules for *ortho-para*-phenolic oxidative coupling which allowed the predication of new natural product systems before they were actually discovered and to the correction of several erroneous structures.

In 1960, his paper on the remarkably short synthesis of the steroid hormone aldosterone (*J. Am. Chem. Soc.*, 1960, **82**, 2641) disclosed the first of many inventions of new reactions—in this case nitrite photolysis—to achieve short, high yielding processes, many of which have been patented and are used worldwide in the pharmaceutical industry.

Moving to 1975, by which time some 500 papers had been published, yet another "Barton reaction" was born—"The Deoxygenation of Secondary Alcohols" (*J. Chem. Soc. Perkin Trans. 1*, 1975, 1574), which has been very widely applied due to its tolerance of quite hostile and complex local environments in carbohydrate and nucleoside chemistry. This reaction is the chemical counterpart to ribonucleotide→ deoxyribonucleotide reductase in biochemistry and, until the arrival of the Barton reaction, was virtually impossible to achieve.

In 1985, "Invention of a New Radical Chain Reaction" involved the generation of carbon radicals from carboxylic acids (*Tetrahedron*, 1985, **41**, 3901). The method is of great synthetic utility and has been used many times by others in the burgeoning area of radicals in organic synthesis.

These recent advances in synthetic methodology were remarkable since his chemistry had virtually no precedent in the work of others. The radical methodology was especially timely in light of the significant recent increase in applications for fine chemical syntheses, and Barton gave the organic community an entrée into what will prove to be one of the most important methods of the twenty-first century. He often said how proud he was, at age 71, to receive the ACS Award for Creativity in Organic Synthesis for work published in the preceding five years.

Much of Barton's more recent work is summarized in the articles "The Invention of Chemical Reactions—The Last 5 Years" (*Tetrahedron*, 1992, **48**, 2529) and "Recent Developments in Gif Chemistry" (*Pure Appl. Chem.*, 1997, **69**, 1941).

Working 12 hours a day, Barton's stamina and creativity remained undiminished to the day of his death. The author of more than 1000 papers in chemical journals, Barton also held many successful patents. In addition to the Nobel Prize he received many honors and awards including the Davy, Copley, and Royal medals of the Royal Society of London, and the Roger Adams and Priestley Medals of the American Chemical Society. He held honorary degrees from 34 universities. He was a Fellow of the Royal Societies of London and Edinburgh, Foreign Associate of the National Academy of Sciences (USA), and Foreign Member of the Russian and Chinese Academies of Sciences. He was knighted by Queen Elizabeth in 1972, received the Légion d'Honneur (Chevalier 1972; Officier 1985) from France, and the Order of the Rising Sun from the Emperor of Japan. In his long career, Sir Derek trained over 300 students and postdoctoral fellows, many of whom now hold major positions throughout the world and include some of today's most distinguished organic chemists.

For those of us who were fortunate to know Sir Derek personally there is no doubt that his genius and work ethic were unique. He gave generously of his time to students and colleagues wherever he traveled and engendered such great respect and loyalty in his students and co-workers, that major symposia accompanied his birthdays every five years beginning with the 60th, ending this year with two celebrations just before his 80th birthday.

With the death of Sir Derek Barton, the world of science has lost a major figure, who together with Sir Robert Robinson and Robert B. Woodward, the cofounders of *Tetrahedron*, changed the face of organic chemistry in the twentieth century.

Professor Barton is survived by his wife, Judy, and by a son, William from his first marriage, and three grandchildren.

A. I. SCOTT
Texas A&M University

Reprinted from *Tetrahedron*, 1998, **54**, 8847
Photograph courtesy of Library and Information Centre, Royal Society of Chemistry. © The Nobel Foundation

Contributors to Volume 4

Dr. G. Bierbaum
Institut für Medizinische Mikrobiologie, Universität Bonn, Sigmund-Freud Strasse 25, D-53105
Bonn, Germany

Dr. A. A. Brakhage
Technische Universität Darmstadt, Institut für Biochemie, Petersenstrasse 22, D-64287 Darmstadt,
Germany

Mr. B. A. Carlson
National Cancer Institute, National Institute of Health, Bethesda, MD 20892, USA

Dr. K.-M. Cheung
Department of Biochemistry and Molecular Biology, Biomedical Sciences Building, University of
Southampton, Basset Crescent East, Southampton, SO16 7PX, UK

Mr. H. S. Chittum
National Cancer Institute, National Institute of Health, Bethesda, MD 20892, USA

Dr. V. de Crécy-Lagard
Institut Pasteur, Unité de Biochimie Cellulaire, 25, rue du Dr Roux, F-75724 Paris, France

Dr. T. Dudler
Imperial Cancer Research Fund, LLMB, Room 113, 44 Lincoln's Inn Fields, London, WC2A 3PX,
UK

Professor M. H. Gelb
Department of Chemistry and Biochemistry, University of Washington, Box 351700, Seattle,
WA 98195-1700, USA

Mr. V. N. Gladyshev
National Cancer Institute, National Institute of Health, Bethesda, MD 20892, USA

Dr. D. L. Hatfield
Chief, Section on the Molecular Biology of Selenium, Laboratory of Basic Research, Building 37,
Room 2D09, National Cancer Institute, National Institute of Health, Bethesda, MD 20892, USA

Mr. J. R. Huh
Seoul National University, Seoul 151-742, Korea

Dr. N. Inoue
Department of Immunoregulation, Research Institute for Microbial Diseases, Osaka University, 3-1
Yamada-oka, Suita, Osaka 565, Japan

Professor J. W. Kelly
The Skaggs Institute for Chemical Biology, The Scripps Research Institute, 10550 North Torrey Pines
Road, La Jolla, CA 92037, USA

Mr. M. Kim
Seoul National University, Seoul 151-742, Korea

Dr. T. Kinoshita
Department of Immunoregulation, Research Institute for Microbial Diseases, Osaka University, 3-1
Yamada-oka, Suita, Osaka 565, Japan

Dr. S. Kumar
Biochemistry and Molecular Biology Division, Central Institute of Medicinal and Aromatic Plants
(CSIR), PO - CIMAP, Picnic Spot Road, Lucknow 226015, India

Dr. B. J. Lee
Associate Professor of Molecular Biology and Genetics, Laboratory of Molecular Genetics, Institute
for Molecular Biology and Genetics, Seoul National University, Seoul 151-742, Korea

Dr. L. Liu
University of Washington, Division of Hematology, School of Medicine, Box 357710, Seattle,
WA 98195-7710, USA

Dr. J. M. Luengo
Departamento de Bioquímica y Biología Molecular, Facultad de Veterinaria, Universidad de León,
E-24007 León, Spain

Dr. R. Luthra
Biochemistry and Molecular Biology Division, Central Institute of Medicinal and Aromatic Plants
(CSIR), PO - CIMAP, Picnic Spot Road, Lucknow 226015, India

Professor M. A. Marahiel
Fachbereich Chemie/Biochemie, Philipps-Universität Marburg, Hans-Meerwein Strasse, D-35032
Marburg, Germany

Dr. N. Misra
Biochemistry and Molecular Biology Division, Central Institute of Medicinal and Aromatic Plants
(CSIR), PO - CIMAP, Picnic Spot Road, Lucknow 226015, India

Mr. M. E. Moustafa
National Cancer Institute, National Institute of Health, Bethesda, MD 20892, USA

Mr. J. M. Park
Seoul National University, Seoul 151-742, Korea

Mr. S. I. Park
National Cancer Institute, National Institute of Health, Bethesda, MD 20892, USA

Dr. N. Rajendran
Department of Civil and Environmental Engineering, Research Engineering Complex, Michigan
State University, East Lansing, MI 48824, USA

Dr. C. A. Roessner
Chemistry Department, Texas A&M University, College Station, TX 77843-3255, USA

Mr. M. Ruppert
Institut für Pharmazie, Johannes Gutenberg-Universität Mainz, Staudinger Weg 5, D-55099 Mainz,
Germany

Dr. P. J. Santander
Chemistry Department, Texas A&M University, College Station, TX 77843-3255, USA

Dr. A. I. Scott
Chemistry Department, Texas A&M University, College Station, TX 77843-3255, USA

Professor P. M. Shoolingin-Jordan
Division of Biochemistry and Molecular Biology, School of Biological Sciences, University of
Southampton, Basset Crescent East, Southampton, SO16 7PX, UK

Dr. K. L. Singh
Biochemistry and Molecular Biology Division, Central Institute of Medicinal and Aromatic Plants
(CSIR), PO - CIMAP, Picnic Spot Road, Lucknow 226015, India

Dr. T. Stein
Technische Universität Berlin, Max-Volmer-Institut für Biophysikalische Chemie und Biochemie,
Fachgebiet Biochemie und Molekulare Biologie, Franklinstrasse 29, D-10587 Berlin, Germany

Professor J. Stöckigt
Institut für Pharmazie, Johannes Gutenberg-Universität Mainz, Staudinger Weg 5, D-55099 Mainz, Germany

Dr. J. Takeda
Department of Environmental Medicine (H3), Osaka University Medical School, 2-2 Yamadaoka, Suita, Osaka 565, Japan

Dr. J. Vater
Technische Universität Berlin, Max-Volmer-Institut für Biophysikalische Chemie und Biochemie, Fachgebiet Biochemie und Molekulare Biologie, Franklinstrasse 29, D-10587 Berlin, Germany

Abbreviations

The most commonly used abbreviations in *Comprehensive Natural Products Chemistry* are listed below. Please note that in some instances these may differ from those used in other branches of chemistry

A	adenine
ABA	abscisic acid
Ac	acetyl
ACAC	acetylacetonate
ACTH	adrenocorticotropic hormone
ADP	adenosine 5'-diphosphate
AIBN	2,2'-azobisisobutyronitrile
Ala	alanine
AMP	adenosine 5'-monophosphate
APS	adenosine 5'-phosphosulfate
Ar	aryl
Arg	arginine
ATP	adenosine 5'-triphosphate
B	nucleoside base (adenine, cylosine, guanine, thymine or uracil)
9-BBN	9-borabicyclo[3.3.1]nonane
BOC	*t*-butoxycarbonyl (or carbo-*t*-butoxy)
BSA	*N,O*-bis(trimethylsilyl)acetamide
BSTFA	*N,O*-bis(trimethylsilyl)trifluoroacetamide
Bu	butyl
Bu^n	*n*-butyl
Bu^i	isobutyl
Bu^s	*s*-butyl
Bu^t	*t*-butyl
Bz	benzoyl
CAN	ceric ammonium nitrate
CD	cyclodextrin
CDP	cytidine 5'-diphosphate
CMP	cytidine 5'-monophosphate
CoA	coenzyme A
COD	cyclooctadiene
COT	cyclooctatetraene
Cp	η^5-cyclopentadiene
Cp*	pentamethylcyclopentadiene
12-Crown-4	1,4,7,10-tetraoxacyclododecane
15-Crown-5	1,4,7,10,13-pentaoxacyclopentadecane
18-Crown-6	1,4,7,10,13,16-hexaoxacyclooctadecane
CSA	camphorsulfonic acid
CSI	chlorosulfonyl isocyanate
CTP	cytidine 5'-triphosphate
cyclic AMP	adenosine 3',5'-cyclic monophosphoric acid
CySH	cysteine
DABCO	1,4-diazabicyclo[2.2.2]octane
DBA	dibenz[*a,h*]anthracene
DBN	1,5-diazabicyclo[4.3.0]non-5-ene

DBU	1,8-diazabicyclo[5.4.0]undec-7-ene
DCC	dicyclohexylcarbodiimide
DEAC	diethylaluminum chloride
DEAD	diethyl azodicarboxylate
DET	diethyl tartrate (+ or -)
DHET	dihydroergotoxine
DIBAH	diisobutylaluminum hydride
Diglyme	diethylene glycol dimethyl ether (or bis(2-methoxyethyl)ether)
DiHPhe	2,5-dihydroxyphenylalanine
Dimsyl Na	sodium methylsulfinylmethide
DIOP	2,3-*O*-isopropylidene-2,3-dihydroxy-1,4-bis(diphenylphosphino)butane
dipt	diisopropyl tartrate (+ or -)
DMA	dimethylacetamide
DMAD	dimethyl acetylenedicarboxylate
DMAP	4-dimethylaminopyridine
DME	1,2-dimethoxyethane (glyme)
DMF	dimethylformamide
DMF-DMA	dimethylformamide dimethyl acetal
DMI	1,3-dimethyl-2-imidazalidinone
DMSO	dimethyl sulfoxide
DMTSF	dimethyl(methylthio)sulfonium fluoroborate
DNA	deoxyribonucleic acid
DOCA	deoxycorticosterone acetate
EADC	ethylaluminum dichloride
EDTA	ethylenediaminetetraacetic acid
EEDQ	*N*-ethoxycarbonyl-2-ethoxy-1,2-dihydroquinoline
Et	ethyl
EVK	ethyl vinyl ketone
FAD	flavin adenine dinucleotide
Fl	flavin
FMN	flavin mononucleotide
G	guanine
GABA	4-aminobutyric acid
GDP	guanosine 5'-diphosphate
GLDH	glutamate dehydrogenase
gln	glutamine
Glu	glutamic acid
Gly	glycine
GMP	guanosine 5'-monophosphate
GOD	glucose oxidase
G-6-P	glucose-6-phosphate
GTP	guanosine 5'-triphosphate
Hb	hemoglobin
His	histidine
HMPA	hexamethylphosphoramide (or hexamethylphosphorous triamide)
Ile	isoleucine
INAH	isonicotinic acid hydrazide
IpcBH	isopinocampheylborane
Ipc$_2$BH	diisopinocampheylborane
KAPA	potassium 3-aminopropylamide
K-Slectride	potassium tri-*s*-butylborohydride

LAH	lithium aluminum hydride
LAP	leucine aminopeptidase
LDA	lithium diisopropylamide
LDH	lactic dehydrogenase
Leu	leucine
LICA	lithium isopropylcyclohexylamide
L-Selectride	lithium tri-*s*-butylborohydride
LTA	lead tetraacetate
Lys	lysine
MCPBA	*m*-chloroperoxybenzoic acid
Me	methyl
MEM	methoxyethoxymethyl
MEM-Cl	ß-methoxyethoxymethyl chloride
Met	methionine
MMA	methyl methacrylate
MMC	methyl magnesium carbonate
MOM	methoxymethyl
Ms	mesyl (or methanesulfonyl)
MSA	methanesulfonic acid
MsCl	methanesulfonyl chloride
MVK	methyl vinyl ketone
NAAD	nicotinic acid adenine dinucleotide
NAD	nicotinamide adenine dinucleotide
NADH	nicotinamide adenine dinucleotide phosphate, reduced
NBS	*N*-bromosuccinimider
NMO	*N*-methylmorpholine *N*-oxide monohydrate
NMP	*N*-methylpyrrolidone
PCBA	*p*-chlorobenzoic acid
PCBC	*p*-chlorobenzyl chloride
PCBN	*p*-chlorobenzonitrile
PCBTF	*p*-chlorobenzotrifluoride
PCC	pyridinium chlorochromate
PDC	pyridinium dichromate
PG	prostaglandin
Ph	phenyl
Phe	phenylalanine
Phth	phthaloyl
PPA	polyphosphoric acid
PPE	polyphosphate ester (or ethyl *m*-phosphate)
Pr	propyl
Pr^i	isopropyl
Pro	proline
Py	pyridine
RNA	ribonucleic acid
Rnase	ribonuclease
Ser	serine
Sia$_2$BH	disiamylborane
TAS	tris(diethylamino)sulfonium
TBAF	tetra-*n*-butylammonium fluoroborate
TBDMS	*t*-butyldimethylsilyl
TBDMS-Cl	*t*-butyldimethylsilyl chloride
TBDPS	*t*-butyldiphenylsilyl
TCNE	tetracyanoethene

TES	triethylsilyl
TFA	trifluoracetic acid
TFAA	trifluoroacetic anhydride
THF	tetrahydrofuran
THF	tetrahydrofolic acid
THP	tetrahydropyran (or tetrahydropyranyl)
Thr	threonine
TMEDA	*N,N,N',N'*,tetramethylethylenediamine[1,2-bis(dimethylamino)ethane]
TMS	trimethylsilyl
TMS-Cl	trimethylsilyl chloride
TMS-CN	trimethylsilyl cyanide
Tol	toluene
TosMIC	tosylmethyl isocyanide
TPP	tetraphenylporphyrin
Tr	trityl (or triphenylmethyl)
Trp	tryptophan
Ts	tosyl (or *p*-toluenesulfonyl)
TTFA	thallium trifluoroacetate
TTN	thallium(III) nitrate
Tyr	tyrosine
Tyr-OMe	tyrosine methyl ester
U	uridine
UDP	uridine 5'-diphosphate
UMP	uridine 5'-monophosphate

Contents of All Volumes

An Historical Perspective of Natural Products Chemistry

KOJI NAKANISHI

Columbia University, New York, USA

To give an account of the rich history of natural products chemistry in a short essay is a daunting task. This brief outline begins with a description of ancient folk medicine and continues with an outline of some of the major conceptual and experimental advances that have been made from the early nineteenth century through to about 1960, the start of the modern era of natural products chemistry. Achievements of living chemists are noted only minimally, usually in the context of related topics within the text. More recent developments are reviewed within the individual chapters of the present volumes, written by experts in each field. The subheadings follow, in part, the sequence of topics presented in Volumes 1–8.

1. ETHNOBOTANY AND "NATURAL PRODUCTS CHEMISTRY"

Except for minerals and synthetic materials our surroundings consist entirely of organic natural products, either of prebiotic organic origins or from microbial, plant, or animal sources. These materials include polyketides, terpenoids, amino acids, proteins, carbohydrates, lipids, nucleic acid bases, RNA and DNA, etc. Natural products chemistry can be thought of as originating from mankind's curiosity about odor, taste, color, and cures for diseases. Folk interest in treatments for pain, for food-poisoning and other maladies, and in hallucinogens appears to go back to the dawn of humanity

For centuries China has led the world in the use of natural products for healing. One of the earliest health science anthologies in China is the Nei Ching, whose authorship is attributed to the legendary Yellow Emperor (thirtieth century BC), although it is said that the dates were backdated from the third century by compilers. Excavation of a Han Dynasty (206 BC–AD 220) tomb in Hunan Province in 1974 unearthed decayed books, written on silk, bamboo, and wood, which filled a critical gap between the dawn of medicine up to the classic Nei Ching; Book 5 of these excavated documents lists 151 medical materials of plant origin. Generally regarded as the oldest compilation of Chinese herbs is Shen Nung Pen Ts'ao Ching (Catalog of Herbs by Shen Nung), which is believed to have been revised during the Han Dynasty; it lists 365 materials. Numerous revisions and enlargements of Pen Ts'ao were undertaken by physicians in subsequent dynasties, the ultimate being the Pen Ts'ao Kang Mu (General Catalog of Herbs) written by Li Shih-Chen over a period of 27 years during the Ming Dynasty (1573–1620), which records 1898 herbal drugs and 8160 prescriptions. This was circulated in Japan around 1620 and translated, and has made a huge impact on subsequent herbal studies in Japan; however, it has not been translated into English. The number of medicinal herbs used in 1979 in China numbered 5267. One of the most famous of the Chinese folk herbs is the ginseng root *Panax ginseng*, used for health maintenance and treatment of various diseases. The active principles were thought to be the saponins called ginsenosides but this is now doubtful; the effects could well be synergistic between saponins, flavonoids, etc. Another popular folk drug, the extract of the Ginkgo tree, *Ginkgo biloba* L., the only surviving species of the Paleozoic era (250 million years ago) family which became extinct during the last few million years, is mentioned in the Chinese Materia Medica to have an effect in improving memory and sharpening mental alertness. The main constituents responsible for this are now understood to be ginkgolides and flavonoids, but again not much else is known. Clarifying the active constituents and mode of (synergistic) bioactivity of Chinese herbs is a challenging task that has yet to be fully addressed.

The Assyrians left 660 clay tablets describing 1000 medicinal plants used around 1900–400 BC, but the best insight into ancient pharmacy is provided by the two scripts left by the ancient Egyptians, who were masters of human anatomy and surgery because of their extensive mummification practices. The Edwin Smith Surgical Papyrus purchased by Smith in 1862 in Luxor (now in the New York Academy of Sciences collection), is one of the most important medicinal documents of the ancient Nile Valley, and describes the healer's involvement in surgery, prescription, and healing practices using plants, animals, and minerals. The Ebers Papyrus, also purchased by Edwin Smith in 1862, and then acquired by Egyptologist George Ebers in 1872, describes 800 remedies using plants, animals, minerals, and magic. Indian medicine also has a long history, possibly dating back to the second millennium BC. The Indian materia medica consisted mainly of vegetable drugs prepared from plants but also used animals, bones, and minerals such as sulfur, arsenic, lead, copper sulfate, and gold. Ancient Greece inherited much from Egypt, India, and China, and underwent a gradual transition from magic to science. Pythagoras (580–500 BC) influenced the medical thinkers of his time, including Aristotle (384–322 BC), who in turn affected the medical practices of another influential Greek physician Galen (129–216). The Iranian physician Avicenna (980–1037) is noted for his contributions to Aristotelian philosophy and medicine, while the German-Swiss physician and alchemist Paracelsus (1493–1541) was an early champion who established the role of chemistry in medicine.

The rainforests in Central and South America and Africa are known to be particularly abundant in various organisms of interest to our lives because of their rich biodiversity, intense competition, and the necessity for self-defense. However, since folk-treatments are transmitted verbally to the next generation via shamans who naturally have a tendency to keep their plant and animal sources confidential, the recipes tend to get lost, particularly with destruction of rainforests and the encroachment of "civilization." Studies on folk medicine, hallucinogens, and shamanism of the Central and South American Indians conducted by Richard Schultes (Harvard Botanical Museum, emeritus) have led to renewed activity by ethnobotanists, recording the knowledge of shamans, assembling herbaria, and transmitting the record of learning to the village.

Extracts of toxic plants and animals have been used throughout the world for thousands of years for hunting and murder. These include the various arrow poisons used all over the world. *Strychnos* and *Chondrodendron* (containing strychnine, etc.) were used in South America and called "curare," *Strophanthus* (strophantidine, etc.) was used in Africa, the latex of the upas tree *Antiaris toxicaria* (cardiac glycosides) was used in Java, while *Aconitum napellus*, which appears in Greek mythology (aconitine) was used in medieval Europe and Hokkaido (by the Ainus). The Colombian arrow poison is from frogs (batrachotoxins; 200 toxins have been isolated from frogs by B. Witkop and J. Daly at NIH). Extracts of *Hyoscyamus niger* and *Atropa belladonna* contain the toxic tropane alkaloids, for example hyoscyamine, belladonnine, and atropine. The belladonna berry juice (atropine) which dilates the eye pupils was used during the Renaissance by ladies to produce doe-like eyes (belladona means beautiful woman). The Efik people in Calabar, southeastern Nigeria, used extracts of the calabar bean known as esere (physostigmine) for unmasking witches. The ancient Egyptians and Chinese knew of the toxic effect of the puffer fish, fugu, which contains the neurotoxin tetrodotoxin (Y. Hirata, K. Tsuda, R. B. Woodward).

When rye is infected by the fungus *Claviceps purpurea*, the toxin ergotamine and a number of ergot alkaloids are produced. These cause ergotism or the "devil's curse," "St. Anthony's fire," which leads to convulsions, miscarriages, loss of arms and legs, dry gangrene, and death. Epidemics of ergotism occurred in medieval times in villages throughout Europe, killing tens of thousands of people and livestock; Julius Caesar's legions were destroyed by ergotism during a campaign in Gaul, while in AD 994 an estimated 50,000 people died in an epidemic in France. As recently as 1926, a total of 11,000 cases of ergotism were reported in a region close to the Urals. It has been suggested that the witch hysteria that occurred in Salem, Massachusetts, might have been due to a mild outbreak of ergotism. Lysergic acid diethylamide (LSD) was first prepared by A. Hofmann, Sandoz Laboratories, Basel, in 1943 during efforts to improve the physiological effects of the ergot alkaloids when he accidentally inhaled it. "On Friday afternoon, April 16, 1943," he wrote, "I was seized by a sensation of restlessness... ." He went home from the laboratory and "perceived an uninterrupted stream of fantastic dreams ... ." (*Helvetica Chimica Acta*).

Numerous psychedelic plants have been used since ancient times, producing visions, mystical fantasies (cats and tigers also seem to have fantasies?, see nepetalactone below), sensations of flying, glorious feelings in warriors before battle, etc. The ethnobotanists Wasson and Schultes identified "ololiqui," an important Aztec concoction, as the seeds of the morning glory *Rivea corymbosa* and gave the seeds to Hofmann who found that they contained lysergic acid amides similar to but less potent than LSD. Iboga, a powerful hallucinogen from the root of the African shrub *Tabernanthe iboga*, is used by the Bwiti cult in Central Africa who chew the roots to obtain relief from fatigue and hunger; it contains the alkaloid ibogamine. The powerful hallucinogen used for thousands of years by the American Indians, the peyote cactus, contains mescaline and other alkaloids. The Indian hemp plant, *Cannabis sativa*, has been used for making rope since 3000 BC, but when it is used for its pleasure-giving effects it is called cannabis and has been known in central Asia, China, India, and the Near East since ancient times. Marijuana, hashish (named after the Persian founder of the Assassins of the eleventh century, Hasan-e Sabbah), charas, ghanja, bhang, kef, and dagga are names given to various preparations of the hemp plant. The constituent responsible for the mind-altering effect is 1-tetrahydrocannabinol (also referred to as 9-THC) contained in 1%. R. Mechoulam (1930–, Hebrew University) has been the principal worker in the cannabinoids, including structure determination and synthesis of 9-THC (1964 to present); the Israeli police have also made a contribution by providing Mechoulam with a constant supply of marijuana. Opium (morphine) is another ancient drug used for a variety of pain-relievers and it is documented that the Sumerians used poppy as early as 4000 BC; the narcotic effect is present only in seeds before they are fully formed. The irritating secretion of the blister beetles, for example *Mylabris* and the European species *Lytta vesicatoria*, commonly called Spanish fly, was used medically as a topical skin irritant to remove warts but was also a major ingredient in so-called love potions (constituent is cantharidin, stereospecific synthesis in 1951, G. Stork, 1921–; prep. scale high-pressure Diels–Alder synthesis in 1985, W. G. Dauben, 1919–1996).

Plants have been used for centuries for the treatment of heart problems, the most important being the foxgloves *Digitalis purpurea* and *D. lanata* (digitalin, diginin) and *Strophanthus gratus* (ouabain). The bark of cinchona *Cinchona officinalis* (called quina-quina by the Indians) has been used widely among the Indians in the Andes against malaria, which is still one of the major infectious diseases; its most important alkaloid is quinine. The British protected themselves against malaria during the occupation of India through gin and tonic (quinine!). The stimulant coca, used by the Incas around the tenth century, was introduced into Europe by the conquistadors; coca beans are also commonly chewed in West Africa. Wine making was already practiced in the Middle East 6000–8000 years ago; Moors made date wines, the Japanese rice wine, the Vikings honey mead, the Incas maize chicha. It is said that the Babylonians made beer using yeast 5000–6000 years ago. As shown above in parentheses, alkaloids are the major constituents of the herbal plants and extracts used for centuries, but it was not until the early nineteenth century that the active principles were isolated in pure form, for example morphine (1816), strychnine (1817), atropine (1819), quinine (1820), and colchicine (1820). It was a century later that the structures of these compounds were finally elucidated.

2. DAWN OF ORGANIC CHEMISTRY, EARLY STRUCTURAL STUDIES, MODERN METHODOLOGY

The term "organic compound" to define compounds made by and isolated from living organisms was coined in 1807 by the Swedish chemist Jons Jacob Berzelius (1779–1848), a founder of today's chemistry, who developed the modern system of symbols and formulas in chemistry, made a remarkably accurate table of atomic weights and analyzed many chemicals. At that time it was considered that organic compounds could not be synthesized from inorganic materials *in vitro*. However, Friedrich Wöhler (1800–1882), a medical doctor from Heidelberg who was starting his chemical career at a technical school in Berlin, attempted in 1828 to make "ammonium cyanate," which had been assigned a wrong structure, by heating the two inorganic salts potassium cyanate and ammonium sulfate; this led to the unexpected isolation of white crystals which were identical to the urea from urine, a typical organic compound. This well-known incident marked the beginning of organic chemistry. With the preparation of acetic acid from inorganic material in 1845 by Hermann Kolbe (1818–1884) at Leipzig, the myth surrounding organic compounds, in which they were associated with some vitalism was brought to an end and organic chemistry became the chemistry of carbon compounds. The same Kolbe was involved in the development of aspirin, one of the earliest and most important success stories in natural products chemistry. Salicylic acid from the leaf of the wintergreen plant had long been used as a pain reliever, especially in treating arthritis and gout. The inexpensive synthesis of salicylic acid from sodium phenolate and carbon dioxide by Kolbe in 1859 led to the industrial production in 1893 by the Bayer Company of acetylsalicylic acid "aspirin," still one of the most popular drugs. Aspirin is less acidic than salicylic acid and therefore causes less irritation in the mouth, throat, and stomach. The remarkable mechanism of the anti-inflammatory effect of aspirin was clarified in 1974 by John Vane (1927–) who showed that it inhibits the biosynthesis of prostaglandins by irreversibly acetylating a serine residue in prostaglandin synthase. Vane shared the 1982 Nobel Prize with Bergström and Samuelsson who determined the structure of prostaglandins (see below).

In the early days, natural products chemistry was focused on isolating the more readily available plant and animal constituents and determining their structures. The course of structure determination in the 1940s was a complex, indirect process, combining evidence from many types of experiments. The first effort was to crystallize the unknown compound or make derivatives such as esters or 2,4-dinitrophenylhydrazones, and to repeat recrystallization until the highest and sharp melting point was reached, since prior to the advent of isolation and purification methods now taken for granted, there was no simple criterion for purity. The only chromatography was through special grade alumina (first used by M. Tswett in 1906, then reintroduced by R. Willstätter). Molecular weight estimation by the Rast method which depended on melting point depression of a sample/camphor mixture, coupled with Pregl elemental microanalysis (see below) gave the molecular formula. Functionalities such as hydroxyl, amino, and carbonyl groups were recognized on the basis of specific derivatization and crystallization, followed by redetermination of molecular formula; the change in molecular composition led to identification of the functionality. Thus, sterically hindered carbonyls, for example the 11-keto group of cortisone, or tertiary hydroxyls, were very difficult to pinpoint, and often had to depend on more searching experiments. Therefore, an entire paper describing the recognition of a single hydroxyl group in a complex natural product would occasionally appear in the literature. An oxygen function suggested from the molecular formula but left unaccounted for would usually be assigned to an ether.

Determination of C-methyl groups depended on Kuhn–Roth oxidation which is performed by drastic oxidation with chromic acid/sulfuric acid, reduction of excess oxidant with hydrazine, neutralization with alkali, addition of phosphoric acid, distillation of the acetic acid originating from the C-methyls, and finally its titration with alkali. However, the results were only approximate, since *gem*-dimethyl groups only yield one equivalent of acetic acid, while primary, secondary, and tertiary methyl groups all give different yields of acetic acid. The skeletal structure of polycyclic compounds were frequently deduced on the basis of dehydrogenation reactions. It is therefore not surprising that the original steroid skeleton put forth by Wieland and Windaus in 1928, which depended a great deal on the production of chrysene upon Pd/C dehydrogenation, had to be revised in 1932 after several discrepancies were found (they received the Nobel prizes in 1927 and 1928 for this "extraordinarily difficult structure determination," see below).

In the following are listed some of the Nobel prizes awarded for the development of methodologies which have contributed critically to the progress in isolation protocols and structure determination. The year in which each prize was awarded is preceded by "Np."

Fritz Pregl, 1869–1930, Graz University, Np 1923. Invention of carbon and hydrogen microanalysis. Improvement of Kuhlmann's microbalance enabled weighing at an accuracy of 1 μg over a 20 g range, and refinement of carbon and hydrogen analytical methods made it possible to perform analysis with 3–4 mg of sample. His microbalance and the monograph *Quantitative Organic Microanalysis* (1916) profoundly influenced subsequent developments in practically all fields of chemistry and medicine.

The Svedberg, 1884–1971, Uppsala, Np 1926. Uppsala was a center for quantitative work on colloids for which the prize was awarded. His extensive study on ultracentrifugation, the first paper of which was published in the year of the award, evolved from a spring visit in 1922 to the University of Wisconsin. The ultracentrifuge together with the electrophoresis technique developed by his student Tiselius, have profoundly influenced subsequent progress in molecular biology and biochemistry.

Arne Tiselius, 1902–1971, Ph.D. Uppsala (T. Svedberg), Uppsala, Np 1948. Assisted by a grant from the Rockefeller Foundation, Tiselius was able to use his early electrophoresis instrument to show four bands in horse blood serum, alpha, beta and gamma globulins in addition to albumin; the first paper published in 1937 brought immediate positive responses.

Archer Martin, 1910–, Ph.D. Cambridge; Medical Research Council, Mill Hill, and Richard Synge, 1914–1994, Ph.D. Cambridge; Rowett Research Institute, Food Research Institute, Np 1952. They developed chromatography using two immiscible phases, gas–liquid, liquid–liquid, and paper chromatography, all of which have profoundly influenced all phases of chemistry.

Frederick Sanger, 1918–, Ph.D. Cambridge (A. Neuberger), Medical Research Council, Cambridge, Np 1958 and 1980. His confrontation with challenging structural problems in proteins and nucleic acids led to the development of two general analytical methods, 1,2,4-fluorodinitrobenzene (DNP) for tagging free amino groups (1945) in connection with insulin sequencing studies, and the dideoxynucleotide method for sequencing DNA (1977) in connection with recombinant DNA. For the latter he received his second Np in chemistry in 1980, which was shared with Paul Berg (1926–, Stanford University) and Walter Gilbert (1932–, Harvard University) for their contributions, respectively, in recombinant DNA and chemical sequencing of DNA. The studies of insulin involved usage of DNP for tagging disulfide bonds as cysteic acid residues (1949), and paper chromatography introduced by Martin and Synge 1944. That it was the first elucidation of any protein structure lowered the barrier for future structure studies of proteins.

Stanford Moore, 1913–1982, Ph.D. Wisconsin (K. P. Link), Rockefeller, Np 1972; and William Stein, 1911–1980, Ph.D. Columbia (E. G. Miller); Rockefeller, Np 1972. Moore and Stein cooperatively developed methods for the rapid quantification of protein hydrolysates by combining partition chromatography, ninhydrin coloration, and drop-counting fraction collector, i.e., the basis for commercial amino acid analyzers, and applied them to analysis of the ribonuclease structure.

Bruce Merrifield, 1921–, Ph.D. UCLA (M. Dunn), Rockefeller, Np 1984. The concept of solid-phase peptide synthesis using porous beads, chromatographic columns, and sequential elongation of peptides and other chains revolutionized the synthesis of biopolymers.

High-performance liquid chromatography (HPLC), introduced around the mid-1960s and now coupled on-line to many analytical instruments, for example UV, FTIR, and MS, is an indispensable daily tool found in all natural products chemistry laboratories.

3. STRUCTURES OF ORGANIC COMPOUNDS, NINETEENTH CENTURY

The discoveries made from 1848 to 1874 by Pasteur, Kekulé, van't Hoff, Le Bel, and others led to a revolution in structural organic chemistry. Louis Pasteur (1822–1895) was puzzled about why the potassium salt of tartaric acid (deposited on wine casks during fermentation) was dextrorotatory while the sodium ammonium salt of racemic acid (also deposited on wine casks) was optically inactive although both tartaric acid and "racemic" acid had identical chemical compositions. In 1848, the 25 year old Pasteur examined the racemic acid salt under the microscope and found two kinds of crystals exhibiting a left- and right-hand relation. Upon separation of the left-handed and right-handed crystals, he found that they rotated the plane of polarized light in opposite directions. He had thus performed his famous resolution of a racemic mixture, and had demonstrated the phenomenon of chirality. Pasteur went on to show that the racemic acid formed two kinds of salts with optically active bases such as quinine; this was the first demonstration of diastereomeric resolution. From this work Pasteur concluded that tartaric acid must have an element of asymmetry within the molecule itself. However, a three-dimensional understanding of the enantiomeric pair was only solved 25 years later (see below). Pasteur's own interest shifted to microbiology where he made the crucial discovery of the involvement of "germs" or microorganisms in various processes and proved that yeast induces alcoholic fermentation, while other microorganisms lead to diseases; he thus saved the wine industries of France, originated the process known as "pasteurization," and later developed vaccines for rabies. He was a genius who made many fundamental discoveries in chemistry and in microbiology.

The structures of organic compounds were still totally mysterious. Although Wöhler had synthesized urea, an isomer of ammonium cyanate, in 1828, the structural difference between these isomers was not known. In 1858 August Kekulé (1829–1896; studied with André Dumas and C. A. Wurtz in Paris, taught at Ghent, Heidelberg, and Bonn) published his famous paper in Liebig's *Annalen der Chemie* on the structure of carbon, in which he proposed that carbon atoms could form C–C bonds with hydrogen and other atoms linked to them; his dream on the top deck of a London bus led him to this concept. It was Butlerov who introduced the term "structure theory" in 1861. Further, in 1865 Kekulé conceived the cyclo-hexa-1:3:5-triene structure for benzene (C_6H_6) from a dream of a snake biting its own tail. In 1874, two young chemists, van't Hoff (1852–1911, Np 1901) in Utrecht, and Le Bel (1847–1930) in Paris, who had met in 1874 as students of C. A. Wurtz, published the revolutionary three-dimensional (3D) structure of the tetrahedral carbon Cabcd to explain the enantiomeric behavior of Pasteur's salts. The model was welcomed by J. Wislicenus (1835–1902, Zürich, Würzburg, Leipzig) who in 1863 had demonstrated the enantiomeric nature of the two lactic acids found by Scheele in sour milk (1780) and by Berzelius in muscle tissue (1807). This model, however, was criticized by Hermann Kolbe (1818–1884, Leipzig) as an "ingenious but in reality trivial and senseless natural philosophy." After 10 years of heated controversy, the idea of tetrahedral carbon was fully accepted, Kolbe had died and Wislicenus succeeded him in Leipzig.

Emil Fischer (1852–1919, Np 1902) was the next to make a critical contribution to stereochemistry. From the work of van't Hoff and Le Bel he reasoned that glucose should have 16 stereoisomers. Fischer's doctorate work on hydrazines under Baeyer (1835–1917, Np 1905) at Strasbourg had led to studies of osazones which culminated in the brilliant establishment, including configurations, of the Fischer sugar tree starting from D-(+)-glyceraldehyde all the way up to the aldohexoses, allose, altrose, glucose, mannose, gulose, idose, galactose, and talose (from 1884 to 1890). Unfortunately Fischer suffered from the toxic effects of phenylhydrazine for 12 years. The arbitrarily but luckily chosen absolute configuration of D-(+)-glyceraldehyde was shown to be correct sixty years later in 1951 (Johannes-Martin Bijvoet, 1892–1980). Fischer's brilliant correlation of the sugars comprising the Fischer sugar tree was performed using the Kiliani (1855–1945)–Fischer method via cyanohydrin intermediates for elongating sugars. Fischer also made remarkable contributions to the chemistry of amino acids and to nucleic acid bases (see below).

4. STRUCTURES OF ORGANIC COMPOUNDS, TWENTIETH CENTURY

The early concept of covalent bonds was provided with a sound theoretical basis by Linus Pauling (1901–1994, Np 1954), one of the greatest intellects of the twentieth century. Pauling's totally interdisciplinary research interests, including proteins and DNA is responsible for our present understanding of molecular structures. His books *Introduction to Quantum Mechanics* (with graduate student E. B. Wilson, 1935) and *The Nature of the Chemical Bond* (1939) have had a profound effect on our understanding of all of chemistry.

The actual 3D shapes of organic molecules which were still unclear in the late 1940s were then brilliantly clarified by Odd Hassel (1897–1981, Oslo University, Np 1969) and Derek Barton (1918–1998, Np 1969). Hassel, an X-ray crystallographer and physical chemist, demonstrated by electron diffraction that cyclohexane adopted the chair form in the gas phase and that it had two kinds of bonds, "standing (axial)" and "reclining (equatorial)" (1943). Because of the German occupation of Norway in 1940, instead of publishing the result in German journals, he published it in a Norwegian journal which was not abstracted in English until 1945. During his 1949 stay at Harvard, Barton attended a seminar by Louis Fieser on steric effects in steroids and showed Fieser that interpretations could be simplified if the shapes ("conformations") of cyclohexane rings were taken into consideration; Barton made these comments because he was familiar with Hassel's study on *cis-* and *trans-*decalins. Following Fieser's suggestion Barton published these ideas in a four-page *Experientia* paper (1950). This led to the joint Nobel prize with Hassel (1969), and established the concept of conformational analysis, which has exerted a profound effect in every field involving organic molecules.

Using conformational analysis, Barton determined the structures of many key terpenoids such as ß-amyrin, cycloartenone, and cycloartenol (Birkbeck College). At Glasgow University (from 1955) he collaborated in a number of cases with Monteath Robertson (1900–1989) and established many challenging structures: limonin, glauconic acid, byssochlamic acid, and nonadrides. Barton was also associated with the Research Institute for Medicine and Chemistry (RIMAC), Cambridge, USA founded by the Schering company, where with J. M. Beaton, he produced 60 g of aldosterone at a time when the world supply of this important hormone was in mg quantities. Aldosterone synthesis ("a good problem") was achieved in 1961 by Beaton ("a good experimentalist") through a nitrite photolysis, which came to be known as the Barton reaction ("a good idea") (quotes from his 1991 autobiography published by the American Chemical Society). From Glasgow, Barton went on to Imperial College, and a year before retirement, in 1977 he moved to France to direct the research at ICSN at Gif-sur-Yvette where he explored the oxidation reaction selectivity for unactivated C–H. After retiring from ICSN he made a further move to Texas A&M University in 1986, and continued his energetic activities, including chairman of the *Tetrahedron* publications. He felt weak during work one evening and died soon after, on March 16, 1998. He was fond of the phrase "gap jumping" by which he meant seeking generalizations between facts that do not seem to be related: "In the conformational analysis story, one had to jump the gap between steroids and chemical physics" (from his autobiography). According to Barton, the three most important qualities for a scientist are "intelligence, motivation, and honesty." His routine at Texas A&M was to wake around 4 a.m., read the literature, go to the office at 7 a.m. and stay there until 7 p.m.; when asked in 1997 whether this was still the routine, his response was that he wanted to wake up earlier because sleep was a waste of time—a remark which characterized this active scientist approaching 80!

Robert B. Woodward (1917–1979, Np 1965), who died prematurely, is regarded by many as the preeminent organic chemist of the twentieth century. He made landmark achievements in spectroscopy, synthesis, structure determination, biogenesis, as well as in theory. His solo papers published in 1941–1942 on empirical rules for estimating the absorption maxima of enones and dienes made the general organic chemical community realize that UV could be used for structural studies, thus launching the beginning of the spectroscopic revolution which soon brought on the applications of IR, NMR, MS, etc. He determined the structures of the following compounds: penicillin in 1945 (through joint UK–USA collaboration, see Hodgkin), strychnine in 1948, patulin in 1949, terramycin, aureomycin, and ferrocene (with G. Wilkinson, Np 1973—shared with E. O. Fischer for sandwich compounds) in 1952, cevine in 1954 (with Barton Np 1966, Jeger and Prelog, Np 1975), magnamycin in 1956, gliotoxin in 1958, oleandomycin in 1960, streptonigrin in 1963, and tetrodotoxin in 1964. He synthesized patulin in 1950, cortisone and cholesterol in 1951, lanosterol, lysergic acid (with Eli Lilly), and strychnine in 1954, reserpine in 1956, chlorophyll in 1960, a tetracycline (with Pfizer) in 1962, cephalosporin in 1965, and vitamin B_{12} in 1972 (with A. Eschenmoser, 1925–, ETH Zürich). He derived biogenetic schemes for steroids in 1953 (with K. Bloch, see below), and for macrolides in 1956, while the Woodward–Hoffmann orbital symmetry rules in 1965 brought order to a large class of seemingly random cyclization reactions.

Another central figure in stereochemistry is Vladimir Prelog (1906–1998, Np 1975), who succeeded Leopold Ruzicka at the ETH Zürich, and continued to build this institution into one of the most active and lively research and discussion centers in the world. The core group of intellectual leaders consisted of P. Plattner (1904–1975), O. Jeger, A. Eschenmoser, J. Dunitz, D. Arigoni, and A. Dreiding (from Zürich University). After completing extensive research on alkaloids, Prelog determined the structures of nonactin, boromycin, ferrioxamins, and rifamycins. His seminal studies in the synthesis and properties of 8–12 membered rings led him into unexplored areas of stereochemisty and chirality. Together with Robert Cahn (1899–1981, London Chemical Society) and Christopher Ingold (1893–1970, University College, London; pioneering mechanistic interpretation of organic reactions), he developed the Cahn–Ingold–Prelog (CIP) sequence rules for the unambiguous specification of stereoisomers. Prelog was an excellent story teller, always had jokes to tell, and was respected and loved by all who knew him.

4.1 Polyketides and Fatty Acids

Arthur Birch (1915–1995) from Sydney University, Ph.D. with Robert Robinson (Oxford University), then professor at Manchester University and Australian National University, was one of the earliest chemists to perform biosynthetic studies using radiolabels; starting with polyketides he studied the biosynthesis of a variety of natural products such as the C_6–C_3–C_6 backbone of plant phenolics, polyene macrolides, terpenoids, and alkaloids. He is especially known for the Birch reduction of aromatic rings, metal–ammonia reductions leading to 19-norsteroid hormones and other important products (1942–) which were of industrial importance. Feodor Lynen (1911–1979, Np 1964) performed studies on the intermediary metabolism of the living cell that led him to the demonstration of the first step in a chain of reactions resulting in the biosynthesis of sterols and fatty acids.

Prostaglandins, a family of 20-carbon, lipid-derived acids discovered in seminal fluids and accessory genital glands of man and sheep by von Euler (1934), have attracted great interest because of their extremely diverse biological activities. They were isolated and their structures elucidated from 1963 by S. Bergström (1916–, Np 1982) and B. Samuelsson (1934–, Np 1982) at the Karolinska Institute, Stockholm. Many syntheses of the natural prostaglandins and their nonnatural analogues have been published.

Tetsuo Nozoe (1902–1996) who studied at Tohoku University, Sendai, with Riko Majima (1874–1962, see below) went to Taiwan where he stayed until 1948 before returning to Tohoku University. At National Taiwan University he isolated hinokitiol from the essential oil of *taiwanhinoki*. Remembering the resonance concept put forward by Pauling just before World War II, he arrived at the seven-membered nonbenzenoid aromatic structure for hinokitiol in 1941, the first of the troponoids. This highly original work remained unknown to the rest of the world until 1951. In the meantime, during 1945–1948, nonbenzenoid aromatic structures had been assigned to stipitatic acid (isolated by H. Raistrick) by Michael J. S. Dewar (1918–) and to the thujaplicins by Holger Erdtman (1902–1989); the term tropolones was coined by Dewar in 1945. Nozoe continued to work on and discuss troponoids, up to the night before his death, without knowing that he had cancer. He was a remarkably focused and warm scientist, working unremittingly. Erdtman (Royal Institute of Technology, Stockholm) was the central figure in Swedish natural products chemistry who, with his wife Gunhild Aulin Erdtman (dynamic General Secretary of the Swedish Chemistry Society), worked in the area of plant phenolics.

As mentioned in the following and in the concluding sections, classical biosynthetic studies using radioactive isotopes for determining the distribution of isotopes has now largely been replaced by the use of various stable isotopes coupled with NMR and MS. The main effort has now shifted to the identification and cloning of genes, or where possible the gene clusters, involved in the biosynthesis of the natural product. In the case of polyketides (acyclic, cyclic, and aromatic), the focus is on the polyketide synthases.

4.2 Isoprenoids, Steroids, and Carotenoids

During his time as an assistant to Kekulé at Bonn, Otto Wallach (1847–1931, Np 1910) had to familiarize himself with the essential oils from plants; many of the components of these oils were compounds for which no structure was known. In 1891 he clarified the relations between 12 different monoterpenes related to pinene. This was summarized together with other terpene chemistry in book form in 1909, and led him to propose the "isoprene rule." These achievements laid the foundation for the future development of terpenoid chemistry and brought order from chaos.

The next period up to around 1950 saw phenomenal advances in natural products chemistry centered on isoprenoids. Many of the best natural products chemists in Europe, including Wieland, Windaus, Karrer, Kuhn, Butenandt, and Ruzicka contributed to this breathtaking pace. Heinrich Wieland (1877–1957) worked on the bile acid structure, which had been studied over a period of 100 years and considered to be one of the most difficult to attack; he received the Nobel Prize in 1927 for these studies. His friend Adolph Windaus (1876–1959) worked on the structure of cholesterol for which he also received the Nobel Prize in 1928. Unfortunately, there were chemical discrepancies in the proposed steroidal skeletal structure, which had a five-membered ring B attached to C-7 and C-9. J. D. Bernal, Minera-logical Museums, Cambridge University, who was examining the X-ray patterns of ergosterol (1932) noted that the dimensions were inconsistent with the Wieland–Windaus formula. A reinterpretation of the production of chrysene from sterols by Pd/C dehydrogenation reported by Diels (see below) in 1927 eventually led Rosenheim and King and Wieland and Dane to deduce the correct structure in 1932. Wieland also worked on the structures of morphine/strychnine alkaloids, phalloidin/amanitin cyclopeptides of toxic mushroom *Amanita phalloides*, and pteridines, the important fluorescent pig-ments of butterfly wings. Windaus determined the structure of ergosterol and continued structural studies of its irradiation product which exhibited antirachitic activity "vitamin D." The mechanistically complex photochemistry of ergosterol leading to the vitamin D group has been investigated in detail by Egbert Havinga (1927–1988, Leiden University), a leading photochemist and excellent tennis player.

Paul Karrer (1889–1971, Np 1937), established the foundations of carotenoid chemistry through structural determinations of lycopene, carotene, vitamin A, etc. and the synthesis of squalene, carotenoids, and others. George Wald (1906–1997, Np 1967) showed that vitamin A was the key compound in vision during his stay in Karrer's laboratory. Vitamin K (K from "Koagulation"), discovered by Henrik Dam (1895–1976, Polytechnic Institute, Copenhagen, Np 1943) and structurally studied by Edward Doisy (1893–1986, St. Louis University, Np 1943), was also synthesized by Karrer. In addition, Karrer synthesized riboflavin (vitamin B$_2$) and determined the structure and role of nicotinamide adenine dinu-cleotide phosphate (NADP$^+$) with Otto Warburg. The research on carotenoids and vitamins of Karrer who was at Zürich University overlapped with that of Richard Kuhn (1900–1967, Np 1938) at the ETH Zürich, and the two were frequently rivals. Richard Kuhn, one of the pioneers in using UV-vis spectroscopy for structural studies, introduced the concept of "atropisomerism" in diphenyls, and stud-ied the spectra of a series of diphenyl polyenes. He determined the structures of many natural carotenoids, proved the structure of riboflavin-5-phosphate (flavin-adenine-dinucleotide-5-phosphate) and showed that the combination of NAD-5-phosphate with the carrier protein yielded the yellow oxidation enzyme, thus providing an understanding of the role of a prosthetic group. He also determined the structures of vitamin B complexes, i.e., pyridoxine, *p*-aminobenzoic acid, pantothenic acid. After World War II he went on to structural studies of nitrogen-containing oligosaccharides in human milk that provide immu-nity for infants, and brain gangliosides. Carotenoid studies in Switzerland were later taken up by Otto Isler (1910–1993), a Ruzicka student at Hoffmann-La Roche, and Conrad Hans Eugster (1921–), a Karrer student at Zürich University.

Adolf Butenandt (1903–1998, Np 1939) initiated and essentially completed isolation and structural studies of the human sex hormones, the insect molting hormone (ecdysone), and the first pheromone, bombykol. With help from industry he was able to obtain large supplies of urine from pregnant women for estrone, sow ovaries for progesterone, and 4,000 gallons of male urine for androsterone (50 mg, crystals). He isolated and determined the structures of two female sex hormones, estrone and progester-one, and the male hormone androsterone all during the period 1934–1939 (!) and was awarded the Nobel prize in 1939. Keen intuition and use of UV data and Pregl's microanalysis all played important roles. He was appointed to a professorship in Danzig at the age of 30. With Peter Karlson he isolated from 500 kg of silkworm larvae 25 mg of α-ecdysone, the prohormone of insect and crustacean molting hormone, and determined its structure as a polyhydroxysteroid (1965); 20-hydroxylation gives the in-sect and crustacean molting hormone or ß-ecdysone (20-hydroxyecdysteroid). He was also the first to isolate an insect pheromone, bombykol, from female silkworm moths (with E. Hecker). As president of the Max Planck Foundation, he strongly influenced the postwar rebuilding of German science.

The successor to Kuhn, who left ETH Zürich for Heidelberg, was Leopold Ruzicka (1887–1967, Np 1939) who established a close relationship with the Swiss pharmaceutical industry. His synthesis of the 17- and 15-membered macrocyclic ketones, civetone and muscone (the constituents of musk) showed that contrary to Baeyer's prediction, large alicyclic rings could be strainless. He reintroduced and refined the isoprene rule proposed by Wallach (1887) and determined the basic structures of many sesqui-, di-, and triterpenes, as well as the structure of lanosterol, the key intermediate in cholesterol biosynthesis. The "biogenetic isoprene rule" of the ETH group, Albert Eschenmoser, Leopold Ruzicka, Oskar Jeger, and Duilio Arigoni, contributed to a concept of terpenoid cyclization (1955), which was consistent with the mechanistic considerations put forward by Stork as early as 1950. Besides making the ETH group into a center of natural products chemistry, Ruzicka bought many seventeenth century Dutch paintings with royalties accumulated during the war from his Swiss and American patents, and donated them to the Zürich Kunsthaus.

Studies in the isolation, structures, and activities of the antiarthritic hormone, cortisone and related compounds from the adrenal cortex were performed in the mid- to late 1940s during World War II by Edward Kendall (1886–1972, Mayo Clinic, Rochester, Np 1950), Tadeus Reichstein (1897–1996, Basel University, Np 1950), Philip Hench (1896–1965, Mayo Clinic, Rochester, Np 1950), Oskar Wintersteiner (1898–1971, Columbia University, Squibb) and others initiated interest as an adjunct to military medicine as well as to supplement the meager supply from beef adrenal glands by synthesis. Lewis Sarett (1917–, Merck & Co., later president) and co-workers completed the cortisone synthesis in 28 steps, one of the first two totally stereocontrolled syntheses of a natural product; the other was cantharidin (Stork 1951) (see above). The multistep cortisone synthesis was put on the production line by Max Tishler (1906–1989, Merck & Co., later president) who made contributions to the synthesis of a number of drugs, including riboflavin. Besides working on steroid reactions/synthesis and antimalarial agents, Louis F. Fieser (1899–1977) and Mary Fieser (1909–1997) of Harvard University made huge contributions to the chemical community through their outstanding books *Natural Products related to Phenanthrene* (1949), *Steroids* (1959), *Advanced Organic Chemistry* (1961), and *Topics in Organic Chemistry* (1963), as well as their textbooks and an important series of books on Organic Reagents. Carl Djerassi (1923–, Stanford University), a prolific chemist, industrialist, and more recently a novelist, started to work at the Syntex laboratories in Mexico City where he directed the work leading to the first oral contraceptive ("the pill") for women.

Takashi Kubota (1909–, Osaka City University), with Teruo Matsuura (1924–, Kyoto University), determined the structure of the furanoid sesquiterpene, ipomeamarone, from the black rotted portion of spoiled sweet potatoes; this research constitutes the first characterization of a phytoallexin, defense substances produced by plants in response to attack by fungi or physical damage. Damaging a plant and characterizing the defense substances produced may lead to new bioactive compounds. The mechanism of induced biosynthesis of phytoallexins, which is not fully understood, is an interesting biological mechanistic topic that deserves further investigation. Another center of high activity in terpenoids and nucleic acids was headed by Frantisek Sorm (1913–1980, Institute of Organic and Biochemistry, Prague), who determined the structures of many sesquiterpenoids and other natural products; he was not only active scientifically but also was a central figure who helped to guide the careers of many Czech chemists.

The key compound in terpenoid biosynthesis is mevalonic acid (MVA) derived from acetyl-CoA, which was discovered fortuitously in 1957 by the Merck team in Rahway, NJ headed by Karl Folkers (1906–1998). They soon realized and proved that this C_6 acid was the precursor of the C_5 isoprenoid unit isopentenyl diphosphate (IPP) that ultimately leads to the biosynthesis of cholesterol. In 1952 Konrad Bloch (1912–, Harvard, Np 1964) with R. B. Woodward published a paper suggesting a mechanism of the cyclization of squalene to lanosterol and the subsequent steps to cholesterol, which turned out to be essentially correct. This biosynthetic path from MVA to cholesterol was experimentally clarified in stereochemical detail by John Cornforth (1917–, Np 1975) and George Popják. In 1932, Harold Urey (1893–1981, Np 1934) of Columbia University discovered heavy hydrogen. Urey showed, contrary to common expectation, that isotope separation could be achieved with deuterium in the form of deuterium oxide by fractional electrolysis of water. Urey's separation of the stable isotope deuterium led to the isotopic tracer methodology that revolutionized the protocols for elucidating biosynthetic processes and reaction mechanisms, as exemplified beautifully by the cholesterol studies. Using MVA labeled chirally with isotopes, including chiral methyl, i.e., -CHDT, Cornforth and Popják clarified the key steps in the intricate biosynthetic conversion of mevalonate to cholesterol in stereochemical detail. The chiral methyl group was also prepared independently by Duilio Arigoni (1928–, ETH, Zürich). Cornforth has had great difficulty in hearing and speech since childhood but has been helped expertly by his chemist wife Rita; he is an excellent tennis and chess player, and is renowned for his speed in composing occasional witty limericks.

Although MVA has long been assumed to be the only natural precursor for IPP, a non-MVA pathway in which IPP is formed via the glyceraldehyde phosphate-pyruvate pathway has been discovered (1995–1996) in the ancient bacteriohopanoids by Michel Rohmer, who started working on them with Guy Ourisson (1926–, University of Strasbourg, terpenoid studies, including prebiotic), and by Duilio Arigoni in the ginkgolides, which are present in the ancient *Ginkgo biloba* tree. It is possible that many other terpenoids are biosynthesized via the non-MVA route. In classical biosynthetic experiments, ^{14}C-labeled acetic acid was incorporated into the microbial or plant product, and location or distribution of the ^{14}C label was deduced by oxidation or degradation to specific fragments including acetic acid; therefore, it was not possible or extremely difficult to map the distribution of all radioactive carbons. The progress in ^{13}C NMR made it possible to incorporate ^{13}C-labeled acetic acid and locate all labeled carbons. This led to the discovery of the nonmevalonate pathway leading to the IPP units. Similarly, NMR and MS have made it possible to use the stable isotopes, e.g., ^{18}O, ^{2}H, ^{15}N, etc., in biosynthetic studies. The current trend of biosynthesis has now shifted to genomic approaches for cloning the genes of various enzyme synthases involved in the biosynthesis.

4.3 Carbohydrates and Cellulose

The most important advance in carbohydrate structures following those made by Emil Fischer was the change from acyclic to the current cyclic structure introduced by Walter Haworth (1883–1937). He noticed the presence of α- and ß-anomers, and determined the structures of important disaccharides including cellobiose, maltose, and lactose. He also determined the basic structural aspects of starch, cellulose, inulin, and other polysaccharides, and accomplished the structure determination and synthesis of vitamin C, a sample of which he had received from Albert von Szent-Györgyi (1893–1986, Np 1937). This first synthesis of a vitamin was significant since it showed that a vitamin could be synthesized in the same way as any other organic compound. There was strong belief among leading scientists in the 1910s that cellulose, starch, protein, and rubber were colloidal aggregates of small molecules. However, Hermann Staudinger (1881–1965, Np 1953) who succeeded R. Willstätter and H. Wieland at the ETH Zürich and Freiburg, respectively, showed through viscosity measurements and various molecular weight measurements that macromolecules do exist, and developed the principles of macromolecular chemistry.

In more modern times, Raymond Lemieux (1920–, Universities of Ottawa and Alberta) has been a leader in carbohydrate research. He introduced the concept of *endo-* and *exo-*anomeric effects, accomplished the challenging synthesis of sucrose (1953), pioneered in the use of NMR coupling constants in configuration studies, and most importantly, starting with syntheses of oligosaccharides responsible for human blood group determinants, he prepared antibodies and clarified fundamental aspects of the binding of oligosaccharides by lectins and antibodies. The periodate–potassium permanganate cleavage of double bonds at room temperature (1955) is called the Lemieux reaction.

4.4 Amino Acids, Peptides, Porphyrins, and Alkaloids

It is fortunate that we have China's record and practice of herbal medicine over the centuries, which is providing us with an indispensable source of knowledge. China is rapidly catching up in terms of infrastructure and equipment in organic and bioorganic chemistry, and work on isolation, structure determination, and synthesis stemming from these valuable sources has picked up momentum. However, as mentioned above, clarification of the active principles and mode of action of these plant extracts will be quite a challenge since in many cases synergistic action is expected. Wang Yu (1910–1997) who headed the well-equipped Shanghai Institute of Organic Chemistry surprised the world with the total synthesis of bovine insulin performed by his group in 1965; the human insulin was synthesized around the same time by P. G. Katsoyannis, A. Tometsko, and C. Zaut of the Brookhaven National Laboratory (1966).

One of the giants in natural products chemistry during the first half of this century was Robert Robinson (1886–1975, Np 1947) at Oxford University. His synthesis of tropinone, a bicyclic amino ketone related to cocaine, from succindialdehyde, methylamine, and acetone dicarboxylic acid under Mannich reaction conditions was the first biomimetic synthesis (1917). It reduced Willstätter's 1903 13-step synthesis starting with suberone into a single step. This achievement demonstrated Robinson's analytical prowess. He was able to dissect complex molecular structures into simple biosynthetic building blocks, which allowed him to propose the biogenesis of all types of alkaloids and other natural products. His laboratory at Oxford, where he developed the well-known Robinson annulation reaction (1937) in connection with his work on the synthesis of steroids became a world center for natural products study. Robinson was a pioneer in the so-called electronic theory of organic reactions, and introduced the use of curly arrows to show the movements of electrons. His analytical power is exemplified in the structural studies of strychnine and brucine around 1946–1952. Barton clarified the biosynthetic route to the morphine alkaloids, which he saw as an extension of his biomimetic synthesis of usnic acid through a one-electron oxidation; this was later extended to a general phenolate coupling scheme. Morphine total synthesis was brilliantly achieved by Marshall Gates (1915–, University of Rochester) in 1952.

The yield of the Robinson tropinone synthesis was low but Clemens Schöpf (1899–1970) , Ph.D. Munich (Wieland), Universität Darmstadt, improved it to 90% by carrying out the reaction in buffer; he also worked on the stereochemistry of morphine and determined the structure of the steroidal alkaloid salamandarine (1961), the toxin secreted from glands behind the eyes of the salamander.

Roger Adams (1889–1971, University of Illinois), was the central figure in organic chemistry in the USA and is credited with contributing to the rapid development of its chemistry in the late 1930s and 1940s, including training of graduate students for both academe and industry. After earning a Ph.D. in 1912 at Harvard University he did postdoctoral studies with Otto Diels (see below) and Richard Willstätter (see below) in 1913; he once said that around those years in Germany he could cover all *Journal of the American Chemical Society* papers published in a year in a single night. His important work include determination of the structures of tetrahydrocannabinol in marijuana, the toxic gossypol in cottonseed oil, chaulmoogric acid used in treatment of leprosy, and the Senecio alkaloids with Nelson Leonard (1916–, University of Illinois, now at Caltech). He also contributed to many fundamental organic reactions and syntheses. The famous Adams platinum catalyst is not only important for reducing double bonds in industry and in the laboratory, but was central for determining the number of double bonds in a structure. He was also one of the founders of the *Organic Synthesis* (started in 1921) and the *Organic Reactions* series. Nelson Leonard switched interests to bioorganic chemistry and biochemistry, where he has worked with nucleic acid bases and nucleotides, coenzymes, dimensional probes, and fluorescent modifications such as ethenoguanine.

The complicated structures of the medieval plant poisons aconitine (from *Aconitum*) and delphinine (from *Delphinium*) were finally characterized in 1959–1960 by Karel Wiesner (1919–1986, University of New Brunswick), Leo Marion (1899–1979, National Research Council, Ottawa), George Büchi (1921–, mycotoxins, aflatoxin/DNA adduct, synthesis of terpenoids and nitrogen-containing bioactive compounds, photochemistry), and Maria Przybylska (1923–, X-ray).

The complex chlorophyll structure was elucidated by Richard Willstätter (1872–1942, Np 1915). Although he could not join Baeyer's group at Munich because the latter had ceased taking students, a close relation developed between the two. During his chlorophyll studies, Willstätter reintroduced the important technique of column chromatography published in Russian by Michael Tswett (1906). Willstätter further demonstrated that magnesium was an integral part of chlorophyll, clarified the relation between chlorophyll and the blood pigment hemin, and found the wide distribution of carotenoids in tomato, egg yolk, and bovine corpus luteum. Willstätter also synthesized cyclooctatetraene and showed its properties to be wholly unlike benzene but close to those of acyclic polyenes (around 1913). He succeeded Baeyer at Munich in 1915, synthesized the anesthetic cocaine, retired early in protest of anti-Semitism, but remained active until the Hitler era, and in 1938 emigrated to Switzerland.

The hemin structure was determined by another German chemist of the same era, Hans Fischer (1881–1945, Np 1930), who succeeded Windaus at Innsbruck and at Munich. He worked on the structure of hemin from the blood pigment hemoglobin, and completed its synthesis in 1929. He continued Willstätter's structural studies of chlorophyll, and further synthesized bilirubin in 1944. Destruction of his institute at Technische Hochschule München, during World War II led him to take his life in March 1945. The biosynthesis of hemin was elucidated largely by David Shemin (1911–1991).

In the mid 1930s the Department of Biochemistry at Columbia Medical School, which had accepted many refugees from the Third Reich, including Erwin Chargaff, Rudolf Schoenheimer, and others on the faculty, and Konrad Bloch (see above) and David Shemin as graduate students, was a great center of research activity. In 1940, Shemin ingested 66 g of ^{15}N-labeled glycine over a period of 66 hours in order to determine the half-life of erythrocytes. David Rittenberg's analysis of the heme moiety with his home-made mass spectrometer showed all four pyrrole nitrogens came from glycine. Using ^{14}C (that had just become available) as a second isotope (see next paragraph), doubly labeled glycine $^{15}NH_2{}^{14}CH_2COOH$ and other precursors, Shemin showed that glycine and succinic acid condensed to yield δ-aminolevulinate, thus elegantly demonstrating the novel biosynthesis of the porphyrin ring (around 1950). At this time, Bloch was working on the other side of the bench.

Melvin Calvin (1911–1997, Np 1961) at University of California, Berkeley, elucidated the complex photosynthetic pathway in which plants reduce carbon dioxide to carbohydrates. The critical $^{14}CO_2$ had just been made available at Berkeley Lawrence Radiation Laboratory as a result of the pioneering research of Martin Kamen (1913–), while paper chromatography also played crucial roles. Kamen produced ^{14}C with Sam Ruben (1940), used ^{18}O to show that oxygen in photosynthesis comes from water and not from carbon dioxide, participated in the *Manhattan* project, testified before the House UnAmerican Activities Committee (1947), won compensatory damages from the US Department of State, and helped build the University of California, La Jolla (1957). The entire structure of the photosynthetic reaction center (>10 000 atoms) from the purple bacterium *Rhodopseudomonas viridis* has been established by X-ray crystallography in the landmark studies performed by Johann Deisenhofer (1943–), Robert Huber (1937–), and Hartmut Michel (1948–) in 1989; this was the first membrane protein structure determined by X-ray, for which they shared the 1988 Nobel prize. The information gained from the full structure of this first membrane protein has been especially rewarding.

The studies on vitamin B_{12}, the structure of which was established by crystallographic studies performed by Dorothy Hodgkin (1910–1994, Np 1964), are fascinating. Hodgkin also determined the structure of penicillin (in a joint effort between UK and US scientists during World War II) and insulin. The formidable total synthesis of vitamin B_{12} was completed in 1972 through collaborative efforts between Woodward and Eschenmoser, involving 100 postdoctoral fellows and extending over 10 years. The biosynthesis of fascinating complexity is almost completely solved through studies performed by Alan Battersby (1925–, Cambridge University), Duilio Arigoni, and Ian Scott (1928–, Texas A&M University) and collaborators where advanced NMR techniques and synthesis of labeled precursors is elegantly combined with cloning of enzymes controlling each biosynthetic step. This work provides a beautiful demonstration of the power of the combination of bioorganic chemistry, spectroscopy and molecular biology, a future direction which will become increasingly important for the creation of new "unnatural" natural products.

4.5 Enzymes and Proteins

In the early days of natural products chemistry, enzymes and viruses were very poorly understood. Thus, the 1926 paper by James Sumner (1887–1955) at Cornell University on crystalline urease was received with ignorance or skepticism, especially by Willstätter who believed that enzymes were small molecules and not proteins. John Northrop (1891–1987) and co-workers at the Rockefeller Institute went on to crystallize pepsin, trypsin, chymotrypsin, ribonuclease, deoyribonuclease, carboxypeptidase, and other enzymes between 1930 and 1935. Despite this, for many years biochemists did not recognize the significance of these findings, and considered enzymes as being low molecular weight compounds adsorbed onto proteins or colloids. Using Northrop's method for crystalline enzyme preparations, Wendell Stanley (1904–1971) at Princeton obtained tobacco mosaic virus as needles from one ton of tobacco leaves (1935). Sumner, Northrop, and Stanley shared the 1946 Nobel prize in chemistry. All these studies opened a new era for biochemistry.

Meanwhile, Linus Pauling, who in mid-1930 became interested in the magnetic properties of hemoglobin, investigated the configurations of proteins and the effects of hydrogen bonds. In 1949 he showed that sickle cell anemia was due to a mutation of a single amino acid in the hemoglobin molecule, the first correlation of a change in molecular structure with a genetic disease. Starting in 1951 he and colleagues published a series of papers describing the alpha helix structure of proteins; a paper published in the early 1950s with R. B. Corey on the structure of DNA played an important role in leading Francis Crick and James Watson to the double helix structure (Np 1962).

A further important achievement in the peptide field was that of Vincent Du Vigneaud (1901–1978, Np 1955), Cornell Medical School, who isolated and determined the structure of oxytocin, a posterior pituitary gland hormone, for which a structure involving a disulfide bond was proposed. He synthesized oxytocin in 1953, thereby completing the first synthesis of a natural peptide hormone.

Progress in isolation, purification, crystallization methods, computers, and instrumentation, including cyclotrons, have made X-ray crystallography the major tool in structural. Numerous structures including those of ligand/receptor complexes are being published at an extremely rapid rate. Some of the past major achievements in protein structures are the following. Max Perutz (1914, Np 1962) and John Kendrew (1914–1997, Np 1962), both at the Laboratory of Molecular Biology, Cambridge University, determined the structures of hemoglobin and myoglobin, respectively. William Lipscomb (1919–, Np 1976), Harvard University, who has trained many of the world's leaders in protein X-ray crystallography has been involved in the structure determination of many enzymes including carboxypeptidase A (1967); in 1965 he determined the structure of the anticancer bisindole alkaloid, vinblastine. Folding of proteins, an important but still enigmatic phenomenon, is attracting increasing attention. Christian Anfinsen (1916–1995, Np 1972), NIH, one of the pioneers in this area, showed that the amino acid residues in ribonuclease interact in an energetically most favorable manner to produce the unique 3D structure of the protein.

4.6 Nucleic Acid Bases, RNA, and DNA

The "Fischer indole synthesis" was first performed in 1886 by Emil Fischer. During the period 1881–1914, he determined the structures of and synthesized uric acid, caffeine, theobromine, xanthine, guanine, hypoxanthine, adenine, guanine, and made theophylline-D-glucoside phosphoric acid, the first synthetic nucleotide. In 1903, he made 5,5-diethylbarbituric acid or Barbital, Dorminal, Veronal, etc. (sedative), and in 1912, phenobarbital or Barbipil, Luminal, Phenobal, etc. (sedative). Many of his syntheses formed the basis of German industrial production of purine bases. In 1912 he showed that tannins are gallates of sugars such as maltose and glucose. Starting in 1899, he synthesized many of the 13 α-amino acids known at that time, including the L- and D-forms, which were separated through fractional crystallization of their salts with optically active bases. He also developed a method for synthesizing fragments of proteins, namely peptides, and made an 18-amino acid peptide. He lost his two sons in World War I, lost his wealth due to postwar inflation, believed he had terminal cancer (a misdiagnosis), and killed himself in July 1919. Fischer was a skilled experimentalist, so that even today, many of the reactions performed by him and his students are so delicately controlled that they are not easy to reproduce. As a result of his suffering by inhaling diethylmercury, and of the poisonous effect of phenylhydrazine, he was one of the first to design fume hoods. He was a superb teacher and was also influential in establishing the Kaiser Wilhelm Institute, which later became the Max Planck Institute. The number and quality of his accomplishments and contributions are hard to believe; he was truly a genius.

Alexander Todd (1907–1997, Np 1957) made critical contributions to the basic chemistry and synthesis of nucleotides. His early experience consisted of an extremely fruitful stay at Oxford in the Robinson group, where he completed the syntheses of many representative anthocyanins, and then at Edinburgh where he worked on the synthesis of vitamin B_1. He also prepared the hexacarboxylate of vitamin B_{12} (1954), which was used by D. Hodgkin's group for their X-ray elucidation of this vitamin (1956). M. Wiewiorowski (1918–), Institute for Bioorganic Chemistry, in Poznan, has headed a famous group in nucleic acid chemistry, and his colleagues are now distributed worldwide.

4.7 Antibiotics, Pigments, and Marine Natural Products

The concept of one microorganism killing another was introduced by Pasteur who coined the term antibiosis in 1877, but it was much later that this concept was realized in the form of an actual antibiotic. The bacteriologist Alexander Fleming (1881–1955, University of London, Np 1945) noticed that an airborne mold, a *Penicillium* strain, contaminated cultures of *Staphylococci* left on the open bench and formed a transparent circle around its colony due to lysis of *Staphylococci*. He published these results in 1929. The discovery did not attract much interest but the work was continued by Fleming until it was taken up further at Oxford University by pathologist Howard Florey (1898–1968, Np 1945) and biochemist Ernst Chain (1906–1979, Np 1945). The bioactivities of purified "penicillin," the first antibiotic, attracted serious interest in the early 1940s in the midst of World War II. A UK/USA team was formed during the war between academe and industry with Oxford University, Harvard University, ICI, Glaxo, Burroughs Wellcome, Merck, Shell, Squibb, and Pfizer as members. This project resulted in the large scale production of penicillin and determination of its structure (finally by X-ray, D. Hodgkin). John Sheehan (1915–1992) at MIT synthesized 6-aminopenicillanic acid in 1959, which opened the route for the synthesis of a number of analogues. Besides being the first antibiotic to be discovered, penicillin is also the first member of a large number of important antibiotics containing the ß-lactam ring, for example cephalosporins, carbapenems, monobactams, and nocardicins. The strained ß-lactam ring of these antibiotics inactivates the transpeptidase by acylating its serine residue at the active site, thus preventing the enzyme from forming the link between the pentaglycine chain and the D-Ala-D-Ala peptide, the essential link in bacterial cell walls. The overuse of ß-lactam antibiotics, which has given rise to the disturbing appearance of microbial resistant strains, is leading to active research in the design of synthetic ß-lactam analogues to counteract these strains. The complex nature of the important penicillin biosynthesis is being elucidated through efforts combining genetic engineering, expression of biosynthetic genes as well as feeding of synthetic precursors, etc. by Jack Baldwin (1938–, Oxford University), José Luengo (Universidad de León, Spain) and many other groups from industry and academe.

Shortly after the penicillin discovery, Selman Waksman (1888–1973, Rutgers University, Np 1952) discovered streptomycin, the second antibiotic and the first active against the dreaded disease tuberculosis. The discovery and development of new antibiotics continued throughout the world at pharmaceutical companies in Europe, Japan, and the USA from soil and various odd sources: cephalosporin from sewage in Sardinia, cyclosporin from Wisconsin and Norway soil which was carried back to Switzerland, avermectin from the soil near a golf course in Shizuoka Prefecture. People involved in antibiotic discovery used to collect soil samples from various sources during their trips but this has now become severely restricted to protect a country's right to its soil. M. M. Shemyakin (1908–1970, Institute of Chemistry of Natural Products, Moscow) was a grand master of Russian natural products who worked on antibiotics, especially of the tetracycline class; he also worked on cyclic antibiotics composed of alternating sequences of amides and esters and coined the term depsipeptide for these in 1953. He died in 1970 of a sudden heart attack in the midst of the 7th IUPAC Natural Products

Symposium held in Riga, Latvia, which he had organized. The Institute he headed was renamed the Shemyakin Institute.

Indigo, an important vat dye known in ancient Asia, Egypt, Greece, Rome, Britain, and Peru, is probably the oldest known coloring material of plant origin, Indigofera and Isatis. The structure was determined in 1883 and a commercially feasible synthesis was performed in 1883 by Adolf von Baeyer (see above, 1835–1917, Np 1905), who founded the German Chemical Society in 1867 following the precedent of the Chemistry Society of London. In 1872 Baeyer was appointed a professor at Strasbourg where E. Fischer was his student, and in 1875 he succeeded J. Liebig in Munich. Tyrian (or Phoenician) purple, the dibromo derivative of indigo which is obtained from the purple snail Murex bundaris, was used as a royal emblem in connection with religious ceremonies because of its rarity; because of the availability of other cheaper dyes with similar color, it has no commercial value today. K. Venkataraman (1901–1981, University of Bombay then National Chemical Laboratory) who worked with R. Robinson on the synthesis of chromones in his early career, continued to study natural and synthetic coloring matters, including synthetic anthraquinone vat dyes, natural quinonoid pigments, etc. T. R. Seshadri (1900–1975) is another Indian natural products chemist who worked mainly in natural pigments, dyes, drugs, insecticides, and especially in polyphenols. He also studied with Robinson, and with Pregl at Graz, and taught at Delhi University. Seshadri and Venkataraman had a huge impact on Indian chemistry. After a 40 year involvement, Toshio Goto (1929–1990) finally succeeded in solving the mysterious identity of commelinin, the deep-blue flower petal pigment of the Commelina communis isolated by Kozo Hayashi (1958) and protocyanin, isolated from the blue cornflower Centaurea cyanus by E. Bayer (1957). His group elucidated the remarkable structure in its entirety which consisted of six unstable anthocyanins, six flavones and two metals, the molecular weight approaching 10 000; complex stacking and hydrogen bonds were also involved. Thus the pigmentation of petals turned out to be far more complex than the theories put forth by Willstätter (1913) and Robinson (1931). Goto suffered a fatal heart attack while inspecting the first X-ray structure of commelinin; commelinin represents a pinnacle of current natural products isolation and structure determination in terms of subtlety in isolation and complexity of structure.

The study of marine natural products is understandably far behind that of compounds of terrestrial origin due to the difficulty in collection and identification of marine organisms. However, it is an area which has great potentialities for new discoveries from every conceivable source. One pioneer in modern marine chemistry is Paul Scheuer (1915–, University of Hawaii) who started his work with quinones of marine origin and has since characterized a very large number of bioactive compounds from mollusks and other sources. Luigi Minale (1936–1997, Napoli) started a strong group working on marine natural products, concentrating mainly on complex saponins. He was a leading natural products chemist who died prematurely. A. Gonzalez Gonzalez (1917–) who headed the Organic Natural Products Institute at the University of La Laguna, Tenerife, was the first to isolate and study polyhalogenated sesquiterpenoids from marine sources. His group has also carried out extensive studies on terrestrial terpenoids from the Canary Islands and South America. Carotenoids are widely distributed in nature and are of importance as food coloring material and as antioxidants (the detailed mechanisms of which still have to be worked out); new carotenoids continue to be discovered from marine sources, for example by the group of Synnove Liaaen-Jensen, Norwegian Institute of Technology). Yoshimasa Hirata (1915–), who started research at Nagoya University, is a champion in the isolation of nontrivial natural products. He characterized the bioluminescent luciferin from the marine ostracod *Cypridina hilgendorfii* in 1966 (with his students, Toshio Goto, Yoshito Kishi, and Osamu Shimomura); tetrodotoxin from the fugu fish in 1964 (with Goto and Kishi and co-workers), the structure of which was announced simultaneously by the group of Kyosuke Tsuda (1907–, tetrodotoxin, matrine) and Woodward; and the very complex palytoxin, $C_{129}H_{223}N_3O_{54}$ in 1981–1987 (with Daisuke Uemura and Kishi). Richard E. Moore, University of Hawaii, also announced the structure of palytoxin independently. Jon Clardy (1943–, Cornell University) has determined the X-ray structures of many unique marine natural products, including brevetoxin B (1981), the first of the group of toxins with contiguous *trans*-fused ether rings constituting a stiff ladder-like skeleton. Maitotoxin, $C_{164}H_{256}O_{68}S_2Na_2$, MW 3422, produced by the dinoflagellate *Gambierdiscus toxicus* is the largest and most toxic of the nonbiopolymeric toxins known; it has 32 alicyclic 6- to 8-membered ethereal rings and acyclic chains. Its isolation (1994) and complete structure determination was accomplished jointly by the groups of Takeshi Yasumoto (Tohoku University), Kazuo Tachibana and Michio Murata (Tokyo University) in 1996. Kishi, Harvard University, also deduced the full structure in 1996.

The well-known excitatory agent for the cat family contained in the volatile oil of catnip, *Nepeta cataria*, is the monoterpene nepetalactone, isolated by S. M. McElvain (1943) and structure determined by Jerrold Meinwald (1954); cats, tigers, and lions start purring and roll on their backs in response to this lactone. Takeo Sakan (1912–1993) investigated the series of monoterpenes neomatatabiols, etc.

from Actinidia, some of which are male lacewing attractants. As little as 1 fg of neomatatabiol attracts lacewings.

The first insect pheromone to be isolated and characterized was bombykol, the sex attractant for the male silkworm, *Bombyx mori* (by Butenandt and co-workers, see above). Numerous pheromones have been isolated, characterized, synthesized, and are playing central roles in insect control and in chemical ecology. The group at Cornell University have long been active in this field: Tom Eisner (1929–, behavior), Jerrold Meinwald (1927–, chemistry), Wendell Roeloff (1939–, electrophysiology, chemistry). Since the available sample is usually minuscule, full structure determination of a pheromone often requires total synthesis; Kenji Mori (1935–, Tokyo University) has been particularly active in this field. Progress in the techniques for handling volatile compounds, including collection, isolation, GC/MS, etc., has started to disclose the extreme complexity of chemical ecology which plays an important role in the lives of all living organisms. In this context, natural products chemistry will be play an increasingly important role in our grasp of the significance of biodiversity.

5. SYNTHESIS

Synthesis has been mentioned often in the preceding sections of this essay. In the following, synthetic methods of more general nature are described. The Grignard reaction of Victor Grignard (1871–1935, Np 1912) and then the Diels–Alder reaction by Otto Diels (1876–1954, Np 1950) and Kurt Alder (1902–1956, Np 1950) are extremely versatile reactions. The Diels–Alder reaction can account for the biosynthesis of several natural products with complex structures, and now an enzyme, a Diels–Alderase involved in biosynthesis has been isolated by Akitami Ichihara, Hokkaido University (1997).

The hydroboration reactions of Herbert Brown (1912–, Purdue University, Np 1979) and the Wittig reactions of Georg Wittig (1897–1987, Np 1979) are extremely versatile synthetic reactions. William S. Johnson (1913–1995, University of Wisconsin, Stanford University) developed efficient methods for the cyclization of acyclic polyolefinic compounds for the synthesis of corticoid and other steroids, while Gilbert Stork (1921–, Columbia University) introduced enamine alkylation, regiospecific enolate formation from enones and their kinetic trapping (called "three component coupling" in some cases), and radical cyclization in regio- and stereospecific constructions. Elias J. Corey (1928–, Harvard University, Np 1990) introduced the concept of retrosynthetic analysis and developed many key synthetic reactions and reagents during his synthesis of bioactive compounds, including prostaglandins and gingkolides. A recent development is the ever-expanding supramolecular chemistry stemming from 1967 studies on crown ethers by Charles Pedersen (1904–1989), 1968 studies on cryptates by Jean-Marie Lehn (1939–), and 1973 studies on host–guest chemistry by Donald Cram (1919–); they shared the chemistry Nobel prize in 1987.

6. NATURAL PRODUCTS STUDIES IN JAPAN

Since the background of natural products study in Japan is quite different from that in other countries, a brief history is given here. Natural products is one of the strongest areas of chemical research in Japan with probably the world's largest number of chemists pursuing structural studies; these are joined by a healthy number of synthetic and bioorganic chemists. An important Symposium on Natural Products was held in 1957 in Nagoya as a joint event between the faculties of science, pharmacy, and agriculture. This was the beginning of a series of annual symposia held in various cities, which has grown into a three-day event with about 50 talks and numerous papers; practically all achievements in this area are presented at this symposium. Japan adopted the early twentieth century German or European academic system where continuity of research can be assured through a permanent staff in addition to the professor, a system which is suited for natural products research which involves isolation and assay, as well as structure determination, all steps requiring delicate skills and much expertise.

The history of Japanese chemistry is short because the country was closed to the outside world up to 1868. This is when the Tokugawa shogunate which had ruled Japan for 264 years was overthrown and the Meiji era (1868–1912) began. Two of the first Japanese organic chemists sent abroad were Shokei Shibata and Nagayoshi Nagai, who joined the laboratory of A. W. von Hoffmann in Berlin. Upon return to Japan, Shibata (Chinese herbs) started a line of distinguished chemists, Keita and Yuji Shibata (flavones) and Shoji Shibata (1915–, lichens, fungal bisanthraquinonoid pigments, ginsenosides); Nagai returned to Tokyo Science University in 1884, studied ephedrine, and left a big mark in the embryonic era of organic chemistry. Modern natural products chemistry really began when three extraordinary organic chemists returned from Europe in the 1910s and started teaching and research at their respective faculties:

Riko Majima, 1874–1962, C. D. Harries (Kiel University); R. Willstätter (Zürich): Faculty of Science, Tohoku University; studied urushiol, the catecholic mixture of poison ivy irritant.

Yasuhiko Asahina, 1881–1975, R. Willstätter: Faculty of pharmacy, Tokyo University; lichens and Chinese herb.

Umetaro Suzuki, 1874–1943, E. Fischer: Faculty of agriculture, Tokyo University; vitamin B_1(thiamine).

Because these three pioneers started research in three different faculties (i.e., science, pharmacy, and agriculture), and because little interfaculty personnel exchange occurred in subsequent years, natural products chemistry in Japan was pursued independently within these three academic domains; the situation has changed now. The three pioneers started lines of first-class successors, but the establishment of a strong infrastructure takes many years, and it was only after the mid-1960s that the general level of science became comparable to that in the rest of the world; the 3rd IUPAC Symposium on the Chemistry of Natural Products, presided over by Munio Kotake (1894–1976, bufotoxins, see below), held in 1964 in Kyoto, was a clear turning point in Japan's role in this area.

Some of the outstanding Japanese chemists not already quoted are the following. Shibasaburo Kitazato (1852–1931), worked with Robert Koch (Np 1905, tuberculosis) and von Behring, antitoxins of diphtheria and tetanus which opened the new field of serology, isolation of microorganism causing dysentery, founder of Kitazato Institute; Chika Kuroda (1884–1968), first female Ph.D., structure of the complex carthamin, important dye in safflower (1930) which was revised in 1979 by Obara *et al.*, although the absolute configuration is still unknown (1998); Munio Kotake (1894–1976), bufotoxins, tryptophan metabolites, nupharidine; Harusada Suginome (1892–1972), aconite alkaloids; Teijiro Yabuta (1888–1977), kojic acid, gibberrelins; Eiji Ochiai (1898–1974), aconite alkaloids; Toshio Hoshino (1899–1979), abrine and other alkaloids; Yusuke Sumiki (1901–1974), gibberrelins; Sankichi Takei (1896–1982), rotenone; Shiro Akabori (1900–1992), peptides, C-terminal hydrazinolysis of amino acid ; Hamao Umezawa (1914–1986), kanamycin, bleomycin, numerous antibiotics; Shojiro Uyeo (1909–1988), lycorine; Tsunematsu Takemoto (1913–1989), inokosterone, kainic acid, domoic acid, quisqualic acid; Tomihide Shimizu (1889–1958), bile acids; Kenichi Takeda (1907–1991), Chinese herbs, sesquiterpenes; Yoshio Ban (1921–1994), alkaloid synthesis; Wataru Nagata (1922–1993), stereocontrolled hydrocyanation.

7. CURRENT AND FUTURE TRENDS IN NATURAL PRODUCTS CHEMISTRY

Spectroscopy and X-ray crystallography has totally changed the process of structure determination, which used to generate the excitement of solving a mystery. The first introduction of spectroscopy to the general organic community was Woodward's 1942–1943 empirical rules for estimating the UV maxima of dienes, trienes, and enones, which were extended by Fieser (1959). However, Butenandt had used UV for correctly determining the structures of the sex hormones as early as the early 1930s, while Karrer and Kuhn also used UV very early in their structural studies of the carotenoids. The Beckman DU instruments were an important factor which made UV spectroscopy a common tool for organic chemists and biochemists. With the availability of commercial instruments in 1950, IR spectroscopy became the next physical tool, making the 1950 Colthup IR correlation chart and the 1954 Bellamy monograph indispensable. The IR fingerprint region was analyzed in detail in attempts to gain as much structural information as possible from the molecular stretching and bending vibrations. Introduction of NMR spectroscopy into organic chemistry, first for protons and then for carbons, has totally changed the picture of structure determination, so that now IR is used much less frequently; however, in biopolymer studies, the techniques of difference FTIR and resonance Raman spectroscopy are indispensable.

The dramatic and rapid advancements in mass spectrometry are now drastically changing the protocol of biomacromolecular structural studies performed in biochemistry and molecular biology. Herbert Hauptman (mathematician, 1917–, Medical Foundation, Buffalo, Np 1985) and Jerome Karle (1918–, US Naval Research Laboratory, Washington, DC, Np 1985) developed direct methods for the determination of crystal structures devoid of disproportionately heavy atoms. The direct method together with modern computers revolutionized the X-ray analysis of molecular structures, which has become routine for crystalline compounds, large as well as small. Fred McLafferty (1923–, Cornell University) and Klaus Biemann (1926–, MIT) have made important contributions in the development of organic and bioorganic mass spectrometry. The development of cyclotron-based facilities for crystallographic biology studies has led to further dramatic advances enabling some protein structures to be determined in a single day, while cryoscopic electron micrography developed in 1975 by Richard Henderson and Nigel Unwin has also become a powerful tool for 3D structural determinations of membrane proteins such as bacteriorhodopsin (25 kd) and the nicotinic acetylcholine receptor (270 kd).

Circular dichroism (c.d.), which was used by French scientists Jean B. Biot (1774–1862) and Aimé Cotton during the nineteenth century "deteriorated" into monochromatic measurements at 589 nm after R.W. Bunsen (1811–1899, Heidelberg) introduced the Bunsen burner into the laboratory which readily emitted a 589 nm light characteristic of sodium. The 589 nm $[\alpha]_D$ values, remote from most chromophoric maxima, simply represent the summation of the low-intensity readings of the decreasing end of multiple Cotton effects. It is therefore very difficult or impossible to deduce structural information from $[\alpha]_D$ readings. Chiroptical spectroscopy was reintroduced to organic chemistry in the 1950s by C. Djerassi at Wayne State University (and later at Stanford University) as optical rotatory dispersion (ORD) and by L. Velluz and M. Legrand at Roussel-Uclaf as c.d. Günther Snatzke (1928–1992, Bonn then Ruhr University Bochum) was a major force in developing the theory and application of organic chiroptical spectroscopy. He investigated the chiroptical properties of a wide variety of natural products, including constituents of indigenous plants collected throughout the world, and established semiempirical sector rules for absolute configurational studies. He also established close collaborations with scientists of the former Eastern bloc countries and had a major impact in increasing the interest in c.d. there.

Chiroptical spectroscopy, nevertheless, remains one of the most underutilized physical measurements. Most organic chemists regard c.d. (more popular than ORD because interpretation is usually less ambiguous) simply as a tool for assigning absolute configurations, and since there are only two possibilities in absolute configurations, c.d. is apparently regarded as not as crucial compared to other spectroscopic methods. Moreover, many of the c.d. correlations with absolute configuration are empirical. For such reasons, chiroptical spectroscopy, with its immense potentialities, is grossly underused. However, c.d. curves can now be calculated nonempirically. Moreover, through-space coupling between the electric transition moments of two or more chromophores gives rise to intense Cotton effects split into opposite signs, exciton-coupled c.d.; fluorescence-detected c.d. further enhances the sensitivity by 50- to 100-fold. This leads to a highly versatile nonempirical microscale solution method for determining absolute configurations, etc.

With the rapid advances in spectroscopy and isolation techniques, most structure determinations in natural products chemistry have become quite routine, shifting the trend gradually towards activity-monitored isolation and structural studies of biologically active principles available only in microgram or submicrogram quantities. This in turn has made it possible for organic chemists to direct their attention towards clarifying the mechanistic and structural aspects of the ligand/biopolymeric receptor interactions on a more well-defined molecular structural basis. Until the 1990s, it was inconceivable and impossible to perform such studies.

Why does sugar taste sweet? This is an extremely challenging problem which at present cannot be answered even with major multidisciplinary efforts. Structural characterization of sweet compounds and elucidation of the amino acid sequences in the receptors are only the starting point. We are confronted with a long list of problems such as cloning of the receptors to produce them in sufficient quantities to investigate the physical fit between the active factor (sugar) and receptor by biophysical methods, and the time-resolved change in this physical contact and subsequent activation of G-protein and enzymes. This would then be followed by neurophysiological and ultimately physiological and psychological studies of sensation. How do the hundreds of taste receptors differ in their structures and their physical contact with molecules, and how do we differentiate the various taste sensations? The same applies to vision and to olfactory processes. What are the functions of the numerous glutamate receptor subtypes in our brain? We are at the starting point of a new field which is filled with exciting possibilities.

Familiarity with molecular biology is becoming essential for natural products chemists to plan research directed towards an understanding of natural products biosynthesis, mechanisms of bioactivity triggered by ligand–receptor interactions, etc. Numerous genes encoding enzymes have been cloned and expressed by the cDNA and/or genomic DNA-polymerase chain reaction protocols. This then leads to the possible production of new molecules by gene shuffling and recombinant biosynthetic techniques. Monoclonal catalytic antibodies using haptens possessing a structure similar to a high-energy intermediate of a proposed reaction are also contributing to the elucidation of biochemical mechanisms and the design of efficient syntheses. The technique of photoaffinity labeling, brilliantly invented by Frank Westheimer (1912–, Harvard University), assisted especially by advances in mass spectrometry, will clearly be playing an increasingly important role in studies of ligand–receptor interactions including enzyme–substrate reactions. The combined and sophisticated use of various spectroscopic means, including difference spectroscopy and fast time-resolved spectroscopy, will also become increasingly central in future studies of ligand–receptor studies.

Organic chemists, especially those involved in structural studies have the techniques, imagination, and knowledge to use these approaches. But it is difficult for organic chemists to identify an exciting and worthwhile topic. In contrast, the biochemists, biologists, and medical doctors are daily facing exciting life-related phenomena, frequently without realizing that the phenomena could be understood or at least clarified on a chemical basis. Broad individual expertise and knowledge coupled with multidisciplinary research collaboration thus becomes essential to investigate many of the more important future targets successfully. This approach may be termed "dynamic," as opposed to a "static" approach, exemplified by isolation and structure determination of a single natural product. Fortunately for scientists, nature is extremely complex and hence all the more challenging. Natural products chemistry will be playing an absolutely indispensable role for the future. Conservation of the alarming number of disappearing species, utilization of biodiversity, and understanding of the intricacies of biodiversity are further difficult, but urgent, problems confronting us.

That natural medicines are attracting renewed attention is encouraging from both practical and scientific viewpoints; their efficacy has often been proven over the centuries. However, to understand the mode of action of folk herbs and related products from nature is even more complex than mechanistic clarification of a single bioactive factor. This is because unfractionated or partly fractionated extracts are used, often containing mixtures of materials, and in many cases synergism is most likely playing an important role. Clarification of the active constituents and their modes of action will be difficult. This is nevertheless a worthwhile subject for serious investigations.

Dedicated to Sir Derek Barton whose amazing insight helped tremendously in the planning of this series, but who passed away just before its completion. It is a pity that he was unable to write this introduction as originally envisaged, since he would have had a masterful overview of the content he wanted, based on his vast experience. I have tried to fulfill his task, but this introduction cannot do justice to his original intention.

ACKNOWLEDGMENT

I am grateful to current research group members for letting me take quite a time off in order to undertake this difficult writing assignment with hardly any preparation. I am grateful to Drs. Nina Berova, Reimar Bruening, Jerrold Meinwald, Yoko Naya, and Tetsuo Shiba for their many suggestions.

8. BIBLIOGRAPHY

"A 100 Year History of Japanese Chemistry," Chemical Society of Japan, Tokyo Kagaku Dojin, 1978.
K. Bloch, *FASEB J.*, 1996, **10**, 802.
"Britannica Online," 1994–1998.
Bull. Oriental Healing Arts Inst. USA, 1980, **5**(7).
L. F. Fieser and M. Fieser, "Advanced Organic Chemistry," Reinhold, New York, 1961.
L. F. Fieser and M. Fieser, "Natural Products Related to Phenanthrene," Reinhold, New York, 1949.
M. Goodman and F. Morehouse, "Organic Molecules in Action," Gordon & Breach, New York, 1973.
L. K. James (ed.), "Nobel Laureates in Chemistry," American Chemical Society and Chemistry Heritage Foundation, 1994.
J. Mann, "Murder, Magic and Medicine," Oxford University Press, New York, 1992.
R. M. Roberts, "Serendipity, Accidental Discoveries in Science," Wiley, New York, 1989.
D. S. Tarbell and T. Tarbell, "The History of Organic Chemistry in the United States, 1875–1955," Folio, Nashville, TN, 1986.

4.01

Overview of the Biosynthesis of Amino Acids, Peptides, Porphyrins, and Alkaloids with a Focus on the Biosynthesis of Aromatic Amino Acids

JEFFERY W. KELLY

The Scripps Research Institute, La Jolla, CA, USA

4.01.1 INTRODUCTION

This volume covers the exciting advances made in the chemistry, biochemistry, and molecular biology of amino acid, peptide, and alkaloid biosynthesis. The 14 chapters in this volume briefly introduce classical approaches to deciphering biosynthetic pathways. These include feeding experiments where the producing organism is supplemented with isotopically labeled substrates to identify the building blocks and assembly mechanism possibilities. Strains of producing cells with disabling mutations in a given enzyme have also proven to be very useful in identifying key intermediates in the pathway. The chapters in this volume largely describe the contemporary approaches used to establish a given biosynthetic pathway, i.e., applying both chemical and genetic methods to uncover the mysteries associated with the synthesis of some of the world's most beautiful molecules.

Two chapters review our current understanding of alkaloid biosynthesis, which provided much of the impetus for the growth of organic chemistry in the twentieth century. Two chapters are devoted to the biosynthesis of heme prosthetic groups, which have attracted the attention of numerous chemists and biologists over the years. Five chapters focus on peptide and antibiotic biosynthesis, in keeping with the fantastic advances that have been made in understanding non-ribosomal peptide synthesis and the unparalleled contribution that this class of molecules has made to modern medicine. A related chapter on the genetics of antibiotic biosynthesis is also very revealing. One chapter on the biosynthesis of the 21st natural amino acid and the means of its selective incorporation into proteins is representative of the beautiful work done on the biosynthesis of amino acids—the building blocks of life. This volume is finished off with two definitive chapters on the biosynthetic modification of proteins with either GPI or palmitoyl groups, utilized to anchor proteins to membranes.

This introductory chapter will focus on the biosynthesis of amino acids, a subject that is under-represented in the chapters that follow. Since the subject of amino acid biosynthesis is very broad, potentially encompassing a whole volume in its own right, I have decided to focus this introduction on a literature review of aromatic amino acid biosynthesis. Rather than writing a comprehensive review, which is outside the scope of an introductory chapter, the subject will be reviewed so as to provide a perspective of how multidisciplinary and exciting studies on biosynthesis of aromatic amino acids are.

4.01.2 BIOSYNTHESIS OF AROMATIC ACIDS

The biosynthesis of the aromatic amino acids tryptophan (Trp), tyrosine (Tyr), and phenylalanine (Phe) is almost twice as costly in terms of ATP than the synthesis of the remaining α-amino acids, requiring 78, 62, and 65 mol of ATP, respectively, to synthesize 1 mol of these residues.[1] Besides being energetically costly, animals lack the ability to make the aromatic amino acids, although some synthesize Tyr by the hydroxylation of Phe provided by their diet.[2] Bacteria, fungi, and plants are capable of synthesizing all three aromatic amino acids.

The biosynthetic pathways of the aromatic amino acids are among the best understood biosynthetic cascades.[3,4] The majority of the enzymes involved have been at least partially characterized, and more and more information is emerging regarding the organization and control of the genes encoding these enzymes, as well as the modulation of the activity of the enzymes themselves once synthesized. The shikimate pathway beginning with a reaction between erythrose-4-phosphate and phosphoenolpyruvate (PEP) leading ultimately to chorismic acid, the precursor for all three aromatic amino acids, is common in all eukaryotic and prokaryotic organisms (Scheme 1).[3–6] This pathway involves seven different enzyme activities which are invariant in the eukaryotic and prokaryotic organisms investigated thus far.[6,7]

The proteins that catalyze the seven steps of the shikimate pathway appear to be more organized in eukaryotic organisms, which have multifunctional enzymes on just a few polypeptide chains which carry out the analogous reactions carried out by individual polypeptides in the prokaryotes.[3–6] In lower eukaryotes, there is one polypeptide that catalyzes five of the seven reactions of the shikimate pathway.[3] In contrast, *Escherichia coli* encodes five separate proteins each having one enzymatic activity, but interestingly the genes coding for these enzymes are grouped into operons unlike the *Saccharomyces cerevisiae* genes which are scattered throughout the genome. The seven enzyme-catalyzed reactions of the shikimate pathway are encoded by four genes in the yeast *S. cerevisiae*. 3-Deoxy-D-arabinoheptulosonate-7-phosphate (DAHP) synthase, the first enzyme on the pathway, orchestrates an unusual coupling reaction between erythrose-4-phosphate and PEP (see Scheme 1). The *E. coli* genome encodes three isoenzymes of DAHP synthase, the activity of each being regulated by one of the three aromatic amino acids.[4,8–11] In yeast, *S. cerevisiae*, two isoenzymes exist, inhibited either by L-Phe or L-Tyr.[12] Not surprisingly, the isoenzymes exhibit a high degree of similarity.

5-Dehydroquinate (DHQ) synthase catalyzes the second reaction towards chorismic acid which involves an oxidation and an elimination to yield DHQ (see Scheme 1, ii). In *E. coli*, DHQ synthase is a monofunctional enzyme[5,13] whereas, in *S. cerevisiae*, DHQ activity is part of the aromatic multifunctional enzyme which catalyzes reactions ii to vi in Scheme 1.[14–16] There is a 36% identity between the sequences of the *E. coli* and *S. cerevisiae* enzymes.[3]

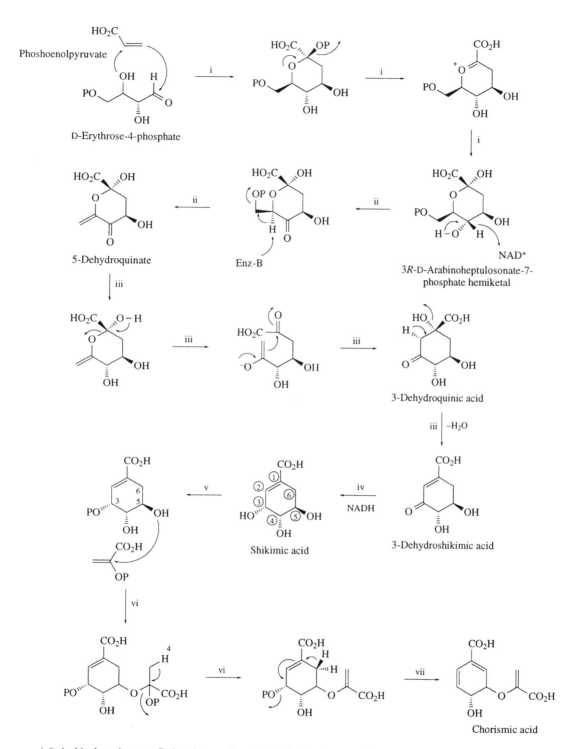

i, D-Arabinoheptulosonate-7-phosphate synthase; ii, 5-Dehydroquinate synthase; iii, 3-Dehydroquinate dehydratase; iv, 3-Dehydroquinate dehydrogenase; v, Shikimate kinase; vi, 5-Enolpyruvyl shikimate 3-phosphate synthase; vii, Chorismate synthase

3-Dehydroquinate (DHQ) dehydratase catalyzes the third reaction in the shikimate pathway that converts DHQ into DHS and introduces the first alkene bond into the carbocyclic ring (Scheme 1, iii). The pentapeptide active site sequences are shared in *E. coli* and *S. cerevisiae*, as well as recognizable sequence similarity between the *E. coli* enzyme and residues 1059–1298 of the *S. cerevisiae* enzyme.[3,17] The fourth step of the pathway is mediated by dehydroshikimate (DHS) dehydrogenase which converts dehydroshikimate into shikimate, the intermediate the pathway was named after, (Scheme 1, iv).[18] The sequence from residues 1306–1588 in *S. cerevisiae* is similar (25% identity) to the *E. coli* DHS dehydrogenase sequence.[3] Shikimate kinase phosphorylates shikimate employing ATP affording shikimate-3-phosphate in the fifth step of the sequence, (Scheme 1, v).[19,20] Residues 886–1059 of the *S. cerevisiae* enzyme share homology to the *E. coli* aro L kinase. The ATP binding site encompasses residues 895–909 in *S. cerevisiae* shikimate kinase. The enzyme 5-enolpyruvyl shikimate 3-phosphate (EPSP) synthase carries out the penultimate reaction, a condensation between PEP and shikimate-3-phosphate, (Scheme 1, vi).[21] The *S. cerevisiae* EPSP synthase domain encompasses residues 404–866 and is very well conserved compared to the *E. coli* EPSP enzyme. These enzymes exhibit 38% identity, except for a spacer region in the *S. cerevisiae* enzyme which has no homology with the *E. coli* enzyme.[3] The EPSP synthase enzyme is the target of glyphosate, the active ingredient of the herbicide Roundup, which has broad spectrum activity against a wide range of plants (see Section 4.01.9).[22] The EPSP enzyme is also one of many enzyme targets of antibacterial agents. Chorismate synthase carries out the final elimination reaction yielding chorismate, which is the key intermediate common to all three aromatic amino acids (Scheme 1, vii).[4–6] The Aro2 gene coding for chorismate synthase is very similar to its *E. coli* counterpart with 47% identity over the entire length of the polypeptide.[23] The *S. cerevisiae* chorismate synthase, like many other chorismate synthases, is a monofunctional enzyme, unlike that of *Neurospora crassa* which is multifunctional.

4.01.3   CHORISMIC ACID—A COMMON INTERMEDIATE

Chorismic acid is an unstable branch point intermediate that serves as the precursor to all the aromatic amino acids in that it can partition into one of two pathways or branches (Scheme 2).[3–6] One path yields Trp, while the other via a Claisen rearrangement yields prephenate with is ultimately transformed into Phe and Tyr (Figure 1). Chorismate is also the precursor to other important aromatic compounds such as ubiquinone, *p*-aminobenzoic acid and vitamin K, which will not be discussed here, but has been reviewed by others.[6]

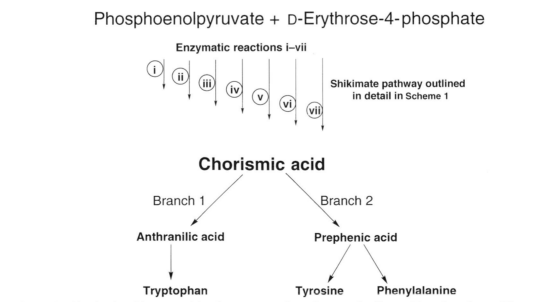

Figure 1   Chorismic acid can partition between two branches, one leading to Trp, the other to Phe and Tyr.

4.01.4 PHENYLALANINE–TYROSINE PATHWAYS

The first committed step in the synthesis of Phe and Tyr in all organisms studied thus far is the pericyclic Claisen rearrangement catalyzed by chorismate mutase.[3,6] This [3,3]-pericyclic reaction likely proceeds, through a six membered chair-like transition state shown in Scheme 2, viii, converting chorismate to prephenic acid.[24] Two separate pathways to Phe and Tyr exist, the exact pathway(s) used depending upon the organism. Phenylalanine can be formed from prephenate proceeding through either arogenate or phenyl pyruvate as intermediates, whereas Tyr may originate from either arogenate or 4-hydroxyphenyl pyruvate (see Scheme 2). In cyanobacteria L-arogenic acid is converted to Phe and in other organisms such as plants arogenate is converted with the help of the coenzyme NAD^+ to Tyr. After surveying a number of organisms, it is clear that all possible biosynthesic pathway combinations to Phe and Tyr are observed in nature.

In *S. cerevisiae* and *E. coli* the Phe/Tyr branch utilizes five enzymatic reactions. The first reaction already alluded to is the pericyclic reaction catalyzed by chorismate mutase that yields prephenate (Scheme 2, viii). In several yeasts, including *S. cerevisiae*, the enzyme is a ∼30 kDa monomeric protein.[16,25,26] The *S. cerevisiae* chorismate mutase is activated by Trp, the end product of the other branch (see Scheme 3), and strongly inhibited by Tyr, one of the amino acids prephenate is ultimately converted into.[25] The exact enzyme regulation scheme varies with the organism being considered, including those organisms which do not appear to regulate chorismate mutases. *E. coli* employs a chorismate mutase that is fused to the second enzyme in either the Phe or the Tyr pathways.[27,28] In both bifunctional enzymes, the N-terminus possesses chorismate mutase activity; however, there is no discernible sequence homology amongst the *S. cerevisiae* and *E. coli* amino acid sequences that encode for chorismate mutase activity.[3]

Prephenate is the substrate for the enzyme prephenate dehydratase which affords phenylpyruvic acid (see Scheme 2, ix). Prephenate also serves as a substrate for prephenate dehydrogenase, leading to the α-ketoacid 4-hydroxyphenyl pyruvic acid (see Scheme 2, x). In *E. coli* the chorismate mutase and the prephenate dehydratase activity leading to the α-keto acid are encoded by a single polypeptide chain.[29] The same is true, for the chorismate mutase and prephenate dehydrogenase activity in *E. coli* that affords the α-keto acid precursor to Tyr.[27,28] In the utilization of arogenate to form Phe and Tyr, the transamination reaction (see next paragraph) on the α-keto group occurs first (Scheme 2 xiii) followed by aromatization.[3,5,6] The α-amino acid arogenate then is enzymatically transformed into Phe by arogenate dehydratase (Scheme 2, xiv). An analogous aromatization reaction, this time eliminating hydride instead of water is mediated by arogenate dehydrogenase and the coenzyme NAD^+ affording Tyr (see Scheme 2, xv). In cyanobacteria, the arogenate pathway is the major pathway utilized for Tyr (Scheme 2, xiii and xv). Interestingly, *Pseudomonas* utilizes all possible pathways to Phe and Tyr.[7]

The transaminases that transform the α-keto acids to Phe and Tyr have historically proven difficult to identify and isolate by the classical approach of knocking out one enzyme at a time with mutagenesis. This is so because there are multiple transaminase enzymes, each of which can process multiple α-keto acids, albeit at different rates and with varying K_m values.[30–32] Significant progress has been made in identification of the aminotransferases that facilitate aromatic amino acid biosynthesis in *E. coli*.[33–42] The major *E. coli* transaminases, coded by the aspC, ilvE, and tyrB genes, can amininate the α-keto acid precursors of Phe, Tyr, and Trp and have a surprisingly broad range of substrate acceptability (Scheme 2, xi and xii). The Tyr B gene codes for the aromatic aminotransferase that is able to complement *E. coli* strains that are deficient in these major aminotransferases. These aminotransferases utilize Asp as a source of NH_2.[38] Expression of the tyrB gene product is controlled at the transcriptional level by the TyrR protein, which when bound to Tyr acts as a corepressor.[42] The greater than 30 aminotransferase sequences, of which the Asp family of aminotransferases is a subset, exhibit 35% sequence identity.[41] The yeast aromatic aminotransferases, encoded by the ARO8 and ARO9 genes of *S. cerevisiae*, were isolated by complementation and code for 500 and 513 amino acid polypeptides, respectively.[30] The expression of the ARO8 gene is subject to the general control of amino acid biosynthesis (see Section 4.01.8). The expression of ARO9 is induced when aromatic amino acids are present in the growth media as well as ammonia.

4.01.5 TRYPTOPHAN BRANCH

The biosynthesis of Trp proceeds by five invariant steps in all organisms studied thus far (Scheme 3). However, as is so often the case, when organisms are compared, the distribution of activities as

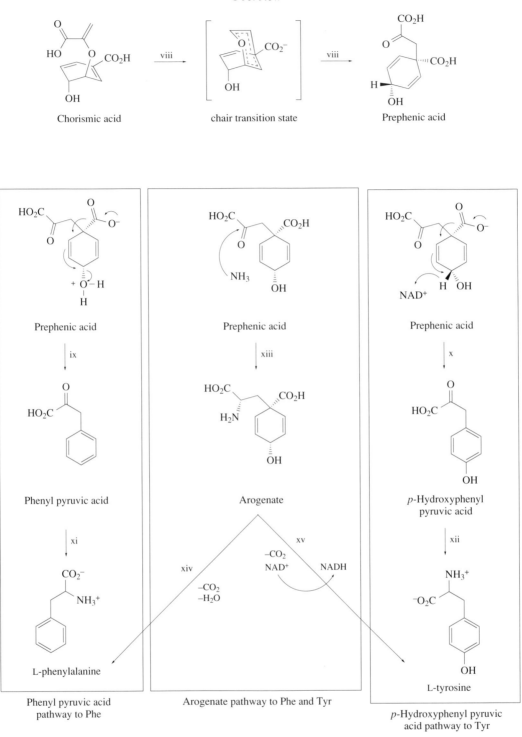

viii, Chorismate mutase; ix, Prephenate dehydratase; x, Prephenate dehydrogenase; xi, Transaminases;
xii, Transaminases; xiii, Transaminases; xiv, Arogenate dehydratase; xv, Arogenate dehydrogenase

Scheme 2

separate proteins is highly variable. The first committed step is the conversion of chorismate to
anthranilate (see Scheme 3, xvi). The anthranilate synthase that carries out this reaction utilizes
glutamine as a source of NH_3 and is feedback inhibited by the ultimate product, tryptophan. The
enzyme in yeast *S. cerevisiae* and in *E. coli* is a heterodimer, one polypeptide chain performing the

amino transferase activity, i.e., producing the nucleophilic NH_3 component from Gln.[43] The other polypeptide chain effects the synchronous conjugate addition of "NH_3" and elimination of H_2O. This polypeptide in *S. cerevisiae* also contains the Trp binding site for feedback inhibition. An amino acid sequence alignment of the *E. coli* and *S. cerevisiae* enzyme exhibits nine short sequences with high homology, with the remainder of the sequences lacking homology. The second step in the biosynthesis of Trp is the nucleophilic attack of the aniline group of anthranilate on the electrophile 5-phosphoribosyl pyrophosphate catalyzed by the enzyme phosphoribosyltransferase yielding phosphoribosylanthranilate (PRA) (see Scheme 3, xvii).[44,45] The yeast *S. cerevisiae* enzyme is a monomer whereas the *E. coli* phosphoribiosyltransferase is a component of a multifunctional protein chain. About half of the residues are identical in two domains of about 50 amino acids each in both the *E. coli* and *S. cerevisiae* proteins. In reaction iii of Trp biosynthesis, PRA isomerase orchestrates the Amadori reaction, a 1,2 hydride shift affording the ketone carboxyphenylamino-1-deoxy ribulose *S*-phosphate (CDRP) (see Scheme 3, xviii).[46] In *E. coli* and other bacteria the PRA isomerase is fused to the next enzyme on the pathway, i.e., GP synthase. However, in the yeast *S. cerevisiae* PRA isomerase is a monofunctional enzyme.[3] A comparison of the PRA isomerase sequences in yeast and *E. coli* reveals a high degree of homology. The structures of several PRA isomerases have been shown crystallographically to be a TIM barrel.[47] The penultimate step in Trp biosyntheses involves the decarboxylation of CDRP and the closure of the second ring to yield indole-3-glycerol-phosphate (InGP), catalyzed by InGP synthase. InGp activity in *S. cerevisiae* is part of the bifunctional enzyme encoded by the TPP3 gene.[45] The InGP synthases also fold into a TIM (triose phosphate isomerase) barrel three-dimensional structure.

In the last catalyzed reaction affording Trp, tryptophan synthase lowers the activation free energy for the aldol cleavage and facilitates a reaction between indole and the Schiff base of dehydroalanine derived from serine to form the rarest amino acid in proteins, Trp. The enzyme has two active sites, one cleaves off the side chain via an aldol cleavage and the other orchestrates the condensation of indole and serine (via dehydroalanine), which affords L-tryptophan. In the yeast *S. cerevisiae* both activities are on a single 76 kDa polypeptide chain, whereas in *E. coli* and most other organisms these activities are on separate polypeptide chains. These single-chain enzymes can, in some cases, associate into an efficient quaternary structural enzyme complex.[48,49]

4.01.6 REGULATION OF AROMATIC AMINO ACID BIOSYNTHESIS

Aromatic amino acid biosynthesis is regulated largely by two mechanisms. First, at the level of transcription and, to a lesser extent, translation of the enzymes involved in aromatic amino acid biosynthesis. Second, amino acid biosynthesis is controlled by modulating the enzyme activities that dictate whether Phe, Tyr, or Trp or some combination thereof are made. In *E. coli* the amino acid biosynthesis genes are organized into operons that are controlled by proteins that bind the operon as a result of an interaction with a given ligand. Binding of a repressor–amino acid complex to an operator prevents the transcription of that operon. In yeast such as *S. cerevisiae*, the genes that code for enzymes of aromatic amino acid biosynthesis are not organized into gene clusters, instead they are scattered through the genome. The GCN4 coiled-coil DNA binding protein regulates the majority of the genes involved in aromatic amino acid biosynthesis. GCN4 is involved in regulation by derepressing at least 30 structural genes involved in multiple amino acid biosynthetic pathways (see Section 4.01.8).

Enzyme function is regulated by changing enzyme activity via binding of small molecules on the aromatic amino acid biosynthetic pathway via an allosteric mechanism. This is accomplished by modulating activity at the first step in the pathway (DAHP synthase) and at the branch points, which differ slightly depending on which organism is being considered. An invariant example is the regulation of anthranilate synthase, the first committed enzyme that partitions carbon flow towards Trp, which is feedback inhibited by the end product of this branch, Trp. The utilization of the end products of each branch to control carbon flow into that branch by modifying enzyme activity is the typical mechanism utilized.

4.01.7 *E. COLI* AROMATIC AMINO ACID BIOSYNTHESIS

The control of carbon flow into the *E. coli* aromatic amino acid biosynthetic pathway starts with three DAHP syntheses, one inhibited by Phe, one by Tyr, and one by Trp. It appears that the Phe

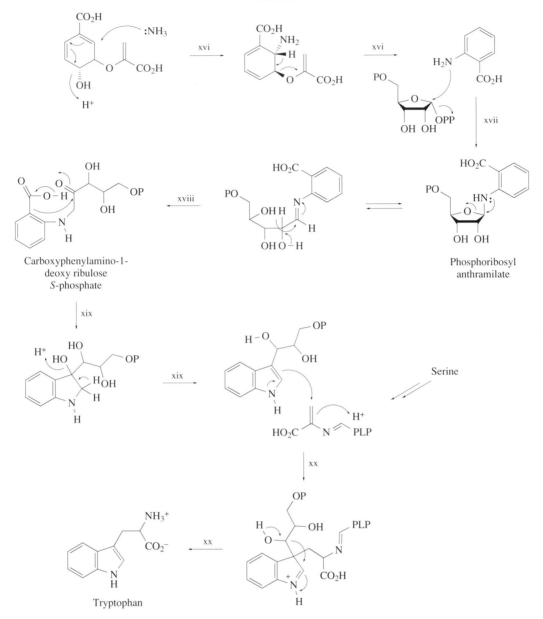

xvi, Anthranilate synthase; xvii, Phosphoribosyl transferase; xviii, Phosphoribosyl anthranilate isomerase; xix, InGP synthase; xx, Tryptophan synthase

Scheme 3

DAHP synthase is sensitive to multivalent repression by Phe and Trp in combination. In *E. coli* the bifunctional enzyme chorismate mutase–prephenate dehydrogenase partitions chorismate to Tyr while the chorismate mutase *p*-prephenate dehydratase commits carbon flow to Phe, (see Scheme 2). The *E. coli* chorismate mutase–prephenate dehydrogenase is feedback inhibited by Tyr whereas the chorismate mutase–prephenate dehydratase is feedback inhibited by Phe.[50] The Tyr repressor is involved in controlling the enzyme concentrations of chorismate mutase T-prephenate dehydrogenase and the Tyr sensitive DAHP synthase by repressing transcription. The Tyr repressor also appears to inhibit transcription of the Phe-modulated DAHP synthase. More on the regulation of the aromatic amino acid synthase by transcriptional control can be found in Section 4.01.8.

The partitioning of metabolites into the Trp pathway is governed by the inhibition of anthranilate synthase by Trp. Genetic studies on Trp biosynthesis in *E. coli* reveals that there is a repressor for the Trp operon that controls the transcription of the five enzymes which orchestrate Trp biosynthesis. The Trp repressor senses the level of Trp and controls transcription accordingly. The Trp repressor upon sensing Trp binds to the Trp operator preventing RNA polymerase from binding, thus inhibiting transcription. Transcriptional termination also controlled by Trp levels is another rheostat of transcriptional control. In addition, there are two general metabolic control systems that have control over Trp operon transcription. One is the rel A locus which controls ribosomal RNA synthesis in response to amino acid starvation. The second kind of metabolic control increases transcription as the growth rate increases, which appears to be affected by a decrease in premature termination.

In *E. coli*, TyrR plays a central role in aromatic amino acid biosynthesis by activating or repressing the expression of at least eight unlinked operons.[51,52] TyrR represses transcription in the presence of Tyr and ATP and activates transcription in the presence of Phe of these operons by binding TyrR boxes, having the palindromic consensus sequence TGTAAANNNNNNTTTACA. Two types of Tyr boxes have been identified, strong boxes which have more than 10 base pairs of consensus sequence and which bind TyrR in the absence of amino acid and weak boxes, which require Tyr and a neighboring strong box for TyrR binding.[52] The weak box resides in the RNA polymerase binding region whereas the strong box lies outside. Only in the presence of Tyr and ATP are both Tyr boxes occupied. The TyrR protein is composed of 513 amino acids and is a homodimer in solution, but becomes hexameric upon binding Tyr and ATP allowing it to bind to multiple TyrR boxes.[53]

4.01.8 THE CONTROL OF AROMATIC AMINO ACID BIOSYNTHESIS IN *S. CEREVISIAE*

The temporal activity of the aromatic amino acid biosynthetic enzymes is regulated by modulation of enzyme activity and by gene expression. The latter is controlled by a variety of parameters such as initiation and termination of transcription, as in the case of operon regulation in *E. coli*. The genes of aromatic amino acid biosynthesis are not organized into operons in *S. cerevisiae*. The aromatic amino acid genes distributed throughout the yeast genome are regulated largely through regulatory protein binding to the 5′ ends of the genes.

In contrast to bacteria where the presence of a given amino acid in the cell generally shuts down transcription of the enzymes in that pathway, eukaryotic cells like *S. cerevisiae* maintain a basal level of gene expression irrespective of the amino acid concentration in the cell. This is affected by the organization of promoters and trans-acting regulators. Deprivation of a given amino acid in a yeast cell results in the upregulation of the transcription, not only of the genes coding for that specific amino acid biosynthetic pathway, but for many of the genes involved in the biosynthesis of a wide variety of amino acids. This type of regulation is known as the general control regulatory system.[54] The starvation response in *S. cerevisiae* is mediated by GCN4, which is a transcriptional activator protein which binds DNA (TGATCT) and is responsible for general control of transcription of the amino acid biosynthetic pathway. GCN4 is also responsible for the basal level control of some of the amino acid biosynthetic genes such as ARO3.[55] The general and basal promoters are differentiated by additional upstream regulatory protein binding sites. While there is little evidence of direct translational control of aromatic amino acid biosynthesis, interestingly the transcriptional activator GCN4 appears to be translational regulated by the small open reading frames in the leader regions where several proteins bind.

The regulation of enzyme activity plays a major role in the control of carbon flow through a given aromatic amino acid pathway. The initiation of aromatic amino acid biosynthesis is facilitated by two DAHP synthes, one inhibited by Tyr and the other by Phe. At the first branch point, Trp biosynthesis is catalyzed by anthranilate synthase which is feedback inhibited by Trp and the chorismate mutase which yields the prephenate intermediate common to both the Phe and Tyr pathways in feedback inhibited by Tyr and strongly activated by Trp binding.

4.01.9 HERBICIDE ACTION AND AROMATIC AMINO ACID BIOSYNTHESIS

Several practical developments have resulted from our current understanding of aromatic amino acid biosynthesis. The product that nearly everyone has encountered in one way or another is the herbicide Roundup. Glyphosate (Roundup) is sold by Monsanto as a broad spectrum herbicide that is very effective in killing a wide variety of plants. Glyphosate inhibits EPSP synthase, (see Scheme 1, vi), severely disrupting the carbon flow through the aromatic amino acid pathway ultimately killing the plant.[56] The inhibitor functions by forming a ternary complex with the synthase and the substrate, shikimate 3-phosphate. Several lines of evidence indicate that glyphosate binds to a subsite that prevents the other component of the enzymatic reaction, PEP from binding (Scheme 1, vi). Interestingly, glyphosate inhibits the EPSP synthase reverse reaction when both substrates are bound (EPSP and P_i).[57] Data indicate that glyphosate and PEP do not bind to overlapping sites, as initially thought. Instead, it appears that glyphosate inhibits EPSP synthase by an allosteric mechanism, i.e., a binding-induced conformational change making the EPSP synthase active site unable to utilize PEP.[56] While the author is a firm supporter of structure-based drug design, it is clear from what is now known, that it is unlikely that Roundup would have been discovered via a structure-based approach. Hundreds of analogues of glyphosate prove less effective, likely because of the complex allosteric inhibition mechanism that may be clarified with a structure of the ternary complex.

4.01.10 SUMMARY

Even though our current understanding of aromatic amino acid biosynthesis is quite impressive, it is clear that the complete genome sequence of several organisms coupled to the ability to get a complete readout of mRNA produced as a result of a given stimulus will significantly increase the sophistication of our understanding of these pathways. The mechanisms of regulation will undoubtedly be more complex than is currently appreciated. Furthermore, a structural and mechanistic understanding of several of the enzymes is lacking. Further structural and mechanistic information will greatly increase our appreciation for these catalysts. This is especially true for the multifunctional enzymes and the complex quaternary structural forms of the biosynthetic enzymes which appear to act as an assembly line.

4.01.11 THE FUTURE OF BIOSYNTHESIS

The late 1990s is a very interesting time for biosynthetic studies. It is absolutely clear that a combination of chemistry, molecular biology, structual biology, enzymology, and functional genomics knowledge is required to effectively study and significantly extend current understanding of biosynthetic pathways in the twenty-first century. The time is ripe to apply all of these modern strategies to the study of biosynthetic pathways, as this will undoubtedly lead to advances in molecular medicine far beyond what we can now envision. Young scientists reading this volume should seriously consider biosynthesis as a futuristic research area. Individuals who think that the heyday of biosynthesis is over will be definitively proven wrong in the twenty-first century.

4.01.12 REFERENCES

1. D. E. Atkinson, "Cellular Energy Metabolism and its Regulation," Academic Press, New York, 1997.
2. E. Haslam, "The Shikimate Pathway," Butterworths, London, 1974.
3. G. H. Braus, *Microbiol. Rev.*, 1991, **55**, 349.
4. H. E. Umbarger, *Ann. Rev. Biochem.*, 1978, **47**, 533.
5. A. J. Pittard, in "Biosynthesis of the Aromatic Amino Acids," *Escherichia coli Salmonella typhimurium*, ed. F. C. Neidhardt, American Chemical Society, 1987.
6. K. B. G. Torssell, "Natural Products Chemistry," Kristianstads Boktryckeri, Stockholm, 1997.
7. K. M. Herman and R. L. Somerville, "Amino Acids: Biosynthesis and Regulation," Addison-Wesley, Reading, MA, 1983.
8. R. Simpson and B. Davidson, *Eur J. Biochem.*, 1976, **70**, 493.
9. I. Dusha and G. Denes, *Biochim. Biophys. Acta.*, 1976, **438**, 563.
10. R. Schoner and K. Herrmann, *J. Biol. Chem.*, 1976, **251**, 5440.
11. H. Camakaris and J. Pittard, *J. Bacteriol.*, 1974, **120**, 590.
12. F. Lingens, W. Goebel, and H. Uesseler, *Eur. J. Biochem.*, 1967, **1**, 363.
13. J. Pittard, in "The regulation of the biosynthesis of aromatic amino acids and vitamins in *Escherichia coli* K-12," Genetics: New Frontiers, Proceedings of the 15th International Congress, Oxford IBH Publishing Co., Oxford, 1984.
14. A. de Leeuw, *Genetics*, 1967, **56**, 554.
15. F. W. Larimer, C. C. Morse, A. K. Beck, K. W. Cole, and F. H. Gaertner, *Mol. Cell. Biol.*, 1983, **3**, 1601.
16. R. K. Mortimer, D. Schild, C. R. Contopoulou, and J. A. Kans, *Yeast*, 1989, **5**, 321.
17. G. Millar and J. R. Coggins, *FEBS Lett.*, 1986, **200**, 11.
18. B. D. Davis, *Adv. Enzymol.*, 1955, **16**, 287.
19. H. G. Griffin and M. J. Gasson, *DNA Sequence*, 1995, **5**, 195.
20. B. Ely and J. Pittard, *J. Bacteriol.*, 1979, **138**, 933.

21. G. R. Welch and F. H. Gartner, *Curr. Top. Cell. Regul.*, 1980, **16**, 113.
22. G. Kishore, D. Shah, S. Padgette, G. Della-Cioppa, C. Gasser, D. Re, C. Hironaka, M. Taylor, J. Wibbenmeyer, *et al.*, *ACS Symp. Ser.*, 1988, **379**, 37.
23. D. G. L. Jones, U. Reusser, and G. H. Braus, *Mol. Microbiol.*, 1991, **5**, 2143.
24. A. Y. Lee, J. D. Stewart, J. Clardy, and B. Ganem, *Chem. Biol.*, 1995, **2**, 195.
25. P. Kradolfer, J. Zeyer, G. Miozzari, and R. Hutter, *FEMS Microbiol Lett.*, 1977, **2**, 211.
26. T. Schmidheini, P. Sperisen, G. Paravicini, R. Hutter, and G. Braus, *J. Bacteriol.*, 1989, **171**, 1245.
27. J. Dayan and D. B. Sprinson, *J. Bacteriol.*, 1971, **108**, 1174.
28. G. L. E. Koch, D. C. Shaw, and F. Gibson, *Biochim. Biophys. Acta*, 1971, **229**, 795.
29. S. W. K. Im and J. Pittard, *J. Bacteriol.*, 1971, **106**, 784.
30. I. Iraqui, S. Vissers, M. Cartiaux, and A. Urrestarazu, *Mol. Gen. Genet.*, 1998, **257**, 238.
31. R. J. Whitaker, C. G. Gaines, and R. A. Jensen, *J. Biol. Chem.*, 1982, **257**, 13550.
32. D. H. Gelfand and R. A. Steinberg, *J. Bacteriol.*, 1977, **130**, 429.
33. R. Marquardt, J. Then, H. M. Deger, G. Woehner, M. Robinson, and A. Doherty, "Cloning of the transaminase gene tyrb of *Escherichia coli* and its use in bacterial manufacture of phenylalanine," *AAA: Eur. Pat. Appl.*, 1988.
34. C. Mavrides and W. Orr, *Biochim. Biophys. Acta*, 1974, **336**, 70.
35. C. Mavrides and W. Orr, *J. Biol. Chem.*, 1975, **250**, 4128.
36. G. Pan and P. Ouyang, *Nanjing Huagong Xueyuan Xuebao*, 1994, **16**, 33.
37. N. Monnier, A. Montmitonnet, S. Chesne, and J. Pelmont, *Biochimie*, 1976, **58**, 663.
38. M. Robinson, R. Marquardt, J. Then, J. McChesney, H. Neatherway, G. Woehner, H. M. Deger, and P. Praeve, *Biotechnol. Lett.*, 1987, **9**, 673.
39. R. F. Senkpeil, D. P. Pantaleone, I. G. Fotheringham, and P. P. Taylor, in "Production of unnatural amino acids via whole-cell bioconversion with transaminases cloned in *E. coli*." Book of Abstracts, 213th ACS National Meeting, San Francisco, April 13–17, 1997.
40. Y. Shi, W. Wu, Y. Zhou, Y. Zhang, and T. Gong, *Yaowu Shengwu Jishu*, 1994, **1**, 8.
41. M. H. Sung, *Saenghwahak Nyusu*, 1991, **11**, 34.
42. J. Yang and J. Pittard, *J. Bacteriol.*, 1987, **169**, 4710.
43. F. A. Prantl, A. Strasser, M. Aebi, R. Furter, P. Neiderberger, K. Kirschner, and R. Hutter, *Eur. J. Biochem.*, 1985, **146**, 95.
44. U. Hommel, A. Lustig, and K. Kirschner, *Eur. J. Biochem.*, 1989, **180**, 33.
45. C. Yanofsky, T. Platt, I. P. Crawford, B. P. Nichols, G. E. Christie, H. Horowitz, M. van Cleemput, and A. M. Wu, *Nucleic Acids Res.*, 1981, **9**, 6647.
46. G. H. Braus, K. Luger, G. Paravicini, T. Schmidheini, K. Kirschner, and R. Hutter, *J. Biol. Chem.*, 1988, **263**, 7868.
47. J. P. Priestle, M. P. Grutter, J. White, M. G. Vincent, M. Kania, E. Wilson, T. S. Jardetzky, K. Kirschner, and J. N. Jansonius, *Proc. Natl. Acad. Sci. USA*, 1987, **84**, 5690.
48. E. W. Miles, R. Bauerle, and S. A. Ahmed, *Methods Enzymol.*, 1987, **142**, 398.
49. M. Dettwiler and K. Kirschner, *Eur. J. Biochem.*, 1979, **102**, 159.
50. R. G. H. Cotton and F. Gibson, *Biochim Biophys. Acta*, 1965, **100**, 76.
51. T. Kwok, J. Yang, A. J. Pittard, T. J. Wilson, and B. E. Davidson, *Mol. Microbiol.*, 1995, **17**, 471.
52. A. J. Pittard and B. E. Davidson, *Mol. Microbiol.*, 1991, **5**, 1585.
53. T. J. Wilson, P. Maroudas, G. J. Howlett, and B. E. Davidson, *J. Mol. Biol.*, 1994, **238**, 309.
54. A. G. Hinnebusch, *Crit. Rev. Biochem.*, 1986, **21**, 277.
55. G. Paravicini, H. U. Moesch, T. Schmidheini, and G. Braus, *Mol. Cell. Biol.*, 1989, **9**, 144.
56. J. A. Sikorski and K. J. Kruys, *Acc. Chem. Res.*, 1997, **30**, 2.
57. M. R. Boocock and J. R. Coggins, *FEBS Lett.*, 1983, **154**, 127.

4.02
Protein Palmitoylation

MICHAEL H. GELB and LI LIU
University of Washington, Seattle, WA, USA

and

THOMAS DUDLER
Imperial Cancer Research Fund, London, UK

4.02.1 INTRODUCTION

Many proteins are modified in cells by covalent attachment of a variety of functional groups. Modifications such as phosphorylation and glycosylation have been known for many years. Lipidation of proteins was discovered in the 1970s and most other discoveries were made in the 1980s and 1990s. This chapter describes proteins that contain palmitoyl groups acyl linked to amino acid residues of proteins. Protein lipidations not covered in this chapter include N-terminal myristoylation, which was discovered in the early 1980s,[1–4] prenylation (see Chapter 2.13), which was discovered in the early to late 1980s,[5,6] attachment of glycosyl phosphatidylinositol anchors to proteins, which was discovered in the 1960s,[7] and other less common protein lipidations.[8,9] Proteins that contain covalently attached palmitoyl groups may contain significant amounts of fatty acyl groups other than palmitoyl (see Section 4.02.2.2). Throughout this chapter, these proteins will be referred to as palmitoylated proteins.

Probably the first reports of palmitoylated proteins are the organic soluble proteolipids found in myelin and sarcoplasmic reticulum.[10,11] Later, viral glycoproteins were shown to be palmitoylated.[12] Covalent incorporation of radiolabeled palmitic acid into proteins in a variety of cells was reported by several groups.[13–20] As will be documented in this chapter, a large number of palmitoylated proteins have been identified, and the functions of the lipidations are becoming better understood in a few cases. In general, the biochemical processes by which proteins are palmitoylated is poorly understood. In some examples, palmitoylation of specific proteins has been shown to be critical for the function of the protein in the intact cell. In other cases, the roles of palmitoylation remain obscure because the nonpalmitoylated proteins retain those functions analyzed by the experimentalist. Earlier reviews on protein palmitoylation have appeared.[21–40]

4.02.2 CHEMICAL STRUCTURES OF PROTEIN PALMITOYL GROUPS

4.02.2.1 Protein–Palmitoyl Linkages

Early studies of palmitoylated proteins including myelin proteolipid and viral envelope glycoproteins showed that the fatty acyl group was released from the protein by treatment with mild alkali, neutral hydroxylamine, or sodium borohydride.[12,41–43] This suggests that the fatty acyl group is ester linked to the protein; amides are usually stable under these conditions. Sodium borohydride and neutral hydroxylamine usually cleave thiolesters and not oxyesters, but it is risky to make a distinction between linkage of the fatty acyl group to cysteine versus serine or threonine based on these reactivities, especially with the latter reagent.[22,44] Fatty acyl groups attached to cysteine residues are cleaved by neutral thioethanol or dithiothreitol,[45] but again this should not be taken as definitive evidence for a thiolester linkage. Perhaps the best one can do with these methods is to employ all of them to see if a consensus for thiol- versus oxy-ester is obtained.[46]

In order to determine the type of linkage between protein and fatty acyl group unambiguously, direct structural data are required. Early direct structural studies could be carried out only in cases where relatively large amounts of palmitoylated protein were available. Virtually unlimited amounts of myelin proteolipid are available, and the entire protein has been sequenced by *N*-terminal degradation of proteolytic fragments.[47] The attached fatty acid was localized to a particular tryptic peptide by analyzing saponified purified peptides for fatty acids by thin layer chromatography.[48] This peptide was subjected to Edman degradation, candidate fatty acylated amino acid phenylthiohydantoins were saponified, and gas chromatography was used to analyze for fatty acids. This analysis showed that threonine198 contains attached palmitic acid, stearic acid, and oleic acid in a 1.3 : 1.0 : 1.0 mole ratio.[47] Clearly this approach is tedious and requires multimilligram amounts of protein. Mack and Kruppa isolated the cytoplasmic domain of vesicular-stomatitis virus glycoprotein, which was known to contain a fatty acyl group, and treated the peptide with radiolabeled iodoacetic acid before and after saponification of the peptide.[49] A single cysteine, which was identified by Edman degradation, became radioalkylated only after the peptide was saponified, which strongly suggests that the fatty acyl group is attached to this cysteine. Likewise, Schmidt *et al.* isolated a radiopalmitoylated chymotryptic peptide from the Semliki Forest virus transmembrane protein E1 and showed that radiolabeled iodoacetamide was incorporated into the peptide only after it was saponified; from the known sequence of this protein, it could be inferred that the acylated peptide has only one cysteine.[44]

Deduction of the location and type of palmitoyl linkage in proteins that are available in relatively small amounts is usually carried out by studies in which the candidate cysteines are individually changed to serines by site-directed mutagenesis. For example, the site of acylation of the transferrin receptor has been inferred from such an approach.[50] However, lack of incorporation of fatty acid into a mutant protein in which a specific cysteine has been replaced with serine is only circumstantial evidence for the acylation of this cysteine residue. The GTP-binding proteins N-Ras and H-Ras were thought for several years to be palmitoylated on cysteine186 because mutation of this residue to serine led to proteins that failed to become palmitoylated in cells and to localize to the membrane. However, it is now clear that these proteins are first farnesylated on cysteine186, and palmitoylation, which requires farnesylation as a prerequisite, occurs on one or more nearby cysteines.[5]

There are a few reports of the use of mass spectrometry to analyze palmitoylated proteolytic fragments of lipidated proteins, and this method is expected to be a powerful tool for analyzing numerous palmitoylated proteins. Tandem mass spectrometry in which the peptide is fragmented at each amide linkage has been used to show that bovine rhodopsin *C*-terminal peptide 241–327 contains adjacent palmitoylated cysteines.[51] Electrospray mass spectrometry has been used to locate the bispalmitoylated-cysteinyl unit in lung surfactant SP-C protein[52] and to locate the palmitoylation site in the transmembrane protein p12E of Friend murine leukemia virus.[53] The ε-amino groups of internal lysine residues of the bacterial proteins adenylate cyclase toxin and hemolysin were shown by mass spectrometry to be palmitoylated.[54,55]

Table 1 gives a listing of most of the reported palmitoylated proteins. Where available, data concerning the type of acyl linkage, the site of acylation, and the fatty acid composition are given.

4.02.2.2 Fatty Acyl Chains

It is often said that protein palmitoylation, unlike myristoylation, is less specific in that other fatty acids such as stearic acid and oleic acid are found acyl linked to proteins in addition to palmitic acid. However, this statement may not be accurate in general because the fatty acid compositions of palmitoylated proteins have been properly analyzed in only a few cases (these are indicated in Table 1). Many studies rely on labeling the protein of interest by adding pure molecular species of radiolabeled fatty acids, and detailed experimental protocols have been reported.[45,129] There are three problems with this method. First, fatty acids are interconvertible in cells by elongation and degradation pathways, and they are also metabolized to amino acids. This leads to partial scrambling of the radiolabel into a mixture of fatty acids.[22,130] Secondly, the pool size of acyl-CoA species in cells depends on the acyl group, and thus the specific radioactivity of different acyl-CoA species in cells can be very different. Thirdly, incorporation of a particular radiolabeled fatty acid into a protein does not give the fatty acid composition of the protein, but only shows that it is possible for the labeled fatty acid to incorporate into the protein. These concerns are not without consequences, for example, src kinase from Rous sarcoma virus was originally thought to be palmitoylated, but is now known to be myristoylated.[22,131]

Here again, direct structural data is needed to determine the fatty acid composition of the palmitoylated protein. Gas chromatography of fatty acid methyl esters derived from methanolysis of purified acylated protein has been the method of choice. A limitation of this method is that 5–10 nmol of highly purified acylated protein is needed, and even then extreme care has to be taken to avoid contamination of the sample by noncovalently bound fatty acids, which are present essentially everywhere.[46] Modern mass spectrometry methods can detect subpicomole to picomole amounts of peptides, and this method may hold promise for fatty acid analysis of palmitoylated proteins, especially if novel methods are developed for the handling and purification of lipidated peptides.

4.02.3 ATTACHMENT OF PALMITOYL GROUPS TO PROTEINS

Very little is known about the enzymology of protein palmitoylation. Since nonenzymatic transfer of the palmitoyl group from a thiolester such as palmitoyl-CoA (the likely donor of protein palmitoyl groups) to a cysteine occurs at a measurable rate *in vitro*, there has been a lively debate about whether protein palmitoylation in cells occurs via a spontaneous process or via an enzyme-catalyzed one. In 1987 it was suggested that the acylation of rat brain myelin proteolipid protein is an enzymatic reaction.[132,133] In their experiment fresh myelin membrane isolated from rat brain was incubated with ATP, $MgCl_2$, CoA, and ^3H-palmitic acid in 20 mM tris buffer pH 7.4 containing 10 mM $MgCl_2$. They found the following: (i) maximum incorporation of ^3H-palmitic acid into proteolipid occurs at physiological temperature and decreases with increasing temperature; (ii) this acylation reaction is severely inhibited by SDS (0.05%); (iii) incorporation of the fatty acid or palmitoyl-CoA into proteolipid was substantially decreased by the process of freezing-and-thawing and by lyophilization of myelin; and (iv) the acylation velocity displays hyperbolic kinetics with respect to the concentration of acyl-CoA (apparent K_M for palmitoyl-CoA is 41 μM). In contrast, in 1988 Ross and Braun reported that the same acylation reaction is nonenzymatic.[134] Their evidence was the following: (i) acylation increases with the increasing pH above 7.5; (ii) heating myelin at 80 °C for 1 h does not inhibit acylation; and (iii) proteolipid could still be acylated after it was extracted with organic solvent or Triton X-100 and even after this material was treated with trypsin to convert it into two fragments.

Slomiany and co-workers studied the palmitoylation of gastric mucus glycoprotein and have provided evidence that the process is enzymatic.[135] The enzymatic activity appears to be associated with a golgi-rich membrane fraction and has been solubilized with Triton X-100. The enzyme has been purified to apparent homogeneity and displays 65 kDa and 67 kDa protein bands on SDS–PAGE (sodium dodecyl sulfate–polyacrylamide gel electrophoresis).[135] Values for K_M for mucus glycoprotein and palmitoyl-CoA are 4.5×10^{-7} M and 3.8×10^{-5} M, respectively. The enzyme is capable of palmitoylating the deglycosylated glycoprotein, but does not act on proteolytically degraded mucus glycoprotein.

In 1984, Berger and Schmidt provided strong evidence that the viral glycoprotein E1 is enzymatically palmitoylated. In contrast, these authors also suggested that the viral glycoprotein E2 is nonenzymatically palmitoylated,[136] since this reaction still occurs after the membranes were boiled. The enzyme that acts on E1 was successfully solubilized with neutral detergents (NP40 or Triton X-100).[137,138] In 1995, the same group reported that human erythrocyte ghosts have palmitoylating

Table 1   Palmitoylated proteins.[a]

Protein	Function of protein	Acyl linkage[b]	Acyl chain[c]	Location of acyl chain(s)	Function(s) of acyl chain	Ref.
Adenylate cyclase toxin (*Bordetella pertussis*)	Virulence factor (cytotoxic and hemolytic)	Amide linkage to the ε-amino group of lysine	16:0 (MS)	Lys983	Required for cytotoxic and hemolytic activities	56
Alpha 2A adrenergic receptor	Signal transduction	TE (SDM)	16:0 (RL)	Cys442	Receptor down regulation	57
Alpha crystallin	Major lens protein	?	16:0 (RL)	?	?	58
Asialoglycoprotein receptor	Cellular uptake of glycoconjugates	TE (HA)	16:0, 18:0 (MS)	?	Receptor reactivation	59
Band 3 protein	Anion exchange in erythrocytes	TE (MS)	14:0, 16:0, 18:0, 18:1 (MS)	Cys843, C-term. intracellular domain	?	60
Beta 2 adrenergic receptor	Signal transduction	TE (HA)	16:0 (RL)	Cys341, C-term. intracellular domain	Agonist-induced desensitization, phosphorylation coupling to adenylate cyclase	61
Caveolin	Forms coat of caveoli	TE (SDM)	16:0 (RL)	Cys133, Cys143, Cys156 on C-terminal side of transmembrane domain	Stabilization of caveolin oligomers	62
CD4	Helper T cell coreceptor	TE (HA, SDM)	16:0 (RL)	Cys394, Cys397 at junction of transmembrane and cytoplasmic domains	?	63
CD44	Lymphocyte migration and adhesion	?	16:0 (RL)	?	Regulates CD44 inhibition of T cell receptor signaling in T cells	64
Cytochrome b561	Electron transfer	Oxyester	?	?	?	65
D2L dopamine receptor	Signal transduction	TE (HA)	16:0 (RL)	?	?	66
ETB endothelin receptor	Signal transduction	?	?	C-term. intracellular domain	Required for calcium signaling	67
Friend murine leukemia virus p12E	Essential part of virus envelope	TE (HA, SDM)	16:0 (RL, MS)	Cy606, C-term. side of transmembrane domain	?	68
G protein alpha subunit	Signal transduction	TE (HA, SDM)	16:0, 20:4 (RL)	Cysteines in the region 3–10, some are also N-terminally myristoylated	Membrane binding, agonist-induced translocation	69

Protein	Function		Fatty acid	Site	Role	Ref.
GAP-43 (neuromodulin)	Neuronal growth, synaptic plasticity, neurotransmitter release	TE (HA, SDM)	16:0 (RL)	Cys3, Cys4	Required for cellular extension and localization of protein growth cones	70
Glucose transporter	Cellular glucose uptake	TE (HA)	16:0 (RL)	?	?	71
GluR6 kainate receptor	Glutamate receptor	TE (HA)	16:0 (RL)	Cys827, Cys840	?	72
mGluR4	Metabotropic glutamate receptor	TE (HA)	16:0 (RL)	?	?	73
gp41 (HIV)	Viral envelope protein	TE (HA, SDM)	16:0 (RL)	?	?	74
GRK6	Desensitization of signal transduction	TE (HA, SDM)	16:0 (RL)	Cys561, Cys562, Cys565	Membrane targeting	75
Hepatitis virus E2 protein	Major viral surface glycoprotein	TE (HA)	16:0 (RL)	?	?	76
Influenza virus hemagglutin	Hemolytic activity	TE (HA, SDM)	16:0 (RL)	Cys211, Cys218, Cys221 at junction of transmembrane and cytoplasmic domain	Membrane fusogenic activity	77
Influenza virus M2 protein	Ion channel	TE (HA)	16:0 (RL, SDM)	Cys50, intracellular domain	?	78
Insulin receptor	Signal transduction	TE (HA)	16:0 (RL)	?	?	79
HCK kinase	Member of the src family of protein kinases	TE (SDM)	16:0 (RL)	Cys3 of p59hck which is N-terminally myristoylated	Localization to caveoli	80
Hemolysin (*E. coli*)	Membrane pore formation	Amide (MS, SDM)	?	Lys564, Lys690	Required for toxin activity	55
Herbicide binding protein	Component of photo-system II	?	16:0 (RL)	?	?	81
Luteinizing hormone, choriogonadotropin, lutrotropin receptor	Signal transduction	TE (HA)	16:0 (RL)	?	?	82
Mannose-6-phosphate receptor	Intracellular protein transport	TE (SDM)	16:0 R(L)	Cys30, Cys34	Trafficking and lysosomal enzyme sorting	83
MHC Class II	Antigen presentation	TE (HA)	16:0 (RL)	Fatty acylation occurs on the α, β chains and on the invariant chain near the membrane spanning segment	?	84, 85

Table 1 (continued)

Protein	Function of protein	Acyl linkage[b]	Acyl chain[c]	Location of acyl chain(s)	Function(s) of acyl chain	Ref.
Mucus glycoprotein	Lubrication of digestive and air passages	TE (HA)	14:0, 14:1, 16:0, 18:0, 18:1 (GC)	?	?	86–88
Myelin p0 glycoprotein	Major protein of myelin	TE (HA, SDM)	16:0 (RL)	Cys153 at the boundary of transmembrane and cytoplasmic domains (SDM)	?	89
Myelin proteolipid	Major protein of myelin	TE (HA, SDM, base hydrolysis, and cysteine alkylation)	Mainly 16:0 (GC)	Cys5, Cys6, Cys9, Cys108, Cys138, Cys140 (SDM, base hydrolysis, and cysteine alkylation)	?	90–93
Nitric oxide synthase (endothelial)	Nitric oxide production	TE (HA)	16:0 (RL)	Cys15, Cys26 (SDM), N-terminal myristoylation required for palmitoylation	Localization to caveolae, also found in Golgi, conflicting reports about the role of palmitoylation in membrane binding	94–100
p56hck	Signal transduction (src kinase homolog)	TE (HA)	16:0 (RL)	Cys3 (SDM), N-terminal myristoylation required for palmitoylation	Binding to GPI-linked proteins in caveolae	101
Pullulanase (*Klebsiella pneumoniae*)	Starch debranching enzyme from bacterium	TE (sequencing)	16:0 (RL)	N-terminal cysteine after the signal peptide is removed (sequencing)	May affect efficiency of pullulanase secretion but not critical	102–104
Rabies virus glycoprotein G	?	TE (HA)	16:0 (RL)	?	?	105
Ras (H-, N-)	Cell proliferation regulator	TE (HA, SDM)	16:0 (RL)	Cys181, Cys184 (H-Ras) Cys181 (N-Ras)	Drives Ras onto membranes	5
Rh protein	Erythrocyte surface antigen	TE (NS)	16:0 (RL)	Multiple cysteines near the cytoplasmic leaflet	?	106, 107
Rhodopsin	Photoreception	TE (MS)	16:0 (MS)	Cys322, Cys323 of cytoplasmic domain	?	108
Semliki Forest virus p62 envelope protein	Major viral surface protein	TE (HA, SDM)	16:0 (RL)	Cys433 in the middle of the putative transmembrane segment	?	44, 109–111
Sindbis virus E1 glycoprotein	Major viral surface glycoprotein	TE (SDM)	16:0 (RL)	Cys416, Cys417 (SDM)	?	111

Protein	Function	Linkage	Fatty acyl	Cysteine	Biological role	Ref.
Sm 25	Major glycoprotein antigen of the parasitic helminth *Schistosoma mansoni*	TE (SDM)	16:0 (RL)	Cys-168 in *C*-terminal hydrophobic domain	Membrane binding	112
SNAP-25	Intracellular vesicle transport	TE (HA)	16:0 (RL)	?	Membrane binding	113–115
SP-C	Lung surfactant protein	TE (MS)	16:0 (MS)	Cys5, Cys6	May be involved in maintaining proper protein structure	116–119
Spectrin (tightly membrane bound fraction)	Cytoskeletal structure	TE (HA)	16:0 (RL)	β-subunit	?	120
Spiralin	Major membrane protein of bacterium *Spiroplasma melliferum*	?	14:0, 16:0 (RL)	?	?	121
Synaptobrevin (yeast)	Membrane recognition and vesicle fusion	TE (HA)	16:0 (RL)	Cys95 (SDM) near transmembrane domain	?	122
Synaptotagmin	Calcium-sensor for synaptic transmission	TE (HA)	16:0 (RL)	?	?	114
Transforming growth factor-α precursor	Precursor to the mitogenic growth factor	TE (HA)	16:0 (RL)	?	May anchor the *C*-terminal fragment in the membrane after proteolytic release of the growth factor	123
Transferrin receptor	Cellular transferrin uptake	TE (HA)	16:0 (RL)	Cys62, Cys67 of cytoplasmic domain (SDM)	One study shows nonlipidated receptor is endocytosed faster, another study shows no effect	124–127
Vesicular stomatitis virus glycoprotein	Major viral surface protein	TE (HA, SDM)	16:0 (RL)	Cys489 in cytoplasmic domain	?	43
YPT proteins	Vesicle secretion	TE (HA, SDM)	16:0 (RL)	*C*-terminal penultimate cysteine	?	128

[a] This table contains only those proteins that have been shown to contain covalently attached palmitoyl groups (and possibly other fatty acyl groups, except *N*-terminal myristoylated proteins, which are not listed). Only proteins that have a known biological function are included. This table contains most of the known palmitoylated proteins, but there is no assurance that it is absolutely comprehensive. [b] TE designates a thiolester linkage to a cysteine. HA designates hydroxylamine sensitivity of the acyl linkage. SDM indicates that the type of acyl linkage is suggested by site-directed mutagenesis of candidate amino acids. [c] Fatty acyl chains are designated as 14:0 (myristoyl), 14:1 (myristoleoyl), 16:0 (palmitoyl), 18:0 (stearoyl), and 18:1 (oleoyl). RL, GC, MS designate, respectively, that the type of acylation linkage was determined by incorporation of radiolabeled fatty acid, gas chromatography analysis of the released fatty acid, or mass spectrometry analysis of the intact acylated protein or acylated proteolytic fragment, respectively.

activity which acylates both endogenous ghost polypeptides and exogenous proteins derived from Semliki Forest virus.[109]

Many members of the src family of protein kinases are palmitoylated on cysteines near the N-terminal myristoyl group, and myristoylation seems to be a prerequisite for palmitoylation (Table 1). The possibility that palmitoylation of lipidated proteins in cells is a nonenzymatic event is suggested by Quesnel and Silvius[139] who reported that the myristoylated peptides, myristoyl-GCX (X = G, L, R, T, or V), and a prenylated peptide, SCRC(S-farnesyl)-CO$_2$Me can accept the palmitoyl group from palmitoyl-CoA when both are incubated in synthetic liposomes at neutral pH. This process may not have the efficiency to account for protein palmitoylation in cells where the concentration of palmitoyl-CoA is probably much lower than the value of 1 mol% used in the model membrane study.

Evidence suggests that palmitoylation of lipidated proteins is enzyme catalyzed. The membrane-dependent acylation of the myristoylated src kinase family member fyn by radioiodinated palmitoyl-CoA is heat sensitive.[140] The activity has been solubilized from membranes with Triton X-100 and partially purified by ion exchange chromatography. Many G protein α-subunits are palmitoylated near their N-terminal myristoyl group (Table 1). Linder and co-workers found heat-, protease-, and SDS-sensitive, G protein palmitoyltransferase activity enriched in plasma membranes.[141] The enzyme could be solubilized with detergent but not with high salt.

In 1996, a protein palmitoyltransferase that acts on Ras proteins was purified to apparent homogeneity (10 000-fold) from rat liver.[142] The enzyme appears as two bands (apparent MW of 30 kDa and 33 kDa) when analyzed by SDS-PAGE. It acts on Ras only if this acceptor is C-terminally farnesylated and methylated. Further work is needed to determine if this enzyme is responsible for Ras palmitoylation *in vivo*. Ras and G proteins undergo multiple cycles of palmitoylation and depalmitoylation, which suggests the existence of a protein thiolesterase. Hofmann and co-workers have purified a palmitoyl protein thiolesterase that removes the palmitoyl group from Ras and G proteins.[143] Molecular cloning of the enzyme reveals that it contains a glycosylation consensus site and a leader sequence, which suggests that this enzyme is located in cellular compartments that do not contain Ras and G proteins.[144] Research in 1996 indicates that the enzyme is located in lysosomes, and its malfunction is associated with infantile neuronal ceroid lipofuscinosis.[145]

4.02.4 FUNCTIONS OF PROTEIN PALMITOYL GROUPS

Apart from the expected effect of favoring protein binding to cellular membranes, the specific function(s) of the palmitoyl-group(s) for many palmitoylated proteins remain, to a large extent, unknown. The primary reasons for the paucity of information are the dynamic nature of protein palmitoylation, the limited information about the palmitoylating and depalmitoylation enzymes, as well as the lack of experimental tools specifically to enhance or decrease the palmitoylation of a given protein. The most revealing insight into the role of palmitoylation for protein functions has come from genetic studies using protein mutants defective in palmitoylation. A few selected examples of such studies are summarized below.

4.02.4.1 G Protein α-Subunits

Heterotrimeric G proteins are involved in a large number of signal transduction pathways and function to couple cell surface receptors to intracellular effectors. Many G protein α-subunits are N-terminally myristoylated and also contain 1–2 palmitoyl groups on cysteine residues close to the myristoylated N-terminus. Knowledge about the biological function of Gα palmitoylation stems mostly from studies using cell lines transfected with Gα subunits carrying mutations in their palmitoylation sites. In the case of Gs-α (which is palmitoylated, but not myristoylated) an intact palmitoylation site is required for its ability to mediate hormonal stimulation of adenylate cyclase.[146] Since palmitoylation of Gs is also required for membrane binding, such studies led to the conclusion that palmitoylation serves as a means to co-localize the heterotrimeric G protein with its receptor and the downstream effector (adenylate cyclase is a membrane-bound enzyme) at the plasma membrane.

Gs-α also undergoes activation-induced depalmitoylation which is accompanied by a membrane-to-cytosol translocation of the activated α-subunit.[147] Since activation-induced depalmitoylation decreases the amount of Gs-α available for interaction with both the receptor as well as the downstream effector, this process has been proposed to serve as a turn-off or desensitization mechanism for G-protein coupled receptor signaling.[147,148]

A distinctly different function has been observed for the Gq-α subunit. Like Gs-α, Gq-α palmitoylation-site mutants are signaling defective. In transfected cell lines palmitoylation-defective mutants are unable to interact productively with the receptor and fail to activate phospholipase C, a downstream effector molecule. Unlike Gs-α mutants, however, nonpalmitoylated Gq mutants are still membrane bound, suggesting that the palmitoylation sites may contribute to protein–protein interaction.[149] An alternative explanation was offered based on *in vitro* studies, suggesting that the palmitoylation sites are functionally important regardless of the state of palmitoylation.[150]

4.02.4.2 Ras Proteins

Mammalian H Ras and yeast RAS2 both require palmitoylation for tight membrane binding. Genetic experiments in yeast revealed that palmitoylation of RAS2 plays a differential role in signal transduction and cell viability.[151] While a single copy of a palmitoylation-deficient RAS2 gene confers cell viability (which is strictly dependent on at least one functional RAS gene), such mutant yeast strains fail to activate adenylate cyclase in response to glucose addition. Since adenylate cyclase is the primary target for RAS in both the glucose response as well as in cell viability, these observations have been interpreted to mean that yeast RAS can couple to both a cytosolic adenylate cyclase required for cell viability and a membrane-bound form implicated in the glucose response. Thus, yeast Ras may couple to different effectors based on the state of palmitoylation.

In mammals, transfection studies indicated that a nonpalmitoylated form of oncogenic H-Ras, even though residing in the cytosol, is competent to transform fibroblasts when overexpressed.[152,153] However, dose-response experiments using microinjection of activated normal H-Ras into *Xenopus leavis* oocytes revealed that the palmitoylated form is 25 to 50 times more efficient in inducing cell cycle progression and activation of mitogen-activated protein kinase, a major downstream signal pathway for Ras in vertebra.[154] Therefore, it is very likely that under physiological conditions (i.e., when Ras is expressed at physiological levels) the membrane bond, palmitoylated form, constitutes the active, signaling-competent pool of H-Ras in cells, a requirement that can be overcome by overexpression.

4.02.5 FUTURE PROSPECTS

Elucidation of the structures of palmitoyl groups attached to proteins is now well established, and several analytical methods have been developed for determining the palmitoylation status of proteins in cells. Most structural studies are incomplete in that the array of fatty acids attached to the protein have not been determined, and the location of the fatty acid is often stated based on circumstantial evidence. One can expect that mass spectrometry will emerge as the method of choice for the direct structural characterization of palmitoylated proteins.

In a few cases, reversible palmitoylation of proteins has been shown to alter the location of the protein in the cell (membrane vs. soluble), and this has functional consequences. Much more information is needed before we can fully understand the function of protein palmitoylation, especially in cases where palmitoylation occurs on integral, transmembrane proteins such as receptors.

Only two protein palmitoyltransferases have been purified as yet; those that act on mucus glycoprotein and Ras proteins. Further work is needed to determine if the isolated enzymes actually have these functions in the intact cell. The existence of other enzymes that fatty acylate other mammalian cell proteins remains to be established, and the molecular mechanisms underlying their regulation remain to be determined.

4.02.6 REFERENCES

1. A. Aitken, P. Cohen, S. Santikarn, D. H. Williams, A. G. Calder, A. Smith, and C. B. Klee, *FEBS Lett.*, 1982, **150**, 314.
2. S. A. Carr, K. Biemann, S. Shoji, D. C. Parmelee, and K. Titani, *Proc. Natl. Acad. Sci. USA*, 1982, **79**, 6128.
3. L. E. Henderson, H. C. Krutzsch, and S. Oroszlan, *Proc. Natl. Acad. Sci. USA*, 1983, **80**, 339.
4. J. I. Gordon, R. J. Duronio, D. A. Rudnick, S. P. Adams, and G. W. Gokel, *J. Biol. Chem.*, 1991, **266**, 8647.

5. J. A. Glomset, M. H. Gelb, and C. C. Farnsworth, *Trends Biochem. Sci.*, 1990, **15**, 139.
6. F. L. Zhang and P. J. Casey, *Annu. Rev. Biochem.*, 1996, **65**, 241.
7. P. T. Englund, *Annu. Rev. Biochem.*, 1993, **62**, 121.
8. S. Yamamoto and J. O. Lampen, *J. Biol. Chem.*, 1976, **251**, 4102.
9. K. Hantke and V. Braun, *Eur. J. Biochem.*, 1973, **34**, 284.
10. P. E. Braun and N. S. Radin, *Biochemistry*, 1969, **8**, 4310.
11. D. H. MacLennan, C. C. Yip, G. H. Isles, and P. Seeman, *Cold Spring Harbor Symp. Quant. Biol.*, 1972, **37**, 469.
12. M. F. Schmidt, M. Bracha, and M. J. Schlesinger, *Proc. Natl. Acad. Sci. USA*, 1979, **76**, 1687.
13. A. I. Magee and S. A. Courtneidge, *EMBO J.*, 1985, **4**, 1137.
14. M. A. Bolanowski, B. J. Earles, and W. J. Lennarz, *J. Biol. Chem.*, 1984, **259**, 4934.
15. C. E. Dahl, J. S. Dahl, and K. Bloch, *J. Biol. Chem.*, 1983, **258**, 11 814.
16. R. A. McIlhinney, S. J. Pelly, J. K. Chadwick, and G. P. Cowley, *EMBO J.*, 1985, **4**, 1145.
17. G. V. Marinetti and K. Cattieu, *Biochim. Biophys. Acta*, 1982, **685**, 109.
18. E. N. Olson, D. A. Towler, and L. Glaser, *J. Biol. Chem.*, 1985, **260**, 3784.
19. D. Riendeau and D. Guertin, *J. Biol. Chem.*, 1986, **261**, 976.
20. D. Wen and M. J. Schlesinger, *Mol. Cell Biol.*, 1984, **4**, 688.
21. M. J. Schlesinger, *Methods Enzymol.*, 1983, **96**, 795.
22. M. F. Schmidt, *Biochim. Biophys. Acta*, 1989, **988**, 411.
23. D. A. Towler, J. I. Gordon, S. P. Adams, and L. Glaser, *Annu. Rev. Biochem.*, 1988, **57**, 69.
24. A. M. Schultz, L. E. Henderson, and S. Oroszlan, *Annu. Rev. Cell Biol.*, 1988, **4**, 611.
25. A. I. Magee, L. Gutierrez, C. J. Marshall, and J. F. Hancock, *J. Cell Sci. Suppl.*, 1989, **11**, 149.
26. R. J. Grand, *Biochem. J.*, 1989, **258**, 625.
27. J. S. Hu, G. James, and E. N. Olson, *Biofactors*, 1988, **1**, 219.
28. D. A. Towler, J. I. Gordon, S. P. Adams, and L. Glaser, *Annu. Rev. Biochem.*, 1988, **57**, 69.
29. A. V. Kabanov, A. V. Levashov, V. Y. Alakhov, K. Martinek, and E. S. Severin, *Biomed. Sci.*, 1990, **1**, 33.
30. C.-S. Jackson, P. Zlatkine, C. Bano, P. Kabouridis, B. Mehul, M. Parenti, G. Milligan, S. C. Ley, and A. I. Magee, *Biochem. Soc. Trans.*, 1995, **23**, 568.
31. R. J. Deschenes, M. D. Resh, and J. R. Broach, *Curr. Opin. Cell Biol.*, 1990, **2**, 1108.
32. J. Blenis and M. D. Resh, *Curr. Opin. Cell Biol.*, 1993, **5**, 984.
33. R. A. McIlhinney, *Trends Biochem. Sci.*, 1990, **15**, 387.
34. G. Milligan, M. A. Grassie, A. Wise, D. J. MacEwan, A. I. Magee, and M. Parenti, *Biochem. Soc. Trans.*, 1995, **23**, 583.
35. D. I. Mundy, *Biochem. Soc. Trans.*, 1995, **23**, 572.
36. S. M. Mumby and K. H. Muntz, *Biochem. Soc. Trans.*, 1995, **23**, 156.
37. M. F. Schmidt and G. R. Burns, *Behring Inst. Mitt.*, 1991, **89**, 185.
38. G. James and E. N. Olson, *Biochemistry*, 1990, **26**, 2623.
39. M. Bouvier, T. P. Loisel, and T. Hebert, *Biochem. Soc. Trans.*, 1995, **23**, 577.
40. M. Bouvier, S. Moffett, T. P. Loisel, B. Mouillac, T. Hebert, and P. Chidiac, *Biochem. Soc. Trans.*, 1995, **23**, 116.
41. J. Folch and M. B. Lees, *J. Biol. Chem.*, 1951, **191**, 807.
42. P. Stoffyn and P. J. Folch, *Biochem. Biophys. Res. Commun.*, 1971, **44**, 157.
43. M. F. G. Schmidt and M. J. Schlesinger, *Cell*, 1979, **17**, 813.
44. M. Schmidt, M. F. G. Schmidt, and R. Rott, *J. Biol. Chem.*, 1988, **263**, 18 635.
45. M. A. Bolanowski, B. J. Earles, and W. J. Lennarz, *J. Biol. Chem.*, 1984, **259**, 4934.
46. O. Bizzozero, *Methods Enzymol.*, 1995, **250**, 361.
47. W. Stoffel, H. Hillen, W. Schroder, and R. Deutzmann, *Hoppe-Seyler's Z. Physiol. Chem.*, 1983, **364**, 1455.
48. J. Jolles, J.-L. Nussbaum, F. Schoentgen, P. Mandel, and P. Jolles, *FEBS Lett.*, 1977, **74**, 190.
49. D. Mack and J. Kruppa, *Biochem. J.*, 1988, **256**, 1021.
50. S. Jing and I. S. Trowbridge, *EMBO J.*, 1987, **6**, 327.
51. D. I. Papac, K. R. Thornburg, E. E. Bullesbach, R. K. Crouch, and D. R. Knapp, *J. Biol. Chem.*, 1992, **267**, 1992.
52. W. Weinmann, C. Maier, K. Baumeister, M. Przybylski, C. E. Parker, and K. B. Tomer, *J. Chromatogr. A*, 1994, **664**, 271.
53. J. Hensel, M. Hintz, M. Karas, D. Linder, B. Stahl, and R. Geyer, *Eur. J. Biochem.*, 1995, **232**, 373.
54. M. Hackett, L. Guo, J. Shabanowitz, D. F. Hunt, and E. L. Hewlett, *Science*, 1994, **266**, 433.
55. P. Stanley, L. C. Packman, V. Koronakis, and C. Hughes, *Science*, 1994, **266**, 1992.
56. M. Hackett, C. B. Walker, L. Guo, M. C. Gray, C. S. Van, A. Ullmann, J. Shabanowitz, D. F. Hunt, E. L. Hewlett, and P. Sebo, *J. Biol. Chem.*, 1995, **270**, 20 250.
57. M. G. Eason, M. T. Jacinto, C. T. Theiss, and S. B. Liggett, *Proc. Natl. Acad. Sci. USA*, 1994, **91**, 11 178.
58. S. Manenti, I. Dunia, and E. L. Benedetti, *FEBS Lett.*, 1990, **262**, 356.
59. F. Y. Zeng, J. A. Oka, and P. H. Weigel, *Biochem. Biophys. Res. Commun.*, 1996, **218**, 325.
60. K. Okubo, N. Hamasaki, K. Hara, and M. Kageura, *J. Biol. Chem.*, 1991, **266**, 16 420.
61. S. Moffett, B. Mouillac, H. Bonin, and M. Bouvier, *EMBO J.*, 1993, **12**, 349.
62. S. Monier, D. J. Dietzen, W. R. Hastings, and D. M. Lublin, *FEBS Lett.*, 1996, **388**, 143.
63. B. Crise and J. K. Rose, *J. Biol. Chem.*, 1992, **267**, 13 593.
64. Y. J. Guo, S. C. Lin, J. H. Wang, M. Bigby, and M. S. Sy, *Int. Immunol.*, 1994, **6**, 213.
65. U. M. Kent and P. J. Fleming, *J. Biol. Chem.*, 1990, **265**, 16 422.
66. G. Y. Ng, B. F. O'Dowd, M. Caron, M. Dennis, M. R. Brann, and S. R. George, *J. Neurochem.*, 1994, **63**, 1589.
67. T. Koshimizu, G. Tsjuimoto, K. Ono, T. Masaki, and A. Sakamoto, *Biochem. Biophys. Res. Commun.*, 1995, **217**, 354.
68. C. Yang and R. W. Compans, *Virology*, 1996, **221**, 87.
69. E. M. Ross, *Curr. Biol.*, 1995, **5**, 107.
70. Y. Liu, D. A. Fisher, and D. R. Storm, *J. Neurosci.*, 1994, **14**, 5807.
71. J. F. Pouliot and R. B'eliveau, *Biochim. Biophys. Acta*, 1995, **1234**, 191.
72. D. S. Pickering, F. A. Taverna, M. W. Salter, and D. R. Hampson, *Proc. Natl. Acad. Sci. USA*, 1995, **92**, 12 090.
73. S. Alaluf, E. R. Mulvihill, and R. A. McIlhinney, *J. Neurochem.*, 1995, **64**, 1548.

74. C. Yang, C. P. Spies, and R. W. Compans, *Proc. Natl. Acad. Sci. USA*, 1995, **92**, 9871.
75. R. H. Stoffel, R. R. Randall, R. T. Premont, R. J. Lefkowitz, and J. Inglese, *J. Biol. Chem.*, 1994, **269**, 27791.
76. M. F. van Berlo, W. J. van den Brink, M. C. Horzinek, and B. A. van der Zeijst, *Arch. Virol.*, 1987, **95**, 123.
77. H. C. Philipp, B. Schroth, M. Veit, M. Krumbiegel, and A. Herrmann, *Virology*, 1995, **210**, 20.
78. L. J. Holsinger, M. A. Shaughnessy, A. Micko, L. H. Pinto, and R. A. Lamb, *J. Virol.*, 1995, **69**, 1219.
79. A. I. Magee and K. Siddle, *J. Cell Biochem.*, 1988, **37**, 347.
80. S. M. Robbins, N. A. Quintrell, and J. M. Bishop, *Mol. Cell. Biol.*, 1995, **15**, 3507.
81. A. K. Mattoo and M. Edelman, *Proc. Natl. Acad. Sci. USA*, 1987, **84**, 1497.
82. H. Zhu, H. Wang, and M. Ascoli, *Mol. Endocrinol.*, 1995, **9**, 141.
83. A. Schweizer, S. Kornfeld, and J. Rohrer, *J. Cell Biol.*, 1996, **132**, 577.
84. S. Simonis and S. E. Cullen, *J. Immunol.*, 1986, **136**, 2962.
85. N. Koch and G. J. Hammerling, *Biochemistry*, 1985, **24**, 6185.
86. Y. H. Liau, V. L. Murty, K. Gwozdzinski, A. Slomiany, and B. L. Slomiany, *Biochim. Biophys. Acta*, 1986, **880**, 108.
87. A. Slomiany, Z. Jozwiak, A. Takagi, and B. L. Slomiany, *Arch. Biochem. Biophys.*, 1984, **229**, 560.
88. A. Slomiany, Y. H. Liau, A. Takagi, W. Laszewicz, and B. L. Slomiany, *J. Biol. Chem.*, 1984, **259**, 13304.
89. O. A. Bizzozero, K. Fridal, and A. Pastuszyn, *J. Neurochem.*, 1994, **62**, 1163.
90. O. A. Bizzozero, L. K. Good, and J. E. Evans, *Biochem. Biophys. Res. Commun.*, 1990, **170**, 375.
91. S. U. Tetzloff and O. A. Bizzozero, *Biochem. Biophys. Res. Commun.*, 1993, **193**, 1304.
92. T. Weimbs and W. Stoffel, *Biochemistry*, 1992, **31**, 12289.
93. O. A. Bizzozero and L. K. Good, *J. Neurochem.*, 1990, 1986.
94. D. T. Hess, S. I. Patterson, D. S. Smith, and J. H. Skene, *Nature*, 1993, **366**, 562.
95. P. W. Shaul, E. J. Smart, L. J. Robinson, Z. German, I. S. Yuhanna, Y. Ying, R. G. Anderson, and T. Michel, *J. Biol. Chem.*, 1996, **271**, 6518.
96. W. C. Sessa, G. Garcia-Cardena, J. Liu, A. Keh, J. S. Pollock, J. Bradley, S. Thiru, I. M. Braverman, and K. M. Desai, *J. Biol. Chem.*, 1995, **270**, 17641.
97. C. G. Garc'ia, P. Oh, J. Liu, J. E. Schnitzer, and W. C. Sessa, *Proc. Natl. Acad. Sci. USA*, 1996, **93**, 6448.
98. J. Liu, C. G. Garc'ia, and W. C. Sessa, *Biochemistry*, 1995, **34**, 12333.
99. L. J. Robinson and T. Michel, *Proc. Natl. Acad. Sci. USA*, 1995, **92**, 11776.
100. L. J. Robinson, L. Busconi, and T. Michel, *J. Biol. Chem.*, 1995, **270**, 995.
101. S. A. M. Shenoy, D. J. Dietzen, J. Kwong, D. C. Link, and D. M. Lublin, *J. Cell Biol.*, 1994, **126**, 353.
102. M. G. Kornacker, D. Faucher, and A. P. Pugsley, *J. Biol. Chem.*, 1991, **266**, 13842.
103. A. P. Pugsley, C. Chapon, and M. Schwartz, *J. Bacteriol.*, 1986, **166**, 1083.
104. I. Poquet, D. Faucher, and A. P. Pugsley, *EMBO J.*, 1993, **12**, 271.
105. Y. Gaudin, C. Tuffereau, A. Benmansour, and A. Flamand, *Virology*, 1991, **184**, 441.
106. M. P. de Vetten and P. Agre, *J. Biol. Chem.*, 1988, **263**, 18193.
107. S. Hartel-Schenk and P. Agre, *J. Biol. Chem.*, 1992, **267**, 5569.
108. Y. A. Ovchinnikov, N. G. Abdulaev, and A. S. Bogachuk, *FEBS Lett.*, 1988, **230**, 1.
109. M. F. G. Schmidt, R. A. J. McIlhinney, and G. R. Burns, *Biochim. Biophys. Acta*, 1995, **1257**, 205.
110. C. G. Scharer, H. Y. Naim, and H. Koblet, *Arch. Virol.*, 1993, **132**, 237.
111. L. Ivanova and M. J. Schlesinger, *J. Virol.*, 1993, **67**, 2546.
112. E. J. Pearce, A. I. Magee, S. R. Smithers, and A. J. Simpson, *EMBO J.*, 1991, **10**, 2741.
113. D. T. Hess, T. M. Slater, M. C. Wilson, and J. H. Skene, *J. Neurosci.*, 1992, **12**, 4634.
114. M. Veit, T. H. Sollner, and J. E. Rothman, *FEBS Lett.*, 1996, **385**, 119.
115. I. C. Bark, *J. Mol. Biol.*, 1993, **233**, 67.
116. T. Curstedt, J. Johansson, P. Persson, A. Eklund, B. Robertson, B. Lowenadler, and H. Jornvall, *Proc. Natl. Acad. Sci. USA*, 1990, **87**, 2985.
117. R. Qanbar and F. Possmayer, *Biochim. Biophys. Acta*, 1995, **1255**, 251.
118. L. A. Creuwels, R. A. Demel, G. L. M. van Golde, B. J. Benson, and H. P. Haagsman, *J. Biol. Chem.*, 1993, **268**, 26752.
119. D. K. Vorbroker, C. Dey, T. E. Weaver, and J. A. Whitsett, *Biochim. Biophys. Acta*, 1992, **1105**, 161.
120. M. Mariani, D. Maretzki, and H. U. Lutz, *J. Biol. Chem.*, 1993, **268**, 12996.
121. X. Foissac, C. Saillard, J. Gandar, L. Zreik, and J. M. Bov'e, *J. Bacteriol.*, 1996, **178**, 2934.
122. A. Couve, V. Protopopov, and J. E. Gerst, *Proc. Natl. Acad. Sci. USA*, 1995, **92**, 5987.
123. T. S. Bringman, P. B. Lindquist, and R. Derynck, *Cell*, 1987, **48**, 429.
124. E. Alvarez, N. Girones, and R. J. Davis, *J. Biol. Chem.*, 1990, **265**, 16644.
125. M. A. C. Turbide and R. M. Johnstone, *Arch. Biochem. Biophys.*, 1988, **264**, 553.
126. S. Jing and I. S. Trowbridge, *EMBO J.*, 1987, **6**, 327.
127. S. Jing and I. S. Trowbridge, *J. Biol. Chem.*, 1990, **265**, 11555.
128. C. M. Newman, T. Giannakouros, J. F. Hancock, E. H. Fawell, J. Armstrong, and A. I. Magee, *J. Biol. Chem.*, 1992, **267**, 11329.
129. S. M. Mumby and J. E. Buss, *Methods*, 1990, **1**, 216.
130. M. F. G. Schmidt, *EMBO J.*, 1984, **3**, 2295.
131. B. M. Sefton, I. S. Trowbridge, J. A. Cooper, and E. M. Scolnick, *Cell*, 1982, **31**, 465.
132. T. Yoshimura, D. Agrawal, and H. C. Agrawal, *Biochem. J.*, 1987, **246**, 611.
133. O. A. Bizzozero, J. F. McGary, and M. B. Lees, *J. Biol. Chem.*, 1987, **262**, 2138.
134. N. W. Ross and P. E. Braun, *J. Neurosci. Res.*, 1989, **21**, 35.
135. C. Kasinathan, E. Grzelinska, K. Okazaki, B. L. Slomiany, and A. Slomiany, *J. Biol. Chem.*, 1990, **265**, 5139.
136. M. Berger and M. F. G. Schmidt, *J. Biol. Chem.*, 1984, **259**, 7245.
137. M. F. Schmidt and G. R. Burns, *Biochem. Soc. Trans.*, 1989, **17**, 625.
138. M. F. G. Schmidt and G. R. Burns, *Biochem. Soc. Trans.*, 1989, **17**, 859.
139. S. Quesnel and J. R. Silvius, *Biochemistry*, 1994, **33**, 13340.
140. L. Berthiaume and M. D. Resh, *J. Biol. Chem.*, 1995, **270**, 22399.
141. J. T. Dunphy, W. K. Greentree, C. L. Manahan, and M. E. Linder, *J. Biol. Chem.*, 1996, **271**, 7154.

142. L. Liu, T. Dudler, and M. H. Gelb, *J. Biol. Chem.*, 1996, **271**, 23 269.
143. L. A. Camp and S. L. Hofmann, *J. Biol. Chem.*, 1993, **268**, 22 566.
144. L. A. Camp, L. A. Verkruyse, S. J. Afendis, C. A. Slaughter, and S. L. Hofmann, *J. Biol. Chem.*, 1994, **269**, 23 212.
145. L. A. Verkruyse and S. L. Hofmann, *J. Biol. Chem.*, 1996, **271**, 584.
146. P. B. Wedegaertner, D. H. Chu, P. T. Wilson, M. J. Levis, and H. R. Bourne, *J. Biol. Chem.*, 1993, **268**, 25 001.
147. P. B. Wedegaertner and H. R. Bourne, *Cell*, 1994, **77**, 1063.
148. P. B. Wedegaertner, P. T. Wilson, and H. R. Bourne, *J. Biol. Chem.*, 1995, **270**, 503.
149. M. D. Edgerton, C. Chabert, A. Chollet, and S. Arkinstall, *FEBS Lett.*, 1994, **354**, 195.
150. J. R. Hepler, G. H. Biddlecome, C. Kleuss, L. A. Camp, S. L. Hofmann, E. M. Ross, and A. G. Gilman, *J. Biol. Chem.*, 1996, **271**, 496.
151. S. Bhattacharya, L. Chen, J. R. Broach, and S. Powers, *Proc. Natl. Acad. Sci. USA*, 1995, **92**, 2984.
152. J. F. Hancock, A. I. Magee, J. E. Childs, and C. J. Marshall, *Cell*, 1989, **57**, 1167.
153. J. F. Hancock, H. Paterson, and C. J. Marshall, *Cell*, 1990, **63**, 133.
154. T. Dudler and M. H. Gelb, *J. Biol. Chem.*, 1996, **271**, 11 541.

4.03
Recent Advances in Biosynthesis of Alkaloids

NEELAM MISRA, RAJESH LUTHRA, KIRAN L. SINGH, and
SUSHIL KUMAR
*Central Institute of Medicinal and Aromatic Plants,
Lucknow, India*

4.03.1 INTRODUCTION

The major source of alkaloids is the angiosperms, but alkaloids are also found in animals, insects, microorganisms, marine organisms, and lower plants. Humans have utilized alkaloids as medicines, poisons, and magical potions. The alkaloids are secondary metabolites that contain secondary, tertiary, or quaternary nitrogen atoms in their structures. They are metabolically active and play an important role in the physiology of plants or organisms. The alkaloids in plants are very important in repelling and discouraging predators and pathogens. Great interest has been shown in the production of alkaloids from plants or plant cells for many years. Over 25% of all prescription alkaloidal medicines are derived from plants.[1,2] Every year more than 1500 new chemical structures are reported, of which a substantial number are found to be useful to humans.[2] The chemical structures of about 130 000 alkaloids are known and their synthesis in plants is one of the most exciting and fascinating subjects. Many biosynthetic pathways of alkaloids have been proposed.

Studies on the biosynthetic pathways of alkaloids involve the administration of the labeled precursors followed by isolation of the alkaloids, which are then degraded in a systematic manner to determine the position of the labeled atoms. Using this approach, many alkaloids, e.g., morphine, pellotine, nicotine, colchicine, gramine, tropane, and indole alkaloids, have been shown to be synthesized from amino acids. Considerable improvements have been made in separation of alkaloids through column chromatographic methods. Spectroscopic and X-ray crystallographic techniques have elucidated many structures and much effort has been directed towards understanding the biosynthesis of alkaloids and the chemical reactions involved. Tracer techniques have opened up new facets of natural product chemistry—for instance, the elucidation of metabolic pathways of alkaloids.

The alkaloids are classified into three widely accepted groups: true alkaloids, protoalkaloids, and pseudoalkaloids.

The true alkaloids are derived from amino acids and show a wide range of physiological activity. The protoalkaloids are simple amines where the amino acid nitrogen group is not found in a heterocyclic ring, e.g., mescaline and ephedrine (Scheme 1). The pseudoalkaloids are not derived from amino acids, e.g., steroidal alkaloids (conessine) and purine alkaloids (caffeine) (Scheme 1).

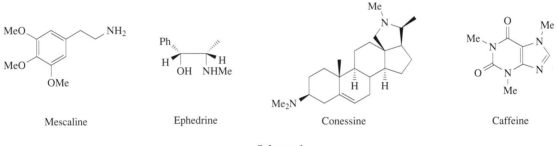

Mescaline   Ephedrine   Conessine   Caffeine

Scheme 1

The majority of alkaloids are produced in covalent combination with acetate and terpenoid units, and in aromatic hydroxylations, methylations, etc. Thus in addition to amino acids, the alkaloids are also derived either from the purine nucleus or by the addition of ammonia or an alkylamine to several terpenoid units. However, in this chapter, emphasis is placed on alkaloids derived from amino acids (Table 1).

The alkaloids are classified into several biogenetically related groups, formed by different biosynthetic routes. The biosynthesis of some important groups of alkaloids has been reviewed.[3–12]

4.03.2 ALKALOIDS DERIVED FROM ORNITHINE

4.03.2.1 Tropane Alkaloids

The tropane alkaloids are a structurally well-defined group of natural products, and have mydriatic and anesthetic properties. The tropane alkaloids mainly occur in the family Solanaceae, and are best known from species of *Datura*, *Atropa*, *Hyoscyamus*, *Scopotia*, and *Duboisia*.

About 200 tropane alkaloids have been isolated and most of the medicinally important tropane alkaloids are obtained from plants. However, tremendous efforts have been made to develop economically feasible methods using tissue or cell culture techniques[13–20] to produce hyoscyamine, scopolamine, anisodamine, and anisodine.[21–26] However, the tropane alkaloid content in cell culture is generally lower than in intact plants.[27] The root cultures of some species, especially when *Agrobacterium*-mediated transformations are used, hold promise for tropane alkaloid production.

Table 1 Some important groups of alkaloids derived from amino acids.

Ornithine
Tropane, pyrrolidine, nicotine, pyrrolizidine

Lysine
Quinolizidine, sedum, lobeline, piperidine, anabasine, pelletierine, lythrine

Tyrosine and phenylalanine
Morphine, isoquinoline, erythrina, Amaryllidaceae alkaloids, securinine, betanidin, melanin, mescaline, ephedrine, pellotine, aporphine, pavine, protopine, protoberberine, rhoeadine, benzophenanthridine, dibenzopyrrocoline, morphinandienone, cularine

Tryptophan
Indole, *Rauwolfia* alkaloids, strychnine, ergot, diketopiperazines, cinchona, yohimbine, pyrrolnitrin, secodine, carbazole, physostigmine, iboga, pandoline, corynanthe, melodinus, camptothecine, ajmalicine, sarpagine, β-carboline

Histidine
Imidazole alkaloids

A review of the biosynthesis of tropane alkaloids has been published.[28] The biochemical pathway of tropane alkaloid synthesis is shown in Schemes 2 and 3.[12,29] The important major aromatic tropeines, hyoscyamine and hyoscine, act as acetylcholine antagonists[30] and have stimulated considerable interest in determining their biosynthetic pathways.[12,28,29]

Hyoscyamine is the most common tropane alkaloid and was first isolated from henbane by Geiger[31] in 1833, and it was subsequently isolated from *Datura* species. Leete[12] proposed that δ-N-methylornithine is not an intermediate in the formation of N-methylputrescine, but rather ornithine is decarboxylated to afford a "bound form" of putrescine, which is then methylated to yield first a "bound form" of N-methylputrescine and then N-methylputrescine itself. Oxidation of N-methylputrescine leads to 4-methylaminobutanal, which then cyclizes to the N-methyl-Δ^1-pyrrolinium salt (Scheme 2). The "bound form" of putrescine may also be first transformed into putrescine and then methylated to N-methylputrescine. The N-methyl-Δ^1-pyrrolinium salt is subsequently formed via 4-methylaminobutanal. Two molecules of acetyl-coenzyme A then condense with the N-methyl-Δ^1-pyrrolinium salt to form an intermediate, which then leads, via the normal pathway (putrescine$\rightarrow$ intermediates I, II, III), first to tropinone and tropine and then to hyoscyamine and related compounds (Scheme 2).

Leete[12] proposed a new biosynthetic pathway for cocaine and suggested that the N-methyl-Δ^1-pyrrolinium salt reacts with one molecule of acetyl-coenzyme A to give the coenzyme A thioester of 1-methylpyrrolidine-2-acetic acid, which reacts with another molecule of acetyl-coenzyme A to yield the thioester, which is first oxidized to the iminium ion and then transformed by a Mannich reaction into the tropinone intermediate. Transesterification to 2-methoxycarbonyl-3-tropinone followed by reduction leads to the corresponding alcohol, methylecgonine. Benzoylation of methylecgonine completes the biosynthetic formation of cocaine (Scheme 3).[32] The β-keto ester was established as a precursor to cocaine by feeding methyl (R,S)-[1,2-$^{13}C_2$,1-^{14}C]-4-methyl-2-pyrrolidinyl-3-oxobutanoate to *Erythroxylum coca*.[33]

It was shown[34–38] that scopolamine, a well known metabolite of hyoscyamine, is not formed via 6,7-dehydrohyoscyamine as previously thought.[39] Rather, the 6β-hydroxy group of the initially formed 6β-hydroxyhyoscyamine (=anisodamine) attacks directly the C-7 position of 6β-hydroxyhyoscyamine, giving rise to the epoxy ring (Scheme 2).

The incorporation of phenylalanine into hyoscyamine was demonstrated in *Datura stramonium* plants by ^{14}C-labeling studies. Hydrolysis of the alkaloid confirmed that the label was in the acidic moiety of the tropane ester.[40,41] 1-Methylpyrrolidine-2-acetic acid, [1′,2′-$^{13}C_2$,1′-^{14}C]-1-methylpyrrolidine-2-acetic acid, and [2′-^{14}C]-1-methylpyrrolidine-2-acetic acid N-acetylcysteamine thioester were not efficient precursors to the tropane alkaloids or cuscohygrine in *E. coca* and *Datura innoxia*.

Putrescine N-methyltransferase (PMT) activity leads to a rapid loss of alkaloid biogenesis in transformed root cultures of *D. stramonium*[42] and *Nicotiana tabacum*[43] when root cultures are treated with a combination of α-naphthaleneacetic acid (NAA) and kinetin (N^6-furfurylaminopurine). However, full biogenesis of the alkaloid is restored in phytohormone-free conditions with the restitution of the rooty phenotype.[42,44] Other enzymes of tropane alkaloid synthesis are also decreased in *D. stramonium* roots treated with phytohormone but their loss from the root cultures is not

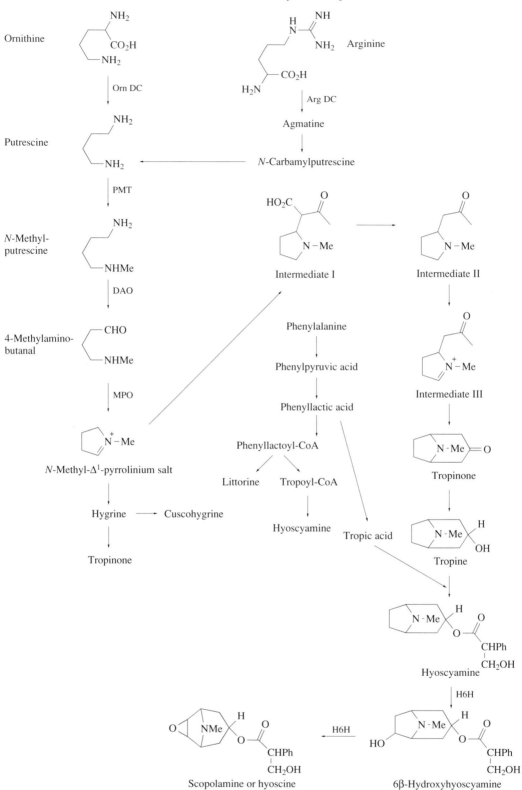

Arg DC = arginine decarboxylase; Orn DC = ornithine decarboxylase; PMT = putrescine *N*-methyltransferase; DAO = diamine oxidase; TR = tropinone reductase; H6H = hyoscyamine 6β-hydroxylase; MPO = N-methyl-Δ'-pyrrolineum oxidase

Scheme 2

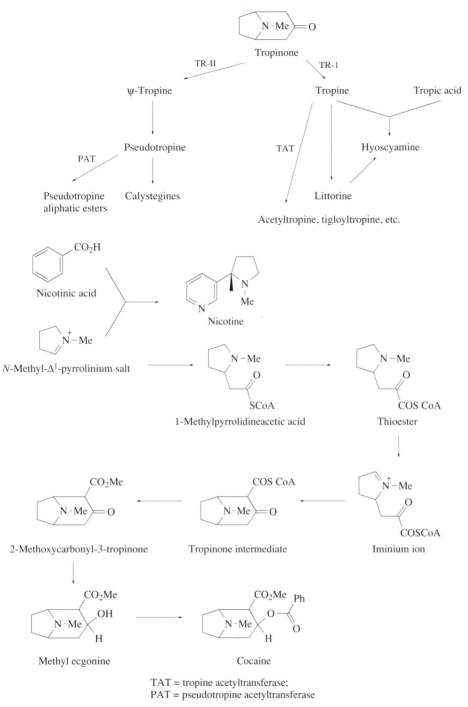

Scheme 3

complete.[42] Some enzyme activity is responsible for the formation of an ester between tropine and phenyllactic acid and the rearrangement of the product, littorine, to hyoscyamine, the major component accumulated by root cultures (Scheme 2). Feeding experiments[45] have suggested that the esterification/rearrangement process plays a key role in determining the rate of hyoscyamine accumulation in the root cultures. *In vivo* ^{15}N NMR spectroscopy has been used to study the metabolism of [^{15}N]tropinone in transformed root cultures of *D. stramonium*.[46] It was shown that dedifferentiated cultures are capable of metabolizing [^{15}N]tropinone to [^{15}N]tropine and [^{15}N]tropine esters but cannot convert [^{15}N]tropinone into [^{15}N]hyoscyamine.[46] Two enzymes of tropinone reduction, namely TR-I (tropine-forming) and TR-II (pseudotropine-forming)[47] were highly active.

Drager *et al.*[48] reported that acetyltropine was found in tropinone-fed cultures, rather than hyoscyamine. Pseudotropine, the major alkaloid of *Convolvulus arvensis*, has been suggested to be the biosynthetic precursor of the calystegines.[49,50] Calystegines (polyhydroxytropanes) have been reported from roots of *C. arvensis*.

Tropane alkaloids are commonly derived from ornithine and/or arginine by the action of ornithine decarboxylase and arginine decarboxylase. A symmetric diamine putrescine, which is found in plants, is metabolized to the polyamines spermidine and spermine. Both amino acids can be metabolized to putrescine, but arginine supplies most of the putrescine for alkaloid biosynthesis.[8] Arginine decarboxylase (ArgDc) genes have been cloned from *Escherichia coli*[51] and barley[52] and a partial cDNA clone enclosing a polypeptide homologous to barley ArgDc has been isolated from a tomato cDNA library.[53] PMT is the first committed enzyme, which drives the flow of primary nitrogen away from polyamine biosynthesis to alkaloid biosynthesis. *N*-Methylputrescine is formed and is oxidatively deaminated to 4-aminobutanal, which cyclizes to the 1-methyl-Δ^1-pyrrolinium cation; the reaction is catalyzed by a diamine oxidase. The *N*-methylputrescine (MP) content in *Hyoscyamus albus* root cultures is five times greater than the putrescine content in alkaloid biosynthesis;[54] putrescine is probably metabolized first to MP and then to the 1-methyl-Δ^1-pyrrolinium cation, rather than being deaminated directly by diamine oxidase. The 1-methyl-Δ^1-pyrrolinium cation is a reactive species. Its imino bond condenses *in vitro* with acetoacetic acid to give hygrine, a precursor of the tropane ring, whereas its coupling with nicotinic acid is presumed to form nicotine. Most tropane alkaloids, such as the hyoscyamine type, are restricted in Solanaceae but many have been found in several other genera.[55]

The first intermediate with a tropane ring is the ketone tropinone, which is located at the branching point in the biosynthetic route. Two different enzymes that reduce the keto group of tropinone have been found: tropinone reductase-I (TR-I) forms tropine, which has a 3α-hydroxy group, whereas tropinone reductase-II (TR-II) gives ψ-tropine, which has a 3β-hydroxy group. Hence the two alcohol products are diastereomers of each other. Tropine is metabolized to hyoscyamine and scopolamine. Hyoscyamine conversion to scopolamine proceeds in two oxidative steps: (i) hydroxylation at the 6β-position of the tropane ring,[56] followed by (ii) intramolecular epoxide formation by removal of the 7β-hydrogen.[57] Expression of a hyoscyamine 6β-hydroxylase (H6H) cDNA clone in *E. coli* has demonstrated that H6H catalyzes both reactions and that the hydroxylase activity is 40 times greater than the epoxidase activity.[58]

4.03.2.2 Pyrrolidine Alkaloids

The pyrrolidine alkaloids are found in nature in various forms. Pyrrolidine and its *N*-methyl derivative occur in tobacco (*N. tabacum* L.), wild carrot (*Daucus carota* L.), and deadly nightshade (*Atropa belladonna* L.). Hygrine and cuscohygrine occur in the Peruvian coca shrub (*Erythroxylon truxillense* Rusby). Many pyrrolidine alkaloids are derived by combination of a cyclized ornithine and an acetoacetate or polyacetate unit. Chromic acid oxidation of hygrine forms hygrinic acid, which could be related to L-proline. A new pyrrolidine alkaloid irnine has been isolated from the Moroccan tuber *Arisarum vulgare* and its structure was demonstrated on the basis of 2-D ^{13}C COSY studies.[59] A new pyrroline–pyrrolidine alkaloid has been isolated from the bulbs of *Lilium candidum* and its structure has been deduced from MS and ^{1}H NMR studies.[60]

Acetoacetate anion reacts with *N*-methylpyrrolinium cation to form an intermediate that undergoes double decarboxylation to form hygrine. Nucleophilic attack on a second molecule of *N*-methyl-Δ^1-pyrrolinium cation gives cuscohygrine (Scheme 4). The *N*-methylpyrrolinium salt is an important intermediate in the biosynthesis of pyrrolidine alkaloids[61] and is formed from *N*-methylputrescine. The enzyme which catalyzes the oxidative conversion has been purified to homogeneity from transformed roots of *N. tabacum* and has been subjected to immunocharacterization.[62]

The first labeled compound to be tested as a precursor of the pyrrolidine alkaloids was ornithine. The ant venom alkaloid and the corresponding pyrroline and nitrone have been utilized in the enantioselective synthesis of L-pyroglutamic acid[62] and the antibiotic (−)-anisomycin,[64] respectively. Racemic pyrrolidine alkaloids such as nurruspoline and ruspolinone have been synthesized[65] from 2-phenylsulfonylpyrrolidine derivatives. Jones and Woo[66] proposed an alternative synthesis of ruspolinone from (2*S*)-proline. The neurotoxic alkaloid (+)-anatoxin-a, isolated from blue–green algae, has been synthesized from L-pyroglutamic acid.[67] Another alkaloid, aphanorphine, a potential analgesic, has been synthesized from the L-pyroglutamic acid ester, also found in blue–green algae.[68]

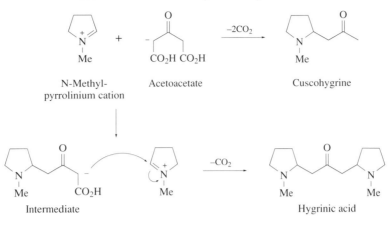

Scheme 4

4.03.2.3 Pyrrolizidine Alkaloids

The pyrrolizidine alkaloids (PAs) are ester alkaloids consisting of a necine base moiety, esterified with one or two necic acids.[69] Tracer studies have shown that the necine base is biosynthesized from two putrescine molecules via homospermidine.[70–72] The pyrrolizidine alkaloids contain one nitrogen at the bridge of two five-membered rings known as 1-azabicyclo[0.3.3]octane. These alkaloids are widely distributed in several families of the plant kingdom, viz., Asteraceae, Boraginaceae, and Fabaceae. About 400 alkaloids are known; many of them are toxic to livestock, hepatotoxic, mutagenic, and carcinogenic,[73] and play a role in insect–plant interactions.[70]

4.03.2.3.1 Synthesis of necines

There are two aspects of PA biosynthesis: (i) the synthesis of the necine group and (ii) the synthesis of the necic acid group. Important members of the necine group found in the PA are supinidine, heliotridine, retronecine, and platynecine, of which the most important is retronecine.

Retronecine is biosynthesized from two molecules of putrescine via homospermidine.[61,72] The enzymatic formation of homospermidine was shown to be catalyzed by homospermidine synthase (EC 2.5.1.44).[72] In an NAD$^+$ dependent reaction, this enzyme combines an aminobutyl group derived either from putrescine or from spermidine with a second mole of putrescine (Scheme 5). Ornithine is a precursor of the retronecine portion of several alkaloids in *Crotalaria* and *Senecio* species. Two four-carbon chains from ornithine enter retronecine nonsymmetrically. Since the levels of activity at the two labeled centers differ, the pathway involves two different intermediates, each derived by a separate path from ornithine. The cycloaddition of a nitroalkene for the synthesis of a wide variety of necines has been described by Denmark and co-workers, e.g., (−)-hastanecine[74] and (−)-rosmarinecine.[75]

Cynoglossum officinale (Boraginaceae) contains high levels of pyrrolizidine alkaloids.[76] *C. officinale* was the first plant species known to be able to synthesize PAs in both shoots and roots. N-Oxides of PAs are trachelanthamine, viridiflorine, 7-angeloylheliotridine, rinderine, echinatine, 3-acetylchinatine, amabitine, and heliosupine.[77] The youngest leaves of *C. officinale* contained PA concentrations up to 190 times higher than the levels found in older leaves.[78] 7-Acetyleuropine has been isolated from *Heliotropium bovei* and its structure determined by one- and two-dimensional NMR spectroscopy.[79]

A number of reviews relating to PAs have been published,[80,81] and the formation of PAs and other ornithine derived alkaloids has been reviewed.[82,83] The role of PAs and other secondary plant metabolites in plant–insect interactions[84] has been discussed. The loline group of PAs appear to be specific to grasses infected by *Acrumonium coenophialum*.[85] The (±)-Geissman–Waiss lactone is

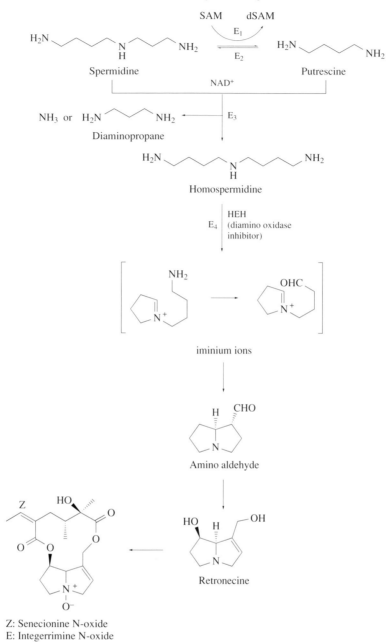

Z: Senecionine N-oxide
E: Integerrimine N-oxide

SAM = *S*-Adenosylmethionine; dSAM = Decarboxylated SAM; MTA = 5'-Methylthioadenosine;
HEH = β-Hydroxyethylhydrazine; E_1 = SAM decarboxylase (EC 4.1.1.50); E_2 = Spermidine synthase (EC 2.5.1.16)
Putrescine producing enzyme activity (HEH insensitive); E_3 = Homospermidine synthase (EC 2.5.1.44);
E_4 = PA biosynthetic oxidase activity (HEH sensitive)

Scheme 5

an important intermediate in the synthesis of necines, including retronecine, platynecine,[40] and croalbinecine.[86–88] The synthesis of ($\pm$)-xenovenine and related alkaloids[43] found in ant venom have been described by Robins[89] and Takahata *et al.*[90]

The most important PAs are retrosine[91] from *Senecio isatideus* (Compositae), rosmarinine[73] from *Senecio pleistocephalus*, monocrotaline[92] from *Crotalaria retusa* (Leguminosae), and echinatine[93] from *Heliotropium indicum* (Boraginaceae).[91]

A rapid method for the determination of PAs in plant extracts involves reduction of the *N*-oxides with an oxygen-absorbing resin followed by ion-exchange purification to carry out GC or TLC detection.[94] Polydimethylsiloxane-coated columns are considered to be the best for comparative studies. Geometrical isomers of PAs in complex mixtures can be determined by this technique.

Proton NMR spectra of more than 350 PAs have been compiled.[95] Rycroft *et al.*[96] reported a detailed ^{1}H and ^{13}C NMR spectroscopic study of 13-*O*-acetyldicrotaline. A selective long-range ^{13}C DEPT NMR technique was used to elucidate the structures of platyphylline and neoplatyphylline from *Senecio fuberi* Hensl.[97] The two PAs, namely intergerrimine and usaramine (18-hydroxy-integerrimine), were identified through their optical, rotational, IR, mass, 2D ^{1}H and ^{13}C NMR spectra and the structure of the latter was verified by COSY and DEPT methods.[97]

The association constants of α- and β-cyclodextrins with many PAs have been determined.[98] Surveillance of natural toxicants in the United Kingdom food supply has indicated that comfrey-containing products are the most important sources of PAs.[99]

Langer and co-workers[100] have developed an immunoassay with antibodies against retrorine to detect the closely related senecionine, the main alkaloid in various Asteraceae. The immunoassay approach is also useful as a screening system for the selection of PAs from plant extracts of *Petasites hybridus*.[100]

Freer *et al.*[101] studied the stereochemistry of the enzymatic route involved in the formation of the otonecine portion of emiline in *Emilia flammea*. The labeling patterns are the same as those obtained for retronecine. Studies of the stereospecificity of the enzymatic process involved the decarboxylases, diamine oxidases, hydroxylases,[102] and homospermidine synthase.[103]

4.03.2.3.2 *Synthesis of necic acids*

A number of alkaloids are known which contain a necic acid moiety, such as senecionine,[104] (+)-yamataimine,[105] stannoxane,[106] axillaridine,[107] methylene acetal,[108] and α-phosphonolactones.[109] The biosynthetic pathway of necic acid is not known. However, the synthesis of senecionine has been reported from a lactone.[104] A coupling reaction involving stannoxane gave the seco acid regio-selectively. This was cyclized under Keck's conditions to give a mixture of a macrocycle and its isomer. This macrocycle was then deprotected to afford the desired senecionine (Scheme 6).

Niwa *et al.*[105] carried out the synthesis of the 12-membered macrocycle (+)-yamataimine using, in part, a similar approach (Scheme 7). In this synthesis the lactone ester was obtained by alkylation of the keto ester, followed by a Baeyer–Villiger oxidation affording the two lactones shown in a 1 : 1 ratio, which were separated chromatographically. Introduction of an ethylidene group into the lactone and reduction of the alkene formed the lactone.

Axillaridine was isolated previously from the seeds of *Crotalaria axillaris* Ait. (Fabaceae).[107] It has eight stereoisomers in the lactones portion and the acid portion has three chiral centers. The eight stereoisomers have been synthesized by Matsumoto *et al.*[110] and the absolute configuration of each stereoisomer has been established. However, apparently the necic acid component of axillaridine has not been isolated.

4.03.3 ALKALOIDS DERIVED FROM LYSINE

4.03.3.1 Quinolizidine Alkaloids

The quinolizidine alkaloids (QAs) are mainly found in the subfamily Papilionoideae of the Fabaceae (Leguminosae) and are especially abundant in the tribes Genisteae, Sophoreae, and Thermopsideae.[111,112] The genus *Lupinus* forms a distinct subtribe within the Genisteae.[113,114] More than 500 lupin species have been described and identified by capillary GC and GC–electron ionization MS.[115] Some QAs exhibit interesting biological activities towards animals, insects, and plants.[116] The value of QAs as chemotaxonomic markers for plants of the Leguminosae has been demonstrated by Greinwald and co-workers.[117] Lupins produce potent defense chemicals affecting a wide array of targets. It has been shown that QAs provide an efficient defense against herbivores.[118,119]

QA esters are mainly distributed in the plants of the genera *Lupinus*,[120] *Cytisus*,[121] *Pearsonia*,[122] and *Rothia*[123] as esters of tigilic acid and *p*-coumaric acid. Some esters of QAs are formed during seedling development.[124] *Lupinus* plants accumulate two types of QAs (Scheme 8): tetracyclic alkaloids (lupanine), such as sparteine and martine, and the bicyclic QAs (lupinine). Some tricyclic

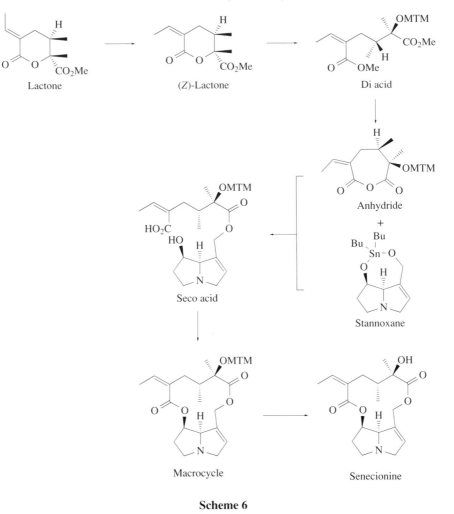

Scheme 6

pyridones such as cytisine are also found in *Lupinus* species. Two ester alkaloids, cineverine[125] and a new compound, cineroctine, have been isolated from twigs of *Genista cinerae* subsp. *cinerae*.[126]

In vivo tracer experiments revealed that QAs are derived from L-lysine via cadaverine as the first detectable intermediate.[127,128] The quinolizidine ring is formed from the cyclization of cadaverine units via an enzyme-bound intermediate.[129,130] Some QAs are modified by intracellular enzymes through dehydrogenation, hydroxylation, esterification, and related reactions.

Biochemical aspects of QA biosynthesis are not well documented compared with the other alkaloids.[6] A review has been published on QAs and their biosynthesis.[131] Saito *et al.*[132] demonstrated the presence of two alkaloid acyltransferases, HMT/HLTase and ECTase, in *Lupinus hirsutus* (Scheme 9). The same group also purified and characterized the enzymes HMT/HLTase from *Lupinus albus* seedlings.[120] Suzuki *et al.*[133] demonstrated the intracellular localization of two alkaloid acyltransferases and their functions in putative metabolic and transport pathways of QAs in *Lupinus* plant cells and in the green seedlings of *L. hirsutus* and *L. albus*.

Robinson[134] suggested that the bicyclic lupinines are formed from two molecules of lysine via cadaverine. The first studies on lupinine biosynthesis were carried out by Schutte.[135] He showed that ^{14}C-labeled lysine and [1,5-14C]cadaverine were incorporated into lupinine. After degradation of the labeled lupinine, one-quarter of the total radioactivity was located in the hydroxymethyl group[135] and half could be accounted for in the C-4 and C-6 atoms by a methylation process and Hofmann degradation of lupinine followed by ozonolysis of the products.[136] Studies by Schutte[135] with ^{14}C-labeled lysine and cadaverine, involving the establishment of partial labeling patterns in alkaloids by degradation, indicated that tetracyclic QAs, such as (−)-sparteine, (±)-lupane, and subsequently matrine, are derived from three units of these precursors. Alkaloid metabolism and gene expression

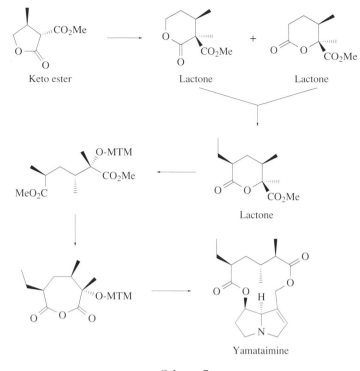

Scheme 7

in cell suspension cultures of *Lupinus* species has been summarized by Wink *et al.*[137] Michael[138,139] published a review on the synthesis of QAs in many Leguminosae plants.

Structurally QAs are constituted from oligomers of cadaverine units, two units for instance being used to construct lupinine. The incorporation of proton-labeled cadaverine into a range of QAs has been examined.[140,141] [2-^{2}H]Cadaverine was incorporated into lupinine in *Lupinus luteus* and three (2,2,4,4-^{2}H$_4$]cadaverines gave lupinine alkaloids.[142]

A novel *O*-trigoyltransferase (acetyltransferase) enzyme associated with the biosynthesis of QAs has been purified to homogeneity and characterized from *Lupinus termis* seedlings.[120] Two isoforms of the enzyme were identified that catalyzed the transfer of tiglic acid from tigloyl-CoA to (−)-13α-hydroxymultiflorine and (+)-13α-hydroxylupanine. Both of these are tetracyclic with an axial hydroxyl group at C-13 and with a 7S,9S configuration.[120] No enzyme activity was detected towards the bicyclic QAs such as epilupinine and 1-lupinine and (7R,9R)-alkaloids, but a new alkaloid, (−)-3β-hydroxy-13α-tigloyloxylupanine, has been isolated from *Cytisus scoparius* seedlings and the enzyme tigloyltransferase was detected in cell-free extracts of this plant.[121] Previously, Saito and co-workers[132,143] described the acyltransferase involved in the biosynthesis of ester-bearing lupin alkaloids from *L. hirsutus* and the enzyme was partially purified and characterized. The production of alkaloids in bitter and sweet forms of *L. luteus* and *L. albus* has been studied.[143] It was concluded that differences between the sweet and bitter forms lie in the suppression of the enzyme which catalyzes the formation of the cyclic alkaloids.[143]

4.03.4 ALKALOIDS DERIVED FROM TYROSINE AND PHENYLALANINE

4.03.4.1 Morphine Alkaloids

The morphine alkaloids are found in opium poppy (*Papaver somniferum* L.), belonging to the family Papaveraceae. They are used as analgesics, antitussives, and antispasmodics. The opium poppy plant produces a number of alkaloids (Scheme 10) that are divided into two groups: (i) phenanthrene alkaloids, e.g., thebaine, morphine, and codeine, and (ii) benzylisoquinoline alkaloids, e.g., papaverine, narcotine, narceine, and reticuline.

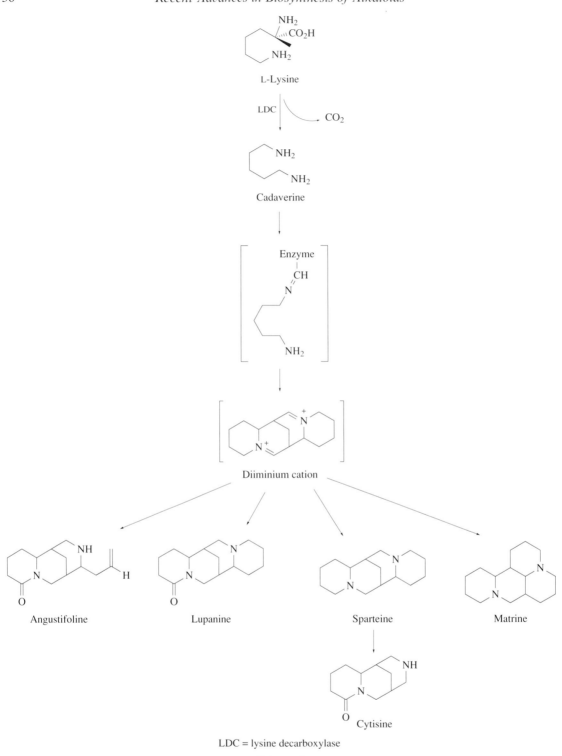

LDC = lysine decarboxylase

Scheme 8

Fairbairn and Djote[144] reported that latex is capable of synthesizing morphine from amino acids such as dihydroxyphenylalanine (DOPA) and tyrosine. Thus, latex carries enzymes required for the biosynthesis of morphine alkaloids. Toth-Lokoes *et al.*[145] reviewed the production of morphine in a variety of strains of *P. somniferum*.

The biosynthesis of opium poppy alkaloids (Scheme 11) involves mainly the following steps:

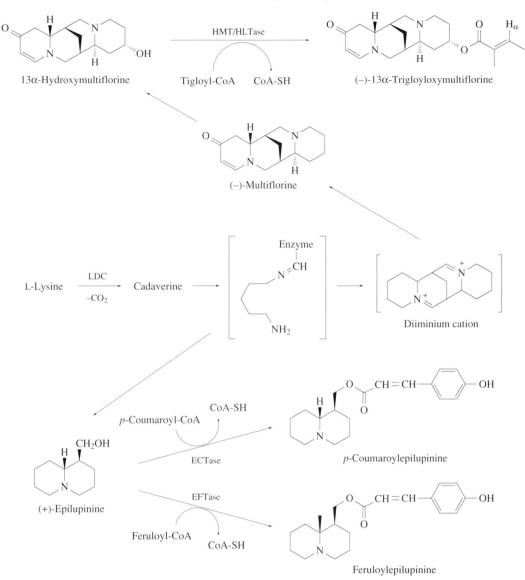

13α-Hydroxymultiflorine

HMT/HLTase

Tigloyl-CoA   CoA-SH

(−)-13α-Trigloyloxymultiflorine

(−)-Multiflorine

L-Lysine $\xrightarrow[-CO_2]{\text{LDC}}$ Cadaverine

Enzyme
N=CH

NH$_2$

Diiminium cation

CoA-SH

p-Coumaroyl-CoA

ECTase

p-Coumaroylepilupinine

CH$_2$OH

(+)-Epilupinine

EFTase

Feruloyl-CoA   CoA-SH

Feruloylepilupinine

HMTase = 13α-hydroxymultiflorine *O*-tiglolyltransferase;
HLTase = 13α-hydroxylupanine *O*-tigloyltransferase; ECTase = epilupinine *O*-coumaroyltransferase;
EFTase = epilupinine *O*-feruloyltransferase

Scheme 9

  (i) decarboxylation and hydroxylation of amino acids such as DOPA, tyrosine, and phenyl-alanine;
  (ii) condensation of two units of amino acid to make the basic structure of the benzylisoquinoline alkaloid and norlaudanosoline;
  (iii) methylation of norlaudanosoline to give reticuline;
  (iv) oxidation and phenol coupling to give phenanthrene and benzylisoquinoline alkaloids; and
  (v) synthesis of demethylated codeine and morphine from thebaine.
  Satutaridinol 7-*O*-acetyltransferase is acetyl-CoA dependent, and the product satutaridinol-7-*O*-acetate is formed, which undergoes ring closure *in vivo* at alkaline pH.[142] The conversion of neopinone into codeinone is a spontaneous reaction.[146] A biosynthetic route to a morphine alkaloid in cell cultures of *P. somniferum* plants has been published.[146–149]

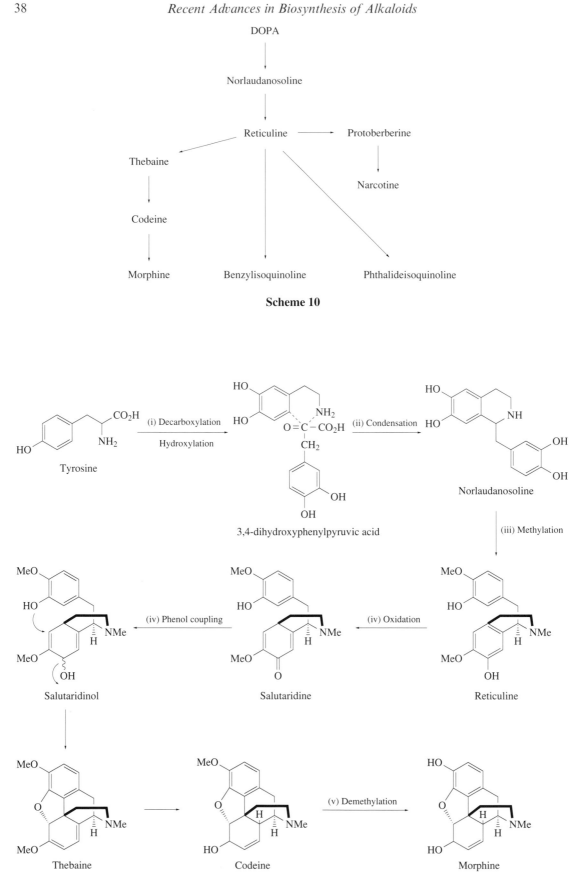

Scheme 10

Scheme 11

4.03.4.1.1 *Biosynthesis of benzylisoquinoline alkaloids*

Reticuline is synthesized from norcoclaurine by a methylation reaction.[61,150] Reticuline is used in thebaine synthesis; the enzyme 1,2-dehydroreticuline reductase catalyzes the reduction of reticuline affording thebaine. The enzyme has been isolated from seedlings of the opium poppy, purified to homogeneity and characterized.[151] A membrane-bound cytochrome P-450 protein from the microsome was involved in the conversion of reticuline into satutaridine,[151] its catalytic action being dependent on O_2 and NADPH. An oxidoreductase enzyme, which catalyzes the reduction of satutaridine to the alcohol satutaridinol and precursors of morphinan alkaloids, has been isolated, purified, and characterized from *P. somniferum*.[152] Gerardy and Zenk[153] reported a cytosolic enzyme satutaridine:NADPH 7-oxidoreductase, which catalyzes the NADPH-dependent reduction of satutaridine to satutaridinol utilizing the *pro-S* hydride (B-type) NADPH in *P. somniferum* tissue cultures.

Hydroxylation of the amino acid tyrosine gives rise to DOPA, which is converted into dopamine by decarboxylation. Dopamine is condensed with 3,4-dihydroxyphenylpyruvic acid (formed from phenylalanine by oxidation) to form the benzylisoquinoline skeleton (Scheme 12).

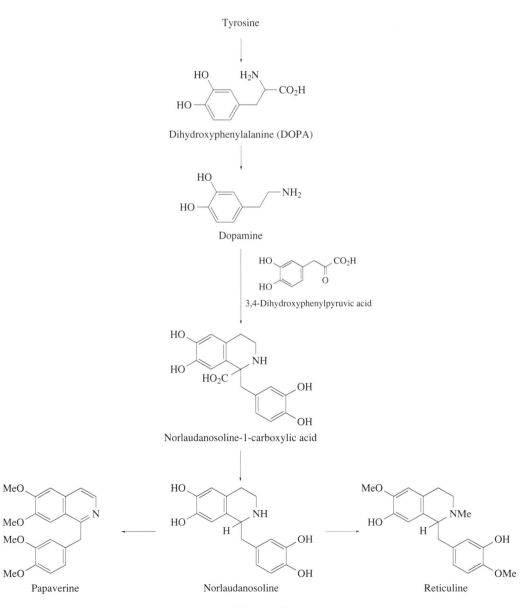

Scheme 12

4.03.4.1.2 Biosynthesis of phenanthrene alkaloids

Reticuline is the precursor to phenanthrene alkaloids. Reticuline gives rise to thebaine by an oxidative coupling reaction, and thebaine undergoes demethylation to afford codeine and morphine. Thebaine conversion to codeinone involves the formation of the intermediate neophinone by demethylation, which on rearrangement gives codeinone, which is ultimately reduced to codeine. The enzyme codeine:NADP oxidoreductase, which catalyzes the reduction of (−)-codeinone and (−)-morphinone into (−)-codeine and (−)-morphine, has been purified to homogeneity from *P. somniferum* cell cultures and differentiated poppy plants.[154] This enzyme also catalyzes the reduction of dihydrocodeinone, dihydromorphinone, naloxone, and naltrexone stereospecifically to the 6α-hydroxy compounds, but does not catalyze the reduction of neopinone or satutaridine.[154]

The aporphine types of alkaloids in poppy plants are synthesized from reticuline by phenol couplings. Protoberberine and orientatine types of alkaloid are also derived from reticuline. The use of radioactive norlaudanosoline[155] revealed the biosynthesis of phthalideisoquinoline alkaloids such as narcotine from reticuline (Scheme 13).

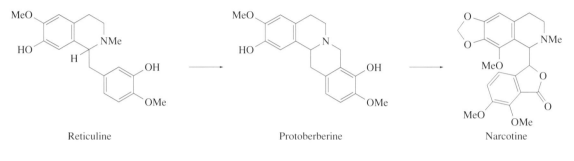

Reticuline   Protoberberine   Narcotine

Scheme 13

4.03.4.2 Isoquinoline Alkaloids

The isoquinoline alkaloids are mainly found in families, viz., Annonaceae, Berberidaceae, Magnoliaceae, Menispermaceae, Monimiaceae, Papaveraceae, and Ranunculaceae and have been reviewed by Bentley.[156,157]

Cularines, isocularines, and berberines form a group of isoquinoline alkaloids and have been isolated from some species of the genera *Cearatocapnos*,[158] *Corydalis*, *Dicentra*, and *Sarcocapnos* (Papaveraceae).[159] These alkaloids are biosynthetically related to crassifoline, a 1-benzylisoquinoline with an unusual 7,8-oxygenation pattern at the isoquinoline nucleus,[160] as shown by incorporation of radioactive crassifoline into cularine.[161]

The production of isoquinoline alkaloids is related to the crassifoline metabolism in callus and cell suspensions of *Ceratocapnos heterocarpa*.[158] Berberine is an isoquinoline alkaloid belonging to the protoberberine type and is used medicinally. The biotransformations of berberine and their tetrahydro derivatives in cell cultures of *Corydalis pallida* var. *tenuis*, which contains the protoberberine alkaloids, were estimated using LC–atmospheric pressure ionization mass spectrometry.[162]

Salsolidine, a simple isoquinoline alkaloid, has been isolated from *Arthrocnemum glaucum*.[163] The new alkaloid tamynine, of uncextamic biogenetic origin, has been isolated from *Murraya paniculata*.[164] (*S*)-Salsolidine has been obtained in a high degree of optical purity by the hydrogenation of the 1-methylenetetrahydroisoquinoline in the presence of diaryl (*S*)-2,2′-bisdiphenyl-phosphine-1,1-bisnaphthylruthenium as a catalyst.[165] A novel approach to the synthesis of pure isoquinoline alkaloids has been reported.[166] Further, a procedure for the synthesis of *N*-methyl-corydaline[167] and (*R*)-(+)-canegine has been developed.[168] The stereospecific synthesis of (*R*)-crypto-styline has been developed based on a previously reported synthesis of salsolidine.[169,170]

Exogenous feeding of a biosynthetic precursor to a culture medium may increase the final yield of alkaloids. The exogenous feeding of precursors, e.g., shikimic acid and L-phenylalanine, influenced the overall biosynthesis of cephaeline in callus cultures of *Cephaelis ipecacuanha*.[171,172] The plant *C. ipecacuanha* contains isoquinoline alkaloids, e.g., emetine, cephaeline, psychotrine, *O*-methyl-psychotrine, and emetamine. Feeding of tropic acid to *Scopolia* and *Datura* callus cultures[173] and L-phenylalanine and tyrosine to the seed callus tissue of *Datura tatula*[174] increased the total alkaloid content.

Thirteen enzymes for the biosynthesis of berberine have been identified, but their regulation is not well understood. The pathway involves several methyl transfer and oxidation reactions. The pathway from tyrosine to (*S*)-norcoclaurine is based on feeding experiments and partial characterization of the enzymes involved.[175] Three cDNA clones and four genomic clones encoding tyrosine decarboxylase (TyrDc) have been isolated from parsley,[176] and these can be used to obtain TyrDc genes from plants that produce isoquinoline alkaloids.

(*S*)-Reticuline is formed from (*S*)-norcoclaurine catalyzed by two *O*-methyltransferases, one *N*-methyltransferase, and one phenolase (Scheme 14). These two enzymes are 3′-hydroxy-*N*-methyl-coclaurine 4′-*O*-methyltransferase (4′-OMT) and noroclaurine 6-*O*-methyltransferase (6-OMT). These 6-methyltransferases have similar physical properties and were purified from *Berberis koetineana* cell cultures.[177] 6-OMT was also purified from *Coptis japonica* cell cultures, composed of 40 and 41 kDa proteins. When a cDNA clone encoded with 41 kDa protein was expressed in *E. coli*, it showed 4′-OMT but not 6-OMT activity, and suggested that 6-OMT and 4′-OMT were co-purified unintentionally.[177]

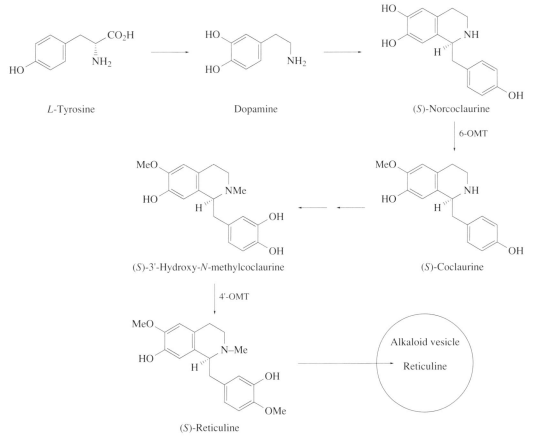

6-OMT = norcoclaurine 6-*O*-methyltransferase;
4′-OMT = 3′-hydroxy-*N*-methylcoclaurine 4′-*O*-methyltransferase

Scheme 14

(*S*)-Reticuline enters in specific vesicles (alkaloid vesicles) and forms the final product berberine (Scheme 15). First the *N*-methyl group of (*S*)-reticuline forms the berberine bridge of (*S*)-scoulerine, catalyzed by berberine bridge enzyme (BBE). BBE cDNA was isolated from *Eschscholtzia californica* cell cultures.[178] The product (*S*)-scoulerine is then methylated to (*S*)-tetrahydrocolumbamine, catalyzed by scoulerine 9-*O*-methyltransferase. A methylenedioxy bridge is formed,[179] giving (*S*)-canadine (Scheme 15). The reaction is catalyzed by canadine synthase. The methylenedioxy bridge was also formed by dimethylation by a peroxidase within an alkaloid vesicle.[180] The methylenedioxy bridge is thought to be synthesized by a cytochrome P-450 enzyme from microsomes of *Berberis thalictrum* and *Coptis* spp. in (*S*)-tetrahydrocolumbamine, but not in the columbamine (quaternary alkaloid).[181] An enzyme system that catalyzes the formation of methylenedioxy bridge at ring A of

(*S*)-canadine from (*S*)-tetrahydrocolumbamine has been isolated from *Thalictrum tuberosum* L. cell line.[181] Owing to this substrate specificity of canadine synthase, the berberine pathway of columbamine operates in *Berberis* spp. only.[182] Finally, (*S*)-canadine is oxidized to the berberine. The oxidase from *Berberis* spp. cell cultures converts several (*S*)-tetrahydroprotoberberines, such as (*S*)-canadine and (*S*)-1-benzylisoquinoline alkaloids, corresponding to quaternary alkaloids.[183] The reaction is catalyzed by (*S*)-tetrahydroprotoberberine oxidase (STOX). In the presence of oxygen, the tetrahydroprotoberberine is first oxidized by STOX to an iminium intermediate that oxidizes spontaneously to the protoberberine, although this spontaneous oxidation does not occur with the 1,2-dehydroiminium ion produced from 1-benzylisoquinoline. Canadine oxidase (COX), which converts (*S*)-canadine to berberine, exhibits biochemical properties that are different from those of STOX in *Berberis* spp. The COX acts on (*S*)-canadine but not on tetrahydroprotoberberines or norreticuline.[184] In *Hydrastis canadensis* L., the formation of canadinic acid is an intermediate step in the oxidative sequence of *Hydrastis* benzylisoquinoline alkaloids towards β-hydrastine.[185]

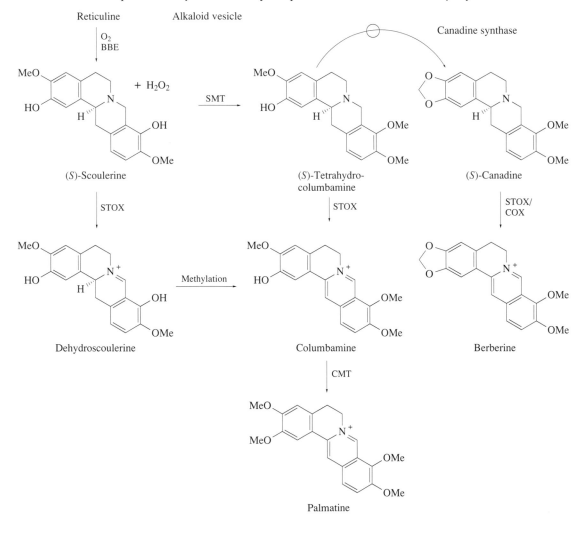

BBE = berberine bridge enzyme; SMT = scoulerine 9-*O*-methyltransferase;
CMT = columbamine *O*-methyltransferase; STOX = (*S*)-tetrahydroproterberberine oxidase;
COX = canadine oxidase

Scheme 15

4.03.4.3 *Erythrina* Alkaloids

These alkaloids are found in *Erythrina* (Leguminosae) and also in Menispermaceae (*Cocculus*, *Pachygone*, and *Hyperbaena*). Their occurrence is limited to tropical areas. The number of *Erythrina* alkaloids that have been found is ~95. The *Erythrina* alkaloids were first assumed to be formed by the oxidative coupling of a C_6—C_2—N—C_2—C_6 type intermediate, which is derived from the two molecules of DOPA. A detailed review on the *Erythrina* alkaloids has been published.[186] The biosynthetic pathway of *Erythrina* alkaloids was established as outlined in Scheme 16. The intramolecular oxidative coupling of the benzylisoquinoline alkaloid (S)-(+)-1-norprotosinomenine proceeds through a symmetrical dibenz[d,f]azonine. Benz[d,f]azonine alkaloids are considered as biosynthetic precursors of the *Erythrina* alkaloids and benz[d,f]azecines are considered as progenitors of the homoerythrinan alkaloids. A variety of *Erythrina* alkaloids have been isolated from the seeds of *Erythrina* species, including dienoid, alkenoid, and lactonic derivatives, which are also found in leaves, stems, bark, roots, pods, and flowers.

The feeding of [4'-methoxy-^{14}C]norprotosinomenine to *E. cristagalli* produced erythaline labeled at the methoxy and the methylenedioxycarbon atoms. The intermediate, namely chiral dibenzazonine, racemizes rapidly and only (5S)-(−)-erysodienone is the precursor of erythraline and erythroidines. These alkaloids are biosynthesized in the growing region of the plant.[187] The biosynthesis of *Erythrina* alkaloids has been explained mainly by Barton and co-workers.[188] [2-^{14}C]Tyrosine labeled the C-8 and C-10 positions of β-erythroidine in *Erythrina berteroana*, and norprotosinomenine and erysodienone were precursors of erythraline in *E. cristagalli*. (+)-Norprotosinomenine is also an effective precursor of cocculine and cocculidine in *Cocculus laurifolius*.[189] The biosynthesis of abnormal *Erythrina* alkaloids starts from norprotosinomenine. Reduction and dienol–benzene rearrangement reactions involved the route through intermediates I → VI → VII → VIII → V shown in Scheme 17, and were established by double radiolabeling with (±)-[1-^{3}H,4'-methoxy-^{14}C]norprotosinomenine and (±)-[1-^{3}H,7-methoxy-^{14}C]norprotosinomenine in *C. laurifolius*.[189] The ^{14}C/^{3}H ratio in the product base, cocculidine, was unchanged. Another route proceeds through intermediates I dibenz[d,f]azonine → II → III → IV → V (Scheme 18). These two routes may be confirmed by comparing the radioactivities of 3-OMe and 14-OMe and/or ^{3}H-8 and ^{3}H-10 in the product. The (±)-isomer of norprotosinomenine was incorporated into cocculidine about 60 times more efficiently than the (−)-isomer. The plant can convert isococculidine into cocculidine with high efficiency.[189] Therefore, feeding the radiolabeled 1,2-alkene isococculidine and the 6-alkene cocculidine into the dienoid alkaloids cocculinine and cocculine revealed that the former was incorporated more efficiently.[190] *O*-Demethylation is the last step in the biosynthesis of phenolic bases, and has been shown by the incorporation method.[189,190] 8-Oxo alkaloids are formed through oxidation at the C-8 position.

Ring D-modified alkaloids, such as erymelanthine, are formed from the aromatic alkaloid by oxidative cleavage and recyclization reactions (Scheme 19). The enzymes involved in the biosynthesis of *Erythrina* alkaloids are not known and very little is known about the pharmacological action of homoerythrinan alkaloids.

4.03.4.4 Amaryllidaceae Alkaloids

The biosynthesis of the Amaryllidaceae alkaloids has been studied beginning in 1957 by Barton *et al.*[191] The most important Amaryllidaceae alkaloids are lycorine,[192–196] galanthine,[194–197] galanthamine,[194,195,197–199] sanguinine,[195–197] and crinine.[200,201] Barton *et al.*[196] postulated that the Amaryllidaceae alkaloids were biogenetically derived from norbelladine. Among these alkaloids, there are three major structural groups, the lycorine, galanthamine, and crinine types, synthesized by various phenolic coupling processes (Scheme 20). Labeled samples of norbelladine and methylnorbelladine are incorporated into alkaloids of the three series. Many new Amaryllidaceae alkaloids have been synthesized and have been reviewed.[202,203]

The methylation of norbelladine is crucial in determining the structural nature of the precursor; thus, *p*-*O*-methylnorbelladine was not incorporated into galanthamine, whereas *N*-methylnorbelladine was incorporated, as demonstrated by Barton and Cohen.[204] *N*-Methylation of norbelladine inhibits entry of the precursor into alkaloids of the lycorine and crinine group (Scheme 20) but does not inhibit its incorporation into the galanthamine group. Thus, methylation of the precursor controls its entry into biosynthetic pathways leading to the different Amaryllidaceae alkaloids. Labeled *p*-*O*-methyl and *N*,*p*'-*O*-dimethylnorbelladine are incorporated into haemanthamine and galanthamine, respectively (Scheme 21). Methylation at the *m*'-O_2 of norbelladine appears to block the phenol coupling reaction and *m*',*o*-methylnorbelladine was not incorporated into haemanthamine. The origin of the C_6—C_2—N—C_1—C_6 system of norbelladine from primary metabolites has been established.[205]

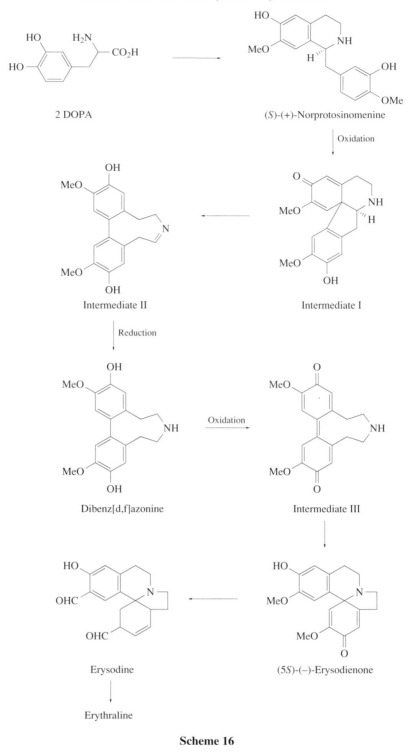

2 DOPA

(S)-(+)-Norprotosinomenine

Oxidation

Intermediate II

Intermediate I

Reduction

Dibenz[d,f]azonine

Intermediate III

Oxidation

Erysodine

(5S)-(−)-Erysodienone

Erythraline

Scheme 16

The C_6—C_2 unit is derived from tyrosine and the C_6—C_1 unit from phenylalanine, but not from DOPA. Through tracer evidence it is known that the norbelladine originates by reduction of the condensation product of tyramine with protocatechuric aldehyde, intermediates which originate from tyrosine and phenylalanine, respectively (Scheme 22). The phenanthridine nucleus can be considered as a catabolic product formed from the cleavage of the "ethano" bridge of crinane precursors. A similar type of catabolic process has been proposed for the biogenesis of ismine and postulated for related oxophenanthridine and phenanthridinium alkaloids.[206]

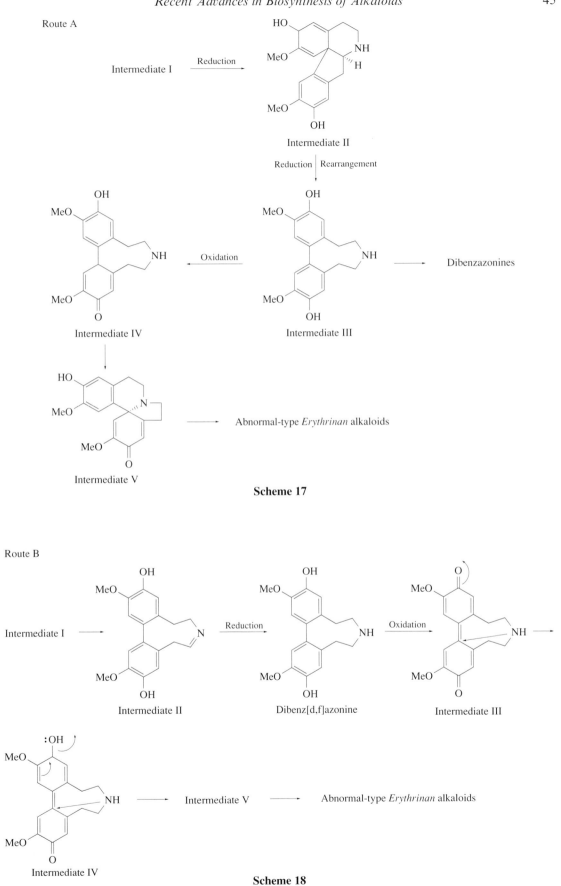

Route A

Intermediate I → Reduction → Intermediate II

Intermediate II → Reduction | Rearrangement → Intermediate III → Dibenzazonines

Intermediate III → Oxidation → Intermediate IV

Intermediate IV → Intermediate V

Intermediate V → Abnormal-type *Erythrinan* alkaloids

Scheme 17

Route B

Intermediate I → Intermediate II → Reduction → Dibenz[d,f]azonine → Oxidation → Intermediate III →

Intermediate IV → Intermediate V → Abnormal-type *Erythrinan* alkaloids

Scheme 18

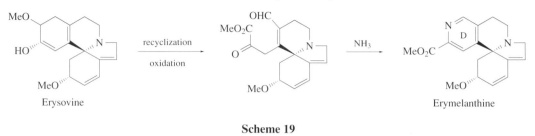

Erysovine   Erymelanthine

Scheme 19

4.03.5 ALKALOIDS DERIVED FROM TRYPTOPHAN

4.03.5.1 Monoterpenoid Indole Alkaloids

The medicinally important terpenoid indole alkaloids are vindoline, catharanthine, ajmalicine, ajmaline, serpentine, β-yohimbine, vinblastine (VLB), and vincristine (VCR). The most important indole alkaloids, VLB and VCR, are clinically useful anticancer agents.[207–209] Ajmalicine is used in the treatment of circulatory diseases. However, VLB accumulates only in trace amounts in the plant and much research has been devoted to the improvement of large-scale production of alkaloids in *Catharanthus roseus*.[3]

The alkaloids are mainly found in Apocynaceae family, i.e., *C. roseus*, *Rauwolfia serpentina*, *Rauwolfia sellowii*, *Leuconotis eugenifolius*, *Tabernaemontana divaricate*, etc. All these medicinal alkaloids are basically indole alkaloids consisting of tryptamine provided by tryptophan and a terpenoid part provided by the iridoid glucoside secologanin. Tryptamine and secologanin are condensed to form strictosidine, which is the common precursor of all indole alkaloids. Strictosidine is deglucosylated and then converted to ajmalicine, catharanthine, vindoline, and other indole alkaloids (Scheme 23). The biosynthesis of indole alkaloids requires a number of enzymatic reactions in plants. A detailed review of the enzymology of the biosynthesis of indole alkaloids in *C. roseus* has been published.[3,4] Vinblastine, a dimeric indole alkaloid, is formed by the coupling of vindoline and catharanthine, catalyzed by horseradish peroxidase.

Cell cultures of *C. roseus* have been developed as a major model system for cell biotechnological studies. Although much effort was put into the production of dimeric alkaloids by means of *C. roseus* cell culture, the cell cultures do not produce dimeric and monomeric indole alkaloids (vindoline). Only catharanthine is produced in considerable amounts[210] in shoot cultures[211] and differentiated tissues[212,213] and occasionally in callus cultures.[214]

Catharanthus seedlings rapidly accumulate catharanthine, vindoline, and tabersonine, reaching a plateau after a few days of seed germination.[215] Jasmonate, a fatty acid-derived plant signaling molecule, and methyl jasmonate (MeJa) enhance the accumulation of monomeric indole alkaloids in seedlings.[215] Jasmonate is produced by the so-called "lipoxygenase pathway" and increases in the activities of some enzymes in this pathway have been observed in germinating seeds.[216–218] The synthesis of lipoxygenase products has been implicated in some defense reactions in plants.[219] Accumulation of endogenous MeJa may regulate alkaloid synthesis during the germination of *Catharanthus*. Indole alkaloid synthesis has been influenced by a precursor of MeJa, 13-hydroperoxylinolenic acid, a fatty acid-derived messenger, traumatic acid,[220] and lysophosphatidylethanolamine.[221] The plant hormones abscisic acid (with effects similar to those of MeJa)[222,223] and indolebutyric acid enhance the activity of the key enzyme tryptophan decarboxylase in *Catharanthus* seedlings.[224] MeJa treatment has been found predominantly to influence regulatory controls leading to monomeric alkaloid synthesis in young *Catharanthus* seedlings.[225] The biosynthesis of these alkaloids, which focused on strictosidine and its synthase including molecular genetics, has been reviewed.[3,226]

4.03.5.2 *Rauwolfia* Alkaloids

The plant *Rauwolfia serpentina* is a major source of biologically active indole alkaloids.[227] About 50 alkaloids have been reported from this plant, of which ajmalicine, ajmaline, reserpine, rescinnamine, yohimbine, and serpentine are of pharmaceutical importance.[227] The plant has been used for the treatment of snake bites and feverish illnesses. Organ and plant cell cultures of this plant have been studied in detail.[228,229] In the normal plant of *R. serpentina*, the roots contain higher levels of indole

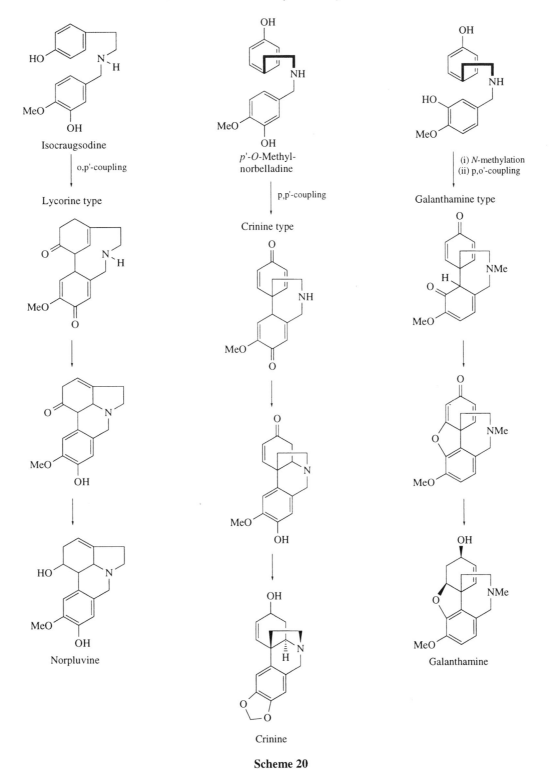

Scheme 20

alkaloids. Callus cultures of *R. serpentina* were able to biosynthesize some of the typical alkaloids, viz., ajmaline, ajmalicine, and serpentine, and their biosynthesis has been studied by tracer experiments.[230]

The *Rauwolfia* alkaloid ajmaline was first studied by Leete.[40] *In vivo* feeding experiments with *Rauwolfia* plants revealed the incorporation of tryptophan into ajmaline, reserpine, and serpentine, and confirmed that tryptophan acts as a precursor amino acid for the indole portion of these alkaloids.

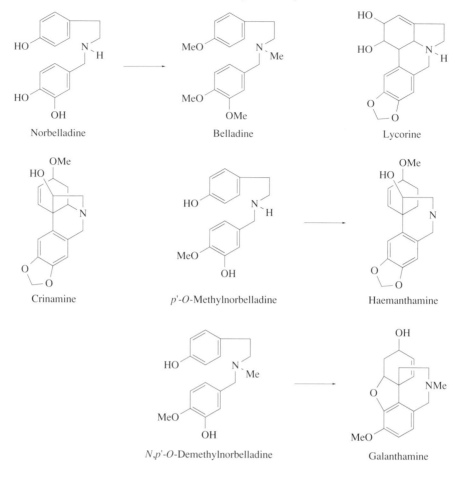

Norbelladine   Belladine   Lycorine

Crinamine   p'-*O*-Methylnorbelladine   Haemanthamine

N,p'-*O*-Demethylnorbelladine   Galanthamine

Scheme 21

Subsequent feeding experiments using *C. roseus* plants confirmed that tryptophan acts as a biosynthetic precursor of monoterpenoid indole alkaloids such as vindoline and catharanthine. The biogenetic precursors tryptamine and secologanin form (*S*)-3α-strictosidine, catalyzed by the enzyme strictosidine synthase (SS). The biosynthetic importance of strictosidine was confirmed, as it serves as the precursor for both the 3α- and 3β-configured alkaloids and also for the *Rauwolfia* alkaloids.[231–235]

The cDNA encoding SS from *R. serpentina* has been heterologously expressed in a cell culture of fall army worm.[236] The complete mRNA sequence of SS from *C. roseus* has been determined.[237] The cDNA from *Rauwolfia* species has been subjected to detailed genetic analysis.[238] The gene encoding SS from *C. roseus* and that encoding TDC are coordinately regulated. An important regulating event in the biosynthesis of terpenoid indole alkaloids is transcription of these genes.[237,239] Analysis of the gene for SS in 10 *Rauwolfia* species revealed high conservation of the mRNA sequence.[240] A detailed enzymatic pathway for indole alkaloid biosynthesis in *Rauwolfia* species has been published.[241] Ten enzymes have been isolated from *Rauwolfia* cell suspension cultures, involved in the biosynthesis of ajmaline from secologanin and tryptamine. Four are membrane-bound proteins whose cellular localization is not known and five reactions of the major biosynthetic route are NADPH$_2$ dependent. Two enzymes depend on the O$_2$ and NADPH$_2$ supplementation as cytochrome P-450 proteins.[242] Metabolism of ajmaline by cell culture of *R. serpentina* has been found to yield raumacline and its two derivatives.[243,244] Cell suspension cultures of plants, which produce terpenoid indole alkaloids, have been examined for the production of alkaloids and activities of chorismate mutase, isochorismate synthase, anthranilate synthase, tryptophan decarboxylase, and strictosidine synthase.[245] Further work on the production of alkaloids in *C. roseus* cell cultures has been reported.[246–248] Genetically transformed hairy root cultures are also known to synthesize a variety of alkaloids characteristic of the parent plant,[249] such as the production of shikonin by *Lithospermum*[250] and indole alkaloids by *C. roseus*.[246]

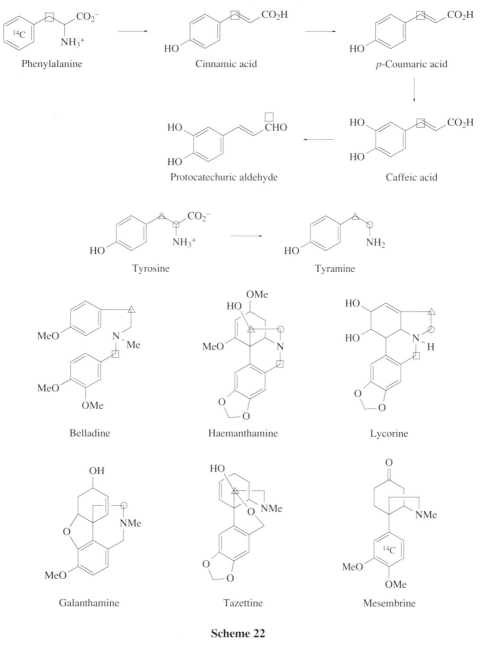

Scheme 22

4.03.5.3 *Strychnos* Alkaloids

The *Strychnos* alkaloids include those alkaloids in which an unrearranged monoterpenoid unit may be identified as being attached to the indole nucleus by C-2—C-16 and C-7—C-3 (or —C-21) bonds.[251] These alkaloids are found only in species of the plant families Apocynaceae and Loganiaceae[252–254] and are formed biogenetically via geissoschizine and preakuammicine, either directly from the latter (strychnan skeletal type; C-7—C-3 bond) or via the intermediacy of stemmadenine-related iminium cations (aspidospermatan skeletal type; C-7—C-21 bond).[254,255]

These alkaloids are derived from tryptophan. Their unrearranged monoterpenoid unit may be identified as an indole nucleus attached by C-2—C-16 and C-7—C-3 (or —C-21) bonds. The alkaloids are widely distributed in the families Aponcynaceae and Loganiaceae. The strychnos alkaloids are biosynthesized via geissoschizine and preakuammicine or via stemmadenine (iminium cations) (Scheme 24). The *Strychnos* alkaloids have received great attention from the synthetic point of view and many alkaloids have been synthesized in the racemic series.[256,257] Mainly three types of *Strychnos* alkaloids have been found, namely aspidospermatan skeletal type,[258] pentacyclic type,[259]

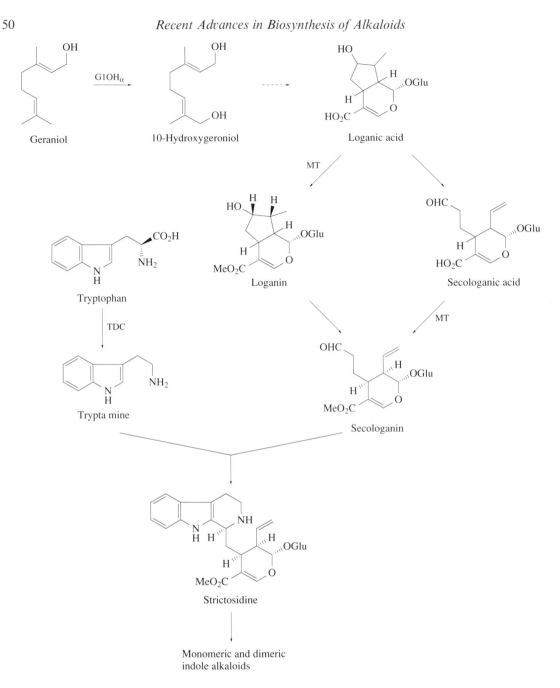

Geraniol

10-Hydroxygeroniol

Loganic acid

Tryptophan

Loganin

Secologanic acid

Tryptamine

Secologanin

Strictosidine

Monomeric and dimeric
indole alkaloids

MT = Methyl transferase; TDC = Tryptophan decarboxylase; G1OH = Geraniol 10 hydroxylase

Scheme 23

and strychnan skeletal type.[260] Some examples of these alkaloid types are stemmadenine, tubotai-wine, geisselline, and dichotine (all aspidospermatan skeletal type), tubifoline, tubifolidine, geis-soschizoline, and norfluorocurarine (pentacyclic type), and akuammicine, rosidiline, diaboline, strychnosilidine, strychnine, icajine, and tsilamine (strychnan skeletal types).

For a long time, *Strychnos* alkaloids have been reported as a group of indole alkaloids. Considerable biosynthetic efforts with *Strychnos* alkaloids have been made. The *Strychnos* alkaloids have been synthesized with either the strychnan or the aspidospermatan skeletal type. Seven novel *Strychnos* alkaloids have been isolated.[261]

In the chemical synthesis of pentacyclic *Strychnos* alkaloids, the pyrrolidine ring is closed by cyclization at the indole 3-position of hexahydro-1,5-methanoazocino[4,3-*b*]indoles. It has been synthesized from either the strychnan or the aspidospermatan skeletal type. The C-6—C-7 bond is

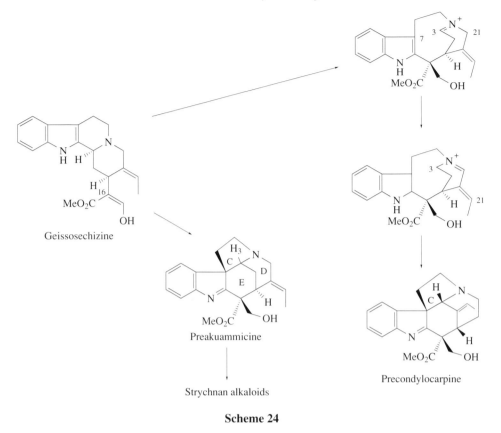

Geissosechizine

Preakuammicine

Strychnan alkaloids

Precondylocarpine

Scheme 24

formed, and generates a quaternary carbon center from the pyrrole nucleus. The cyclization procedure was accomplished by generation of thionium ion by treating dithioacetal with dimethylsulfonium fluoroborate (DMTSF).[262] The pentacyclic indolenine was converted into tubifoline[263] and 19,20-dihydroakuammicine.[262,264]

The required tetracyclic intermediate was prepared from 2-cyanotetrahydropyrine through condensation with indole, addition of an ethyl group and cyclization of a 2-(3-indolyl)piperidine-4-carboxylic acid.[262] DMTSF-induced cyclizations of dithioacetals on 2,3-indoles have proved that the indole nucleus is not deactivated and the piperidine nitrogen is not included in an amide function. Hence dithioacetals were reluctant to undergo cyclization when treated with DMTSF.[265,266]

4.03.5.4 Ergot Alkaloids

The ergot alkaloids are generally derived from L-tryptophan and mevalonic acid. These alkaloids are among the most important group of pharmaceuticals. Reviews appeared on these alkaloids in 1965,[267] 1990,[268] and 1994.[269] Ergot is the dried sclerotium of the filamentous fungus *Claviceps purpurea* (Fries), Tulasne (Hypocreaceae), which grows parasitically on rye and graminaceous crop plants. There are three main groups of ergot alkaloids, the syntheses of which are biogenetically related: clavine derivatives, ergolene derivatives, and lysergic acid derivatives.

A chemical total synthesis of five-membered D-ring ergot analogues was performed starting from indole-3-propionic acid. However, the removal of the *N*-benzoyl group was unsuccessful.[270]

4.03.5.4.1 Biosynthesis of clavine alkaloids

The clavine alkaloids are distributed only in *Claviceps* species, but three clavine-related alkaloids were obtained from *Aspergillus fumigatus*, namely agroclavine, elymoclavine, and chanoclavine, and also from several *Penicillium* and *Aspergillus* species.

4.03.5.4.2   *Biosynthesis of ergolenes and lysergic acid*

Ergolene-type alkaloids are biosynthesized from dimethylallyltryptophan (DMAT), the precursors of which are L-tryptophan and mevalonic acid. There are three steps involved in the biosynthesis of these alkaloids (Scheme 25). DMAT synthetase catalyzes the first step in the formation of DMAT from mevalonic acid and tryptophan. This enzyme has been isolated and purified from a protoplast suspension of mycelia of *Claviceps* species strain SD58.[271] Lee *et al.*[271] showed that tryptophan serves as a biosynthetic precursor of the ergoline ring system and induces alkaloid synthesis. In the second step of the biosynthesis of ergot alkaloids, *N*-methylation of DMAT or decarboxylation occurs. However, two tryptophan derivatives synthesized in cell-free extracts of *Claviceps* species showed that derivative I was incorporated into the elymoclavine whereas II was not incorporated (Scheme 26(a)). These studies show that the *N*-methylation of DMAT precedes decarboxylation and that tryptophan derivative II is not used in the formation of ring C in the ergolene. Anderson and Saini[272] isolated 4-[(*E*)-4′-hydroxy-3′-methylbut-2′-enyl]-L-tryptophan (E-HODMAT) from the culture broth of *C. purpurea* PRL, which is formed from DMAT in the ammonium sulfate fraction in the presence of NADPH molecules.[272,273] Both DMAT and E-HODMAT are also obtained from cell-free extracts of *Claviceps paspali* following treatment with L-tryptophan and isopentenyl pyrophosphate.[274] However, E-HODMAT was incorporated into elymoclavine but not into agroclavine or chanoclavine, and it was concluded that a biosynthetic route via E-HODMAT to elymoclavine could not be the main route in the biosynthesis of ergot alkaloids (Scheme 26(b)).

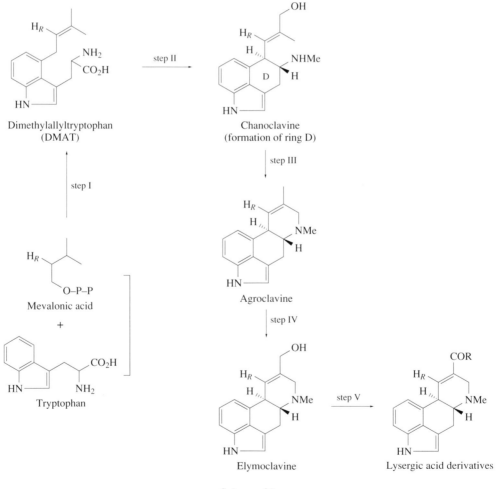

Scheme 25

The biosynthetic route from chanoclavine to lysergic acid derivatives presents various problems. The (*Z*)-methyl group was known to be isomerized to an (*E*)-methyl group and chanoclavine I

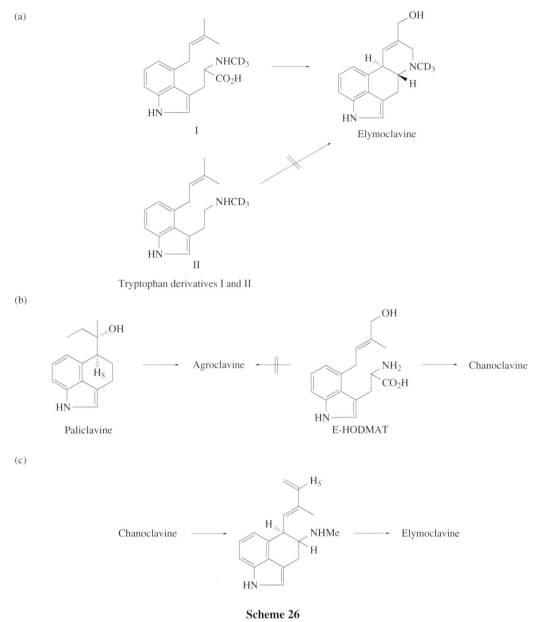

(a)

I

Elymoclavine

II

Tryptophan derivatives I and II

(b)

Paliclavine

Agroclavine

E-HODMAT

Chanoclavine

(c)

Chanoclavine

Elymoclavine

Scheme 26

aldehyde was isolated from a culture broth of *C. purpurea*,[275,276] and construction of ring D would occur via route I.[277] Acklin *et al.*[278] proposed the biosynthetic pathway of paliclavine in cultures of *C. paspali* and agroclavine derived from the paliclavine route (Scheme 26(c)).

4.03.5.4.3 Biosynthesis of peptide ergot alkaloids

Peptide ergot alkaloids are derived from L-proline, which forms a tripeptide that couples to the ergolene moiety to form the lactam ring, e.g., clavicipitic acid. Bajwa *et al.*[279] investigated the biosynthesis of clavicipitic acid. DMAT was bioconverted to clavicipitic acid in cell-free extracts of *Claviceps* sp.[215]

An efficient chemical synthesis of clavicipitic acid has been found via palladium-catalyzed reactions.[280–282] A proline-free ergopeptine-type alkaloid, ergobalansinine, has been isolated from the seeds of *Ipomoea piurensis*.[283] A new natural ergopeptine alkaloid hydroxylated at C-8 of the lysergic acid moiety has been isolated from sclerotia of the field-growing parasitic fungus *C.*

purpurea.[267] Ergopeptines are usually considered to be the final step in the synthesis of peptide ergot alkaloids.[284] A new peptide, an ergot alkaloid, ergolodinine, has been isolated from the sclerotia of parasitic fungus *C. purpurea*.[285]

4.03.5.5 Carbazole Alkaloids

The carbazole alkaloids include several important drugs. In addition, a large number of carbazole alkaloids are found in indole alkaloids. Carbazole alkaloids have been found in the leaf plant *Murraya koenigii* (Rutaceae) and the roots of *Gylcosmis pentaphylla* and *Clausena heptaphylla*. Five new carbazole alkaloids, clausine B, E, H, I, and K, were isolated from the stem bark of *Clausena excavata*.[286] The isolation and pharmacological evaluation of some carbazole and carbazolequinone alkaloids obtained from this species have been published.[287–289] *C. excavata* is used as a folk medicine for the treatment of snakebite and abdominal pain and as a detoxicant. Five carbazole alkaloids, murrayamine-M, -J, -N, -I, and -K, were isolated from the leaves of *Murraya euchrestifolia*.[290] Two other carbazole alkaloids, murrayamine-D and -E, were also isolated from the leaves of *M. euchrestifolia*[291] and from a stem bark extract of *M. koenigii*. Two carbazole alkaloids, 2-methoxycarbazole-3-methylcarboxylate and 1-hydroxy-3-methylcarbazole, were identified using spectroscopic and chemical evidence.[292]

4.03.5.6 β-Carboline Alkaloids

The biosynthesis of β-carboline alkaloids, e.g., harman from tryptophan in *Passiflora edulis* and *Eleagneus angustifolia*, has been discussed.[61,293] A close relationship between tryptophan decarboxylase activity and the synthesis of serotonin in *Perganum harmala* root cultures has been observed.[294] The content of serotonin was enhanced but not that of β-carbolines by adding tryptamine to the cell cultures. An attempt was made to identify the intermediates of β-carboline biosynthesis and the oxidation of dihydro-β-carbolines to aromatic alkaloids was observed.

4.03.6 ALKALOIDS DERIVED FROM HISTIDINE

4.03.6.1 Imidazole Alkaloids

Imidazole alkaloids contain an imidazole nucleus and are derived from histidine. They are found in various *Pilocarpus* species (Rutaceae) and fungi, e.g., ergothioneine and hercynine. A number of imidazole alkaloids found in marine sponges have been reviewed.[202,203] Pilocarpine, the only naturally occurring imidazole alkaloid, is used in clinical medicines for the treatment of glaucoma. It is isolated from the leaves of *Pilocarpus microphyllus*. The medicinally important "jaborandis" are species of *Pilocarpus* but the vernacular name is also applied to the other rutaceous and piperaceous plants.[295] A review on jaborandi alkaloids has been published.[296] Boit,[297] Leete,[298] and Nunes and co-workers[299,300] have reported a number of pilocarpine biogenetic and biosynthetic pathways.

4.03.7 CONCLUSION AND FUTURE PROSPECTS

Studies on alkaloids and their biosynthesis were originally initiated because of their pharmacological characteristics and perplexing phytochemical diversity. Each group of alkaloids has received the attention of various groups with backgrounds in biochemistry, plant physiology, organic chemistry, and plant cell culture. As a result, several exciting aspects of alkaloid biosynthesis or biogenesis and their medicinal properties have come to be better understood. There are two major subjects of considerable contemporary interest. The first concerns the structural groups and biogenesis of the alkaloids, determined by incorporation studies in intact plants. The second centers around the involvement of the various enzymes and their regulation, compartmentalization of alkaloid biosynthesis, and identification and isolation of genes that regulate alkaloid biosynthesis. Molecular cloning of some specific master regulatory genes of enzymes for putative alkaloids, however, requires some ingenious work. Any approach to improve alkaloid production requires a thorough knowledge of the alkaloid biosynthetic pathways and their structures, enzymology, and genetic expression. Considerable achievements have been made in understanding the pathways of alkaloid biosynthesis and the enzymology of indole, tropane, morphine, and isoquinoline alkaloids. It is expected that in the years to come the combined efforts of chemists and biochemists will unravel the regulation of various facets of alkaloid biosynthesis.

4.03.8 REFERENCES

1. N. R. Farnsworth, in "Natural Product and Drug Development," eds. P. Krogsgaard-Larsen, S. Brogger Christenssen, and H. Kofud, Musksgaard, Copenhagen, 1984, p. 17.
2. A. Stafford, P. Morris, and M. W. Fowler, *Enzyme Microb. Technol.*, 1986, **8**, 578.
3. N. Misra, R. Luthra, and S. Kumar, *Indian J. Biochem. Biophys.*, 1996, **33**, 261.
4. R. Verpoorte, R. van der Heijden, and P. R. H. Moreno, in "The Alkaloids," ed. G. A. Cordell, Academic Press, San Diego, CA, 1997, vol. 49, p. 227.
5. R. H. Moreno Paulo, R. van der Herjden, and R. Verpoorte, *Plant Cell Tissue Organ Cult.*, 1995, **42**, 1.
6. T. Hashimoto and Y. Yamada, *Plant Physiol. Mol. Biol.*, 1994, **45**, 257.
7. A. H. Meijer, R. Verpoorte, and J. H. Hoge, *J. Plant Res.*, 1993, **3**, Special Issue, 145.
8. T. Hashimoto and Y. Yamada, *Methods Plant Biochem.*, 1993, **9**, 369.
9. T. M. Kutchan and M. H. Zenk, *J. Plant Res.*, 1993, **3**, Special Issue, 165.
10. V. de Luca, *Methods Plant Biochem.*, 1993, **9**, 345.
11. T. Hashimoto and Y. Yamada, in "Proceedings of the 7th Annual Pennsylvania State Symposium on Plant Physiology," Society of Plant Physiology, Rockville, MD, 1992, p. 122.
12. E. Leete, *Planta Med.*, 1990, **56**, 339.
13. Y. Kitamura, in "Biotechnology in Agriculture and Forestry," ed. Y. P. S. Bajaj, Springer, Berlin, 1988, vol. 4, p. 419.
14. G.-Z. Zheng, in "Biotechnology in Agriculture and Forestry," ed. Y. P. S. Bajaj, Springer, Berlin, 1989, vol. 7, p. 23.
15. G. Petriand and Y. P. S. Bajaj, in "Biotechnology in Agriculture and Forestry," ed. Y. P. S. Bajaj, Springer, Berlin, 1988, vol. 7, p. 135.
16. A. Straus, in "Biotechnology in Agriculture and Forestry," ed. Y. P. S. Bajaj, Springer, Berlin, 1989, vol. 7, p. 286.
17. Y. P. S. Bajaj and L. K. Simola, in "Biotechnology in Agriculture and Forestry," ed. Y. P. S. Bajaj, Springer, Berlin, 1991, vol. 15, p. 1.
18. Y. P. S. Bajaj, in "Biotechnology in Agriculture and Forestry," ed. Y. P. S. Bajaj, Springer, Berlin, 1989, vol. 9, p. 1.
19. R. Verpoorte, R. van der Heijden, W. M. van Gullik, and H. J. G. ten Hoopen, in "The Alkaloids," ed. A. Brossi, Academic Press, San Diego, CA, 1990, vol. 40, p. 52.
20. W. Gritsawapan and W. J. Griffin, *Phytochemistry*, 1992, **31**, 3069.
21. Y. Yamada and T. Endo, *Plant Cell Rep.*, 1984, **3**, 186.
22. T. Hashimoto and Y. Yamada, *Agric. Biol. Chem.*, 1987, **51**, 2769.
23. L. K. Simola, R. Parviainen, A. Martinsen, A. Huhtikangas, R. Jakela, and M. Lounasmaa, *Acta Chem. Scand.*, 1989, **43**, 702.
24. L. K. Simola, R. Parviainen, A. Martinsen, A. Huhtikangas, R. Jakela, and M. Lounasmaa, *Phytochemistry*, 1990, **29**, 3517.
25. K. Yoshimatsu, T. Halano, M. Katayama, S. Muramo, H. Kamada, and K. Shimomura, *Phytochemistry*, 1990, **29**, 3525.
26. Y. Kitamura, A. Taura, Y. Kajiya, and H. Miura, *J. Plant Physiol.*, 1992, **40**, 141.
27. T. Hashimoto and Y. Yamada, *Planta Med.*, 1983, **47**, 195.
28. M. Lounasmaa and T. Tamminen, in "The Alkaloids," ed. G. A. Cordell, Academic Press, New York, 1993, vol. 44, p. 1.
29. E. Leete, *Planta Med.*, 1979, **36**, 97.
30. W. C. Evans, "Trease and Evans' Pharmacognosy," 13th edn., Baillière Tindall, London, 1989, p. 832.
31. L. Geiger, *Liebigs Ann. Chem.*, 1833, **7**, 269.
32. E. Leete, J. A. Bjorklund, and S. H. Kim, *Phytochemistry*, 1988, **27**, 2563.
33. E. Leete, J. A. Bjorklund, M. M. Couladis, and S. H. Kim, *J. Am. Chem. Soc.*, 1991, **113**, 9286.
34. T. Hashimoto and Y. Yamada, *Agric. Biol. Chem.*, 1989, **53**, 863.
35. Y. Yamada and T. Hashimoto, *Proc. Jpn. Acad. Sci.*, 1989, **B65**, 156.
36. K. Cheng, H. Fang, W. Zhu, C. Mang, and L. Li, *Planta Med.*, 1989, **55**, 391.
37. Y. Yamada, T. Hashimoto, T. Endo, Y. Yukimune, J. Kohno, N. Hamaguchi, and B. Drager, in "Secondary Products from Plant Tissue Culture," eds. B. V. Charlwood and M. J. C. Rhodes, Oxford University Press, Oxford, 1990, p. 227.
38. T. Hashimoto, J. Matsuda, S. Okabe, Y. Amano, D. J. Yun, A. Hayashi, and Y. Yamada, in "Progress in Plant Cellular and Molecular Biology," eds. H. J. J. Nijkamp, L. H. W. van der Plass, and J. van Aartrijk, Kluwer, Dordrecht, 1990, p. 775.
39. M. Lounasmaa, in "The Alkaloids," ed. A. Brossi, Academic Press, New York, 1988, vol. 33, p. 1.
40. E. Leete, *J. Am. Chem. Soc.*, 1960, **82**, 612.
41. E. Leete and M. L. Louden, *Chem. Ind. (London)*, 1961, 1405.
42. R. J. Robins, E. G. Bent, and M. J. C. Rhodes, *Planta*, 1991, **185**, 385.
43. M. J. C. Rhodes, R. J. Robins, E. L. H. Aird, J. Payne, A. J. Parr, and N. J. Walton, in "Primary and Secondary Metabolism of Plant Cell Cultures II," ed. W. G. W. Kurg, Springer, Berlin, 1989, p. 58.
44. E. L. H. Aird, J. D. Hamill, R. J. Robins, and M. J. C. Rhodes, in "Manipulating Secondary Metabolism in Culture," eds. R. J. Robins and M. J. C. Rhodes, Cambridge University Press, Cambridge, 1988, p. 137.
45. R. J. Robins, A. J. Parr, E. G. Bent, and M. J. C. Rhodes, *Planta*, 1991, **183**, 185.
46. Y. Y. Ford, R. G. Ratcliffe, and R. J. Robins, *Phytochemistry*, 1996, **43**, 115.

47. R. J. Robins and N. J. Walton, in "The Alkaloids," ed. G. A. Cordell, Academic Press, San Diego, CA, 1993, vol. 44, p. 116.
48. B. Drager, A. Portsteffen, A. Schaal, P. H. McCabe, A. C. J. Peerless, and R. J. Robins, *Planta*, 1992, **188**, 581.
49. A. Goldman, M. L. Milat, P. H. Ducrot, J. Y. Lallen, M. Maille, A. Lepingle, I. Charpin, and D. Tepfer, *Phytochemistry*, 1990, **29**, 2125.
50. B. Drager, *Planta Med.*, 1993, **59**, A584.
51. R. C. Moore and S. M. Boyle, *J. Bacteriol.*, 1990, **172**, 4631.
52. E. Bell and R. L. Malmberg, *Mol. Gen. Genet.*, 1990, **224**, 431.
53. A. J. Fleming, T. Mandal, I. Roth, and C. Kuhlemeier, *Plant Cell*, 1993, **5**, 297.
54. T. Hashimoto, Y. Yukimune, and Y. Yamada, *Planta*, 1989, **178**, 123.
55. A. Romeike, *Bot. Not.*, 1978, **131**, 85.
56. T. Hashimoto and Y. Yamada, *Plant Physiol.*, 1986, **81**, 619.
57. T. Hashimoto, J. Kohno, and Y. Yamada, *Phytochemistry*, 1989, **28**, 1077.
58. T. Hashimoto, J. Matsuda, and Y. Yamada, *FEBS Lett.*, 1993, **329**, 35.
59. A. Melhaou, A. Jossang, and B. Bodo, *J. Nat. Prod.*, 1992, **55**, 950.
60. E. Eisenrecichova, M. Haladora, A. Buckora, J. Tomko, D. Uhrin, and K. Ubik, *Phytochemistry*, 1992, **31**, 1085.
61. R. B. Herbert, in "The Biosynthesis of Secondary Metabolites," 2nd edn., Chapman and Hall, London, 1989.
62. W. R. McLaughlin, R. A. McKee, and D. M. Evans, *Planta*, 1993, **191**, 440.
63. S. Rosset, J. Cilirier, and G. Lhommet, *Tetrahedron Lett.*, 1991, **32**, 7521.
64. R. Ballini, E. Marcantoni, and M. Petrini, *J. Org. Chem.*, 1992, **57**, 1316.
65. D. S. Brown, P. Charreau, T. Hansson, and S. V. Ley, *Tetrahedron*, 1991, **47**, 1311.
66. K. Jones and K. Woo, *Tetrahedron*, 1991, **47**, 479.
67. P. Somfai and J. Ahman, *Tetrahedron Lett.*, 1992, **33**, 3791.
68. T. Honda, A. Yamamoto, Y. Cui, and M. Tsubuki, *J. Chem. Soc., Perkins Trans. 1*, 1992, 531.
69. L. Witte, P. Rubiolo, C. Bicchi, and T. Hartmann, *Phytochemistry*, 1993, **32**, 187.
70. T. Hartmann and L. Witte, in "Alkaloids: Chemical and Biological Perspectives," ed. S. W. Pelletier, Pergamon Press, New York, 1994, p. 155.
71. D. J. Robins, *Chem. Soc. Rev.*, 1989, **18**, 375.
72. G. Graser, L. Witte, D. J. Robins, and T. Hartmann, *Phytochemistry*, 1998, **47**, 1017.
73. A. R. Mattocks, in "Chemistry and Toxicology of Pyrrolizidine Alkaloids," Academic Press, London, 1986, p. 191.
74. S. E. Denmark and A. Thorarensen, *J. Org. Chem.*, 1994, **59**, 5672.
75. S. E. Denmark, A. Thorarensen, and D. S. Middleton, *J. Org. Chem.*, 1995, **60**, 35.
76. E. Pedersen, *Arch. Pharm. Chem. Sci.*, 1975, **3**, 375.
77. N. M. van Dam, L. Witte, C. Theuring, and T. Hartmann, *Phytochemistry*, 1995, **39**, 287.
78. N. M. van Dam, R. Verpoorte, and E. van der Meijden, *Phytochemistry*, 1994, **37**, 1013.
79. M. Riena, A. H. Merichi, R. Cabrera, and A. Gonzalez-Coloma, 1995, **38**, 355.
80. E. Roeder, *Pharmazie*, 1995, **50**, 83.
81. H. Ishibashi, N. Uemura, H. Nakatani, M. Okazaki, T. Sato, N. Nakamura, and M. Ikeda, *J. Org. Chem.*, 1993, **58**, 2360.
82. H.-R. Schutte, *Prog. Bot.*, 1994, **55**, 96.
83. D. J. Robins, in "The Alkaloids," ed. G. A. Cordell, Academic Press, San Diego, CA, 1995, vol. 46, p. 1.
84. M. Wink, *Bioforum*, 1993, **16**, 360.
85. J. K. Porter, *J. Anim. Sci.*, 1995, **73**, 871.
86. D. J. Robins, *Nat. Prod. Rep.*, 1989, **6**, 581.
87. J. Cooper, P. T. Gallagher, and D. W. Knight, *J. Chem. Soc., Perkin Trans. 1*, 1993, 1313.
88. A. R. de Faria, C. R. R. Motos, and C. R. D. Correia, *Tetrahedron Lett.*, 1993, **34**, 27.
89. D. J. Robins, *Nat. Prod. Rep.*, 1992, **9**, 316.
90. H. Takahata, H. Bandoh, and T. Momose, *J. Org. Chem.*, 1992, **57**, 4401.
91. D. J. Robins, *Fortschr. Chem. Org. Naturst.*, 1982, **41**, 115.
92. M. Suffness and G. A. Cordell, in "The Alkaloids," ed. A. Brossi, Academic Press, Orlando, FL, 1985, vol. 25, p. 1.
93. E. Nowacki and R. U. Byerrum, *Life Sci.*, 1962, **1**, 157.
94. R. Chizzola, *J. Chromatogr.*, 1994, **668**, 427.
95. C. G. Logie, M. R. Grue, and J. R. Liddell, *Phytochemistry*, 1994, **37**, 43.
96. D. S. Rycroft, I. R. Stirling, and D. J. Robins, *Magn. Reson. Chem.*, 1992, **30**, S42.
97. Z. Miao and D. Cai, *Youji Huaxue*, 1992, **12**, 503.
98. N. Anderton, J. J. Gosper, and C. May, *Aust. J. Chem.*, 1994, **47**, 853.
99. Minist. Agric. Fish. Food (GB), *Food Surveil. Pap.*, 1994, **42**, 68.
100. T. Langer, E. Mostl, R. Chizzola, and R. Gutleb, *Planta Med.*, 1996, **62**, 267.
101. K. A. Freer, J. R. Matheson, M. Rodgers, and D. J. Robins, *J. Chem. Res. (S)*, 1991, 46.
102. D. J. Robins, *Experientia*, 1991, **47**, 1118.
103. F. Bottcher, R.-D. Adolph, and T. Hartmann, *Phytochemistry*, 1993, **32**, 679.
104. H. Niwa, T. Sakata, and K. Yamada, *Bull. Chem. Soc. Jpn.*, 1994, **67**, 1990.
105. H. Niwa, K. Kunitani, T. Nagoya, and K. Yamada, *Bull. Chem. Soc. Jpn.*, 1994, **67**, 3094.
106. H. Niwa, Y. Miyachi, O. Uosaki, A. Kuroda, H. Ishiwata, and K. Yamada, *Tetrahedron*, 1992, **48**, 393.
107. D. H. G. Crout, *J. Chem. Soc. C*, 1969, 1379.
108. T. Honda, K. Tomitsuka, and M. Tsubuki, *J. Org. Chem.*, 1993, **58**, 4274.
109. K. Lee, J. A. Jackson, and D. R. Weimer, *J. Org. Chem.*, 1993, **58**, 5969.
110. T. Matsumoto, H. Terao, N. Ishizuka, S. Usui, and H. Nozaki, *Bull. Chem. Soc. Jpn.*, 1992, **65**, 1761.
111. A. D. Kinghorn and M. F. Balandrin, in "Alkaloids: Chemical and Biological Perspectives," ed. W. S. Pelletier, Wiley, Chichester, 1984, vol. 2, p. 105.
112. M. Wink, *Methods Plant Biochem.*, 1993, **8**, 197.
113. J. A. Mears and T. J. Mabry, in "Chemotaxonomy: the Leguminosae," eds. J. B. Harborne, D. Boulter, and B. L. Turner, Academic Press, London, 1971, p. 73.

114. E. Kass and M. Wink, *Planta Med.*, 1993, **59**, A594.
115. M. Wink, C. Meibnev, and L. Witte, *Phytochemistry*, 1995, **38**, 139.
116. S. Ohmiya, K. Saito, and I. Murakoshi, in "The Alkaloids—Lupine Alkaloids," ed. G. A. Cordell, Academic Press, San Diego, CA, 1995, vol. 47, p. 1.
117. R. Greinwald, J. H. Ross, L. Witte, and F. C. Czygan, *Biochem. Syst. Ecol.*, 1993, **21**, 405.
118. M. Wink, in "Insect–Plant Interactions," ed. E. A. Bernays, CRC Press, Boca Raton, FL, 1992, vol. IV, p. 137.
119. M. Wink, in "The Alkaloids," ed. J. Cordell, Academic Press, New York, 1993, vol. 43, p. 1.
120. H. Suzuki, I. Murakoshi, and K. Saito, *J. Biol. Chem.*, 1994, **269**, 15 853.
121. K. Saito, H. Suzuki, Y. Yamashita, and I. Murakoshi, *Phytochemistry*, 1994, **36**, 309.
122. G. H. Verdoorn and B. E. van Wyk, *Phytochemistry*, 1990, **29**, 1297.
123. R. A. Hussain, A. D. Kinghorn, and R. J. Molyneux, *J. Nat. Prod.*, 1988, **51**, 809.
124. S. Takamatsu, K. Saito, and I. Murakoshi, *J. Nat. Prod.*, 1991, **54**, 477.
125. R. Greinwald, P. Canto, P. Bachmann, L. Witte, and F.-C. Czygan, *Biochem. Syst. Ecol.*, 1992, **20**, 75.
126. I. V. van Rensen, V. Wray, L. Witte, P. Canto, R. Greinwald, G. Veen, M. Veit, and F.-C. Czygan, *Phytochemistry*, 1994, **35**, 421.
127. W. M. Golebiewski and I. D. Spenser, *J. Am. Chem. Soc.*, 1984, **106**, 7925.
128. A. M. Fraser and D. J. Robins, *J. Chem. Soc., Chem. Commun.*, 1987, **14**, 77.
129. M. Wink, T. Hartmann, and H. M. Schiebel, *Z. Naturforsch., Teil C*, 1978, **34**, 704.
130. K. Saito and I. Murakoshi, in "Studies in Natural Products Chemistry. Chemistry, Biochemistry and Chemotaxonomy of Lupin Alkaloids in the Leguminosae," ed. Atta-ur-Rahman, Elsevier, Amsterdam, 1995, vol. 15, p. 519.
131. D. J. Robins, in "The Alkaloids," ed. G. A. Cordell, Academic Press, San Diego, CA, 1995, vol. 46, p. 1.
132. K. Saito, H. Suzuki, S. Takamatsu, and I. Murakoshi, *Phytochemistry*, 1993, **32**, 87.
133. H. Suzuki, Y. Koike, I. Murakoshi, and K. Saito, *Phytochemistry*, 1996, **42**, 1557.
134. R. Robinson, "The Structural Relations of Natural Products," Oxford University Press, Oxford, 1955.
135. H. R. Schutte, *Arch. Pharm. (Weinheim, Ger.)*, 1960, **293**, 1006.
136. M. Soucek and H. R. Schutte, *Angew. Chem., Int. Ed. Engl.*, 1962, **1**, 597.
137. M. Wink, R. Perrey, M. Schneider, U. Warskulat, K. van Borstel, and P. Menede, in "Plant Tissue. Culture and Gene Manipulation for Breeding," ed. K. Oono, NIAR, Tokyo, 1992, p. 99.
138. J. P. Michael, *Nat. Prod. Rep.*, 1994, **11**, 639.
139. J. P. Michael, *Nat. Prod. Rep.*, 1995, **12**, 535.
140. R. B. Herbert, *Nat. Prod. Rep.*, 1988, **5**, 525.
141. R. B. Herbert, *Nat. Prod. Rep.*, 1990, **7**, 107.
142. R. B. Herbert, *Nat. Prod. Rep.*, 1996, **13**, 45.
143. K. Saito, Y. Koike, H. Suzuki, and I. Murakoshi, *Phytochemistry*, 1993, **34**, 1041.
144. J. W. Fairbairn and M. Djote, *Phytochemistry*, 1970, **9**, 739.
145. K. Toth-Lokoes, L. Helzkey, T. Hoeroempoeli, and R. Fleschmann, *Novenytermeles*, 1994, **43**, 279.
146. R. B. Herbert, *Nat. Prod. Rep.*, 1995, **12**, 445.
147. R. Gerardy and M. H. Zenk, *Phytochemistry*, 1993, **32**, 79.
148. R. B. Herbert, *Nat. Prod. Rep.*, 1995, **12**, 55.
149. K. W. Bentley, *Nat. Prod. Rep.*, 1996, **13**, 129.
150. R. B. Herbert, *Nat. Prod. Rep.*, 1993, **10**, 575.
151. W. De-Ekanamkul and M. H. Zenk, *Phytochemistry*, 1992, **31**, 813.
152. D. H. R. Barton, D. S. Bhakuni, R. James, and G. W. Kirby, *J. Chem. Soc.*, 1967, 128.
153. R. Gerardy and M. H. Zenk, *Phytochemistry*, 1993, **34**, 125.
154. R. Lenz and M. Zenk, *Tetrahedron Lett.*, 1995, **36**, 2449.
155. A. R. Battersby and M. Hirst, *Tetrahedron Lett.*, 1965, 669.
156. K. W. Bentley, *Nat. Prod. Rep.*, 1995, **12**, 419.
157. K. W. Bentley, *Nat. Prod. Rep.*, 1996, **13**, 127.
158. M. Valpuesta, N. Posadas, I. Ruiz, M. V. Silva, A. I. Gomez, R. Suau, B. Perez, F. Pliego, and B. Cabezudo, *Phytochemistry*, 1995, **38**, 113.
159. V. Preininger, in 'The Alkaloids," ed. A. Brossi, Academic Press, Orlando, FL, 1986, vol. 29, p. 1.
160. L. Castedo and R. Suau, in "The Alkaloids," ed. A. Brossi, Academic Press, Orlando, FL, 1986, vol. 29. p. 287.
161. G. Blaschke and G. Scriba, *Phytochemistry*, 1986, **25**, 111.
162. K. Iwasa, Y. Kondo, M. Kamigauchi, and M. Tokao, *Planta Med.*, 1994, **60**, 290.
163. A.-T. Khalil, *Mansoura J. Pharm. Sci.*, 1994, **10**, 96.
164. M. A. Khan, S. S. Nizami, T. Oamar, T. Rasheed, M. N. I. Khan, and S. W. Asif, *Heterocycles*, 1994, **38**, 2005.
165. M. Kitamura, Y. Hsiao, M. Ohta, T. Ohta, and J. Takaya, *J. Org. Chem.*, 1994, **59**, 297.
166. A. W.-M. Lee, W. H. Chan, and Y. K. Lee, *Tetrahedron Lett.*, 1991, **32**, 3861.
167. A. W.-M. Lee, W. H. Chan, and E. T. T. Chan, *J. Chem. Soc., Perkin Trans. 1*, 1992, 309.
168. A. W.-M. Lee, W. H. Chan, Y. Tao, and Y. K. Lee, *J. Chem. Soc., Perkin Trans. 1*, 1994, 477.
169. Y. Yamamoto, K. Hashigaki, S. Ishikawa, and H. Quais, *Tetrahedron Lett.*, 1988, **29**, 6439.
170. A. C. Carbonelle, V. Gott, and G. Roussi, *Heterocycles*, 1993, **36**, 1763.
171. C. Veeresham, C. Kokata, and V. Venkateshwarlu, *Phytochemistry*, 1994, **35**, 947.
172. M. Tabata, H. Yamamoto, N. Hiraoka, and M. Konoshima, *Phytochemistry*, 1972, **2**, 949.
173. C. Veeresham, C. Kokata, and V. Venkateshwarlu, *Indian Drugs*, 1991, **29**, 1.
174. T. V. Sairam and P. Khanna, *Lloydia*, 1971, **36**, 170.
175. M. Rueffer and M. H. Zenk, *Z. Naturforsch., Teil C*, 1987, **42**, 319.
176. P. Kawalleck, H. Keller, K. Hahlbrock, D. Schell, and I. E. Somssich, *J. Biol. Chem.*, 1993, **268**, 2189.
177. T. Frenzel and M. H. Zenk, *Phytochemistry*, 1990, **29**, 3505.
178. H. Dittrich and T. M. Kutchan, *Proc. Natl. Acad. Sci. USA*, 1991, **88**, 9969.
179. M. Rueffer and M. H. Zenk, *Tetrahedron Lett.*, 1985, **26**, 201.
180. W. Bauer, R. Stadler, and M. H. Zenk, *Bot. Acta*, 1992, **105**, 370.
181. M. Rueffer and M. H. Zenk, *Phytochemistry*, 1994, **36**, 1219.

182. M. H. Zenk, M. Rueffer, T. M. Kutchan, and E. Galneder, in "Applications of Plant Cell and Tissue Culture," ed. Y. Yamada, Wiley, Chichester, 1988, p. 213.
183. M. Amann, N. Nagakura, and M. H. Zenk, *Eur. J. Biochem.*, 1988, **175**, 17.
184. E. Galneder and M. H. Zenk, in "Progress in Plant Cellular and Molecular Biology," ed. H. J. J. Nijkamp, L. H. W. van der Plas, and J. van Artrijk, Kluwer, Dordrecht, 1990, p. 754.
185. G. Corrado, M. F. Cometa, L. Tomassini, and M. Nicoletti, *Planta Med.*, 1997, **63**, 194.
186. Y. Tsuda and T. Sano, in "The Alkaloids," ed. G. A. Cordell, Academic Press, San Diego, CA, 1996, vol. 48, p. 249.
187. P.-G. Mantel and M. J. Coleman, *Phytochemistry*, 1984, **23**, 1617.
188. D. H. R. Barton, D. J. Lester, W. B. Motherwell, and M. T. B. Papoula, *J. Chem. Soc., Chem. Commun.*, 1979, 705.
189. D. S. Bhakuni and A. N. Singh, *J. Chem. Soc., Perkin Trans. 1*, 1978, 618.
190. D. S. Bhakuni and S. Jain, *Tetrahedron*, 1980, **36**, 2153.
191. D. H. R. Barton, G. W. Kirby, J. B. Taylor, and G. M. Thomas, *J. Chem. Soc.*, 1962, 4545.
192. F. Viladomat, J. Bastida, C. Codina, W. E. Campbell, and S. Mathee, *Phytochemistry*, 1994, **35**, 809.
193. J. Razafimbelo, M. Andriantsiferana, G. Baudouin, and F. Tillequin, *Phytochemistry*, 1996, **41**, 323.
194. C. Codina, J. Bastida, F. Viladomat, J.-M. Fernandez, S. Bergonon, M. Rubiralta, and J. C. Quirion, *Phytochemistry*, 1993, **32**, 1354.
195. J. Bastida, C. Codina, F. Viladomat, S. Bergonon, J.-M. Fernandez, M. Rubiralta, and J. C. Quirion, *Phytochemistry*, 1993, **34**, 1656.
196. A. Latvala, M. A. Onur, T. Gozler, A. Linden, B. Kivcak, and M. Hesse, *Phytochemistry*, 1995, **39**, 1229.
197. M. Kreh and R. Matusch, *Phytochemistry*, 1995, **38**, 1533.
198. O. M. Abdallah, *Phytochemistry*, 1995, **39**, 477.
199. O. M. Abdallah, *Phytochemistry*, 1993, **34**, 1447.
200. Y. Ma, Y. Ito, E. Sokolsky, and H. M. Fales, *J. Chromatogr.*, 1994, **685**, 259.
201. R. J. Tang, N. J. Bi, and G. E. Ma, *Chin. Chem. Lett.*, 1994, **5**, 855.
202. J. R. Lewis, *Nat. Prod. Rep.*, 1996, **13**, 171.
203. J. R. Lewis, *Nat. Prod. Rep.*, 1995, **12**, 339.
204. D. H. R. Barton and T. Cohen, in "Festschrift Arthur Stoll," Birkhauser, Basel, 1957, p. 117.
205. A. R. Battersby, R. Birks, S. W. Brener, N. M. Fales, W. C. Wildman, and R. J. Highet, *J. Chem. Soc.*, 1964, 1595.
206. R. Suau, A.-I. Gomez, and R. Rico, *Phytochemistry*, 1990, **29**, 1710.
207. N. Neuss and M. N. Neuss, in "The Alkaloids: Antitumor Bisindole Alkaloids from *Catharanthus roseus* (L)," eds. A. Brossi and M. Suffness, Academic Press, San Diego, CA, 1990, vol. 37, p. 229.
208. O. van Tellingen, J. H. M. Sips, J. H. Beijnen, A. Butt, and W. J. Nooijen, *Anticancer Res.*, 1992, **12**, 1699.
209. V. DeLuca, N. Brisson, J. Balsevich, and W. G. W. Kurz, in "Primary and Secondary Metabolism of Plant Cell Cultures," ed. W. G. W. Kurz, Springer, Berlin, 1989, vol. 2, p. 154.
210. W. G. W. Kurz, K. B. Chatson, and F. Constabel, in "Primary and Secondary Metabolism of Plant Cell Cultures," eds. K. H. Neumann, W. H. Barz, and E. Reinhard, Springer, Berlin, 1985, p. 143.
211. T. Hirata, Y. Tang, K. Okano, and T. Sugu, *Phytochemistry*, 1989, **28**, 333.
212. V. DeLuca, J. A. Fernandez, D. Campbell, and W. G. W. Kurz, *Plant Physiol.*, 1988, **86**, 447.
213. V. DeLuca, C. Marineau, and M. Brisson, *Proc. Natl. Acad. Sci. USA*, 1989, **86**, 2582.
214. K. Hirata, M. Kobayashi, K. Miyamoto, T. Hoshi, and M. Okazaki, *Planta Med.*, 1989, **55**, 262.
215. R. J. Aerts, D. Gisi, E. De Carolia, V. DeLuca, and T. W. Baumann, *Plant J.*, 1994, **5**, 635.
216. B. A. Vick and D. C. Zimmerman, *Plant Physiol.*, 1984, **75**, 458.
217. B. A. Vick and D. C. Zimmerman, in "The Biochemistry of Plants," eds. P. K. Stumpf and E. E. Conn, Academic Press, Orlando, FL, 1987, vol. 19, p. 53.
218. J. N. Siedow, *Plant Mol. Biol.*, 1994, **42**, 145.
219. E. E. Farmer and C. A. Ryan, *Plant Cell*, 1992, **4**, 129.
220. D. C. Zimmerman and C. A. Coudron, *Plant Physiol.*, 1979, **63**, 536.
221. K. M. Farag and J. P. Palta, *Physiol. Plant.*, 1993, **87**, 515.
222. T. Hildmann, M. Ebneth, H. Pena-Cortés, J. J. Scanchez-Serrano, L. Willmitzer, and S. Prat, *Plant Cell*, 1992, **4**, 1157.
223. C. A. Ryan, *Plant Mol. Biol.*, 1992, **19**, 123.
224. R. J. Aerts, A.-M. Alarco, and V. DeLuca, *Plant Physiol.*, 1992, **100**, 1014.
225. R. J. Aerts, A. Schafer, M. Hesse, T. W. Baumann, and A. Slusarenko, *Phytochemistry*, 1996, **42**, 417.
226. T. M. Kutchan, *Phytochemistry*, 1993, **32**, 493.
227. V. E. Tyler, L. R. Brady, and J. E. Robbers, in "Pharmacognosy," 8th edn., Lea & Febiger, Philadelphia, PA, 1981, p. 235.
228. P. C. Roja, A. T. Sipahimalani, M. R. Heble, and M. S. Chadha, *J. Nat. Prod.*, 1987, **50**, 872.
229. J. Stockigt, A. Pfitzner, and J. Firl, *Plant Cell Rep.*, 1981, **1**, 36.
230. S. Ohta and M. Yatazawa, *Agric. Biol. Chem.*, 1979, **43**, 2297.
231. J. Stockigt and M. H. Zenk, *J. Chem. Soc., Chem. Commun.*, 1977, 646.
232. A. I. Scott, S. L. Lee, P. De Capite, M. G. Culver, and C. R. Hutchinson, *Tetrahedron Lett.*, 1977, 979.
233. A. H. Heckendef and C. R. Hutchinson, *Tetrahedron Lett.*, 1977, 4153.
234. R. T. Brown, J. Leonard, and S. K. Sleigh, *Phytochemistry*, 1978, **7**, 899.
235. M. Rueffer, N. Nagakura, and M. H. Zenk, *Tetrahedron Lett.*, 1978, 1593.
236. T. M. Kutchan, A. Bock, and H. Dittrich, *Phytochemistry*, 1994, **35**, 353.
237. G. Pasquali, O. J. M. Goddijn, A. De Waal, R. Verpoorte, R. A. Schilperoort, J. H. C. Hoge, and J. Memelink, *Plant Mol. Biol.*, 1992, **18**, 1121.
238. D. Bracher and T. M. Kutchan, *Arch. Biochem. Biophys.*, 1992, **294**, 717.
239. I. A. Roewer, N. Cloutier, C. L. Nessler, and V. DeLuca, *Plant Cell Rep.*, 1992, **11**, 86.
240. D. Bracher and T. M. Kutchan, *Plant Cell Rep.*, 1992, **11**, 179.
241. J. Stockigt, in "The Alkaloids," ed. G. A. Cordell, Academic Press, San Diego, CA, 1995, vol. 47, p. 115.
242. J. Stockigt, A. Pfitzner, and P. J. Keller, *Tetrahedron Lett.*, 1983, **24**, 2485.
243. H. Takayama, M. Kitajima, S. Suda, N. Aimi, S. Sakai, S. Endress, and J. Stockigt, *Tetrahedron*, 1992, **48**, 2627.
244. L. Polz, J. Stockigt, H. Takayama, N. Uchida, N. Aimi, and S. Sakai, *Tetrahedron Lett.*, 1990, **31**, 6693.

245. C. Poulsen and R. Verpoorte, *Plant Physiol. Biochem.*, 1992, **30**, 105.

246. L. Toivonen, S. Laakso, and H. Rosenqvist, *Plant Cell Rep.*, 1992, **11**, 390.

247. A. Decendit, D. Liu, L. Ouelhazi, P. Doireau, J. M. Merillon, and M. Rideau, *Plant Cell Rep.*, 1992, **11**, 400.

248. J. M. Merillon, D. Liu, Y. Laurent, M. Rideau, and C. Viel, *Phytochemistry*, 1992, **31**, 1609.

249. H. E. Flores, *Chem. Ind. (London)*, 1992, 374.

250. K. Shimomura, H. Saga, and H. Kamada, *Plant Cell Rep.*, 1991, **10**, 282.

251. K. S. J. Stapleford, in "Rodd's Chemistry of Carbon Compounds," ed. S. Coffey, 2nd edn., Elsevier, Amsterdam, 1977, vol. 43, p. 111.

252. M. V. Kisakurek and M. Hesse, in "Indole and Biogenetically Related Alkaloids," eds. J. D. Phillipson and M. H. Zenk, Academic Press, London, 1980, p. 11.

253. M. V. Kisakurek, A. J. M. Leeuwenberg, and M. Hesse, in "Alkaloids: Chemical and Biological Perspectives," ed. S. W. Pelletier, Wiley, New York, 1983, vol. 1, p. 211.

254. N. G. Bisset, in "Indole and Biogenetically Related Alkaloids," eds. J. D. Phillipson and M. H. Zenk, Academic Press, London, 1980, p. 27.

255. Atta-ur-Rahman and A. Basha, "Biosynthesis of Indole Alkaloids," Oxford University Press, Oxford, 1983.

256. J. Bosch, J. Bonjoch, and M. Amat, in "The Alkaloids," ed. G. A. Cordell, Academic Press, San Diego, CA, 1996, vol. 48, p. 75.

257. J. Sapi and G. Massiot, in "The Chemistry of Heterocyclic Compounds," ed. E. C. Taylor, Wiley, Chichester, 1994, vol. 25, Suppl. to Part 4, p. 279.

258. M. Lounasmaa and P. Somersato, in "Progress in the Chemistry of Organic Natural Products," eds. W. Herz, H. Grisebach, G. W. Kirby, and C. Tamm, Springer, Vienna, 1986, vol. 50, p. 28.

259. J. Bosch and J. Bonjoch, in "Studies in Natural Products," ed. Atta-ur-Rahman, Elsevier, Amsterdam, 1988, p. 31.

260. U. Beifuss, *Angew. Chem., Int. Ed. Engl.*, 1994, **33**, 1144.

261. J.-M. Nuzillard, P. Thepeneir, M.-J. Jacquier, G. Massiot, L. L. Men-Oliver and C. Delaude, *Phytochemistry*, 1996, **43**, 897.

262. M. Amat, A. Linares, and J. Bosch, *J. Org. Chem.*, 1990, **55**, 6295.

263. M. Amat, A. Linares, J. Munoz, and J. Bosch, *Tetrahedron Lett.*, 1988, **29**, 6573.

264. M. Amat, A. Linares, and J. Bosch, *Tetrahedron Lett.*, 1989, **30**, 2293.

265. M. Amat, A. Linares, and J. Bosch, *J. Org. Chem.*, 1990, **55**, 6299.

266. M. Amat, A. Linares, M.-L. Salas, M. Alvarez, and J. Bosch, *J. Chem. Soc., Chem. Commun.*, 1988, 420.

267. A. Stoll and A. Holfmann, in "The Alkaloids," eds. R. H. F. Manske and H. L. Holmes, Academic Press, New York, 1965, vol. 8, p. 725.

268. I. Ninomiya and T. Kiguschi, in "The Alkaloids," ed. A. Brossi, Academic Press, San Diego, CA, 1990, vol. 38, p. 1.

269. E. Eich and H. Pertz, *Pharmazie*, 1994, **49**, 867.

270. G. B. Okide, *Tetrahedron Lett.*, 1993, **35**, 1135.

271. S. L. Lee, H. G. Floss, and P. Heinsteine, *Arch. Biochem. Biophys.*, 1976, **177**, 84.

272. J. A. Anderson and M. S. Saini, *Tetrahedron Lett.*, 1974, 2107.

273. M. S. Saine and J. A. Anderson, *Phytochemistry*, 1978, **17**, 799.

274. R. J. Petroski and W. J. Kellcher, *Lloydia*, 1978, **41**, 332.

275. W. Maier, D. Erge, J. Schmidt, and D. Groger, *Experientia*, 1980, **36**, 1353.

276. W. Maier, E. Erge, and D. Groger, *Planta Med.*, 1980, **40**, 104.

277. S. B. Hassam and H. G. Floss, *J. Nat. Prod.*, 1981, **44**, 756.

278. W. Acklin, T. Fehr, and P. A. Stadlev, *Helv. Chim. Acta*, 1975, **58**, 2492.

279. R. S. Bajwa, R. D. Kohler, M. S. Saini, M. Cheng, and J. A. Anderson, *Phytochemistry*, 1975, **14**, 735.

280. Y. Yokoyawa, T. Matsumoto, and Y. Murakami, *J. Org. Chem.*, 1995, **60**, 1486.

281. M. Ihara and K. Fukumoto, *Nat. Prod. Rep.*, 1996, **13**, 241.

282. G. Galambos, C. Szantay, J. Tamas, Jr., and C. Szantay, *Heterocycles*, 1993, **36**, 2241.

283. K. Jenett-Siems, M. Kaloga, and E. Eich, *J. Nat. Prod.*, 1994, **57**, 1304.

284. L. Cvak, J. Minar, S. Pakhomova, J. Ondracek, B. Kratochvil, P. Sedmera, V. Havlicek, and A. Jegorov, *Phytochemistry*, 1996, **42**, 237.

285. L. Cvak, A. Jegorov, S. Pakhomova, B. Kratochvil, P. Sedmera, V. Havlicek, and J. Minar, *Phytochemistry*, 1997, **44**, 365.

286. T. S. Wu, S. C. Huang, P. L. Wu, and C. M. Teng, *Phytochemistry*, 1996, **43**, 133.

287. T. S. Wu and S. C. Huang, *Chem. Pharm. Bull.*, 1992, **40**, 1069.

288. T. S. Wu, S. C. Huang, and J. S. Lai, *J. Chim. Chem. Soc.*, 1992, **40**, 319.

289. T. S. Wu, S. C. Huang, P. L. Wu, and K. H. Lee, *Bioorg. Med. Chem. Lett.*, 1994, **4**, 2395.

290. T. S. Wu, M. L. Wang, P. L. Wu, C. Ito, and H. Furukawa, *Phytochemistry*, 1996, **41**, 1433.

291. T. S. Wu, S. C. Huang, P. L. Wu, and T. T. Jong, *Phytochemistry*, 1995, **40**, 1817.

292. P. Bhattacharyya, A. K. Maiti, K. Basu, and B. K. Chowdhury, *Phytochemistry*, 1994, **35**, 1085.

293. R. B. Herbert and J. Mann, *J. Chem. Soc., Perkin Trans. 1*, 1982, 1523.

294. J. Berlin, C. Rugenhagen, N. Greidziak, I. N. Kuzovikina, L. Witte, and V. Wray, *Phytochemistry*, 1993, **33**, 593.

295. B. Holmstedt, S. H. Wassen, and R. E. Schultes, *J. Ethnopharmacol.*, 1979, **1**, 3.

296. L. Maat and H. C. Beyerman, in "The Alkaloids," ed. A. Brossi, Academic Press, San Diego, CA, 1983, vol. 22, p. 282.

297. H. G. Boit, "Ergebnisse der Alkaloid Chemie bis 1960," Akademie Verlag, Berlin, 1961, p. 753.

298. E. Leete, in "Biogenesis of Natural Compounds," ed. P. Bernfeld, Pergamon, Oxford, 1963, vol. 63, p. 791.

299. M. A. Nunes, Ph.D. Thesis, University of California at San Francisco, 1974.

300. C. Brochmann, E. Hanssen, M. A. Nunes, and C. K. Olah, *Planta Med.*, 1975, **28**, 1.

4.04
Biosynthesis of Heme

PETER M. SHOOLINGIN-JORDAN and KWAI-MING CHEUNG
University of Southampton, UK

4.04.1   INTRODUCTION

The function of tetrapyrroles in biological systems is, with few exceptions, to chelate metal ions with the pyrrole nitrogen ligands of the macrocyclic ring system and to modulate the properties of the metal ion for the purposes of catalyzing a variety of redox processes, novel bioinorganic chemistry, and electron transfer reactions. Thus, tetrapyrroles act as prosthetic groups for a wide number of proteins, many of which are involved in vital energy generating reactions in life processes such as photosynthesis, respiration, methanogenesis, etc.[1] The metal ion chemistry is finely tuned by variations in the range of substituents at the β-positions of the pyrrole rings, the redox state of the tetrapyrrole, the nature of axial ligands attached to the metal ion and, most importantly, the unique environment provided to the metallotetrapyrrole by interaction with the protein. This chapter is confined to the biosynthesis pathway for heme, one of the most widely used tetrapyrroles in biological systems, although reference is made to the fact that other tetrapyrroles arise from the heme pathway intermediate, uroporphyrinogen III, notably siroheme, factor F_{430}, and vitamin B_{12}, as well as the magnesium derivatives such as the chlorophylls and bacteriochlorophylls (Scheme 1).

It is convenient to divide the multistep heme biosynthesis pathway into three main stages: first, the biosynthesis of the ubiquitous 5-carbon precursor, 5-aminolevulinic acid; second, the transformation of eight molecules of 5-aminolevulinic acid into the macrocyclic tetrapyrrole precursor, uroporphyrinogen III; and finally, the conversion of uroporphyrinogen III into heme by decarboxylation and oxidation of the tetrapyrrole macrocycle to the conjugated porphyrin system and chelation of ferrous iron (Scheme 2). Texts are available concerning the biosynthesis of heme, chlorophylls, and other tetrapyrroles.[2–5]

4.04.2   THE BIOSYNTHESIS OF 5-AMINOLEVULINIC ACID

5-Aminolevulinic acid is biosynthesized by two completely different routes, according to the organism. Animals, fungi, facultative aerobic bacteria, and photosynthetic bacteria employ a single enzymic step in which glycine and succinyl-CoA are condensed in a reaction catalyzed by 5-aminolevulinic acid synthase. Higher plants, algae, anaerobic bacteria, and some oxygenic bacteria utilize the intact carbon skeleton of glutamate and generate 5-aminolevulinic acid through the C_5 pathway.[6]

4.04.2.1   5-Aminolevulinic Acid Synthase

In animals, fungi, and photosynthetic bacteria, 5-aminolevulinic acid is formed by the condensation of glycine and succinyl-CoA (Scheme 3).

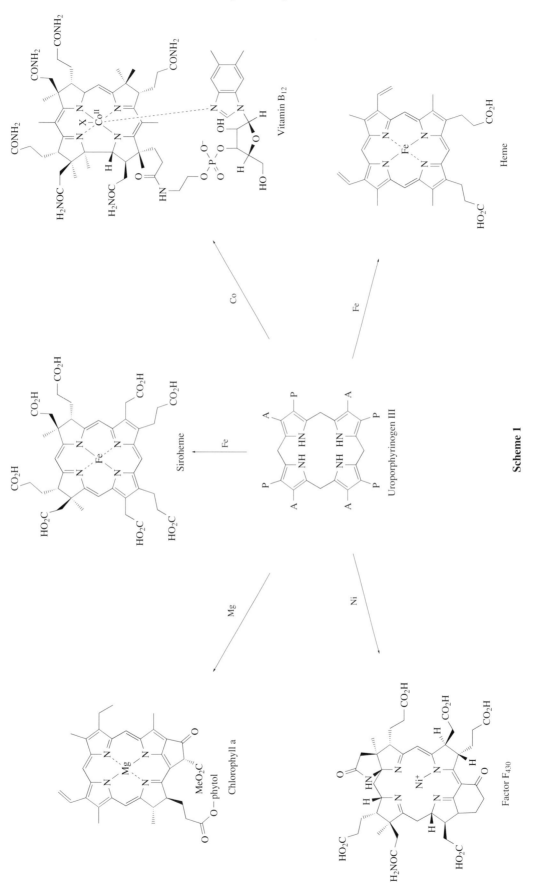

Scheme 1

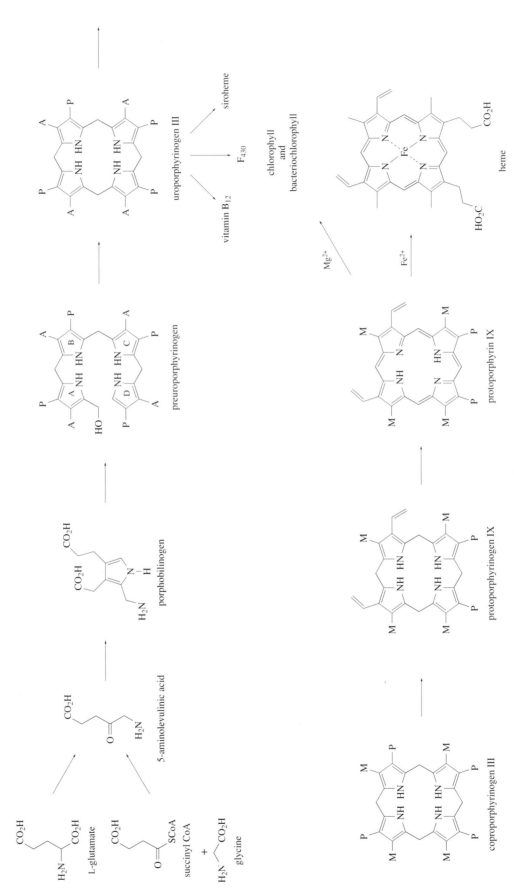

Scheme 2

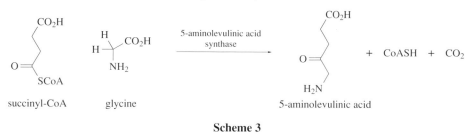

Scheme 3

4.04.2.1.1 Occurrence of the enzyme and its properties

The enzyme 5-aminolevulinic acid synthase was first described in the photosynthetic bacterium *Rhodobacter spheroides* and has since been isolated from several eukaryote sources including human, mouse, and avian erythrocytes and yeast. In mammals there are two forms of the enzyme, a ubiquitous form occurring in all cells, termed ALAS1 or ALAS-H, and a form expressed only in erythrocytes, ALAS2 or ALAS-E.[7] Interestingly, two forms also exist in *Rhodobacter spheroides*,[8] although their individual roles are not known. In eukaryotes the enzyme is located in the mitochondria, where ready access to succinyl-CoA is available. 5-Aminolevulinic acid synthase is the rate-controlling enzyme for heme synthesis and, because of the cytotoxic photodynamic properties of oxidized forms of subsequent tetrapyrrole intermediates (uroporphyrin III, coproporphyrin III, and protoporphyrin IX), the cellular level of the enzyme is always tightly regulated.[9] Consequently, 5-aminolevulinic acid synthase is rather low in abundance, although under conditions where intracellular heme levels are depressed, the enzyme level increases several fold to restore the heme concentration. The mechanism by which heme regulates its own biosynthesis is a classical example of end-product regulation.

Accessibility to 5-aminolevulinic acid synthase has been dramatically influenced as a result of the isolation of a number of genes/cDNAs encoding the enzyme and both prokaryotic and eukaryotic synthases have been overexpressed in recombinant strains of *E. coli*. The gene/cDNA sequences have allowed the primary sequence of the 5-aminolevulinic acid synthase proteins to be derived from several sources.[7] Comparison of the sequences suggests that the enzyme consists of a C-terminal catalytic core, common to all organisms, with a variable N-terminal region that is involved in mitochondrial import and regulatory control mechanisms. The prokaryote enzymes are therefore smaller and possess only the core component. The mRNA specifying the eukaryote protein contains iron-responsive elements important in the regulation of translation and the coordination of enzyme expression with iron metabolism.[10]

4.04.2.1.2 Substrate specificity and kinetics

5-Aminolevulinic acid synthase accepts no other amino acid substrate except glycine. Despite this rigid specificity, the K_m for glycine is in the millimolar range. The K_m for succinyl-CoA is, by contrast, in the lower micromolar range but the enzyme is far less specific for this substrate, accepting several other acyl-CoA derivatives.[11] Acetoacetyl-CoA functions as a substrate at 10% the rate of succinyl-CoA, whereas acetyl-CoA and valeryl-CoA are particularly poor substrates. Prokaryote and eukaryote enzymes show remarkably similar acyl-CoA specificities,[11] suggesting that the active site architecture has been conserved over the course of evolution. Steady-state kinetic analysis indicates an ordered reaction with glycine binding first and 5-aminolevulinic acid dissociating last.[7,12]

4.04.2.1.3 The mechanism and steric course of the 5-aminolevulinic acid synthase reaction

The mechanism of 5-aminolevulinic acid synthase involves reaction of the C-2 atom of glycine with the electrophilic carbonyl carbon of succinyl-CoA. In common with other pyridoxal 5′-phosphate dependent enzymes, the initial stage in the reaction is the formation of a Schiff base between the amino group of glycine and the carbonyl group of pyridoxal 5′-phosphate by trans-aldimination with the active site lysine–pyridoxal 5′-phosphate Schiff base (Scheme 4). To react with succinyl-CoA, the C-2 position of glycine in the enzyme-bound glycine–pyridoxal-5′-phosphate complex must be converted into a stabilized carbanion or equivalent nucleophilic species. This can, in principle, be accomplished by either a decarboxylation or proton abstraction mechanism. These two possibilities have been investigated by following the fate of the hydrogen atoms at C-2 when $2RS[^3H_2]$-glycine is transformed into 5-aminolevulinic acid. A mechanism involving initial decarboxylation would leave the two labeled hydrogen atoms of glycine undisturbed, whereas a mechanism involving initial deprotonation would lead to loss of half of the tritium label. The finding that half of the tritium label is lost from the 2-position of glycine in the enzymic conversion to 5-aminolevulinic acid indicated a mechanism in which the initial stabilized carbanion is generated by the loss of a proton.[13,14] This study was confirmed and extended by the synthesis of the two stereospecifically tritiated glycines, $2R[^3H_2]$-glycine and $2S[^3H_2]$-glycine,[13,15] establishing that it is the glycine hydrogen atom with the *proR* configuration which is removed in the 5-aminolevulinic acid synthase reaction (Scheme 4).

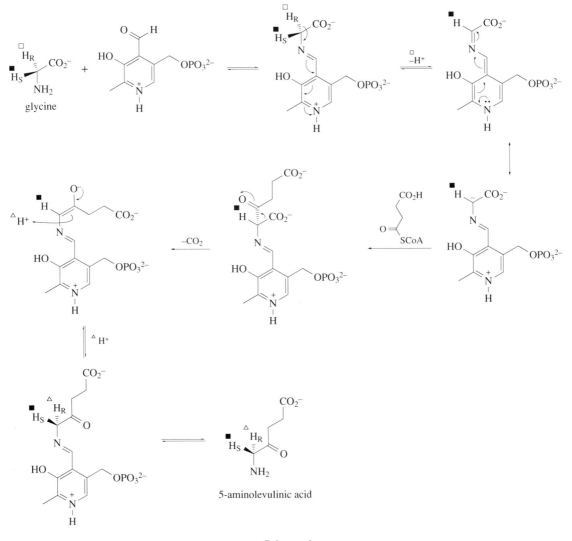

Scheme 4

This finding implicated 2-amino-3-ketoadipic acid in the form of a Schiff base linked to the enzyme-bound pyridoxal 5′-phosphate as an intermediate. There are two broad routes by which this intermediate may be transformed into 5-aminolevulinic acid. Either 2-amino-3-ketoadipic acid could be released from the enzyme and, as an unstable 3-ketoacid, would decarboxylate nonenzymically. Alternatively, the enzyme itself could catalyze the decarboxylation and reprotonation to form 5-aminolevulinic acid. Since the *proS* hydrogen atom of glycine is stereospecifically retained as the *proS* hydrogen at the 5-position of 5-aminolevulinic acid, the decarboxylation of 2-amino-3-ketoadipic acid must be an enzymically mediated process (Scheme 4). The experimental findings also indicate that the mechanism of 5-aminolevulinic acid formation from glycine proceeds with overall inversion.[16]

Whether it is the condensation step or the decarboxylation that occurs with inversion is not known since there are no examples of either substitution or decarboxylation reactions catalyzed by pyridoxal 5′-phosphate-requiring enzymes occurring with inversion.[17] It can be argued reasonably, however, that if decarboxylation of the 2-amino-3-ketoadipic acid intermediate is facilitated by the newly generated keto group rather than by the pyridine ring of the coenzyme then it could proceed by inversion without contravening the accepted retention mechanism followed by most pyridoxal 5′-phosphate-dependent enzymes[17,18] (Scheme 4). However, argument can also be made for inversion in the initial condensation if the bulky acyl-CoA moiety cannot approach the deprotonated glycine species from the side originally occupied by the *proR* hydrogen. Rotation of the bound 3-ketoacid intermediate so that the final decarboxylation/reprotonation occurs with retention would explain the stereochemical findings. The latter mechanism is also consistent with the unimpressive affinity of glycine for the enzyme and the observation that methylamine is a good substrate, suggesting that the interaction of the glycine carboxyl group with the enzyme is relatively weak.

4.04.2.1.4 *Enzymic structure and catalytic groups*

In the absence of an X-ray structure for 5-aminolevulinic acid synthase, the identity of catalytic groups at the active site of 5-aminolevulinic acid synthase has been investigated by site-directed mutagenesis and by comparison between the amino acid sequences of a number of other pyridoxal 5′-phosphate-dependent proteins, especially those catalyzing related reactions to form 2-amino-ketones.[19] Of paramount importance is an invariant lysine that forms the Schiff base with pyridoxal 5′-phosphate. This lysine has been identified as Lys313 in the mouse ALAS2 sequence GTLGKAFGCVGG.[20] Mutation leads to a complete loss of enzyme activity. It is likely that the hydrogen atom initially removed from the *proR* orientation of glycine in the initial deprotonation and the *proR* hydrogen added finally to the 5-position of 5-aminolevulinic acid is transferred by this lysine. The *proR* hydrogen atom of both glycine and 5-aminolevulinic acid undergoes enzyme-catalyzed stereospecific exchange reactions with protons of the medium.[21,22] Assuming that the same enzymic base catalyzes both the reactions, this requires that the carboxyl group of glycine and the succinyl group of 5-aminolevulinic acid occupy the same relative positions at the active site. The structure of 5-aminolevulinic acid synthase derived from a model of the enzyme based on the structure of dialkylglycine decarboxylase[23] indicates that Lys313 plays a key role in catalysis. Conserved histidines at positions 208, 281, and 446 are also likely to play an important part. The invariant Arg439, a conserved residue involved in binding the carboxyl group of amino acids in other pyridoxal 5′-phosphate-dependent enzymes, is likely to be involved in binding the glycine carboxyl group. Mutagenesis of this residue in the mouse ALAS2 enzyme has dramatic effects on the activity.[24] The sequences of all 5-aminolevulinic acid synthases also contain an invariant aspartic acid essential for charge stabilization of the positive pyridine nitrogen in all pyridoxal 5′-phosphate-dependent enzymes. Site-directed mutagenesis of this aspartate, Asp279 in the mouse ALAS2 sequence, leads to an enzyme with very low activity.[25] The final determination of the X-ray structure of the enzyme is awaited with the expectation that many of the questions regarding the active site geometry may have already been answered from model studies.

4.04.2.2 The Biosynthesis of 5-Aminolevulinic Acid by the C₅ Pathway

4.04.2.2.1 *Discovery of the C₅ pathway*

The unsuccessful search for 5-aminolevulinic acid synthase in plants over a period of nearly 20 years was explained when an alternative route for 5-aminolevulinic acid was discovered. The first indications came from experiments in which greening plant tissue was incubated with labeled glutamate and 2-ketoglutamate precursors in the presence of levulinic acid, an inhibitor of 5-aminolevulinic acid utilization. In these experiments a higher incorporation of label was incorporated into 5-aminolevulinic acid from glutamate and related compounds than from glycine.[26–28] Three enzymes are involved in the conversion of glutamate into 5-aminolevulinic acid in this pathway: glutamyl-tRNA[Glu] synthetase (ligase), glutamyl-tRNA[Glu] reductase, and glutamate 1-semialdehyde aminotransferase (Scheme 5). All three enzymes were first isolated from greening barley[29] and subsequently tRNA[Glu] was established as a cofactor in the conversion.[30,31] Glutamyl-tRNA[Glu] synthetase plays a dual role in biological systems in that it also provides glutamyl-tRNA[Glu] for protein synthesis. The first dedicated enzyme to tetrapyrrole biosynthesis in organisms that use the C_5 pathway is therefore glutamyl-tRNA^{Glu} reductase. This enzyme reduces the activated glutamyl-tRNA^{Glu} ester to the novel 2-aminoaldehyde, glutamate 1-semialdehyde. Glutamate 1-semialdehyde is finally converted into 5-aminolevulinic acid by glutamate 1-semialdehyde aminotransferase. This subject area has been reviewed elsewhere.[32,33]

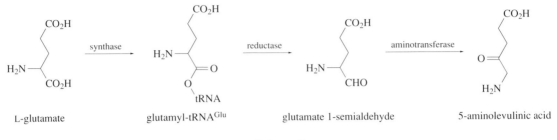

L-glutamate   glutamyl-tRNA^{Glu} glutamate 1-semialdehyde   5-aminolevulinic acid

Scheme 5

4.04.2.2.2   *Glutamyl-tRNA^{Glu} synthetase*

There are two glutamyl-tRNA^{Glu} synthetases in higher plants, one in the cytoplasm and the other in the chloroplast, although both are encoded by the nuclear genome. It is the chloroplast enzyme, however, which is involved in the synthesis of glutamyl-tRNA^{Glu} for use in chlorophyll biosynthesis, although the enzyme is unusual in that it is also able to charge tRNA^{Gln} to form glutamyl-tRNA^{Gln}.[34] The enzyme reaction proceeds through a two-stage mechanism, in common with other aminoacyl-tRNA synthases (ligases), involving an initial reaction between the enzyme, glutamate, ATP, and Mg^{2+} to give enzyme-bound aminoacyl-AMP with the release of pyrophosphate. The presence of tRNA^{Glu} is necessary for the first stage. In the second stage, tRNA^{Glu} reacts with the aminoacyl-AMP forming glutamyl-tRNA^{Glu} and AMP is displaced.

The tRNA^{Glu} utilized in tetrapyrrole biosynthesis in barley is encoded by chloroplast DNA and contains the modified base 5-methylaminomethyl-2-thiouridine at the first position of the anticodon UUC.[30] The same base is found in the analogous position in *E. coli*. It has been proposed that this unstable base is involved in the regulation of tetrapyrrole biosynthesis since, in its oxidized form, the tRNA^{Glu} can no longer function in the reaction to give glutamyl-tRNA^{Glu}. In the light, the reduced base can be reformed, allowing the restoration of chlorophyll biosynthesis.[33]

4.04.2.2.3   *Glutamyl-tRNA^{Glu} reductase*

This enzyme catalyzes the NADPH-dependent reduction of glutamyl-tRNA^{Glu} to give the novel intermediate, glutamate 1-semialdehyde. The reductase enzyme from various sources appears to be very different in size and properties with aggregation and complex formation being used to explain these observations. Glutamyl-tRNA^{Glu} reductases appear to be inhibited by heme at micromolar concentrations and their levels are tightly regulated to prevent the overproduction of tetrapyrroles.[35] The reaction is considered to be the rate-determining step in chlorophyll biosyntheses. The low abundance of the reductase has made it difficult to obtain sufficient amounts of enzyme for detailed biochemical studies, although the cloning of the *hemA*[5] gene encoding the *E. coli* enzyme[36] and the *HEMA*[5] gene encoding *Arabidopsis thaliana* enzyme[37] will permit the development of overexpression systems.

An interesting property of glutamyl-tRNA^{Glu} reductases is their specificity for the tRNA^{Glu} moiety, not only between different organisms but even within subcellular compartments of the same organism.[38] Thus the barley reductase can recognize glutamyl-tRNA^{Glu} from chloroplasts but not from barley cytoplasm or from *E. coli*. On the other hand, the *E. coli* glutamyl-tRNA^{Glu} is recognized by the reductase from *Chlamydomonas reinhardtii* and *A. thaliana*. There is no obvious pattern regarding the nature of the recognition process although there are clues. For instance, chloroplast genes encoding tRNA^{Glu} all contain the DNA sequence that specifies the UUC anticodon and possess the A53:U61 base pair in the Tψ-stem of the molecule, instead of a G:C base pair found in tRNA^{Glu} involved in protein synthesis. Mutations in this base pair in *Euglena gracilis* lead to poor recognition of the glutamyl-tRNA^Glu by the reductase. Despite the lack of precise detail, it is clear that different regions of tRNA^Glu are involved with recognition of the synthetase, the reductase, and the ribosome.

The reaction mechanism has not been established but is expected to involve either the reduction of the aminoacyl-tRNA ester to the hemiacetal that would eliminate tRNA^Glu to yield the semialdehyde product (Scheme 6), or, alternatively, a mechanism involving nucleophilic attack on glutamyl-tRNA^Glu by an enzyme thiol to give a thioester followed by reduction to the thiohemiacetal and release of the aldehyde. The latter has precedents for the transformation of carboxylic acids into aldehydes, notably in the glycolytic/photosynthesis pathways in which 1,3-diphosphoglyceric acid is reduced to glyceraldehyde 3-phosphate. Whether glutamyl-tRNA^Glu ester is sufficiently reactive to form a thioester in this way is, however, debatable. In this context, it may be significant that an invariant thiol, essential for activity, is present in all glutamyl-tRNA^Glu reductases. The involvement of glutamate 1-semialdehyde in tetrapyrrole biosynthesis is somewhat unexpected since 2-amino-aldehydes are notoriously unstable in their unprotected forms. Two suggestions have been put forward to explain the anomalous stability of glutamate 1-semialdehyde, one being its existence in the form of a hydrate[39] and the other as a cyclic tetrahydropyranone, 5-amino-3,4,5,6-tetrahydro-6-hydroxy-2H-pyran-2-one.[40] The surprising stability of glucosamine, also a 2-aminoaldehyde, is also explained by a cyclic structure. If the product of the reductase is a cyclic form of glutamate 1-semialdehyde then any mechanism must take into consideration the role of the enzyme in this process, as well as the possible role of glutamate 1-semialdehyde aminotransferase in ring opening.

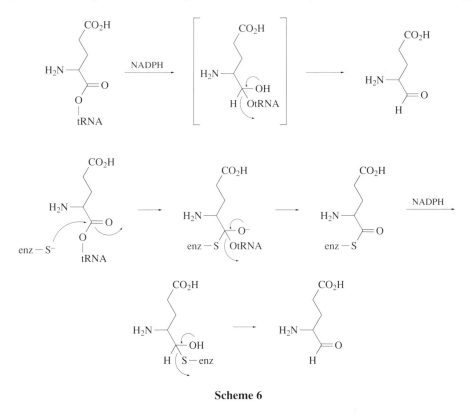

Scheme 6

4.04.2.2.4 *Glutamate 1-semialdehyde aminotransferase*

The final step of the C_5 pathway is catalyzed by the pyridoxal 5'-phosphate-dependent enzyme glutamate 1-semialdehyde aminotransferase and involves the apparent switching of the carbonyl and amino functions of glutamate 1-semialdehyde to give 5-aminolevulinic acid. The enzyme, on first consideration, appears to be a mutase, involving 4,5-intramolecular transfer of the amino group; however, labeling experiments[41] have shown that this is not the case and that the enzyme reaction proceeds in two stages with intermolecular transfer of the amino group. Aminotransferase, rather than mutase, is therefore the appropriate name for the enzyme.

The enzyme is particularly abundant *in vivo*, and glutamate 1-semialdehyde aminotransferases have been isolated from a number of bacteria and plants.[5,6,33] All appear to exist as dimeric enzymes with subunit M_r values in the 44–46 kDa range. Glutamate 1-semialdehyde aminotransferases contain either pyridoxal 5'-phosphate or pyridoxamine 5'-phosphate, or a mixture of the two.[33] The genes specifying the enzyme from higher plants (*gsa*) and the equivalent in bacteria (*hemL*) have been isolated and several have been overexpressed in recombinant bacterial strains.

(i) The mechanism of transamination and inhibition of the enzyme

Unlike other pyridoxal 5'-phosphate-dependent enzymes that catalyze transamination, glutamate 1-semialdehyde aminotransferase is unique in that it can use the substrate L-glutamate 1-semi-aldehyde as an amino donor as well as an amino acceptor. Thus, the apoenzyme can bind either pyridoxal 5'-phosphate or pyridoxamine 5'-phosphate. The enzyme mechanism can conveniently be studied by spectroscopic methods.[42] In the former case, the enzyme is converted into the pyrid-oxamine form by transamination with glutamate 1-semialdehyde to 4,5-dioxovaleric acid. In the latter case, the pyridoxamine form is converted into the pyridoxal form and 4,5-diaminovaleric acid. The 4,5-diaminovalerate intermediate is tightly bound although it can dissociate partially to yield the free enzyme–pyridoxal-5'-phosphate complex. Enzymes from different sources appear to have preferential pathways,[43] although the enzyme most well understood uses the 4,5-diaminovalerate reaction sequence shown in Scheme 7. This involves the initial formation of a Schiff base between the enzyme-bound pyridoxamine 5'-phosphate and the aldehyde carbonyl followed by proton transfer using Lys272 as the catalytic group and hydrolysis to yield the enzyme–pyridoxal-5'-phosphate and 4,5-diaminovalerate. Formation of the isomeric Schiff base with the 4-amino group, followed by analogous proton transfer, restores the enzyme to its pyridoxamine 5'-phosphate form and yields 5-aminolevulinic acid. The equilibrium favors the formation of 5-aminolevulinic acid and analogues of glutamate 1-semialdehyde, such as aspartate 1-semialdehyde and 4-amino-5-hydroxyvaleric acid, are transaminating substrates, whereas the 3,4-diaminobutyrate and 4-hydroxy-5-aminovaleric acid are not[44] (Scheme 8). D-Glutamate 1-semialdehyde can participate in the initial transamination but the reaction is slow. The binding site thus appears to discriminate against substrates or analogues with a terminal amino group, such as found in 5-aminolevulinic acid. Glutamate-406 is thought to play an important role in this interaction.[45]

Glutamate 1-semialdehyde aminotransferases are susceptible to suicide inactivation by the micro-bial toxin gabaculine that reacts with the pyridoxal 5'-phosphate form of the enzyme. Initially the amino group of gabaculine forms a Schiff base and, after proton transfer, an enzyme base is probably involved in generating a tightly bound anthranilate intermediate that causes inactivation of the enzyme (Scheme 9). An interesting gabaculine-resistant mutant aminotransferase from barley has been isolated that contains isoleucine instead of methionine at position 248 in the active site region.[46] The enzyme is also susceptible to other classical pyridoxal 5'-phosphate inhibitors.

(ii) The structure of glutamate 1-semialdehyde aminotransferase

The abundance of the enzyme and the ease with which it can be isolated in milligram amounts from recombinant sources has permitted crystallization and determination of the X-ray structure by molecular replacement.[45] The structure of the enzyme from *Synochococcus* reveals a dimeric enzyme with many of the features expected for a pyridoxal 5'-phosphate-dependent enzyme (Figure 1). One notable exception is the absence of the key arginine residue that binds the carboxyl carbon of aminoacids in other aminotransferases. Lys272 appears to be the only major catalytic residue, not only interacting with the coenzyme but also acting as the proton transfer group at all stages of the reaction.[47] Crystallization of the enzyme in the presence of gabaculine suggests that Arg32 plays an important role in the binding of the carboxyl group of the substrate. This is in a different position to Arg292 that interacts with the glutamate C-5 position in aspartate aminotransferase.[48] The crystal structure shows an interesting asymmetry within the dimer so that only one subunit appears to be functional at any one time.

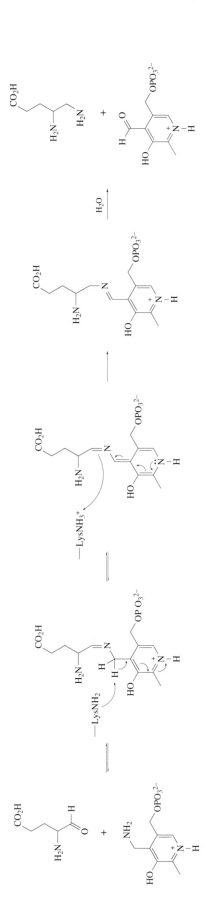
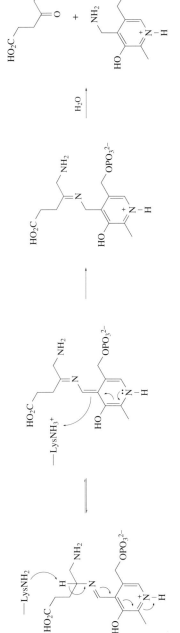
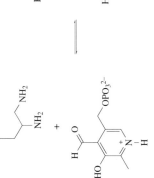

Scheme 7

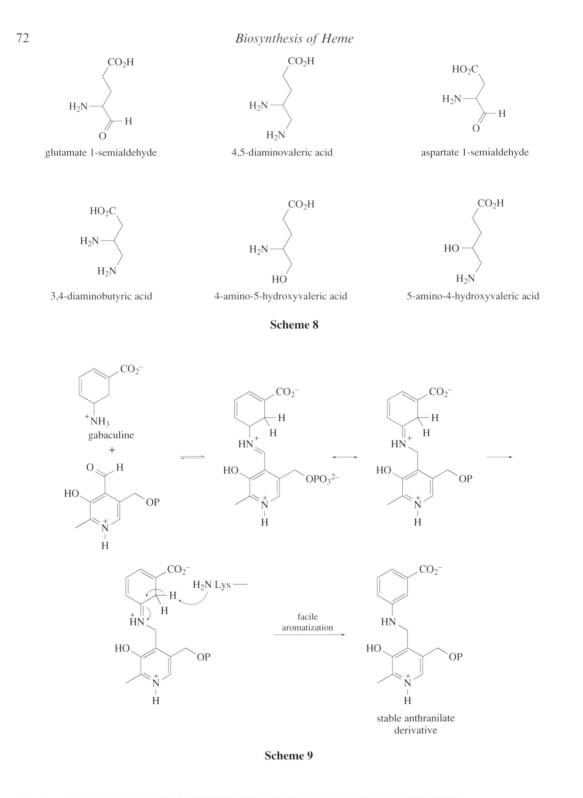

glutamate 1-semialdehyde   4,5-diaminovaleric acid   aspartate 1-semialdehyde

3,4-diaminobutyric acid   4-amino-5-hydroxyvaleric acid   5-amino-4-hydroxyvaleric acid

Scheme 8

gabaculine

facile
aromatization

stable anthranilate
derivative

Scheme 9

4.04.3 THE TRANSFORMATION OF 5-AMINOLEVULINIC ACID INTO UROPORPHYRINOGEN III

The second part of the heme biosynthesis pathway is common to all organisms that biosynthesize tetrapyrroles and is employed for the biosynthesis of corrinoids, siroheme, the nickel cofactor of methanogenic bacteria, chlorophylls, bacteriochlorophylls, and noncyclic tetrapyrroles.[2,3,49] This part of the pathway is macromolecular synthesis evolved to perfection since in only three enzymic steps (Scheme 10) eight molecules of 5-aminolevulinic acid are transformed into the tetrapyrrole macrocycle, uroporphyrinogen III. 5-Aminolevulinic acid dehydratase first dimerizes two molecules of 5-aminolevulinic acid to yield the ubiquitous pyrrole precursor, porphobilinogen. Next, four

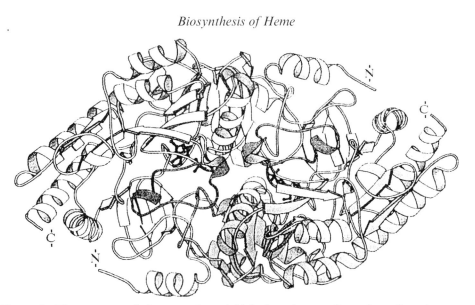

Figure 1 The structure of glutamate 1-semialdehyde aminotransferase from *Synechococcus*.

molecules of porphobilinogen are tetrapolymerized to preuroporphyrinogen by the enzyme porphobilinogen deaminase. Finally, the linear preuroporphyrinogen is cyclized with rearrangement of the D ring to give uroporphyrinogen III. Preuroporphyrinogen is a highly unstable 1-hydroxymethylbilane and, in the absence of uroporphyrinogen III synthase, cyclizes rapidly without rearrangement to yield the nonphysiological isomer, uroporphyrinogen I (Scheme 10).

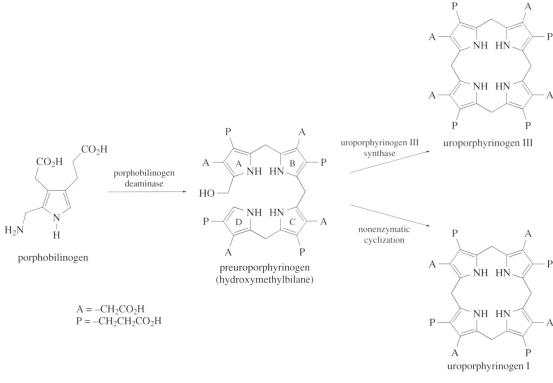

Scheme 10

4.04.3.1 5-Aminolevulinic Acid Dehydratase

5-Aminolevulinic acid dehydratase (also named porphobilinogen synthase) catalyzes the dimerization of two molecules of 5-aminolevulinic acid to give the monopyrrole intermediate, porphobilinogen.[5,50,51] This reaction is essentially a Knorr pyrrole synthesis involving an aldol condensation and the formation of a Schiff base (although not necessarily in that order) and is accompanied by elimination of two molecules of water (Scheme 11). Since 5-aminolevulinic acid, like other 2-aminoketones, is an intrinsically reactive compound, one of the most important contributions of the enzyme is to ensure that the two 5-aminolevulinic acid molecules condense in a "parallel" fashion to form the desired pyrrole product rather in an antiparallel manner to give the dihydropyrazine, as occurs in the nonenzymic dimerization. To achieve this the enzyme has two different substrate binding sites: the "A-site" which interacts with the 5-aminolevulinic acid that forms the acetic acid side chain of porphobilinogen and the "P-site" which binds the molecule contributing to the propionic acid side chain. The chemical and enzymic synthesis of porphobilinogen has been reviewed extensively.[51]

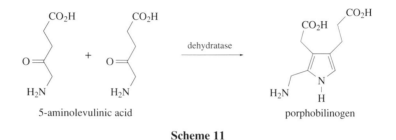

5-aminolevulinic acid   porphobilinogen

Scheme 11

4.04.3.1.1   *5-Aminolevulinic acid dehydratase and its properties*

5-Aminolevulinic acid dehydratase was first described by the pioneering groups of Shemin and Neuberger[52] and since then the enzyme has been isolated from a large number of sources including animals, higher plants, yeast, and bacteria.[5] The genes/cDNAs encoding several 5-aminolevulinic acid dehydratases have been isolated and sequenced and, from a comparison of the derived amino acid sequences,[50] it is clear that the structure of the enzyme is highly conserved throughout the biosphere. The majority of dehydratases are homooctamers with D_4 symmetry.[53] 5-Aminolevulinic acid dehydratases show two important characteristics, each relating to the binding of one of the substrates. First, they all employ covalent catalysis forming a Schiff base with one of the two substrate molecules and, second, they are all metalloenzymes requiring a divalent metal ion for activity.[5,50,51]

4.04.3.2   The Schiff Base Linkage Between the Enzyme and Substrate at the P-site

The first indication that a Schiff base is formed between the enzyme and the substrate in the 5-aminolevulinic acid dehydratase mechanism came from studies on the enzyme from *Rhodopseudomonas spheroides*.[54] Other dehydratases have since been shown to form a Schiff base linkage with 5-aminolevulinic acid.[55,56] Treatment of dehydratases with sodium borohydride in the presence of 5-aminolevulinic acid leads to reduction of the enzyme-substrate Schiff base and irreversible inactivation of the enzyme. Reduction in the presence of labeled substrate leads to the incorporation of label into the enzyme protein. By isolation and sequencing of the labeled peptide, Lys252 in the human enzyme[55] and Lys247 in the *E. coli* enzyme[56] have been identified as the Schiff base site in the invariant sequence -Val-Lys-Pro-, suggesting that all dehydratases interact with the substrate in a similar way. Acid hydrolysis of the enzyme reduced with sodium borohydride in the presence of 5-amino-[¹⁴C]levulinic acid yields the reduced adduct between 5-aminolevulinic acid and lysine.[55] Single turnover experiments using either [¹⁴C]- or [¹³C]labeled substrate have established that, out of the two 5-aminolevulinic acid molecules employed by the enzyme, it is the one that initially binds as a Schiff base which contributes to the propionic acid side chain of porphobilinogen.[57,58] Site-directed mutation of the Schiff base lysine confirms its crucial importance for the normal function of the enzyme with substitutions all causing a dramatic loss of enzymic activity. An additional invariant lysine at position 195 in the *E. coli* enzyme has a dramatic effect on the ability of Lys247 to form a Schiff base with the substrate, indicating that Lys195 is close enough to have a major influence on the reactivity of Lys247. The immediate proximity of these two lysines to one another has been verified by the X-ray structure of the enzyme, as discussed below.

4.04.3.2.1 *The metalloenzyme nature of 5-aminolevulinic acid dehydratase and the substrate binding A site*

All 5-aminolevulinic acid dehydratases require divalent metal ions for activity although, surprisingly, there are two distinct enzymic classes, one requiring zinc, found predominantly in animals and fungi, and another requiring magnesium, present in plants. Bacteria have representatives of both zinc- and magnesium-dependent classes.[5,50,51] In addition, several of the zinc-dependent enzymes are activated by magnesium. The mammalian enzymes are also active in the presence of cadmium and the enzyme from *E. coli* can function with cobalt.

The zinc dehydratases exhibit markedly different properties from the magnesium-dependent enzymes, reflecting the presence of several reactive thiol groups which act as metal ligands. The presence of thiol groups makes the zinc-dependent enzymes exceptionally sensitive to oxidation by air and particularly reactive to thiophilic reagents such as iodoacetic acid, 5,5′-dithio*bis*(2-nitrobenzoic acid) (DTNB),[59] and methylmethanethiosulfonate.[60] The cysteine most reactive to thiophilic reagents in the bovine enzyme is Cys223 which is replaced by either serine or threonine in bacterial and plant enzymes.

In addition to this highly reactive cysteine, there is a short cysteine-rich sequence in mammalian enzymes important in metal binding, extending from residues 120–135 in the human enzyme.[61] The reaction of mammalian enzymes with thiophilic reagents is accompanied by the loss of zinc, although this does not prevent the ability of the enzyme to bind the substrate at the P-site.[62] The same is true for the enzyme from *E. coli*.[63] The thiophilic reagent DTNB promotes oxidative cross-linking between the four cysteines in the cysteine-rich sequence of the *E. coli* enzyme, implying their close proximity to one another.[63]

The cysteine-rich sequence is also found in other zinc-dependent dehydratases but is absent in the dehydratases from plants in which the cysteine residues are substituted by aspartic acid and an alanine residue (Figure 2). Because the differences in metal requirements between animals and plants relate to whether cysteine or aspartic acid residues are present, it was suggested[64] that this region of the protein is of crucial importance in determining the metalloenzyme nature of the enzyme. This has been borne out both by mutagenesis experiments[65] and, more recently, from the crystal structure of the dehydratase.

Human	PNLLVACDV**C**L**C**PYTSHGH**C**G
Yeast	PELYIICDV**C**L**C**EYTSHGH**C**G
E. coli	PEMIVMSDT**C**F**C**EYTSHGH**C**G
Pea	PDLIIYTDV**A**L**D**PYSSDGH**D**G
Photosynthetic bacteria	PEVAIMTDI**A**L**D**PYNANGH**D**G

Figure 2 Metal binding region of 5-aminolevulinic acid dehydratases showing the three invarient cysteines (emboldened) in the zinc-dependent dehydratases and their substitution by aspartic acid and alanine in the magnesium-dependent enzymes.

The nature of the metal binding sites in the bovine enzyme has been investigated by a number of workers and evidence for half-site reactivity has been obtained for the bovine enzyme. Four zinc ions, termed ZnA, appear to be required per octamer for catalysis.[66] On the basis of studies with EXAFS, ZnA was predicted to have the five ligands, tyrosine, histidine, aspartic acid, cysteine, and a water molecule. A second zinc, ZnB, bound by four cysteine ligands has also been predicted on the basis of EXAFS studies.[67] In the *E. coli* enzyme, two metal binding sites, termed α- and β-, have been characterized by spectroscopic studies in which zinc was substituted by cobalt, taking advantage of the cobalt–thiol charge transfer complexes.[68] A complex model involving three metal sites has also been suggested.[50] The X-ray structures both of prokaryote and eukaryote 5-aminolevulinic acid dehydratases (see Figure 3 below) have placed doubt on many of the above predictions and consequently the nature of the metal binding centers, their stoichiometry, and the role of the metal ion(s) needs to be re-evaluated in the light of this structural information.

4.04.3.2.2   *Inhibitor studies with 5-aminolevulinic acid dehydratases*

The mammalian 5-aminolevulinic acid dehydratases are exquisitely sensitive to inhibition by heavy metals such as lead and mercury and these almost certainly exhibit their effects by exchanging with one or both of the metals at the active site. The human enzyme is inhibited by micromolar amounts of lead and its inhibition has been used as a marker for lead poisoning.[69] Preincubation with zinc prevents inhibition with lead as long as added thiols are present.[70]

In addition to inhibitors that affect the metal binding properties of 5-aminolevulinic acid dehydratases, a great number of compounds have been synthesized in attempts to unravel the nature of the enzyme mechanism, or to inhibit the enzyme in a biospecific manner. Initial studies established that 4-keto acids, such as levulinic acid (4-ketopentanoic acid), interact with the enzyme to form a Schiff base[54] and a great many inhibitors based on this structure have since been synthesized, one of the most notable being succinyl acetone.[71] Such 4-keto acids normally act as competitive inhibitors by forming a Schiff base at the P-site and can be linked irreversibly to the enzyme by reduction with sodium borohydride.[51,68,72]

Several inhibitors have higher affinity for the magnesium-dependent plant enzyme compared with the zinc-dependent enzymes,[72,73] possibly reflecting their ability to form both a Schiff base linkage and to provide oxygen as a magnesium ligand. One particularly interesting inhibitor, in which the C-3 of 5-aminolevulinic acid has been substituted by sulfur, is able to acylate the enzyme by reaction with the Schiff base lysine[74] (Scheme 12). A novel homologue of 5-aminolevulinic acid, 4-amino-3-ketobutyrate, has been synthesized and shown to interact specifically at the A-site, but not at the P-site.[72]

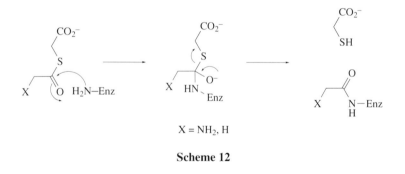

$X = NH_2$, H

Scheme 12

Chemical modification by the putative active site-directed inhibitors 3-chlorolevulinic acid and 5-chlorolevulinic acid[75] showed interesting differences in their reaction with the bovine enzyme, implying the existence of half-site reactivity. These studies have been extended[76] to show that 3-chlorolevulinic acid preferentially modifies Cys234 in the bovine enzyme (equivalent to Ser218 in the *E. coli* dehydratase) but either Cys120 or Cys130 in the *E. coli* enzyme, possibly reflecting the nonselective nature of this inhibitor in alkylating the most reactive thiol. Other detailed studies of inhibitors of the enzyme have been carried out[74,77] and the synthesis and action of a number of inhibitors have been comprehensively reviewed.[51]

4.04.3.2.3   *The X-ray structure of prokaryote and eukaryote 5-aminolevulinic acid dehydratases*

The X-ray structures of *E. coli* and yeast enzymes have been determined[78,79] and these show a number of interesting features that explain many of the discrepancies about the number of active sites, the number of metal ions per active site, and the coordination geometry of the metal ions. The bacterial and yeast enzymes are both homooctamers with each of the eight subunits made up from an (α/β_8)-barrel with an N-terminal arm of approximately 40 amino acids (Figure 3).

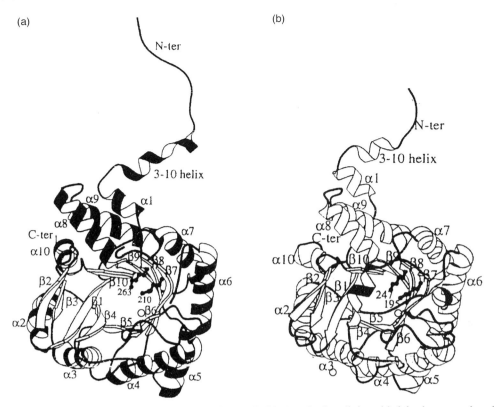

Figure 3 The X-ray structures of yeast (a) and *E. coli* (b) 5-aminolevulinic acid dehydratases showing the (α/β_8) barrel. Lys210 and Lys263 (yeast) and Lys195 and Lys247 (*E. coli*) are indicated in the center of the barrel. The zinc ion is shown as a circle.

The subunits are arranged as dimers with the two arms interacting with the β-barrel region of the partner in a "69" arrangement. The subunits of the four dimers are arranged in D_4 symmetry with the active site of each monomer facing the exterior of the octamer and set independently in the central cleft of the β-barrel. The catalytic Lys247 in *E. coli* (Lys263 in yeast and Lys252 in mammals) is located in a hydrophobic environment within the central cleft of the β-barrel, close to Lys195 in *E. coli* (Lys210 in yeast and Lys198 in mammals) (Figure 4). The proximity of the Lys195 to Lys247 suggests that the latter would be unprotonated to engage in the initial nucleophilic reaction to form the Schiff base with the P-site substrate. The structure of the *E. coli* enzyme in the presence of levulinic acid[79] suggests that the carboxyl group of the P-site substrate interacts with the invariant Tyr312 and Ser273 residues.

The X-ray structure of yeast 5-aminolevulinic acid dehydratase delineates a two-metal center per subunit with scope for binding either two zinc ions or a zinc ion and a magnesium ion. One zinc ion is coordinated to Cys133, Cys135, and Cys143 (bold in Figure 2) and a water molecule. This metal ion is near the Schiff base lysine residue and is likely to play a vital catalytic role. It is this zinc which is exchangeable for lead[78] that causes inactivation of the enzyme[69] in lead poisoning. The other metal ion is more than 7Å distant, coordinated to His142 and Cys234. In the *E. coli* dehydratase, Cys120, Cys122, and Cys130 form the triple cysteine site and a further metal site is present at a subunit–subunit contact that could bind the activating magnesium ion.[79]

4.04.3.2.4 *The enzyme mechanism*

The first mechanism proposed for the 5-aminolevulinic acid dehydratase reaction (Scheme 13) involved the formation of a Schiff base between the substrate that gives rise to the acetic acid side of porphobilinogen.[54] This proposal was based on the observation that levulinic acid not only forms a Schiff base with the enzyme but that together with 5-aminolevulinic acid the enzyme also catalyzes

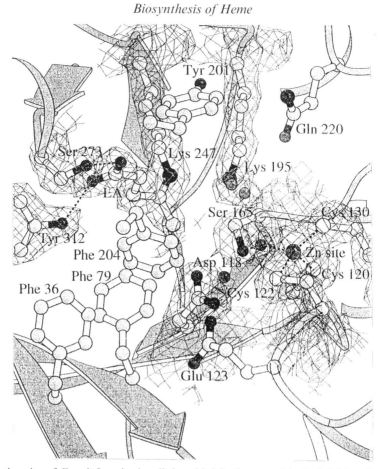

Figure 4   The active site of *E. coli* 5-aminolevulinic acid dehydratase with the inhibitor levulinic acid bound
as a Schiff base to Lys247.

the formation of a mixed pyrrole (see also Scheme 13). However, single turnover experiments[57,58]
pointed to a different mechanism in which the first substrate to bind to the enzyme not only does
so through a Schiff base but also provides the propionic acid side of porphobilinogen. Thus, levulinic
acid can inhibit by forming a Schiff base at the P-site, but to form a mixed pyrrole with 5-
aminolevulinic acid it would have to bind to the A-site.

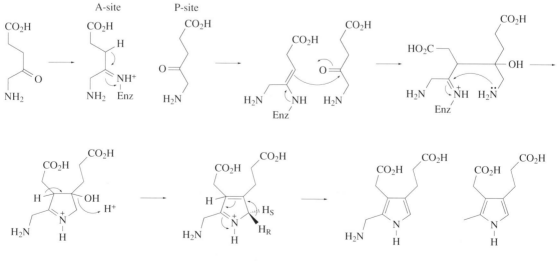

Scheme 13

Two broad mechanisms may be considered for the formation of porphobilinogen: mechanism (a) in which the formation of the C=N bond precedes the aldol condensation (Scheme 14a); and mechanism (b) in which the aldol condensation precedes the formation of the C=N bond (Scheme 14b). In the light of the determination of the X-ray structure of the yeast and *E. coli* enzymes, the mechanism can be reconsidered, taking advantage of the newly found information about possible catalytic groups.

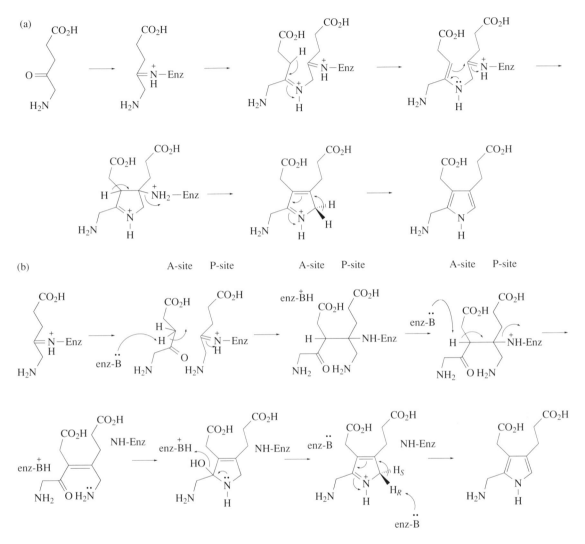

Scheme 14

The initial event in the mechanism of porphobilinogen formation is the binding of 5-amino-levulinic acid at the P-site to form a Schiff base with Lys247 (*E. coli* numbering). The nucleophilicity of this lysine would be enhanced by Lys195 through electrostatic effects. Once formed, the Schiff base is likely to be stabilized by the nearby Asp118. The binding of the second molecule of substrate at the A-site is dependent on the presence of bound substrate at the P-site. Equilibrium dialysis of enzyme, with 5-aminolevulinic acid reduced at the P-site, has shown that 5-aminolevulinic acid or levulinic acid can bind at the A-site. However, if levulinic acid, which lacks an amino group, is reduced at the P-site, then neither substrate nor levulinic acid can bind at the A-site. This suggests that the 5-amino group of the P-site substrate is important for binding of the A-site substrate. It has been suggested from elegant NMR studies that the environment in the vicinity of the amino group of the P-site substrate is hydrophobic and that the P-site substrate is bound with its amino group in an unprotonated state.[80] The close proximity of Lys195 is likely to play a key role in endowing the amino group of the P-substrate with nucleophilic character, again through electrostatic effects. The binding of the second molecule of 5-aminolevulinic acid to the A-site is also dependent on the metal center. The X-ray structure suggests that the triple cysteine metal site interacts with the A-site substrate (Figure 4).

The subsequent order of bond formation between the two bound substrate molecules is less clearly understood and whether C=N bond formation precedes C—C bond formation (Scheme 14a) or vice versa (Scheme 14b) has not been resolved. If C=N bond formation occurs first then the protonated Lys195 is likely to play a key role in promoting nucleophilic attack of the P-site substrate amino group on the carbonyl carbon of the A-site substrate. The resulting imine could then facilitate loss of a proton from the 3-position, followed by C—C bond formation. The positively charged Schiff base nitrogen of Lys247 linked to the P-site substrate may be stabilized by the close proximity of Asp118 to assist this reaction. Having formed the five-membered ring, two deprotonation reactions are necessary: one to remove stereospecifically the *proR* hydrogen from the 5-position of the P-site substrate[81,82] and the other to remove the remaining hydrogen atom from the 3-position of the A-site substrate. In the mechanism shown in Scheme 14a the final tautomerization is the latter, deprotonation being carried out by Lys247 as the catalytic base. The alternative mechanism (Scheme 14b) involves the formation of the C—C bond prior to the C=N bond and is equally plausible.

The precise nature of participation of the metal ion in the mechanism is of some debate since, in principle, zinc could act either as a Lewis acid to polarize the carbonyl group of the A-site substrate facilitating deprotonation at the 3-position or, alternatively, could function in the form of a zinc hydroxide as the base employed in deprotonation at the 3-position. The X-ray structure indicates the presence of a water molecule as the fourth ligand to the triple cysteine metal ion and that this is close to the A-site substrate. It would be unusual for a zinc metal ion bound to three cysteines to act as a Lewis acid and a metal hydroxide mechanism is thus favored. The fact that cadmium, when exchanged for zinc, also supports catalysis is also consistent with a metal hydroxide mechanism. Whether an analogous reaction occurs with the magnesium-dependent dehydratases needs further investigation.

4.04.3.3 Porphobilinogen Deaminase

Originally this enzyme was called uroporphyrinogen I synthase because, in purified form, it appeared to catalyze the formation of the nonphysiological uroporphyrinogen I isomer. The discovery that its true product is an unstable intermediate, called preuroporphyrinogen,[83] which acts as the substrate for the next enzyme in the heme pathway, uroporphyrinogen synthase[84] (previously called cosynthetase), solved many of the longstanding puzzles about the biosynthesis of uroporphyrinogen III and proved that the two enzymes function independently. These studies were confirmed by the chemical synthesis of preuroporphyrinogen, showing unambiguously that it is a 1-hydroxymethylbilane.[85] The reaction catalyzed by porphobilinogen deaminase (also called hydroxymethylbilane synthase) is shown in Scheme 15.

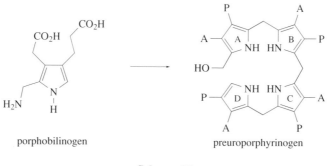

porphobilinogen   preuroporphyrinogen

Scheme 15

4.04.3.3.1 Occurrence and properties of porphobilinogen deaminases

Porphobilinogen deaminases have been isolated and characterized from a variety of sources and in all cases have M_r values ranging from 34 to 44 kDa, consistent with the enzyme being a monomer.

The first gene encoding a porphobilinogen deaminase was isolated from *E. coli*. Since then a number of genes/cDNAs specifying deaminases from plants, animals, and bacteria have been isolated and cloned and several of the enzymes have been overexpressed from recombinant bacterial strains (reviewed by Shoolingin-Jordan[86]). Comparison of the deaminase primary protein structures, derived from the nucleotide sequences, reveals that there has been a considerable degree of conservation of the enzyme during evolution. For instance, the human and *E. coli* protein sequences show 60% similarity, suggesting that the three-dimensional structure and mechanism of action are likely to be very similar throughout the biosphere.

4.04.3.3.2 The X-ray structure of porphobilinogen deaminase

Of all the enzymes of heme synthesis, porphobilinogen deaminase is one of the best understood since its X-ray structure was the first of all the enzymes of the heme synthesis pathway to be solved.[87,88] The structure of the deaminase from *E. coli* (Figure 5) reveals a protein with three domains, each of approximately 100 amino acids, that are linked to one another by flexible strands. The active site is highlighted between domains 1 and 2 by the presence of a dipyrromethane cofactor covalently linked to domain 3. There are relatively few contacts between the three domains, suggesting that substantial conformational changes may be possible during the polymerization reaction to allow the enzyme structure to adapt as the tetrapyrrole chain is assembled during the elongation process (see Figure 5).

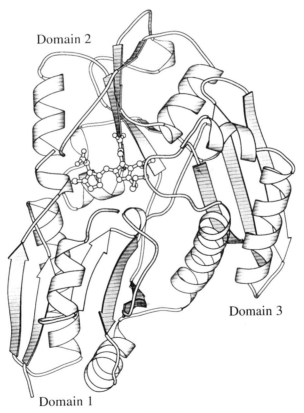

Figure 5 The X-ray structure of *E. coli* porphobilinogen deaminase.

4.04.3.3.3 The dipyrromethane cofactor and its interactions with the enzyme

The most striking aspect of the enzyme active site is the presence of a novel prosthetic group, named the dipyrromethane cofactor,[89] that is attached covalently to the enzyme by a thioether linkage to Cys242 in the *E. coli* enzyme (Scheme 16). The cofactor provides a covalent attachment point for the four substrate molecules that form the product[89,90] and acts as a primer that is elongated in a stepwise mechanism[91,92] to give a chain of six pyrrole residues via the enzyme intermediate complexes ES, ES$_2$, ES$_3$, and ES$_4$, as shown in Scheme 17. Only then is the tetrapyrrole product preuroporphyrinogen released from the "hexapyrrole" by hydrolysis, regenerating the dipyrromethane cofactor intact. Despite arising from two porphobilinogen units, the dipyrromethane cofactor, once formed, remains permanently and covalently bound to the enzyme.[89] In this context, porphobilinogen deaminase resembles other polymerases, such as DNA polymerase and glycogen synthase, both of which use primers in their reaction mechanisms. The dipyrromethane cofactor is a unique primer, however, in that it remains permanently and covalently bound to the enzyme during catalytic turnover and does not contribute to the product.[89] The most important amino acids for binding the acetic acid (A) and propionic acid (P) side chains of the cofactor, substrate, and intermediate complexes are the positively charged side chains from several conserved arginine residues. Site-directed mutagenesis of these arginines in the *E. coli* enzyme[93,94] lead to dramatic effects on cofactor assembly and on enzyme activity. The roles of individual residues were confirmed by the X-ray structure which revealed that the side chains of Arg131 and Arg132 interact with the acetic and propionic acid side chains of the two cofactor rings (Figure 6). Arg149, Arg155, and other active site residues, such as Asp106 and Lys83, play key roles in a complex hydrogen bonding network with the dipyrromethane cofactor. The cofactor–protein interactions have a substantial effect on the deaminase with respect to increasing stability to heat, thus the apoenzyme is readily denatured at 40 °C whereas the holoenzyme is stable at 70 °C.

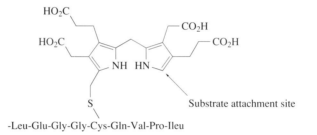

-Leu-Glu-Gly-Gly-Cys-Gln-Val-Pro-Ileu

Scheme 16

4.04.3.3.4   *The porphobilinogen deaminase substrate binding site*

The C-1 and C-2 rings of the dipyrromethane cofactor (Scheme 17) occupy two pyrrole binding sites in the catalytic cleft (Figure 6); however, if allowed to oxidize, the C-2 ring of the cofactor adopts an alternative position that is likely to resemble the substrate binding site.[95] In the reduced, active form of the enzyme this binding site is vacant, with Arg11 and Arg149 available to interact with the A and P side chains. Mutation of Arg11 results in an enzyme unable to bind substrate and which is consequently devoid of catalytic activity. Mutation of Arg155 has similar drastic effects.[93,94] The X-ray structure also highlights the importance of Phe62, that occupies a central position, with the potential to "stack" against the pyrrole ring of porphobilinogen at the substrate binding site, and to increase the pK_a of Asp84.

4.04.3.3.5   *Catalytic groups involved in substrate deamination and product release*

The α-position of the dipyrromethane cofactor is the site for initiating substrate polymerization and thus any potential catalytic group(s) are expected to be in this vicinity. The X-ray structure of *E. coli* porphobilinogen deaminase has highlighted Asp84 as a prime candidate for a catalytic group since the two oxygen atoms of its side chain interact with the hydrogen atoms attached to the nitrogens of the C-1 and C-2 cofactor rings. Most significantly, the oxygen atoms of Asp84 can also interact with the pyrrole NH and amino group of the substrate when bound at the putative active

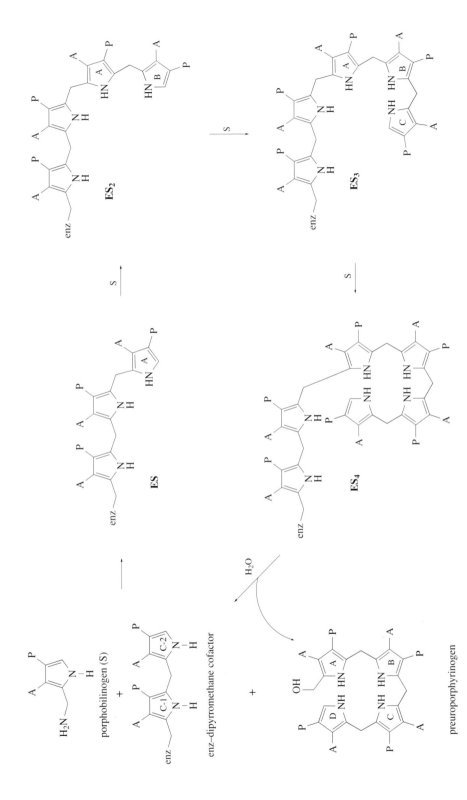

Scheme 17

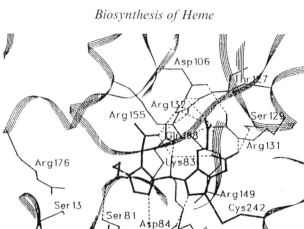

Figure 6 Structure of the catalytic cleft and the cofactor binding groups of *E. coli* porphobilinogen deaminase.

site. Asp84 is invariant in all porphobilinogen deaminases and site-directed mutagenesis to the closely related glutamic acid[96] results in an enzyme with a greatly reduced activity (0.5% of the wild type activity). The only major effect of the mutation is on k_{cat}, the mutant enzyme still forming the enzyme intermediate complexes but at a far slower rate. Not only do the enzyme intermediate complexes form slowly in the mutant but they are also slow to hydrolyze, ES, ES_2, and ES_3 being largely unaffected by incubation at 37 °C, whereas equivalent complexes of the wild type enzyme hydrolyze rapidly under such conditions.[89] The X-ray structure of Asp-84-Glu has been determined[95] and the properties of the mutant can readily be explained from the lack of any interaction between the pyrrole NH of the C-2 ring of the cofactor and the longer glutamate side chain.

The properties of two other site-directed mutants of Asp84, Asp-84-Ala, and Asp-84-Asn are of special interest since both mutant enzymes are totally devoid of enzyme activity. However, these two mutant enzymes not only contain the dipyrromethane cofactor but exist largely as ES_2 complexes. How a catalytically inactive mutant can exist predominantly as an ES_2 complex is explained when the mechanism of cofactor assembly is considered in the following section.

4.04.3.3.6 *The mechanism of dipyrromethane cofactor assembly*

The ability of the apoenzyme to form holoenzyme in the presence of porphobilinogen is well documented.[91,97,98] However, the rate of holoenzyme formation from porphobilinogen always appeared too slow and inefficient to account for the *in vivo* rate with an appreciable lag phase in *in vitro* experiments. A clue to the mechanism of cofactor assembly came from the observation that the regeneration of holoenzyme from apoenzyme and porphobilinogen was greatly accelerated if catalytic amounts of active holodeaminase were included, suggesting that preuroporphyrinogen may be involved. This was confirmed when preuroporphyrinogen was prepared independently and shown to transform apoenzyme into the holoenzyme rapidly and far more efficiently than porphobilinogen.[99] The initial product of the reaction of apoenzyme with preuroporphyrinogen was found to be "ES_2" in which the four pyrrole units of the preuroporphyrinogen had been incorporated intact and covalently linked to the protein. To convert the ES_2 intermediate into the dipyrromethane cofactor, the enzyme is required to complete the catalytic cycle by the addition of two molecules of substrate (Scheme 17), leaving two of the pyrrole rings of the original preuroporphyrinogen as the C-1 and C-2 units of the dipyrromethane cofactor.[99] The reactive nature of preuroporphyrinogen, a 1-hydroxymethylbilane, would readily allow the formation of an azafulvene intermediate that could alkylate Cys242 (Scheme 18). This mechanism for cofactor assembly explains how the catalytically inactive Asp-84-Ala and Asp-84-Asn mutants can exist as ES_2 complexes. Furthermore, such an assembly mechanism also overcomes the perplexing question of why the cofactor is permanently bound to the protein and the apoenzyme → holoenzyme reaction is irreversible under physiological conditions. Thus, the use of a preformed tetrapyrrole ensures that the catalytic machinery can never gain access to the cofactor in such a way that can lead to loss of the C-1 and C-2 rings.

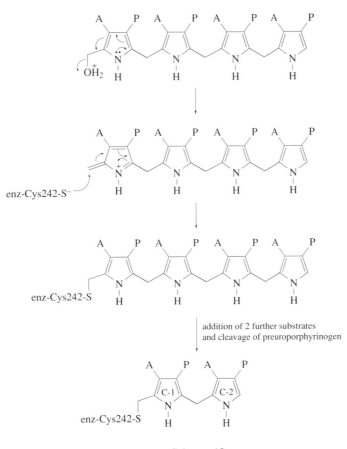

Scheme 18

Mutants of Cys242 have provided further evidence that an azafulvene intermediate is involved in cofactor attachment since a mutant with serine in place of cysteine is also able to form active holoenzyme.[100] Electrospray mass spectrometry of the Cys-242-Ser mutant confirms the existence of a novel holoenzyme with an M_r 16 less than the wild-type deaminase, consistent with the cofactor linkage being an oxygen ether. Although the substitution of O for S at position 242 reduces the efficiency of cofactor assembly, the mutant holoenzyme, once formed, has similar enzymic properties to those of the wild-type enzyme.

4.04.3.3.7 *Inhibitors of porphobilinogen deaminase*

Porphobilinogen deaminase accepts only porphobilinogen or the hydroxymethyl analogue as a substrate, although several porphobilinogen analogues bind to the enzyme and form inhibitory complexes.[86,101] The A and P side chains are particularly important for recognition by the enzyme and a range of potential compounds have been investigated. Opsopyrroledicarboxylic acid (porphobilinogen without the aminomethyl side chain) acts as a competitive inhibitor (K_i = 10 mM).[86] The porphobilinogen analogue, α-bromoporphobilinogen (Br), in which the reactive α-position has been substituted by a bromine atom, acts as a suicide substrate and adds to any of the catalytic intermediates forming chain termination complexes such as EBr, ESBr, and ES_2Br.[91]

4.04.3.3.8 *The mechanism of the porphobilinogen deaminase reaction*

The nonenzymic polymerization of porphobilinogen is extremely facile in dilute acid at 100 °C[102] and results in almost quantitative yields of uroporphyrinogens I, II, III, and IV. Acid conditions are required to protonate the product ammonia to prevent the reverse reaction. At the active site of porphobilinogen deaminase, Asp84 is located close to the predicted position of the substrate amino group and could act as a proton source, since in its hydrophobic environment its pK_a is likely to be around 8. After deamination, the resulting azafulvene would react with the free α-position of the cofactor C-2 ring to make the new C—C bond. Asp84, in its deprotonated form, is ideally placed to stabilize the positive charges that develop on the pyrrole nitrogens during condensation (Scheme 19). After loss of the α-proton, formation of the ES complex is complete. The complete inability of the Asp-84-Ala and Asp-84-Asn mutants to catalyze the reaction in either direction is consistent with this central role for Asp84. The remaining three reactions are likely to follow a similar pattern through ES_2 and ES_3, until ES_4 is formed. At this point the "hexapyrrole" chain is hydrolyzed to release preuroporphyrinogen leaving the cofactor attached to the enzyme. Any of the ES complexes can be cleaved by hydrolysis, although the enzyme carries this out preferentially at the ES_4 stage. The reason for this is unclear at present. Porphobilinogen deaminase is known to catalyze loss of a hydroxy group equally efficiently since the porphobilinogen analogue, hydroxyporphobilinogen, is a good substrate.[103]

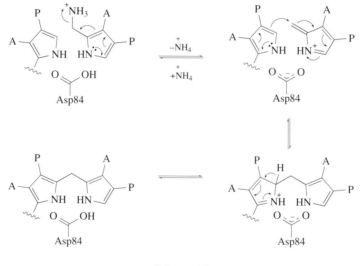

Scheme 19

At high concentrations of NH_4Cl, porphobilinogen deaminase can catalyze the reverse of the deamination reaction in which ES, ES_2, and ES_3 lose pyrrole units as porphobilinogen.[104] Related nitrogenous bases such as NH_2OH or NH_2OMe are also able to displace the terminal unit of the polypyrrole chain with the formation of a porphobilinogen analogues.[105] The nitrogenous bases also react with ES_4 to give the equivalent 1-aminomethylbilane. Thus, the enzyme is able to catalyze either the loss of NH_3 or H_2O from a substrate as well as the addition of NH_3 or H_2O to enzyme-bound ES complexes. These findings are all consistent with the central involvement of an azafulvene species (Scheme 19). An extensive analysis with inhibitors has resulted in a detailed discussion about the mechanism of deamination.[101,106]

4.04.3.3.9 *Conformational changes to the enzyme during polymerization*

The X-ray structure of *E. coli* porphobilinogen deaminase indicates a protein in which all three domains appear to be largely independent with few direct interdomain interactions.[87,88] The structure suggests that during chain elongation, in which the dipyrromethane cofactor is extended to give the "hexapyrrole" enzyme intermediate complex ES$_4$, substantial conformation changes are likely to occur in order to accommodate the four substrate pyrrole units into the catalytic site. Evidence for such conformational changes have been obtained from observations that during the catalytic cycle E → ES → ES$_2$ → ES$_3$ → ES$_4$, the enzyme becomes progressively more susceptible to alkylation by thiophilic reagents such as *N*-ethylmaleimide.[91] Thus, E is largely inert to inactivation by the reagent, whereas ES$_3$ is rapidly modified. The site of modification has been identified[100] as Cys134 which is located in a buried position between domains 2 and 3. It is thus highly probable that during the catalytic cycle the distance between these two domains increases significantly to expose Cys134 to the *N*-ethylmaleimide reagent. Interestingly, enzymes where serine rather than cysteine is present at position 134 are insensitive to *N*-ethylmaleimide.[107]

The precise mechanism by which porphobilinogen deaminase carries out the sequential manipulation of four substrates during the formation of the tetrapyrrole is not completely understood. The X-ray structure shows no evidence for multiple catalytic sites, indicating that the same catalytic machinery is used for all reactions. The question arises therefore as to whether the elongating polypyrrole chain is "pulled through" the catalytic site or, alternatively, whether the growing chain is accommodated in the large active site cleft that is present between domains 1 and 2.[87,88] The inactivation by *N*-ethylmaleimide, described above, supports the former possibility since this would result in the progressive movement of domain 3 away from domain 2. Furthermore, the fact that the substrate binding residue (Arg11) and catalytic group (Asp84) are located on domain 1, whereas the binding sites for the cofactor (and presumably the enzyme intermediate complexes) are largely on domain 2, could permit the catalytic machinery to "slide" along, inserting substrates one by one, until the chain is completed. Once ES$_4$ is formed, steric hindrance or an unacceptable build up of negatively charged groups may cause the polypyrrole to reposition allowing the regiospecific hydrolysis of ES$_4$ to occur in order to liberate the product.

4.04.3.4 Uroporphyrinogen III Synthase

The formation of preuroporphyrinogen by porphobilinogen deaminase marks the first of the two stages in uroporphyrinogen III synthesis from porphobilinogen. The macrocycle is finally completed by joining the A and D rings of preuroporphyrinogen with the elimination of water in a reaction catalyzed by uroporphyrinogen III synthase (Scheme 20). During the condensation, the D ring is turned around so that the linear repeating pattern of acetic acid and propionic acid side chains at the *β*-positions is interrupted and the macrocycle assumes the asymmetry that characterizes all tetrapyrroles and related compounds within the biosphere.

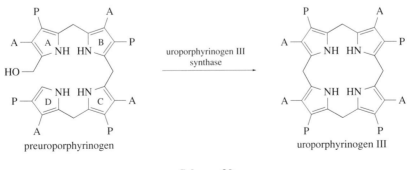

Scheme 20

4.04.3.4.1 *Properties of uroporphyrinogen III synthase*

Relatively little is known about the enzyme, since uroporphyrinogen III synthases are notoriously temperature sensitive and difficult to purify. In addition, the substrate preuroporphyrinogen is exceptionally unstable and rapidly cyclizes nonenzymically to give uroporphyrinogen I (Scheme 10) with a half-life of less than 5 min at pH 8.[83,84] The chemical synthesis of the substrate is complex[85] and, in practice, it is more convenient to generate preuroporphyrinogen enzymically from porphobilinogen using purified porphobilinogen deaminase and to couple the two enzymes together. This approach forms the basis of a rapid assay[108] since, although the rate of nonenzymic cyclization of preuroporphyrinogen to uroporphyrinogen I is fast, enzymic conversion to uroporphyrinogen III is much faster. Thus, subtraction of the uroporphyrinogen I formed in a nonenzymic control from the uroporphyrinogen III synthesized in the presence of the synthase gives a reasonable measure of uroporphyrinogen III synthase activity. The assay has assisted in the isolation of uroporphyrinogen synthases from several sources.[5,86]

The genes/cDNAs specifying uroporphyrinogen III synthases have been isolated from several animals and bacteria.[86] Whereas the other enzymes of the heme pathway show remarkable cross-species similarity, the derived protein structures of the synthases show little, although the hydrophobicity indices indicate some structural relationship. The absence of a crystal structure for uroporphyrinogen III synthase and the lack a conserved primary structure, except for an invariant tyrosine, serine, and threonine, gives little clue to the enzyme groups involved in substrate recognition and binding or to the nature of the catalytic process.

4.04.3.4.2 *The nature of the rearrangement process*

Previous investigators had established the labeling pattern of the heme macrocycle from early precursors such as glycine and 5-aminolevulinic acid and it was clear that during the formation of uroporphyrinogen III it was the D-ring which was turned around.[5] More recently, experiments using ^{13}C NMR confirmed this supposition and, using 1-aminomethylbilanes doubly labeled with ^{13}C (Scheme 21), proved that rearrangement of the D-ring occurs by an intramolecular process.[109] At the time these experiments were carried out, a 1-aminomethylbilane was thought to be the product of the porphobilinogen deaminase.[110] When preuroporphyrinogen was discovered,[83,84] it then became clear that porphobilinogen deaminase, present in the incubation, was able to convert the 1-aminomethylbilane into the equivalent 1-hydroxymethylbilane, preuroporphyrinogen.[84]

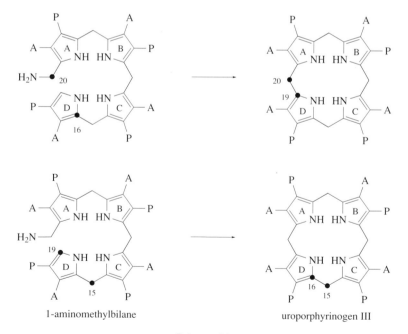

1-aminomethylbilane   uroporphyrinogen III

Scheme 21

Important information about the substrate specificity of the synthase has been obtained from a study with chemically synthesized hydroxymethylbilanes related to preuroporphyrinogen[111] (see also

Shoolingin-Jordan[86] for a review). The findings indicate that the A- and B-ring acetic acid and propionic acid side chains are essential for recognition by the synthase. The orientation of the D ring is less important. Thus, the hydroxymethylbilane with the D ring "rearranged" by chemical synthesis, surprisingly, is transformed into uroporphyrinogen I by the synthase, although the yield is only 45%. The remainder is cyclized directly to the unrearranged product uroporphyrinogen III. The lack of fidelity of this reaction when compared to the natural substrate is probably due to the fact that the enzyme cannot bind the "rearranged" D-ring of the analogue well enough to prevent C-19 from reacting with C-20.

4.04.3.4.3 The spiro-mechanism for uroporphyrinogen III synthase

One of the most enduring mechanisms to explain the formation of uroporphyrinogen III, suggested originally in the 1960s,[112] involves the formation of a *spiro*-pyrrolenine. This intermediate is then cleaved to generate an azafulvene that finally cyclizes to uroporphyrinogen III (Scheme 22). The key aspect of this mechanism is that the two bonds linking C-16 are approximately equivalent although, with the natural substrate, the C-15—C-16 bond is cleaved quantitatively. This is probably achieved because the asymmetric enzyme active site directs the process by binding preuroporphyrinogen in a "uroporphyrinogen III conformation" so that the D-ring of the natural substrate is placed to ensure the initial reaction of C-16 with C-20 to form the *spiro* intermediate. Strategically placed binding groups on the enzyme thus only allow a unidirectional process. Because of the high reactivity of preuroporphyrinogen, all that is necessary for the reaction to proceed is for the enzyme to bind the substrate in the correct geometry and to provide a proton source to facilitate loss of the hydroxyl group. The input of the enzyme to catalysis is thus likely to be minimal apart from providing a suitable binding site.

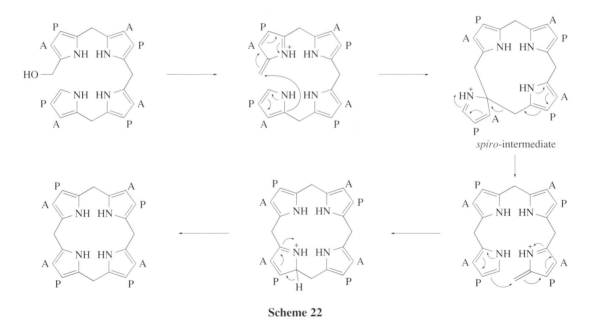

Scheme 22

A possible criticism of the *spiro* mechanism has been raised because the hydroxymethylbilane related to the penultimate intermediate azafulvene (Scheme 22) is not recognized as a substrate. This can be explained because the A and B rings of this hydroxymethylbilane are unlikely to be recognized by the enzyme so that the azafulvene is never generated at the active site. A computer modeling study of the enzyme mechanism is helpful in explaining this and other aspects.[113]

4.04.3.4.4 Experiments with inhibitors

Attempts to synthesize the *spiro* intermediate have proved unsuccessful, although two closely related *spiro*-lactams have been prepared.[114,115] Interestingly, out of the two isomers obtained, the one with *R*-stereochemistry at the *spiro* center is a far better competitive inhibitor of the synthase than the other (Scheme 23). The effect of other analogues and inhibitors on the synthase has been reviewed.[4,5,86,116]

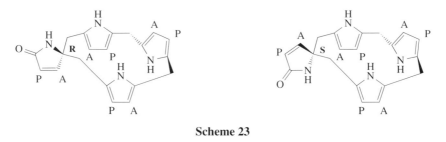

Scheme 23

The availability of recombinant *E. coli*[117] and *Bacillus subtilis*[118] genes specifying uroporphyrinogen III synthases has allowed isolation of the proteins in milligram amounts. The availability of the protein will assist crystallization of the enzyme in the presence of the *spiro*-lactam inhibitor and, it is hoped, will permit the enzymic groups responsible for substrate recognition and catalysis to be identified, as has been carried out for the two preceding enzymes of the pathway.

4.04.3.4.5   *Further conversions of uroporphyrinogen III to siroheme, factor F$_{430}$, and vitamin B$_{12}$*

Uroporphyrinogen III is a truly versatile intermediate since all other tetrapyrrole biosynthetic pathways originate from its template. If the acetic acid side chains are decarboxylated to copro-porphyrinogen III, uroporphyrinogen III follows the heme synthesis pathway. However, if methylation at C-2 and C-7 occurs to yield dihydrosirohydrochlorin,[119] this can then be transformed into siroheme[119] or factor F430.[120] Dihydrosirohydrochlorin, also named precorrin-2, is also an early precursor for the biosynthesis of vitamin B$_{12}$.[121-123] Details of these pathways are outside the scope of this chapter and the final sections cover only the transformation of uroporphyrinogen III into heme.

4.04.4   THE TRANSFORMATION OF UROPORPHYRINOGEN III INTO HEME

The final part of the heme biosynthetic pathway (Scheme 2) involves modification of the uro-porphyrinogen III macrocycle for its future role as a metalloporphyrin prosthetic group[1] (Scheme 1). This requires the regioselective decarboxylation of all but two of the eight β-substituent carboxylic acids to yield a macrocycle with the necessary amphipathic properties to allow the ring system to orientate within the heme binding pockets of hemoproteins. Two of the decarboxylations generate vinyl groups to allow covalent attachment of the heme macrocycle to enzyme thiols. Most important is oxidation of the porphyrinogen ring system to provide a planar coordination geometry for the central ferrous ion.

The initial reaction of this sequence involves the decarboxylation of all four acetic acid side chains of uroporphyrinogen III to form coproporphyrinogen III in a reaction catalyzed by uro-porphyrinogen decarboxylase. Regiospecific oxidative decarboxylation of the propionic acid side chains on rings A and B by coproporphyrinogen III oxidase follows, yielding protoporphyrinogen IX and only then does oxidation of the macrocycle occur in a reaction catalyzed by pro-toporphyrinogen IX oxidase. Finally, the insertion of ferrous iron is carried out by ferrochelatase. Several reviews are available covering this part of the pathway.[124-126]

4.04.4.1   Uroporphyrinogen Decarboxylase

4.04.4.1.1   *Substrate specificity*

The conversion of uroporphyrinogen III into coproporphyrinogen III by decarboxylation of all four acetic acid side chains is catalyzed by a single enzyme uroporphyrinogen decarboxylase (Scheme 2). The enzyme also catalyzes the decarboxylation of uroporphyrinogen I to coproporphyrinogen I and can also decarboxylate uroporphyrinogens II and IV.[127] Early investigators in the field had identified porphyrins with polarities between those of uroporphyrin III and coproporphyrin III and hypothesized that they may be intermediates in heme synthesis. The first attempts to demonstrate enzymic decarboxylation, however, were frustrated because at that time it was not known that the reduced, porphyrinogen oxidation state was required.

4.04.4.1.2 *Occurrence of the enzyme and its properties*

Uroporphyrinogen decarboxylases have been isolated from a variety of sources.[124–126,128] The enzymes all have an M_r of approximately 42 kDa and appear to exist as dimers. Characterization of the enzymes indicates that no coenzyme or metal is required. The enzyme is encoded by a single gene in mammals and other organisms and, unlike several of the other enzymes of the heme biosynthetic pathway, appears to exist in a single form with no tissue-specific isozymes. The genes/cDNAs have been cloned and sequenced from both prokaryotes and eukaryote sources.[128] Comparison of the amino acid primary structures derived from the nucleotide sequences indicate a similarity in excess of 60%.[128] Several highly conserved regions are present in the primary structures, suggesting that the enzymes all have similar three-dimensional structures.

4.04.4.1.3 *A preferential order of decarboxylation?*

There are theoretically 14 possible intermediates between uroporphyrinogen III and copro-prophyrinogen III: four heptacarboxylic acid porphyrinogens, six hexacarboxylic acid porphyrinogens, and four pentacarboxylic acid porphyrinogens. Whether the four decarboxylations involved in the transformation of uroporphyrinogen III into coproporphyrinogen III occur by a random or sequential route has been the subject of controversy for several years.[125,126] Previous analysis of the accumulated decarboxylation products from patients suffering from porphyrias cutanea tarda or from rats poisoned with hexachlorobenzene[129,130] suggested an ordered mechanism in which the acetic acid group on the D ring was initially decarboxylated followed by the A, B, and C rings (Scheme 24). This study was underpinned by the heroic chemical synthesis of all 14 possible intermediates, as their porphyrins.[131] Whether the porphyrins that accumulate *in vivo* reflect the formation of actual intermediates or merely result from the accumulation of the poorest substrates has been a source of continued debate.

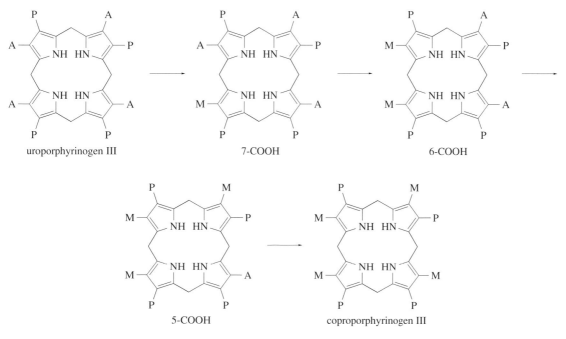

Scheme 24

In contrast to the above observations, studies with the decarboxylase isolated from human erythrocytes suggested that the decarboxylation followed a random route under some conditions.[132,133] Similar observations were made with the enzyme from *Rhodobacter spheroides*.[134] There are several possible explanations for these discrepancies: first, the decarboxylases from different sources may follow different reaction courses; second, the decarboxylases from patients with porphyria cutanea tarda or from rats treated with hexachlorobenzene may possess modified decarboxylases with altered substrate specificities;[135] or third, differences may result from variations in the enzyme:substrate ratio during the assays. There is evidence that the enzyme:substrate ratio can influence the decarboxylation route[133,135–137] with high substrate levels reducing the selectivity of the enzyme for the individual pyrrole rings of uroporphyrinogen III and resulting in a random pathway.[138] *In vivo* the lower levels of the substrate may favor a greater ring selectivity, leading to an ordered decarboxylation. It has not been excluded that under such conditions, intermediates may remain in the vicinity of the catalytic site and may be channeled in the dimeric enzyme from one subunit to the other, possibly accounting for an ordered decarboxylation route. Whether or not there is a preferential sequence of decarboxylation is of no physiological importance since all 14 possible carboxylic acid porphyrinogens between uroporphyrinogen III and coproporphyrinogen III are substrates for the decarboxylase and are all converted to coproporphyrinogen III anyhow.

4.04.4.1.4   Aspects of recognition of the acetic acid side chains

Uroporphyrinogen decarboxylase is highly specific for the acetic acid side chain and requires that a propionic acid side chain is also present on the adjacent β-position. The enzyme clearly has less specificity towards the nature of the substituents of the adjacent rings, since close scrutiny of the uroporphyrinogen III structure (Scheme 24) reveals that the substituents on the neighboring rings are all different (A and A bordering ring A; P and A bordering ring B; P and P bordering ring C; A and P bordering ring D). These differences are accentuated by the appearance of methyl groups when decarboxylation begins. Since enzymes normally exhibit supreme specificity in their recognition of substrates, it is therefore most likely that there are few significant interactions at the active site between the side chains of the pyrrole rings either side of the ring interacting with the catalytic groups. Given the conformation of porphyrinogens, it is more likely that the active site recognizes the more conserved features of the substrate in the vicinity of the pyrrole nitrogen and *meso*-carbon atoms, leaving the substituents of the adjoining rings oriented towards the solvent.

If the stepwise sequence of decarboxylation proposed[131] is operative, then after decarboxylation of the D ring, the heptacarboxylic acid porphyrinogen would have to flip by 180° and turn 90° to allow the A ring to enter the active site.[126] Further decarboxylations of the B and C rings would require only rotation of the substrate by 90° although, because of the puckered nature of porphyrinogens, there would also have to be adjustment to the substrate conformation. Whether the dimeric nature of the enzyme plays a role in substrate channeling is an interesting question that may be answered by the X-ray structure of the enzyme (Figure 7) which has now been determined.[139]

4.04.4.1.5   Stereochemical analysis of the uroporphyrinogen decarboxylase reaction

The steric course of the reaction catalyzed by uroporphyrinogen decarboxylase has been investigated in an elegant study using succinic acid stereospecifically labeled with deuterium and tritium in the C-2 position.[140,141] The basis of the study centered on the generation of acetic acid side chains in uroporphyrinogen III that, on decarboxylation, would yield chiral methyl groups which could be isolated by chemical degradation for analysis as chiral acetate (Scheme 25). Thus, doubly labeled $2R[2-{}^2H,{}^3H]$succinate was incorporated into uroporphyrinogen III and thence into coproporphyrinogen III by conversion via succinyl-CoA though the biosynthesis pathway to heme using chicken erythrocytes. Degradation of the heme to acetic acid and chiral analysis established that the acetic acid has an *S*-configuration,[141] indicating that decarboxylation must have occurred by retention of stereochemistry.

4.04.4.1.6   A mechanism for the reaction catalyzed by uroporphyrinogen decarboxylase

In the light of the above stereochemical findings, a mechanism has been proposed for the decarboxylation.[140,141] This takes into account the fact that the two labeled hydrogen atoms, orig-

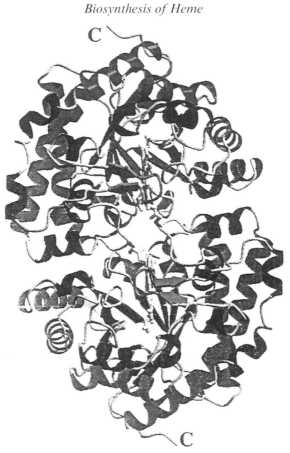

Figure 7 Structure of human uroporphyrinogen decarboxylase.[139]

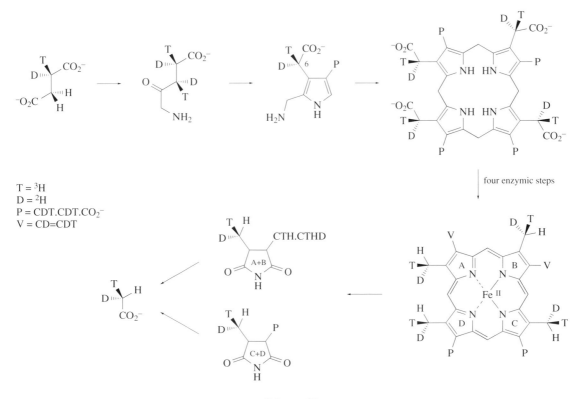

T = ^{3}H
D = ^{2}H
P = CDT.CDT.CO$_2^-$
V = CD=CDT

Scheme 25

inally present in succinate, remain undisturbed throughout the transformation and that the reaction occurs with retention of stereochemistry. A plausible mechanism consistent with these findings is shown in Scheme 26. In this, the pyrrole ring acts as an electron sink in a manner similar to the protonated pyridine ring in pyridoxal 5'-phosphate-dependent decarboxylases. Such reactions also follow a retention mechanism.[17] The initial stage in this mechanism involves a tautomeric shift to form a positively charged ring nitrogen. This promotes the decarboxylation, assisted presumably by a suitably placed basic group on the enzyme. After decarboxylation, reprotonation of the double bond must occur from the same side as the departed CO_2, followed by the tautomeric shift in reverse. It is significant that uroporphyrin III cannot function as a substrate, highlighting the requirement for porphyrinogen chemistry in the decarboxylation.

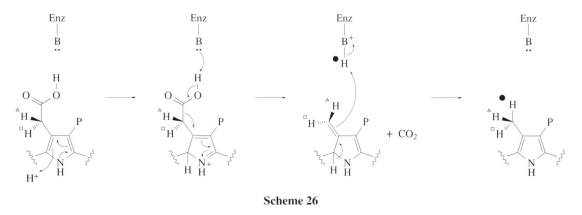

Scheme 26

4.04.4.1.7   *The active site of uroporphyrinogen decarboxylase*

Examination of the amino acid sequences derived from a number of cloned genes/cDNAs[128] reveals several invariant amino acids including arginine, aspartic acid, lysine, histidine, and tyrosine residues, all of which are potential candidates for either binding or catalytic groups. Of particular note are two invariant arginine residues, near the N-terminus, in a highly conserved region, PVWCMRQAGR, that could be involved with substrate carboxyl group binding. The enzyme is inactivated by the arginine modifying reagent phenylglyoxal.[134,135] Arginine residues have been shown to be central to binding the acetic acid and propionic acid side chains of porphobilinogen at the active site of porphobilinogen deaminase[93,94] (see also in Section 4.04.3.3).

By analogy with pyridoxal 5'-phosphate-dependent enzymes, it is possible that an aspartic acid may be necessary to stabilize the positive pyrrole ring nitrogen during the reaction (assuming the mechanism in Scheme 26). There are three invariant aspartates as possible candidates. The mammalian enzymes are particularly sensitive to reaction with thiophilic reagents[136,142] and heavy metals;[143] however, there are no invariant cysteines and it is unlikely that the cysteines in the mammalian enzymes play a specific role. Mammalian uroporphyrinogen decarboxylase is also sensitive to diethylpyrocarbonate, implicating histidine as an important residue.[137] The nature of the catalytic group postulated to promote the decarboxylation and reprotonation steps is more difficult to predict and its identity will have to await three-dimensional structural determination of the enzyme.

One intriguing aspect of the enzyme is its sensitivity in mammalian systems to inactivation by certain chemicals such as chlorinated phenyls. There is evidence that the process of inactivation requires iron and that an inhibitory analogue of uroporphyrinogen may be involved. This aspect is reviewed elsewhere.[128]

4.04.4.2   Coproporphyrinogen III Oxidase

Coproporphyrinogen III oxidase catalyzes oxidative decarboxylation of the propionic acid side chains of rings A and B to the vinyl groups of protoporphyrinogen IX[144] (Scheme 2). Whereas uroporphyrinogen decarboxylase accepts uroporphyrinogen I as a good substrate, copro-porphyrinogen III oxidase is specific only for coproporphyrinogen III. Building on pioneering studies carried out in the 1960s,[145] it is now generally accepted that two classes of enzyme exist, one class found in aerobic organisms and another in anaerobic cells. Many organisms have both aerobic and anaerobic coproporphyrinogen III oxidases.

4.04.4.2.1 Aerobic coproporphyrinogen III oxidase

Aerobic coproporphyrinogen III oxidase has been extensively studied in animals and yeast, where it is present as a soluble mitochondrial enzyme. The rat[146] and bovine[147,148] liver enzymes have been purified, the latter to homogeneity, and the protein characterized. The aerobic enzyme uses oxygen although there is no evidence for the requirement of a reducing agent such as NADPH. Furthermore, there is no evidence for the requirement for, or presence of, any chromophore, coenzyme or metal ion. The bovine enzyme has an M_r of 72×10^3 and appears to be a dimer. It contains two tyrosine residues that are essential for enzyme activity and can be modified by tetranitromethane.[148] The yeast enzyme has also been isolated and is also a dimer, comprising two identical subunits each with 328 amino acids and an M_r of 37 673.[149] It harbors an iron atom in each subunit, although there is no direct evidence to suggest that iron plays a role in catalysis.

Genes/cDNAs for coproporphyrinogen III oxidase from human,[150] soybean,[151] yeast[152] (*HEM 13*), and *Salmonella typhimurium*[153] (*hem F*) encode proteins with astonishing similarities in their amino acid sequences. Apart from the N-terminal regions, where there are differences reflecting signal peptides for import into the mitochondrial intermembrane space, the conservation is over 30% of invariant residues. Some of the genes/cDNAs have been expressed to generate recombinant enzymes.[152,154]

4.04.4.2.2 Anaerobic coproporphyrinogen III oxidase

Early reports based on experiments with *Rhodobacter spheroides*, a bacterium able to grow aerobically in the dark and anaerobically in the light, provided the first evidence for the presence of an anaerobic coproporphyrinogen III oxidase, which did not require oxygen but instead could transform coproporphyrinogen III into protoporphyrinogen IX in the presence of NADP, ATP, and L-methionine.[145] Attempts to extend this work have proved difficult although a gene encoding a putative anaerobic coproporphyrinogen III oxidase has been identified which complements an *R. spheroides* mutant that accumulates large amounts of coproporphyrinogen III. The gene has been mapped and sequenced[155] and encodes a protein completely different in sequence from the aerobic enzyme. Other members of this new class, termed *hemN* rather than *hemF*, have been identified in *Salmonella typhimurium*[156] as well as in other organisms, although attempts to demonstrate its activity *in vitro* have proved problematical, possibly because the enzyme requires an as yet unknown electron acceptor.

4.04.4.2.3 The order of side chain oxidation

The isolation of harderoporphyrinogen, an intermediate in which position 3 of the macrocycle bears a vinyl group, provided the first evidence that the propionic acid side chain of ring A is oxidatively decarboxylated prior to the propionic acid side chain on ring B[157,158] (Scheme 27). Although isoharderoporphyrinogen, which bears the vinyl group on ring B at the 8 position, is also a substrate for the enzyme, the rate of protoporphyrinogen formation is much slower than with harderoporphyrinogen, indicating a preferred pathway. Further experiments indicated that harderoporphyrinogen was not released as a free intermediate, since exogenous harderoporphyrinogen could not be equilibrated with labeled enzyme-bound intermediate,[159,160] leading to the proposal that substrate channeling occurs with the "rotation" occurring at the active site, after the first decarboxylation.

Other porphyrinogens are accepted as poor substrates, notably the pentacarboxylic acid porphyrinogen (Scheme 24), which results in dehydroisocoproporphyrinogen.[161,162] This intermediate is then converted into harderoporphyrinogen by uroporphyrinogen decarboxylase with decarboxylation of the ring C active acid side chain. The unnatural coproporphyrinogen IV isomer is also a poor substrate, being transformed into protoporphyrinogen XIII.[163]

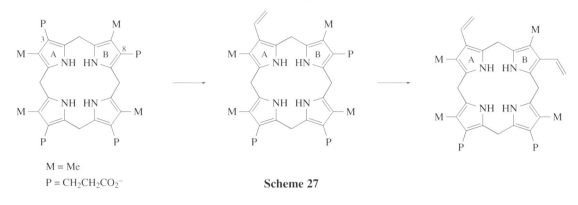

M = Me
P = CH₂CH₂CO₂⁻

Scheme 27

4.04.4.2.4 Incorporation of chiral succinate to investigate the steric course of coproporphyrinogen III oxidase

Succinates labeled stereospecifically with tritium have been employed to investigate the status of the two pairs of methylene hydrogens of the propionic acid side chains of coproporphyrinogen III during formation of the vinyl groups of protoporphyrinogen IX[164–166] using a similar approach to that described for the investigation of uroporphyrinogen decarboxylase. Thus, $2S[^3H]$succinate was first converted into succinyl-CoA and incorporated by enzymic methods via $6S,8S,9S\text{-}[^3H_3]$por-phobilinogen into heme using avian erythrocytes. The biosynthetic heme was degraded to the maleimides (as shown in Scheme 25) and the tritium label in ethyl methyl maleimide (from rings A and B) was compared with hematinic acid (from rings C and D). Scheme 28 summarizes the findings which conclusively demonstrate that only the β hydrogen atom occupying the H_{si} configuration is removed during the transformation of the propionic acid into the vinyl group.[167] These results were confirmed by introducing the deuterium label into the H_{si} orientation at the β-position by another approach.[168]

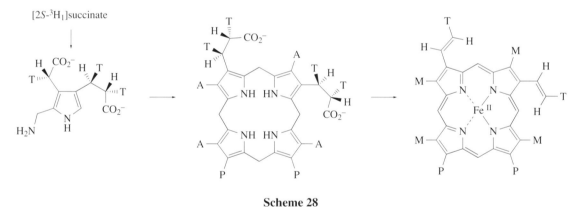

Scheme 28

Further information about the steric course of the reaction was obtained from experiments in which deuterium labels were introduced into the propionic acid side chain of porphobilinogen by chemical synthesis.[169] Incorporation of the diastereomeric mixture of 8,9-bisdeuterated por-phobilinogen (Scheme 29) into heme, followed by product analysis, established that conversion of the propionic acid side chains into vinyl groups occurs by an antiperiplanar elimination[170] (the bonds cleaved by the enzyme are arrowed in Scheme 29).

4.04.4.2.5 The mechanism of coproporphyrinogen III oxidase

The loss of only a single hydrogen atom eliminates several possible mechanisms, including the involvement of a β-keto acid intermediate. This would have required the loss of an α-hydrogen and both β-hydrogen atoms. The participation of an acrylic acid intermediate is also prohibited since such a course would have required the loss of one of the α-hydrogen atoms in addition to the β-hydrogen.

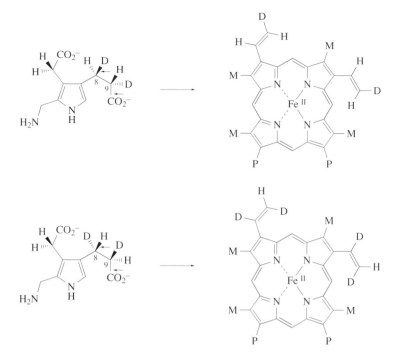

Scheme 29

In the light of these stereochemical studies the mechanism for both aerobic and anaerobic coproporphyrinogen III oxidase may be considered, taking due notice of the requirement for oxygen in the former case. Mechanism 1 proposes that the H_{si} β-hydrogen atom is replaced by a hydroxyl group arising from molecular oxygen to give a β-hydroxypropionate intermediate (Scheme 30). This, in itself, is uncontroversial since a synthetic 3-(β-hydroxypropionate), 8-propionate porphyrinogen, is efficiently transformed into protoporphyrinogen IX in the presence of oxygen, both with avian erythrocyte preparations and purified bovine liver coproporphyrinogen III oxidase.[147,148] Furthermore, incubation of this substrate under anaerobic conditions led to harderoporphyrinogen containing a vinyl group at C-3 but an unchanged propionate side chain at C-8. These observations were supported by the detection of the same 3-(β hydroxypropionate), 8-propionate porphyrinogen, during aerobic incubations of coproporphyrinogen III with a cell-free preparation from avian erythrocytes.[171] Thus, mechanism 1 (Scheme 30) is consistent with the experimental results but represents a novel type of reaction in that it cannot involve hydroxylation, in the accepted sense, because of the absence of a heme P450 chromophore and a reducing agent. In this context, the presence of iron in the aerobic yeast enzyme may point to a novel oxygen insertion mechanism.

mechanism 1

mechanism 2

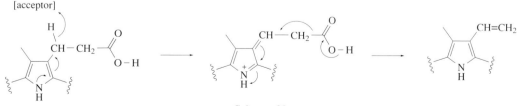

Scheme 30

In the case of the anaerobic mechanism, an electron acceptor other than molecular oxygen is required, although all attempts have failed to identify such a species. In mechanism 2 the advantage of using the porphyrinogen pyrrole ring to facilitate electron transfer is indicated only for the anaerobic process, although there is no reason why the oxygen-dependent mechanism cannot also use the vinylogous pyrrole nitrogen in an analogous way. The oxidized intermediate in mechanism 2 can readily decarboxylate to generate the vinyl group.

Comparative studies with the aerobic and anaerobic enzyme systems from *R. spheroides* have established that the H_{si} from the β-position of coproporphyrinogen III is lost in both cases.[172] Thus, even if the *hemF* and *hemN* genes encode proteins of very different sequence[173] (and presumably structure), the basic chemistry has been largely conserved. If the aerobic mechanism is unclear, the anaerobic process is even more unclear, although the now classical studies with *R. spheroides*, which implicated *S*-adenosylmethionine, remain the only clue to understanding the mechanism. In the light of better understanding about the role of *S*-adenosylmethionine and iron in enzyme mechanisms,[174] it is tempting to speculate that both of these cofactors could be playing a combined role in the decarboxylation. It is also necessary to implicate an alternative to oxygen and there is every likelihood that the reaction may be coupled to an electron transfer chain. Few ideas are given by the emerging sequences of *hemN* genes.[175]

4.04.4.3   Protoporphyrinogen IX Oxidase

Only now that the side chains of uroporphyrinogen and coproporphyrinogen have been appropriately modified, by employing the chemistry of the unconjugated pyrrole rings, can oxidation of the macrocycle take place so that a planar porphyrin ring is established in readiness for iron chelation. Oxidation is accompanied by dramatic changes in physical and chemical properties, with the colorless porphyrinogen becoming an intense purple/red chromophore with a large Soret peak at 400 ($E = 500\,000$ $L^{-1}M^{-1}$) and red fluorescence. Most importantly, oxidation to the porphyrin places the four pyrrole nitrogen atoms of the conjugated macrocycle in perfect position to chelate ferrous iron and to impart to the metal the unique properties associated with metalloporphyrins.[176]

4.04.4.3.1   *Properties of protoporphyrinogen IX oxidase*

Although the chemical oxidation of protoporphyrinogen is known to occur, the existence of an enzyme was proposed to account for the accumulation of protoporphyrinogen under some conditions.[177] Observations that the oxidation of *meso*-tritiated protoporphyrinogen IX in cell-free extracts led to the retention of exactly 50% of ^3H label, as opposed to 95% found in the nonenzymic reaction,[178] further supported the involvement of an enzyme. The existence of protoporphyrinogen IX oxidase was confirmed finally when it was isolated from yeast.[179] Further studies established the substrate specificity, protoporphyrinogen IX being the only major substrate, with protoporphyrinogen XIII, *meso*-porphyrinogen IX, harderoporphyrinogen, and isoharderoporphyrinogen, all active as poor substrates.[160] Eukaryotic enzymes are located in mitochondria and detergents are necessary for their solubilization.[179–183] The k_m for protoporphyrinogen IX is in the 5–10 µM range and oxygen is the electron acceptor with a k_m of 125 µM. Studies on the purified enzyme from bovine liver have provided evidence that a tightly bound flavin is present, suggesting a plausible oxidative mechanism.[184] The involvement of FMN has been supported by studies with the murine enzyme.[185] The bovine and murine enzymes are monomers with M_r values around 65 kDa, somewhat larger than the yeast enzyme. The M_r of the enzyme from *B subtilis*[186] as judged from the gene sequence is somewhat smaller. Further nucleotide sequences for protoporphyrinogen oxidases are available.[187–189] A comprehensive review describes the properties of several enzymes from other sources.[124] One particularly interesting facet of the enzyme is its sensitivity to diphenyl ether herbicides, not only in plants[190] but in animals and yeast.[191] The inhibition is competitive with respect to protoporphyrinogen IX.

4.04.4.3.2   *The mechanism and steric course of the protoporphyrinogen IX oxidase reaction*

The observation that [5,10,15,20-$^3H_8$]coproporphyrinogen III is converted into protoporphyrin IX with retention of 50% of the ^3H label[178,192] prompted a more detailed study in which the bridging methylenes were labeled stereospecifically (Scheme 31). This was accomplished by the incubation of 11S[^{3}H]porphobilinogen (prepared from 2RS[^{3}H$_2$]glycine using 5-aminolevulinic acid synthase coupled to 5-aminolevulinic acid dehydratase) with an avian erythrocyte preparation containing all the enzymes of heme synthesis.[193,194]

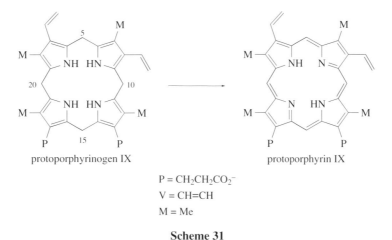

protoporphyrinogen IX   protoporphyrin IX

$P = CH_2CH_2CO_2^-$
$V = CH=CH$
$M = Me$

Scheme 31

Following the status of the labeled porphobilinogen would provide insight into the subsequent stages of heme biosynthesis. The incorporation of ^{3}H label was followed by isolating heme and comparing the ^{3}H:^{14}C label with the original isotopic finding that 75% of the ^{3}H label was lost with 25% remaining in the heme macrocycle suggested that one of the four *meso*-positions had retained the ^{3}H label. The location of this remaining ^{3}H was established by degradation of the macrocycle to the α-, β-, γ-, and δ-biliverdins by ascorbic acid oxidation. The lone label was found to be at C-10 (Scheme 32). Control experiments with 11RS[^{3}H$_2$]porphobilinogen, as expected, led to retention of 50% of the label, one at each of the *meso*-positions. Taken cumulatively, these results suggest a mechanistic scheme for protoporphyrinogen oxidase in which three hydrogen atoms are lost as H$^-$ (or an equivalent) during the oxidation and that these are specifically removed by the enzyme from positions 5, 15, and 20. With hindsight, this is not an unexpected result since the oxidation of protoporphyrinogen involves the loss of 3H$_2$ (6H). The hydrogen atom removed from C-10 would originate from the other side of the macrocycle and be removed as a proton.

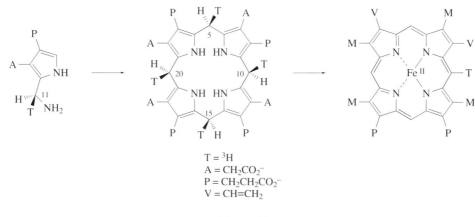

$T = {}^3H$
$A = CH_2CO_2^-$
$P = CH_2CH_2CO_2^-$
$V = CH=CH_2$

Scheme 32

The mechanism shown (Scheme 33) makes a reasonable assumption that the elaboration of all four *meso*-positions occurred by identical processes and that all four ^{3}H labels are located on the same "side" of the porphyrinogen macrocycle. The absolute stereochemistry at the *meso*-positions of protoporphyrinogen IX biosynthesized from 11S[^{3}H]porphobilinogen is not known although, for the sake of argument, the transformation has been shown to proceed by retention mechanisms. What is quite clear from the results is that the reactions which establish the *meso*-bridges from porphobilinogen, catalyzed by porphobilinogen deaminase and uroporphyrinogen III synthase, both proceed by tightly controlled stereochemical processes. The mechanism shown in Scheme 33 makes use of flavin (F) detected in the aerobic enzymes[184,185] for the removal of hydrogen atoms (H_a) from the 5, 15, and 20 positions as "hydride" to give FH_2. This is then oxidized by molecular oxygen to regenerate F. The loss of H_b from the 10 position as H^+ is shown as a tautomeric shift promoted by an enzymic base.

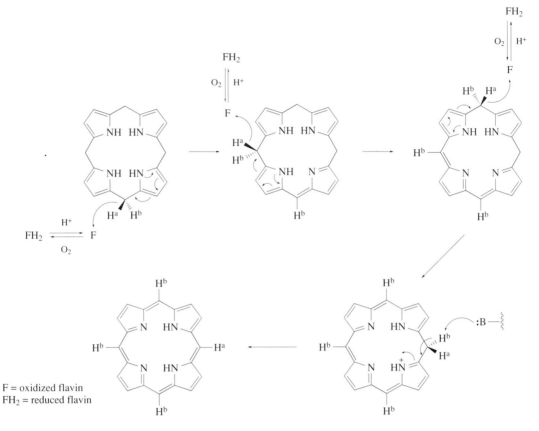

F = oxidized flavin
FH_2 = reduced flavin

Scheme 33

In microorganisms, protoporphyrinogen IX oxidase is thought to be linked to the respiratory chain components[195] and attempts to separate the enzyme from membrane fractions has proved difficult in many cases, although isolation of the oxidase from *Desulfovibrio gigas* has been reported.[196]

4.04.4.4   Ferrochelatase

Ferrochelatase catalyzes the final stage of heme biosynthesis involving the insertion of ferrous iron into the planar protoporphyrinogen IX macrocycle (Scheme 34). This represents an important branch in plants and photosynthetic bacteria since it is at this point that the magnesium pathways to chlorophylls and bacteriochlorophylls diverge.[28]

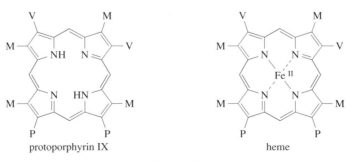

protoporphyrin IX   heme

Scheme 34

4.04.4.4.1 The occurrence and properties of ferrochelatase

Ferrochelatase has been detected in all organisms that biosynthesize heme and purification has been achieved from mammalian sources,[197] yeast,[198] and bacteria.[199,200] In the large majority of organisms, ferrochelatase is membrane-associated, although the enzyme from *Bacillus subtilis*[200] is soluble. The M_r values for bacterial ferrochelatases are lower than those from eukaryotes, being of the order of 35×10^3. The eukaryote M_r values are variable, reflecting a larger protein which has the tendency to aggregate.

Several studies have established the metal specificity for ferrochelatase with Fe^{2+}, Co^{2+}, and Zn^{2+} all being accepted as substrates.[201,202] The most important requirement is for paired propionic acid side chains at positions 13 and 17 of the macrocycle ring, although uncharged groups on the A and B rings are also important.

Cysteine residues were originally implicated in the mechanism for metal binding,[203] but since there are no invariant cysteines in the ferrochelatases, this is unlikely. Arginine residues are suggested as groups with potential for binding and orienting the carboxylic acid side chains.[204]

4.04.4.4.2 Molecular biology of ferrochelatase and information from protein sequence comparisons

The gene/cDNA sequences for several ferrochelatases have been determined and, like most of the heme synthesis enzymes, show substantial conservation.[205] Invariant histidine, serine phenylalanine, and glutamate residues suggest that they play a key role in the enzyme function. This has been borne out by the structure of the enzyme (see below).

The gene/cDNA sequences show that the eukaryote enzymes both have N- and C-terminal extensions and are expressed as larger precursors that are difficult to purify without detergents. The N-termini involved with mitochondria import,[198,206] whereas the C-terminal region of the enzyme is thought to specify the final location of the enzyme to the inner surface of the mitochondrial membrane.[201] A novel feature of the mammalian ferrochelatases is the presence of a cysteine-rich sequence that contributes to an iron–sulfur cluster. EPR and Mössbauer spectroscopy are consistent with a high-spin iron(III) structure of a tetrahedrally coordinated iron linked to four sulfurs as -2S_{cys}-Fe-2S_{inorg}-Fe-2S_{cys}-. The cysteine-containing sequence (-C-X_7-C-X_2-C-X_4-C-) provides the ligands for the [2Fe2S] cluster.[207] It is unlikely that the [2Fe2S] cluster has any mechanistic significance, although its destruction leads to a loss of enzyme activity.

Removal of the N- and C-termini regions to mirror the *B. subtilis* enzyme has permitted expression of the *E. coli*[208] and human enzymes[209] in soluble form.

4.04.4.4.3 The structure of ferrochelatase from B. subtilis

The three-dimensional structural determination of the ferrochelatase from *B. subtilis*[210] has been a major breakthrough in our understanding of the way the enzyme may function (Figure 8). The protein is made up from two topologically related domains held together by two protein strands and with a deep cleft that is considered to represent the catalytic site. This cleft is lined with invariant residues, some of which are structural but others have the potential to act in the enzyme mechanism, especially Ser118, Ser181, His183, Asp194, Tyr196, Gln221, Ser222, Trp230, Phe258, and Glu264 (Figure 9). A model with protoporphyrin IX bound has also been constructed that may be useful in predicting the way that the substrate interacts with the protein during catalysis.

4.04.4.4.4 Site-directed mutagenesis of ferrochelatases and structural implications

The results from site-directed mutagenesis experiments, together with the three-dimensional structure of the ferrochelatase from *B. subtilis*, with the substrate modeled at the active site,[209] have suggested a role for most of the invariant residues in the enzyme in the mechanism.

Site-directed mutagenesis of the invariant His263 in human ferrochelatase[211] (His183 in *B. subtilis*) affected V_{max} and k_m for the divalent metals Fe^{2+} and Zn^{2+}, although the binding of protoporphyrin IX was unaffected, implicating this histidine in metal binding. Histidine-183 is located close to the pyrrole nitrogens of the substrate modeled into the catalytic cleft. Site-directed mutagenesis of the invariant Phe417 in human ferrochelatase[212] (Phe248 in *B. subtilis*) resulted in a dramatic loss of

Figure 8   The structure of ferrochelatase from *Bacillus subtilis.*[210]

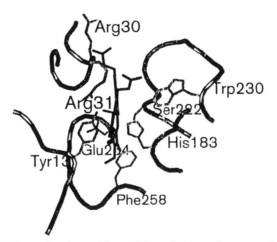

Figure 9   The active site residues of ferrochelatase from *Bacillus subtilis.*

activity, suggesting a key role in catalysis. Phe258 is close to an invariant glutamate (Glu264), located near to the pyrrole nitrogen atoms on the other side of the substrate.

Consideration should also be given to the possible role of structural changes in the enzyme during the reaction. Ferrochelatase belongs to the well-characterized family of two-domain binding proteins[213] which function by oscillating from open to closed conformations on ligand binding. This is facilitated by a hinge region linking two topologically related domains. The two domains of ferrochelatase each present a surface to the active site cleft containing the sequence Gln-Ser-Glu-Gly-Asn-Thr-Pro-Asp-Pro-Trp-Leu extending from 221–231 in the *B. subtilis* enzyme and containing the invariant Gln-Ser and Trp-Leu residues. This region forms key contacts with the pyrrole ring of the modeled macrocycle.[210] The C and D propionic acid side chains of the substrate are likely to interact with the paired Arg30 and Arg31 residues. Although not invariant, their counterparts appear in other ferrochelatases at topologically significant locations.

4.04.4.4.5   *The mechanism of action of ferrochelatase*

Kinetic studies have suggested that ferrochelatase follows an ordered bi–bi mechanism with ferrous iron binding before protoporphyrin IX[214] and the release of the product occurring prior to the two protons. Other studies in contrast suggest a random mechanism.[215]

The ferrochelatase mechanism involves the exchange of Fe^{2+} for two of the hydrogen atoms attached to the pyrrole nitrogens of protoporphyrin IX. Two key elements are likely to be required: first, a basic group to accept the protons from the pyrrole nitrogens after reaction with the ferrous ion; and second, a means to deliver the metal ion simultaneously. The most attractive mechanism would be the addition of Fe^{2+} to one side of the macrocycle as the protons are extruded from the other side. In view of the structure of ferrochelatase, it is envisaged that on protoporphyrin IX binding, a conformational change occurs which leads to the active site becoming closed to the solvent. Distortion of the tetrapyrrole ring system in the transition state could be accomplished by conformational changes in the two protein domains. It has been suggested that the dramatic inhibition of ferrochelatases by *N*-methyl porphyrin, in which one of the four rings is twisted because of the *N*-methylation, represents the conformation of the transition state.[216] Account also needs to be taken of additional conserved acid side chains which may play a role in removing protons from the active site or in delivering the ferrous ion to the hydrophobic pocket prior to chelation.

4.04.5   EPILOGUE

The biosynthesis of heme represents only a small part of the field of tetrapyrrole biosynthesis and other pathways fall outside the scope of this chapter. Reviews covering many aspects of tetrapyrrole biosynthesis are to be found in Jordan,[2] Chadwick and Ackrill,[3] and Dailey.[124] There are also many derivatives of heme, each with new pathways emerging for their biosynthesis.

There is remarkable evidence accumulating to suggest that the heme in cytochrome c_3 in *Desulfovibrio vulgaris* may arise from precorrin-2 since the methyl groups at positions C-2 and C7 arise, not from the carbon atoms of 5-aminolevulinic acid, but from the methyl group of L-methionine[217] (Scheme 35). Thus, a novel mechanism would have to be proposed to account for the loss of the acetic acid side chains in rings A and B. This would indeed open up a completely new chapter in heme biosynthesis and the suggestions need further work to substantiate the evidence and to identify intermediates,[218] enzymes, and genes. Such a pathway may throw light on the biosynthesis of heme d_1 (Scheme 35) and the possibility exists that an even more ancient pathway existed before the one with which we are most familiar.

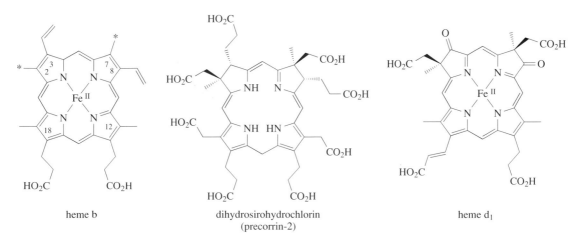

heme b   dihydrosirohydrochlorin   heme d_1
                          (precorrin-2)

Scheme 35

4.04.6 REFERENCES

1. M. J. Warren and A. I. Scott, *Trends Biochem. Sci.*, 1990, **18**, 486.
2. P. M. Jordan (ed.), "Biosynthesis of Tetrapyrroles," Elsevier, Amsterdam, 1991.
3. D. J. Chadwick and K. Ackrill (eds.), "The Biosynthesis of the Tetrapyrrole Pigments," Wiley, Chichester, 1994.
4. A. R. Battersby and F. J. Leeper, *Top. Curr. Chem.*, 1998, **195**, 140.
5. P. M. Jordan (ed.), in "Biosynthesis of Tetrapyrroles," Elsevier, Amsterdam, p. 1.
6. S. I. Beale and J. D. Weinstein, in "Biosynthesis of Tetrapyrroles," ed. P. M. Jordan, Elsevier, Amsterdam, 1991, p. 155.
7. G. C. Ferreira and J. Gong, *J. Bioenerg. Biomembr.*, 1995, **27**, 151.
8. E. L. Neidle and S. Kaplan, *J. Bacteriol.*, 1993, **175**, 2292.
9. P. Dierks, in "*Biosynthesis of Hemes and Chlorophylls*," ed. H. A. Dailey, McGraw-Hill, New York, p. 201.
10. T. C. Cox, M. J. Bawden, A. Martin, and B. K. May, *EMBO. J.*, 1991, **10**, 1891.
11. J. E. Lelean, A. Lloyd, and P. M. Shoolingin-Jordan, *Methods Enzymol.*, 1997, **218**, 309.
12. M. Fanica-Gaignier and J. Clement-Metral, *Eur. J. Biochem.*, 1973, **268**, 584.
13. M. Akhtar and P. M. Jordan, *J. Chem. Soc., Chem. Commun.*, 1968, 1691.
14. Z. Zaman, P. M. Jordan, and M. Akhtar, *Biochem. J.*, 1973, **135**, 257.
15. P. M. Jordan and M. Akhtar, *Tetrahedron Lett.*, 1969, **11**, 875.
16. M. M. Abboud, P. M. Jordan, and M. Akhtar, *J. Chem. Soc., Chem. Commun.*, 1974, 643.
17. V. C. Emery and M. Akhtar, in "Enzyme Mechanisms," eds. M. I. Page and A. Williams, Royal Society of Chemistry, London, p. 345.
18. H. C. Dunathan, "Advances in Enzymology," ed. F. Nord, Wiley, New York, 1971, p. 79.
19. O. Ploux and A. Marquet, *Eur. J. Biochem.*, 1996, **236**, 301.
20. G. C. Ferreira, P. J. Neame, and H. A. Dailey, *Protein Sci.*, 1993, **2**, 1959.
21. A. Laghai and P. M. Jordan, *Biochem. Soc. Trans.*, 1976, **4**, 52.
22. A. Laghai and P. M. Jordan, *Biochem. Soc. Trans.*, 1977, **5**, 299.
23. M. D. Toney, E. Hohenester, J. W. Keller, and J. N. Jansonius, *J. Mol. Biol.*, 1995, **241**, 151.
24. D. Tan, T. Harrison, G. A. Hunter, and G. C. Ferreira, *Biochemistry*, 1998, **37**, 1478.
25. J. Gong, G. A. Hunter, and G. C. Ferreira, *Biochemistry*, 1998, **37**, 3509.
26. S. I. Beale and P. A. Castelfranco, *Plant Physiol.*, 1974, **53**, 297.
27. S. I. Beale, S. P. Gough, and S. Granick, *Proc. Natl. Acad. Sci. USA*, 1975, **72**, 2719.
28. S. I. Beale and J. D. Weinstein, in "Biosynthesis of Hemes and Chlorophylls," ed. H. A. Dailey, McGraw-Hill, New York, 1990, p. 287.
29. W.-Y. Wang, S. P. Gough, and C. G. Kannangara, *Carlsberg Res. Commun.*, 1981, **46**, 243.
30. A. Schön, G. Krupp, S. P. Gough, S. Berry-Lowe, C. G. Kannangara, and D. Söll, *Nature (London)*, 1986, **322**, 281.
31. D. D. Huang and W.-Y. Wang, *J. Biol. Chem.*, 1986, **261**, 13 451.
32. C. G. Kannangara, S. P. Gough, P. Bruyant, J. K. Hoober, A. Kahn, and D. von Wettstein, *Trends Biochem. Sci.*, 1988, **13**, 139.
33. C. G. Kannangara, R. V. Andersen, B. Pontoppidan, R. Willows, and D. von Wettstein, in "The Biosynthesis of the Tetrapyrrole Pigments," eds. J. Chadwick and K. Ackrill, Wiley, Chichester, 1994, p. 3.
34. A. Schön, C. G. Kannangara, S. P. Gough, and D. Söll, *Nature (London)*, 1986, **331**, 187.
35. S. Reinbothe and C. Reinbothe, *Eur. J. Biochem.*, 1996, **237**, 323.
36. D. Jahn, E. Verkamp, and D. Söll, *Trends Biochem. Sci.*, 1992, **17**, 215.
37. L. I. Ilag, A. M. Kumar, and D. Söll, *Plant Cell*, 1994, **6**, 265.
38. E. Verkamp, A. M. Kumar, A. Lloyd, O. Martins, N. Stange-Thomann, and D. Söll, in "tRNA: Structure Biosynthesis and Function," eds. D. Söll and U. Raj Bhandary, American Society of Microbiology, Washington, DC, chap. 27, p. 545.
39. J. K. Hoober, A. Kahn, D. E. Ash, S. P. Gough, and C. G. Kannangara, *Carlsberg Res. Commun.*, 1988, **53**, 11–25.
40. P. M. Jordan, K.-M. Cheung, R. P. Sharma, and M. J. Warren, *Tetrahedron Lett.*, 1993, **34**, 1177,
41. S. Mayer, E. Gawlita, Y. Avissar, V. Anderson, and S. I. Beale, *Plant Physiol.*, 1993, **101**, 1029.
42. C. Pugh, J. Harwood, and R. A. John, *J. Biol. Chem.*, 1992, **267**, 1584.
43. V. Breu and D. Dornemann, *Biochem. Biophys. Acta*, 1988, **967**, 135.
44. P. M. Shoolingin-Jordan, J. F. A. Callaghan, K.-M. Cheung, P. Spencer, and M. J. Warren, in "Phytochemical Diversity, a Source of New Industrial Products," eds. S. Wrigley, M. Hayes, R. Thomas, and E. Chrystal, Royal Society of Chemistry London, 1997, p. 115.
45. M. Hennig, B. Grimm, R. Contestabile, R. A. John, and J. N. Jansonius, *Proc. Natl. Acad. Sci. USA*, 1997, **94**, 4866.
46. B. Grimm, A. J. Smith, C. G. Kannangara, and M. Smith, *J. Biol. Chem.*, 1991, **296**, 12 495.
47. B. Grimm, M. Smith, and D. von Wettstein, *Eur. J. Biochem.*, 1992, **206**, 579.
48. R. A. John, in "Comprehensive Biological Catalysis," Academic Press, New York, 1998, vol. 2, p. 173.
49. P. M. Jordan, in "Biosynthesis of Heme and Chlorophylls," ed. H. A. Dailey, McGraw-Hill, New York, 1990, p. 55.
50. E. K. Jaffe, *J. Bioenerg. Biomembr.*, 1995, **27**, 165.
51. R. Neier, *Adv. Nitrogen Heterocycles*, 1996, **2**, 35.
52. D. Shemin, in "The Enzymes," 3rd edn., ed. P. D. Boyer, Academic Press, New York, 1972, p. 323.
53. W. Wu, D. Shemin, K. E. Richards, and R. C. Williams, *Proc. Natl. Acad. Sci. USA*, 1974, **69**, 2585.
54. D. L. Nandi and D. Shemin, *J. Biol. Chem.*, 1968, **243**, 1231.
55. P. N. B. Gibbs and P. M. Jordan, *Biochem. J.*, 1986, **236**, 447.
56. P. Spencer and P. M. Jordan, *Biochem. J.*, 1995, **305**, 151.
57. P. M. Jordan and J. S. Seehra, *J. Chem. Soc., Chem. Commun.*, 1980, 240.
58. P. M. Jordan and J. S. Seehra, *FEBS Lett.*, 1980, **114**, 283.
59. G. F. Barnard, R. Itoh, L. H. Hohberger, and D. Shemin, *J. Biol. Chem.*, 1977, **252**, 8965.
60. E. K. Jaffe, S. P. Salowe, N. T. Chen, and P. A. DeHaven, *J. Biol. Chem.*, 1984, **259**, 5032.
61. J. G. Wetmer, D. F. Bishop, C. Cantelmo, and R. J. Desnick, *Proc. Natl. Acad. Sci. USA*, 1986, **83**, 7703.
62. E. K. Jaffe and D. Hanes, *J. Biol. Chem.*, 1986, **261**, 9348.

63. P. Spencer and P. M. Jordan, *Biochem. J.*, 1993, **290**, 279.
64. Q. F. Boese, A. J. Spano, J. M. Li, and M. P. Timko, *J. Biol. Chem.*, 1991, **266**, 17 060.
65. S. Chauhan and M. R. O'Brian, *J. Biol. Chem.*, 1995, **270**, 19 823.
66. D. R. Bevan, P. Bodlaender, and D. Shemin, *J. Biol. Chem.*, 1980, **255**, 2030.
67. A. J. Dent, D. Beyersmann, C. Block, and S. S. Hasnain, *Biochemistry*, 1990, **29**, 7822.
68. P. Spencer and P. M. Jordan, *Biochem. J.*, 1994, **300**, 373.
69. J. J. Chisholm *Sci. Am.*, 1971, **224** (2), 15.
70. P. N. B. Gibbs, M. G. Gore, and P. M. Jordan, *Biochem. J.*, 1985, **225**, 573.
71. P. J. Brumm and H. C. Friedmann, *Biochem. Biophys. Res. Commun.*, 1981, 240.
72. K.-M. Cheung, P. Spencer, M. P. Timko, and P. M. Shoolingin-Jordan, *Biochemistry*, 1997, **36**, 1148.
73. N. Senior, P. G. Thomas, J. B. Cooper, S. P. Wood, P. T Erskine, P. M. Shoolingin-Jordan, and M. J. Warren, *Biochem. J.*, 1996, **320**, 401.
74. D. Appleton, A. B. Duguid, S.-K. Lee, H.-J. Ha, H.-J. Ha, and F. J. Leeper, *J. Chem. Soc., Perkin. Trans. 1*, 1998, 89.
75. J. S. Seehra and P. M. Jordan, *Eur. J. Biochem.*, 1981, **113**, 435.
76. E. K. Jaffe, W. R. Abrams, K. X. Kaempfen, and K. A. Harris, *Biochemistry*, 1992, **31**, 2113.
77. H.-J. Ha, J.-W. Park, S.-J. Oh, J.-C. Lee, and C. E. Song, *Bull. Korean Chem. Soc.*, 1993, **14**, 652.
78. P. T. Erskine, N. Senior, S. Awan, R. Lambert, G. Lewis, I. J. Tickle, M. Sarwar, P. Spencer, P. Thomas, M. J. Warren, P. M. Shoolingin-Jordan, S. P. Wood, and J. B. Cooper, *Nature Struct. Biol.*, 1997, **4**, 1025.
79. G. Lewis, P. T. Erskine, E. Norton, J. B. Cooper, R. Lambert, P. Spencer, M. Sarwar, S. P. Wood, M. J. Warren, and P. M. Shoolingin-Jordan, *Biochemistry*, 1998, in press.
80. E. K. Jaffe, G. D. Markham, and J. S. Rajagopalan, *Biochemistry*, 1990, **29**, 8345.
81. M. M. Abboud and M. Akhtar, *J. Chem. Soc., Chem. Commun.*, 1976, 1007.
82. A.-G. Chaudhry and P. M. Jordan, *Biochem. Soc. Trans.*, 1976, **4**, 760.
83. G. Burton, P. E. Fagerness, S. Hosazawa, P. M. Jordan, and A. I. Scott, *J. Chem. Soc., Chem. Commun.*, 1979, 202.
84. P. M. Jordan, G. Burton, H. Nordlov, M. Schneider, L. Pryde, and A. I. Scott, *J. Chem. Soc., Chem. Commun.*, 1979, 204.
85. A. R. Battersby, C. J. R. Fookes, K. E. Gustafson-Potter, G. W. J. Matcham, and E. McDonald, *J. Chem. Soc., Chem. Commun.*, 1979, 1155.
86. P. M. Shoolingin-Jordan, *J. Bioenerg. Biomembr.*, 1995, **27**, 181.
87. G. V. Louie, P. D. Brownlie, R. Lambert, J. B. Cooper, T. L. Blundell, S. P. Wood, M. J. Warren, S. C. Woodcock, and P. M. Jordan, *Nature (London)*, 1992, **359**, 33.
88. G. V. Louie, P. D. Brownlie, R. Lambert, J. B. Cooper, T. L. Blundell, S. P. Wood, V. N. Malashkevich, A. Hadener, M. J. Warren, and P. M. Shoolingin-Jordan, "Proteins: Structure, Function and Genetics," 1996, **25**, 48.
89. P. M. Jordan and M. J. Warren, *FEBS Lett.*, 1987, **225**, 87.
90. G. J. Hart, A. D. Miller, F. J. Leeper, and A. R. Battersby, *J. Chem. Soc., Chem. Commun.*, 1987, 762.
91. M. J. Warren and P. M. Jordan, *Biochemistry*, 1988, **27**, 9020.
92. R. T. Aplin, J. R. Baldwin, C. Pichon, C. A. Roessner, A. I. Scott, C. J. Schofield, N. J. Stolowich, and M. J. Warren, *Bioorg. Med. Chem. Lett.*, 1991, **1**, 503.
93. P. M. Jordan and S. C. Woodcock, *Biochem. J.*, 1991, **280**, 445.
94. M. Lander, A. R. Pitts, P. R. Alefounder, D. Bardy, C. Abell, and A. R. Battersby, *Biochem. J.*, 1991, **275**, 447.
95. R. Lambert, P. D. Brownlie, S. C. Woodcock, G. V. Louie, J. B. Cooper, M. J. Warren, P. M. Jordan, T. L. Blundell, and S. P. Wood, in "The Biosynthesis of Tetrapyrrole Pigments," eds. D. J. Chadwick and K. Ackrill, Wiley, Chichester, 1994, p. 97.
96. S. C. Woodcock, and P. M. Jordan, *Biochemistry*, 1994, **33**, 2688.
97. A. I. Scott, K. R. Clemens, N. J. Stolowich, P. J. Santander, M. D. Gonzalez, and C. A. Roessner, *FEBS Lett.*, 1989, **242**, 319.
98. A. D. Miller, G. J. Hart, L. C. Packman, and A. R. Battersby, *Biochem. J.*, 1988, **254**, 915.
99. P. M. Shoolingin-Jordan, M. J. Warren, and S. J. Awan, *Biochem. J.*, 1996, **316**, 373.
100. M. J. Warren, S. Gul, R. T. Aplin, A. I. Scott, C. A. Roessner, P. I. O'Grady, and P. M. Shoolingin-Jordan, *Biochemistry*, 1996, **34**, 11 288.
101. K. R. Clemens, C. Pichon, A. R. Jacobson, P. Yon-Hin, M. D. Gonzalez, and A. I. Scott, *Bioorg. Med. Chem. Lett.*, 1994, **4**, 521.
102. D. Mauzerall, *J. Am. Chem. Soc.*, 1960, **82**, 2605.
103. A. R. Battersby, C. J. R. Fookes, G. W. J. Matcham, E. McDonald, and K. E. Gustafson-Potter, *J. Chem. Soc., Chem. Commun.*, 1979, 316.
104. R. Radmer and L. Bogorad, *Biochemistry*, 1972, **11**, 904.
105. R. C. Davies and A. Neuberger, *Biochem. J.*, 1973, **133**, 471.
106. C. Pichon, K. R. Clemens, A. R. Jacobson, and A. I. Scott, *Tetrahedron*, 1992, **48**, 4687.
107. R. M. Jones and P. M. Jordan, *Biochem. J.*, 1994, **299**, 895.
108. P. M. Jordan, *Enzyme*, 1982, **28**, 158.
109. A. R. Battersby, C. J. R. Fookes, E. McDonald, and M. Megan, *J. Chem. Soc., Chem Commun.*, 1978, 185.
110. A. R. Battersby, C. J. R. Fookes, G. W. J. Matcham, and E. McDonald, *J. Chem. Soc., Chem. Commun.*, 1978, 1064.
111. A. R. Battersby, C. J. R. Fookes, G. W. J. Matcham, and P. S. Pandy, *Angew. Chem., Int. Ed. Engl.*, 1981, **20**, 293.
112. J. H. Mathewson and A. H. Corwin, *J. Am. Chem Soc.*, 1961, **83**, 135.
113. F. J. Leeper, in "The Biosynthesis of Tetrapyrrole Pigments," eds. D. J. Chadwick and K. Ackrill, Wiley, Chichester, 1994, p. 111.
114. W. M. Stark, G. J. Hart, and A. R. Battersby, *J. Chem. Soc., Chem. Commun.*, 1986, 465.
115. A. C. Spivey, A Capretta, C. S. Frampton, F. J. Leeper, and A. R. Battersby, *J. Chem. Soc., Perkin Trans. 1*, 1996, 2091.
116. A. R. Battersby and F. J. Leeper, *Chem. Rev.*, 1990, **90**, 1261.
117. A. F. Alwan, B. I. A. Mgbeje, and P. M. Jordan, *Biochem. J.*, 1989, **264**, 397.
118. N. J. P. Stamford, A. Capretta, and A. R. Battersby, *Eur. J. Biochem.*, 1996, 236.
119. M. J. Warren, E. Bolt, and S. C. Woodcock, in "The Biosynthesis of Tetrapyrrole Pigments," eds. D. J. Chadwick and K. Ackrill, Wiley, Chichester, 1994, p. 26.

120. R. K. Thauer and L. G. Bonacker, in "The Biosynthesis of Tetrapyrrole Pigments," eds. D. J. Chadwick and K. Ackrill, Wiley, Chichester, 1994, p. 210.
121. F. Blanche, D. Thibaut, L. Debussche, R. Hertle, F. Zipfel, and G. Muller, *Angew. Chem., Int. Ed. Engl.*, 1993, **32**, 1651.
122. F. Blanche, B. Cameron, J. Crouzet, L. Debussche, D. Thibaut, M. Vuilhorgue, F. J. Leeper, and A. R. Battersby, *Angew. Chem., Int. Ed. Engl.*, 1995, **34**, 383.
123. A. I. Scott, *Tetrahedron*, 1994, **50**, 13 315.
124. H. A. Dailey, in "Biosynthesis of Heme and Chlorophylls," ed. H. A. Dailey, McGraw-Hill, New York, 1990, p. 123.
125. M. Akhtar, in "Biosynthesis of Tetrapyrroles," ed. P. M. Jordan, Elsevier, Amsterdam, 1991, p. 67.
126. M. Akhtar, in "The Biosynthesis of Tetrapyrrole Pigments," eds. D. J. Chadwick and K. Ackrill, Wiley, Chichester, 1994, p. 131.
127. A. G. Smith and J. E. Francis, *Biochem. J.*, 1979, **183**, 455.
128. G. H. Elder and A. G. Roberts, *J. Bioenerg. Biomembr.*, 1995, **27**, 207.
129. L. C. San Martin de Vialle and M. Grinstein, *Biochim. Biophys. Acta*, 1968, **158**, 79.
130. L. C. San Martin de Vialle, A. A. Vialle, S. Nacht, and M. Grinstein, *Clin. Chem. Acta*, 1970, **28**, 13.
131. A. H. Jackson, H. A. Sancovich, A. M. Ferramola, N. Evans, D. E. Games, S. A. Matlin, G. H. Elder, and S. Smith, *Philos. Trans. R. Soc. London Ser. B*, 1976, **273**, 191.
132. J. Luo and C. K. Lim, *Biochem. J.*, 1990, **268**, 513.
133. J. Luo and C. K. Lim, *Biochem. J.*, 1993, **289**, 529.
134. R. M. Jones and P. M. Jordan, *Biochem. J.*, 1993, **293**, 703.
135. S. Billi de Catabbi, M. C. Rios de Molina, and L. C. San Martin de Vialle, *Int. J. Biochem.*, 1991, **23**, 675.
136. H. de Verneuil, S. Sassa, and A. Kappas, *J. Biol. Chem.*, 1983, **258**, 2454.
137. S. Kawanishi, Y. Seki, and S. Sano, *J. Biol. Chem.*, 1983, **258**, 4285.
138. T. D. Lash, *Biochem. J.*, 1991, **278**, 901.
139. F. G. Whitby, J. D. Phillips, J. P. Kushner, and C. P. Hill, *EMBO J.*, 1998, **17**, 2463.
140. G. F. Barnard and M. Akhtar, *J. Chem. Soc., Chem. Commun.*, 1975, 494.
141. G. F. Barnard and M. Akhtar, *J. Chem. Soc., Perkin Trans. 1*, 1979, 2354.
142. G. H. Elder, J. Tovey, and D. M. Sheppard, *Biochem. J.*, 1983, **215**, 45.
143. J. P. Kushner, G. R. Lee, and S. Nacht, *J. Clin. Invest.*, 1972, **51**, 3044.
144. S. Sano and S. Granick, *J. Biol. Chem.*, 1961, **236**, 1173.
145. G. H. Tait, *Biochem. J.*, 1972, **128**, 1159.
146. A. M. Del, C. Batlle, A. Benson, and C. Rimington, *Biochem. J.*, 1965, **97**, 731.
147. T. Yoshinaga and S. Sano, *J. Biol. Chem.*, 1980, **255**, 4722.
148. T. Yoshinaga and S. Sano, *J. Biol. Chem.*, 1980, **255**, 4727.
149. J. M. Camadro, H. Chambon, J. Jolles, and P. Labbe, *Eur. J. Biochem.*, 1986, **156**, 579.
150. S. Taketani, H. Kohno, T. Furukawa, T. Yoshinaga, and R. Tokunaga, *Biochim. Biophys. Acta*, 1994, **1183**, 547.
151. O. Madsen, N. N. Sandal, and K. A. Marcker, *Plant Mol. Biol.*, 1993, **23**, 35.
152. M. Zagorec, J. M. Buhler, I. Treich, T. Keng, L. Guarente, and R. Labbe-Bois, *J. Biol. Chem.*, 1988, **263**, 9718.
153. K. Xu and T. Elliott, *J. Bacteriol.*, 1993, **75**, 4990.
154. P. Martasek, J. M. Camadro, M. H. Delfaularue, J. B. Dumas, J. J. Montagne, H. de Verneuil, P. Labbe, and B. Grandchamp, *Proc. Natl. Acad. Sci. USA*, 1994, **91**, 3024.
155. S. A. Coomber, R. M. Jones, P. M. Jordan, and C. N. Hunter, *Mol. Microbiol.*, 1992, **6**, 3155.
156. K. Xu, J. Delling, and T. Elliott, *J. Bacteriol.*, 1992, **174**, 3953.
157. G. Y. Kennedy, A. H. Jackson, G. W. Kenner, and C. J. Suckling, *FEBS Lett.*, 1970 **6**, 9.
158. J. A. S. Cavaleiro, G. W. Kenner, and K. M. Smith, *J. Chem. Soc., Perkin Trans. 1*, 1974, 1188.
159. G. H. Elder, J. O. Evans, J. R. Jackson, and A. H. Jackson, *Biochem. J.*, 1978, **169**, 215.
160. A. H. Jackson, G. H. Elder, and S. G. Smith, *Int. J. Biochem.*, 1978, **9**, 877.
161. G. H. Elder and J. O. Evans, *Biochem. J.*, 1978, **169**, 205.
162. A. H. Jackson, T. D. Lash, and D. J. Ryder, *Int. J. Biochem.*, 1980, **12**, 775.
163. R. B. Frydman and B. Frydman, *FEBS Lett.*, 1975, **52**, 317.
164. Z. Zaman, M. M. Abboud, and M. Akhtar, *J. Chem. Soc., Chem. Commun.*, 1972, 1263.
165. M. M. Abboud and M. Akhtar, *Nouv. J. Chim.*, 1978, **2**, 419.
166. M. Akhtar and C. Jones, *Methods Enzymol.*, 1986, **123**, 375.
167. Z. Zaman and M. Akhtar, *Eur. J. Biochem.*, 1976, **61**, 215.
168. A. R. Battersby, *Experentia*, 1978, **34**, 1.
169. A. R. Battersby, J. Baldas, J. Collins, D. H. Grayson, K. M. James, and E. McDonald, *J. Chem. Soc., Chem. Commun.*, 1972, 1265.
170. A. R. Battersby and E. McDonald, in "Porphyrins and Metalloporphyrins," ed. K. M. Smith, Elsevier, Amsterdam, 1975, p. 61.
171. A. H. Jackson, D. M. Jones, G. Phillip, T. D. Lash, A. M. Del, and C. Batlle, *Int. J. Biochem.*, 1980, **12**, 681.
172. J. S. Seehra, P. M. Jordan, and M. Akhtar, *Biochem. J.*, 1983, **209**, 709.
173. B. Grandchamp, J. Lamoril, and H. Puy, *J. Bioenerg. Biomembr.*, 1995, **27**, 215.
174. P. Frey, *Biochem. Soc. Trans.*, 1998, **26**, 326.
175. B. Hippler, G. Homuth, T. Hoffman, C. Hungerer, W. Schumann, and D. Jahn, *J. Bacteriol.*, 1997, **179**, 7181.
176. K. M. Smith (ed.), "Porphyrins and Metalloporphyrins," Elsevier, Amsterdam, 1975.
177. R. J. Porra and J. E. Falk, *Biochem. J.*, 1964, **90**, 69.
178. A. H. Jackson, D. E. Games, P. W. Couch, J. R. Jackson, R. V. Belcher, and S. G. Smith, *Enzyme*, 1974, **17**, 81.
179. R. Poulson and W. J. Polglase, *J. Biol. Chem.*, 1975, **250**, 1269.
180. R. Poulson, *J. Biol. Chem.*, 1976, **251**, 3730.
181. J. M. Jacobs and N. J. Jacobs, *Biochem. J.*, 1987, **244**, 219.
182. J. M. Camadro, N. G. Ibraham, and R. D. Levere, *Arch. Biochem. Biophys.*, 1985, **242**, 206.
183. H. A. Dailey and S. W. Karr, *Biochemistry*, 1987, **26**, 2697.
184. L. J. Siepker, M. Ford, R. de Kock, and S. Kramer, *Biochim. Biophys. Acta*, 1987, **913**, 349.

185. K. L. Proulx and H. A. Dailey, *Protein Sci.*, 1992, **1**, 801.
186. M. Hansson and L. Hederstedt, *J. Bacteriol.*, 1992, **174**, 8081.
187. J. Frank, H. Lam, and A. M. Christiano, *Am. J. Hum. Genet.*, 1997, **61**, 1952.
188. S. Narita, R. Tanka, T. Ito, K. Okada, S. Taketani, and M. Inokuchi, *Gene*, 1996, **182**, 169.
189. J. M. Camadro and P. Labbe, *J. Biol. Chem.*, 1996, **271**, 9120.
190. D. A. Witkowski and B. P. Halling, *Plant Physiol.*, 1989, **90**, 1239.
191. M. Matringe, J. M. Camadro, P. Labbe, and R. Scalla, *Biochem. J.*, 1989, **260**, 231.
192. A. R. Battersby, J. Staunton, E. McDonald, J. R. Redfern, and R. H. Wightman, *J. Chem. Soc., Perkin Trans. 1*, 1976, 266.
193. C. Jones, P. M. Jordan, A. G. Chaudry, and M. Akhtar, *J. Chem. Soc., Chem. Commun.*, 1979, 96.
194. C. Jones, P. M. Jordan, and M. Akhtar, *J. Chem. Soc., Perkin Trans. 1*, 1984, 2625.
195. N. J. Jacobs and J. N. Jacobs, *Arch. Biochem. Biophys.*, 1981, **211**, 305.
196. D. J. Klemm and L. L. Barton, *J. Bacteriol.*, 1987, **169**, 5209.
197. S. Taketani and R. Tokunaga, *J. Biol. Chem.*, 1981, **256**, 12 748.
198. J. M. Camadro and P. Labbe, *J. Biol. Chem.*, 1988, **263**, 11 675.
199. H. A. Dailey, *J. Biol. Chem.*, 1982, **257**, 14 714.
200. M. Hansson and L. Hederstedt, *Eur. J. Biochem.*, 1994, **220**, 201.
201. M. S. Jones and O. T. G. Jones, *Biochem. J.*, 1969, **113**, 207.
202. J. M. Camadro and P. Labbe, *Biochim. Biophys. Acta*, 1982, **707**, 280.
203. H. A. Dailey, *J. Biol. Chem.*, 1984, **259**, 2711.
204. H. A. Dailey and J. E. Fleming, *J. Biol. Chem.*, 1986, **261**, 7902.
205. G. C. Ferreira, R. Franco, S. G. Lloyd, I. Mourna, J. J. G. Mourna, and B. H. Huynh, *J. Bioenerg. Biomembr.*, 1995, **27**, 221.
206. S. R. Karr and H. A. Dailey, *J. Biol. Chem.*, 1988, **254**, 799.
207. G. C. Ferreira, *J. Biol. Chem.*, 1994, **269**, 4396.
208. K. Miyamoto, S Kanaya, K. Morikawa, and H. Inokuchi, *J. Biochem.*, 1994, **115**, 545.
209. M. Okuda, H. Kohno, F. Furukawa, R. Tokunaga, and S. Taketani, *Biochim. Biophys. Acta*, 1994, **1200**, 123.
210. S. Al-Karadaghi, M. Hansson, S. Nikolov, B. Jönsson, and L. Hederstedt, *Structure*, 1997, **5**, 1501.
211. H. Kohno, T. Furukawa, T. Yoshinaga, R. Tokunaga, and S. Taketani, *J. Biol. Chem.*, 1993, **268**, 21 359.
212. D. A. Brenner, J. M. Didlier, F. Frusier, S. R. Christensen, G. A. Evans, and H. A. Dailey, *Am. J. Hum. Genet.*, 1992, **50**, 1203.
213. G. V. Louie, *Curr. Opin. Struct. Biol.*, 1993, **3**, 401.
214. H. A. Dailey, C. S. Jones, and S. W. Karr, *Biochem. Biophys. Acta*, 1989, **999**, 7.
215. R. Labbe-Bois and J. M. Camadro, in "Metal Ions in Fungi," eds. G. Winkelmann and Dwinge, Marcel Dekker, New York, 1994, p. 413.
216. D. K. Lavallee, in "Mechanistic Principles of Enzyme Activity," eds. J. F. Webman and A. Greenberg, VCH, Weinheim, 1988, p. 279.
217. H. Atkutsu, J.-S. Park, and S. Sano, *J. Am. Chem. Soc.*, 1993, **115**, 12 185.
218. T. Ishida, L. Yu, H. Akutsu, K. Ozawa, S. Kawanishi, A. Seto, T. Inubushi, and S. Sano, *Proc. Natl. Acad. Sci. USA*, 1998, **95**, 4853.

4.05

Strictosidine—The Biosynthetic Key to Monoterpenoid Indole Alkaloids

JOACHIM STÖCKIGT and MARTIN RUPPERT
Johannes Gutenberg-Universität Mainz, Germany

4.05.1 INTRODUCTION

Strictosidine (**5**) belongs to the large group of plant indole alkaloids derived from tryptamine (**2**), which is the decarboxylation product of the amino acid tryptophan (**1**), and from the monoterpene loganin (3) leading to the seco-iridoid secologanin (4). Strictosidine (5) is a glucoside which is an intermediate in the biosynthesis of monoterpenoid indole alkaloids and some biosynthetically related alkaloid groups. It is estimated that more than 2000 such indolic compounds occur in higher plants.[1] Many of these indole alkaloids (Scheme 1) are of considerable pharmacological and therapeutic interest, including the following:

(i) the highly toxic strychnine (6),[2] the study of which led to the discovery of the glycine receptor;[3]

(ii) the antihypertensive reserpine (7),[4] interesting because of its antipsychotic activity;[5]

(iii) the structurally complex ajmaline (8),[6] showing antiarrhythmic activity and applied in the therapy of heart diseases;[7]

(iv) the vasodilator yohimbine (10),[8] exhibiting a complex spectrum of pharmacological effects previously reviewed in great detail;[9]

(v) the antihypertensive ajmalicine (9),[10] employed in the treatment of post-ischemic damage as an admixture with almitrine;[11] and

(vi) the dimeric (bisindole) alkaloids vincaleucoblastine (11)[12] and vincristine (12),[13] important cytotoxic drugs in clinical use for cancer chemotherapy.[14]

Strictosidine (5) is the biogenetic precursor of these natural products and therefore is of extraordinary significance, which was recognized particularly from the 1960s onwards. A large number of research papers have appeared concerning strictosidine (5) and its role in alkaloid formation was frequently discussed at the end of the 1960s.[15,16] An excellent overview of this important intermediate of indole alkaloid biosynthesis[17] has been published. This chapter summarizes our present knowledge of this intriguing alkaloid.

4.05.2 OCCURRENCE OF STRICTOSIDINE

Especially during the 1960s and 1970s, the isolation and structural elucidation of indole alkaloids were one of the greatest challenges in organic chemistry. At that time, physical and spectroscopic methods of structure elucidation were under development; for example, high-field NMR spectroscopy (>240 MHz) was not yet routine and mass spectrometry (MS) was still in the process of becoming established as a routine method for structure determination. Work on the isolation and identification of monoterpenoid indole alkaloids led to an understanding of MS fragmentation processes, the development of general fragmentation rules[18] and application to the identification of natural products.

The identification of these alkaloids and their structural features permitted their classification into three main groups:

(i) the Corynanthé type (13);

(ii) the Aspidosperma type (14); and

(iii) the Iboga type (15), all having the monoterpenoid C_{10} unit.

Scheme 2 shows the main skeletons of the iridoid part of monoterpenoid indole alkaloids.

The question of how the plant synthesizes such structurally complicated natural products soon became one of the major points of interest in the field of biosynthesis of secondary products. The step from the development of biogenetic hypotheses during the 1950s[19–21] to their experimental verification became possible within a few years after the first *in vivo* feeding experiments with radioactively labeled precursors.[22–24]

Evidence accumulated indicating that two biosynthetic precursors are involved. The one delivering the indole part of the molecule is tryptophan (1), leading to tryptamine (2), which was shown to be incorporated into the *Rauwolfia* alkaloids ajmaline (8), reserpine (7), and serpentine.[22] The second direct progenitor, which was suggested to be a monoterpene, was much more difficult to identify in the early days of unraveling the indole alkaloid pathways. For other groups of alkaloids an aldehydic monoterpene was the most probable building block and this was later shown to be secologanin (4). Many tracer experiments were undertaken in order to delineate the major steps in the biosynthesis of this secoiridoid derived from loganin (3). According to *in vivo* feeding experiments, geraniol (16) and its isomer nerol (17) were both precursors of loganin (3) (Scheme 3),[25] as were the appropriate 10-hydroxy derivatives which were incorporated into monoterpenoid indole alkaloids such as ajmalicine (9).[26,27] These derivatives were incorporated into iridodial (20), as were the appropriate aldehydic derivatives such as 10-oxogeraniol, 10-hydroxygeraniol (18), and 10-oxogeranial, as discerned by cell-free experiments.[28] The sequence which follows the 10-hydroxy compounds affording loganin (3) and secologanin (4) involves a number of steps, the mechanisms of which are mostly unknown. The formation of the cyclopentane unit of loganin (3) and its transformation into the

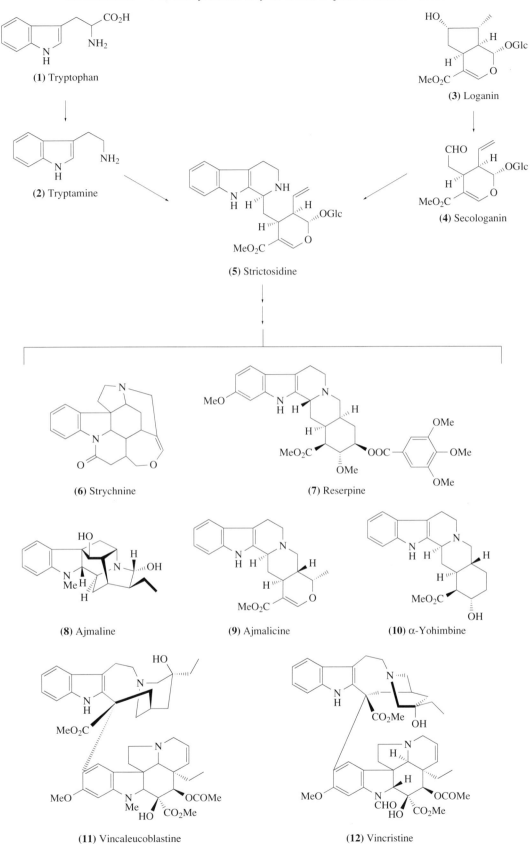

(1) Tryptophan

(2) Tryptamine

(3) Loganin

(4) Secologanin

(5) Strictosidine

(6) Strychnine

(7) Reserpine

(8) Ajmaline

(9) Ajmalicine

(10) α-Yohimbine

(11) Vincaleucoblastine

(12) Vincristine

Scheme 1

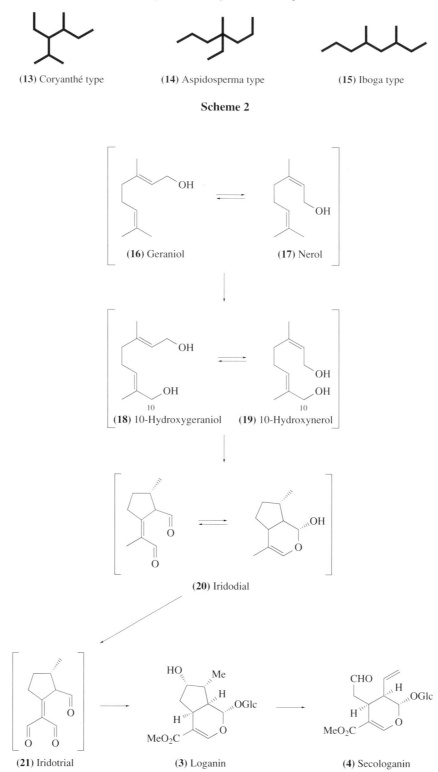

(13) Coryanthé type   **(14)** Aspidosperma type   **(15)** Iboga type

Scheme 2

(16) Geraniol   **(17)** Nerol

(18) 10-Hydroxygeraniol   **(19)** 10-Hydroxynerol

(20) Iridodial

(21) Iridotrial   **(3)** Loganin   **(4)** Secologanin

Scheme 3

seco-iridoid secologanin (**4**) by ring opening is still unknown. Nevertheless, the earlier steps on this pathway of the formation of iridodial (**20**) and iridotrial (**21**) and the significance for loganin (**3**) biosynthesis have been investigated by detailed *in vivo* and cell-free experiments. Taking into account tracer experiments and enzymatic studies, such as studies of loganic acid methyltransferase with

Catharanthus roseus seedlings,[29,30] the results suggest the sequential formation of 7-deoxyloganic acid → loganic acid → loganin (**3**) → secologanin (**4**). Of these reactions, the ring cleavage of loganin (**3**) giving rise to the formation of secologanin (**4**) seems to be of particular interest because this cleavage has no precedent in biochemistry or organic chemistry. From several feeding experiments, only two mechanisms could be deduced. On the one hand there appears to be involvement of 10-hydroxyloganin[31] and on the other the direct ring fission of loganin (**3**).[32] Extensive feeding and enzymatic experiments with cell suspension cultures, especially of *Rauwolfia serpentina* Benth. *ex* Kurz as an indole alkaloid-producing system, and also investigations on several loganin (**3**)- and secologanin (**4**)-forming cultures favor the second mechanistic possibility.[28,33–35] The isolation, purification, and characterization of the particular enzymes involved in that step are important to gain insight into the mechanism of this exceptional ring cleavage reaction.

Although at this stage of the biosynthesis studies the formation of both principal alkaloid precursors tryptamine (**2**) and secologanin (**4**) had been sufficiently explained, the coupling reaction of (**2**) and (**4**) to give the first monoterpenoid indole alkaloid required clarification.

In the course of phytochemical investigations concentrated especially on the plant families Apocynaceae, Rubiaceae, and Loganiaceae, the genus *Rhazya* attracted much interest. *Rhazya* species were found to be rich in water-soluble indole alkaloids containing a glucose unit, the so-called amino acid glycosides. Investigation of this plant material yielded a number of alkaloidal glucosides, a rare group of alkaloids of which strictosidine (**5**) is probably the most important.

4.05.2.1 Occurrence of Strictosidine in Differentiated Plants

In 1968, the detailed analysis of the leaves of the plant *Rhazya stricta* Decaisne performed by Smith[15] at Manchester revealed glucoalkaloids. The most interesting glucoalkaloid in *Rhazya* was found as an amorphous material in a yield of about 0.01%. This alkaloid was named strictosidine (**5**) because of its first isolation from *R. stricta* (Scheme 4). Its significance in the biosynthesis of monoterpenoid indole alkaloids was proposed at the same time. The analytical method of countercurrent distribution allowed the isolation of (**5**) because it is relatively unstable, resulting in the formation of vallesiachotamines on mild hydrolysis. In addition, the occurrence of strictosidine (**5**) could be demonstrated by isotope dilution analysis after feeding [aryl-³H]tryptophan and [O-methyl-³H]loganin to *Vinca rosea* plants.[16] At this point it should be noted that, for taxonomic reasons, *V. roseus* became reclassified in an independent genus named *Catharanthus* and the correct taxonomic name of the plant used for the feeding experiments is not *Vinca rosea* but *Catharanthus roseus* (L.) G. Don. After the feeding experiment, the glucoalkaloid (**5**) was purified to constant specific activity through its pentaacetate, clearly indicating the presence of strictosidine (**5**) in the plant material. The biogenetic function of (**5**) had already been suggested at that time.[15,16] In similar experiments using [U-³H]tryptamine (**22**) and [O-methyl-³H]secologanin (**23**), an epimeric mixture of strictosidine (=isovincoside) (**5**) and vincoside (**24**) was prepared and administered to *Vinca rosea* plants and evidence was obtained of incorporation into some of the *Vinca* (=*Catharanthus*) alkaloids (Scheme 4). Moreover, both glucoalkaloids were detected in the plant material by isotope dilution analysis.[36] These sets of experiments together with the isolation of appropriate labeled *Catharanthus* alkaloids pointed to the biogenetic intermediacy of such a glucoalkaloid (**5, 24**) for monoterpenoid indole alkaloid biosynthesis.

From *in vivo* feeding experiments, it was claimed that the 3β-glucoalkaloid vincoside (**24**) was the biogenetic precursor for the different classes of monoterpenoid indole alkaloids and that 3α-strictosidine (**5**) was not a progenitor.[36] Because the monoterpenoid indole alkaloids usually possess the 3α-configuration, the discrepancy of incorporation of 3β-vincoside (**24**) needed to be explained. Cell-free experiments finally solved this issue.

4.05.3 CELL-FREE FORMATION OF STRICTOSIDINE

Detection of the enzyme proceeded as follows. Employing seedlings of *C. roseus* and, with the advent of cultured tissue of the same plant, cell-free preparations of that material were shown to catalyze the formation of the alkaloids ajmalicine (**9**) and geissoschizine from [2-¹⁴C]tryptamine and the monoterpene glucoside secologanin (**4**) (Scheme 5).[37] Cell suspension cultures of *C. roseus*, which were more efficient in producing monoterpenoid indole alkaloids,[38,39] were also superior to the above callus system. Cell-free extracts of the latter material revealed NADPH-dependent formation of the

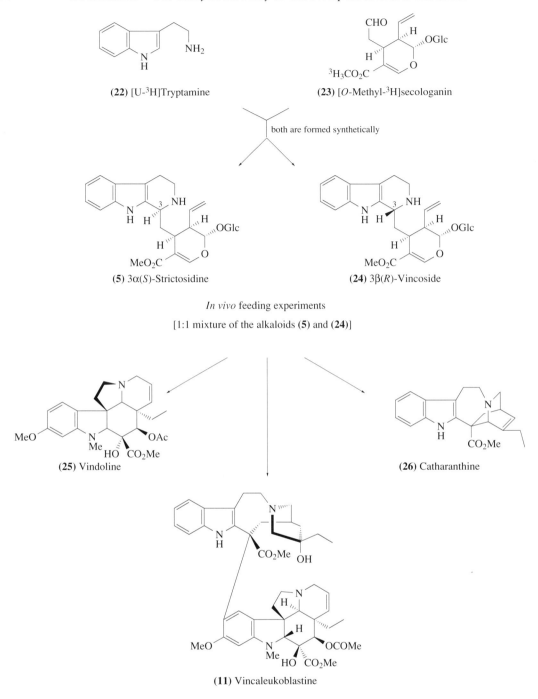

(22) [U-³H]Tryptamine

(23) [*O*-Methyl-³H]secologanin

both are formed synthetically

(5) 3α(*S*)-Strictosidine

(24) 3β(*R*)-Vincoside

In vivo feeding experiments

[1:1 mixture of the alkaloids **(5)** and **(24)**]

(25) Vindoline

(26) Catharanthine

(11) Vincaleukoblastine

Scheme 4

three heteroyohimbine alkaloids ajmalicine (**9**), its epimer 19-epiajmalicine (**34**) and tetra-hydroalstonine (**36**)[40] in excellent yields, indicating remarkable enzyme activities of the cultured *Catharanthus* cells. When the reducing cofactor NADPH was omitted from the cell-free incubation mixtures, several transformation products appeared together with the major alkaloidal component cathenamine (**30**).[41,42]

In the course of the optimized formation of the unknown compounds, two sets of experiments resulted in the formation of a polar compound in high yield: (i) performing the cell-free experiments at low pH (4.1); and (ii) incubation of the precursors tryptamine (**2**) and secologanin (**4**) in the presence of the glucosidase inhibitor δ-gluconolactone.

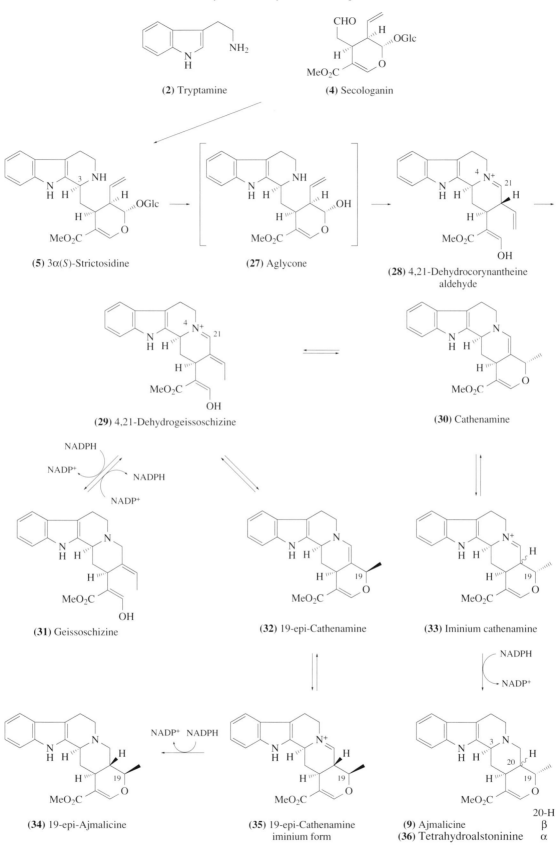

(2) Tryptamine

(4) Secologanin

(5) 3α(*S*)-Strictosidine

(27) Aglycone

(28) 4,21-Dehydrocorynantheine aldehyde

(29) 4,21-Dehydrogeissoschizine

(30) Cathenamine

(31) Geissoschizine

(32) 19-epi-Cathenamine

(33) Iminium cathenamine

(34) 19-epi-Ajmalicine

(35) 19-epi-Cathenamine iminium form

(9) Ajmalicine   20-H   β
(36) Tetrahydroalstoninine   α

Scheme 5

The expectations concerning both of these experiments were the inhibition of some enzymes of the crude protein mixture at a relatively low pH.[43,44] Such inhibition could lead to interference with the overall reaction to ajmalicine (**9**), leading to the accumulation of new biosynthetic intermediates. Thin-layer chromatography (TLC) revealed that the same polar compound accumulated when the experiments (i) and (ii) were performed and large incubation mixtures of tryptamine (**2**) and secologanin (**4**) provided enough material for further investigation. Since accumulation of the unidentified product was due to inhibition of an involved glucosidase, the unknown compound could be expected to be an alkaloidal glucoside. Determination of its structure was necessary after its intermediacy in the cell-free biosynthesis of the heteroyohimbines ajmalicine (**9**) and its isomers (**34**, **36**) had been demonstrated indirectly by (i) careful cell-free time-course experiments showing accumulation followed by transformation of the unknown compound; and (ii) incubating the isolated glucoalkaloid with the *C. roseus* enzyme mixture leading to (**9**), (**34**) and (**36**).

The same result was obtained when this polar glucoalkaloid was enzymatically produced by a crude protein mixture isolated from *Rhazya* cells. It was this last result which made a rigorous structure determination of the accumulated polar intermediate necessary because of the discrepancy described above concerning the function and the stereochemistry at C-3 of strictosidine (**5**), vincoside (**24**), and isovincoside (= strictosidine) (**5**), compounds which could be identical with the accumulated intermediate.

4.05.3.1   Structure of Enzymatically Formed Strictosidine

For the identification of the polar intermediate of the heteroyohimbine alkaloids (**9**), (**34**), and (**36**), the pH-dependent cell-free synthesis was applied, generating the product in labeled form from [^{14}C]tryptamine or unlabeled in a semipreparative way in milligram amounts. On TLC the labeled polar compound comigrated in a neutral solvent system with a chemically synthesized mixture of strictosidine (**5**) and vincoside (**24**), indicating chromatographically its glucoalkaloid behavior. In a basic TLC solvent system, vincoside (**24**) is, in contrast to strictosidine (**5**), converted into its lactam, which allows a simple separation of both epimers. In fact, the cell-free product obtained comigrated under these conditions with authentic strictosidine (**5**). After saponification of the unknown compound it showed the same R_f value as strictosidine lactam (**37**) and was clearly separated from vincoside lactam, which finally disproved the identity of the enzymatically formed polar product as vincoside (**24**).

Moreover, the pentaacetate of the unknown exhibited the same R_f as 3α-pentaacetylstrictosidine and was well separated chromatographically from 3β-pentaacetylvincoside. The same result was obtained when four different TLC systems were applied.

However, because of the earlier assumed role of 3β-vincoside (**24**) as the biogenetic precursor of ajmalicine (**9**), a clear observation was necessary to be certain that the isolated enzymatic product was, indeed, strictosidine (**5**). The direct evidence for its structure came from c.d. and NMR data.[43–45] The c.d. curves of authentic pentaacetylstrictosidine and pentaacetylvincoside can be easily distinguished in the region at 217, 271, and 290 nm. The c.d. curve of the cell-free product is superimposable on that of authentic acetylated strictosidine (**5**). In addition, the unknown compound, after conversion to the tetraacetyl lactam, gave identical ^{1}H NMR data when compared with authentic strictosidine lactam (**37**). The latter can be distinguished from the tetraacetylvincoside lactam by the significant shift of one acetyl signal. All these data were in complete agreement with the structure of strictosidine (**5**) for the enzymatically formed product. At that time it became clear that the enzyme combining tryptamine (**2**) and secologanin (**4**) was the newly detected protein strictosidine synthase. The results, however, did not exclude any additional formation of the 3β-epimer vincoside (**24**) by the cell-free system. Investigation of this point would also allow judgement on the specificity of the enzyme involved (strictosidine synthase) in the crude protein mixture. Three sets of isotope dilution analyses with concomitant preparation of several derivatives demonstrated without any doubt the exclusive formation of the 3α-epimer (**5**)[44,45] and were also evidence for high specificity of strictosidine synthase. This was also necessary for demonstration of the purified enzyme later.

These observations strongly suggested strictosidine (**5**) as the key biosynthetic intermediate for the biogenesis of heteroyohimbine alkaloids with a 3α-configuration such as ajmalicine (**9**), but they did not disprove at this stage a biosynthetic role for its isomer vincoside (**24**). Experiments on feeding both epimers to appropriate plants or cell systems would be expected to resolve this point.

4.05.4 SIGNIFICANCE OF STRICTOSIDINE

4.05.4.1 Biosynthetic Function of Strictosidine

In order to clarify the function of the glucoalkaloid (**5**) in the biosynthesis of different classes of monoterpenoid indole alkaloids and to compare it with the 3β-epimer vincoside (**24**), both radio-active compounds were administered to young *C. roseus* plants under identical conditions. Extraction of the plant material and isolation and purification of the major alkaloids representing the Corynanthé/Strychnos (**13**), Aspidosperma (**14**) and Iboga (**15**) types revealed very clearly that the exclusive precursor for these groups of alkaloids was strictosidine (**5**) and not the 3β-epimer vincoside (**24**),[43,44,46] which was in agreement with the earlier assumption by Smith.[15]

Additional work on this subject soon confirmed the biogenetic role of (**5**) in tissue systems and in differentiated plants[47–49] and ruled out a possible biogenetic function for vincoside (**24**). Scheme 6 shows strictosidine (**5**) as the sole biogenetic precursor of the different alkaloid types Corynanthé (**13**), Aspidosperma (**14**), and Iboga (**15**).

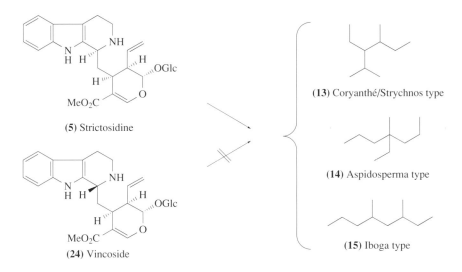

(5) Strictosidine

(24) Vincoside

(13) Corynanthé/Strychnos type

(14) Aspidosperma type

(15) Iboga type

Scheme 6

In a similar series of experiments, the lactam of strictosidine (=strictosamide) (**37**) was shown to occupy a role as precursor of the alkaloid camptothecin (**38**) in *Camptotheca acuminata* plants[50,51] (Scheme 7).

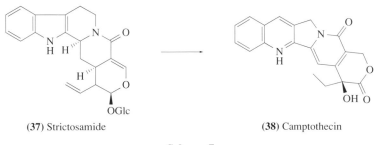

(37) Strictosamide

(38) Camptothecin

Scheme 7

The crucial position of strictosidine (**5**) in the biosynthesis of the major classes of monoterpenoid indole alkaloids and related alkaloid groups became evident after more extensive *in vivo* feeding experiments. Application of strictosidine (**5**) to plants belonging to *Amsonia tabernaemontana*, *Cinchona pubescens*, *Rhazya orientalis*, *Rhazya stricta*, *Uncaria gambir*, and *Vallesia glabra* resulted in high incorporation into the corresponding indole alkaloids, whereas vincoside (**24**) was biosyn-thetically inert as far as alkaloid synthesis was concerned.[52]

In the class of Corynanthé alkaloids, several examples exist in which the C-3 hydrogen is in the β-configuration, not α as in strictosidine (**5**). The question was whether the 3β-alkaloids derived from the 3β-vincoside (**24**) instead of its 3α-epimer (**5**). Such a postulate[48] would make sense but does not fit with the general role of 3α-strictosidine (**5**) as a precursor.

It was demonstrated in an elegant series of feeding experiments that the 3α-epimer strictosidine (**5**) also serves as a precursor of Corynanthé alkaloids (**13**) with a 3β-configuration.[52,53] The incorporation of the 3α-glucoalkaloid (**5**) into reserpiline (**44**), isoreserpiline (**42**), and speciociliatine was monitored. This result is in sharp contrast to the observation that the 3β-akuammigine (**40**) was not labeled in *C. roseus* plants and led to the assumption that the β-glucoalkaloid vincoside (**24**) might be the progenitor.[48] It is noteworthy that the *in vivo* incorporation of strictosidine (**5**) into 3β-alkaloids occurs with loss of the C-3 proton whereas it is retained during the course of the formation of 3α-alkaloids. An epimerization process which was postulated in the past to explain the transformation of 3β-vincoside (**24**) into 3α-alkaloids was excluded on the basis of these experiments. The incorporation of strictosidine (**5**) into ajmaline (**8**), sarpagine (**55**), and gelsemine (**39**) indicated the role of strictosidine (**5**) as the sole and central biosynthetic intermediate of a large variety of alkaloids of monoterpenoid origin.[52]

The precursor role of strictosidine (**5**) for the ajmaline- and sarpagine-type alkaloids is also of major interest. For mechanistic reasons, 5-carboxystrictosidine (**54**) and 5-carboxyvincoside had been assumed originally as possible progenitors.[54,55] Indeed, the importance of these carboxyalkaloids in the biomimetic synthesis of ajmaline (**8**) had been demonstrated. However, extensive feeding experiments with both carboxyglucoalkaloids, 5-carboxystrictosidine (**54**) and 5-carboxyvincoside, to three different plant species, *Rauwolfia vomitoria*, *Vallesia glabra*, and *Voacanga africana*, species synthesizing ajmaline- and/or sarpagine-type alkaloids, indicated that the carboxy compounds were not involved in the alkaloid biosynthesis.[56] These findings were perfectly in accord with the key biosynthetic role of strictosidine (**5**).

This role is illustrated in Scheme 8, summarizing the feeding experiments with labeled strictosidine (**5**) and its incorporation into the various monoterpenoid indole alkaloids with 3α- and 3β-configurations.

4.05.4.2   Ecological Significance of Strictosidine

A possible ecological function for strictosidine (**5**) has been discussed.[57] The involvement of the glucoalkaloid (**5**) in antimicrobial and antifeedant activities was investigated using leaves of *C. roseus*. It was demonstrated that strictosidine (**5**), and also its deglucosylation product formed specifically by the enzyme strictosidine glucosidase, were active against various microorganisms. In fact, one could already expect from the high reactivity and the instability of these compounds that some biological activities such as antimicrobial activity would be a characteristic. When larvae of *Spodoptera exigua* were feeding on intact *C. roseus* leaves, the extracts of the leaves showed an antifeedant activity against the organisms. In contrast to these results, it was observed that neither strictosidine (**5**) nor the aglycone product(s) exhibited any antifeedant effect on the larvae. It was suggested that a rather non-polar compound derived later in the biosynthetic pathway, presumably an alkaloid, might be involved in the observed ecological effects. This compound has not been identified.

4.05.5   PURIFICATION OF STRICTOSIDINE SYNTHASE FROM *C. ROSEUS*

All the experiments described thus far clearly indicate the presence of an enzyme in monoterpenoid-bearing plant cells, which catalyzes the following reaction:

tryptamine (**2**) + secologanin (**4**) → strictosidine (**5**)

The appropriate enzyme combining tryptamine (**2**) and secologanin (**4**) was named strictosidine synthase and attracted much interest. To gain more knowledge about this new plant catalyst, isolation, purification, and characterization of this protein became necessary.

One of the most important prerequisites for the detection, purification, and enrichment of an enzyme activity is the development of a fast and highly sensitive enzyme assay. For strictosidine synthase purification, because TLC scanning of the [^{14}C]strictosidine formed[43,44] is an inconvenient and time-consuming procedure, a simple and highly efficient assay was established based on ^{3}H release during condensation of secologanin (**4**) and [2-^{3}H]tryptamine.[58] Meanwhile, two additional

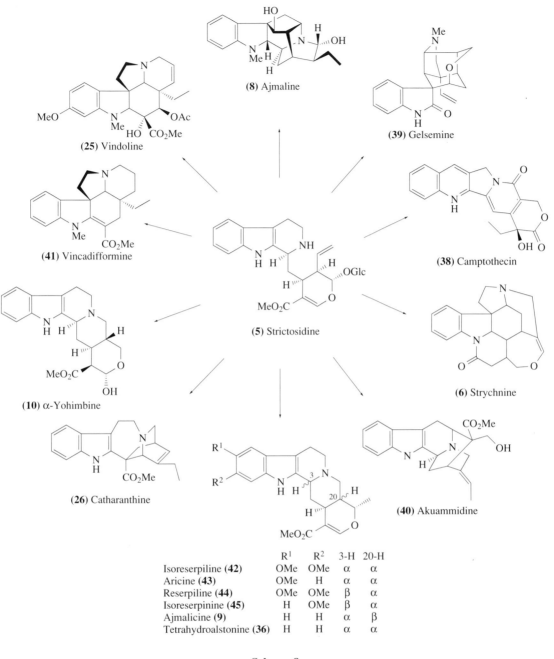

(8) Ajmaline

(39) Gelsemine

(25) Vindoline

(41) Vincadifformine

(38) Camptothecin

(5) Strictosidine

(10) α-Yohimbine

(6) Strychnine

(26) Catharanthine

(40) Akuammidine

	R^1	R^2	3-H	20-H
Isoreserpiline **(42)**	OMe	OMe	α	α
Aricine **(43)**	OMe	H	α	α
Reserpiline **(44)**	OMe	OMe	β	α
Isoreserpinine **(45)**	H	OMe	β	α
Ajmalicine **(9)**	H	H	α	β
Tetrahydroalstonine **(36)**	H	H	α	α

Scheme 8

tests have been published based on spectrophotometric measurements[59] and HPLC separation and quantitation of the product,[60] but these methods are not competitive with the high-throughput ^{3}H-release assay to measure strictosidine synthase activities.[58] Many of the properties of the synthase were determined in 1979 when, as a result of these assays, the enzyme activity could be highly enriched from *C. roseus* cells and other systems.[58,61,62] Employing the above-mentioned ^{3}H-releasing assay, a broad variety of cell suspension systems was investigated for the occurrence and activity of the synthase.[58] As depicted in Table 1, the enzyme activities varied greatly, depending on the cell culture analyzed. Whereas excellent enzyme activities were measured in cells of *Rauwolfia serpentina* and *Voacanga africana*, significantly lower activities were found in *C. roseus* and *Rhazya orientalis* cells. Low enzyme activity (<20%) was measured in *Amsonia*, *Stemmadenia*, *Vinca*, and *Ochrosia* cell systems.

The data also showed that the enzyme is present exclusively in monoterpenoid indole alkaloid-producing cells of the Apocynaceae family. When cells of other plant families devoid of such

Table 1 Distribution of strictosidine synthase activity in different genera of Apocynaceae and unrelated species.

Cell material	Enzyme activity (%)	Plant family
Rauwolfia vomitoria	100.0	Apocynaceae
Voacanga africana	66.5	Apocynaceae
Catharanthus roseus	35.4	Apocynaceae
Rhazia orientalis	35.0	Apocynaceae
Amsonia salicifolia	17.9	Apocynaceae
Stemmadenia tomentosa	10.9	Apocynaceae
Vinca minor	10.9	Apocyanaceae
Ochrosia elliptica	10.4	Apocynaceae
Trifolium pratense[a]	0.0	Leguminosae
Nicotiana tabacum[a]	0.0	Solanaceae

[a]These plants do not synthesize monoterpenoid indole alkaloids.

alkaloids were tested, e.g., cell material of the Leguminosae or Solanaceae, no enzyme activity could be measured; for example, cells of *Nicotina*, which synthesize nicotine alkaloids, did not contain strictosidine synthase. The data proved for the first time that this novel enzyme was of common occurrence in plant cells synthesizing indole alkaloids of monoterpenoid origin.[58] This investigation characterized this enzyme for the first time.

Because of the intense interest in the *Catharanthus* alkaloids[25] in particular, the purification and characterization of the synthase from *C. roseus* cell cultures[62,63] was soon described.

Since many examples exist supporting the observation that the first enzyme in a given pathway serves as a site of control in primary[64] and secondary metabolism,[65] it seemed vitally necessary to investigate the strictosidine synthase at a homogeneous stage for the determination of its role in the regulation of alkaloid biosynthesis and also to establish its physical and kinetic properties. In the first publication[62] on this topic, an apparently homogeneous synthase was obtained by a five-step purification procedure involving ammonium sulfate precipitation, chromatography on DEAE-cellulose and Hydroxyapatite, Sephadex G-75 gel filtration, and preparative isoelectric focusing. In an overall yield of 10%, 740-fold enrichment of the enzyme could be achieved. At that stage of purification, the protein showed a single band on analytical polyacrylamide gel electrophoresis and analytical isoelectric focusing, which indeed suggested a pure enzyme.[62] Later work,[63] however, was not in agreement with these findings. Taking advantage of the above-mentioned fast, sensitive strictosidine synthase assay, the enzyme(s) were obtained in pure form after only a four-step purification protocol. Using ammonium sulfate precipitation, AcA 44 gel filtration, Matrex Gel Red A, and PBE 94 chromatofocusing, 2 kg of fresh *C. roseus* suspended cells delivered at least four multiple forms of the strictosidine synthase (I–IV) in surprisingly high total yield (40%).[63] These isoforms were detected for the first time when a chromatofocusing column was used. This pattern of synthase isoforms was confirmed by a completely different purification procedure based on chromatography on an immuno matrix which was prepared by coupling monospecific strictosidine synthase antibodies to agarose. The 1000-fold purified enzyme so obtained also showed four isoforms on polyacrylamide gel electrophoresis. The same immuno-absorption purification was applied to an enzyme extract of *C. roseus* leaves and finally showed the identical isoform pattern after an excellent 1700-fold enzyme enrichment.[63]

It can be concluded that strictosidine synthase from *C. roseus* cell culture systems and from differentiated plants consists of multiple forms, up to seven isoforms having been detected, and not of a single protein as suggested earlier.[62] These isoenzymes catalyze exactly the same reaction. Careful control experiments were performed in order to ascertain that these different forms are not due to proteolysis or disruption of an oligomeric protein.

The physical and catalytic properties of the major isoenzymes (I–IV) were determined. As pointed out in Table 2, all four isoforms exhibit an identical pH optimum but they can be easily distinguished by their isoelectric points and by their K_M values (shown for tryptamine).

From comparison of the final specific activity of the major isoform III, which showed a value of 104 nKat mg^{-1} protein (1292-fold purified) with that of the pure enzyme described before[62] showing only 5.85 nKat mg^{-1} at a stage of 740-fold purification, it was argued that the latter enzyme preparation was not homogeneous. It is worth noting that the determination of the molecular mass of the synthase III by different methods also did not lead to the same value (Table 3), and this fact probably needs to be considered in general in enzyme characterization work.

Table 2 Comparison of the physicochemical and catalytic properties of different strictosidine synthase isoforms I–IV isolated from cultured *C. roseus* cell suspensions.

Parameter	Isoform			
	I	*II*	*III*	*IV*
pH optimum	6.7	6.7	6.7	6.7
Isoelectric point	4.8	4.6	4.5	4.3
K_M for tryptamine (mmol L^{-1})	0.9	6.6	1.9	2.2
V_{max} for tryptamine (pKat)	28.9	93.6	48.0	46.3

Table 3 Comparison of the molecular masses of different isolates of strictosidine synthase from *C. roseus* determined by gel filtration and SDS–PAGE.

Enzyme	Gel filtration (kDa)	SDS–PAGE (kDA)	Ref.
Synthase III	31	41.5	63
Synthase	38		62
Cloned synthase		38.119	66

Further determination of the substrate specificity of the enzyme was interesting because some derivatives of tryptamine (**2**) and secologanin (**4**) were detected which could also serve as substrates.[61,63] For instances, 5- or 6-hydroxylated or -fluorinated tryptamines and 7-methyltryptamine were accepted by the enzyme, although the activity towards the substrates was much lower with tryptamine (**2**) itself. Similar observations were made for the second partner of the condensation reaction, secologanin (**4**) (Table 4).

Table 4 Substrate acceptance of strictosidine synthase isolated from *C. roseus*.

Tested compound	Accepted as substrate
Amine components	
Tryptamine (**2**)	Yes
5- and 6-Hydroxytryptamine	Yes
5- and 6-Fluorotryptamine	Yes
7-Methyltryptamine	Yes
N-Methyltryptamine	No
L-Tryptophan (**1**)	No
Dopamine	No
Veratrylamine	No
Monoterpenoid components	
Secologanin (**4**)	Yes
9,10-Dihydrosecologanin	Yes
2′-*O*-Methylsecologanin	Yes
3′-*O*-Methylsecologanin	Yes
Secologanin acid	No
Loganin (**3**)	No

From the biosynthetic point of view, important conclusions can be drawn from the observed substrate specificities of the *Catharanthus roseus* strictosidine synthase. The most efficient substrates are tryptamine (**2**) and secologanin (**4**). However, it remains uncertain whether the other substrates tested might also be of biogenetic significance. Acceptance of 6-hydroxytryptamine leads to 9-hydroxystrictosidine. There are several examples of monoterpenoid indole alkaloids displaying a hydroxy or methoxy group at that position;[67] thus, 11-hydroxytabersonine (**46**), 11-hydroxypleiocarpamine (**47**), vincanidine (**49**), vindoline (**25**), 11-methoxytabersonine (**48**), horhammerinine (**53**), tetraphylline (**50**), vincine (**51**), and vincaminoridine (**52**) could in principle all arise biosynthetically from 9-hydroxystrictosidine (Scheme 9). Because 6-hydroxytryptamine has not been isolated from *C. roseus* and strictosidine (**5**) itself was efficiently incorporated into vindoline (**25**),[43] the introduction of OH and methylation at that position appears to occur after the formation of strictosidine (**5**), because the pathway leading from strictosidine (**5**) to vindoline (**25**) involves the

11-dehydroxy compound tabersonine.[68] Work on the enzymatic biosynthesis of vindoline (**25**) clearly supports this line and rules out the involvement of the hydroxylated strictosidine derivative in *C. roseus* cell suspensions and in the plant.[69–71]

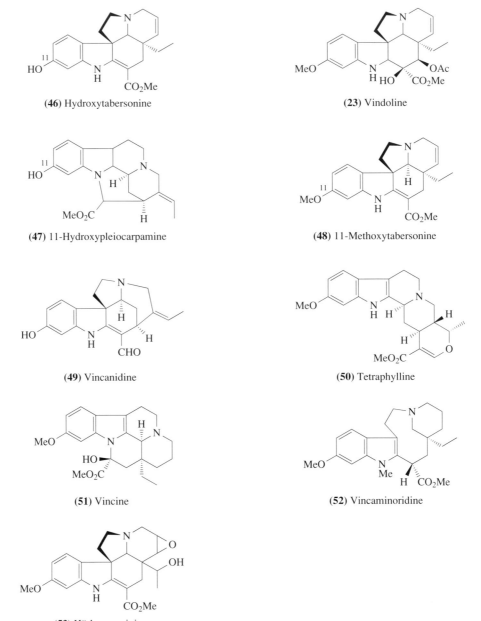

(**46**) Hydroxytabersonine

(**23**) Vindoline

(**47**) 11-Hydroxypleiocarpamine

(**48**) 11-Methoxytabersonine

(**49**) Vincanidine

(**50**) Tetraphylline

(**51**) Vincine

(**52**) Vincaminoridine

(**53**) Hörhammerinine

Scheme 9

Strictosidine synthase does not accept *N*-methyltryptamine as a substrate. Again, there are many occurrences of N^α-methylated indole alkaloids in plants, for example in the Loganiaceae family (e.g., *Strychnos*) and in Apocynaceous plants (e.g., *Rauwolfia*). Although *N*-methylstrictosidine (=dolichantoside)[72,73] occurs in *Strychnos gossweileri* plants, it is unlikely that a special enzyme from *Strychnos* would combine secologanin (**4**) with *N*-methyltryptamine. It seems more probable that a subsequent *N*-methylation takes place after the stage of strictosidine (**5**), as *N*-methylation is a characteristic enzyme reaction in alkaloid biosynthesis that is observed for many alkaloid classes. In *Rauwolfia serpentina* plants and cell cultures, N^α-methylated alkaloids are also common[74,75] and the *Rauwolfia* alkaloid ajmaline (**8**) is a typical example of this group of compounds.

Since nearly all the enzymes involved in the ajmaline biosynthetic pathway are known[76] starting from tryptamine (**2**) and secologanin (**4**), the participation of *N*-methylstrictosidine can be ruled out with certainty. Enough evidence has been accumulated showing that the N^{α}-methylation takes place at the end, not at the beginning of ajmaline (**8**) formation.[77,78]

Strictosidine synthase is also completely inactive towards L-tryptophan (**1**). The above-mentioned *Rhazya* and also *Adina* alkaloids bearing a carboxy group at the C-5 position, including 5-carboxy-strictosidine (**53**), which co-occur with strictosidine (**5**), cannot be expected to be formed directly by the synthase described here. Obviously another enzyme must be postulated for the biosynthesis of such alkaloids[63,79] or carboxylation at C-5 must occur after the formation of strictosidine (**5**) or later pathway intermediates. Both possibilities are still speculative and need to be proven experimentally. Scheme 10 illustrates the non-involvement of 5-carboxystrictosidine (**54**) in the biosyntheses of 5-carboxymonoterpenoid indole alkaloids.

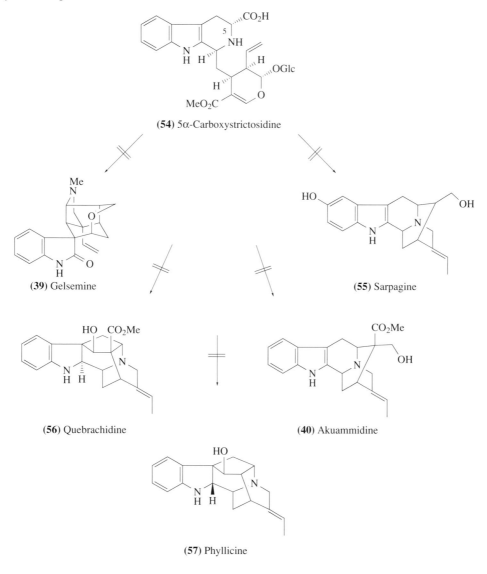

(**54**) 5α-Carboxystrictosidine

(**39**) Gelsemine

(**55**) Sarpagine

(**56**) Quebrachidine

(**40**) Akuammidine

(**57**) Phyllicine

Scheme 10

It also should be pointed out that lack of acceptance of dopamine rules out the proposition that this enzyme is involved in the biosynthesis of alkaloids of the ipecac group, e.g., formation of deacetylipecoside and deacetylisoipecoside.[80,81]

4.05.5.1 Further Purification of Strictosidine Synthase Isoforms

Repetition of the work on *C. roseus* strictosidine synthase was undertaken several years later.[82] Starting also with cell suspension culture material of *C. roseus* and applying Phenyl-Sepharose

CL-4B chromatography, followed by Sephacryl S-200 and Mono Q anion exchange and Mono P chromatofocusing, respectively, the separation of the synthase into several isoforms was confirmed. In this series of experiments, six isoforms of the enzyme were detected with specific activities between 400 and 600 nKat mg^{-1} protein, suggesting high enzyme purity. The number of these multiple forms nearly correlates with the seven isoforms reported several years earlier.[63,83] Using this protocol for enzyme enrichment, the purification of the single isoforms could be enhanced to values between 3000- and 4000-fold. At this stage of extensive purification only microgram amounts of protein (12–100 μg) remained and were obviously homogeneous (with one exception), as proved by sodium dodecylsulfate polyacrylamide gel electrophoresis (SDS–PAGE) and silver staining. In contrast, native PAGE revealed that the isoforms were not as homogeneous as SDS–PAGE has indicated. Because limited amounts of enzyme isoforms were available, only a few investigations were undertaken to characterize these forms in more detail, e.g., the pH dependency was only determined for one form and no isoelectric points were described which would have allowed a better comparison and an assignment of the different isolated forms. However, the K_M value determined for tryptamine (**2**) was much lower (9 μmol L^{-1}) and did not vary much, in contrast to the earlier determination, and V_{max} values were measured between 150 and 210 nKat mg^{-1}. The authors believe that the last-mentioned variation is not significant and that all isoforms probably exhibit the same V_{max} values. Moreover, weak product inhibition was observed, but the substrate inhibition (by tryptamine) described above was not found for these particular enzyme preparations. Although it was known that strictosidine synthase is a glycoprotein, its separation into the six forms allowed the investigation of the glycosylation of each single form. All isoforms reacted in a lectin-based glycan-differentiation kit with *Galanthus nivalis* agglutinin and it was concluded that all forms represent glycoproteins. Reaction with the agglutinin indicated the presence of terminal bound mannose residues, which could not be verified by experiments with endoglycosidase H performed by the same group.[82]

Isolation of larger amounts of the single enzyme should allow more detailed analyses concerning this point and also allow one to obtain more knowledge of the amino acid sequences. Only limited data on the N-terminal sequence of two of the multiple forms are available, indicating that at least the residues 2–9 are identical. At this experimental stage, it seems that more detailed knowledge of the properties and especially of the physical and chemical characteristics of these isoforms is necessary; such data can only be obtained after complete purification of the enzyme to real homogeneity.

When the multiple forms were detected for the first time in cultured *C. roseus* cells, the question was posed concerning their distribution in other material. As mentioned before, the isoforms were also present in leaves of *Catharanthus*, not being a characteristic of the dedifferentiated cell suspension only.[63,83] The occurrence of the aforementioned six isoforms was later investigated and compared in different plant sources, cell suspension cultures, leaves, seedlings, roots, and hairy roots, and all of them showed the presence of the six isoforms. Hence the physiological significance of multiple forms of strictosidine synthase is important, but only speculation is possible. Careful experiments which are difficult to perform are necessary to yield reproducible enzyme activities. There were significant variations in the isolated activities of single isoforms when the same material (cell suspensions) was used in two sets of experiments; this makes the interpretation of data difficult. A discussion regarding the ratio of the different synthases is therefore not possible. That specific enzyme forms occur in specific tissues or during different developmental stages can probably be ruled out on the basis of the data presented. More work is needed to clarify the physiological function of expression of strictosidine synthase isoforms.

4.05.5.2 Subcellular Localization of Strictosidine Synthase

Relatively little experimental work has been carried out on the subcellular localization of strictosidine synthase, strictosidine glucosidase(s), and tryptophan decarboxylase. Isolated and purified protoplasts and vacuoles of cell suspension cultures of *C. roseus* were used to determine the activities of the first two enzymes, respectively. In these experiments, strictosidine synthase was not detected inside the vacuoles and only very small amounts of the appropriate specific glucosidase activity ($\sim$0.1% of the total activity that could arise from this enzyme of protoplast preparations) was associated with the vacuoles.[84] The activity of acid phosphates could be used in these experiments as a vacuolar-marker protein. The results suggested that the synthase and strictosidine glucosidase should have an extravacuolar localization as it has the acid phosphatase and unspecific glucosidases, respectively. Linear sucrose density gradient fractionation of lysed protoplasts of *Catharanthus* leaves showed that both tryptophan decarboxylase and strictosidine synthase could be cytoplasmic enzymes.[85] Both enzymes have also been detected in roots, hypocotyls, and cotyledons of developing seedlings of the same plant and both enzyme activities appeared not to be regulated by light during the process of germination.[86] Similar work has been published yielding information on the compartmentation of these enzymes in suspensions of cultured cells of *C. roseus* and the indole alkaloid-delivering plant *Tabernaemontana divaricata*.[87]

Again, comparison of enzyme activities with intra- and extravacuolar marker enzymes isolated from protoplasts and vacuoles, respectively, were made. α-Mannosidase was used as a vacuolar marker and malate dehydrogenase as a cytoplasmic enzyme.[88,89] From the ratios of enzyme activities measured, it was concluded that tryptophan decarboxylase is a cytoplasmic enzyme and strictosidine synthase is active inside the vacuoles of *C. roseus* and *T. divaricata* cells.[87] Unfortunately, for strictosidine glucosidase the enzyme activity was not compared with that of unspecific glucosidase activities and it seems that it can be detected vacuole-associated as well as outside this compartment. The vacuole-associated enzyme activity is suggested to be bound to the exterior of the tonoplast. From simple centrifugal fractionation experiments it was concluded that on the one hand strictosidine synthase is a soluble enzyme, as demonstrated during the 1980s. On the other hand, the authors believe that strictosidine glucosidase occurs at least in *C. roseus*, in membrane-bound and soluble forms. In *T. divaricata* cells three different soluble strictosidine glucosidases exist[90] but their subcellular localizations are not yet known.

4.05.5.3   Application of Strictosidine for Biomimetic Syntheses

The substrate specificity of strictosidine synthase is interesting from a different point of view. Strictosidine (**5**) not only serves as a biosynthetic precursor of a broad range of pharmaceutically valuable alkaloids, but it also should be an excellent synthon for the generation, chemically or by biomimetic syntheses, of new derivatives some of which may possess outstanding biological activities. If such syntheses can be optimized, such as the biomimetic synthesis of heteroyohimbine alkaloids described by Brown's group in the late 1970s,[91] then strictosidine (**5**) will be needed in large amounts. Using the other substrates for the synthase discussed previously, novel hydroxylated or fluorinated alkaloids could become available for pharmacological testing.

Scheme 11 summarizes some of the approaches already developed for biomimetic strictosidine (**5**) transformations into various (a) new and (b) known alkaloids:

(i)   the generation of the heteroyohimbine isomers ajmalicine (**9**), tetrahydroalstonine (**36**), and 19-epi-ajmalicine (**34**) as reported many years ago;[91–93]
(ii) formation of the N-analogues of heteroyohimbines[93–95] (e.g., (**58**));
(iii) formation of substituted N-analogues of the heteroyohimbine type;[96] and
(iv) generation of the nacycline skeleton[96,97] (**59**).

4.05.6   PREPARATIVE BIOSYNTHESIS OF STRICTOSIDINE

Immobilization of the enzyme proceeded as follows. For preparative transformations of strictosidine (**5**) into both novel and known alkaloids, large amounts of the glucoalkaloid would be required. One interesting approach yielding gram quantities of strictosidine (**5**) is the immobilization of the synthase.[83,98] For such an experiment, an enzyme preparation which was enriched only 15-fold by ammonium sulfate precipitation and AcA 44 gel filtration (with an 80% recovery) proved to be adequate. This crude protein was immobilized on cyanobromide-activated Sepharose with a 40% yield based on synthase activity. Investigation of the pH tolerance of the coupled enzyme showed this preparation to be more stable than the soluble enzyme at pH 4 and 8.

Other properties of the matrix-bound enzyme were also remarkable. Although the differences in the affinity of the synthases towards the substrates were less pronounced (Table 5), the stability of the immobilized enzyme increased dramatically. It has been calculated that a 325-fold higher stability of strictosidine synthase can be achieved by the immobilization procedure described.[98] Such an enhancement of stability is important for the large-scale preparation of strictosidine (**5**). Employing only a small column (1 × 6 cm, 2 g of Sepharose, 40 nKat of bound enzyme), 6 g of pure strictosidine

(a)

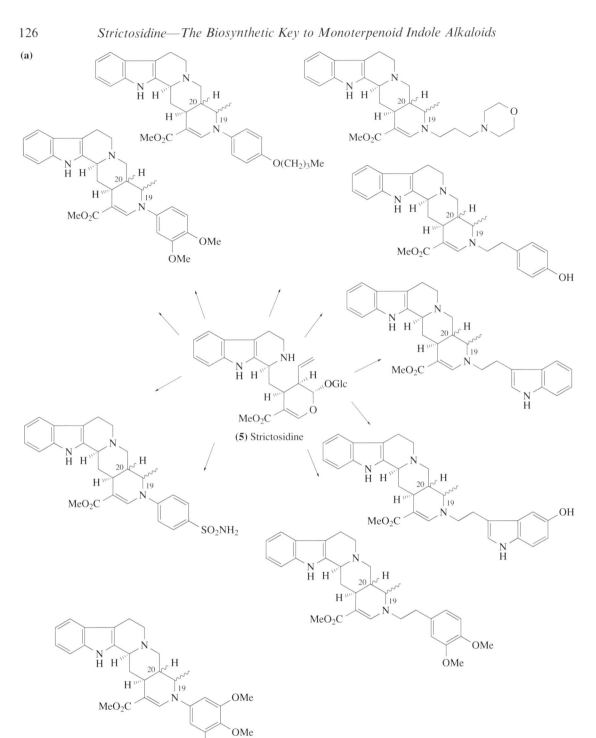

Scheme 11

(**5**) were prepared from tryptamine (**2**) and secologanin (**4**), an amount which already permits a large variety of experiments in the field of biomimetic syntheses. The overall yield of strictosidine formation by this procedure was highly impressive and amounted to 95% of the pure glucoalkaloid. Strictosidine synthase is the first enzyme of the secondary metabolism of plants that has been successfully immobilized and applied to the enzymatic synthesis of an alkaloid in gram quantities.

This is still the best way of preparing large amounts of strictosidine (**5**) in either labeled or unlabeled form. However, at least two other approaches to a sufficient supply of strictosidine (**5**) are available: the chemical synthesis of strictosidine (**5**) and the enzymatic synthesis of strictosidine (**5**) by heterologous expressed strictosidine synthase.

(b)

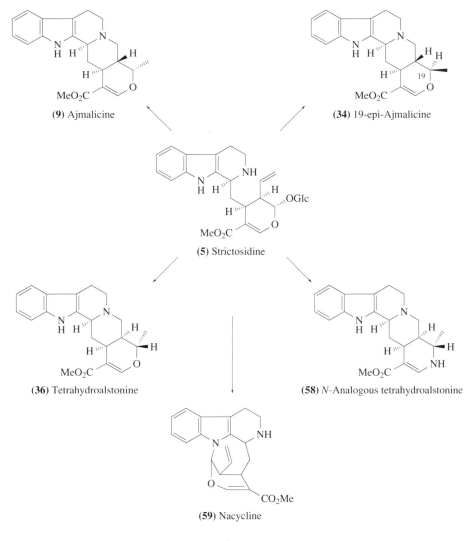

Scheme 11 (cont.)

Table 5 Comparison of the properties of soluble and immobilized strictosidine synthase.

Substrate/property	Soluble synthase	Immobilized synthase
Tryptamine (2)		
K_M (mmol L^{-1})	2.3	0.9
V_{max} (pKat)	130.0	18.9
Secologanin (4)		
K_M (mmol L^{-1})	3.4	2.1
V_{max} (pKat)	55.0	9.1
Inhibition by 2.5 mmol L^{-1} tryptamine (2) (%)	46	17
Stability at 37 °C	$t_{1/2} = 5$ h	$t_{1/2} = 68$ d
Stability after 2 years at 4 °C		45% of activity retained
Enzyme product	Pure strictosidine	Pure strictosidine

4.05.7 CHEMICAL SYNTHESIS OF STRICTOSIDINE

As far as the chemical formation of strictosidine (**5**) is concerned, the classical procedure using the Pictet–Spengler-like condensation of tryptamine (**2**) and secologanin (**4**) reported in 1969 by Battersby *et al.*[99] is a very simple and useful reaction. For analytical amounts it might be applicable, but it has the great disadvantage of the ∼1:1 formation of an epimeric mixture of 3β-vincoside

(24) and 3α-strictosidine (5). For preparative purposes, this non-stereoselective condensation of tryptamine (2) and secologanin (4) is of little value.

This also seems to be true for another optimized, chemical synthesis, which followed the same principle.[100,101] Taking advantage of the lower stability of vincoside (24), which is most easily transformed into its lactam, the authors reported the preparation of pure strictosidine (5). (This property of instability of vincoside (24) has been used previously to separate radioactive strictosidine (5) from its epimer during studies of the enzymatic formation of (5); see Section 4.05.3.1.)

In this "novel" chemical synthesis of strictosidine (5), the reaction conditions were changed (condensation at 100 °C, 6 h, dilute acetic acid) and the condensation took place with a yield of 89%. Separation of vincoside lactam and chromatographic purification of strictosidine (5) resulted in a final synthesis of 390 mg of the glucoalkaloid, corresponding to a total yield of 36%. Two-dimensional NMR methods were used to determine all ^{1}H and ^{13}C NMR chemical shifts, the ^{1}H–^{1}H and ^{1}H–^{13}C coupling constants, and the ^{1}H NOE interactions for strictosidine (5).[101] The 3α S-configuration at the C-3 chiral center has thus been confirmed independently by spectroscopic data. Obviously this success in synthetic strictosidine formation cannot compete with the enzymatic synthesis catalyzed by the immobilized enzyme from *C. roseus* cell culture, a technique that resulted in a 95% formation of pure strictosidine (5) compared with a 36% yield in the chemical synthesis. Nevertheless, the available amounts of strictosidine synthase from that particular cell culture might be still insufficient if economically acceptable amounts of enzyme activity are required (e.g., if new biologically valuable alkaloids or their analogues are needed in large amounts).

The isolation of strictosidine synthase from physiologically highly active *Rauwolfia* cell cultures, the sequencing of the enzyme, detection of the responsible gene, and its heterologous expression could lead to the desired amounts of strictosidine (5) and so to its final biotechnological applications.

4.05.8   STRICTOSIDINE SYNTHASE FROM *RAUWOLFIA SERPENTINA* CELL SUSPENSION CULTURES

Cell suspension cultures of the Indian medicinal plant *Rauwolfia serpentina* were known to produce not only a great variety[75] of monoterpenoid indole alkaloids of different classes, but also extremely high concentrations of particular alkaloids, which have not so far been observed in plant cell suspensions.[102,103]

Moreover, work on the biosynthesis of *Rauwolfia* alkaloids performed during the 1980s revealed excellent enzyme activities for various enzymes of ajmaline biosynthesis. The impressive growth characteristics of these *Rauwolfia* cultures allowed the generation of fresh cell material in amounts up to about 500 g (48 g dry weight) per liter of nutrition medium, thus providing a rich source for enzyme isolation. These cells were also expected to be efficient for the isolation and characterization of strictosidine synthase, although many reports had followed the original ones of synthase isolation from plant cell cultures[104–107] also yielding low enzyme activity levels.

C. roseus produced multiple enzyme forms, which were not easy to handle and to purify. From a range of cell suspension culture systems the *Rauwolfia* cells were finally found to be the most promising for preparative enzyme production and isolation. Compared with the previously reported activities of strictosidine synthase in *C. roseus* cultured cells, *Rauwolfia* suspensions contained about 20-fold enzyme activity. In order to optimize the system, its growth parameter and time course for strictosidine synthase formation were studied in detail.

Cells of *R. serpentina* were grown in Linsmaier and Skoog medium,[108] resulting in a maximum cell yield after 10 d of cultivation. Optimum strictosidine synthase activity appeared, however, after 12 d of cultivation, indicating that maximal enzyme activity is reached in the stationary phase. This observation has also been described for other enzymes involved in the biosynthesis of *Rauwolfia* alkaloids,[109] especially in the biosynthetic pathway of ajmaline (8).

Starting with 2.5 kg of fresh cells (corresponding to 4.8 g of crude protein), the synthase was obtained after five steps of purification as a pure enzyme in an impressive overall yield of 3.5% (1.9 mg of enzyme).[110] This enzyme preparation was rigorously proven for its purity. When the synthase was investigated by isoelectric focusing in the pH range 4–5 it showed a single band at pH 4.5, suggesting a homogeneous enzyme. Under these conditions, proteins with as little as 1% difference in pH would have been resolved. In fact, analytical gel permeation chromatography using TSK G 3000 SM also gave a single peak, which indicated that the described purification procedure yields an apparently pure protein.

It is obvious that strictosidine synthase is present in the *Rauwolfia* cells, forming about 0.1% of the total soluble protein. In sharp contrast to the enzyme from *C. roseus* cell suspensions, the enzyme from *Rauwolfia* does not split into multiple forms or isoforms. For the first time strictosidine-forming enzymes were completely purified from two different sources, allowing a direct comparison of the properties of both proteins (Table 6).

Table 6 Comparison of strictosidine synthase properties of homogeneous preparations from *C. roseus* (isoform III) and from *Rauwolfia serpentina*.[110]

	Strictosidine synthase	
Property	From C. roseus (*isoform III*)	From R. serpentina
pH optimum	6.8	6.5
Temperature optimum ($^\circ$C)		45
Isoelectric point	4.5	4.5
Relative molecular mass (kDa)		
By gel filtration	31	30
By SDS–PAGE	41.5	35
K_M (*mmol* L^{-1})		
Tryptamine	1.9	4
Secologanin		4
K_{cat} (s^{-1})	3.2	
Specific activity (nKat mg^{-1})	5.9[62]	184[110]
Stability, $t_{1/2}$ at 37 $^\circ$C	5 h	36 d
Inhibition by tryptamine	Yes	Yes
Substrate specificity	High	High
Carbohydrate content	Yes	5.3%

Considering the purification of strictosidine synthase from *Rauwolfia* described in this work, it can be seen from Table 6 that the enzymes from both sources exhibit very similar properties. The isoelectric points are identical and the pH optimum differs only slightly by 0.3 units. The relative molecular mass determined by gel filtration is practically the same. The synthases from *C. roseus* and *R. serpentina* both display the pronounced high substrate specificity and at high tryptamine concentrations strong substrate inhibition can be observed. It was this substrate inhibition that did not allow the determination of the real K_M value of the second substrate, secologanin (**4**). However, when the tryptamine concentration was fixed at 0.9 mmol L^{-1}, the K_M of secologanin (**4**) corresponded to 4 mmol L^{-1}. As one would expect, the sole product of catalysis is the 3α-configuration glucoalkaloid strictosidine (**5**). Importantly, the formation of the appropriate 3β-isomer vincoside (**24**) has never been observed with any enzyme preparation so far isolated. A careful investigation of both pure proteins revealed a carbohydrate content showing clearly that both enzymes are glycoproteins. In the case of the *Rauwolfia* enzyme hydrolysis, sugar derivatization and gas chromatography were applied to determine the sugar content. When both enzymes were subjected to periodate oxidation their catalytic activity was retained at 100%, suggesting that the intact carbohydrate moiety is not necessary for the catalytic activity of the synthase. This experiment has important implications for further work on the synthase and especially for the expression of the enzyme in heterologous systems. Both enzymes must be similar in structure because antibodies directed against the enzyme from *C. roseus* cells cross-reacted well with that from *R. serpentina*. Synthase isolated from Rubiaceae sources did not cross-react.[106]

The turnover number of the *Rauwolfia* synthase is $\sim$270 and in an immobilized stage it still showed greater stability than the *Catharanthus* synthase. Obviously the *Rauwolfia* enzyme is more suitable for the preparative synthesis of strictosidine (**5**) and therefore seems to be the enzyme of choice for the study of the biotechnological application of alkaloid biosynthesis and secondary metabolism.[111]

The purification of the enzyme to homogeneity was a breakthrough in the investigation of the structure of this important protein, leading to its molecular application by isolating the corresponding gene and to elucidating its heterologous expression. Establishing the protocol for enzyme purification was a prerequisite for the studies of molecular genetic analyses of this important protein carried out by the groups of Zenk and Kutchan.[111]

4.05.9   GENETIC ANALYSES OF STRICTOSIDINE SYNTHASE

After isolation of homogeneous strictosidine synthase, the determination of its partial amino acid sequences became possible for the first time. Because the N-terminus of the homogeneous synthase was blocked, the enzyme was digested with TPCK-trypsin.[112] Eight tryptic peptides were isolated, purified, and sequenced and were the prerequisites for the subsequent cloning procedure of strictosidine synthase.[66] From the sequence segment of one of these peptides, an oligodeoxynucleotide 17 nucleotides in length was synthesized in three subsets. Indirect RNA sequencing primed by these subsets of oligodeoxynucleotide mixtures was performed which finally yielded a hybridization probe with 52 nucleotides with the structure 5′-TCCATAATT/CTGT/CTGGACACCTCTGTCATCA TATAAGGTGCTAACATCGGTGA-3′.

4.05.9.1   Cloning of the Strictosidine Synthase Gene

Using this oligonucleotide for hybridization with poly (A$^+$) RNA from *R. serpentina* produced an ~1200–1300 base nucleotide which migrated as a single band, suggesting the specificity of this particular oligonucleotide. Screening of the established λgt 11 cDNA library of *R. serpentina* cells gave a positive clone from which the λDNA could be isolated. From that, an insert ~1200 base pairs in length was isolated and subcloned in pUC18. In some of the pUC18 subclones the strictosidine synthase cDNA could be clearly identified. The identified sequence was 1177 base pairs in length. It contained an open reading frame of 1032 base pairs, which corresponds to 344 amino acids, the appropriate polypeptide having a molecular mass of 38.118 kDa. When the nucleotides were translated into the amino acid sequence the tryptic peptides originally obtained from the sequencing of strictosidine synthase could be detected as illustrated in Figure 1.

This finding strongly suggested that the nucleotide sequence obtained was that of the cDNA clone of the synthase. At the nucleotide positions 331–339 a putative carbohydrate attachment site (Asn–Ser–Thr) could be recognized. As discussed earlier, strictosidine synthase contains 5.3% carbohydrates and the detection of this attachment site is in line with that observation.

As proof of the identity of the clone being that of strictosidine synthase, the expression of active enzyme would be the final evidence. In a first attempt, the clone was expressed in the *Escherichia coli* strain DH5. Analysis of the protein extracts of the bacteria by immunoblotting with antibodies directed against strictosidine synthase revealed a slightly smaller protein than expected for the synthase. Since bacterially produced proteins do not contain carbohydrates, the protein detected here might correspond to the appropriate carbohydrate-deficient enzyme. This protein was absent in bacteria containing, for example, empty vectors, so indicating its specific origin from the cloned sequence. As a next step, the expression of enzymatically active enzyme in a heterologous system, such as the *E. coli* DH5 strain, was investigated[113] starting with the construction of an efficient expression vector.

4.05.9.2   Heterologous Expression of the Enzyme

The described cDNA of strictosidine synthase contained, in addition to the 1032 base pair open reading frame, 120 bases at the 3′-end and 60 bases at the 5′-flanking region. In order to increase the total amount of enzyme produced in the bacterium the 5′ *Bal I*/*Eco RI* fragment was replaced with a synthetic oligodesoxynucleotide. This nucleotide contained the ribosome-binding site of the β-galactosidase gene together with the ATG sequence of the reading frame, which was lost during *Bal I* digestion. Then the 3′-flanking region was digested with exonuclease III to within 32 base pairs of the stop codon in order to remove the poly A tail for stabilization of the mRNA. As depicted in Figure 2, the cDNA and the synthetic adapter were ligated into the *Eco RI*/*Hind III* restricted pKK 223-3 vector, which finally resulted in the construction of the expression vector pKSS1.[113]

A whole series of experiments based on different vector constructs, bacterial strains, and culture conditions were then performed to optimize the expression of active strictosidine synthase. Measurement of the synthase activity in the soluble protein fraction of the different *E. coli* strains gave different results depending on the applied vector and culture growth. As can be seen from Table 7, pKSS1 vector in *E. coli* DH5 produced about half the amounts of enzyme compared with pUC18-

```
   1 CTA CTA GTT GAG CTC CAC TCA TCG TTA AAT CCC AGT CCT TCG TCC   45

  46 GTC TGT CCC CCT ACT ATG GCC AAA CTT TCT GAT TCG CAA ACT ATG   90
                  Met Ala Lys Leu Ser Asp Ser Gln Thr Met
  91 GCA CTG TTC ACC GTC TTC CTT CTT TTC CTC TCC TCT TCG CTC GCT  135
     Ala Leu Phe Thr Val Phe Leu Leu Phe Leu Ser Ser Ser Leu Ala
 136 CTC TCC TCT CCA ATC TTG AAA GAG ATT TTG ATT GAG GCT CCT TCC  180
     Leu Ser Ser Pro Ile Leu Lys Glu Ile Leu Ile Glu Ala Pro Ser
 181 TAT GCC CCC AAT TCC TTC ACC TTC GAC TCA ACC AAC AAA GGG TTC  225
     Tyr Ala Pro Asn Ser Phe Thr Phe Asp Ser Thr Asn Lys Gly Phe
 226 TAC ACC TCC GTC CAA GAT GGC CGA GTT ATC AAG TAC GAA GGA CCC  270
     Tyr Thr Ser Val Gln Asp Gly Arg Val Ile Lys Tyr Glu Gly Pro
 271 AAC TCC GGT TTC GTC GAC TTC GCC TAT GCA TCT CCC TAC TGG AAC  315
     Asn Ser Gly Phe Val Asp Phe Ala Tyr Ala Ser Pro Tyr Trp Asn
 316 AAA GCG TTC TGT GAG AAC AGC ACA GAT GCA GAG AAA AGA CCC TTG  360
     Lys Ala Phe Cys Glu Asn Ser Thr Asp Ala Glu Lys Arg Pro Leu
 361 TGT GGG AGG ACA TAT GAT ATT TCA TAT AAC TTG CAA AAC AAC CAG  405
     Cys Gly Arg Thr Tyr Asp Ile Ser Tyr Asn Leu Gln Asn Asn Gln
 406 CTT TAC ATT GTT GAT TGC TAT TAT CAT CTT TCT GTG GTT GGT TCT  450
     Leu Tyr Ile Val Asp Cys Tyr Tyr His Leu Ser Val Val Gly Ser
 451 GAA GGT GGG CAT GCT ACC CAA CTC GCC ACC AGC GTT GAT GGA GTG  495
     Glu Gly Gly His Ala Thr Gln Leu Ala Thr Ser Val Asp Gly Val
 496 CCA TTC AAG TGG CTC TAT GCA GTA ACA GTT GAT CAG AGA ACT GGG  540
     Pro Phe Lys Trp Leu Tyr Ala Val Thr Val Asp Gln Arg Thr Gly
 541 ATT GTT TAC TTC ACC GAT GTT AGC ACC TTA TAT GAT GAC AGA GGT  585
     Ile Val Tyr Phe Thr Asp Val Ser Thr Leu Tyr Asp Asp Arg Gly
 586 GTC CAA CAA ATT ATG GAT ACA AGC GAT AAA ACA GGA AGA CTA ATA  631
     Val Gln Gln Ile Met Asp Thr Ser Asp Lys Thr Gly Arg Leu Ile
 631 AAG TAT GAT CCC TCC ACC AAA GAA ACA ACA CTA CTG TTG AAA GAG  675
     Lys Tyr Asp Pro Ser Thr Lys Glu Thr Thr Leu Leu Leu Lys Glu
 676 CTA CAC GTT CCA GGT GGC GCA GAA GTC AGT GCA GAT AGC TCC TTT  720
     Leu His Val Pro Gly Gly Ala Glu Val Ser Ala Asp Ser Ser Phe
 721 GTT CTT GTG GCT GAG TTT TTG AGC CAT CAA ATT GTC AAA TAT TGG  765
     Val Leu Val Ala Glu Phe Leu Ser His Gln Ile Val Lys Tyr Trp
 766 CTA GAA GGG CCT AAG AAG GGC ACT GCG GAG GTT TTA GTG AAA ATC  810
     Leu Glu Gly Pro Lys Lys Gly Thr Ala Glu Val Leu Val Lys Ile
 811 CCA AAC CCA GGA AAT ATA AAG AGG AAC GCT GAT GGA CAT TTT TGG  855
     Pro Asn Pro Gly Asn Ile Lys Arg Asn Ala Asp Gly His Phe Trp
 856 GTT TCC TCA AGT GAA GAA TTA GAT GGA AAT ATG CAC GGA AGA GTT  900
     Val Ser Ser Ser Glu Glu Leu Asp Gly Asn Met His Gly Arg Val
 901 GAT CCT AAA GGA ATA AAA TTT GAT GAG TTT GGG AAC ATT CTT GAA  945
     Asp Pro Lys Gly Ile Lys Phe Asp Glu Phe Gly Asn Ile Leu Glu
 946 GTT ATC CCA CTC CCA CCA CCA TTT GCA GGT GAA CAC TTC GAA CAA  990
     Val Ile Pro Leu Pro Pro Pro Phe Ala Gly Glu His Phe Glu Gln
 991 ATT CAA GAG CAT GAT GGT TTG CTG TAC ATT GGA ACC CTG TTC CAT 1035
     Ile Gln Glu His Asp Gly Leu Leu Tyr Ile Gly Tyr Leu Phe His
1036 GGC TCT GTG GGC ATA TTA GTA TAT GAT AAG AAG GGA AAT TCT TTT 1080
     Gly Ser Val Gly Ile Leu Val Tyr Asp Lys Lys Gly Asn Ser Phe
1081 GTT TCA AGT CAT TAA ATT TTC CAC GAA CCA GAT GGG TTT TGT TTT 1125
     Val Ser Ser His
1126 TGA TTT GTA TAA CAC TCT TTA AAG GTT TTG TAT TCG AAT CAC GTC 1170

1171 ATC TCA GCC TCC AGG AAT AAG AAA AAA GCA GCA GAA TAA CTT CTC 1215
```

Figure 1 Nucleotide sequence of the cDNA clone of strictosidine synthase from *R. serpentina*. The underlined sequences are those obtained from the amino acid sequencing of tryptic peptides from strictosidine synthase (source Kutchan *et al*[66]).

4 expressed in *E. coli* SG935. When pKSS1 was introduced into *E. coli* SG935, strictosidine synthase production was enhanced about fivefold as observed with the same vector in *E. coli* DH5. Under optimum growth conditions (absorbance at $\lambda = 590$ nm (A_{590}) to 12 instead of 2), a 69-fold enzyme activity was determined (Table 7).

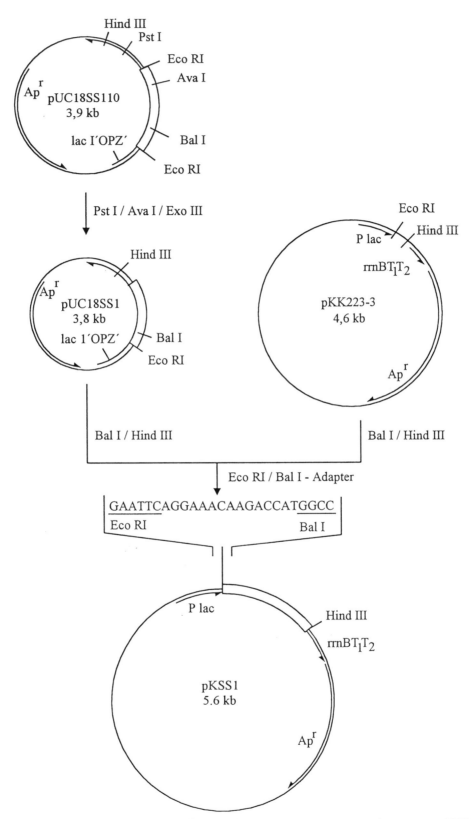

Figure 2 Strategy for the construction of the strictosidine synthase expression vector pKSS1 (source Kutchan[113]).

Table 7　Heterologous expression of strictosidine synthase in *E. coli.*

Vector system	Enzyme activity (nKat L^{-1})	Strictosidine synthase (% of soluble protein)
pKSS1/DH5	1.93	0.0008
pUC18-4/SG935	4.91	0.0020
pKSS1/SG935	9.72	0.0040
pKSS1/SG935[a]	138.00	0.0630

[a]Grown to absorbance of $A_{590} \approx 12.0$ instead of 2.0.

When tryptamine (**2**) and secologanin (**4**) were incubated in the presence of isolated *E. coli* protein or even with the intact *E. coli* cells, strictosidine (**5**) was synthesized and showed chromatographic and spectroscopic properties which were in complete agreement with those of a reference sample. Thus, the gene from *R. serpentina* cell suspension cultures responsible for the formation of strictosidine synthase has been isolated, identified, and expressed in *E. coli* systems delivering the active enzyme. When the enzyme production (per hour, per liter) of the *Rauwolfia* culture and the *E. coli* were compared, the latter produced about 20 times more strictosidine synthase, indicating the high efficiency of the genetically altered bacterial system.

Whole *E. coli* cells containing the pKSS1 vector were also tested for their capacity to catalyze the condensation of tryptamine (**2**) and secologanin (**4**). In fact, in equimolar concentrations (15 mmol L^{-1}), both precursors were quantitatively converted into strictosidine (**5**). Analysis of the soluble bacteria protein, the membrane-bound protein, and the surrounding medium of the bacterial suspension demonstrated that the entire enzyme amounts were present within the cells (soluble enzyme fraction) and that strictosidine (**5**) accumulated within the nutrition medium. Addition of the substrates might be done either during the exponential growth or during the stationary phase of *E. coli*. During accumulation, the glucoalkaloid (**5**) is not modified and obviously not degraded. Because of these results it seemed certain that with this particular system, unlimited quantities of strictosidine (**5**) could be synthesized rapidly for the first time. However, the continuous effort and success of Kutchan's group[114] led to a further most significant optimization of strictosidine synthase production by heterologous systems.

Three additional systems for heterologous expression were developed based on bacteria, yeast, and insect cell systems. Although the *E. coli* bacteria system was already very efficient in taking up both precursors, converting them into strictosidine (**5**), followed by secretion of the glucoalkaloid, the system was further modified. In this case the vector pJUB5, which carries the gene for ubiquitin[115] in front of the strictosidine synthase gene, was used. Under these conditions, 3% of the total soluble protein of the *E. coli* bacteria strain SG 935 was represented by the synthase. As shown in Table 8, as much as 6600 nKat L^{-1} of enzyme activity was produced. In contrast, the yeast (*Saccharomyces cerevisiae*) system with the vector construct pEVP11 was less productive. Only 0.005% of its soluble protein comprised active strictosidine synthase. The insect cells *Spodoptera frugiperda* were much more efficient and the best results with the latter were obtained when the AcMNPV vector was introduced. More than 2% of the soluble protein of the cells consisted of enzymatically active strictosidine synthase (Table 8).

Table 8　Efficiency of three of the best systems described for the heterologous expression of enzymatically active strictosidine synthase.

Organism	Vector construct	Enzyme activity (nKat L^{-1})	Strictosidine synthase (% of soluble protein)
Escherichia coli	pJUB5/SG935	6600	3
Saccharomyces cerevisiae	pEVP11/DH484	90	0.005
Spodoptera frugiperda	AcMNPV/Sf9	900	>2

With the advent of these systems for heterologous expression of active strictosidine synthase, production of this key enzyme in large amounts became attainable for the first time. On the basis of enzyme activity, the above optimized *E. coli* strain products 6600 nKat L^{-1}. The enormous advantage of the *E. coli* system would now allow the unlimited production of the enzyme. Applying the developed procedures for enzyme immobilization, the glucoalkaloid strictosidine (**5**) would also be available in almost any quantities desired.

The cloning of the synthase gene and its heterologous expression in microorganisms and eukaryote cell culture represents a major breakthrough for further physical and biochemical characterization of the enzyme and for its final biotechnological application. For instance, the enzyme mechanism can be investigated in great detail and biomimetic syntheses starting with strictosidine (5) and leading to novel alkaloid structures or pharmacologically interesting alkaloids have become attainable for the first time. Gene analyses of the strictosidine synthase gene in different plants and plant cell cultures can be performed, the gene structures will be comparable in the future and a taxonomic application based on this gene should also become possible.

Meanwhile, the gene for strictosidine synthase, *str 1*, from the *Rauwolfia* species *R. serpentina* and *R. mannii* has been isolated and partly characterized.[116] Analysis of *str 1* demonstrated that it does not contain introns. Therefore, it belongs to the 17% of plant genes known which are intronless.[117] Moreover, *str 1* showed a 100% homology over 1180 base pairs when compared with both *Rauwolfia* species. This gene could also be used for taxonomic purposes (Figure 3).

```
  1 MAKLSDSQTM  ALFTVFLLFL  SSSLALSSPI  LKEILIEAPS  YAPNSFTFDS
      MMAV      FFMFFL LSS    SSS          K F S        A
 51 TNKGFYTSVQ  DGRVIKYEGP  NSGFVDFAYA  SPYWNKAFCE  NSTDAEKRPL
      D                         T           F           P
101 CGRTYDISYN  LQNNQLYIVD  CYYHLSVVGS  EGGHATQLAT  SVDGVPFKWL
      D         YK S M       GH   C  K      Y          Q
151 YAVTVDQRTG  IVYFTDVSTL  YDDR  GVQQIM DTSDKTGRLI  KYDPSTKETT
                     SI      H   SPEGVEE  N    R   M
201 LLLKELHVPG  GAEVSADSSF  VLVAEFLSHQ  IVKYWLEGPK  KGTAEVLVKI
                  I    G      V     NR                S F   T
251 PNPGNIKRNA  DGHFWVSSSE  ELDGNMHGRV  DPKGIKFDEF  GNILEVIPLP
       S                      GQ  S      VSR   G                Q
301 PPFAGEHFEQ  IQEHDGLLYI  GTLFHGSVGI  LVYDKKGNSF  VSSH*
       YE                    S  S  S       DHD KG   N YVSQLVIN*
```

Figure 3 Comparison of the amino acid sequences as single-letter code for strictosidine synthases from *R. serpentina* (top sequence) and from *C. roseus* (bottom sequence). Sequences were generated by translation of the nucleotide sequences of the appropriate cDNAs. For the sequence of *C. roseus* only those parts which differ from that of *R. serpentina* are illustrated.

4.05.10   CHEMOTAXONOMIC CONSIDERATIONS BASED ON THE STRICTOSIDINE SYNTHASE GENE *Str 1*

Applying the polymerase chain reaction (PCR), a comparison of the strictosidine synthase gene, *str 1*, from 10 *Rauwolfia* species has been published.[118] No differences between the 10 genes were detectable, indicating an extremely high conservation of *str 1*. Even when the cDNA was compared with that of *C. roseus*,[119] an impressively high homology of 76% at the nucleotide level was observed. Although extensive chemotaxonomic studies of indole alkaloid-bearing plant species have been made in the past considering the different skeletal types, especially of the families Apocynaceae, Loganiaceae, and Rubiaceae,[120–122] or the iridoid components,[123] it was a challenge to include comparisons at the gene level. Therefore, the cDNA of strictosidine synthase from *R. serpentina* has been used for hybridization experiments with genomic DNA isolated from 34 plant species belonging to eight families.[17] Part of this work is summarized in Table 9.

All *Rauwolfia* species tested showed a high stringency hybridization. In *C. roseus*, however, which of course expresses the gene for strictosidine synthase, no strong hybridization could be observed. In contrast, the genomic DNA from *Vinca minor*, which also contains the strictosidine synthase, did hybridize. The plant *Glycine max* does not biosynthesize monoterpenoid indole alkaloids and certainly has no strictosidine synthase activity, but it also showed strong hybridization. It is therefore not certain that the gene of strictosidine synthase is an acceptable tool for further taxonomic classification of plants producing monoterpenoid indole alkaloids.

4.05.11   DOES STRICTOSIDINE SYNTHASE HAVE A REGULATORY FUNCTION IN ALKALOID BIOSYNTHESIS?

As the precursor of the ~2000 monoterpenoid indole alkaloids, strictosidine (5) occupies a central biosynthetic role. Glucosides or glycosides of plant secondary metabolites are usually final products

Table 9 Survey of plant species from which genomic DNA was used for hybridization experiments with *R. serpentina* strictosidine synthase cDNA (after Kutchan[17]).

Species	Family	Tribe	Hybridization
Rauwolfia canescens	Apocynaceae	*Rauwolfieae*	+
R. chinensis	Apocynaceae	*Rauwolfieae*	+
R. mannii	Apocynaceae	*Rauwolfieae*	+
R. serpentina	Apocynaceae	*Rauwolfieae*	+
R. verticillata	Apocynaceae	*Rauwolfieae*	+
Alstonia scholaris	Apocynaceae	*Plumerieae*	−
Catharanthus ovalis	Apocynaceae	*Plumerieae*	−
C. pusillus	Apocynaceae	*Plumerieae*	−
C. roseus	Apocynaceae	*Plumerieae*	−
C. trichophyllus	Apocynaceae	*Plumerieae*	−
Plumeria sericifolia	Apocynaceae	*Plumerieae*	−
Vinca minor	Apocynaceae	*Plumerieae*	+
Trachomitum venetum	Apocynaceae	*Apocyneae*	−
Carissa carandas	Apocynaceae	*Carisseae*	−
Thevetia ahouai	Apocynaceae	*Cerbereae*	+
Strophanthus candatus	Apocynaceae	*Nerieae*	−
Tabernaemontana divaricata	Apocynaceae	*Tabernaemontaneae*	−
Voacanga africana	Apocynaceae	*Tabernaemontaneae*	+
Gelsemium sempervirens	Loganiaceae	*Gelsemieae*	−
Strychnos colubrina	Loganiaceae	*Strychneae*	+
Cinchona ledgeriana	Rubiaceae	*Cinchoneae*	−
Pausinystalia yohimbe	Rubiaceae	*Cinchoneae*	+
Chiococca bermudiana	Rubiaceae	*Chiococceae*	+
Morinda citrifolia	Rubiaceae	*Morindeae*	+
Mitragyna speciosa	Rubiaceae	*Naucleeae*	−
Nauclea latifolia	Rubiaceae	*Naucleeae*	−
Psychotria orophilia	Rubiaceae	*Psychotrieae*	−
Glycine max	Fabaceae	*Phaseoleae*	+

of biosynthetic pathways, which become water-soluble storage compounds by the glucosylation process.

In sharp contrast, strictosidine (**5**) as a glucoalkaloid is a biogenetic progenitor, not an end product. It is relatively stable but becomes a highly reactive intermediate after removal of the glucose moiety, catalyzed by the appropriate strictosidine glucosidases.[46,90,124,125] At this stage different biogenetic pathways leading to the various alkaloid skeletons are initiated. Because of this key position of the condensing enzyme strictosidine synthase, it can be argued that this enzyme has a regulatory function or at least could be a point of control of alkaloid production and accumulation. Such control could, however, also occur at the preceding enzyme, which is tryptophan decarboxylase, generating tryptamine (**2**) from the amino acid tryptophan (**1**).[126,127] For that reason, several studies were performed to obtain information on regulatory aspects of these enzymes. In such investigations, the time course of alkaloid biosynthesis and tryptophan decarboxylase and strictosidine synthase activities were compared in different plant systems.[104–106]

In a *C. roseus* strain, an increased formation of indole alkaloids following the transfer to a "production medium" is accompanied by increases in tryptophan decarboxylase. Under these conditions, no changes in strictosidine synthase activity have been observed.[126] These results lead to the suggestion that tryptophan decarboxylase is an important factor in the regulation of indole alkaloid biosynthesis. Another study compared different *C. roseus* strains grown in the same nutrition medium and their contents of tryptophan (**1**), tryptamine (**2**), the alkaloids ajmalicine (**9**), and serpentine, and the enzyme activities of tryptophan decarboxylase and strictosidine synthase. By comparing the values obtained from all of the cell strains, no correlation could be observed between enzyme activities and accumulation of alkaloids.[106] When one strain was cultured in three different nutrition media, again no correlation was found between the decarboxylase activity and alkaloid formation.[128] These observations are in contrast with those in which a 12-fold increase of the decarboxylase was measured[126] after the cells had been transferred to an "alkaloid production" medium.[38] This effect led to the assumption that tryptophan decarboxylase is indeed an enzyme which has regulatory function.[126] Also, work on other cell systems, such as *Cinchona ledgeriana* cells transformed by *Agrobacterium tumefaciens*,[105] was concerned with the comparison of the mentioned enzyme activities and alkaloid contents. However, in all these cases the strictosidine synthase activity did not vary very much.

Considerably less research was concentrated on finding a regulatory function in the pathway responsible for the formation of the second substrate of strictosidine synthase, the monoterpene secologanin (**4**). The geraniol-10-hydroxylase, which is the first cytochrome P450-dependent enzyme in that particular biosynthetic line, has been well characterized.[129,130] The activity of this hydroxylase was found to increase in *C. roseus* cells upon transfer to the production medium[131] and addition of secologanin (**4**) to the nutrition medium increased the alkaloid level.[128] From these limited experiments, it can be concluded that the site of regulation of monoterpenoid indole alkaloid biosynthesis is difficult to determine and that the terpenoid pathway leading to secologanin (**4**) might also be an important factor in controlling the alkaloid biosynthesis.

It seems certain that strictosidine synthase, the activities of which are mostly similar in time-course experiments performed with different cell systems, is not a regulatory enzyme. Further research must therefore concentrate on the delineation of regulatory mechanisms of alkaloid biosynthesis in order finally to understand cellular alkaloid formation and to develop opportunities to enhance further the productivity of the plant cell.

ACKNOWLEDGMENTS

The continuous and generous support of the Deutsche Forschungsgemeinschaft (Bonn, Bad-Godesberg, Germany) and of the Fonds der Chemischen Industrie (Frankfurt/Main, Germany) is gratefully acknowledged. We also appreciate very much the help of Heide Höhle and Christina Bolle for typing the text and drawing the structural formulae. We thank Professor W. E. Court (Mold, UK) for linguistic advice in preparing this chapter.

4.05.12 REFERENCES

1. T. W. Southon and J. Buckingham, "Dictionary of Alkaloids," Chapman & Hall, London, 1979, vol. 3.
2. J. B. Hendrickson, in "The Alkaloids," eds. R. H. F. Manske and H. L. Holmes, Academic Press, New York, 1960, vol. 6, p. 179.
3. L. M. Pullan and R. J. Powel, *Neuroscience*, 1992, **148**, 199.
4. J. M. Müller, E. Schlittler, and H. J. Bein, *Experientia*, 1952, **8**, 338.
5. E. Mutschler, "Arzneimittl ewirkungen," 7. Auflage Wissenschaftliche Verlagsgesellschaft, Stuttgart, 1996, p. 152.
6. M. Hesse, "Alkaloid Chemistry," Wiley, New York, 1981, p. 21.
7. W. A. Creasey, in "The Monoterponoid Indole Alkaloids," Supplement to part 4, vol. 24 in the series "The Chemistry of Heterocyclic Compounds," ed. J. E. Saxton, Wiley, Chichester, 1994, p. 715, and references cited therein.
8. W. Steglich, B. Fugmann, and S. Lang-Fugmann (eds.), "RÖMPP Lexikon Naturstoffe," Thieme, Stuttgart, 1997, p. 702.
9. W. A. Creasey, in "Indoles, Part 4, The Monoterpenoid Indole Alkaloids," ed. J. E. Saxton, Wiley–Interscience, Chichester, 1983, p. 723.
10. N. Neuss, in "Indole and Biogenetically Related Alkaloids," eds. J. D. Phillipson and M. H. Zenk, Academic Press, London, 1980, p. 293.
11. P. Chan, Y. Li, Z. Zheng, P. Hu, S. D. Sarbach, and D. Guez, *Eur. J. Pharmacol.*, 1993, **231**, 175.
12. A. van Camp, *J. Am. Pharm. Assoc., Sci. Ed.*, 1960, **49**, 681.
13. N. Neuss, M. Gorman, W. Hargrove, N. J. Cone, K. Biemann, G. Büchi, and R. E. Manning, *J. Am. Chem. Soc.*, 1964, **86**, 1440.
14. G. H. Svoboda, N. Neuss, and M. Gorman, *J. Am. Pharm. Assoc., Sci. Ed.*, 1959, **48**, 659.
15. G. N. Smith, *J. Chem. Soc., Chem. Commun.*, 1968, 912.
16. R. T. Brown, G. N. Smith, and K. S. J. Stableford, *Tetrahedron Lett.*, 1968, **41**, 4349.
17. T. M. Kutchan, *Phytochemistry*, 1993, **32**, 493.
18. M. Hesse, "Indolalkaloide Teil 1, Fortschritte der Massenspektrometrie," Verlag Chemie, Weinheim, 1974.
19. R. B. Woodward, *Angew. Chem.*, 1956, **68**, 13.
20. D. H. R. Barton and T. Cohen, "Festschrift A. Stoll," Birkhauser, Basle, 1957, p. 144.
21. E. Wenkert and N. V. Bringi, *J. Am. Chem. Soc.*, 1959, **81**, 1474.
22. E. Leete, *J. Am. Chem. Soc.*, 1960, **82**, 6338.
23. A. R. Battersby, *Proc. Chem. Soc.*, 1963, 189.
24. D. H. R. Barton, *Proc. Chem. Soc.*, 1963, 293.
25. G. A. Cordell, *Lloydia*, 1974, **37**, 219.
26. S. Escher, P. Loew, and D. Arigoni, *J. Chem. Soc., Chem. Commun.*, 1970, 823.
27. A. R. Battersby, S. H. Brown, and T. G. Payne, *J. Chem. Soc., Chem. Commun.*, 1970, 827.
28. S. Uesato, Y. Ogawa, H. Inouye, K. Saiki, and M. H. Zenk, *Tetrahedron Lett.*, 1986, **27**, 2893.
29. K. M. Madyastha, R. Guarnaccia, and C. J. Coscia, *FEBS Lett.*, 1971, **14**, 175.
30. K. M. Madyastha, R. Guarnaccia, and C. J. Coscia, *Biochem. J.*, 1972, **128**, 34P.
31. A. R. Battersby, *Pure Appl. Chem.*, 1967, **14**, 117.
32. H. Inouye, *Planta Med.*, 1978, **33**, 193.
33. S. Uesato, S. Kanomi, A. Iida, H. Inouye, and M. H. Zenk, *Phytochemistry*, 1986, **25**, 839.

34. S. Uesato, H. Ikeda, T. Fujita, H. Inouye, and M. H. Zenk, *Tetrahedron Lett.*, 1987, **28**, 4431.
35. T. Tanahashi, N. Nagakura, H. Inouye, and M. H. Zenk, *Phytochemistry*, 1984, **23**, 1917.
36. A. R. Battersby, A. R. Burnett, and P. G. Parsons, *J. Chem. Soc., Chem. Commun.*, 1968, 1280.
37. A. I. Scott and S.-L. Lee, *J. Am. Chem. Soc.*, 1975, **97**, 6906.
38. M. H. Zenk, H. El-Shagi, H. Arens, J. Stöckigt, E. W. Weiler, and B. Deus, in "Plant Tissue Culture and its Biotechnological Applications," eds. W. Barz, E. Reinhard, and M. H. Zenk, Springer, Berlin, 1977, p. 27.
39. J. Stöckigt and H. J. Soll, *Planta Med.*, 1980, **40**, 22.
40. J. Stöckigt, J. Treimer, and M. H. Zenk, *FEBS Lett.*, 1976, **70**, 267.
41. J. Stöckigt, H.-P. Husson, C. Kan-Fan, and M. H. Zenk, *J. Chem. Soc., Chem. Commun.*, 1977, 164.
42. J. F. Treimer and M. H. Zenk, *Phytochemistry*, 1978, **17**, 227.
43. J. Stöckigt and M. H. Zenk, *J. Chem. Soc., Chem. Commun.*, 1977, 646.
44. J. Stöckigt and M. H. Zenk, *FEBS Lett.*, 1977, **79**, 233.
45. J. Stöckigt, *Phytochemistry*, 1979, **18**, 965.
46. M. H. Zenk, *J. Nat. Prod.*, 1980, **43**, 438.
47. A. I. Scott, S.-L. Lee, P. De Capite, M. G. Culver, and C. R. Hutchinson, *Heterocycles*, 1977, **7**, 979.
48. R. T. Brown, J. Leonard, and S. K. Sleigh, *Phytochemistry*, 1978, **17**, 899.
49. A. R. Battersby, N. G. Lewis, and J. M. Tippet, *Tetrahedron Lett.*, 1978, 4849.
50. A. H. Heckendorf and C. R. Hutchinson, *Tetrahedron Lett.*, 1977, 4153.
51. C. R. Hutchinson, A. H. Heckendorf, J. L. Staughn, P. E. Daddona, and D. E. Cane, *J. Am. Chem. Soc.*, 1979, **101**, 3358.
52. N. Nagakura, M. Rüffer, and M. H. Zenk, *J. Chem. Soc., Perkin Trans. 2*, 1979, 2308.
53. M. Rüffer, N. Nagakura, and M. H. Zenk, *Tetrahedron Lett.*, 1978, 1593.
54. E. E. von Tamelen, V. R. Haarstadt, and R. L. Orvis, *Tetrahedron Lett.*, 1968, **24**, 687.
55. E. E. van Tamelen and L. K. Oliver, *Bioorg. Chem.*, 1976, **5**, 309.
56. J. Stöckigt, *Tetrahedron Lett.*, 1979, **28**, 2615.
57. T. J. C. Luijendijk, E. van der Meijden, and R. Verpoorte, *J. Chem. Ecol.*, 1996, **22**, 1355, 48a.
58. J. F. Treimer and M. H. Zenk, *FEBS Lett.*, 1979, **97**, 159.
59. N. J. Walton, S. E. Skinner, R. J. Robins, and M. J. C. Rhodes, *Anal. Biochem.*, 1987, **163**, 482.
60. E. J. M. Pennings, R. A. van den Bosch, R. van der Heijden, L. H. Stevens, J. A. Duine, and R. Verpoorte, *Anal. Biochem.*, 1989, **176**, 412.
61. J. F. Treimer and M. H. Zenk, *Eur. J. Biochem.*, 1979, **101**, 225.
62. H. Mizukami, H. Nordlöv, S.-L. Lee, and A. I. Scott, *Biochemistry*, 1979, **18**, 3760.
63. U. Pfitzner and M. H. Zenk, *Planta Med.*, 1989, **55**, 525.
64. R. A. Yates and A. B. Pardee, *J. Biol. Chem.*, 1956, **221**, 757.
65. P. F. Heinstein, S.-L. Lee, and H. G. Floss, *Biochem. Biophys. Res. Commun.*, 1971, **44**, 1244.
66. T. M. Kutchan, N. Hampp, F. Lottspeich, K. Beyreuther, and M. H. Zenk, *FEBS Lett.*, 1988, **237**, 40.
67. B. Gabetta and G. Mustich, in "Spectral Data of Indole Alkaloids," Inverni della Beffa, Milan, 1975.
68. R. B. Herbert, in "Indoles, Part 4, The Monoterpenoid Indole Alkaloids," ed. J. E. Saxton, Wiley–Interscience, Chichester, 1983, p. 1.
69. W. Fahn, E. Laussermair, B. Deus-Neumann, and J. Stöckigt, *Plant Cell Rep.*, 1985, **4**, 333.
70. M. Dethier and V. De Luca, *Phytochemistry*, 1993, **32**, 673.
71. E. De Carolis and V. De Luca, *J. Biol. Chem.*, 1993, **268**, 5504.
72. C. Coune and L. Angenot, *Planta Med.*, 1978, **34**, 53.
73. M. Tits, V. Brandt, J.-N. Wauters, C. Delaude, G. Llabres, and L. Angenot, *Planta Med.*, 1996, **62**, 73.
74. A. M. G. Nasser and W. E. Court, *J. Ethnopharmacol.*, 1984, **11**, 99.
75. J. Stöckigt, A. Pfitzner and J. Firl, *Plant Cell Rep.*, 1981, **1**, 36.
76. J. Stöckigt, in "The Alkaloids," ed. G. A. Cordell, Academic Press, San Diego, 1995, vol. 47, p. 115.
77. J. Stöckigt, A. Pfitzner, and P. J. Keller, *Tetrahedron Lett.*, 1983, **24**, 2485.
78. J. Stöckigt, in "Pharmazie und Drogenkunde," GIT Verlag, Darmstadt, 1988, p. 29.
79. K. T. D. De Silva, D. King, and G. N. Smith, *J. Chem. Soc., Chem. Commun.*, 1971, 908.
80. N. Nagakura, G. Höfle, and M. H. Zenk, *J. Chem. Soc., Chem. Commun.*, 1978, 896.
81. N. Nagakura, G. Höfle, D. Coggiola, and M. H. Zenk, *Planta Med.*, 1978, **34**, 381.
82. A. De Wal, A. H. Meijer, and R. Verpoorte, *Biochem. J.*, 1995, **306**, 571.
83. U. Pfitzner and M. H. Zenk, *Methods Enzymol.*, 1987, **136**, 342.
84. B. Deus Neumann and M. H. Zenk, *Planta*, 1984, **162**, 250.
85. V. De Luca and A. J. Cutler, *Plant Physiol.*, 1987, **85**, 1099.
86. V. De Luca, J. A. Fernandez, D. Campbell, and G. W. Kurz, *Plant Physiol.*, 1988, **86**, 447.
87. H. Stevens, T. J. M. Blom, and R. Verpoorte, *Plant Cell Rep.*, 1993, **12**, 573.
88. H. Masuda, S. Narita, and S. Sugawara, *Phytochemistry*, 1990, **29**, 393.
89. M. Betz, E. Martinoia, D. K. Hincha, J. M. Schmitt, and K.-J. Dietz, *Phytochemistry*, 1992, **31**, 433.
90. T. J. C. Luijendijk, A. Nowak, and R. Verpoorte, *Phytochemistry*, 1996, **41**, 1451.
91. R. T. Brown, J. Leonard, and S. K. Sleigh, *J. Chem. Soc., Chem. Commun.*, 1977, 636.
92. R. T. Brown and J. Leonard, *J. Chem. Soc., Chem. Commun.*, 1979, 877.
93. R. T. Brown, in "Stereoselectic Synthesis of Natural Products," eds. W. Bartmann and E. Winterfeldt, Excerpta Medica, Amsterdam, 1979, p. 62.
94. P. Heinstein, G. Höfle, and J. Stöckigt, *Planta Med.*, 1979, **37**, 349.
95. H. P. Husson, C. Kan-Fan, T. Sevenet, and J.-P. Vidal, *Tetrahedron Lett.*, 1977, **22**, 1889.
96. H. Schübel and J. Stöckigt, Ph.D. Thesis, Ludwig-Maximilians-Universität, Munich, 1986.
97. R. T. Brown, C. L. Chapple, and A. G. Lashford, *J. Chem. Soc., Chem. Commun.*, 1975, 295.
98. U. Pfitzner and M. H. Zenk, *Planta Med.*, 1982, **46**, 10.
99. A. R. Battersby, A. R. Burnett, and P. G. Parsons, *J. Chem. Soc. C*, 1969, 1193.
100. A. Koksis, Z. Pál, Z. Szabó, P. Tetenyi, and M. Varga-Balazs, *Eur. Pat. Appl.*, EP 156267, 1985; *Chem. Abstr.*, 1986, **104**, P149345q.

101. A. Patthy-Lukats, L. Karolyházy, L. F. Szabó, and B. Podanyi, *J. Nat. Prod.*, 1997, **60**, 69.
102. H. Schübel and J. Stöckigt, *Plant Cell Rep.*, 1984, **3**, 72.
103. H. Schübel, C. M. Ruyter, and J. Stöckigt, *Phytochemistry*, 1989, **28**, 491.
104. W. Noe and J. Berlin, *Planta*, 1985, **166**, 500.
105. S. E. Skinner, N. J. Walton, R. J. Robins, and M. J. C. Rhodes, *Phytochemistry*, 1987, **26**, 721.
106. P. Doireau, J. M. Merillon, A. Guillot, M. Rideau, J. C. Chenieux, and M. Brillard, *Planta Med.*, 1987, **53**, 364.
107. U. Eilert, V. De Luca, W. G. W. Kurz, and F. Constabel, *Plant Cell Rep.*, 1987, **6**, 271.
108. E. M. Linsmaier and F. Skoog, *Physiol. Plant.*, 1965, **18**, 110.
109. A. Pfitzner, Ph.D. Thesis, Ludwig-Maximilians-Universität, Munich, 1984.
110. N. Hampp and M. H. Zenk, *Phytochemistry*, 1988, **27**, 3811.
111. T. M. Kutchan, H. Dittrich, D. Bracher, and M. H. Zenk, *Tetrahedron*, 1991, **47**, 5945.
112. K. Beyreuther, K. Adler, E. Fanning, C. Murray, A. Klemm, and N. Geisler, *Eur. J. Biochem.*, 1975, **59**, 491.
113. T. M. Kutchan, *FEBS Lett.*, 1989, **257**, 127.
114. T. M. Kutchan, A. Bock, and H. Dittrich, *Phytochemistry*, 1994, **35**, 353.
115. T. R. Butt, S. Jonnalagadda, P. B. Monia, E. J. Sternberg, J. A. Marsh, J. M. Stadel, D. J. Ecker, and S. T. Crooke, *Proc. Natl. Acad. Sci. USA*, 1989, **86**, 2540.
116. D. Bracher and T. M. Kutchan, *Arch. Biochem. Biophys.*, 1992, **294**, 717.
117. J. D. Hawkins, *Nucleic Acid Res.*, 1988, **16**, 9893.
118. D. Bracher and T. M. Kutchan, *Plant Cell Rep.*, 1992, **11**, 179.
119. T. D. McKnight, C. A. Roessner, R. Devagupta, A. I. Scott, and C. L. Nessler, *Nucleic Acids Res.*, 1990, **18**, 4939.
120. A. J. M. Leeuwenberg, in "Indole and Biogenetically Related Alkaloids," eds. J. D. Phillipson and M. H. Zenk, Academic Press, London, 1980, p. 1.
121. M. V. Kisakürek and M. Hesse, in "Indole and Biogenetically Related Alkaloids," eds. J. D. Phillipson and M. H. Zenk, Academic Press, London, 1980, p. 11.
122. M. V. Kisakürek, A. J. M. Leeuwenberg, and M. Hesse, in "Alkaloids—Chemical and Biological Perspectives," ed. S. W. Pelletier, Wiley, New York, 1983, vol. 1, p. 211.
123. K. Inoue, T. Tanahashi, H. Inouye, H. Kawajima, and K. Takaishi, *Phytochemistry*, 1989, **28**, 2971.
124. T. Hemscheidt and M. H. Zenk, *FEBS Lett.*, 1980, **110**, 187.
125. D. Schmidt and J. Stöckigt, *Planta Med.*, 1995, **61**, 254.
126. K. H. Knobloch, B. Hansen, and J. Berlin, *Z. Naturforsch., Teil C*, 1981, **36**, 40.
127. O. Schiel and J. Berlin, *Planta Med.*, 1986, **51**, 422.
128. J. M. Merillon, P. Doreau, A. Guillot, J. C. Chenieux, and M. Rideau, *Plant Cell Rep.*, 1986, **5**, 23.
129. K. M. Madyastha, T. D. Meehan, and C. J. Coscia, *Biochemistry*, 1976, **15**, 1097.
130. K. M. Madyastha and C. J. Coscia, *J. Biol. Chem.*, 1979, **254**, 2419.
131. O. Schiel, L. Witte, and J. Berlin, *Z. Naturforsch., Teil C*, 1987, **42**, 1075.

4.06
Enzymatically Controlled Steps in Vitamin B$_{12}$ Biosynthesis

A. IAN SCOTT, CHARLES A. ROESSNER, and
PATRICIO J. SANTANDER
Texas A&M University, College Station, TX, USA

4.06.1 INTRODUCTION

In the 25 years prior to 1989, much of the effort in delineating the vitamin B$_{12}$ biosynthetic pathway was expended on the search for intermediates in those organisms that make the cobalamin cofactor. This major direction was changed by the advent in 1989[1] of a genetic and molecular biological approach to the isolation and recombinant expression of genes required for vitamin B$_{12}$ biosynthesis, which ushered in a new era in the discovery of the vitamin B$_{12}$ pathway. Application of these techniques to the aerobic bacterium *Pseudomonas denitrificans*, along with the enzymatic synthesis and NMR analysis of the intermediate structures, led by the end of 1993 to the complete elucidation of the pathway encompassing over 20 genes and their enzymatic products. Since this pathway requires molecular oxygen, it is obvious, however, that organisms that synthesize vitamin B$_{12}$ under strict anaerobic conditions, such as *Clostridium thermoaceticum* and the methanogens, must have an alternative pathway, at least at those steps requiring oxygen. Genetic and biochemical studies have established that there is a similar, but distinct, pathway which does not require molecular oxygen that exists in strict anaerobes and some other aerotolerant organisms such as *Propionibacterium shermanii* and *Salmonella typhimurium*. In this chapter, we will first outline the enzymatic steps of the oxygen-requiring pathway and then discuss the similarities and differences among the corresponding processes of the anoxic pathway leading to Nature's most beautiful cofactor, adenosyl cobalamin.

4.06.2 THE OXYGEN-REQUIRING PATHWAY TO VITAMIN B$_{12}$ AS EXEMPLIFIED BY *PSEUDOMONAS DENITRIFICANS*

4.06.2.1 Synthesis of Aminolevulinic Acid (ALA)

In *P. denitrificans*, ALA is synthesized by ALA synthase from glycine and succinyl CoA via the Shemin pathway (Figure 1). However, the other two bacteria in which vitamin B$_{12}$ synthesis has been studied in detail and reviewed here (*S. typhimurium* and *P. shermanii*) utilize the C-5 pathway starting with glutamyl-tRNA for the synthesis of ALA (Figure 1). This pathway requires glutamyl-tRNA synthase and the products of the *hemA* (glutamyl-tRNA reductase) and *hemL* (glutamate semi-aldehyde aminomutase) genes of these organisms. (In this chapter, genes are named in

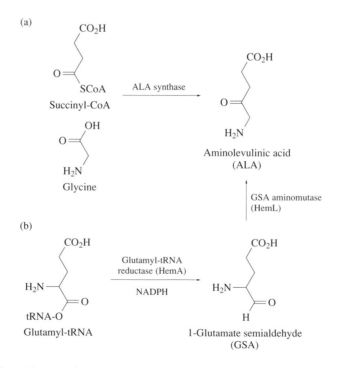

Figure 1 The two biosynthetic pathways to aminolevulinic acid (ALA): (a) the Shemin pathway found in *P. denitrificans*; (b) the C-5 pathway found in *P. shermanii* and *S. typhimurium*.

italics and beginning with a lower case letter, e.g., *hemA*, whereas the protein encoded by a given gene is written in normal style and begins with an upper case letter, e.g., HemA.)

4.06.2.2 Enzymes for the Conversion of ALA to Hydrogenobyrinic Acid (HBA)

The conversion of ALA to HBA requires the action of the twelve enzymes described in this section. Several of these enzymes perform more than one step as shown in Table 1 and Figure 2. The genes encoding the enzymes in *P. denitrificans* and the corresponding genes from *S. typhimurium* and *P. shermanii* are also listed in Table 1. The first three enzymes (ALA dehydratase, PBG deaminase, and uro'gen III synthase), also needed for the biosynthesis of the other "pigments of life" (heme, siroheme, chlorophyll, and factor F$_{430}$), are reviewed in depth elsewhere in this volume and thus will be mentioned only briefly here. Since the genes encoding these three enzymes have not been isolated from *P. denitrificans*, all characterization and mechanistic studies of the corresponding enzymes have been done with enzymes isolated from other sources. The first five enzymes have been overproduced in recombinant strains of *E. coli*.[4-7] The remaining seven enzymes have all been overproduced in recombinant strains of *P. denitrificans* (see references for the individual enzymes described below) and in recombinant strains of *E. coli*.[4]

Table 1 The enzymes and corresponding genes required for the conversion of ALA to adenosyl cobalamin.

Enzyme	Function (product)	Gene		
		P. denitrificans	S. typhimurium	P. shermanii
ALA dehydratase	(PBG)	*hemB*	*hemB*	*hemB*
PBG deaminase	(HMB)		*hemC*	
Urogen III synthase	(Urogen III)		*hemD*	
Urogen III methyltransferase	C-2, C-7 meth. (precorrin 2)	*cobA*	*cysG*	*cobA*
Precorrin 2 methyltransferase	C-20 meth. (precorrin 3)	*cobI*	*cbiL*	*cbiL*
Precorrin 3 oxidase	C-20 hydroxylation (precorrin 3 hydroxylactone)	*cobG*		
Precorrin 3 hydroxylactone methyltransferase	C-17 meth., ring contraction (precorrin 4)	*cobJ*	*cbiH*	*cbiH*
Precorrin 4 methyltransferase	C-11 meth. (precorrin 5)	*cobM*	*cbiF*	*cbiF*
Precorrin 5 methyltransferase	C-1 meth., acetate extrusion (precorrin 6)	*cobF*		
Precorrin 6 reductase	(dihydroprecorrin 6)	*cobK*	*cbiJ*	*cbiJ*
Precorrin 6 methyltransferase	C-5, C-10 meth.	*cobL*	*cbiE*	*cbiE*
Precorrin 6 decarboxylase	(precorrin 8)	*cobL*	*cbiT*	*cbiT*
Precorrin 8 mutase	C-11 → C-12 methyl shift (hydrogenobyrinic acid, HBA)	*cobH*	*cbiC*	*cbiC*
HBA *a,c*-diamide synthase	(HBA *a,c*-diamide)	*cobB*	*cbiA*	
Cobaltochelatase	(cobyrinic acid *a,c*-diamide)	*cobNST*	*cysG, cbiK*	
Cob(II)yrinic acid *a,c*-diamide reductase	(cob(I)yrinic acid *a,c*-diamide)			
Cob(I)adenosyl transferase	(adenosyl cobyrinic acid *a,c*-diamide)	*cobO*	*cobA*	
Cobyric acid synthase	(adenosyl cobyric acid)	*cobQ*	*cbiP*	*cbiP*
Cobinamide synthase	(adenosyl cobidamide)	*cobCD*	*cbiB*	
Cobinamide kinase/cobinamide phosphate guanylyl transferase	(adenosyl GDP-cobinamide)	*cobP*	*cobU*	
Cobalamin (5′-phosphate) synthase	(adenosyl cobalamin)	*cobV*	*cobS*	
DMBI phosphoribosyltransferase	(α-ribazole phosphate)	*cobU*	*cobT*	
Cobalt transport system			*cbiMNOQ*	*cbiMNOQ*
Uncharacterized function		*cobEW*	*cbiDG*	*cbiDG*

4.06.2.2.1 ALA dehydratase

ALA dehydratase catalyzes the condensation of two molecules of ALA to afford the pyrrole, porphobilinogen (PBG). This enzyme, encoded by the *hemB* gene, is a homooctamer with a mass of 280 kDa and a subunit M_r of 35 000.

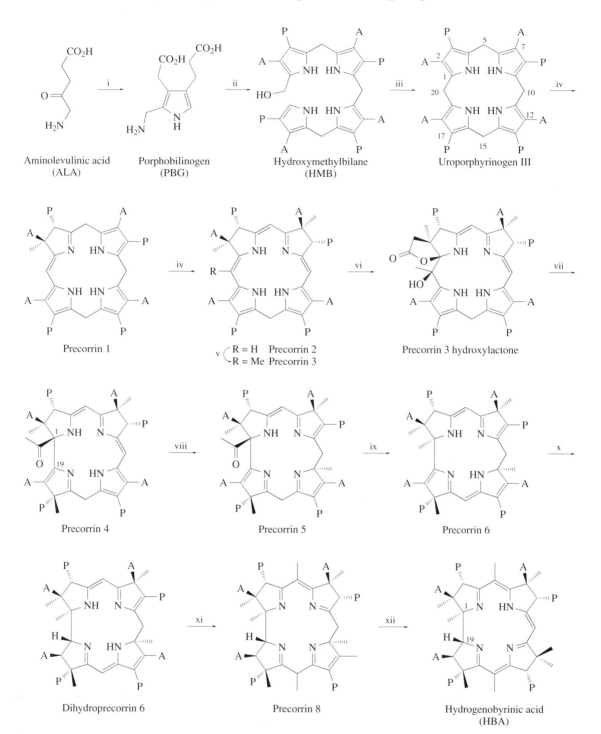

Figure 2 The oxygen-requiring biosynthetic pathway from ALA to hydrogenobyrinic acid (HBA) as determined in *P. denitrificans*. The required enzymes are: (i) ALA dehydratase; (ii) porphobilinogen (PBG) deaminase; (iii) uroporphyrinogen III (uro'gen III) synthase; (iv) uro'gen III methyltransferase; (v) precorrin 2 methyltransferase; (vi) precorrin 3 oxidase; (vii) precorrin 3 hydroxylactone methyltransferase; (viii) precorrin 4 methyltransferase; (ix) precorrin 5 methyltransferase; (x) precorrin 6 reductase; (xi) dihydroprecorrin 6 methyltransferase and decarboxylase; and (xii) precorrin 8 mutase. Most of the intermediates shown in this figure can exist in several tautomeric forms. In some cases the most stable form, e.g., precorrin 8, which can be analyzed by NMR, is shown as a tautomer of the true intermediate (A = CH$_2$CO$_2$H; P = CH$_2$CH$_2$CO$_2$H).

4.06.2.2.2   PBG deaminase

PBG deaminase catalyzes the condensation of four molecules of PBG to form the tetrapyrrole, hydroxymethylbilane (HMB). The enzyme is a monomer with an M_r of 34 000 and contains at its active site a unique dipyrromethane cofactor upon which the tetrapyrrole is built. The structure of the enzyme from *E. coli* has been determined by X-ray crystallography.[8]

4.06.2.2.3   Uroporphyrinogen (uro'gen) III synthase (cosynthetase)

Uro'gen III synthase, encoded by the *hemD* gene, catalyzes the inversion of the D-ring of HMB and ring closure to yield uro'gen III.

4.06.2.2.4   Uro'gen III methyltransferase (precorrin 2 synthase)

Uro'gen III methyltransferase, encoded by the *cobA* gene of *P. denitrificans*[9] adds two methyl groups from *S*-adenosyl-L-methionine to uro'gen III. The first methyl is added at C-2 (the numbering system used throughout is shown within the structure of uro'gen III in Figure 2) to form precorrin 1 which is then further methylated at C-7 to form precorrin 2. The enzyme was purified to homogeneity from a recombinant strain of *P. denitrificans*. The purified enzyme, a homodimer with a subunit M_s of 30 000, forms equal amounts of precorrin 1 and precorrin 2 and exhibits substrate inhibition at concentrations > 2 mM, a property which may play a regulatory role in cobalamin biosynthesis.[10]

4.06.2.2.5   Precorrin 2 methyltransferase (precorrin 3 synthase)

Precorrin 2 methyltransferase, encoded by the *cobI* gene,[11] catalyzes the SAM-dependent methylation of precorrin 2 to yield precorrin 3. Since precorrin 2 is also the precursor of both siroheme and, in methanogenic bacteria, factor F$_{430}$, the biosynthesis of precorrin 3 by this enzyme is the first step committed solely to the vitamin B$_{12}$ pathway. The enzyme has been purified to homogeneity from a recombinant strain of *P. denitrificans* and shown to be a homodimer with a subunit of M_r of 26 000.[10]

4.06.2.2.6   Precorrin 3 oxidase (precorrin 3 hydroxylactone synthase)

Precorrin 3 oxidase, encoded by the *cobG* gene,[12] catalyzes the addition of a hydroxy group to C-20 of precorrin 3, with the subsequent formation of a γ-lactone between C-1 and the acetate of ring A to afford precorrin 3 hydroxylactone.[12,13] In this chapter we have chosen to give precorrin isomers containing the same number of methyl groups descriptive chemical names, e.g., precorrin 3 and precorrin 3 hydroxylactone, in lieu of confusing, less meaningful descriptors, e.g., precorrin 3A, precorrin 3B, precorrin 3x, etc. This becomes essential in view of the second pathway (see below). The purified enzyme has a M_r of 46 000 and is a nonheme iron–sulfur protein containing 4–5 mol of iron per mole of protein suggesting either an [4Fe–4S] or two [2Fe–2S] complexes.[12] Whether the iron–sulfur complex is acting as an electron carrier or is involved in the active center has not been determined. Contrary to the original reports[12,14] that the formation of intermediates beyond precorrin 3 proceeds anaerobically, the hydroxy group of precorrin 3 hydroxylactone has been shown conclusively to be derived from molecular oxygen,[13,15] which has been added *cis* to the oxygen terminus of the γ-lactone at C-1.[16] Based on these findings, two mechanisms have been proposed for addition of the hydroxyl functional group (Figure 3). The first (a), which is analogous to the nonenzymatic oxidation of the C-5 position of corrins with Udenfriend's reagent[17] (Fe^{II}, ascorbate, O$_2$), involves the direct insertion of oxygen from the upper face of precorrin 3 from an Fe^{III}–O$^+$

species ($\equiv$OH$^+$). An alternative mechanism (b) involves formation of a 1,20 epoxide which, for steric reasons, requires opening through participation of the nitrogen lone electron pair rather than by direct attack by the carboxylate of the C-2 acetate side chain.

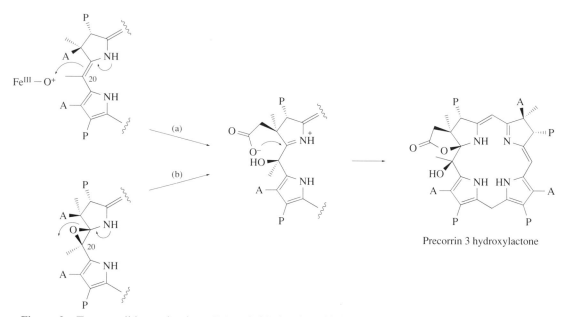

Figure 3 Two possible mechanisms ((a) and (b)) for the addition of the hydroxy group to precorrin 3.

4.06.2.2.7   Precorrin 3 hydroxylactone methyltransferase (precorrin 4 synthase)

Precorrin 3 hydroxylactone methyltransferase, encoded by the *cobJ* gene,[11] is a typical methyl-transferase with a subunit M_r of 27 000 that catalyzes the SAM dependent addition of the fourth methyl group, at C-17.[12,13] Surprisingly, ring contraction is triggered by this addition, *without the requirement of any other enzyme*, and is accompanied by the genesis of a new methyl ketone function pendant from C-1 that has been derived by a pinacol type rearrangement[13] as illustrated in Figure 4. In the absence of SAM, no alteration of the substrate is observed, suggesting that methylation of C-17 is the sole function of the enzyme.

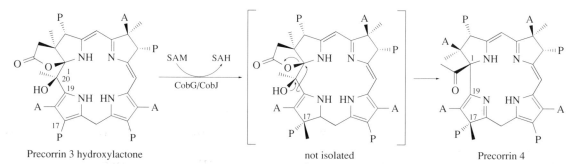

Figure 4 Mechanism of ring contraction of precorrin 3 hydroxylactone via a pinacol type rearrangement.

4.06.2.2.8   Precorrin 4 methyltransferase (precorrin 5 synthase)

Precorrin 4 methyltransferase (M_r 31 000), encoded by the *cobM* gene,[12] catalyzes the SAM-dependent methylation of C-11 to give precorrin 5.[12,13,18] This methyl group later migrates to become the C-12-*re*-methyl found in vitamin B$_{12}$ (see Section 4.06.2.2.12).

4.06.2.2.9 Precorrin 5 methyltransferase (precorrin 6 synthase)

Precorrin 5 methyltransferase (M_r 31 000), encoded by the *cobF* gene,[12] is a bifunctional enzyme that catalyzes removal of the methyl ketone moiety of precorrin 5 (deacylation) and the subsequent SAM-dependent methylation at C-1 to afford precorrin 6.[12,19] In the mechanistic model depicted in Figure 5, deacylation provides a proposed short-lived intermediate (shown in brackets) which then undergoes C-1 methylation. The model features the bis-imino chromophore in the intermediate extended from C-19 as an electron sink created by reaction with the enzyme. This structure, in which C-11 methylation insulates rings C and D electronically from rings A and B, not only facilitates deacetylation but also allows the return of electron density to C-1, followed by C-methylation mediated by the enzyme and SAM. The process is completed by prototropic shift from C-18 in the kinetic product to give precorrin 6. The extruded two carbon species has been identified as acetic acid.[12]

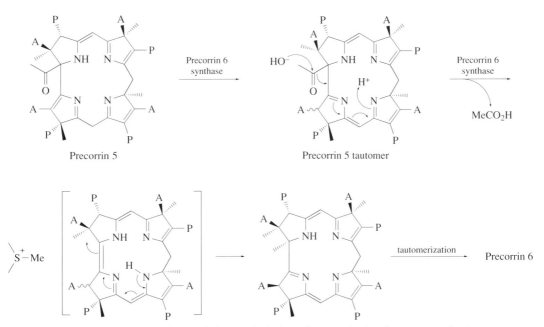

Figure 5 Deacylation and C-1 methylation of precorrin 5 to form precorrin 6.

4.06.2.2.10 Precorrin 6 reductase

Precorrin 6 reductase (M_r 31 000), encoded by the *cobK* gene, catalyzes the NADPH-dependent reduction of precorrin 6[20] to afford dihydroprecorrin 6. The reduction proceeds via the mechanism shown in Figure 6 in which C-18 is protonated from the medium and the hydrogen at C-19 is derived by enzymatic transfer of H$_R$ from NADPH.[21,22] Dihydroprecorrin 6 now matches the oxidation level of hydrogenobyrinic acid.

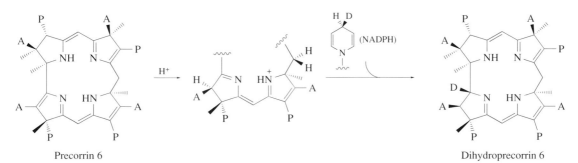

Figure 6 Mechanism of reduction of precorrin 6 with NADPH as cofactor.

4.06.2.2.11 Dihydroprecorrin 6 methyltransferase and decarboxylase (precorrin 8 synthase)

Precorrin 8 synthase (M_r 43 000), encoded by the *cobL* gene,[12] is a trifunctional enzyme that catalyzes not only the SAM-dependent methylation of both C-5 and C-15 but is also responsible for decarboxylation of the C-12 acetate (to form the *si*-methyl found at C-12) resulting in precorrin 8.[23] The amino-terminal half of the enzyme shows homology to the other methyltransferases, so the remaining carboxyl-portion ($M_r \sim 20\,000$) probably contains the decarboxylase activity. In the absence of SAM, no decarboxylation is observed, suggesting that SAM binding or at least one methylation must occur prior to decarboxylation. A mechanism for the decarboxylation of the C-12 acetate is depicted in Figure 7. Precorrin 8 exists in several tautomeric forms with the one shown in Figure 2 being the most stable form and the only one observed so far by NMR. However, it is *not* a substrate for the next enzyme and thus does not depict the true structure of the intermediate.

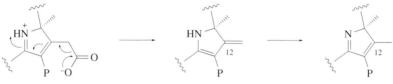

Figure 7 Mechanism of decarboxylation of the C-12 acetate.

4.06.2.2.12 Precorrin 8 mutase (hydrogenobyrinic acid (HBA) synthase)

Precorrin 8 mutase (M_r 22 000), encoded by the *cobH* gene,[12] catalyzes the final step in hydrogenobyrinic acid synthesis, the migration of the methyl group previously added at C-11 to C-12.[24] The enzyme exhibits product inhibition, a property that allowed the accumulation, isolation, and subsequent characterization of precorrin 8. The rearrangement, a typical suprafacial [1,5]-sigmatropic shift, may occur as shown in Figure 8. This rearrangement could occur spontaneously in the stable structure of precorrin 8, but does not. The purpose of the enzyme may be to change and hold precorrin 8 in a form favorable for rearrangement.

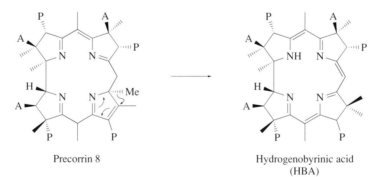

Precorrin 8   Hydrogenobyrinic acid
                                        (HBA)

Figure 8 The rearrangement of the C-11 methyl group to C-12 via a 1,5 sigmatropic shift to afford hydrogenobyrinic acid.

4.06.2.3 Enzymes for the Conversion of HBA to Cobalamin

The following enzymes, required for the biosynthetic pathway from HBA to cobalamin, are depicted in Figure 9.

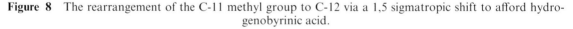

Figure 9 The biosynthetic pathway from hydrogenobyrinic acid to cobalamin. The enzymes required are: (i) HBA *a,c*-diamide synthase; (ii) cobaltochelatase; (iii) cob(II)yrinic acid *a,c*-diamide reductase; (iv) cobyrinic acid *a,c*-diamide adenosyltransferase; (v) ado-cobyric acid synthase; (vi) cobinamide synthase; (vii) cobinamide kinase; and (viii) cobalamin synthase.

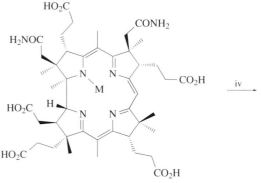

i $\left\{\begin{array}{l} R^1 = OH \quad \text{Hydrogenobyrinic acid} \\ R^1 = NH_2 \quad \text{Hydrogenobyrinic acid} \\ \qquad\qquad a,c\text{-diamide} \end{array}\right.$

iii $\left\{\begin{array}{l} M = Co^+ \quad \text{Cob(II)byrinic acid } a,c\text{-diamide} \\ M = Co \quad \text{Cob(I)byrinic acid } a,c\text{-diamide} \end{array}\right.$

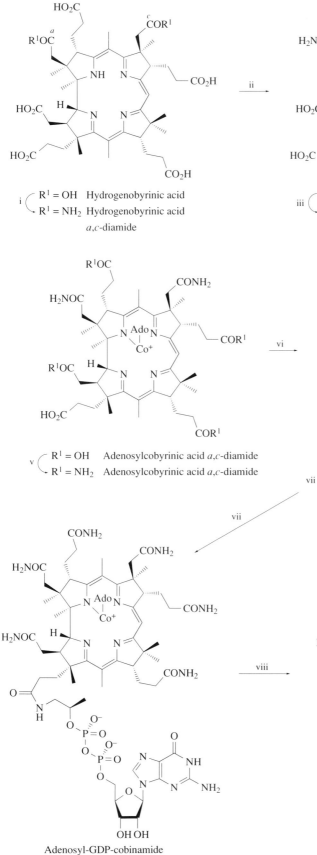

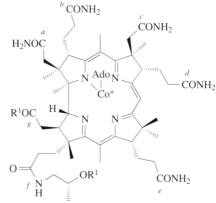

v $\left\{\begin{array}{l} R^1 = OH \quad \text{Adenosylcobyrinic acid } a,c\text{-diamide} \\ R^1 = NH_2 \quad \text{Adenosylcobyrinic acid } a,c\text{-diamide} \end{array}\right.$

vii $\left\{\begin{array}{l} R^1 = H \qquad \text{Adenosylcobinamide} \\ R^1 = PO_3^{2-} \quad \text{Adenosylcobinamide phosphate} \end{array}\right.$

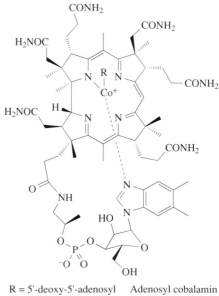

Adenosyl-GDP-cobinamide

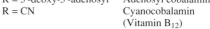

R = 5'-deoxy-5'-adenosyl   Adenosyl cobalamin
R = CN   Cyanocobalamin
                            (Vitamin B₁₂)

4.06.2.3.1   HBA a,c-diamide synthase

The first committed step linking HBA with the coenzyme is the side chain amidation of the *a,c* acetate side chains, a reaction requiring ATP/Mg^{2+} and glutamine, catalyzed by HBA *a,c* diamide synthase (M_r 45 000), encoded by the *cobB* gene.[25,26] Substrate specificity showed that the *c* side chain is amidated before the *a* acetate. This sequence is also believed to operate in *P. shermanii* and *C. tetanomorphum*[27] although the substrates in these organisms are cobalt complexes (see Section 4.06.3.2). Homology comparison of the enzyme suggested that the "amino transfer domain" is at the carboxy terminus, with Cys326 being a likely candidate for the prosthetic group involved in the transfer.

4.06.2.3.2   Cobaltochelatase (cob[II]yrinic acid a,c diamide synthase)

The cobaltochelatase activity of *P. denitrificans* contains no less than three subunits encoded by the *cobN, S,* and *T* genes.[28] Whereas CobN (M_r 140 000) appears to be the key component, CobS (M_r 38 000) and CobT (M_r 80 000) are certainly required. There is a strict dependence of cobaltochelatase activity on hydrolyzable ATP and Mg^{2+} and specificity for Co^{2+}. CobN shows 28% and 22% identity with Oli and BchH, respectively, both of which are implicated in Mg^{2+} chelation during bacteriochlorophyll biosynthesis.[29,30] HBA diamide was shown to be the physiological substrate although cobyrinic acid is a poor substrate. It was also established spectrophotometrically that the product is the Co^{II} complex (rather than Co^{III}) so no overall redox change is taking place during cobalt insertion.

4.06.2.3.3   Cob(II)yrinic acid a,c-diamide reductase

Prior to adenosylation, Cob(II)yrinic acid *a,c*-diamide is reduced to the Co^I complex. Based on the nucleophilicity of Co^I, this reducing activity was sought in *P. denitrificans* using both a chemical trap and coupling to cob(I)alamin adenosyl transferase.[31] Although the encoding gene has yet to be identified, the enzymatic activity (M_r 16 000) was purified 6300-fold and was found to require FMN and NADH. The reductase from *C. tetanomorphum* behaves similarly.[32]

4.06.2.3.4   Cobyrinic acid a,c-diamide adenosyltransferase

The adenosylating enzyme that forms adenosyl (ado-)cobyrinic acid *a,c*-diamide was shown to be encoded by *cobO*[33] and is a homodimer (subunit M_r 27 000). There is a fairly rigid substrate specificity requiring the presence of the *a,c*-diamide functionality.

4.06.2.3.5   Adenosyl cobyric acid synthase

Amidation of the *b, d, e,* and *g* (but not *f*) carboxylates results in ado-cobyric acid formation. Extracts of engineered *P. denitrificans* harboring the *cobQ* gene by itself supported the amidation of all four carboxyls demonstrating that its product is sufficient.[34] There is a requirement in the substrate for the adenosyl group. Comparison of this amidase (CobQ) with the first amidase (CobB) shows 22% amino acid identity and a tentative allocation of the conserved Cys333 in CobQ to the active site residue.

4.06.2.3.6   Cobinamide synthase

The attachment of (*R*)-1-amino-2-propanol to ado-cobyric acid affords ado-cobinamide. Two proteins (*α* and *β*) are involved (as shown by HPLC assay) and the reaction is dependent on ATP/Mg^{2+} and (*R*)-1-amino-2-propanol (K_m 20 mM). It was concluded that *cobC* and *cobD* encode protein *β* and that protein *α* is encoded by a gene yet to be identified.[35] The source of the isopropanolamine is believed to be threonine but this has yet to be established genetically and enzymologically. *O*-phospho-L-threonine may be an intermediate since it has been shown to be a better substrate than propanolamine itself in complementation studies of a mutant of *P. denitrificans* that requires one of these compounds to make vitamin B$_{12}$.[36] Attachment of isopropanolamine in *S. typhimurium* similarly appears to involve three gene products.[37]

4.06.2.3.7 Cobinamide kinase

Ado-cobinamide is converted to Ado-cobalamin in a three step process. The first two of these, ATP-dependent phosphorylation and transfer of GMP from GTP, to form ado-cobinamide phosphate and ado-GDP-cobinamide, respectively, are mediated by a single bifunctional enzyme, called cobinamide kinase (M_r 19 442) encoded by the *cobP* gene.[38] Remarkably, GTP functions both as a substrate for the second step and as a regulator for the first, increasing the affinity for the first substrate, ado-cobinamide, by lowering the K_m from 11.5 μM to 0.4 μM.[38] The homologous enzymes in *S. typhimurium* perform the same functions.[37,39]

4.06.2.3.8 Cobalamin synthase

The final step of the pathway to coenzyme B$_{12}$ is the substitution of the GMP of ado-GDP-cobinamide with either α-ribazole or α-ribazole-phosphate to afford ado-cobalamin or its 5′ phosphate. The responsible enzyme, cobalamin (5′-phosphate) synthase, appears to be encoded by the *cobV* gene[40] although it has not yet been isolated for comparison with the predicted sequence, since the expression system afforded a mixture of six proteins in a high molecular weight complex. The α-ribazole-phosphate required for this reaction is derived from dimethylbenzimidazole (DBI, see below for its biosynthesis) and nicotinic acid mononucleotide (NAMN) as shown in Figure 10. This reaction is performed by the enzyme encoded by the *cobU* gene[38] which exists as a homodimer with a monomeric M_r of 34 640.

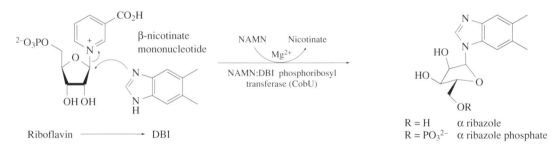

Figure 10 The biosynthesis of α-ribazole phosphate from dimethylbenzimidazole and nicotinic acid mononucleotide.

The complete pathway to adenosylcobalamin in aerobic bacteria such as *P. denitrificans* can now be observed by combining Figure 2 with Figure 9. Adenosylcobalamin is converted to vitamin B$_{12}$ (cyanocobalamin) by substitution of the adenosyl moiety with cyanide (Figure 9).

4.06.2.4 Formation of the 5,6-Dimethylbenzimidazole (DBI) Moiety

Although the biosynthesis of the corrin ring follows a quite distinct oxygen-dependent route in the aerobic *P. denitrificans*, as opposed to the anoxic route found in the aerotolerant bacteria *P. shermanii* and *S. typhimurium* (see Section 4.06.3), the DBI unit apparently can be formed in all three organisms via the same oxygen-dependent route. In strict anaerobes, however, the DBI moiety must be formed by a completely different pathway due to the nature of the organisms. Pathways for both the oxygen-dependent and the anoxic synthesis of DBI have been proposed based on the extensive work of Renz.[41]

4.06.2.4.1   The oxygen-dependent route to DBI

It has been shown[42] that the precursor is riboflavin-5′-phosphate (FMN), and that C-1′ of the ribityl side chain is transformed into C-2 of DBI[43] (Figure 11). During this transformation the pro *S* hydrogen at C-1′ of riboflavin is retained, appearing as the C-2H of DBI. Carbons 3′, 4′, and 5′ of the riboflavin are excised as a C-3 fragment (presumably glyceraldehyde-3-phosphate), which is further processed to hexoses. Another byproduct is urea, whose carbon is derived from 2-^{14}C-riboflavin. A working hypothesis for aerobic DBI formation is shown in Figure 11.

4.06.2.4.2   Anoxic formation of DBI

It has been shown that, in the strict anaerobe *Eubacterium limosum*, the DBI structure is synthesized from glycine, formate, glutamine, erythrose, and methionine. Regiospecific incorporation of C-1, C-2, and N of glycine into C-9, C-8 and N-1 of DBI (●) has been demonstrated (Figure 12). Formate (▲) is clearly incorporated into C-2 and N-3 (■) is derived from the amide nitrogen of glutamine. The remaining four carbons of the ring system (C-4, -5, -6, and -7) are provided by erythrose which is introduced regiospecifically, C-1 of erythrose being incorporated into C-4 of DBI, as shown by NMR spectroscopy using [1-^{13}C]erythrose as precursor to DBI in *E. limosum*. Finally, the two methyl groups are derived from methionione (^{13}C). Interestingly, the labeling pattern of DBI is "inverted" from that found in the purine nucleotides, such as AMP. The anaerobic pathway to DBI has not been established at the enzymatic level, but could follow purine biosynthesis as far as 5-amino imidazole β-D-ribofuranoside. In fact, amino imidazole is taken up by *E. limosum* to form 7-azabenzimidazolylcobamide and its 5,6-dimethyl homologue. The DBI moiety of B$_{12}$ was not labeled in this experiment, however, indicating that 5-amino imidazole is not a DBI precursor *per se*. Also of note is the occurrence of 5-methoxy-6-methyl-benzimidazolyl cobamide in *Clostridium formicoaceticum* hinting that 5-hydroxy-benzimidazole is the precursor of 5-hydroxy-6-methyl benzimidazole, which in turn can be transformed to DBI. This scenario was clearly demonstrated when the 5-hydroxy bases were labeled with ^{13}C at C-2 and with ^{15}N at N-1 leading to complete B$_{12}$ with the DBI moiety labeled as predicted, both in *E. limosum* and *Clostridium barkeri*. The hypothetical anaerobic route to DBI is shown in Figure 13. Implicit in this scheme is the desymmetrization of DBI itself, for if the ribose-5-phosphate of nicotinic acid mononucleotide is transferred to DBI in solution, N-1 and N-3 are substituted at equal rates. However, the regiospecific incorporation of the amide-N of glutamine and C-1 of erythrose require that either DBI is enzyme bound until ribosylation takes place, or that the α-glucosidic bond is formed on N-1 before completion of DBI synthesis.

4.06.2.4.3   Other benzimidazole bases found in B$_{12}$ analogues

Several analogues of DBI occur in nature[41] including 5-hydroxy benzimidazolyl cobamide (Factor III) in methanogens, the 5-methoxy analogue in *C. thermoaceticum*, and, as mentioned above, 5-methoxy-6-methyl benzimidazolyl cobamide in *C. formicoaceticum*. Benzimidazole and its 5-methyl analogue occur as the corrinoid bases in anaerobic sewage sludge along with a naphthimidazole analogue. Little is known about the biosynthesis of these analogues except that erythrose is not a precursor of 5-hydroxy benzimidazole in methanogenic bacteria, carbons 4–7 being derived from acetate and pyruvate (two from C-1). Another ligand that can replace DBI is *p*-cresolyl to form *p*-cresolyl cobamide necessary for methanol metabolism in the acetogenic bacterium, *Sporumosa ovata*.[44]

4.06.3   THE ANOXIC PATHWAY TO VITAMIN B$_{12}$ AS EXEMPLIFIED BY *S. TYPHIMURIUM* AND *P. SHERMANII*

There is no doubt that there are at least two pathways to vitamin B$_{12}$ in nature, the pathway described above requiring O$_2$, and an alternate anoxic pathway that must exist in organisms such

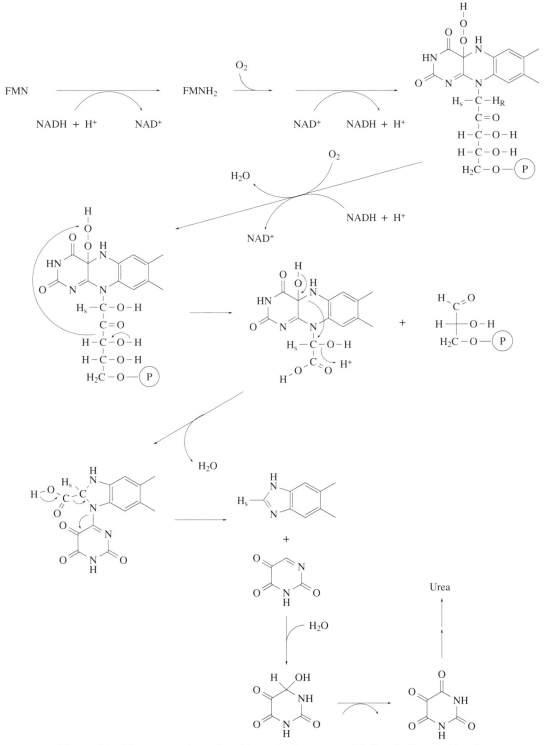

Figure 11 The oxygen-dependent biosynthetic route to 5,6-dimethylbenzimidazole.

as *Methanococcus jannaschii*[45] that grow only under strict anaerobic conditions. Based on homology studies (see below), the anoxic pathway is not absolutely restricted to obligate anaerobes since it apparently is used by aerotolerant bacteria such as *S. typhimurium*[46] and *P. shermanii*[47] that synthesize B₁₂ under microaerophilic conditions (and require O₂ for DBI synthesis) and *Bacillus megaterium*,[48] an obligate aerobe.

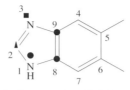

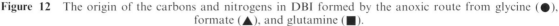

Figure 12 The origin of the carbons and nitrogens in DBI formed by the anoxic route from glycine (●), formate (▲), and glutamine (■).

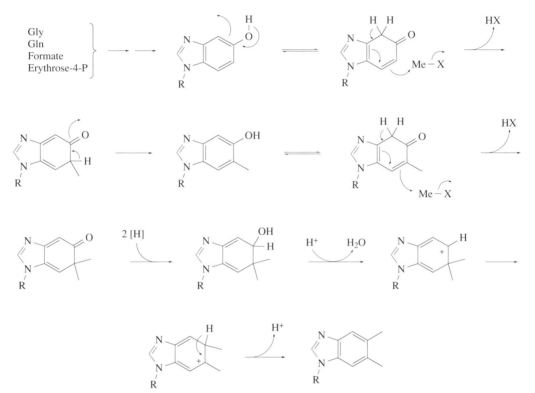

Figure 13 The hypothetical oxygen-free route to DBI.

4.06.3.1 Comparison of the Enzymes of the Oxygen-requiring Pathway of *P. denitrificans* to Those Encoded by Genes of Other Organisms

Vitamin B₁₂ biosynthetic genes have been isolated from several other organisms including complete sets from *S. typhimurium*,[37] *M. jannaschii*,[45] and *B. megaterium*[48] and a partial set from *P. shermanii*.[47] Genetic studies with *S. typhimurium* have established three regions associated with B₁₂ biosynthesis, CobI, II, and III located at 41 minutes of the chromosomal map. Homology comparisons of the proteins encoded by the *S. typhimurium cbi* genes found in the CobI region with those encoded by the previously characterized *cob* genes from *P. denitrificans* has led to tentative functional assignments for the Cbi proteins (Figure 14, Table 1). The genes (*cbiA–cbiL*) correspond approximately to the *P. denitrificans cobF–cobM* genes responsible for the conversion of precorrin-2 to hydrogenobyrinic acid but there are sufficient differences to indicate that at an early stage the pathways are quite distinct. Although homologies exist between *cbiC* and *cobH*; *cbiE + T* and *cobL*; *cbiF* and *cobM*; *cbiH* and *cobJ*; *cbiJ* and *cobK*; and *cbiL* and *cobI*, there are no gene sequences in *P. denitrificans* corresponding to *cbiD*, *cbiG*, or *cbiK* in *Salmonella*, while *cobF* and *cobG* in *P. denitrificans* are not found in *Salmonella*. Although most of the *cbi* genes *A–L* have been expressed,[4] only two have so far been assigned specific functions—*cbiL* (*cobI*) is the C-20 methyltransferase and *cbiF* is known to have C-11 methylase activity. As might be expected for an anoxic pathway, no protein in *S. typhimurium* is homologous to that encoded by *cobG*, the oxygen-requiring precorrin 3 hydroxylactone synthase. Also missing is a homologue to precorrin 6 synthase for deacylation and C-1 methylation. Presumably one of the other methyltransferases assumes the functions of this enzyme. In addition, no role has been assigned to three of the other *cbi* gene products, CbiD, CbiG, and CbiK. However, evidence has suggested that CbiK may be involved in cobalt insertion (see below), and CbiG shows partial homology to the *cobE* gene product whose function is unknown but which is not required for the cell-free synthesis of hydrogenobyrinic acid.[5,12,19] CbiD and CbiK show no homology to any other known vitamin B$_{12}$ biosynthetic enzymes.

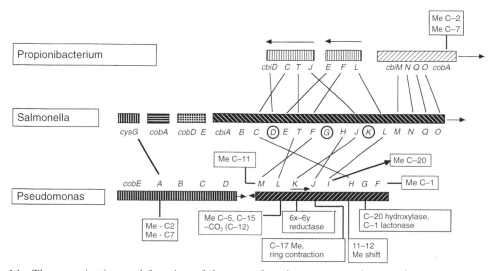

Figure 14 The organization and function of the genes found on two complementation groups required for the biosynthesis of HBA in *P. denitrificans*. The known homologues in *S. typhimurium* and *P. shermanii* are shown.

The sequence of the complete genome of *M. jannaschii* has been determined.[45] A homology search with the *S. typhimurium cbi* gene products has found proteins homologous to all of them except CbiK. In addition, genes for the biosynthesis of cobyrinic acid corresponding to *cbiC*, *cbiD*, *cbiE*, *cbiF*, *cbiJ*, *cbiH*, *cbiL*, and *cbiT* have been isolated from *P. shermanii*.[47]

4.06.3.2   Evidence for Early Cobalt Insertion in the Anoxic Pathway

As described in Section 4.06.2.3.2, cobalt insertion occurs at a fairly late stage into HBA *a,c*-diamide in the oxygen-dependent pathway. Studies using [14]C- or ^{60}Co-labeled precursors,[49,50] however, have demonstrated that cobalt insertion in *P. shermanii* occurs as early as the precorrin 2 or precorrin 3 stage. Both genetic and biochemical evidence have been provided to show that, in *S. typhimurium*, the *cysG* gene product has not only the methyltransferase activity responsible for the synthesis of precorrin 2 and the oxidase and ferrochelatase activities for the conversion of precorrin 2 to siroheme,[51,52] but also is a cobalt-inserting enzyme in the B$_{12}$ pathway.[51,53] In addition, an unusual function for *cbiK* was promulgated when it was found[54] that, in a *cysG* mutant auxotrophic for B$_{12}$, the function of *cysG* could be restored by a plasmid bearing both *cobA* and *cbiK*. Since neither gene by itself restored activity, and CobA is the known urogen III methyltransferase in *P. denitrificans*, by implication the activity associated with CbiK is that of a surrogate cobalto-chelatase, which is recruited in mutants lacking *cysG*. The cobalt-inserting enzyme of *P. shermanii* has not yet been described. However, urogen III methyltransferase (CobA) has been isolated from this organism[55] and is known to be unlike CysG in that it does not contain the amino-terminal portion believed to be responsible for metal chelation. When whole cells or cell-free extracts of *P. shermanii* are incubated in the absence of cobalt, precorrins 1–3 accumulate and are isolated in the oxidized (factor) forms. The structure of the monomethyl chlorin and the dimethyl and trimethylisobacteriochlorins (factors I, II, and III) were established as indicated in Figure 15 based both on NMR spectroscopy studies and incorporation experiments.[56-64] During the incubation and/or the isolation process of these pigments oxidations and epimerizations at C-3 and/or C-8 can occur so that lactones and epimers such as those shown in structures 1–9 can be formed.[65-68]

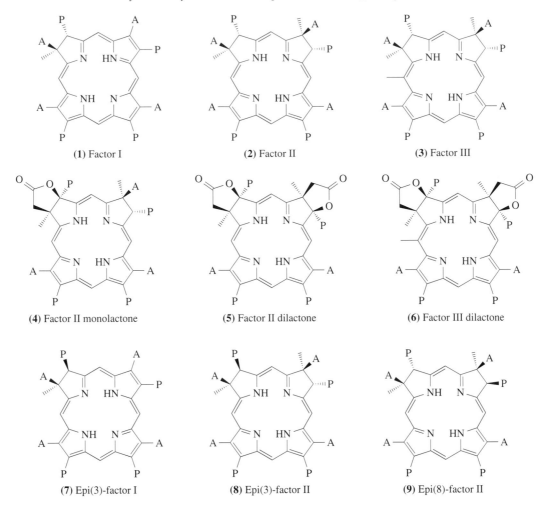

(**1**) Factor I   (**2**) Factor II   (**3**) Factor III

(**4**) Factor II monolactone   (**5**) Factor II dilactone   (**6**) Factor III dilactone

(**7**) Epi(3)-factor I   (**8**) Epi(3)-factor II   (**9**) Epi(8)-factor II

4.06.3.3   Exchange of Oxygen at C-27 in the Anoxic Pathway

During the synthesis of HBA in *P. denitrificans*, no exchange of oxygen occurs at any of the carbonyl centers of the acetate or propionate side chains.[16] In sharp contrast, a unique and substantial exchange has been observed between the aqueous medium and the carbonyl oxygens at the ring A acetamido group (C-27) during the anoxic biosynthesis of vitamin B$_{12}$ in *P. shermanii*.[69,70] This exchange has been shown to occur prior to the cobyrinic acid stage.[71] The formation of a mixed anhydride during the removal of the acetate at C-1 of precorrin 5 was invoked[72] to explain the observed transfer of ^{18}O from ring A acetate to the acetic acid liberated in *P. shermanii*. As described below (Section 4.06.3.4), however, the 2-carbon fragment in *P. shermanii* is not acetic acid, but rather acetaldehyde (which is subsequently oxidized to acetic acid) thus requiring revision of the earlier mechanistic proposals.

4.06.3.4   Isolation and Structure of Factor 4: Extrusion of Acetaldehyde

Consistent with the evidence of early cobalt insertion is the structure of factor IV, an intermediate from *P. shermanii* that was isolated by incubating a mixture of 3-epi; 8-epi factor II (the oxidized form of precorrin 2) in cell-free lysates of *P. shermanii*.[73] In this experiment, a ring-contracted, tetramethylated, cobalt-containing, epimeric derivative accumulated, presumably because the epimer(s) could act as substrate(s) up to this point in the pathway. A low conversion (1.0%) of the 3-epi; 8-epi factor II to cobyrinic acid was observed, and factor IV was converted to cobyrinic acid at 0.6% efficiency based on radio labeling experiments. The structure of factor IV suggested that, in *P. shermanii*, C-20 and its associated methyl should be extruded as acetaldehyde rather than as

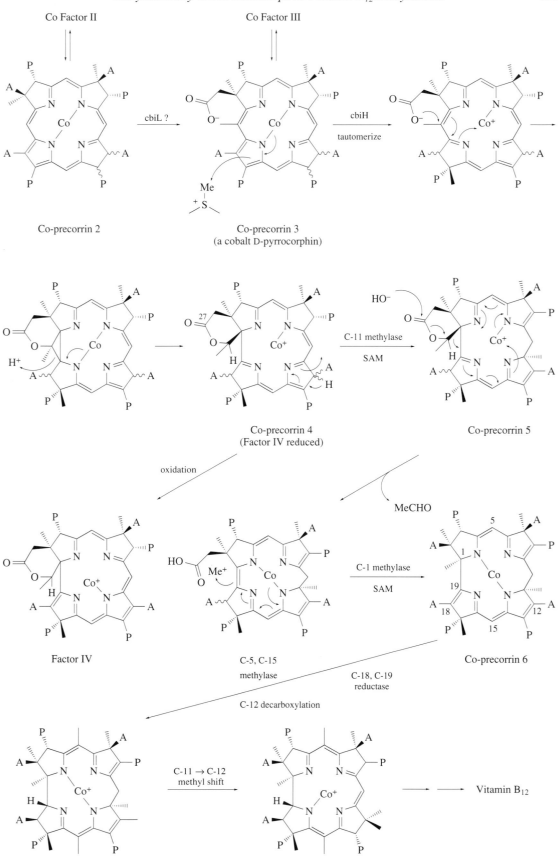

Figure 15 The proposed anoxic route for the biosynthesis of cobyrinic acid from precorrin 2.

acetic acid as in *P. denitrificans*. Indeed, experiments using dimedone as a trapping agent have confirmed this postulate,[74] and it has been shown that acetaldehyde is converted to acetic acid in cell-free extracts of *P. shermanii*.

4.06.3.5 A Working Hypothesis for the Anoxic Pathway

With the above information in hand, the complete cobalt-mediated anoxic route to cobyrinic acid shown in Figure 15 can now be proposed. The enzymes responsible for the conversion of precorrin 2 to cobalt-precorrin 3 have not yet been identified but are thought to be CysG and CbiL. Although tetrapyrroles are natural metal chelating agents and cobalt insertion may occur spontaneously, the process is probably catalyzed by CysG or CbiK *in vivo*. Experiments by the authors' group[75] have shown that CbiH, the homologue of CobJ (the C-17 methyltransferase and ring-contracting enzyme in *P. denitrificans*), is necessary and sufficient for the synthesis of cobalt-precorrin 4 from cobalt-precorrin 3. The C-11 methyltransferase that catalyzes the formation of cobalt-precorrin 5 is likely to be the *cbiF* gene product since it has previously been shown to have C-11 methyltransferase activity and is homologous to precorrin 5 synthase of the *P. denitrificans* series. The only link now missing from the anoxic series is the enzyme responsible for the conversion of cobalt-precorrin 5 to cobalt-precorrin 6 with concomitant release of acetaldehyde. The analogous step in *P. denitrificans* is performed by the CobF enzyme. However, there has as yet been no homologous enzyme discovered for the anoxic series. Before release of the two carbon fragment, the δ-lactone must first be opened, a process which would involve exchange of ^{18}O with the medium at C-27 and which may be carried out by CbiD or CbiG. One of the other methyltransferases, most probably CbiF or CysG (CobA) could then come back to methylate C-1 with release of acetaldehyde. Since precorrin 6 has been shown to be converted (albeit at low efficiency) to cobyrinic acid by cell-free extracts of *P. shermanii*,[76] it may be assumed that the oxygen-dependent and anoxic pathways converge at this stage and that the sequence cobalt-precorrin 6 → cobalt-precorrin 8 → cobyrinic acid → cobyrinic acid *a,c*-diamide → ado-cobyrinic acid *a,c*-diamide is followed in the anoxic pathway. Studies on the origin and exchangeability of hydrogen atoms on the periphery of the macrocycle[77,78] indicate possible existence of intermediates with a double bond at C-8 to C-9. Rigorous proof for this scenario awaits experimental verification.

In summary, the complex aerobic pathway to B$_{12}$ has been elucidated and considerable progress has been made in defining the steps of the anoxic route. It is unlikely that more than two pathways have evolved (and been conserved) for B$_{12}$ biosynthesis. In fact, it is remarkable that Nature has set in place two completely different mechanisms for the ring contraction step, the ancient version requiring cobalt, the modern one utilizing molecular oxygen.

4.06.4 REFERENCES

1. B. Cameron, K. Briggs, S. Pridmore, G. Brefort, and J. Crouzet, *J. Bacteriol.*, 1989, **171**, 547.
2. P. M. Jordan, Key Features of Heme Biosynthesis, in "Comprehensive Natural Products Chemistry, vol. 4, Amino-acids, Peptides, Porphyrins, and Alkaloids."
3. S. I. Beale, Heme Biosynthesis, in "Comprehensive Natural Products Chemistry, vol. 4, Amino-acids, Peptides, Porphyrins, and Alkaloids."
4. C. A. Roessner, M. J. Warren, P. J. Santander, B. P. Atshaves, S.-I. Ozaki, N. J. Stolowich, K. Iida, and A. I. Scott, *FEBS Lett.*, 1992, **301**, 73.
5. C. A. Roessner, J. B. Spencer, S. Ozaki, C. Min, B. A. Atshaves, P. Nayar, N. Anousis, N. J. Stolowich, M. T. Holderman, and A. I. Scott, *Prot. Expr. Purif.*, 1995, **6**, 155.
6. N. P. J. Stamford, J. Crouzet, B. Cameron, A. I. D. Alanine, A. R. Pitt, A. A. Yeliseev, and A. R. Battersby, *Biochem. J.*, 1996, **313**, 335.
7. A. F. Alwan, B. O. Mgbeje, and P. M. Jordan, *Biochem. J.*, 1989, **264**, 397.
8. P. M. Jordan, M. J. Warren, B. I. Mgbeje, S. P. Wood, J. B. Cooper, G. Louie, P. Brownlie, R. Lambert, and T. L. Blundell, *J. Mol. Biol.*, 1992, **224**, 269.
9. J. Crouzet, L. Cauchois, F. Blanche, L. Debussche, D. Thibaut, M.-C. Rouyez, S. Rigault, J.-F. Mayaux, and B. Cameron, *J. Bacteriol.*, 1990, **172**, 5968.
10. F. Blanche, L. Debussche, D. Thibaut, J. Crouzet, and B. Cameron, *J. Bacteriol.*, 1989, **171**, 4222.
11. J. Crouzet, B. Cameron, L. Cauchois, F. Blanche, S. Rigault, M.-C. Rouyez, D. Thibaut, and L. Debussche, *J. Bacteriol.*, 1990, **172**, 5980.
12. L. Debussche, D. Thibaut, B. Cameron, J. Crouzet, and F. Blanche, *J. Bacteriol.*, 1993, **175**, 7430.
13. A. I. Scott, C. A. Roessner, N. J. Stolowich, J. B. Spencer, C. Min, and S. Ozaki, *FEBS Lett.*, 1993, **331**, 105.
14. D. Thibaut, L. Debussche, and F. Blanche, *Proc. Natl. Acad. Sci. USA*, 1990, **87**, 8795.
15. J. B. Spencer, N. J. Stolowich, C. A. Roessner, C. Min, and A. I. Scott, *J. Am. Chem. Soc.*, 1993, **115**, 11 610.

16. N. J. Stolowich, J. Wang, J. B. Spencer, P. J. Santander, C. A. Roessner, and A. I. Scott, *J. Am. Chem. Soc.*, 1996, **118**, 1657.
17. A. Gossauer, B. Grüning, L. Ernst, W. Becker, and W. S. Sheldrick, *Angew. Chem., Int. Ed. Engl.*, 1977, **16**, 481.
18. C. Min, B. A. Atshaves, C. A. Roessner, N. J. Stolowich, J. B. Spencer, and A. I. Scott, *J. Am. Chem. Soc.*, 1993, **115**, 10 380.
19. C. A. Roessner, J. B. Spencer, N. J. Stolowich, J. Wang, N. C. Parmesh, P. J. Santander, C. Pichon, C. Min, P. Nayar, S. Ozaki, B. A. Atshaves, N. Anousis, M. T. Holderman, and A. I. Scott, *Chem. Biol.*, 1994, **1**, 119.
20. F. Blanche, D. Thibaut, A. Famechon, L. Debussche, B. Cameron, and J. Crouzet, *J. Bacteriol.*, 1992, **174**, 1036.
21. G. W. Weaver, F. J. Leeper, A. R. Battersby, F. Blanche, D. Thibaut, and L. Debussche, *J. Chem. Soc., Chem. Commun.*, 1991, 976.
22. F. Kiuchi, D. Thibaut, L. Debussche, F. J. Leeper, F. Blanche, and A. R. Battersby, *J. Chem. Soc., Chem. Commun.*, 1992, 306.
23. F. Blanche, A. Famechon, D. Thibaut, L. Debussche, B. Cameron, and J. Crouzet, *J. Bacteriol.*, 1992, **174**, 1050.
24. D. Thibaut, M. Couder, A. Famechon, L. Debussche, B. Cameron, J. Crouzet, and F. Blanche, *J. Bacteriol.*, 1992, **174**, 1043.
25. J. Crouzet, L. Cauchois, F. Blanche, L. Debussche, D. Thibaut, M.-C. Rouyez, S. Rigault,, J. F. Mayaux, and B. Cameron, *J. Bacteriol.*, 1990, **172**, 5968.
26. L. Debussche, D. Thibaut, B. Cameron, J. Crouzet, and F. Blanche, *J. Bacteriol.*, 1990, **172**, 6239.
27. H. C. Friedman and L. M. Cagen, *Annu. Rev. Biochem.*, 1970, **24**, 159.
28. L. Debussche, M. Couder, D. Thibaut, B. Cameron, J. Crouzet, and F. Blanche, *J. Bacteriol.*, 1992, **174**, 7445.
29. A. Hudson, R. Carpenter, S. Doyle, and E. Coen, *EMBO J.*, 1993, **12**, 3711.
30. D. Burke, M. Alberti, G. Armstrong, and J. Hearst, *Genbank database*, 1991, accession no. Z11165.
31. F. Blanche, L. Maton, L. Debussche, and D. Thibaut, *J. Bacteriol.*, 1992, **174**, 7452.
32. F. M. Huennekens, K. S. Vitols, K. Fujii, and D. W. Jacobsen, in "B$_{12}$," ed. D. Dolphin, Wiley, New York, 1982, vol. 1, p. 145.
33. L. Debussche, M. Couder, D. Thibaut, B. Cameron, J. Crouzet, and F. Blanche, *J. Bacteriol.*, 1991, **173**, 6300.
34. F. Blanche, M. Couder, L. Debussche, D. Thibaut, B. Cameron, and J. Crouzet, *J. Bacteriol.*, 1991, **173**, 6046.
35. F. Blanche, B. Cameron, J. Crouzet, L. Debussche, D. Thibaut, M. Vuilhorgne, F. J. Leeper, and A. R. Battersby, *Angew. Chem., Int. Ed. Engl.*, 1995, **34**, 383.
36. E. Remy, L. Debussche, and D. Thibaut, *Biofactors*, 1996, **5**, 248.
37. J. R. Roth, J. G. Lawrence, M. Rubenfield, S. Kieffer-Higgens, and G. M. Church, *J. Bacteriol.*, 1993, **175**, 3303.
38. F. Blanche, L. Debussche, A. Famechon, D. Thibaut, B. Cameron, and J. Crouzet, *J. Bacteriol.*, 1991, **173**, 6052.
39. G. A. O'Toole, M. R. Rondon, and J. C. Escalante-Semerana, *J. Bacteriol.*, 1993, **175**, 3317.
40. B. Cameron, F. Blanche, M.-C. Rouyez, D. Bisch, A. Famechon, M. Couder, L. Cauchois, D. Thibaut, L. Debussche, and J. Crouzet, *J. Bacteriol.*, 1991, **173**, 6066.
41. P. Renz, in "Vitamin B$_{12}$ and B$_{12}$-Proteins," eds. B. Kräutler, D. Arigoni, and B. T. Golding, Wiley, Weinheim, 1998, p. 119.
42. P. Renz, *FEBS Lett.*, 1970, **6**, 187.
43. P. Renz and R. Weyhenmeyer, *FEBS Lett.*, 1972, **22**, 124.
44. E. Stupperich, H. J. Eisinger, and B. Kräutler, *Eur. J. Biochem.*, 1988, **172**, 459.
45. C. J. Bult, O. White, G. J. Olson, L. Zhou, R. D. F. Leishmann, G. G. Sutton, J. A. Blake, L. M. Fitzgerald, and R. A. Clayton, *Science*, 1996, **273**, 1043.
46. R. M. Jeter, B. M. Olivera, and J. R. Roth, *J. Bacteriol.*, 1984, **159**, 206.
47. C. A. Roessner, K. Huang, and A. I. Scott, *Biofactors*, 1996, **5**, 208.
48. E. Raux, M. J. Warren, A. Lanois, A. Rambach, and C. Thermes, *Biofactors*, 1996, **5**, 247.
49. G. Muller, F. Zipfel, K. Hlineny, R. Savvidis, R. Hertle, U. Traub-Eberhard, A. I. Scott, H. J. Williams, N. J. Stolowich, P. J. Santander, M. J. Warren, F. Blanche, and D. Thibaut, *J. Am. Chem. Soc.*, 1991, **113**, 9893.
50. S. Balchandran, R. A. Vishwakarma, S. M. Monaghan, A. Prelle, N. P. J. Stamford, F. J. Leeper, and A. R. Battersby, *J. Chem. Soc., Perkin Trans. 1*, 1994, 487.
51. J. B. Spencer, N. J. Stolowich, C. A. Roessner, and A. I. Scott, *FEBS Lett.*, 1993, **335**, 57.
52. M. J. Warren, E. L. Bolt, C. A. Roessner, A. I. Scott, J. B. Spencer, and S. C. Woodcock, *Biochem. J.*, 1994, **302**, 837.
53. T. G. Fazzio and J. R. Roth, *J. Bacteriol.*, 1996, **178**, 6952.
54. E. Raux, R. Beck, F. Levillayer, A. Rambach, C. Thermes, and M. J. Warren, *J. Bacteriol.*, 1997, **179**, 3202.
55. I. Sattler, C. A. Roessner, N. J. Stolowich, S. H. Hardin, L. W. Harris-Haller, N. T. Yokubaitis, Y. Murooka, Y. Hashimoto, and A. I. Scott, *J. Bacteriol.*, 1995, **177**, 1564.
56. A. I. Scott, A. J. Irwin, L. M. Siegel, and J. N. Shoolery, *J. Am. Chem. Soc.*, 1978, **100**, 7987.
57. A. R. Battersby, E. McDonald, H. Morris, M. Thompson, D. C. Williams, V. Y. Bykhovsky, N. Zaitseva, and V. Bukin, *Tetrahedron Lett.*, 1977, 2217.
58. A. R. Battersby, E. McDonald, M. Thompson, and V. Y. Bykhovsky, *J. Chem. Soc., Chem. Commun.*, 1978, 150.
59. G. Müller, K. D. Gneuss, H.-P. Kriemler, A. I. Scott, and A. J. Irwin, *J. Am. Chem. Soc.*, 1979, **101**, 3655.
60. G. Müller, K. D. Gneuss, H.-P. Kriemler, A. J. Irwin, and A. I. Scott, *Tetrahedron*, Suppl., 1981, **37**, 81.
61. A. R. Battersby, G. W. J. Matcham, E. McDonald, R. Neier, M. Thompson, W.-D. Woggon, V. Y. Bykhovsky, and R. Morris, *J. Chem. Soc., Chem. Commun.*, 1979, 185.
62. N. G. Lewis, R. Neier, G. W. J. Matcham, E. McDonald, and A. R. Battersby, *J. Chem. Soc., Chem. Commun.*, 1979, 541.
63. R. Deeg, H.-P. Kriemler, K.-H. Bergman, and G. Müller, *Hoppe-Seyler's Z. Physiol. Chem.*, 1977, **358**, 339.
64. M. Imfeld, D. Arigoni, R. Deeg, and G. Müller, in "Vitamin B$_{12}$," eds. B. J. Zagalak and W. Friedrich, de Gruyter, Berlin, 1979, p. 315.
65. A. R. Battersby, in "B$_{12}$," ed. D. Dolphin, Wiley, New York, 1982, vol. 1, p. 107.
66. A. R. Battersby, K. Frobel, F. Hammerschmidth, and C. Jones, *J. Chem. Soc., Chem. Commun.*, 1982, 455.
67. A. R. Battersby and S. Seo, *J. Chem. Soc., Perkin Trans. 1*, 1983, 3049.
68. A. R. Battersby, E. McDonald, R. Neier, and M. Thompson, *J. Chem. Soc., Chem. Commun.*, 1979, 960.
69. K. Kurumaya, T. Okazaki, and M. Kajiwara, *Chem. Pharm. Bull.* (*Tokyo*), 1989, **37**, 1151.

70. A. I. Scott, N. J. Stolowich, B. P. Atshaves, P. Karuso, M. J. Warren, M. Kajiwara, K. Kurumaya, and T. Okazaki, *J. Am. Chem. Soc.*, 1991, **113**, 9891.

71. R. A. Vishwakarma, S. Balachandran, A. I. D. Alanine, W. P. J. Stamford, F. Kiuchi, F. J. Leeper, and A. R. Battersby, *J. Chem. Soc., Perkin Trans. 1*, 1993, 2893.

72. D. Arigoni, in "Biosynthesis of the Tetrapyrrole Pigments," CIBA Foundation Symposium 180, Wiley, Chichester, 1994, p. 281.

73. A. I. Scott, N. J. Stolowich, J. Wang, O. Gawatz, E. Fridrich, and G. Müller, *Proc. Natl. Acad. Sci. USA*, 1996, **93**, 14316.

74. J. Wang, N. J. Stolowich, P. J. Santander, J.-H. Park, and A. I. Scott, *Proc. Natl. Acad. Sci. USA*, 1996, **93**, 14320.

75. P. J. Santander, C. A. Roessner, N. J. Stolowich, M. T. Holderman, and A. I. Scott, *Chem. and Biol.*, 1997, **4**, 659.

76. F. Blanche, D. Thibaut, L. Debussche, R. Hertle, F. Zipfel, and G. Müller, *Angew. Chem., Int. Ed. Engl.*, 1993, **32**, 1651.

77. A. I. Scott, M. Kajiwara, and P. J. Santander, *Proc. Natl. Acad. Sci. USA*, 1987, **84**, 6616.

78. M. Kodera, F. J. Leeper, and A. R. Battersby, *J. Chem. Soc., Chem. Commun.*, 1992, 835.

4.07
Biosynthesis of β-Lactam Compounds in Microorganisms

AXEL A. BRAKHAGE
Technische Universität Darmstadt, Germany

4.07.1   INTRODUCTION

Modern antibiotic therapy started with the discovery of β-lactam antibiotics in 1929, when Alexander Fleming published his observation about the inhibition of growth of *Staphylococcus aureus* on an agar plate contaminated with *Penicillium notatum*.[1] Three years later, it was shown that the growth inhibition was due to penicillin.[2] Because of its instability, the authors abandoned further work with this compound. About seven years later, however, investigations on penicillin were reopened by scientists around Chain and Florey in Oxford. Based on their research efforts the first clinical trials with penicillin were undertaken in 1941.[3] The further development of this β-lactam is the first and one of the most successful examples of the application of a compound isolated from microorganisms to infectious disease therapy. The success of β-lactams, most notably penicillins and cephalosporins, in the treatment of infectious diseases is due to their high specificity and their low toxicity. Despite a growing number of antibiotics and the incidence of penicillin-resistant isolates, the β-lactams are still by far the most frequently utilized.[4]

Reviews on the biosynthesis and structures of β-lactam compounds,[5-7] and on the biosynthesis, enzymology, and genetics of fungal penicillin and cephalosporin biosyntheses have been published.[4,7-10] Nonribosomal peptide synthesis was reviewed by Kleinkauf and von Döhren.[11] Here, the author focuses on the biosynthesis and the producing microorganisms, as well as some aspects of the molecular genetics of β-lactams.

4.07.2   GENERAL ASPECTS OF β-LACTAM COMPOUNDS

According to their chemical structures, β-lactams can be classified into five groups (Table 1). All of these compounds have in common the four-membered β-lactam ring. With the exception of monolactams which have a single ring only, β-lactams typically comprise a bicyclic ring system. The ability to synthesize β-lactams is widespread in nature. β-lactams have been found in fungi, as well as in Gram-positive and Gram-negative bacteria (Table 1). The production of hydrophilic cephalosporins is accomplished by organisms belonging to all three groups, whereas hydrophobic penicillins are only produced as end-products by filamentous fungi. For the remaining groups of β-lactams listed in Table 1, only bacterial producers have been reported. The number of pro-karyotic and eukaryotic microorganisms able to synthesize β-lactam antibiotics is continuously increasing.[7]

4.07.3   PENICILLINS, CEPHALOSPORINS, AND CEPHAMYCINS

4.07.3.1   General Aspects

In the penams (penicillins) and ceph-3-ems, the four-membered β-lactam ring is fused to a five-membered thiazolidine ring and a six-membered dihydrothiazine ring, respectively (Table 1). Ceph-3-ems are divided into four groups (Table 1). These include the cephamycins and cephabacins, which have a 7-α-methoxy substituent, in contrast to the cephalosporins. In chitinovorins, making up the remaining group, there is a 7-α-formamide residue (Table 2). These substructures impart resistance against some β-lactamases comparable with that exhibited by cephalosporin C.[5,6]

The early penicillins showed susceptibility to β-lactamase cleavage, whereas cephalosporin C, as well as advanced penicillins, proved to be resistant to *S. aureus* β-lactamase and displayed activity against Gram-negative bacteria. A wide spectrum of different derivatives is available exhibiting resistance against β-lactamases and activity against Gram-positive and Gram-negative bacteria.[5,6]

Table 1 Naturally occurring classes of β-lactams, examples of producing microorganisms are given.

Classes of naturally occurring β-lactams	Antibiotics (examples)	Producing microorganisms (examples)		
		Fungi	Bacteria	
			Gram⁺	Gram⁻
Penam	penicillins	Penicillium chrysogenum, Penicillium notatum, Aspergillus nidulans, Trichophyton, Sartorya		
Ceph-3-em	cephalosporins, cephamycins, cephabacins, chitinovorins	Acremonium chrysogenum (syn. Cephalosporium acremonium), Paecilomyces persicinus	Streptomyces clavuligerus, Streptomyces lipmanii, Nocardia lactamdurans	Flavobacterium sp., Lysobacter lactamgenus
Clavam	clavulanic acid, 2(2-hydroxyethyl) clavam		Streptomyces clavuligerus, Streptomyces antibioticus	
Carbapenem	thienamycins, olivanic acid, epithienamycins		Streptomyces cattleya, Streptomyces olivaceus, Streptomyces paracidomyceticus	Erwinia carotovora, Erwinia herbicola, Serratia sp.
Monolactam	nocardicins		Nocardia uniformis subsp. tsuyamanensis	
	monobactams			Agrobacterium radiobacter, Pseudomonas acidophila, Gluconobacter sp., Chromobacterium violaceum

After O'Sullivan and Sykes,[5] Gräfe,[6] and Aharonowitz et al.[8]

Table 2 Structures and producers of some ceph-3-ems.

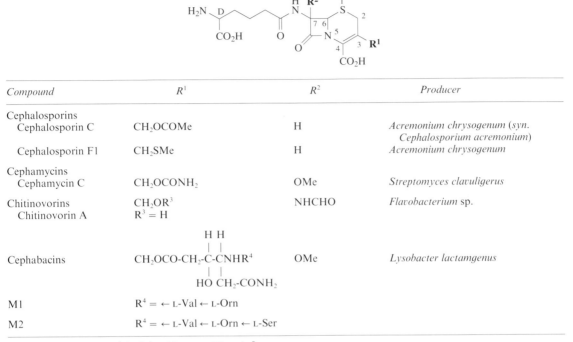

Compound	R^1	R^2	Producer
Cephalosporins			
Cephalosporin C	CH_2OCOMe	H	*Acremonium chrysogenum (syn.*
			Cephalosporium acremonium)
Cephalosporin F1	CH_2SMe	H	*Acremonium chrysogenum*
Cephamycins			
Cephamycin C	CH_2OCONH_2	OMe	*Streptomyces clavuligerus*
Chitinovorins	CH_2OR^3	NHCHO	*Flavobacterium* sp.
Chitinovorin A	$R^3 = H$		
Cephabacins	$CH_2OCO\text{-}CH_2\text{-}\overset{\text{H}}{\underset{\text{HO}}{C}}\text{-}\overset{\text{H}}{\underset{CH_2\text{-}CONH_2}{C}}NHR^4$	OMe	*Lysobacter lactamgenus*
M1	$R^4 = \leftarrow$ L-Val $\leftarrow$ L-Orn		
M2	$R^4 = \leftarrow$ L-Val $\leftarrow$ L-Orn $\leftarrow$ L-Ser		

After O'Sullivan and Sykes,[5] Gräfe,[6] and Jensen and Demain.[7]

4.07.3.2   A Brief Summary of Biosynthesis

To provide a brief overview of the biosynthetic pathways affording β-lactams, references have been omitted, but they are given in the following sections.

The biosyntheses of both penicillins and cephalosporins have the first two steps in common (Scheme 1).[10] All naturally occurring penicillins and cephalosporins/cephamycins are formed from the same three amino acids: L-α-aminoadipic acid (AAA), L-cysteine, and L-valine. In fungi, the nonproteinogenic amino acid AAA is derived from the fungus-specific aminoadipate pathway that leads to formation of lysine. It can also be provided by catabolic degradation of lysine although the contribution of this pathway to penicillin biosynthesis has not yet been clarified. In bacteria, a specific pathway for formation of AAA for β-lactam biosynthesis has been found (see Section 4.07.3.3).

In the first reaction cycle, the amino acid precursors are condensed to the tripeptide δ-(L-α-aminoadipyl)-L-cysteinyl-D-valine (ACV). All reactions that are required for formation of this tripeptide are catalyzed by a single multifunctional enzyme, designated according to the product formed δ-(L-α-aminoadipyl)-L-cysteinyl-D-valine synthetase (ACVS; Scheme 1). ACVS is encoded by a single structural gene designated *acvA* (*pcbAB*) (Scheme 1, Figure 1; see Section 4.07.3.4.1).

In the second step, oxidative ring closure of the linear tripeptide leads to formation of a bicyclic ring, i.e., the four-membered β-lactam ring fused to the five-membered thiazolidine ring, which is characteristic of all penicillins. The resulting compound isopenicillin N (IPN) possesses weak antibiotic activity and is thus the first bioactive intermediate of both penicillin and cephalosporin pathways. This reaction is catalyzed by isopenicillin N synthase (IPNS) encoded by the *ipnA* (*pcbC*) gene (see Section 4.07.3.4.2). IPN is the branch point of penicillin and cephalosporin biosyntheses (Scheme 1).

In the third and final step of penicillin biosynthesis, the hydrophilic AAA side-chain of IPN is exchanged for a hydrophobic acyl group catalyzed by acyl coenzyme A:isopenicillin N acyltransferase (IAT). The corresponding gene was designated *aat* (*penDE*) (see Section 4.07.3.5). In natural habitats, penicillins such as penicillin F and K, which contain Δ3-hexenoic acid and octenoic acid as side-chains, respectively, are synthesized (Table 3). By supplying the cultivation medium

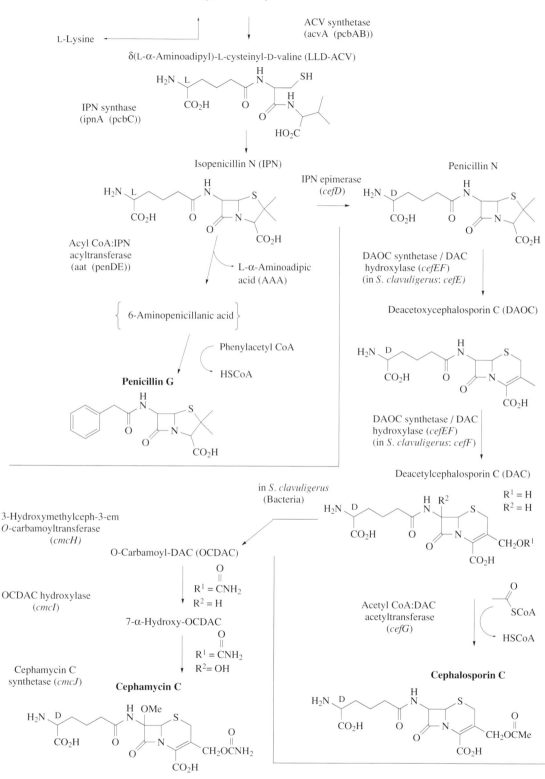

Scheme 1

with phenylacetic or phenoxyacetic acid, the synthesis can be directed mainly towards penicillin G and V, respectively (Table 3). The side-chain precursors have to be activated before they become substrates for the IAT. It is generally believed that the activated forms of the side-chains consist of their CoA-thioesters, but the mechanism behind this activation is still not fully elucidated.

The first step that commits the pathway to the production of cephalosporins is the isomerization of the AAA side-chain of IPN to the D-enantiomer to give penicillin N. This reaction is catalyzed by IPN epimerase (Scheme 1; see Section 4.07.3.6.1). Penicillin N is the precursor for antibiotics containing the cephem nucleus. It is converted to deacetoxycephalosporin C (DAOC) by DAOC synthetase (expandase) activity (Scheme 1). This ring expansion step involves the oxidative opening of the penam thiazolidine ring to give, upon reclosure, the six-membered dihydrothiazine ring, which is characteristic of all cephems (Table 2). In the next step, the methyl group at the C-3 atom of DAOC is hydroxylated/oxidized to form deacetylcephalosporin C (DAC; Scheme 1). In *Acremonium chrysogenum* both reactions are catalyzed by a single enzyme, DAOC synthetase (expandase)/DAC hydroxylase encoded by the *cefEF* gene, whereas in the bacterial cephalosporin-producer *Streptomyces clavuligerus*, one enzyme has been found for each reaction, encoded by the two genes *cefE* and *cefF* (see Section 4.07.3.6.2).

Table 3 Structures of some penicillins produced by fungi such as *P. chrysogenum* and *A. nidulans*.

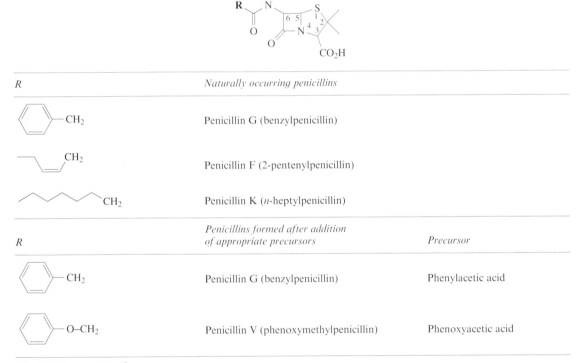

R	Naturally occurring penicillins	
⬡—CH$_2$	Penicillin G (benzylpenicillin)	
⌇CH$_2$	Penicillin F (2-pentenylpenicillin)	
⌇CH$_2$	Penicillin K (*n*-heptylpenicillin)	
R	*Penicillins formed after addition of appropriate precursors*	*Precursor*
⬡—CH$_2$	Penicillin G (benzylpenicillin)	Phenylacetic acid
⬡—O–CH$_2$	Penicillin V (phenoxymethylpenicillin)	Phenoxyacetic acid

After O'Sullivan and Sykes.[5]

In the last step of cephalosporin C biosynthesis, which is best studied in the fungus *A. chrysogenum*, an acetyl moiety from acetyl coenzyme A is transferred to the hydroxy group of DAC catalyzed by the product of *cefG*, acetyl coenzyme A:DAC acetyltransferase (Scheme 1; see Section 4.07.3.7). Several cephalosporins have been isolated from a variety of microorganisms that differ from cephalosporin C in the substituent attached to the 3′C oxygen (Table 2).

Cephamycin C biosynthesis, which has been studied best in *S. clavuligerus*, starts from the intermediate DAC. A carbamoyl group is attached to DAC to give *O*-carbamoyl-DAC (OCDAC). This reaction is catalyzed by 3-hydroxymethyl ceph-3-em *O*-carbamoyltransferase, which is encoded by the *cmcH* gene (Scheme 1; see Section 4.07.3.8.1). Then, the C-7 is hydroxylated by the action of OCDAC hydroxylase encoded by *cmcI* (Scheme 1; see Section 4.07.3.8.2). In the final step of cephamycin biosynthesis, the hydroxy group at C-7 is methylated to form cephamycin C

(7-methoxycephalosporin) catalyzed by cephamycin C synthetase. The corresponding gene is designated *cmcJ* (see Section 4.07.3.8.3).

4.07.3.3   Precursor Amino Acids

Radiochemical tracer studies during the 1940s–1950s revealed that penicillins and cephalosporins are naturally synthesized from the three amino acid precursors AAA, L-cysteine, and L-valine. Although only cysteine and valine are finally incorporated into the β-lactam nucleus, the requirement for AAA has long been established.[12]

Cysteine and valine are ubiquitous, proteinogenic amino acids, whereas AAA is an intermediate of the lysine biosynthetic pathway, which only occurs in this form in higher fungi (Scheme 2).[7,9,10,13] The pathway starts from α-ketoglutarate and is named according to its key intermediate α-aminoadipate pathway.[13] Hence, AAA is a branch point between lysine and penicillin/cephalosporin biosynthesis (Scheme 2). Genetic data suggested, however, that in *Aspergillus nidulans* AAA for penicillin biosynthesis is also provided by catabolic conversion of L-lysine by an as yet unidentified pathway. For *Penicillium chrysogenum*, a similar observation was made. AAA can also be obtained for penicillin synthesis by reversal of the last steps of the L-lysine biosynthetic pathway (Scheme 2). In addition, a high lysine:α-ketoglutarate-ε-aminotransferase activity was found in *P. chrysogenum*, which converts lysine into piperideine-6-carboxylic acid (Scheme 2). This finding suggests the existence of at least two different pathways involved in the degradation of lysine. In fungi, AAA can thus be provided by both the lysine biosynthetic pathway and degradation of lysine. The contribution of the latter catabolic pathway to β-lactam production, however, remains to be shown.[7,9,10]

Bacteria synthesize lysine starting from aspartic acid via a different route, designated diaminopimelic acid pathway. Since in this pathway AAA is not an intermediate, the question was addressed how cephalosporin-producing bacteria obtain the required AAA. In *S. clavuligerus* and *Nocardia lactamdurans* a specific enzyme activity, lysine ε-aminotransferase (LAT) encoded by the *lat* gene (Figure 1), was discovered removing an amino group of lysine to give cyclic 1-piperideine-6-carboxylic acid which is then oxidized to α-aminoadipic acid (Scheme 2).[7] Cephamycin-producing actinomycetes such as *S. clavuligerus* and *N. lactamdurans*, can metabolize lysine by two different pathways.[7] For growth lysine is degraded via the cadaverine pathway, whereas to generate AAA for cephamycin biosynthesis the LAT pathway is used (Scheme 2). The LAT activity seems to be present only in actinomycetes producing β-lactams and can thus be regarded as specific for a secondary metabolic product.[7] Consistent with this view is the notion that the *lat* gene encoding this activity is located in the β-lactam biosynthesis gene cluster (Figure 1).[7,9]

The synthesis of cysteine is another crucial factor in the economics of β-lactam formation (Scheme 1). Genetic and biochemical analyses have shown that the pathway in the β-lactam producing *A. nidulans* requires a sulfate permease (encoded by the *sB* gene) to transport sulfate into the cell, ATP sulfurylase (*sC* gene), adenosine-5-phosphosulfate (APS) kinase (*sD* gene), and 3-phosphoadenosine-5-phosphosulfate (PAPS) reductase. Sulfite is reduced to sulfide by the multienzyme complex sulfite reductase which involves the transfer of six electrons from NADPH to the sulfur atom.[21] From β-lactam producing fungi, some of the enzymes involved have been characterized, and the first genes cloned. The regulation of both cysteine and methionine biosynthesis appears to be complex and awaits further investigations.[10]

Starting from sulfide, three different routes leading to cysteine biosynthesis have been described.[10,22,23] In the "direct sulfhydrylation" pathway, reduced sulfur is incorporated by *O*-acetylserine sulfhydrylase into *O*-acetyl-L-serine to give cysteine. In a different pathway, "transsulfuration," sulfide incorporation is catalyzed by *O*-acetylhomoserine sulfhydrylase. The third possibility is the "reverse transsulfuration," in which the sulfur of methionine is transferred to cysteine. Interestingly, different routes leading to formation of cysteine are used by the various β-lactam producing fungi. This is apparently genetically determined. Although all pathways seem to exist in *A. chrysogenum*, the fungus prefers conversion of methionine to cysteine (reverse transsulfuration) for optimal cephalosporin C synthesis. In contrast, *P. chrysogenum* and *A. nidulans* synthesize cysteine mainly from direct sulfhydrylation.[23] The direct sulfhydrylation pathway is energetically more favorable than the transsulfuration pathway.[24]

The biosynthesis of valine is closely connected to the biosynthesis of the other two branched-chain amino acids leucine and isoleucine. Valine biosynthesis proceeds by four enzymatic steps with two moles of pyruvate as precursor metabolites.[10,23]

In fungi, pools of these three amino acids specific for β-lactam antibiotic biosynthesis, separate from the main pool of each amino acid have been proposed.[10]

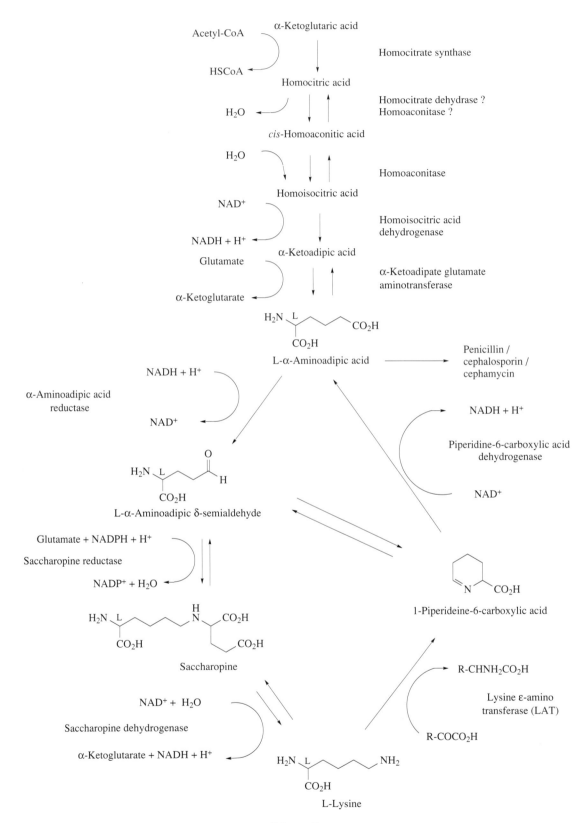

Scheme 2

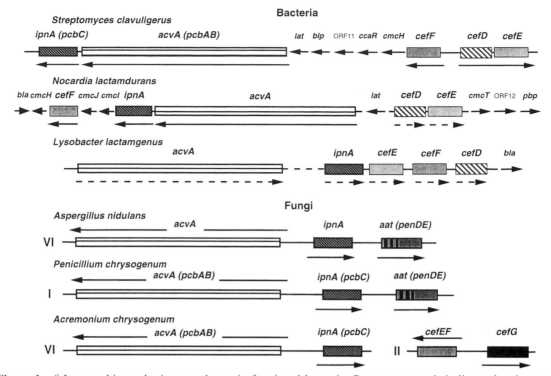

Figure 1 β-Lactam biosynthesis gene cluster in fungi and bacteria. Roman numerals indicate the chromosomes (in fungi) on which the genes are localized. Bacterial genes with fungal homologues are boxed. Abbreviations not mentioned in the text or Figure 2: *cmcT*, transmembrane protein; *pbp*, penicillin-binding protein; *bla*, β-lactamase; *blp*, showing similarity to the extracellular β-lactamase inhibitory protein BLIP. (After Brakhage.[10] The organization of the *Lysobacter lactamgenus* gene cluster was after Kimura *et al.*[14] The whole gene cluster of *Nocardia lactamdurans* has been published: Coque *et al.*[15–19] and Péres-Llarena *et al.*[20])

4.07.3.4 Steps Common to Both Penicillins and Cephalosporins

4.07.3.4.1 *Synthesis of the tripeptide δ-(L-α-aminoadipyl)-L-cysteinyl-D-valine*

The first reaction that has been shown for the biosyntheses of penicillin and cephalosporin/cephamycin is the formation of the ACV tripeptide. All of the reactions required for synthesis of the tripeptide are catalyzed by a single enzyme, ACVS, which is encoded by the *acvA* (*pcbAB*) gene (Scheme 1). Thus, the ACV tripeptide is formed via a nonribosomal enzyme thiotemplate mechanism from its amino acid precursors. This is similar in many aspects to the synthesis of other microbial peptides.[11,25,26]

ACVS activity was demonstrated in cell-free systems of *P. chrysogenum*, *A. chrysogenum*, and *S. clavuligerus*.[7] The first isolation of an ACVS protein was achieved by van Liempt *et al.*[27] who partially purified ACVS of *A. nidulans* 118-fold. Since then, ACVS enzymes have been purified from different organisms, including *S. clavuligerus*, *A. chrysogenum*, and *N. lactamdurans*. Attempts to purify ACVS from *P. chrysogenum* have thus far been unsuccessful because the enzyme seems to be rapidly degraded during chromatographic purification.[11,25]

Although not entirely clarified, it is believed that ACVS multienzymes are monomers with molecular masses of around 420 kDa composed of more than 3700 amino acids. They exhibit different catalytic activities, such as the specific recognition of the three amino acid precursors and their activation, peptide bond formation, isomerization of the L-valine moiety to the D-form, etc. As in ribosomal peptide biosynthesis, the carboxy function of the amino acid is activated by the formation of a mixed anhydride with the α-phosphate of ATP, resulting in the release of pyrophosphate. This has been used to develop assays based on amino acid-dependent exchange of ^{32}P between pyrophosphate and ATP.[27]

After activation of an amino acid, the formed aminoacyl adenylate is cleaved by the action of a thiol group present on the enzyme, resulting in formation of a thioester bond between the enzyme (at an appropriate location on the enzyme) and the amino acid, and in the release of AMP. These thioesterified amino acids play the same role as the tRNA-bound amino acids in the ribosomal peptide biosynthesis. They are high-energy intermediates that are the targets for nucleophilic attack by the amino group of a second amino acid, resulting in the formation of a peptide bond. As in the ribosome, the nascent peptide grows from the amino terminus to the carboxy terminus and the intermediate peptides remain bound (as thioesters) to the enzyme. Substrate specificity is less strict than in protein synthesis, since a variety of tripeptide analogues are known.[10,25]

Assuming three independent activation sites, the dissociation constants for the *S. clavuligerus* ACVS have been estimated to be 1.25 mM and 1.5 mM for cysteine and ATP, respectively, and 2.4 mM and 0.25 mM for valine and ATP. No AAA-dependent ATP/PPi exchange was detected with the enzyme preparation used, although the amino acid was bound to the enzyme in an ATP-dependent fashion.[28] This seems to distinguish the bacterial enzyme from fungal ACV synthetases (from both *A. nidulans* and *A. chrysogenum*), which drove radioactivity exchange in dependence of all three amino acids. For aminoacyl tRNA synthetases, dissociation constants are much lower when compared with *S. clavuligerus* ACVS, usually below 100 μM for their respective amino acids. This may be a way of guaranteeing the supply of amino acids to the primary metabolism, and avoiding the depletion of vital cellular components by secondary metabolism. L-Valine is apparently epimerized to the D-form at the tripeptide stage since no D-valine intermediate has been detected (Scheme 1).[10,11,28]

Each ACVS is encoded by a single structural gene (designated *acvA* or *pcbAB*) with a size of more than 11 kbp (Figure 1). The first *acvA* gene was cloned and sequenced from *P. chrysogenum* independently by Smith *et al.*[29] and Diez *et al.*,[30] based on the assumption that biosynthesis genes for antibiotics are clustered and information had accumulated about *ipnA* genes from several organisms. Subsequently, the corresponding genes have been cloned and sequenced from *A. nidulans*, *A. chrysogenum*, and bacterial cephamycin producers such as *N. lactamdurans*, *S. clavuligerus*, and *Lysobacter lactamgenus*.[7,9,10,14] Even in fungi, the open reading frame (ORF) does not seem to be interrupted by introns. Fungal *acvA* genes are divergently oriented to the *ipnA* genes (Figure 1). The genes are separated by about 1 kbp. Sizes of the intergenic regions between both genes vary slightly among the different fungi (Figure 1). In both *A. nidulans* and *P. chrysogenum*, it is known that the *acvA* mRNA starts within the intergenic region between *acvA* and *ipnA*.[7,10]

Amino acid sequences of ACVS proteins of all fungal and bacterial species so far identified contain three homologous regions of about 1000 amino acids. These contain repeated domains sharing extensive amino acid sequence similarities with each other, with the corresponding regions of ACVS protein of other fungi and bacteria, and with the repeated domains identified for *Bacillus brevis* gramicidin S synthetase 1 and 2, and tyrocidine synthetase I.[11] Since all of these enzymes specifically recognize amino acids and form adenylates, it is most likely that the respective adenylate forming domains (for nomenclature of ACVS domains, see Kleinkauf and von Döhren[11]) recognize and adenylate one of the constituent amino acids. The order of the biosynthesis of the AAA-Cys-Val tripeptide is believed to reflect the linear organization of the ACVS in AAA-, Cys-, and Val-activating domains.[11] A surprising result, however, was the observation of formation of *O*-methyl-seryl-D,L-valine by ACVS upon replacement of cysteine by *O*-methylserine.[31] This finding suggested that the second peptide bond is formed initially. Consequently, an order of peptide formation starting with Cys-Val and, subsequently, addition of AAA would thus be conceivable. Additional experiments, however, make this suggestion unlikely.[10]

Based on a microbiological assay for detection of pantothenic acid, it was observed that one mole of pantothenic acid was liberated per mole of purified *A. chrysogenum* ACVS. This implied the presence of one phosphopantetheine per ACVS molecule. Sequencing of the ACVS structural genes revealed, however, that in the three repeated regions of each ACVS some similarity to 4'-phosphopantetheine attachment sites described for polyketide synthases (i.e., Asp-Ser-Leu) is evident. This may reflect the attachment of multiple cofactors to ACVS. Because a single phosphopantetheine arm is sufficient for activity of fatty acid synthases, the finding of several phosphopantetheine attachment sites suggests a modified mechanism for the thiotemplate pathways to polypeptides. The relevance of all three pantetheine attachment sites of ACVSs, however, has not yet been proven experimentally. In the carboxy terminal region of these enzymes, sequence similarities to the thioesterase active site region, GXSXG, have been found, which would be required to release the generated tripeptide from the enzyme.[10,11,25,26] The current view of the thiotemplate mechanism of ACVS catalysis is summarized in detail by Kleinkauf and von Döhren,[11] and Zhang and Demain.[26]

ACV synthetases are of special interest since they represent a route for peptide bond formation independent of the ribosome and allow the incorporation of many nonproteinogenic amino acids. This was shown by Baldwin *et al.*,[32] who demonstrated that *S*-carboxymethylcysteine was an effective substitute for AAA and both allylglycine and vinylglycine could substitute for cysteine, indicating that the thiol group of cysteine is not essential for peptide formation. L-Alloisoleucine substituted effectively for valine. These results indicated that ACVS has a broad substrate specificity. Furthermore, since different parts of peptide synthetases are specific for certain amino acids, this can be used to engineer genetically new peptide synthetases producing new compounds, possibly with new pharmacological activities. This approach has been successfully used by Stachelhaus *et al.* (see Chapter 8).[33]

4.07.3.4.2 Ring formation reaction by isopenicillin N synthase

The second step of the penicillin/cephalosporin/cephamycin biosynthesis, i.e., cyclization of the linear ACV tripeptide to the bicycle IPN, is catalyzed by IPNS ("cyclase"), a nonheme Fe^{II}-dependent oxidase (Scheme 1). The enzyme has a molecular mass of approximately 38 kDa and formally catalyzes the removal of four hydrogen equivalents of the ACV tripeptide in a desaturative ring closure with concomitant reduction of dioxygen to water.[7,8,34,35] The IPNS reaction requires ferrous iron, molecular oxygen as cosubstrate, and ascorbate as electron donor to form the β-lactam and thiazolidine ring of IPN.[23]

IPNS was purified to homogeneity from *A. chrysogenum*[36–38] and has subsequently been obtained from *P. chrysogenum*, *A. nidulans*, several actinomycetes such as *Streptomyces clavuligerus*, *S. lipmanii*, *N. lactamdurans*, and the Gram-negative bacterium *Flavobacterium* sp.[7] It was shown that two interconvertible forms of the enzyme exist: an oxidized state with a disulfide linkage and a reduced state.[38] For *P. chrysogenum* IPNS, it was found that it is strongly inhibited by glutathione and also sensitive to cobalt inhibition.[9]

Only the free thiol form of the tripeptide ACV serves as a substrate, the bisdisulfide dimer, which is spontaneously formed, being inactive.[34] *S. clavuligerus* possesses a disulfide reductase that recognizes bis-ACV as a substrate.[39] In *P. chrysogenum*, a broad-range disulfide reductase belonging to the thioredoxin family of oxidoreductases was found which efficiently reduced bis-ACV to the thiol monomer. When coupled to IPNS *in vitro*, it converted bis-ACV to IPN and was therefore suggested to play a role in penicillin biosynthesis.[40] In enzyme assays *in vitro*, the thiol groups of both the ACV tripeptide and the IPNS enzyme are kept in a reduced state by addition of ascorbate and dithiothreitol. In these assays, the appearance of antibiotic activity due to formation of IPNS from the antibiotically inactive ACV is measured with an indicator organism that is sufficiently sensitive. Alternatively, IPN can be monitored by HPLC.[7,34]

Mössbauer, EPR, and optical spectroscopies suggested that ACV binds directly to the active-site iron of IPNS via the cysteinyl thiol of ACV. A six-coordinate metal centre at the active site was proposed, with two or three endogenous histidine ligands, an aspartate and sites for the thiolate of ACV, oxygen, and solvent.[34,35] The ACV sulfur atom was shown to bind to the active-site iron of the enzyme.[41,42] The crystal structure of the *A. nidulans* IPNS was solved at a resolution of 2.5 Å and 1.3 Å complexed with manganese,[35] and with Fe^{2+} and substrate,[43] respectively. The secondary nature of IPNS was found to consist of ten helices and 16 β-strands. Eight of the β-strands fold to give a jelly-roll motif. The active structure shows the manganese ion attached to four protein ligands (His214, Asp216, His270, and Gln330) and bears two water molecules occupying coordination sites directed into a hydrophobic cavity within the protein.[35] The Fe^{II}:ACV:IPNS structure has one protein molecule with ferrous ion and ACV bound at the active site. The side-chain of Gln330, which coordinates the metal in the absence of substrate, is replaced by the ACV thiolate.[43] In the substrate complex, three of the five coordination sites are filled with protein ligands: His214, His270, and Asp216.[44] The remaining two sites are occupied by a water molecule (at position 398) and the ACV thiolate.[43] Such a structural characteristic (an iron-binding site within an unreactive hydrophobic substrate binding cavity) is probably a requirement for this class of enzyme, as it results in the isolation of the reactive complex and subsequent intermediates from the external environment. Thus, the reaction can be channeled along a single path, avoiding the many side reactions potentially open to the highly reactive species resulting from the reduction of dioxygen at the metal.[35] The role of enzymes in such processes has been designated negative catalysis.[45] IPNS catalyzes a unique enzyme reaction with no precedent in chemistry.[35]

All intact IPNS enzymes cloned to date have proline at position 285 in a highly conserved region. This Pro residue seems to be essential for activity because a mutant (N2) of *A. chrysogenum*, which did not produce cephalosporin, encodes a defective *ipnA* gene, probably due to a single base-pair change, resulting in a change of Pro285 to Leu.[10,46]

Baldwin and Wan[47] proposed a catalytic mechanism for IPNS which involves formation of an intermediate carbon radical of the LLD-ACV, but complete details of the reaction are yet to be determined. Additional data on the mechanism of the IPNS reaction suggest that initial formation of the β-lactam ring is followed by closure of the thiazolidine ring.[48] The current model of the catalytic mechanism can be found in Roach *et al.*[35,43]

The IPNS shows a broad substrate specificity in particular with alterations in the AAA moiety and the valine residue of ACV. This finding has an ingenious use in creating novel penicillins from ACV analogues, although cyclization of unnatural tripeptides occurs at lower efficiency.[49] Nevertheless, many new penicillins have been produced biosynthetically via this route which is thus very promising for the generation of new β-lactam compounds *in vivo.*[50]

The genes encoding IPNS enzymes are designated *ipnA* (*pcbC*) and have a size of approximately 1 kbp. The *ipnA* (*pcbC*) gene from *A. chrysogenum* was the first gene encoding an enzyme of β-lactam biosynthesis that was cloned and sequenced.[51] Thereafter, *ipnA* (*pcbC*) genes have been isolated from different fungi and bacteria such as *A. nidulans, P. chrysogenum, S. clavuligerus, Streptomyces griseus, S. lipmanii, Flavobacterium* sp., *N. lactamdurans*, etc. (Figure 1).[7,15]

In contrast to bacteria, in fungi *ipnA* and *acvA* are bidirectionally oriented (Figure 1). For the fungal genes, it has been demonstrated that the *ipnA* transcripts start in the corresponding intergenic regions. Fungal IPNS genes identified up until now do not possess introns.[34]

4.07.3.5 Step Specific for Penicillin Biosynthesis: Acyl Coenzyme A:Isopenicillin N Acyltransferase

The third and final reaction of penicillin biosynthesis, which does not occur in cephalosporin biosynthesis and has been found in fungi only, is catalyzed by acyl coenzyme A:IAT.[9,10,52] The hydrophilic AAA side-chain is exchanged for a hydrophobic acyl group, e.g., phenylacetyl in penicillin G (Scheme 1, Table 3). IAT shows a broad substrate specificity.[9,52,53] By addition of appropriate precursor molecules, the fermentation can be directed towards a specific penicillin; e.g., for production of penicillin G, phenylacetic acid is added, or for production of penicillin V, phenoxyacetic acid (Table 3). Once the precursor has been taken up, it must be activated to its CoA thioester. This reaction seems to be carried out by phenylacetyl-CoA ligase.[52] It is unclear, however, whether a specific enzyme is needed for this reaction because a possible candidate is the acetyl-CoA synthase (ACS), which was purified from *P. chrysogenum* and its structural gene *acuA* cloned. It was shown that the ACS enzymes of both *P. chrysogenum* and *A. nidulans* have the capability to catalyze *in vitro* the activation (to their CoA thioesters) of some of the side-chain precursors required for the production of several penicillins by these fungi.[52]

A two-step enzymatic process for conversion of IPN to penicillin G has been proposed.[54] In the first step, IPN is deacylated to 6-aminopenicillanic acid (6-APA), which in the second step is acylated to penicillin G through addition of a phenylacetyl group from its CoA derivative (Scheme 1). Thus, two enzymatic functions are required, an amidohydrolase and an acyl-CoA:6-aminopenicillanic acid acyltransferase function. For many years, it remained unclear whether two enzymes and, in addition, 6-APA as an intermediate are involved in this step of penicillin biosynthesis. The cloning and sequencing of the gene helped to solve this question. It was found that IAT of *P. chrysogenum* has isopenicillin N amidohydrolase, 6-aminopenicillanic acid acyltransferase and penicillin amidase activities, all of which are encoded by the single *aat* (*penDE*) gene.[53] 6-APA remains bound to IAT when the enzyme is saturated with appropriate acyl-CoA substrates but is released in their absence. Significant amounts of 6-APA are produced when exogenous side-chain precursors are not fed to *P. chrysogenum.*[55]

Purification of *P. chrysogenum* IAT led to the assumption that the active enzyme is a monomeric protein with a molecular mass of 29 kDa.[56,57] In several partially purified enzyme preparations, however, three proteins of about 40, 29, and 11 kDa were present in ratios that differed between experiments.[9,52,55]

The cloning and sequencing of the corresponding *P. chrysogenum* and *A. nidulans aat* (*penDE*) genes with sizes of about 1.2 kbp were independently achieved by Barredo *et al.*[58] and Tobin *et al.*[59] (Figure 1). The deduced amino acid sequences of the ORFs (357 amino acids for both proteins) encoded a protein of about 40 kDa. In contrast to the other penicillin biosynthesis genes (*acvA* and *ipnA*), the *aat* genes contain three introns in both organisms at similar positions.[58–60] No DNA sequence homologous to the *aat* gene of *P. chrysogenum* was found in the genome of three different strains of *A. chrysogenum* and actinomycetes.[9] This finding is consistent with the notion that 6-aminopenicillanic acid:acyltransferase activity which is also carried out by IAT is lacking in *A.*

chrysogenum and other cephalosporin producers.[57] Therefore, these organisms do not produce penicillin G or any other penicillins with a hydrophobic side-chain.

The active form of the IAT enzyme results from processing of the 40 kDa monomeric precursor to a heterodimer containing subunits of 11 kDa and 29 kDa.[9,52,55] When both subunits from *P. chrysogenum* were expressed from two different plasmids in *Escherichia coli*, production of either subunit in the absence of the other did not form active recombinant IAT. However, cotransformation of *E. coli* with two plasmids, each encoding a different IAT subunit, produced recombinant IAT having acyl-coenzyme A:6-aminopenicillanic acid acyltransferase activity, providing evidence that both subunits are required for activity.[61]

To investigate the mechanism of IAT proteolysis in detail, *P. chrysogenum* IAT was overexpressed from the *aat* gene in *E. coli* and recombinant active IAT isolated. When purified to homogeneity, recombinant IAT was an α, β-heterodimer, comprising 11 kDa (α) and 29 kDa (β) subunits, derived from a 40 kDa precursor polypeptide by a posttranslational cleavage. In *P. chrysogenum*, it was shown that the processing event that generated the two subunits from the 40 kDa precursor polypeptide occurred between Gly102 and Cys103.[62] Additional investigations suggest that the formation of recombinant IAT involves cooperative folding events between the subunits and that IAT hydrolysis is an autocatalytic event.[61]

Site-directed mutagenesis of the *aat* gene and expression in *E. coli* revealed that Cys103 is required for IAT proenzyme cleavage. Whether this requirement reflects a direct participation of Cys103 in cleavage or as part of a cleavage recognition site has not yet been clarified. However, it cannot be entirely excluded that Cys103 is involved in IAT enzyme activity because all of these experiments were based on the detection of enzyme specific activity.[63] The encoded amino acid sequence in the cleavage site is identical in *P. chrysogenum* and *A. nidulans* (Arg-Asp-Gly ... Cys-Thr-Thr).[62–64]

IAT enzymes seem to be similar to bacterial penicillin and cephalosporin acylases that catalyze the deacylation of acyl side-chains of penicillins and cephalosporins, to yield 6-APA and 7-aminocephalosporanic acid, respectively. The penicillin acylase from *E. coli* ATCC11105 is a periplasmic enzyme which consists of two nonidentical subunits that are produced by posttranslational processing from a precursor protein. The functional and structural similarity between fungal IATs and bacterial acylases is striking, although there is only very low sequence similarity (approximately 11% identical amino acids).[10]

4.07.3.6 Steps Common to Cephalosporin C/Cephamycin C Biosynthesis

4.07.3.6.1 Isopenicillin N epimerase

Isopenicillin N epimerase catalyzes the epimerization of the AAA side-chain of IPN to the D-enantiomer to give penicillin N which is the precursor to antibiotics containing the cephem nucleus, i.e., cephalosporins and cephamycins (7-methoxycephalosporins, Scheme 1).[7,10] The purification of *A. chrysogenum* IPN epimerase proved to be difficult because the enzyme was extremely labile in cell-free preparations.[65] No further data on the fungal protein or gene are available.[7]

The *S. clavuligerus* IPN epimerase was found to be more stable and was characterized biochemically.[66] The corresponding gene (*cefD*) was cloned and sequenced, and is located immediately upstream of the gene for the deacetoxycephalosporin C synthetase (expandase; *cefE*; Figure 1). *cefD* encodes 398 amino acid residues which give a protein with a deduced molecular mass of 43 497 Da.[67] The cloning of the *cefD* genes of *S. lipmanii*, *N. lactamdurans*, and *L. lactamgenus* YK90 was reported (Figure 1).[10]

P. chrysogenum expresses no isopenicillin N epimerase activity and a probe of the *S. lipmanii cefD* gene did not hybridize to the DNA of *P. chrysogenum*.[68] However, some minor isopenicillin N epimerase activity possibly due to some unspecific amino acid racemase exists in *P. chrysogenum*, because recombinant strains of *P. chrysogenum* expressing *S. clavuligerus* deacetoxycephalosporin C synthetase (expandase; *cefE*) were found to produce deacetoxycephalosporin C.[69,70]

4.07.3.6.2 Ring expansion by deacetoxycephalosporin C synthetase (expandase)/deacetylcephalosporin C hydroxylase

In cephalosporin biosynthesis penicillin N is converted to DAOC by expansion of the five-membered thiazolidine ring to give the six-membered dihydrothiazine ring of DAOC (Scheme 1).

This reaction is catalyzed by DAOC synthetase which possesses the required expandase function.[23] The enzyme was purified from *A. chrysogenum*, *S. clavuligerus* and *N. lactamdurans*.[10] Fungal and bacterial expandases are stimulated by reducing agents, like DTT or glutathione, and show an absolute requirement for Fe^{2+}, molecular oxygen, α-ketoglutarate, ascorbate, and possibly ATP. These unusual cofactor requirements are characteristic of the class of intermolecular dioxygenases which activate oxygen in the decomposition of equimolar amounts of α-ketoglutarate to form carbon dioxide and succinate.[7,23] In the following reaction step, the methyl group at the C-3 atom of DAOC is hydroxylated/oxidized to give DAC.[71,72] This reaction is catalyzed by DAC hydroxylase, which is very similar to DAOC synthetase. The enzyme also belongs to the group of α-ketoglutarate-linked dioxygenases.

Ring expansion by DAOC synthetase and hydroxylation reaction both reside on the same peptide in *A. chrysogenum*.[71–73] Expression of the corresponding intronless *cefEF* gene with a size of 996 bp encoding 332 amino acids (deduced M_r 36 462) in *E. coli* established unequivocally that, in this fungus, one protein is responsible for the ring expansion of penicillin N to DAOC and the 3′-hydroxylation of DAOC to DAC (Scheme 1).[73] In contrast, in *S. clavuligerus* both enzymes can be separated by anion-exchange chromatography[74] and were later found to be encoded by two genes, *cefE* and *cefF*. *cefE* consists of 933 bp encoding 311 amino acids with a deduced M_r of 34 519, *cefF* of 954 bp coding for 318 amino acids with a deduced M_r of 34 584.[75,76]

The *cefEF* gene of *A. chrysogenum* is located on chromosome II. It is closely linked to the *cefG* gene but transcribed in the opposite direction (Figure 1). The intergenic region of about 1 kbp is believed to contain the promoters for both genes. In *S. clavuligerus*, the *cefF* gene is closely linked to *cefD* and *cefE*, but is transcribed in the opposite orientation (Figure 1).[76] In *L. lactamgenus* and *N. lactamdurans* different orders of genes were found (Figure 1).[10,14–19]

4.07.3.7 Step Specific for Cephalosporin C Biosynthesis in Fungi: Acetyl Coenzyme A: Deacetylcephalosporin C Acetyltransferase

In the last step of the cephalosporin C biosynthesis which is best studied in the fungus *A. chrysogenum*, an acetyl moiety from acetyl coenzyme A is transferred to the hydroxy group of DAC catalyzed by acetyl coenzyme A:DAC acetyltransferase (Scheme 1).[7] The corresponding structural gene (*cefG*) of *A. chrysogenum* was cloned and sequenced independently by three groups who obtained slightly different results.[10] It is composed of about 1300 bp and contains two introns. The protein consists of about 400 amino acids giving a deduced molecular mass of about 40 kDa. It is transcribed in the opposite orientation of *cefEF*, separated by an intergenic region of about 1 kbp (Figure 1).[10]

4.07.3.8 Steps Specific for Cephamycin Biosynthesis in Actinomycetes

4.07.3.8.1 3-Hydroxymethyl ceph-3-em O-carbamoyltransferase

Starting from the intermediate DAC, a carbamoyl group is attached to DAC to give *O*-carbamoyl-DAC (OCDAC). This reaction is catalyzed by 3-hydroxymethyl ceph-3-em *O*-carbamoyl-transferase[77] which is encoded by the *cmcH* gene[19] (Scheme 1). The enzyme has been partially purified from *S. clavuligerus*; it transfers the carbamoyl group from carbamoyl phosphate to DAC in a reaction stimulated by Mg^{2+}, Mn^{2+}, and ATP.[7,77] The gene has been characterized from *N. lactamdurans*, and in part from *S. clavuligerus*.[19] The corresponding genes are present within the gene clusters (Figure 1). The *N. lactamdurans* gene encodes a protein with a deduced molecular mass of 57 149 Da. It contains a carbamoylphosphate-binding specific sequence.

4.07.3.8.2 O-Carbamoyl deacetylcephalosporin C (OCDAC) hydroxylase

The C-7 is hydroxylated by the action of OCDAC hydroxylase. It is also a dioxygenase which requires oxygen, iron, ascorbate, and 2-ketoglutarate as a cosubstrate. The enzyme has been purified to near homogeneity and migrates in SDS-PAGE with a molecular mass of 32 kDa.[7] The corresponding gene was designated *cmcI* and found to be located on a 1.5 kbp *Sal*I DNA fragment isolated from *S. clavuligerus*.[78] The *cmcI* gene was also cloned from *N. lactamdurans*.[18]

4.07.3.8.3 *Cephamycin C synthetase*

In the final step of cephamycin biosynthesis, the hydroxy group at C-7 is methylated to form cephamycin C (7-methoxycephalosporin) catalyzed by cephamycin C synthetase. The corresponding gene is named *cmcJ*. It has been cloned and sequenced from *N. lactamdurans*.[18] It consists of 876 bp coding for 292 amino acids which give a protein with a deduced M_r of 32 090.

4.07.3.9 Genetics of Cephalosporin/Penicillin Biosynthesis

4.07.3.9.1 *Structural genes are organized in clusters*

In bacteria and fungi all structural genes of β-lactam biosynthesis are clustered (Figure 1). The penicillin biosynthesis genes in *P. chrysogenum* and *A. nidulans* were found to be tightly clustered. In *A. chrysogenum*, two clusters containing the cephalosporin biosynthesis genes have been identified. In cephamycin C producing bacteria the genes are organized into a single cluster which, in *S. clavuligerus*, together with the clavulanic acid biosynthesis gene cluster forms a "super-cluster" (Figures 1 and 2).[8–10,79]

The linkage of antibiotic-biosynthesis genes is a well-known phenomenon in many antibiotic-producing organisms. It has been speculated that linkage has occurred during evolution owing to an ecological selective advantage.[77] It was suggested that coordinated regulation of antibiotic-biosynthesis genes could be achieved by organizing the genes into large operons controlled by a single promoter.[80] For example, genes of the actinorhodin biosynthetic pathway in *Streptomyces coelicolor* are clustered and expressed in several polycistronic messages.[81] In eukaryotic fungi, however, β-lactam biosynthesis genes are transcribed separately, and are expressed from different promoters.[34] Hence, in fungi, there is no obvious need for clustering and it thus seems more likely that linkage reflects a common ancestral origin (see Section 4.07.3.9.6). However, there is no evidence that the acyl coenzyme A:isopenicillin N acyltransferase gene (*aat*) has a close relative in modern prokaryotes, even though it is part of the cluster. This fact supports the hypothesis that linkage might also confer an ecological advantage to the eukaryotic fungi, although the reason for this is not yet understood.[8,10]

4.07.3.9.2 *Regulation in fungi*

(i) General aspects

Much of the knowledge on the regulation of β-lactam biosynthesis genes was obtained in fungi. Industrial production of penicillin and cephalosporin was achieved with *P. chrysogenum* and *A. chrysogenum* (syn. *Cephalosporium acremonium*), respectively. These fungi, however, belong to the deuteromycetes which are in general difficult to analyze genetically. Therefore, up to now, the greatest progress in elucidation of the molecular regulation of biosynthesis of β-lactams has been made in the penicillin-producer *A. (Emericella) nidulans*, because this fungus is an ascomycete with a sexual cycle. Hence, classical genetic techniques can be applied to *A. nidulans*, which, together with molecular techniques, facilitated a thorough analysis of the genetic regulation of metabolic pathways, including that of penicillin biosynthesis.[82]

Within the last few years, several studies have indicated that the β-lactam biosynthesis genes are controlled by a complex regulatory network.[10] These studies led to the finding that the promoter strengths of penicillin biosynthesis genes are rather different. In all three fungi investigated, i.e., *A. nidulans*, *P. chrysogenum*, and *A. chrysogenum acvA* had much lower expression than *ipnA*.[10] In addition, in *A. nidulans*, it was shown that *aat* had lower expression than *ipnA*.[83] The intergenic regions between *acvA* and *ipnA* thus seem to contain the information required for the remarkable difference in expression levels between *acvA* and *ipnA*. The low expression of *acvA* is, at least in wild-type strains of *A. nidulans*, rate-limiting for penicillin production because overexpression of *acvA* led to drastically increased production of penicillin,[84] while similar overexpression of *ipnA* and *aat* did not.[60]

Previous studies of the formation of secondary metabolites in batch fermentations led to the definition of two phases. The growth phase or trophophase, and the period of secondary metabolite production or idiophase. A strict separation of both phases, with production of secondary metabolites virtually restricted to the idiophase, has been observed in antibiotic-producing bacterial cultures. In *A. nidulans*, when lactose is used as the sole carbon source, there is no sharp separation of tropho- and idiophase with respect to penicillin production or the expression of penicillin biosynthesis genes.[10] For some secondary metabolites, it seems likely that their production in a clear-cut idiophase reflects the inhibiting effects of certain compounds in the medium, e.g., glucose,[85,86] rather than an intrinsic temporal delay in the product formation pattern. This view is consistent with the results showing that in *P. chrysogenum* wild-type (NRRL1951) and high-penicillin producing strain P2, the highest steady-state level of mRNAs of all penicillin biosynthesis genes was observed during maximal growth in both shake flasks and 2-1 fermenters.[87] However, some observations imply that in *P. chrysogenum* there might be a temporal expression of β-lactam biosynthesis genes.[10,88,89]

(ii) Carbon source regulation and implications for posttranscriptional regulation

Industrial production of penicillin with *P. chrysogenum* was usually carried out by using lactose as the carbon source (C-source), which gave the highest penicillin titer. The use of excess glucose leads to a drastic reduction of the penicillin titer. This problem is partially overcome by feeding subrepressing doses of glucose and by the use of lactose as C-source.[90] Since, in general, the fungus grows better with glucose than with lactose,[85] the production of penicillin appears to be favored by suboptimal growth conditions.

C-source regulation seems to act at several points of the penicillin biosynthesis: (i) flux of AAA (3.3), (ii) activation of side-chain precursors, (iii) transcription of penicillin biosynthesis genes, and (iv) posttranscriptional regulation of penicillin biosynthesis genes.[10]

Glucose represses the formation of the enzymes ACVS and IPNS in *P. chrysogenum*. In agreement with these findings was the observation that the expression of both the *P. chrysogenum* Q176 *acvA* (*pcbAB*) and *ipnA* (*pcbC*) gene was repressed by glucose.[10,34,89] In *A. nidulans*, the use of repressing carbon sources such as glucose or sucrose instead of lactose in the fermentation medium reduced the amount of penicillin produced and the expression of the *ipnA* gene (Figure 1).[85,86] This was further supported by the finding that the IPNS specific activity was drastically reduced in glucose-grown mycelia.[85] The repression of *ipnA* expression by repressing C-sources occurs, at least in part, at the transcriptional level.[86]

Unexpectedly, the expression of both *A. nidulans* *acvA* and *aat* reporter gene fusions was only slightly, if at all, repressed by glucose in fermentation medium. However, the specific activity of the *aat* gene product, IAT, was reduced in mycelia grown with glucose instead of lactose.[83,85] This suggests that the glucose regulation of IAT takes place, at least in part, posttranscriptionally. The effect of glucose on IAT specific activity could be reversed by washing cells and their reincubation in lactose-containing medium.[83]

Cephalosporin C formation by *A. chrysogenum* also depends on the C-source used. C-sources that support the most rapid growth, like glucose or glycerol, had a negative effect on β-lactam production.[7,10,12] The ACVS level was not affected by higher concentrations of glucose, making a major repressive effect of glucose at the level of gene expression unlikely. ACVS specific activity measured in crude protein extracts was severely inhibited by glucose and glycerol. The inhibition could be reversed by the addition of ATP. Interestingly, the purified enzyme was not inhibited by glucose, but was inhibited by glyceraldehyde-3-phosphate and glyceraldehyde.[91] From these observations the authors concluded that inhibition of ACVS specific activity in crude extracts, by glucose and some other sugars, is due to depletion of the cofactor ATP via sugar metabolism.

In contrast to the penicillin production strain AS-P-78 of *P. chrysogenum*, the IAT-specific activity of both *A. nidulans* and the *P. chrysogenum* wild-type strain NRRL1951 was clearly reduced in glucose-grown cultures.[34] Glucose was also found to cause inactivation of *P. chrysogenum* acetyl-CoA synthetase, which has the capability to catalyze the activation (to their CoA thioesters) of some of the side-chain precursors required for the production of several penicillins *in vitro*.[92]

Previously, it was reported that the uptake of side-chain precursors of phenylacetic acid was regulated by glucose.[93] It has been shown, however, that phenylacetic acid passes through the plasma membrane via passive diffusion of the protonated species, thus excluding the possibility that the uptake could be regulated by the available C-source.[94]

The molecular mechanisms of the C-source regulation of *ipnA* expression are not understood, although the glucose repression of several genes of the primary metabolism has been intensively investigated in *A. nidulans*.[10]

(iii) pH regulation mediated by the transcriptional regulator PACC

Penicillin production is subject to regulation by ambient pH.[95,96] There was markedly more penicillin in the culture broth when the pH of the medium was kept constant at 8.1 than at 6.5 or 5.1.[95] Control experiments excluded the possibility that the difference in penicillin yield measured was due to the higher stability of penicillin G at alkaline pH.[96] It was found that the penicillin biosynthesis is part of the pH regulatory system. Extracellular enzymes and permeases, for example, are not protected by the intracellular pH homeostatic system and, therefore, their synthesis is controlled by external pH. This mode of regulation avoids the synthesis of, for example, secreted alkaline phosphatase in an acidic pH environment. The central regulator is the transcriptional factor PACC. PACC consists of 678 amino acid residues that form three putative Cys$_2$His$_2$ zinc fingers required for DNA binding.[97] The full-length form of PACC, which predominates at acidic ambient pH, is not functional.[98] It must be specifically proteolyzed to yield the functional version (for both positive and negative roles) containing the N-terminal 40% of the protein. The processed form is functional as both activator and repressor.[98,99]

At alkaline ambient pH, PACC activates transcription of alkaline-expressed genes, e.g., of the alkaline phosphatase and protease genes *palD* and *prtA*, respectively, and also of the penicillin biosynthesis genes *ipnA*[97] and very likely of *acvA*.[10] The intergenic region between *acvA* and *ipnA* was found to contain four *in vitro* PACC binding sites designated ipnA1, ipnA2, ipnA3, and ipnA4AB, recognized by a GST::PACC (amino acids 31–195) fusion protein. The fusion protein was demonstrated to bind to the core consensus GCCARG.[97] A mutation analysis of each of these sites revealed that, *in vivo*, the binding site ipnA3 was most important for PACC dependent *ipnA* expression, whereas sites ipnA2 and ipnA4AB were less important, although site ipnA2 was bound with highest affinity by PACC *in vitro*. Binding site ipnA1 was apparently not required for PACC dependent *ipnA* expression.[100] PacC mutants mimicking the effect of growth at alkaline pH produced about twofold more penicillin than wild-type strains, whereas strains carrying *pal* mutations mimicking growth at acidic pH showed lowest penicillin titers. These findings are consistent with the observation that penicillin titers were higher when wild-type strains were grown at alkaline pH.[95]

The signal triggered by alkaline ambient pH is transduced via the *pal* gene products. This has been concluded from genetic experiments. So far, there is a total of six discrete regulatory *pal* genes (*palA*, *-B*, *-C*, *-F*, *-H*, and *-I*). Pal mutations impair pH regulation of the syntheses of extracellular enzymes, like phosphatases and permeases. PACC proteolysis occurs in response to a signal provided by the six *pal* gene products in alkaline environments.[97–99] Thus, the products of these *pal* genes constitute an alkaline ambient pH signal transduction pathway, which is required for conversion of PACC to its functional form.[101]

The *P. chrysogenum pacC* gene encodes a predicted 641-residue protein which exhibits most of the features described for the *A. nidulans* PACC protein, including three zinc fingers of the Cys$_2$His$_2$ class.[88] The *pacC* genes of both *P. chrysogenum* and *A. nidulans* seem to be autogenously regulated.[88,97] A fusion protein of glutathione S-transferase with amino acids 46–154 of *P. chrysogenum* PACC (GST::PACC (46–154)), bound *in vitro* to the intergenic region between *P. chrysogenum acvA* and *ipnA*. By computer analysis, seven PACC binding consensus sites (5′-GCCARG-3′) were found in the intergenic region. This is consistent with the finding that steady state *ipnA* mRNA levels were increased at alkaline pH.[88]

pacC genes are not confined to β-lactam producing fungi, as the cloning of the *Aspergillus niger pacC* gene showed.[102] Hence, it seems to represent a wide-domain regulator that regulates, with other genes, the penicillin biosynthesis genes.

Because the use of glucose or sucrose as the C-source leads to acidification of the medium[85,86,96] it was conceivable that the glucose/sucrose effect was due to pH regulation.[96] However, neither acidic external pH nor mutations *palA1*, *palB7*, and *palF5* mimicking the effects of growth at acidic pH, prevented C-source derepression.[96] Furthermore, the PACC binding sites determined *in vitro* by the use of a fusion polypeptide containing the PACC DNA-binding domain are not located in the *cis*-acting region which was shown to mediate C-source repression of *ipnA-lacZ* expression.[96,97] Taken together, these data support the model of Espeso *et al.*[96] and Tilburn *et al.*[97] of independent regulatory mechanisms, one mediating carbon regulation and another mediating pH regulation through the *pacC*-encoded transcriptional regulator. Since alkaline pH values *per se* seem to derepress *ipnA* transcription, it was proposed that internal alkalinity represents a physiological signal which triggers penicillin biosynthesis.[96]

In contrast to the situation in *A. nidulans*, alkaline ambient pH did not seem to override the negative effect of repressing C-source on *ipnA* (*pcbC*) transcription in *P. chrysogenum*.[88] The reason for the pH mediated regulation of penicillin biosynthesis is unclear. It was suggested that it might be connected to the observation that β-lactams exhibit increased toxicity on at least some bacterial species at alkaline pH. Furthermore, bacterial competition with fungi may be more intense at alkaline pH.[99]

(iv) Nitrogen regulation

The effect of the availability of nitrogen source on penicillin biosynthesis has been discussed for a long time. The penicillin biosynthesis in *P. chrysogenum* is inhibited by high levels of ammonium. In *A. chrysogenum*, it was found that ammonium concentrations ($(NH_4)_2SO_4$) higher than 100 mM also strongly interfered with cephalosporin C production.[7]

It was demonstrated that in *P. chrysogenum* ammonium (40 mM NH_4Cl in lactose-grown mycelia) directly repressed the expression of penicillin biosynthesis genes *ipnA* (*pcbC*) and *acvA* (*pcbAB*).[89]

In *A. nidulans* and *Neurospora crassa*, global nitrogen repression/derepression of several genes of the primary metabolism is mediated by the major positive control genes *areA* and *nit-2*, respectively.[103] In addition, the homologous gene of *P. chrysogenum*, *nre*, was cloned.[104] Each of these three genes encodes regulatory factors with a single Cys-X2-Cys-X17-Cys-X2-Cys-type zinc finger which in combination with an immediate downstream basic region constitutes a DNA-binding domain. A similar zinc finger motif has been found in a wide range of eukaryotic organisms including yeast, plants, chicken, mice, and humans. All of these transcription factors recognize the consensus sequence GATA and can be grouped together into a GATA protein family.[103]

A protein consisting of 181 amino acid residues of (the 835-residue) *P. chrysogenum* NRE, containing its zinc-finger domain, was fused to the N terminus of *E. coli* β-GAL.[105] The hybrid β-GAL-NRE protein bound with high affinity to a DNA fragment derived from the intergenic region between *acvA* (*pcbAB*) and *ipnA* (*pcbC*) of *P. chrysogenum*. Although there are six GATA sequences found in the intergenic region, the β-GAL-NRE fusion protein strongly interacted only with a site that contains two of these GATA sequences.[105] In this binding site, the two GATA core sequences are arranged in a head-to-head fashion and separated by 27 bp which resembles the optimal binding site for NIT-2. Therefore, it appears very likely that nitrogen metabolite regulation of penicillin biosynthesis genes is mediated through NRE and that these structural genes for secondary metabolism are regulated as members of a nitrogen control circuit, although *in vivo* studies are clearly needed to relate NRE binding to potential regulatory functions.[105] Furthermore, this suggests that the availability of favored nitrogen sources and thus good growth conditions leads to reduced penicillin synthesis by the fungus.

(v) Regulation by amino acids

Because penicillin and cephalosporin are synthesized from amino acid precursors it was conceivable that amino acids play a role in the regulation of their biosynthesis. This was supported by the observation that in both *A. nidulans* and *P. chrysogenum* the addition of L-lysine to fermentation medium led to reduced penicillin titers. High levels of lysine also interfere with cephalosporin production in *A. chrysogenum*.[7,10]

Since AAA is a branch point between lysine and penicillin/cephalosporin biosynthetic pathways, lysine inhibition of penicillin biosynthesis seems to act by inhibiting several enzymes of the pathway (α-aminoadipate reductase, homocitrate synthase, and homoaconitase) and/or by repressing their synthesis (Scheme 2).[7,10] This might lead to a reduction of the AAA pool available for penicillin production.

Interestingly, when the AAA reductase activity (Scheme 2) of three penicillin-producing strains of *P. chrysogenum* was compared, AAA reductase from the lowest penicillin producer was found to be least sensitive towards L-lysine inhibition. This might explain the ability of high-producing strains to accumulate increased amounts of AAA, and consequently more penicillin.[106]

In addition, a more direct effect of lysine on the expression of penicillin biosynthesis genes *acvA* and *ipnA* was found in *A. nidulans*.[10] Biochemical data obtained with *S. cerevisiae* suggest that the reactions of the first part of the pathway up to formation of AAA take place in the mitochondria, whereas those of the second part occur in the cytosol.[13] Hence, AAA apparently has to leave the mitochondria. In addition, the major part of intracellular lysine is sequestered in the vacuole.[13,107] Hence, compartmentation might be an important aspect as sell (see Section 4.07.3.10).[10]

In *A. chrysogenum*, it was reported that the addition of D,L-methionine to the medium (final concentration 20 mM) led to a three- to fourfold increase in production of cephalosporin C.[7,108] This methionine effect was found to act, at least in part, at the gene expression level because the increased production of cephalosporin C triggered by addition of methionine to the medium was paralleled by increased steady state levels of mRNAs of cephalosporin biosynthesis genes *pcbAB*, *pcbC*, *cefEF* and, to a slight extent, *cefG*.[108]

In *A. nidulans*, differential effects due to various amino acids on the expression of penicillin biosynthesis genes *acvA* and *ipnA*, and penicillin production were measured. All of the amino acids showing a major negative effect on the expression of *acvA-uidA* and *ipnA-lacZ* gene fusions, i.e., His, Val, Lys, and Met (only at concentrations greater than 10 mM), led to decreased ambient pH during cultivation of the fungus.[109] Further analysis revealed that the negative effects at least of His and Val were due to reduced activation of the *acvA* gene fusion by the pH regulator PACC[99] under acidic conditions caused by these amino acids.[109] These data also implied that PACC regulates the expression of *acvA*, predominantly by using PACC binding site ipnA3. The repressing effect caused by Lys and Met on *acvA* expression, however, was suggested to act independently of PACC by unknown mechanisms which even seem to differ between lysine and methionine. It is interesting to note that lysine and methionine are closely related to the precursor amino acids AAA and cysteine, respectively. The effects due to their presence in the medium even seem to be mediated by specific regulatory mechanisms.[10,109]

(vi) Influence of phosphate and oxygen

Excess phosphate was found to exert a negative effect on cephalosporin production in *A. chrysogenum* strain W53253. The basis of this effect is still a matter of debate and the molecular mechanism is not yet understood.[7,9,10]

The availability of oxygen is clearly very important for β-lactam production, at least in fungi. Good aeration of mycelia with oxygen is a prerequisite for high β-lactam titers.[10,90,110] Since several enzymatic reactions require oxygen, it is conceivable that this is the reason for oxygen requirement. The importance of oxygen is also supported by the possibility of increasing cephalosporin production by introducing a bacterial oxygen binding protein in *A. chrysogenum* (see Section 4.07.3.9.5.2).[111]

(vii) The CCAAT-box binding protein complex PENR1

Based on results of a moving window analysis of the intergenic region between the *A. nidulans* *acvA* and *ipnA* genes and together with band shift and methyl interference assays, a CCAAT-containing DNA motif (box I) located 409 bp upstream of the ATG initiation codon of the *A. nidulans* *acvA* gene was identified. This is bound by a protein complex designated PENR1, for penicillin regulator 1. A 4 bp-deletion of this site (ΔCCA-G) led to an eightfold increase of *acvA* expression and, simultaneously, to a reduction of *ipnA* expression to about 30%.[112] Furthermore, an additional CCAAT-containing DNA element (box II) located about 250 bp upstream of the transcriptional start sites of *aat* was specifically bound by the same PENR1 regulator. Substitution of the CCAAT core sequence by GATCC led to a fourfold reduction of expression of an *aat-lacZ* gene fusion, indicating that the identified binding site positively influenced *aat* expression.[10,113]

Biochemical analysis indicated that PENR1 is a HAP-like transcriptional complex. Although in higher eukaryotes several different CCAAT-binding proteins have been described,[10,114] so far, in lower eukaryotes, the only CCAAT-box binding factor characterized in detail, both genetically and biochemically, is the *S. cerevisiae* HAP complex. It consists of at least four subunits: HAP2, HAP3 and HAP5 form a heterotrimeric complex, which is essential for DNA binding; HAP4 is an acidic protein that acts as the transcriptional activation domain.[115,116]

Papagiannopoulos *et al.*[117] reported the cloning of a gene from *A. nidulans* designated *hapC*, which exhibits significant similarity to the *Saccharomyces cerevisiae HAP3* gene. The putative complex analogous to the yeast HAP complex was designated *A. nidulans* CCAAT binding factor (AnCF). It was shown that HAPC is involved in the regulation of genes that have been shown to contain CCAAT sequences in their promoters, such as the *amdS* gene, which is required for the use of acetamide as the nitrogen and C-source.[117] Hence, it seems very likely that HAPC is a member of the proposed AnCF complex. Genetic and biochemical data suggest that HAPC is also part of the PENR1 complex. Hence, PENR1 might be a HAP-like regulatory complex. The question whether PENR1 corresponds to AnCF remains to be answered.[10,118]

Interestingly, PENR1 only binds with high affinity to certain CCAAT-containing DNA elements. The CCAAT sequence alone is not sufficient for binding.[10]

Deletion or mutagenesis of the PENR1 binding sites had opposite effects; the expression of *acvA* was increased eightfold, while the expression of *ipnA* and *aat* was reduced.[112,113] Formally, the CCAAT box I mediated a negative effect on *acvA*, but a positive effect on *ipnA*, and box II a positive effect on *aat* expression. The analysis of the effect of the deletion of *hapC in vivo* showed that PENR1 is of major importance as a positive acting factor of both *ipnA* and *aat* expression, whereas it only slightly affects the regulation of the *acvA* gene. This finding suggests that, in addition to PENR1, a repressor acts on the expression of *acvA*, which binds closely to or overlaps the PENR1 binding site.[10,118]

In *S. cerevisiae*, the HAP complex activates the expression of genes whose products are required for respiration. Hence, Hap mutants are not able to grow on nonfermentable C-sources.[115,116] However, because *A. nidulans* is an aerobically growing fungus, *S. cerevisiae* may not be a good model to understand the role of HAP complexes in this fungus and, in general, in aerobically growing eukaryotes. In *A. nidulans*, HAP-like PENR1 also regulates β-lactam biosynthesis genes. Hence, it will be interesting to understand whether PENR1 activates genes belonging, for reasons not yet understood, to a common regulatory circuit.

It is interesting to note that deletion of 4 bp of the PENR1 binding site led to increased *acvA* expression and reduced *ipnA* expression.[112] It has not yet been clarified whether this increase of *acvA* and decrease of *ipnA* expression is of physiological relevance. It is tempting to speculate, however, that under certain physiological conditions, an increase of *acvA* expression and simultaneously, reduced expression of *ipnA* and *aat*, could lead to higher cellular amounts of ACV. The tripeptide could have, apart from being the precursor of β-lactams, additional functions, e.g., in amino acid transport.[119] ACV is structurally very similar to glutathione (GSH), which consists of the tripeptide γ-(L-glutamyl)-L-cysteinyl-glycine. In higher eukaryotes, GSH is a part of the γ-glutamyl cycle which among other functions serves to transport extracellular amino acids across the plasma membrane.[10,120]

Since HAPC is part of PENR1, it seems very likely that the PENR1 complex is conserved among the industrially important β-lactam producing fungi. This assumption is supported by biochemical data of *P. chrysogenum* and *A. chrysogenum*,[112,113] suggesting that the corresponding genes of these fungi are regulated by PENR1 analogous complexes.[10]

(viii) Mutations affecting putative regulatory genes

To identify additional regulatory genes, mutants of *A. nidulans* were isolated carrying mutations that are specifically involved in *trans* regulation of the penicillin biosynthesis genes. These mutations were designated *prg* (for penicillin regulation)[121] and *npeE1* (impaired in penicillin biosynthesis).[122] In penicillin production medium, the mutants exhibited only 5–50% of *ipnA* expression, and produced only 10–30% of penicillin compared with the wild-type strain. For mutants Prg-1 (*prgA1*) and Prg-6 (*prgB1*), it was demonstrated that they also differed in *acvA* expression levels from the wild-type. The Prg mutants led to the identification of two different genes (*prgA1* and *prgB1*) whose defect caused the Prg phenotype.[121]

The *npeE1* gene is located on linkage group I.[122] To date, it has not been clarified whether *npeE1* differs from *prgA1* and *prgB1*. The results obtained by genetic and biochemical analyses indicated that the mutants isolated most likely carry mutations in positively acting regulatory genes, which, in the case of *prgA1* and *prgB1*, specifically affect both *acvA* and *ipnA*, and in the case of *npeE1*, at least *ipnA* expression.[10]

Nine mutants of the *P. chrysogenum* Wis54-1255 strain impaired in penicillin production were isolated.[123] Biochemical and genetic analyses suggested that two of these (mutants Npe2 and Npe3) carry mutations in regulatory genes affecting the expression of the entire penicillin biosynthesis gene cluster.

4.07.3.9.3 Regulation of β-lactam biosynthesis in fungal production strains

Because β-lactam compounds are still the most sold antibiotics in the world market, there is a strong interest to analyze high-producing production strains that are highly mutated and have been derived from several different strain development programmes.[9] This will help to elucidate both the molecular basis of deregulation and thus high production and also any remaining bottlenecks.

Industrial penicillin and cephalosporin production is carried out mainly with *P. chrysogenum* and *A. chrysogenum*, respectively. Most of the strains have been produced by mutagenesis followed by screening or selection. In 1972, the initial Panlabs Inc. *P. chrysogenum* strain made 20 000 units of penicillin ml^{-1} in 7 days (an activity equivalent to 12 mg pure penicillin G, Na salt).[90] In 1990, the improved strain made 70 000 units ml^{-1} in 7 days. Penicillin titers in industry in 1993 were as high as 100 000 units ml^{-1}.[9,124]

Two important genetic features of *P. chrysogenum* production strains have been identified: (i) amplification of structural genes and (ii) their massively increased steady-state mRNA levels. Between six and 16 copies of the penicillin biosynthesis gene cluster have been found in various penicillin production strains.[125,126] In the high titer *P. chrysogenum* strains E1 and AS-P-78, the amplifications are organized in tandem repeats.[127] A conserved TTTACA hexanucleotide sequence may be involved in their generation. This TTTACA sequence borders the 106.5 kb long penicillin biosynthesis gene cluster in the wild-type strain NRRL 1951 and also the *P. notatum* strain ATCC9478 (Fleming's isolate).[9,127]

Transcript analysis established that the *ipnA* mRNA steady state level of strain BW 1890 was 32 to 64-fold that of NRRL1951, an increase too great to be due to the amplification alone.[125] Since sequence analysis of promoter regions has shown that no mutations have been generated within the promoter regions of the penicillin biosynthesis structural genes, the increased penicillin production in amplified strains was suggested to be due to altered regulation of the biosynthesis pathway through changes in *trans*-acting regulatory factors.[10]

Other factors, such as stability of enzymes at later stages of fermentations, loss of glucose repression/inhibition might be important for *P. chrysogenum* production strains.[10]

In contrast, in the cephalosporin C production strain *A. chrysogenum* LU4-79-6, the β-lactam biosynthesis genes seem only to be present in single copy.[4,128] It was shown that *A. chrysogenum* strain DSM produces a considerably higher amount of *ipnA* transcript and of cephalosporin C (tenfold higher) than the wild-type strain ATCC140553. Hence, in *A. chrysogenum* increased transcription and consequently expression of biosynthesis genes seems also to be an important feature of production strains.[129]

There are certainly numerous other mutations involved that lead to a high-producing phenotype, such as deregulation of enzymes involved in amino acid biosynthetic pathways, and hence the amount of precursor amino acids produced. It was shown, for instance, that improved *P. chrysogenum* penicillin production strains have reduced catabolism of AAA and flux of AAA to lysine. This agrees well with the finding that AAA reductase activity from the lowest penicillin producer of *P. chrysogenum* was least sensitive towards L-lysine inhibition (see Section 4.07.3.9.2.5). A high-producing *P. chrysogenum* strain had twice the acetohydroxy acid synthase activity of an ancestral strain and the enzyme in the superior strain was deregulated to valine feedback inhibition. A similar observation was reported from an *A. chrysogenum* high-producing mutant in which acetohydroxy acid synthase was found to be partially desensitized to valine feedback inhibition. These alterations seem to be required to provide the high amounts of valine needed for penicillin production. Furthermore, the transport of intermediates of the penicillin biosynthesis between organelles, and the number of organelles can be predicted also to be important for penicillin production strains.[10]

4.07.3.9.4 *Regulation of cephamycin biosynthesis in* S. clavuligerus

There is not much known about the regulation of cephamycin biosynthesis in bacteria. The *ipnA* gene of both *S. clavuligerus* and *S. lipmanii* seems to be transcribed from its own promoter leading to a 1.2 kb transcript (Figure 1).[7] In both species the promoter lies within the upstream *acvA* gene. In addition, in *S. clavuligerus* a second longer *ipnA* transcript was found, which seems to extend throughout *lat*, *acvA*, and *ipnA*.[130] The *cefE* gene is transcriptionally linked to *cefD* while *cefF* is transcribed separately. Northern analysis has indicated that the *cefD-cefE* transcript is greater than 10 kb in length, which still leaves space for other yet unidentified genes (Figure 1).[67]

A regulatory gene (*ccaR*) of both the cephamycin and clavulanic acid biosynthesis was found to be located in the cephamycin biosynthesis gene cluster of *S. clavuligerus* (Figure 1). This gene shows high similarity to regulatory genes involved in the biosynthesis of other secondary metabolites in Streptomycetes, such as actinorhodin.[20] Cell density was proposed to stimulate β-lactam production in Streptomycetes,[131] as it was previously found to trigger carbapenem production in *Erwinia carotovora* (see Section 4.07.5.3).

The reaction catalyzed by lysine-ε-aminotransferase (encoded by *lat*) may be the rate-limiting reaction for cephamycin biosyntheses in *S. clavuligerus* because insertion of an extra copy of *lat* into the chromosome led to a twofold increase in cephamycin production.[132]

S. clavuligerus produces cephamycin C under numerous nutrient limitations, whilst clavulanic acid was produced under phosphate and carbon limitations. Biochemical investigations point to a possible role of guanosine 5′-diphosphate 3′-diphosphate (ppGpp) in the regulation of cephamycin production in *S. clavuligerus*.[133]

4.07.3.9.5   *Applications*

The increasing knowledge of the molecular genetics of β-lactam biosynthesis has opened up new possibilities to improve rationally β-lactam production strains and to engineer new biosynthetic pathways. Cephalosporin production with *A. chrysogenum* is well below the productivity reached with *P. chrysogenum* for penicillin. For penicillin, there are estimates that it should still be possible to improve the current yields by a factor of 4–5, which would have a dramatic effect on the economy of the process.[10,24]

Several molecular strategies have been followed to improve or alter β-lactam production: (i) introducing additional copies or overexpression using strong promoters of β-lactam biosynthesis genes; (ii) metabolic engineering of β-lactam biosynthetic pathways by expression of heterologous genes, e.g., production of cephalosporin precursors in *P. chrysogenum*; (iii) use of the increasing knowledge of peptide synthetase genes such as those for ACVS enzymes to produce novel compounds by genetic engineering (see Chapter 8);[33] (iv) manipulation of regulatory genes, which have not been reported because results on the identification of regulatory genes are just being accumulated.

(i)   *Increase of expression of β-lactam biosynthesis genes*

In *A. nidulans*, overexpression of the *acvA* gene led to a 30-fold increase of penicillin yield,[84] whereas overexpression of *ipnA* had only a modest effect on penicillin production (increase of about 25%). Forced expression of the *aat* gene even reduced penicillin production (decrease of about 10–30%).[60] This demonstrated that *acvA* expression is rate-limiting for penicillin production in *A. nidulans*.

Consistent with data obtained with *A. nidulans*, attempts to increase cephalosporin C yields in *A. chrysogenum* strain 394-4 (improved cephalosporin C production strain) by inserting multiple copies of the *ipnA* gene were unsuccessful.[4] However, when *P. chrysogenum* Wis 54-1255 (an early strain of the strain improvement series) was transformed with a DNA fragment containing *P. chrysogenum* Wis 54-1255 *ipnA* and *aat* genes, some of the resulting transformants produced up to 40% higher amounts of penicillin V.[134]

It was demonstrated that transformants of *A. chrysogenum* wild-type strain, carrying additional copies of the *cefG* gene only, showed a direct relationship between *cefG* copy number, *cefG* mRNA levels, and cephalosporin C titers, suggesting that this enzyme might be a rate-limiting step in cephalosporin C production.[10,124,135]

(ii)   *Genetic engineering of β-lactam biosynthetic pathways*

Cloned β-lactam biosynthesis genes can already be used for a rational improvement of β-lactam production. DAOC and cephalosporin C can be enzymatically deacylated to form 7-aminodeacetoxycephalosporanic acid (7-ADCA) and 7-aminocephalosporanic acid (7-ACA), respectively, important intermediates in the manufacturing of oral cephalosporin antibiotics. Medically important oral cephalosporins, e.g., cephalexin and cephradine, are synthesized by derivatizing the 7-amino group of 7-ADCA or 7-ACA with appropriate side-chain moieties.[136] Although cephalosporins are superior antibiotics compared to penicillins, their production is limited because the processes for producing 7-ADCA and 7-ACA are complex. Removal of the natural D-α-aminoadipyl side-chain from cephalosporins is inefficient. Hence, there have been efforts in both directions: isolation of superior enzymes to remove the side-chain directly and development of alternative biosynthetic routes to 7-ACA and/or 7-ADCA.[70] The latter approach was followed by expressing the *S. clavuligerus cefE* (expandase) or *A. chrysogenum cefEF* (expandase-hydroxylase) gene, with and without the acetyltransferase gene (*cefG*), in *P. chrysogenum*.[70] Feeding of such transformants with adipic acid led to production of various cephalosporins having an adipyl side-chain. The adipyl side-chain was easily removed with a *Pseudomonas*-derived glutaryl amidase to yield the cephalosporin intermediates. Hence, by these measures, an important step in cephalosporin manufacturing could be improved.

Cephalosporin C biosynthesis is regulated by the oxygen content of the medium (see Section 4.07.3.9.2.6).[110] Possibly, the requirement for oxygen is connected to the oxygen requiring reactions involved in cephalosporin biosynthesis.

DeModena *et al.*[111] improved the aerobic metabolism in *A. chrysogenum* by using the oxygen-binding heme protein hemoglobin from the bacterium *Vitreoscilla*. Its structural gene (*VHb*) was expressed in *A. chrysogenum* strain C10. Several transformants produced significantly higher yields of cephalosporin C than control strains.

(iii) Generation of novel compounds by genetic engineering of peptide synthetases

In general, peptide synthetases such as ACVS enzymes possess a highly conserved domain structure. The arrangement of these domains determines the number and order of the amino acid constituents of the peptide product.[11] Hence, the shuffling of domains should result in the synthesis of new peptides exhibiting novel amino acid orders. Stachelhaus *et al.*[33] exchanged domain-coding regions of bacterial and fungal origin. The authors developed a system allowing targeted substitution of amino acid activating domains within the *srfA* operon, which encodes the protein templates for the synthesis of the lipopeptide antibiotic surfactin in *Bacillus subtilis*. By this method, new hybrid genes were produced whose products showed altered amino acid specificities. The genes were expressed in *B. subtilis* and led to the production of novel peptides (see Chapter 8).[33]

4.07.3.9.6 Evolution of fungal β-lactam biosynthesis genes

The availability of sequence information regarding bacterial and fungal penicillin/ceph-3-em biosynthesis genes led to speculations about their evolutionary relationship. Based on several observations, a horizontal transfer of β-lactam biosynthesis genes from bacteria to fungi has been proposed.[8,137–140] This hypothesis has been questioned.[141] The arguments in favor of a horizontal gene transfer are as follows.[10] (i) *ipnA* genes of fungi and bacteria show high sequence similarities. More than 60% of the nucleotide bases and 50% of the deduced amino acids are identical. (ii) Bacterial as well as fungal β-lactam genes are organized in clusters. This finding led to the assumption that the β-lactam biosynthesis genes were transferred as a single cluster from an ancestral prokaryote to a common ancestor of the β-lactam synthesizing fungi. In the eukaryotic ancestor, the biosynthesis genes were split onto two chromosomes (Figure 1). One part encodes the early genes of cephalosporin biosynthesis, the other the late genes. Later in the lineage an ancestor of *A. nidulans* and *P. chrysogenum* diverged from *A. chrysogenum* and has presumably lost the cluster with the genes for the late stage of cephalosporin biosynthesis.[4] (iii) Fungal *acvA* and *ipnA* genes do not contain introns indicating a bacterial origin of the genes. In addition, it was proposed that during the evolution of β-lactam biosynthesis genes *Streptomyces* sp. must have evolved specific resistance mechanisms to avoid self-toxification. If the transfer had occurred from fungi to bacteria, it would have been lethal for bacteria.[8]

In contrast to the other penicillin biosynthesis genes, the *aat* genes contain introns. On the basis of linkage of the *aat* to *ipnA* genes (Figure 1), Skatrud[4] suggested that a sequence functionally related to *aat* was transferred together with the β-lactam genes and was later modified to its current functional form. Since IAT possesses amidohydrolase activity to deacylate IPN to 6-APA,[53] which shows a weak antibiotic activity only, an ancestral amidohydrolase activity in the prokaryote ancestor might have had a resistance function and was fused in fungi with a eukaryotic gene.[4]

Based on the DNA sequences of *ipnA* genes from Gram-positive Streptomycetes and fungi and a rate of nucleotide substitution of 10^{-9} nucleotide changes per site per year, it was proposed that the transfer occurred 370 million years ago.[138] The cloning and sequencing of an *ipnA* gene from a Gram-negative bacterium, *Flavobacterium* sp., however, led to an extension/modification of the hypothesis of horizontal gene transfer. The *ipnA* gene of *Flavobacterium* sp. shares 69% sequence identity with the Streptomycetes gene and 64–65% with the fungal genes (*A. chrysogenum*, *P. chrysogenum*).[8] Based on 5 S rRNA analysis Gram-positive and Gram-negative bacteria split about 1–1.5 billion years ago, and prokaryotes and eukaryotes split about 2 billion years ago.[142] If the gene transfer had occurred only 370 million years ago from Streptomycetes to fungi as proposed by Weigel *et al.*,[138] it could be expected that the fungal and Streptomycete genes show a greater homology than the Gram-positive (Streptomycetes) and Gram-negative genes (*Flavobacterium* sp.).[8,10] As outlined above, this is not the case. Hence, it was suggested that multiple gene transfer events might have occurred from bacteria to fungi.[8]

However, Smith *et al.*[141] criticized this hypothesis because it was made solely on assumptions about rates of change. The authors rooted the tree with two distantly related β-lactam biosynthetic enzymes. They compared the similarity of both IPNS of *A. nidulans*, *P. chrysogenum*, *A. chrysogenum*, *S. clavuligerus*, *Streptomyces anulatus*, and *Flavobacterium* sp., and deacteoxycephalosporin C synthetase of *S. clavuligerus* and *A. chrysogenum*. Based on these similarities, a tree arose with conventional evolutionary descent. The simplest interpretation is that the genes for the two enzymes are the result of a duplication that occurred before the prokaryote/eukaryote divergence. The topology of the tree rooted with the duplicated enzymes, the depth of the bacterial branches, and the different orientations of genes in fungi and eubacteria all appear to be consistent with an ordinary evolution for isopenicillin N synthase. However, the authors do not discuss why most of the eukaryotes and fungi have lost the gene cluster.

4.07.3.10   Compartmentation of Penicillin Biosynthesis Enzymes in Fungi

For *P. chrysogenum*, the penicillin biosynthesis pathway was shown to occur in different compartments of the cell.[10] The ACVS is located either within or bound to the vacuolar membrane.[143] In addition, a large portion of cellular AAA which is most likely used for β-lactam synthesis is also contained in the vacuoles and thus sequestered from the cytosol.[107,144]

The IPNS protein from *P. chrysogenum* was found in the cytoplasm whereas IAT was detected in organelles with a diameter of 200–800 nm, which were assumed to be microbodies.[145] The latter result has been supported by the finding that the *P. chrysogenum* IAT contains a putative targeting signal sequence, a C-terminal alanine-arginine-leucine which is important to direct the enzyme to the microbodies.[9,10,146] Furthermore, a positive correlation between the capacity for penicillin production and the number of organelles per cell was observed when comparing different *P. chrysogenum* strains.[145] Several transport steps are thus required to bring intermediates of the penicillin biosynthesis together with the later enzymes. Phenylacetic acid required for production of penicillin G passes through the plasma membrane via passive diffusion of the protonated species.[94]

4.07.4   CLAVAMS

4.07.4.1   General Aspects

Clavams are composed of a β-lactam ring condensed with an oxazoline ring giving rise to a penicillin nucleus with an oxygen substituting for the sulfur atom (Tables 1 and 4).[5] The first clavam described was clavulanic acid which was isolated from *S. clavuligerus* ATCC 27064 in a screening program designed to detect inhibitors of β-lactamases produced by *Klebsiella aerogenes*.[148] Clavulanic acid was subsequently detected in other Streptomycetes such as *Streptomyces jumonjinensis* and *Streptomyces katsurakamanus*.[5] It has a quite broad antibacterial spectrum but with marginal activity.[148] It penetrates well into Gram-negative bacteria.[149] Clavulanic acid is a powerful, irreversible inhibitor (suicide inhibitor) of many β-lactamases (e.g., from *Staphylococcus*, but also from Gram-negative bacteria such as *E. coli*, *Proteus*, and *Haemophilus*). It is used clinically in combination with β-lactamase-sensitive antibiotics to combat infections that would be resistant to the β-lactam antibiotic alone.[7]

There are other clavams which have been found in different *Streptomyces* species (Table 4). Interestingly, in addition to cephamycins, *S. clavuligerus* was found to have the capability to synthesize several clavams, most notably clavulanic acid (Table 4).[5,6] The clavams valclavam[150,151] and 2-hydroxyethylclavam[152] were isolated from *S. antibioticus*, which does not produce clavulanic

Table 4 Clavam compounds.

Compound	R^1	R^2	Producer
Clavulanic acid	HC⌢OH	CO_2H	*Streptomyces clavuligerus* *Streptomyces jumonjinensis*
Alanylclavam	H_2C⌄CO_2H / NH_2	H	*S. clavuligerus*
Clavam 2-carboxylate	CO_2H	H	*S. clavuligerus*
2-Hydroxyethylclavam	H_2C⌄OH	H	*Streptomyces antibioticus*
2-Formyloxymethylclavam	H_2C⌄O–C(=O)–O	H	*S. clavuligerus*
2-Hydroxymethylclavam	CH_2OH	H	*S. clavuligerus*
Valclavam	H_2C⌄ CO_2H / OH — N(H)—C(=O)—CH(NH_2)	H	*S. antibioticus*

After O'Sullivan and Sykes,[5] Gräfe,[6] and Janc *et al.*[147]

acid.[147] For instance, 2-hydroxymethylclavam, 2-formyloxymethylclavam, and clavam 2-carboxylate[153] have antifungal activities, and alanylclavam has both antifungal and antibacterial properties (Table 4).[154]

4.07.4.2 Biosynthesis and Genetics

The only data regarding clavam biosynthesis focuses on the biosynthesis of clavulanic acid. The current knowledge has been summarized by Janc *et al.*[147] and Paradkar and Jensen[155] and is given in Scheme 3. Feeding experiments with isotopically labeled substrates suggested that the biosynthesis of clavulanic acid begins with the condensation of arginine, a five-carbon moiety, and a three-carbon glycolytic intermediate, such as D-glycerate, which is followed by a series of reactions to form proclavaminic acid.[156–161] Experiments with mutants demonstrated that arginine specifically initiated the biosynthetic pathway.[158]

Two enzymatic activities involved in the clavulanic acid biosynthesis have been characterized in detail (Scheme 3). The proclavaminic acid amidinohydrolase (PAH) was discovered as a contaminating protein copurified in the course of purification of *S. clavuligerus* ACVS. The protein was named ACVS-related protein (ACVSR). When cloned, the encoding gene, *acvsr* (= *cla*), was discovered to lie 5.7 kbp from the *ipnA* (*pcbcC*) gene, which is located in the cephamycin cluster, and directly upstream of *cs2* (Figure 2).[162,166] Insertional inactivation of this gene reduced clavulanic acid levels but left cephamycin production essentially unchanged. Because of its clear relation to clavulanic acid biosynthesis the gene was renamed *cla*.[162] Its deduced gene product shows a high level of homology to amidinohydrolases and was suggested to encode PAH.[157,162] The same gene was cloned by Wu *et al.*[163] and designated *pah* for proclavaminate amidinohydrolase. In addition, PAH was purified and determined to have a molecular mass of 33 kDa by SDS-PAGE.[158] The *cla/pah* gene consists of 939 bp coding for an ORF of 33 368 Da and thus a protein of this size.[162,163] PAH is a member of the evolutionary divergent arginase family of enzymes.[163] It catalyzes the hydrolysis of the guanyl moiety of guanidinoproclavaminic acid to produce proclavaminic acid.[157]

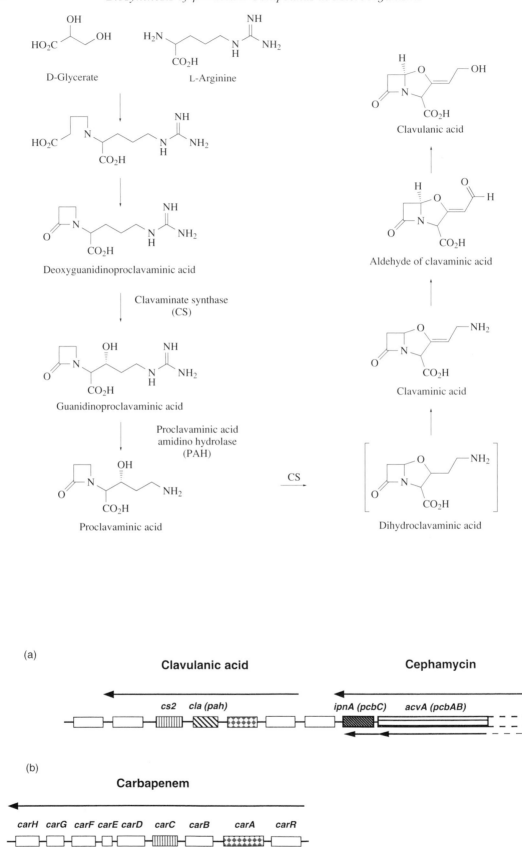

Figure 2 Clavulanic acid (a) and carbapenem (b) biosynthesis gene cluster of *S. clavuligerus*[78] and *Erwinia carotovora*, respectively. Same shading indicates high similarity (after Ward and Hodgson,[79] Paradkar and Jensen,[155] Aidoo *et al.*,[162] Wu *et al.*,[163] Hodgson *et al.*,[164] and McGowan *et al.*[165]).

The other enzymatic activity analyzed in detail is clavaminate synthase (CS), a nonheme iron, α-ketoglutarate-dependent oxygenase, which catalyzes three distinct transformations in the biosynthetic pathway (Scheme 3).[167,168] The first of these reactions is the hydroxylation of deoxyguanidinoproclavaminic acid to produce guanidinoproclavaminic acid.[157,169] Before CS can perform its second oxidative function, PAH has to hydrolyze guanidinoproclavaminic acid to produce proclavaminic acid.[157] CS then catalyzes the oxidative cyclization of proclavaminic acid to give the transient intermediate dihydroclavaminic acid (Scheme 3).[170,171] In a third oxidative process, the latter is desaturated to yield clavaminic acid.[168] Clavaminic acid was suggested to undergo oxidative deamination[172] and an unusual stereochemical inversion to yield the aldehyde, which is then reduced to give clavulanic acid.[172–174] These steps, however, have not yet been fully characterized.

Initial isolation of CS from *S. clavuligerus* revealed the unexpected presence of two isoenzymes, CS1 and CS2,[168] whose genes *cs1* (*cas1*) and *cs2* (*cas2*) have been cloned, sequenced,[166] and overexpressed in *E. coli*.[175] The genes are 87% identical and the proteins differ in length by a single amino acid residue. CS1 and CS2 have a deduced M_r of 35 347 and 35 774, respectively.[166] The two isoenzymes have the same cofactor requirements and possess nearly identical kinetic properties.[168]

The *cs2* gene has been located within a cluster of genes encoding clavulanic acid biosynthetic enzymes, located adjacent to the cephamycin gene cluster (Figure 2(a)).[79,162] However, the location of *cs1* in relation to this cluster remains unknown. Analysis of cosmid clones indicated that *cs1* and *cs2* are separated by at least 28 kb in the genome.[166] Moreover, the rationale for the presence of two CS isoenzymes and the role played by each in clavulanic acid biosynthesis are elusive. Paradkar and Jensen[155] reported the isolation of a mutant blocked specifically in CS2, by disrupting the *cs2* gene, which resulted in a clavulanic acid-nonproducing phenotype in starch-asparagine (SA) medium. Characterization of the Cs2 mutant suggested that both CS1 and CS2 can contribute to clavulanic acid formation. CS2 is indispensable for clavulanic acid formation under some nutritional conditions. In soy medium, however, which was used to demonstrate the presence of the two isoenzymes,[168] the Cs2 mutant was able to produce clavulanic acid. This finding correlated well with the presence of *cs1* transcript in cultures grown in soy medium and the absence of such transcript in those grown in SA medium. *cs2* is transcribed both as a 1.2 kb monocistronic transcript from a promoter which lies within the upstream *cla* gene and as part of a 5.3 kb polycistronic transcript. The size of the longer transcript suggests that it could accommodate *cs2*, *cla*, and two open reading frames further upstream.[155,176,177] It was concluded that *cs1* and *cs2* represent redundant genes encoding two proteins which can be used for clavulanic acid biosynthesis, but there are significant differences in their transcriptional organization and regulation. *cs2* is clustered with other genes involved in clavulanic acid biosynthesis and is transcribed as both mono- and multicistronic transcripts, in both SA and soy media. *cs1* transcript is transcribed as a monocistronic transcript only, and then in cultures grown in soy medium but not in those grown in SA medium.[155]

Marsh *et al.*[166] suggested that the conversion of proclavaminic acid to clavaminic acid is a rate-limiting step, which would explain the need for gene duplication. Alternatively, it has been postulated that the biosynthetic pathways for clavulanic acid and other clavams share reactions up to proclavaminic acid or clavaminic acid.[178,179] The two CS isoenzymes may simply function to provide enough precursors to be channeled into these pathways.[155]

In agreement with the latter proposal is the finding that *S. antibioticus*, which does not produce clavulanic acid, synthesizes proclavaminic acid. This common intermediacy suggests that a shared biosynthetic pathway indeed exists to clavulanic acid and other clavams.[179] It was proposed that the branch point of these two pathways may lie beyond clavaminic acid.[178,179] This was supported by the purification and characterization of an *S. antibioticus* CS designated CS3 with a molecular mass of ~36 kDa.[152] The enzyme shares many features with CS1 and CS2 of *S. clavuligerus*. All three enzymes require the same cofactors, namely Fe^{2+}, α-ketoglutarate and oxygen. The biosynthesis of all clavam metabolites, including clavulanic acid, are therefore thought to follow a common pathway, at least up to proclavaminic acid and perhaps to clavaminic acid.[155]

4.07.4.3 Organization of Clavulanic Acid Biosynthesis Genes in *S. clavuligerus*

An early report described the cloning of the *dclC* locus of *S. clavuligerus*, which is involved in clavulanic acid biosynthesis, by complementation of a clavulanic acid low-producing mutant.[180] However, the nature of the genetic defect complemented by the cloned DNA fragment has not yet been identified. Using this DNA as a probe, cosmid clones were isolated, which encoded additional biosynthetic genes including the clavaminic acid synthase gene *cs2*.[7,166] Using DNA probes from both the cephamycin (e.g., derived from the *ipnA* (*pcbC*) gene) and the clavulanic acid gene cluster to hybridize gene libraries in *E. coli* of *S. clavuligerus*, *S. jumonjinensis*, and *S. katsurahamanus*, it was shown that both clusters are closely linked on the chromosome in a super-cluster (Figure 2(a)).[79] A 12 kbp segment of the super-cluster carrying the clavulanic acid part has been cloned, sequenced and shown to confer upon *Streptomyces lividans* the ability to form clavulanic acid.[162,177] Analysis of the sequenced DNA indicated that eight open reading frames are present. One of them corresponds to the gene *cs2*, previously identified,[166] another to *cla/pah* encoding proclavaminic acid amidinohydrolase, but none shows similarity to genes encoding peptide synthetase enzymes.[177] These observations highlight the differences between penicillin/cephalosporin/cephamycin and clavulanic acid biosynthesis.[7]

4.07.5   CARBAPENEMS

4.07.5.1   General Aspects

The carbapenems resemble penicillins in having a β-lactam ring fused to a five-membered ring. The five-membered ring, however, is unsaturated and does not contain sulfur (Table 5).[5] The first naturally occurring β-lactam antibiotic containing the carbapenem ring system was reported in 1976.[181] This antibiotic, thienamycin, was the first of a series of carbapenems[182] that are produced by *Streptomyces cattleya* and other Streptomycetes as well as by unicellular bacteria such as *Erwinia* and *Serratia* (Table 5).[7] The carbapenem group includes a variety of naturally occurring products such as thienamycins, epithienamycins, and olivanic acid derivatives.[7] Carbapenems tend to occur in families and despite close structural similarity, a wide variety of names are applied to them. *S. cattleya* produces several thienamycins and *Streptomyces olivaceus* different olivanic acid derivatives.[5] Besides the compounds shown in Table 5, additional structurally related carbapenems can be found in the reference by O'Sullivan and Sykes.[5]

Table 5   Structures of some carbapenem compounds.

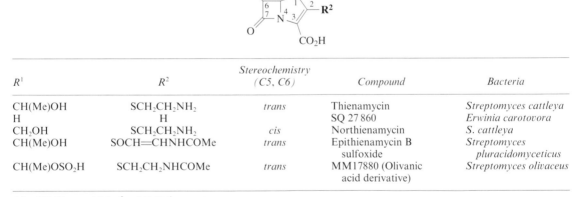

R^1	R^2	Stereochemistry (C5, C6)	Compound	Bacteria
CH(Me)OH	SCH₂CH₂NH₂	*trans*	Thienamycin	*Streptomyces cattleya*
H	H		SQ 27 860	*Erwinia carotovora*
CH₂OH	SCH₂CH₂NH₂	*cis*	Northienamycin	*S. cattleya*
CH(Me)OH	SOCH=CHNHCOMe	*trans*	Epithienamycin B sulfoxide	*Streptomyces pluracidomyceticus*
CH(Me)OSO₂H	SCH₂CH₂NHCOMe	*trans*	MM17880 (Olivanic acid derivative)	*Streptomyces olivaceus*

After O'Sullivan and Sykes[5] and Gräfe.[6]

Carbapenems are broad-spectrum antibiotics which exhibit intrinsic resistance to most of the clinically encountered β-lactamases.[182] The desirable characteristics have spurred continued research to overcome the inherent instability that has limited the clinical usefulness of the compound.[7] A more stable and more active derivative used in medicine is *N*-formidoylthienamycin, also known as imipenem.[183] Although the carbapenems exhibit a high degree of stability to β-lactamases, unlike other β-lactam-containing antibiotics described to date, they are susceptible to hydrolytic cleavage by mammalian dipeptidase such that only a small percentage of the administered dose is recovered in the urine.[184,185]

4.07.5.2 Biosynthesis

Radioactive and stable isotope studies with *S. cattleya* demonstrated that, in thienamycin, the C-6 and C-7 of the β-lactam ring are derived from acetate, the cysteinamyl side-chain from cysteine and both carbon atoms of the hydroxyethyl substituent at C-6 from the methyl of methionine (Table 5).[186–188] Possible biosynthetic routes were suggested, which involve the condensation of acetyl CoA with γ-glutamyl phosphate, formation of a monocyclic intermediate, followed by ring closure to give carbapenem and finally carbapenem compounds.[6,189] Detailed analysis of the pathway has not been reported.[7]

4.07.5.3 Genetics

Studies on the molecular genetics of carbapenem synthesis in Streptomycetes have been difficult because of the poor genetic tractability of the producer organisms. In contrast, a few strains of the Gram-negative bacteria *Erwinia* and *Serratia* make the simple 1-carbapen-2-em-3-carboxylic acid (SQ27 860), which contains only the bicyclic ring system, in contrast to the more complex carbapenem thienamycin (Table 5).[190,191] Because the Gram-negative bacteria are amenable to sophisticated genetic methods, they represent the most advanced systems to understand the genetic basis of carbapenem biosynthesis.[165]

The genes encoding both the carbapenem biosynthetic enzymes and regulators have been cloned and sequenced from *Erwinia carotovora* and designated *car*.[165,192,193] A putative operon of eight genes (*carA–H*) responsible for production of carbapenem was identified and sequenced (Figure 2(b)). Their products are unrelated to enzymes involved in the synthesis of the sulfur-containing β-lactams, namely penicillins, cephalosporins, and cephamycins.[165] Therefore, the carbapenem nucleus must be constructed by new enzymes that constitute a novel biosynthetic pathway.[165] This notion is consistent with the concept of independently evolved routes to synthesize the many β-lactam-containing structures seen in nature.[165,194]

The carbapenem biosynthesis genes were activated by CARR, a transcriptional activator.[192] The *carR* gene is part of the cluster (Figure 2(b)). CARR activity requires the presence of the bacterial pheromone signaling molecule *N*-(3-oxohexanoyl)-L-homoserine lactone (OHHL). Another gene designated *carI*, encodes a protein responsible for synthesis of this pheromone. Synthesis of OHHL is cell-density dependent, and because it is required for activation of carbapenem production, synthesis of the carbapenem antibiotic is thus cell-density dependent.[193] The *carI* gene seems to be unlinked to the *car* biosynthetic locus (Figure 2(b)).[165]

Two of the genes, *carA* and *carC*, encode proteins which show significant similarity to the functionally uncharacterized ORF and the CS2 proteins, respectively, encoded by the *S. clavuligerus* gene cluster responsible for the production of clavulanic acid (Figure 2). Hence, a degree of similarity in the structure of these two gene clusters is apparent, suggesting that they are evolutionarily related.[165]

4.07.6 MONOLACTAMS

4.07.6.1 General Aspects

Monolactams are a group of monocyclic β-lactams produced by bacteria (Table 1). They include monobactams such as sulfazecin and related compounds from *Pseudomonas*, *Gluconobacter*, etc. (Table 6), and nocardicins produced by *Nocardia uniformis* ssp. *tsuyamanensis* (Table 7).[7] The monobactams are a diverse group of compounds. They are *N*-acyl derivatives of (*S*)-3-amino-monobactamic acid having different side-chains (Table 6).[195] The simplest monobactam is that produced by *Chromobacterium violaceum* ATCC31532, having an *N*-acetyl side-chain and a methoxy group at the 3α-position of the β-lactam ring (Table 6). Monobactams have been isolated from different bacteria (Table 6). Some of these compounds have good activity against Gram-negative, but poor activity against Gram-positive pathogens. Aztreonam (SQ 26 776), a synthetic monobactam, is used clinically for Gram-negative infections.[5,7] Nocardicins are a group of closely related compounds (Table 7). They have a peptide-based structure derived from methionine, serine and tyrosine.[196] No member of the nocardicins has been utilized clinically.[7]

Table 6 Structures of some monobactams.

R^2	R^1	SQ^a	*Bacteria*
	OMe	26 180	*Chromobacterium violaceum*
	OMe	26 445 (Sulfazecin)	*Pseudomonas acidophila Gluconobacter* sp.
	OMe	26 823	*Agrobacterium radiobacter*
	OMe	26 970	*Agrobacterium radiobacter*
Oligopeptide (M_r 1462)	H	28 502	*Flexibacter* sp.

After O'Sullivan and Sykes.[5]
[a]SQ, Squibb Chemical Number.

4.07.6.2 Biosynthesis

Some information is available on the biosynthesis of nocardicin A. Scheme 4 shows the likely origin of its component parts.[5-7,197] Biosynthetic studies show that nocardicin A production was stimulated by L-tyrosine, *p*-hydroxyphenylpyruvic acid, D,L-*p*-hydroxymandelic acid, and L-*p*-hydroxyphenylglycine, and that radiolabeled L-tyrosine, L-serine, glycine, and L-homoserine were incorporated in nocardicin A. The two aromatic rings were derived from tyrosine via L-*p*-hydroxyphenylglycine and the carbon atoms of the β-lactam ring from serine. Later, it was shown that L-methionine was incorporated into the homoserine unit to a far higher level than L-homoserine and that the serine was incorporated into the β-lactam ring without a change in the oxidation state of the β-carbon atom.[5]

Less is known about the biosynthesis of monobactams.[5,7] Using a radiolabeled mixture of [3-^{14}C]serine and [3-^{3}H]serine in feeding experiments, it was established that the carbon atoms of the β-lactam ring in SQ 26 180, SQ 26 445 (Sulfazecin), and SQ 26 812 were derived from serine (Table 6).[198] It thus appeared likely that a single mechanism of ring closure applied in three monobactam producers.[5] The side-chains of monobactams vary in origin. Only inorganic sulfur was used as the sulfur source by the investigation mentioned above. The methoxymethyl group, if present, comes from methionine.[7,199]

Table 7   Structures of some nocardicins.

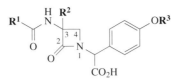

R^1		R^2	R^3	Nocardicin	Bacteria
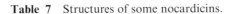		H	H	A	*Nocardia uniformis* ssp. *tsuyamanensis*
		H	H	C	*Nocardia uniformis* ssp. *tsuyamanensis*
		H	H	D	*Nocardia uniformis* ssp. *tsuyamanensis*
		H	H	F	*Nocardia uniformis* ssp. *tsuyamanensis*
		NHCHO $R^4 = OH$	D-Glucuronic acid	Formadicin A	*Flexibacter alginoliquefaciens*

After O'Sullivan and Sykes[5] and Gräfe.[6]

L-Tyrosine ⟶⟶ *p*-Hydroxyphenylpyruvate ⟶

L-*p*-Hydroxyphenylglyoxylic acid

L-Methionine   L-*p*-Hydroxyphenylglycine   L-Serine

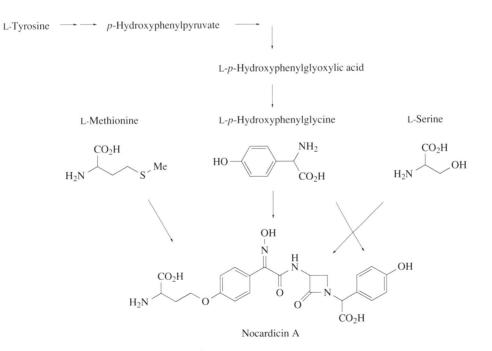

Nocardicin A

Scheme 4

ACKNOWLEDGMENTS

Research in the author's laboratory was supported by the Deutsche Forschungsgemeinschaft (Sonderforschungsbereich 369) and the European Union (EUROFUNG).

4.07.7 REFERENCES

1. A. Fleming, *Br. J. Exp. Pathol.*, 1929, **10**, 226.
2. P. W. Clutterbuck, R. Lovell, and R. Raistrick, *Biochem. J.*, 1932, **26**, 1907.
3. E. P. Abraham, *BioEssays*, 1990, **12**, 601.
4. P. L. Skatrud, in "More Gene Manipulations in Fungi," eds. J. W. Bennett and L. L. Lasure, Academic Press, New York, 1991, p. 364.
5. J. O'Sullivan and R. B. Sykes, in "Biotechnology, A Comprehensive Treatise in 8 Volumes," eds. H. Pape and H.-J. Rehm, VCH Verlagsgesellschaft, Weinheim, 1986, vol. 4, p. 247.
6. U. Gräfe, "Biochemie der Antibiotika," Spektrum Akademischer Verlag, Heidelberg, Berlin, New York, 1992.
7. S. E. Jensen and A. L. Demain, in "Genetics and Biochemistry of Antibiotic Production," eds. L. C. Vining and C. Stuttard, Butterworth-Heinemann, Newton, MA, 1995, p. 239.
8. Y. Aharonowitz, G. Cohen, and J. F. Martin, *Annu. Rev. Microbiol.*, 1992, **46**, 461.
9. J. F. Martin, S. Gutiérrez, and A. L. Demain, in "Fungal Biotechnology. Antibiotics," ed. T. Anke, Chapman and Hall, Weinheim, 1997, p. 91.
10. A. A. Brakhage, *Microbiol. Mol. Biol. Rev.*, 1998, **62**, in press.
11. H. Kleinkauf and H. von Döhren, *Eur. J. Biochem.*, 1996, **236**, 335.
12. A. L. Demain, in "Antibiotics Containing the β-Lactam Structure I," eds. A. L. Demain and N. A. Solomon, Springer Verlag, New York, 1983, p. 189.
13. J. K. Bhattacharjee, in "The Evolution of Metabolic Function," ed. R. P. Mortlock, CRC Press, Boca Raton, FL, 1992, p. 47.
14. H. Kimura, M. Izawa, and Y. Sumino, *Appl. Microbiol. Biotechnol.*, 1996, **44**, 589.
15. J.-J. R. Coque, J. F. Martin, J. G. Calzada, and P. Liras, *Mol. Microbiol.*, 1991, **5**, 1125.
16. J.-J. R. Coque, P. Liras, and J. F. Martin, *EMBO J.*, 1993, **12**, 631.
17. J.-J. R. Coque, P. Liras, and J. F. Martin, *Mol. Gen. Genet.*, 1993, **236**, 453.
18. J.-J. R. Coque, F. J. Enguita, J. F. Martin, and P. Liras, *J. Bacteriol.*, 1995, **177**, 2230.
19. J.-J. R. Coque, F. J. Pérez-Llarena, F. J. Enguita, J. L. Fuente, J. F. Martin, and P. Liras, *Gene*, 1995, **162**, 21.
20. F. J. Pérez-Llarena, P. Liras, A. Rodriguez-Garcia, and J. F. Martin, *J. Bacteriol.*, 1997, **179**, 2053.
21. G. A. Marzluf, *Annu. Rev. Microbiol.*, 1993, **47**, 31.
22. J. F. Martin and Y. Aharonowitz, in "Antibiotics Containing the β-Lactam Structure I," eds. A. L. Demain and N. A. Solomon, Springer Verlag, 1983. p. 229.
23. J. Nüesch, J. Heim, and H.-J. Treichler, *Annu. Rev. Microbiol.*, 1987, **41**, 51.
24. J. Nielsen, "Physiological Engineering Aspects of *Penicillium chrysogenum*," Polyteknisk Forlag, Lyngby, 1995.
25. Y. Aharonowitz, J. Bergmeyer, J. M. Cantoral, G. Cohen, A. L. Demain, U. Fink, J. Kinghorn, H. Kleinkauf, A. MacCabe, H. Palissa, E. Pfeifer, T. Schwecke, H. van Liempt, H. von Döhren, S. Wolfe, and J. Zhang, *Bio/Technology*, 1993, **11**, 807.
26. J. Zhang and A. L. Demain, *Crit. Rev. Biotechnol.*, 1992, **12**, 245.
27. H. van Liempt, H. von Döhren, and H. Kleinkauf, *J. Biol. Chem.*, 1989, **264**, 3680.
28. T. Schwecke, Y. Aharonowitz, H. Palissa, H. von Döhren, H. Kleinkauf, and H. van Liempt, *Eur. J. Biochem.*, 1992, **205**, 687.
29. D. J. Smith, A. J. Earl, and G. Turner, *EMBO J.*, 1990, **9**, 2743.
30. B. S. Diez, S. Gutiérrez, J. L. Barredo, P. van Solingen, L. H. M. van der Voort, and J. F. Martin, *J. Biol. Chem.*, 1990, **265**, 16358.
31. C.-Y. Shiau, J. E. Baldwin, M. F. Byford, W. J. Sobey, and C. J. Schofield, *FEBS Lett.*, 1995, **358**, 97.
32. J. E. Baldwin, C.-Y. Shiau, M. F. Byford, and C. J. Schofield, *Biochem. J.*, 1994, **301**, 367.
33. T. Stachelhaus, A. Schneider, and M. A. Marahiel, *Science*, 1995, **269**, 69.
34. A. A. Brakhage and G. Turner, in "The Mycota II. Genetics and Biotechnology," ed. U. Kück, Springer Verlag, Berlin, 1995, p. 263.
35. P. L. Roach, I. J. Clifton, V. Fülöp, K. Harlos, G. J. Barton, J. Hajdu, I. Andersson, C. J. Schofield, and J. E. Baldwin, *Nature (London)*, 1995, **375**, 700.
36. C. P. Pang, B. Chakravarti, R. M. Adlington, H. H. Ting, R. L. White, G. S. Jayatilake, J. E. Baldwin, and E. P. Abraham, *Biochem. J.*, 1984, **222**, 789.
37. I. J. Hollander, V. Q. Shen, J. Heim, A. L. Demain, and S. Wolfe, *Science*, 1984, **224**, 610.
38. J. E. Baldwin, J. Gagnon, and H. Ting, *FEBS Lett.*, 1985, **188**, 253.
39. Y. Aharonowitz, Y. Av-Gay, R. Schreiber, and G. Cohen, *J. Bacteriol.*, 1993, **175**, 623.
40. G. Cohen, A. Argaman, R. Schreiber, M. Mislovati, and Y. Aharonowitz, *J. Bacteriol.*, 1994, **176**, 973.
41. A. M. Orville, V. J. Chen, A. Kriauciunas, M. R. Harpel, B. G. Fox, E. Munck, and J. D. Lipscomb, *Biochemistry*, 1992, **31**, 4602.
42. R. C. Randall, Y. Zang, A. E. True, L. Que, Jr., J. M. Charnock, C. D. Garner, Y. Fujishima, C. J. Schofield, and J. E. Baldwin, *Biochemistry*, 1993, **32**, 6664.
43. P. L. Roach, I. J. Clifton, C. M. H. Hensgens, N. Shibata, C. J. Schofield, J. Hajdu, and J. E. Baldwin, *Nature (London)*, 1997, **387**, 827.
44. I. Borovok, O. Landman, R. Kreisberg-Zakarin, Y. Aharonowitz, and G. Cohen, *Biochemistry*, 1996, **35**, 1981.
45. J. Retey, *Angew. Chem.*, 1990, **102**, 373 (*Angew. Chem., Int. Ed. Engl.*, **29**, 355).
46. M. Ramsden, B. A. McQuade, K. Saunders, M. K. Turner, and S. Harford, *Gene*, 1989, **85**, 267.

47. J. E. Baldwin and T. E. Wan, *Tetrahedron*, 1981, **37**, 1589.
48. J. E. Baldwin, R. M. Adlington, S. E. Moroney, L. D. Field, and H.-H. Ting, *J. Chem. Soc., Chem. Commun.*, 1984, 984.
49. S. Wolfe, A. L. Demain, S. E. Jensen, and D. W. S. Westlake, *Science*, 1984, **226**, 1386.
50. J. E. Baldwin and E. P. Abraham, *Nat. Prod. Rep.*, 1988, **5**, 129.
51. S. M. Samson, R. Belagaje, D. T. Blankenship, J. L. Chapman, D. Perry, P. L. Skatrud, R. M. van Frank, E. P. Abraham, J. E. Baldwin, S. W. Queener, and T. D. Ingolia, *Nature (London)*, 1985, **318**, 191.
52. J. M. Luengo, *J. Antibiot.*, 1995, **48**, 1195.
53. E. Alvarez, B. Meesschaert, E. Montenegro, S. Gutiérrez, B. Diez, J. L. Barredo, and J. F. Martin, *Eur. J. Biochem.*, 1993, **215**, 323.
54. S. W. Queener and N. Neuss, in "Chemistry and Biology of β-Lactam Antibiotics," eds. R. B. Morin and M. Gorman, Academic Press, London, 1982, vol. 3, p. 1.
55. S. W. Queener, *Antimicrob. Agents Chemother.*, 1990, **34**, 943.
56. J. M. Luengo, J. L. Iriso, and M. J. Lopez-Nieto, *J. Antibiot.*, 1986, **34**, 1565.
57. E. Alvarez, J. M. Cantoral, J. L. Barredo, B. Diez, and J. F. Martin, *Antimicrob. Agents Chemother.*, 1987, **31**, 1675.
58. J. L. Barredo, P. van Solingen, B. Diez, E. Alvarez, J. M. Cantoral, A. Kattevilder, E. B. Smaal, M. A. M. Groenen, A. E. Venstra, and J. F. Martin, *Gene*, 1989, **83**, 291.
59. M. B. Tobin, M. D. Fleming, P. L. Skatrud, and J. R. Miller, *J. Bacteriol.*, 1990, **172**, 5908.
60. J. M. Fernández-Cañón and M. A. Peñalva, *Mol. Gen. Genet.*, 1995, **246**, 110.
61. M. B. Tobin, J. E. Baldwin, S. C. J. Cole, J. R. Miller, P. L. Skatrud, and J. D. Sutherland, *Gene*, 1993, **132**, 199.
62. R. T. Aplin, J. E. Baldwin, S. C. Cole, J. D. Sutherland, and M. B. Tobin, *FEBS Lett.*, 1993, **319**, 166.
63. M. B. Tobin, S. C. J. Cole, J. R. Miller, J. E. Baldwin, and J. D. Sutherland, *Gene*, 1995, **162**, 29.
64. R. T. Aplin, J. E. Baldwin, P. L. Roach, C. V. Robinson, and C. J. Schofield, *Biochem. J.*, 1993, **294**, 357.
65. T. Konomi, S. Herchen, J. E. Baldwin, M. Yoshida, N. A. Hunt, and A. L. Demain, *Biochem. J.*, 1979, **184**, 427.
66. S. E. Jensen, D. W. S. Westlake, and S. Wolfe, *Can. J. Microbiol.*, 1983, **29**, 1526.
67. S. Kovacevic, M. B. Tobin, and J. R. Miller, *J. Bacteriol.*, 1990, **172**, 3952.
68. C. Cantwell, R. Beckmann, P. Whiteman, S. W. Queener, and E. P. Abraham, *Proc. R. Soc. Lond. (Biol.)*, 1992, **248**, 283.
69. K. A. Alvi, C. D. Reeves, J. Peterson, and J. Lein, *J. Antibiot.*, 1995, **48**, 338.
70. L. Crawford, A. M. Stepan, P. C. McAda, J. A. Rambosek, M. J. Conder, V. A. Vinci, and C. D. Reeves, *Bio/Technology*, 1995, **13**, 58.
71. A. Scheidegger, M. T. Küenzi, and J. Nüesch, *J. Antibiot.*, 1984, **37**, 522.
72. J. E. Dotzlaf and W.-K. Yeh, *J. Bacteriol.*, 1987, **169**, 1611.
73. S. M. Samson, J. E. Dotzlaf, M. L. Slisz, G. W. Becker, R. M. van Frank, L. E. Veal, W.-K. Yeh, J. R. Miller, S. W. Queener, and T. D. Ingolia, *Bio/Technology*, 1987, **5**, 1207.
74. S. E. Jensen, D. W. S. Westlake, and S. Wolfe, *J. Antibiot.*, 1985, **38**, 263.
75. S. Kovacevic, B. J. Weigel, M. B. Tobin, T. D. Ingolia, and J. R. Miller, *J. Bacteriol.*, 1989, **171**, 754.
76. S. Kovacevic and J. R. Miller, *J. Bacteriol.*, 1991, **173**, 398.
77. J. F. Martin and P. Liras, *Annu. Rev. Microbiol.*, 1989, **43**, 173.
78. X. Xiao, G. Hintermann, A. Hausler, P. J. Barker, F. Foor, A. L. Demain, and J. Piret, *Antimicrob. Agents Chemother.*, 1993, **37**, 84.
79. J. M. Ward and J. E. Hodgson, *FEMS Microbiol. Lett.*, 1993, **110**, 239.
80. E. T. Seno and R. H. Baltz, in "Regulation of Secondary Metabolism in Actinomycetes," ed. S. Shapiro, CRC Press, Boca Raton, FL, 1989, p. 1.
81. F. Malpartida and D. A. Hopwood, *Mol. Gen. Genet.*, 1986, **205**, 66.
82. K. D. MacDonald and G. Holt, *Sci. Prog., Oxf.*, 1976, **63**, 547.
83. O. Litzka, K. Then Bergh, and A. A. Brakhage, *Mol. Gen. Genet.*, 1995, **249**, 557.
84. J. Kennedy and G. Turner, *Mol. Gen. Genet.*, 1996, **253**, 189.
85. A. A. Brakhage, P. Browne, and G. Turner, *J. Bacteriol.*, 1992, **174**, 3789.
86. E. A. Espeso and M. A. Peñalva, *Mol. Microbiol.*, 1992, **6**, 1457.
87. D. V. Renno, G. Saunders, A. T. Bull, and G. Holt, *Curr. Genet.*, 1992, **21**, 49.
88. T. Suárez and M. A. Peñalva, *Mol. Microbiol.*, 1996, **20**, 529.
89. B. Feng, E. Friedlin, and G. A. Marzluf, *Appl. Environ. Microbiol.*, 1994, **60**, 4432.
90. R. W. Swartz, in "Comprehensive Biotechnology. The Principles, Applications and Regulations of Biotechnology in Industry, Agriculture and Medicine," eds. H. W. Blanch, S. Drew, and D. I. C. Wang, Pergamon Press, Oxford, 1985, vol. 3, p. 7.
91. J. Zhang and A. L. Demain, *Arch. Microbiol.*, 1992, **158**, 364.
92. H. Martinez-Blanco, A. Reglero, M. Fernández-Valverde, M. A. Ferrero, M. A. Moreno, M. A. Peñalva, and J. M. Luengo, *J. Biol. Chem.*, 1992, **267**, 5474.
93. J. M. Fernández-Cañón, A. Reglero, H. Martinez-Blanco, and J. M. Luengo, *J. Antibiot.*, 1989, **42**, 1389.
94. D. J. Hillenga, H. J. M. Versantvoort, S. van der Molen, A. J. M. Driessen, and W. N. Konings, *Appl. Environ. Microbiol.*, 1995, **61**, 2589.
95. A. J. Shah, J. Tilburn, M. W. Adlard, and H. N. Arst, Jr., *FEMS Microbiol. Lett.*, 1991, **77**, 209.
96. E. A. Espeso, J. Tilburn, H. N. Arst, Jr., and M. A. Peñalva, *EMBO J.*, 1993, **12**, 3947.
97. J. Tilburn, S. Sarkar, D. A. Widdick, E. A. Espeso, M. Orejas, J. Mungroo, M. A. Peñalva, and H. N. Arst, Jr., *EMBO J.*, 1995, **14**, 779.
98. M. Orejas, E. A. Espeso, J. Tilburn, S. Sarkar, H. N. Arst, Jr., and M. A. Peñalva, *Genes Dev.*, 1995, **9**, 1622.
99. H. N. Arst, Jr., in "The Mycota III, Biochemistry and Molecular Biology," eds. R. Brambl and G. A. Marzluf, Springer Verlag, Berlin, 1996. p. 235.
100. E. A. Espeso and M. A. Peñalva, *J. Biol. Chem.*, 1996, **271**, 28 825.
101. S. Negrete-Urtasun, S. H. Denison, and H. N. Arst, Jr., *J. Bacteriol.*, 1997, **179**, 1832.
102. A. P. MacCabe, J. P. T. W. van den Hombergh, J. Tilburn, H. N. Arst, Jr., and J. Visser, *Mol. Gen. Genet.*, 1996, **250**, 367.

103. G. A. Marzluf, *Microbiol. Mol. Biol. Rev.*, 1997, **61**, 17.
104. H. Haas, B. Bauer, B. Redl, G. Stöffler, and G. A. Marzluf, *Curr. Genet.*, 1995, **27**, 150.
105. H. Haas and G. A. Marzluf, *Curr. Genet.*, 1995, **28**, 177.
106. Y. Lu, R. L. Mach, K. Affenzeller, and C. P. Kubicek, *Can. J. Microbiol.*, 1992, **38**, 758.
107. C. Hönlinger and C. P. Kubicek, *Biochim. Biophys. Acta*, 1989, **993**, 204.
108. J. Velasco, S. Gutiérrez, F. J. Fernandez, A. T. Marcos, C. Arenos, and J. F. Martin, *J. Bacteriol.*, 1994, **176**, 985.
109. K. Then Bergh and A. A. Brakhage, *Appl. Environ. Microbiol.*, 1998, **64**, 843.
110. P. Hilgendorf, V. Heiser, H. Diekmann, and M. Thoma, *Appl. Microbiol. Biotechnol.*, 1987, **27**, 247.
111. J. A. DeModena, S. Gutiérrez, J. Velasco, F. J. Fernández, R. A. Fachini, J. L. Galazzo, D. E. Hughes, and J. F. Martin, *Bio/Technology*, 1993, **11**, 926.
112. K. Then Bergh, O. Litzka, and A. A. Brakhage, *J. Bacteriol.*, 1996, **178**, 3908.
113. O. Litzka, K. Then Bergh, and A. A. Brakhage, *Eur. J. Biochem.*, 1996, **238**, 675.
114. P. J. Johnson and S. L. McKnight, *Annu. Rev. Biochem.*, 1989, **58**, 799.
115. L. Guarente, in "The Molecular and Cellular Biology of the Yeast *Saccharomyces cerevisiae*. Gene Expression," eds. E. W. Jones, J. R. Pringle, and J. R. Broach, Cold Spring Harbor Laboratory Press, Cold Spring Harbor, NY, 1992, vol. 2, p. 49.
116. D. S. McNabb, Y. Xing, and L. Guarente, *Genes Dev.*, 1995, **9**, 47.
117. P. Papagiannopoulos, A. Andrianopoulos, J. A. Sharp, M. A. Davis, and M. J. Hynes, *Mol. Gen. Genet.*, 1996, **251**, 412.
118. O. Litzka, P. Papagiannopoulos, M. A. Davis, M. J. Hynes, and A. A. Brakhage, *Eur. J. Biochem.*, 1998, **251**, 758.
119. R. Del Carmen Mateos and S. Sanchez, *J. Gen. Microbiol.*, 1990, **136**, 1713.
120. A. Meister and M. E. Anderson, *Annu. Rev. Biochem.*, 1983, **52**, 711.
121. A. A. Brakhage and J. Van den Brulle, *J. Bacteriol.*, 1995, **177**, 2781.
122. B. Pérez-Esteban, E. Gómez-Pardo, and M. A. Peñalva, *J. Bacteriol.*, 1995, **177**, 6069.
123. J. M. Cantoral, S. Gutiérrez, F. Fierro, S. Gil-Espinosa, H. van Liempt, and J. F. Martin, *J. Biol. Chem.*, 1993, **268**, 737.
124. L. Mathison, C. Soliday, T. Stepan, T. Aldrich, and J. Rambosek, *Curr. Genet.*, 1993, **23**, 33.
125. D. J. Smith, J. H. Bull, J. Edwards, and G. Turner, *Mol. Gen. Genet.*, 1989, **216**, 492.
126. J. L. Barredo, B. Diez, E. Alvarez, and J. F. Martin, *Curr. Genet.*, 1989, **16**, 453.
127. F. Fierro, J. L. Barredo, B. Diez, S. Gutiérrez, F. J. Fernandez, and J. F. Martin, *Proc. Natl. Acad. Sci. USA*, 1995, **92**, 6200.
128. P. L. Skatrud and S. W. Queener, *Gene*, 1989, **79**, 331.
129. U. Kück, M. Walz, G. Mohr, and M. Mracek, *Appl. Microbiol. Biotechnol.*, 1989, **31**, 358.
130. A. K. Petrich, B. K. Leskiw, A. S. Paradkar, and S. E. Jensen, *Gene*, 1994, **142**, 41.
131. L. Sánchez and A. F. Braña, *Microbiology*, 1996, **142**, 1209.
132. L. Malmberg, W. Hu, and D. H. Sherman, *J. Bacteriol.*, 1993, **175**, 6916.
133. C. Jones, A. Thompson, and R. Englan, *Microbiology*, 1996, **142**, 1789.
134. A. E. Veenstra, P. van Solingen, R. A. L. Bovenberg, and L. H. M. van der Voort, *J. Biotechnol.*, 1991, **17**, 81.
135. A. Matsuda, H. Sugiura, K. Matsuyama, H. Matsumoto, S. Ichikawa, and K.-I. Komatsu, *Biochem. Biophys. Res. Commun.*, 1992, **186**, 40.
136. C. A. Bunnel, W. D. Luke, and F. M. Perry, Jr., in "β-Lactam Antibiotics for Clinical Use," eds. S. F. Queener, J. A. Webber, and S. W. Queener, Marcel Dekker, New York, 1986, p. 255.
137. L. G. Carr, P. L. Skatrud, M. E. Scheetz, II, S. W. Queener, and T. D. Ingolia, *Gene*, 1986, **48**, 257.
138. B. J. Weigel, S. G. Burgett, V. J. Chen, P. L. Skatrud, C. A. Frolik, S. W. Queener, and T. D. Ingolia, *J. Bacteriol.*, 1988, **170**, 3817.
139. G. Landan, G. Cohen, Y. Aharonowitz, Y. Shuali, D. Graur, and D. Shiffman, *Mol. Biol. Evol.*, 1990, **7**, 399.
140. M. A. Peñalva, A. Moya, J. Dopazo, and D. Ramon, *Proc. R. Soc. Lond. (Biol.)*, 1990, **241**, 164.
141. M. W. Smith, D.-F. Feng, and R. F. Doolittle, *Trends Biochem. Sci.*, 1992, **17**, 489.
142. H. Hori and S. Osawa, *Mol. Biol. Evol.*, 1987, **4**, 445.
143. T. Lendenfeld, D. Ghali, M. Wolschek, E. M. Kubicek-Pranz, and C. P. Kubicek, *J. Biol. Chem.*, 1993, **268**, 665.
144. K. Affenzeller and C. P. Kubicek, *J. Gen. Microbiol.*, 1991, **137**, 1653.
145. W. H. Müller, T. P. van der Krift, A. J. J. Krouwer, H. A. B. Wösten, L. H. M. van der Voort, E. B. Smaal, and A. J. Verkleij, *EMBO J.*, 1991, **10**, 489.
146. W. H. Müller, R. A. L. Bovenberg, M. H. Groothuis, F. Kattevilder, E. B. Smaal, L. H. M. van der Voort, and A. J. Verkleij, *Biochim. Biophys. Acta*, 1992, **1116**, 210.
147. J. W. Janc, L. A. Egan, and C. A. Townsend, *J. Biol. Chem.*, 1995, **270**, 5399.
148. A. G. Brown, D. Butterworth, M. Cole, G. Honscomb, J. D. Hood, C. Reading, and G. N. Robinson, *J. Antibiot.*, 1976, **29**, 668.
149. G. Lancini, F. Parenti, and G. G. Gallo, "Antibiotics. A Multidisciplinary Approach," Plenum Press, New York, 1995.
150. F. Röhl, J. Rabenhorst, and H. Zähner, *Arch. Microbiol.*, 1987, **147**, 315.
151. J. E. Baldwin, T. D. W. Claridge, K.-C. Goh, J. W. Keeping, and C. Schofield, *Tetrahedron Lett.*, 1993, **34**, 5645.
152. M. Wanning, H. Zähner, B. Krone, and A. Zeeck, *Tetrahedron Lett.*, 1981, **22**, 2539.
153. D. Brown, J. R. Evans, and R. A. Fletton, *J. Chem. Soc., Chem. Commun.*, 1979, 282.
154. D. L. Pruess and M. Kellett, *J. Antibiot.*, 1983, **36**, 208.
155. A. S. Paradkar and S. E. Jensen, *J. Bacteriol.*, 1995, **177**, 1307.
156. S. W. Elson and R. S. Oliver, *J. Antibiot.*, 1978, **31**, 586.
157. S. W. Elson, K. H. Baggaley, M. Davidson, M. Fulstone, N. H. Nicholson, G. D. Risbridger, and J. W. Tyler, *J. Chem. Soc., Chem. Commun.*, 1993, 1212.
158. S. W. Elson, K. H. Baggaley, M. Davidson, M. Fulstone, N. H. Nicholson, G. D. Risbridger, and J. W. Tyler, *J. Chem. Soc., Chem. Commun.*, 1993, 1211.
159. C. A. Townsend and M.-F. Ho, *J. Am. Chem. Soc.*, 1985, **107**, 1066.
160. B. P. Valentine, C. R. Bailey, A. Doherty, J. Morris, S. W. Elson, K. H. J. Baggaley, and N. H. Nicholson, *J. Chem. Soc., Chem. Commun.*, 1993, 1210.

161. J. Romero, P. Liras, and J. F. Martin, *Appl. Environ. Microbiol.*, 1986, **52**, 892.
162. K. A. Aidoo, A. S. Paradkar, D. C. Alexander, and S. E. Jensen, in "Proceedings of the Third International Conference on the Biotechnology of Microbial Products: Novel Pharmacological and Agrobiological Activities," eds. V. P. Gullo, J. C. Hunter-Cevera, R. Cooper, and R. K. Johnson, Society for Industrial Microbiology, Annandale, VA, 1994.
163. T.-K. Wu, R. W. Busby, T. A. Houston, D. B. McIlwaine, L. A. Egan, and C. A. Townsend, *J. Bacteriol.*, 1995, **177**, 3714.
164. J. E. Hodgson, A. P. Fosberry, N. S. Rawlinson, H. N. M. Ross, R. J. Neal, I. C. Amell, A. J. Earl, and E. J. Lawlor, *Gene*, 1995, **166**, 49.
165. S. J. McGowan, M. Sebaihia, L. E. Porter, G. S. A. B. Stewart, P. Williams, B. W. Bycroft, and G. P. C. Salmond, *Mol. Microbiol.*, 1996, **22**, 415.
166. E. N. Marsh, M. D. T. Chang, and C. A. Townsend, *Biochemistry*, 1992, **31**, 12648.
167. S. W. Elson, K. H. Baggaley, J. Gillett, S. Holland, N. H. Nicholson, J. T. Sime, and S. R. Woroniecki, *J. Chem. Soc., Chem. Commun.*, 1987, 1736.
168. S. P. Salowe, E. N. Marsh, and C. A. Townsend, *Biochemistry*, 1990, **29**, 6499.
169. J. E. Baldwin, M. D. Lloyd, B. Wha-Son, C. J. Schofield, S. W. Elson, K. H. Baggaley, and N. H. Nicholson, *J. Chem. Soc., Chem. Commun.*, 1993, 500.
170. J. E. Baldwin, R. M. Adlington, J. S. Bryans, A. O. Bringhen, J. B. Coates, N. P. Crouch, M. D. Lloyd, C. J. Schofield, S. W. Elson, K. H. Baggaley, R. Cassels, and N. H. Nicholson, *J. Chem. Soc., Chem. Commun.*, 1990, 617.
171. S. P. Salowe, W. J. Krol, D. Iwata-Reuyl, and C. A. Townsend, *Biochemistry*, 1991, **30**, 2281.
172. C. A. Townsend and W. J. Kroll, *J. Chem. Soc., Chem. Commun.*, 1988, 1234.
173. N. H. Nicholson, K. H. Baggaley, R. Cassels, M. Davison, S. W. Elson, M. Fulston, J. W. Tyler, and S. R. Woroniecki, *J. Chem. Soc., Chem. Commun.*, 1994, 1281.
174. J. E. Baldwin, R. M. Adlington, J. S. Bryans, A. O. Bringhen, J. B. Coates, N. P. Crouch, M. D. Lloyd, C. J. Schofield, S. W. Elson, K. H. Baggaley, R. Cassels, and N. H. Nicholson, *Tetrahedron*, 1991, **47**, 4089.
175. E. J. Lawlor, S. W. Elson, S. Holland, R. Cassels, J. E. Hodgson, M. D. Lloyd, J. E. Baldwin, and C. J. Schofield, *Tetrahedron*, 1994, **50**, 8737.
176. K. A. Aidoo, A. S. Paradkar, D. C. Alexander, and S. E. Jensen, in "Developments in Industrial Microbiology Series," eds. V. P. Gullo, J. C. Hunter-Cevera, R. Cooper, and R. K. Johnson, Society for Industrial Microbiology, Fredericksburg, VA, 1993, vol. 35, p. 219.
177. S. E. Jensen, D. C. Alexander, A. S. Paradkar, and K. A. Aidoo, in "Industrial Microorganisms: Basic and Applied Molecular Genetics," eds. R. H. Baltz, G. D. Hegeman, and P. L. Skatrud, American Society for Microbiology, Washington, DC, 1993, p. 169.
178. D. Iwata-Reuyl and C. A. Townsend, *J. Am. Chem. Soc.*, 1992, **114**, 2762.
179. J. W. Janc, L. A. Egan, and C. A. Townsend, *Bioorg. Med. Chem. Lett.*, 1993, **3**, 2313.
180. C. R. Bailey, M. J. Butler, I. D. Normansell, R. T. Rowlands, and D. J. Winstanley, *Bio/Technology*, 1984, **2**, 808.
181. J. S. Kahan, F. M. Kahan, R. Goegelman, S. A. Currie, M. Jackson, E. O. Stapley, T. W. Miller, A. K. Miller, D. Hendlin, S. Mochales, S. Hernandez, and H. B. Woodruff, in "16th Interscience Conference on Antimicrobial Agents and Chemotherapy," Chicago, 1976, Abstract No. 227.
182. R. C. Moellering, G. M. Eliopoulos, and D. E. Sentochnik, *J. Antimicrob. Chemother.*, 1989, **24**(Suppl. A), 1.
183. H. Kropp, J. G. Sundelof, J. S. D. Kahan, F. M. Kahan, and J. Birnbaum, *Antimicrob. Agents Chemother.*, 1990, **17**, 993.
184. H. Kropp, J. G. Sundelof, R. Hajdu, and J. S. D. Kahan, *Antimicrob. Agents Chemother.*, 1982, **22**, 62.
185. S. R. Norrby, K. Alestig, B. Bjornegaard, L. A. Burman, F. Ferber, J. L. Huber, K. H. Jones, F. M. Kahan, J. S. Kahan, H. Kropp, M. A. P. Meisinger, and J. G. Sundelof, *J. Antimicrob. Agents Chemother.*, 1983, **23**, 300.
186. J. M. Williamson, E. Inamine, K. E. Wilson, A. W. Douglas, J. M. Liesch, and G. Albers-Schönberg, *J. Biol. Chem.*, 1985, **260**, 4637.
187. K. Kubo, T. Ishikura, and Y. Fukagawa, *J. Antibiot.*, 1985, **38**, 904.
188. D. R. Houck, K. Kobayashi, J. M. Williamson, and H. G. Floss, *J. Am. Chem. Soc.*, 1986, **108**, 5365.
189. B. W. Bycroft, C. Maslen, S. J. Box, A. Brown, and J. W. Tyler, *J. Antibiot.*, 1988, **41**, 1231.
190. W. L. Parker, M. L. Rathnum, J. S. Wells, W. H. Trejo, P. H. Principe, and R. B. Sykes, *J. Antibiot.*, 1982, **35**, 653.
191. B. W. Bycroft, C. Maslen, S. J. Box, A. Brown, and J. W. Tyler, *J. Chem. Soc., Chem. Commun.*, 1987, **21**, 1623.
192. S. J. McGowan, M. Sebaihia, S. Jones, B. Yu, N. Bainton, P. F. Chan, B. W. Bycroft, G. S. A. B. Stewart, P. Williams, and G. P. C. Salmond, *Microbiology*, 1995, **141**, 541.
193. N. J. Bainton, P. Stead, S. R. Chabra, B. W. Bycroft, G. P. C. Salmond, G. S. A. B. Stewart, and P. Williams, *Biochem. J.*, 1992, **288**, 997.
194. C. A. Townsend, *Biochem. Soc. Trans.*, 1993, **21**, 208.
195. R. B. Sykes, C. M. Cimarusti, D. P. Bonner, K. Bush, D. M. Floyd, N. H. Georgopapadakou, W. M. Koster, W. C. Liu, W. L. Parker, P. A. Principe, M. L. Rathnum, W. A. Slusarchyk, W. H. Trejo, and J. S. Wells, *Nature (London)*, 1981, **291**, 489.
196. C. A. Townsend and A. M. Brown, *J. Am. Chem. Soc.*, 1983, **105**, 913.
197. N. Katayama, Y. Nozaki, K. Okonogi, H. Ono, S. Harada, and H. Okazaki, *J. Antibiot.*, 1985, **38**, 1117.
198. J. O'Sullivan, M. L. Souser, C. C. Kao, and C. A. Aklonic, *Antimicrob. Agents Chemother.*, 1982, **21**, 558.
199. J. O'Sullivan, M. L. Souser, C. C. Kao, and C. A. Aklonic, *Antimicrob. Agents Chemother.*, 1983, **23**, 598.

4.08
Multifunctional Peptide Synthetases Required for Nonribosomal Biosynthesis of Peptide Antibiotics

NARAYANAN RAJENDRAN

Michigan State University, East Lansing, MI, USA

and

MOHAMED A. MARAHIEL

Philipps-Universität Marburg, Germany

4.08.1   INTRODUCTION

Many microorganisms are pathogenic to humans or animals and cause various diseases resulting in extensive morbidity and mortality. Other microbial species synthesize and release an array of special low molecular weight peptide products with antibiotic activities, which are beneficial to mankind. In both cases, it is essential for the microorganisms to protect themselves in a complex ecological niche in which an enormous diversity of other products or toxins exist. A single microbe, for example *Pseudomonas aeruginosa*, can either act as a pathogen to humans[1] or can be a beneficial antibiotic producer. However, the metabolites produced must not be self-toxic to the producer organism. A nonautointoxication mechanism, often called a self-resistance process,[2] associated with antibiotic production through a different mode of biosynthesis, helps the survival of the producer organisms.[3]

Special metabolites,[4] often called secondary metabolites, such as tetracycline, tyrocidine, gramicidin S, bacitracin, tylosine, etc., are produced through multistep pathways.[5] Their production takes place at synthetase enzymes,[6] leading from precursors — usually intermediates of primary metabolism — to the specific moieties of the secondary metabolites.[7] These metabolites may be produced at various stages of growth,[8] such as on changing from the vegetative to the stationary phase,[9] or at specific intervals of various life processes, for example doubling of chromosomes during the process of cell division, or at intermittent levels of primary metabolic activities.[10] The component moieties of some of these special metabolites are activated in the form of adenylated, phosphorylated, or coenzyme A derivatives, and finally linked together to form the end products,[11] such as polypeptides or polyketides.

These special metabolites represent an enormously diverse collection of compounds, including aminoglycosides, β-lactams, polyketides, and small polypeptides. Their production depends on the rate of synthesis of enzymes[10] involved in the biosynthesis process.[12] The activities of these enzymes are fairly high during the transition state[8] from the fast growing phase to the stationary phase of the cells.[13] The synthesis of enzymes is a short-term process in secondary metabolism. When the synthesis of enzyme stops or the entire enzyme system is inactivated,[14] depending on various growth and differentiation processes, the production of special metabolites decreases.[10]

4.08.2   PHYSIOLOGICAL ROLE OF METABOLITE SYNTHESIS

Special metabolites are often not essential for growth and proliferation[15] of many producer microorganisms. Their production could be abandoned without any deleterious effect to the producer organisms under laboratory conditions.[8] However, another view suggests that as these compounds are complex in nature and derived by complex pathways, they must serve certain functions for the producer organisms[16] during growth in the natural habitat.

Microorganisms are sensitive to various environmental factors. Selection pressure seems to have a greater influence on producer organisms to induce some developmental pathways through which they can synthesize special metabolites.[17] Therefore, the production of various antimicrobial antibiotics[18] may be necessary for the organisms in the competitive environment,[5] enabling them to differentiate their phenotypic characteristics and escape from deleterious factors. Various studies on characterization of secreted metabolites have provided clues for the so-called purpose of biosynthesis of peptide metabolites.[5,10,19,20]

Microorganisms live in different habitats, spanning the spectrum from aquatic to terrestrial environments. They need to be adaptive towards various extreme environmental conditions where plants and other organisms are producing various toxic factors such as phenolics and activated oxygen-containing radicals. Moreover, several predators such as bacteriophages, bacteria like bdellovibrios and protozoa may be present in their ecosystem.[21] Also, the complex nature of the competitive environment in the soil, in which other parameters such as nonavailability of carbon or nitrogen sources, as well as depletion of micro or macro elements,[22] may affect the production of secondary metabolites.

Under conditions of nutritional stress,[23,24] microorganisms have evolved schemes that include irreversible cellular differentiation, as in the cases of widely-studied genera such as *Streptomyces* of Actinomycetes, *Bacillus* of Gram-positive bacteria, and *Pseudomonas* of Gram-negative bacteria.[25] This leads to the development of individual pathways[26] for the synthesis of structurally diverse groups of special metabolites[27] to cope with these adversities. Some of these metabolites are endowed with bioactive properties,[28] which include toxins[29] bacteriolytic enzymes such as lysostaphin,[30] phospholipase A,[31] and hemolysins[32] as well as peptide antibiotics like bacitracin, valinomycin,[33] penicillin,[34] and cephalosporin.[35]

4.08.3 MECHANISMS FOR PREVENTION OF AUTOINTOXICATION

A universal discussion on antibiotic resistance is beyond the scope of this chapter. However, a few comments regarding developments in microbial resistance may be useful for understanding how antibiotic-producing species have developed suicide-avoiding mechanisms,[3] which ultimately help to induce the production of one or more metabolite(s) with antibiotic properties towards the same strain.[36]

4.08.3.1 Microbial Resistance

It is now increasingly obvious that many pathogens have become resistant to various drugs, due to the frequent and widespread use of antibiotics.[10] Numerous microorganisms have acquired plasmids or unique resistant R-factors, which provide the organism with the ability to synthesize enzymes that modify the antimicrobial substances.[10] The plasmids may also cause changes in the organism's ability to accumulate the antimicrobial agent and inactivate the compound through metabolic enzymes.[2] Microbial resistance may occur as the result of chromosomal changes such as the expression of a latent chromosomal gene or the exchange of genetic material via plasmids or transposons.[3] Resistance can also be acquired by a microorganism through a spontaneous or induced mutation. Due to mutations in a locus that controls sensitivity to a given antimicrobial agent, microbes may also lose their susceptibility. Alteration in the permeability of the peptidoglycan unit of the bacterial envelope also results in drug resistance.[10]

Pathogenic microbes that cause diseases in human beings, especially lung diseases such as tuberculosis, pneumonia, and others such as cancer and gastrointestinal diseases, have evolved potent mechanisms of drug resistance. For example, methicillin-resistant *Staphylococcus aureus* (MRSA) is resistant to all the currently used antibiotics except vancomycin.[37] However, vancomycin may not be very effective for deep-rooted infections,[35] because the minimum inhibitory concentration (MIC) has been increasing against some MRSA isolates.

The combined use of more than one antibiotic retards the appearance of resistant organisms; however, their toxicological side-effects on the host organisms may restrict the further use of a combinational therapeutic strategy. Due to the emergence of such cross-restraint microbial groups, certain valuable antibiotics have only restricted uses, as in the case of vancomycin.[10] Strategies such as antibiotic control programs, monitoring drug doses, better hygiene, engineering of agents with refined antimicrobial activities, and development of novel drugs, need to be adopted in order to limit microbial resistance.[10]

Many diseases and the causative agents, such as manifold pathogenic bacteria, fungi, and viruses, are becoming more and more resistant to traditional drug therapy. There is an urgent need for innovative concepts to develop commercially viable and pharmaceutically useful new drugs in order to face the resistance problems. Research and development has therefore turned to advanced molecular biology methods to generate novel therapeutic substances. Pharmaceutical companies are well equipped to face modern technical developments and utilize engineering techniques in their routine screening strategies.

Such developments have unveiled new classes of antibiotics which interfere with the microbes' ability to synthesize proteins in a novel way.[38] Nearly a dozen new antibiotics that show promise in controlling drug resistant organisms have been developed. For example, oleandoymycin (antibacterial, by Pfizer), josamycin (antibacterial, by Yamanouchi), chromomycin (anticancer, by Takeda), and griseofulvin (antifungal, by Schering Plough) are some of the polyketides which show significant therapeutic activities.[38]

Although toxicological tests must be carried out to make use of many antibiotic discoveries for clinical purposes, reports from research expose optimism for treating infectious diseases. While research groups are studying the defense mechanisms of microorganisms against a given antimicrobial agent, others are trying to utilize the significant clues obtained from the resistance mechanisms of microorganisms to explain the physiological role of prevention mechanisms of autointoxication.[10]

4.08.3.2   Suicide-avoiding Mechanism

The mechanism of prevention of autointoxication is a complex process. It is still not fully understood how the organism producing antibiotic prevents autointoxication.[39] However, possible scenarios for the presence of suicide-avoiding mechanisms have been studied in *Bacillus*.[40]

4.08.3.2.1   Modification of autogenous antibiotics

Despite the structural diversity of the antibiotic substrates, two modes of enzymatic modifications exist for detoxification of host antibiotics, namely *N*-acetylation of amino groups and *O*-phosphorylation of hydroxyl groups, for which the donor cofactors are acetyl coenzyme A and ATP, respectively.[3] The toxicity of the antibiotic produced by the host organism can be attenuated by adapting one or both modes of enzymatic modification by the producer organisms that possess enzymes capable of modifying the autogenous drugs.[41] For example, many species of *Streptomyces* have such mechanisms[42] to prevent autointoxication.

4.08.3.2.2   Drug target modifications to protect host organisms

Host self-defense in the case of aminoglycoside antibiotics employs ribosomal modification whereas in antibiotics that do not target RNA, the protein targets of cell membranes are modified in the host cell. In ribosomal target sites, for example in *Micromonospora purpurea*[43] or in *Streptomyces fradiae*,[44] the post-transcriptional methylation of ribosomal RNA is a critical parameter, responsible for nonautointoxication. In cases where RNA is not the target it is not known whether the antibiotic toxicity is diminished owing to post-translational modification of an otherwise-sensitive target enzyme or whether a resistant version of the enzyme is synthesized *de novo* in a nonribosomal site, as in the case of *Cephalosporium caerulens*.[45]

4.08.3.2.3   Extracellular mode of antibiotic secretions

Every antibiotic-producing organism must have a specific mechanism by which the produced antibiotics are secreted out of the cells. Many peptide antibiotics, such as surfactin of *Bacillus subtilis*[46,47] or saframycin of *Myxococcus xanthus*[48] are secreted outside of the cells by an extracellular efflux mechanism. Also for *Tolypocladium niveum*, the cyclosporin A producer,[49] such an efflux mechanism to overcome the possible autointoxication has been described.

4.08.3.2.4   Intracellular mode of antibiotic secretions

Some microorganisms synthesize special metabolites including antibiotics within the cells, but not secreting out of the cells. For example, in *Bacillus sphaericus* and *B. thuringiensis* the endotoxins are produced within the cells.[50] These protein crystals are toxic against many insect larvae when they enter into the mid-gut of the larvae. Generally, the synthesized cellular products are kept stored inside the cells until needed or until the rate of cellular antibiotic production is under control.[51] In *B. licheniformis*, it is essential for the producer organism to keep the cellular product within the cell at a certain level for their cell differentiation.[20] In *B. brevis*, the secreted antibiotics may have some physiological role[52] either in sporulation[53, 54] or in germination.[55]

4.08.3.2.5 *Alteration of membrane permeability barriers*

Modification of bacterial cell wall permeability is a known mechanism for many microorganisms to overcome autointoxication.[3] The secreted antibiotics, available by any means of secretion, whether intracellular or extracellular, are accumulated near the cell membrane region, where cell permeability is taking place. Due to environmental stimulation, physiological stress, or urgent need of the synthesized metabolites, the produced antibiotics are permeated through cell barriers.

The best example of alteration of membrane permeability can be derived from the peptide antibiotic bacitracin of *B. licheniformis*.[56] Initially, a complex of a metal ion with isoprenyl pyrophosphate (IPP) is formed near the cell membrane and later it tries to cross the membrane. Bacitracin, by binding with IPP, prevents its dephosphorylation and thus interrupts the permeability of this complex. Hence, bacitracin passes freely through the cell wall.[57]

The nucleotide sequence of the resistance determinant of bacitracin of *B. licheniformis* has revealed the presence of three open reading frames, *bcrA*, *bcrB*, and *bcrC*. The *bcrA* shows similarity to the ABC family of transporter proteins, whereas *bcrB* and *bcrC* genes may function as membrane components of permease 39 and confirms the presence of permeability mechanisms in bacitracin producer organisms, to regulate the level of produced antibiotics in the cell and avoid autointoxication.

4.08.4 BIOACTIVE PEPTIDE PRODUCTS

Many efforts have been made towards the isolation of bioactive substances, including antibacterial, antifungal, antiviral, and antitumor compounds, from various microorganisms.[58,59] These heterogeneous biochemical compounds, which possess a diverse range of structural characteristics, inhibit the growth of other microorganisms and tumors.[60]

The effective antagonistic activities of secondary metabolites against various microorganisms were first observed and systematically recorded in 1877, by Pasteur and Joubert.[61] Following their observations, many intensive searches for microbial organisms possessing nonpathogenic effects against humans and at the same time having high antagonistic potential against pathogens were made.[62,63] Although an enormous amount of information was gathered on antimicrobial effects, only a fraction of metabolites possessing antibiotic properties were studied in depth for clinical uses.

Since the 1950s, significant progress has been made towards understanding the biosynthesis of the bioactive peptide metabolites, such as depsipeptides, peptidolactones, and lipopeptides,[64] which are of ribosomal and nonribosomal origin with unusual amino acid compositions and structures.[65–67] Some of these special metabolites are widely used in food products for human consumption,[68,69] for example nisin, a polypeptide bacteriocin produced ribosomally by *Lactococcus lactis* subsp. *lactis*,[70] whereas many others such as cyclosporin[71] and penicillin of *Penicillium chrysogenum* are of pharmacological grade and used for clinical purposes.[72]

In 1902, when peptides were first isolated from natural materials by Gulewitsch and Amiradzibi, much effort was applied towards isolation of many peptides.[73] Since the first clinical trial of the antibiotic gramicidin S, a cyclic decapeptide discovered in 1939 by Rene Dubos,[74] several peptide antibiotics have found widespread acceptance in medicine, as well as in animal care, and in industrial and agricultural programmes. The majority of them had antimicrobial activities, like penicillin, or antibacterial activities, like polymyxin.[75]

While some of them have been exploited as immunomodulators, like cyclosporin, or specific enzyme inhibitors, for example polypeptin, many others have specific applications such as extracellular surfactant like surfactin, or siderophores such as enterobactin and ferrichrome, or feed additives like parvulin.[76] Many peptide antibiotics exhibit valuable properties qualifying them for biotechnological and medical uses as antivirals,[76] such as actinomycin, and antitumor agents[64] like blasticidin S.

Microbial peptides considered to be nontoxic act as extracellular regulators that mediate inter-cellular communications between producer organisms.[12] Some peptides are virulence determinants produced by plant or animal pathogens whereas a few others are peptide toxins, such as microcystins, cyclic heptatoxins produced by *Microcystis aeruginosa*,[77] which cause the death of animals and illnesses in humans.

4.08.5   TOWARDS COMMERCIAL ANTIBIOTICS

After the discovery of penicillin[78] by Fleming in 1928 and after its value became known in the mid-1930s by Howard Florey and Ernst Chain,[74] attention turned towards β-lactam antibiotics produced by bacteria and fungi. After the isolation of antibiotics such as actinomycin from *Strep-tomyces* in 1940, streptomycin in 1942, steptomycin in 1944 by Waksman,[79] bacitracin in 1945, polymyxin in 1947, and cephalosporin in 1948 by Brotzu,[80] the search was on for novel antibiotics in the United States, Germany, France, Japan, and England.

In 1948, aureomycin from *Streptomyces aureofaciens* and terramycin from *S. rimosus* were isolated and later the parent structure of tetracyline was established.[81] Two years later in 1950, rhodomycin was isolated by Brockman from *S. purpurascens*. The broad-spectrum antimicrobial antibiotic chloramphenicol, used primarily in the treatment of typhoid and other gastrointestinal diseases, was isolated from *S. venezuelae*.[82] Because of its simple chemical structure, this antibiotic is produced commercially by an entirely chemical synthesis.

At the time of discovery of many other antibiotics like cordycepin in 1950, erythromycin in 1952, and kanamycin in 1957 by Umezawa,[83] many macrolitic and aminoglycoside antibiotics such as spectinomycin, ribostamycin, and puromycin were discovered and clinically used as antibacterial and antineoplastic agents.[82] It was in the 1950s that Birch proposed that polyketides are constructed from acetyl- and malonyl thioesters[84] and it was only in the 1970s that Staunton, Simpson and others using the stable deuterium and ^{13}C isotopes in conjunction with nuclear magnetic resonance, revealed the importance of the new class of aromatic polyketides.[84]

Polyketides are a large family of structurally diverse natural products possessing a wide range of biological activities including antibiotic properties. For example, erythromycin (antibacterial), idarubicin (anticancer), candicidin (antifungal), spiramycin (antirickettsial) are examples of impor-tant therapeutic polyketides.[38]

Since the multienzymatic activities involved in the formation of polyketides were suggested, attention has been concentrated on polyketide synthases. Some are multifunctional proteins similar to type I fatty acid synthases, others including the predominant type II fatty acid synthase are dissociable multi-subunit enzyme complexes. Because of the close functional relationship between fatty acid and polyketide metabolism, it was found that bacterial polyketide synthases could have a similar organization.[85]

Because of the commercial success of polyketides as drugs, flavoring agents, and pigments, as well as the shared mechanism of biosynthesis and the high degree of conservation that exists among polyketide synthase (PKS) gene clusters, a considerable interest has developed in the molecular biology of polyketides, to generate novel ones. Some polyketides, such as tetracyline (antibiotics), anthracyclines (antitumor), daunomycin, aclacinomycin A and doxorubicin,[86] formerly called adri-amycin (anticancer), FK 506 (immunosuppressant), monensin, and avermectin (antiparasitics), were all isolated from Actinomycetes.[87]

Followed by the clinical success of polyketides as water-soluble and heat-stable antibiotics, particular attention was diverted towards economically viable antibiotics,[10] those composed in part of short polypeptides, grouped as peptide antibiotics. The new discoveries and genetic analysis of the multi-enzymatic (nonribosomal) biosynthesis systems that catalyze the polymerization, epi-merization, and tailoring of polypeptides to a specific function have occupied most of the intervening decades to give birth to many commercially viable antibiotics.

The abundance of polypeptide antibiotics accounts for the enormous diversity (see Table 1 for important peptide antibiotics produced nonribosomally), with more than 550 different constituent compounds[72] now identified. It is reasonable to speculate that peptide antibiotics may dominate as commercially successful candidates in the future.

Table 1 Nonribosomally produced peptide antibiotics.

Producers and antibiotics	Therapeutic area
Bacillus brevis Gramicidin S	antibacterial, biosurfactant, nucleotide binding
Bacillus brevis Tyrocidine	antibacterial, hemolytic
Bacillus circulans Octapeptin	antibacterial, antimycobacterial, antifungal, antiprotozoal
Bacillus licheniformis Bacitracin	membrane acting, antibacterial, metal-ion binding, inhibition of cell wall synthesis
Bacillus pumilis, *Micrococcus* spp. Micrococcin	ribosome binding, antibacterial, antimycobacterial
Bacillus subtilis Iturin	antifungal, clinically used
Bacillus subtilis Mycosubtilin	antifungal
Bacillus subtilis Surfactin	antimycobacterial, biosurfactant, membrane acting, hemolytic
Cochliobolus carbonum HC-toxin	antifungal, phytotoxic
Fusarium scripi Enniatin	antifungal
Myxococcus xanthus, *Streptomyces lavendulae* Saframycin	inhibition of DNA/RNA synthesis, antimicrobial, antitumor
Pseudomonas tolaasii Tolaasin	antifungal
Streptomyces canus Amphomycin	inhibition of cell-wall synthesis, antibacterial
Streptomyces griseus Albomycin	clinically used, iron binding, membrane acting
Streptomyces hygroscopicus, *Streptomyces viridochromogenes* Bialaphos	antifungal, antibacterial
Tolypocladium niveum Cyclosporin A	immunosuppressant, clinically used
Aspergillus nidulans, *Penicillin chrysogenum*, *Streptomyces clavuligerus*, *Norcardia lactamdurans*, *Flavobacterium* SC12 β-Lactams	clinically used, inhibition of cell-wall synthesis, precursor of cephalosporins, precursor of penicillins

For other compounds, see references 65 and 88.

4.08.6 CHARACTERISTICS OF PEPTIDE PRODUCTS

The structural diversity exhibited by the peptide antibiotics known today may not have similar biosynthetic origins. Due to proteolytic processing of gene encoded polypeptide precursors,[89] some peptides are multicyclic[88] in structure. However, they often undergo extensive post-translational processing and modification[90] before being secreted outside of the cell,[76] whereas other peptides, composed of between two and 20 amino acids, are made on protein templates,[91] for example precursors of penicillin and cephalosporin. They possess remarkable structural diversities such as linear, cyclic or branched cyclic structures[76] and are synthesized through a series of reactions catalyzed by large multisubunit enzyme complexes[92] called peptide synthetases.[93] The simplest representative of this class of enzymes is L-D-(α-aminoadipoyl)-L-cysteinyl-D-valine (ACV) synthetase, the key initial enzyme in the biosynthesis of both penicillin and cephalosporin antibiotics.[94]

Although polypeptides and polyketides have remarkable structural differences, most of them have some basic analogies in their biosynthesis.[76] For example, in the biosynthesis of polypeptide antibiotics, the polymerization of amino acids takes place on the protein template, similar to polymerization of acetate units during the biosynthesis of polyketides, macrolites, and polyethers.

4.08.7 TWO TYPES OF PEPTIDE SYNTHESIS

Many groups of a special class of metabolites, peptide antibiotics, are produced by an abundance of species of pathogenic as well as nonpathogenic prokaryotes such as unicellular bacteria and actinomycetes and many lower eukaryotes such as filamentous fungi.[88,95,96] The unicellular bacteria and filamentous fungi, including a number of nonpathogens that inhabit soil and aquatic environments, have remarkably complex biosynthetic pathways,[97] and are thought to be synthesized independently of genetic translations, through metabolic pathways.[98]

In contrast to this synthetic pathway, many organisms utilize an alternative pathway, in which the amino acid sequence in the peptide product is not determined directly at the gene level through nucleic acid interactions, but rather by the protein structure of the respective enzyme system.[97] Thus, amino acids are incorporated into peptides during the process of biosynthesis of peptides, and the catalysis of peptide bond formation takes place by either one of the two peptide-forming strategies,[99] that is, the ribosomal system or on a nonribosomal protein template.[10] In both cases, the common feature is the activation of amino acids by the energy derived from hydrolysis of an ATP α–β linkage.[100]

4.08.7.1   Ribosomal Antibiotic Synthesis

In biological systems, the majority of cellular peptides and all proteins are of ribosomal origin.[101] In the ribosomal system, amino acids are activated by aminoacyl-tRNA synthetases as tRNA esters and the peptide bond formation is initiated on the ribosome.[100] The total number of 20 proteinogenic amino acids (21, including selenocysteine)[102,103] is limited by the number of tRNA ligases which specifically recognize the amino acid and esterify the constituent amino acids to their cognate tRNAs.

For example, the bioactive lantibiotics represent highly stable multicyclic structures resulting from the processing of gene-encoded peptide precursors that have undergone extensive post-translational changes.[65] The gene-coded antibiotic subtilin, which is synthesized as a precursor through transcription and translation, is a well-known example of ribosomal peptide antibiotic synthesis.[5] Colicins and microcins are also plasmid-encoded and are of ribosomal origin.[76] The type A lantibiotics, such as nisin of *L. lactis*, subtilisin of *B. subtilis*, epidermin of *Staphylococcus epidermidis*, gallidermin of *S. gallinarum* and type B lantibiotics, such as duramycins (including cinnamycin and ancovenin) of Acinomycetes are all ribosomally synthesized.

4.08.7.2   Nonribosomal Antibiotic Synthesis

In nonribosomal peptide antibiotic synthesis, amino acids are activated employing a multienzyme. Peptide bond formation takes place without ribosomal involvement.[100] The synthesis of gramicidin S, tyrocidine, surfactin, and bacitracin are some of the best characterized examples of a nonribosomal biosynthetic system.

It is now accepted that nonribosomal peptide analysis, as shown in Figure 1, is an alternative means of manufacturing polypeptides, playing an important role in the production of small molecules of quite specialized function.[10] A nonribosomally produced peptide can be composed of linear, cyclic and/or branched peptide chains, often containing amino acids not found in ribosomally synthesized products. Although structurally diverse, most of the nonribosomal peptides share a common mode of synthesis, the multienzyme thiotemplate mechanism facilitated by peptide synthetases. This alternative machinery for peptide synthesis, which depends on a protein template, seems to be necessary for introducing the required variability into the peptides synthesized.

4.08.8   OVERVIEW OF THE NONRIBOSOMAL SYSTEM

Many drugs in use today are derived from natural products. Investigations reveal that many of these produced by various microorganisms[16] are of peptide origin and are structurally complex in nature. Such compounds are composed of an oligopeptide skeleton having extensive modifications, such as the incorporation of D-amino acids, hydroxy acids, and other unusual pseudo amino acids, by use of peptide synthetases.

The majority of peptides originate from Gram-positive bacteria and filamentous fungi. Among them Bacilli and Actinomycetes are leading producers. However, bouvardin from the medicinal

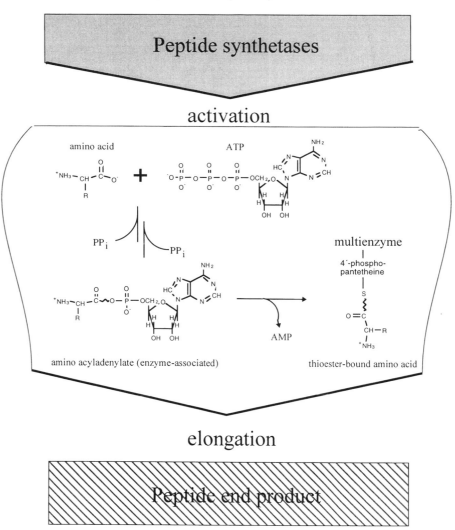

Figure 1 Substrate activation in multienzymatic peptide synthesis system. Peptide synthetases are composed of highly conserved modules. They activate and covalently bind an amino acid specifically by the two-step-reaction shown in the box. Similarly to the ribosomal machinery, the multienzymatic system activates the cognate amino acids under ATP hydrolysis as amino acyl-adenylates. These relatively unstable intermediates are stabilized by thioesterification on the prosthetic 4′-phosphopantetheine groups of the synthetases. According to the multicarrier model, the growing peptide chain is transferred from one domain-linked cofactor to the next.

plant *Bouvardia terniflora*, lophyrotomin, a benzoylated octapeptide toxin from the sawfly larvae *Lophyrotoma interrupta* and *Arge pullata* are promising candidates to prove that some higher organisms contain peptide compounds.[94,95] However, lack of genes encoding peptide synthetase within the determined DNA sequence of *Haemophilus influenzae* has proved that not every organism contains such an enzyme system.[95] Most of these enzymatically synthesized peptides[102,103] are composed of peptide chains that may be modified by acylation or glycosylation.[91]

Many Gram-negative bacteria, for example *Pseudomonas*, produce secondary metabolites in this manner,[104,105] Some of the well-known models are syringotoxin from *P. syringae*,[106] phaseolotoxin from *P. syringae*, var. *phaseolica*[107], the iron siderophores enterobactin[108,109] and pyoverdin[110–112] from *P. aeruginosa*. The fresh-water cyanobacterium *M. aeruginosa* produces mycrocystin,[113–115] and *Zolarian arboricola* produces pneumocandin.[116] Table 1 describes other significant peptide products and their producing strains in detail (for further details about peptide products see refs. 65, 88). Nonribosomally produced peptides thus contribute to the remarkable structural diversity of low molecular weight peptide products.[117]

4.08.9 PROPOSED MODE OF NONRIBOSOMAL PEPTIDE SYNTHESIS

Since many biochemical investigations on peptide biosynthesis have already been carried out, a considerable amount of information on the peptide products and their mode of biosynthesis has accumulated. In the 1950s the first literature appeared describing polypeptide synthesis proceeding without ribosomes or RNA in the presence of protein synthesis inhibitors such as puromycin.[118] In 1954, Lipman[119] proposed a possible mechanism to explain the pathway of nonribosomal peptide biosynthesis. Data in this paper demonstrate that large multienzyme complexes catalyze elongation of a peptide covalently linked to a phosphopantetheinyl arm by a thioester bond.[33]

Although the above studies on polypeptide synthesis first appeared in the early 1950s, in the 1960s special attention was given to so-called particle-free polypeptide synthesis, and further investigations were directed towards understanding nonribosomal peptide synthesis. A revised multienzymatic model proposed by several groups in the early 1970s has provided a general picture regarding the multienzymatic mode of nonribosomal peptide synthesis.[119–121]

The regulation of peptide synthetase-catalyzed reactions and the production of peptide synthetase subunits have also drawn considerable attention from molecular biologists and biochemists in the past, which ultimately yielded a universally accepted model for nonribosomal peptide synthesis.

4.08.10 MULTI-CARRIER THIOTEMPLATE MODEL

Investigations on the characterization of peptide synthetases have provided valuable insights into the molecular structure and *modus operandi* of these enzymes.[72,117] According to these studies, peptide synthetases consist of as many repeated units as the number of amino acids incorporated into the oligopeptide. Each of these repeated units (domains), regardless of the amino acid composition of the molecule synthesized, comprises approximately 1000–1500 amino acid residues and contains conserved motifs involved in substrate activation and thioester formation.[122]

Sequence alignments of peptide synthetases clearly indicate that the peptide synthetases are a related group of enzymes, responsible for the binding, activation, and modification of the individual substrate amino acids.[123] These enzymes are proposed to act by a common mechanism,[124] but their individual units/modules seem to act as independent enzymes,[125] which invoke transfer of the carboxyl group of the amino acid by formation of the respective aminoacyl adenylate. Although the thiol groups were originally proposed to be provided by cysteines, this has been revised to invoke multiple 4′-phosphopantetheine moieties attached to conserved serines which are found in common motifs, since there are no conserved cysteines.[93] The mode of action, in which the activated amino acids are presented as thioesters on a protein template via distinctive 4′-phosphopantetheine groups, was thus recognized as the multiple carrier thiotemplate mechanism.[125]

According to the revised model suggested by Schlumbohm and Wittmann-Liebold[126] in 1991 a two-step process is involved in the mechanism, namely ATP-dependent aminoacyladenylation and subsequent thioesterificaton.[92,122] In this model, shown schematically in Figure 1, each substrate amino acid is first activated as an aminoacyl adenylate (via ATP hydrolysis), as in the case of the amino acid activation step catalyzed by tRNA ligases.[98]

This activated aminoacyl residue is then linked to the active site or substrate-activating domain of the modular peptide synthetase enzyme by a thioester linkage of an individual cofactor, 4′-phosphopantetheine group. Within a single amino acid activating domain, three specific interaction positions for the prosthetic 4′-phosphopantetheine group have been postulated:[72] a charging site for the covalent binding of the cognate amino acid (thioester formation) a peptidyl acceptor site (interaction with preceding adenylation and the following elongation domain) and a peptidyl donor position (interaction with consecutive module).

Thioesterified substrate amino acids thus include a step-by-step elongation of peptide product, in a series of transpeptidation reactions,[63] which occurs by transfer of the activated carboxyl to the amino group of the next amino acid, thus affecting nitrogen to carbon step-wise condensation. This determines the amino acid sequence of the peptide.[126] Basically, this mode of transpeptidation resembles the ribosomal peptidyl transfer from the A (aminoacyl) to P (peptidyl) site. With regard to the initial ATP-dependent substrate activation as shown in Figure 1, this represents an additional similarity between ribosomal and nonribosomal peptide biosynthesis.[92] In conclusion, the cofactors facilitate the ordered shift of the carboxy thioester-activated substrates between the modules that constitute the peptide synthetases, resulting in formation of a defined peptide.[122]

Based on defining the cofactor content of the amino acid activating module of gramicidin S synthetase, and the structures of all thio template sites of the peptide-forming multienzyme, Stein *et al.*[125] have provided firm evidence to support the hypothesized multicarrier model of nonribosomal peptide biosynthesis on multifunctional protein templates.

4.08.11 MODULAR ARRANGEMENT OF PEPTIDE SYNTHETASE

The multifunctional peptide synthetases possess a highly conserved and ordered structure of semi-autonomous modules as depicted in Figure 2. Studies such as site-directed mutagenesis and photoaffinity labeling with ATP analogues, suggest that such modules have a linear arrangement of conserved domains, which are involved in the catalysis of substrate activation, modification, and condensation.[72] Within some intermodule regions, the location of individual domains involved in substrate recognition and adenylation, thioester formation, and racemization[46,122] have been identified. These investigations, which were carried out through specific deletion experiments, enabled the amplification of specific domain-coding regions from a diverse group of bacterial and fungal genes encoding peptide synthetases that were shown to be enzymatically active and clearly supported the model of modular arrangement of peptide synthetases.[117]

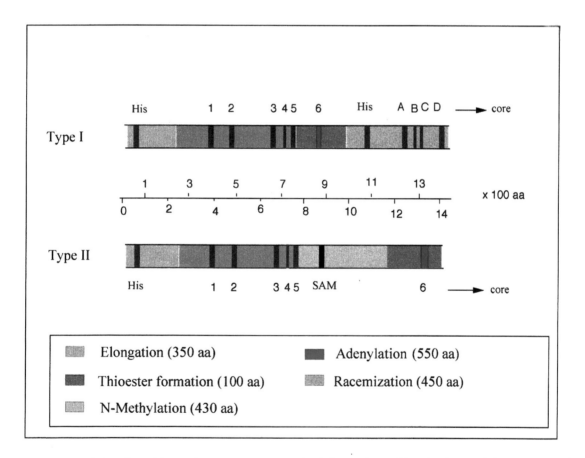

Figure 2 Modules of peptide synthetases. Two types of existing amino acid activating modules, Type I and II, in peptide synthetases are shown. The relative location, extension (size in amino acid residues), and organization of specialized domains such as elongation, adenylation, thioester formation, racemization, and *N*-methylation are schematically depicted. The numbers 1 to 5 indicate the location of conserved core sequences within the adenylation domain such as LKAGGAYVPID, YSGTTGXPKGV, GELCIGGXGXARGYL, YXTGD and VKIRGXRIELGEIE, respectively. The green strip numbered is core 6 (DNFYXLGGHSL) of the domain required for thioester formation and the serine residue(s) acts as cofactor (4'-phosphopantetheinyl) binding site. Putative *N*-methylation (SAM, VLE/DXGXGXG), peptide elongation (His, HHILXDGW), and optional racemization (His, HHILXDGW; A, AYXTEXNDILLT AXG; B, EGHGREXIIE; C, RTVGWFTS-MYPXXLD; and D, FNYLGQFD) core sequences are shown.

The modules of Type I and Type II, shown in Figure 2 for example, are found in peptide synthetases, which catalyze both amino acid adenylation and thioester formation. In Type II modules of N-methyl peptide synthetases, such as cyclosporin synthetase, the adenylating and the thioester-forming domains are separated by an insertion of about 400 amino acids that catalyzes N-methylation of the cognate amino acid.[122] In some Type I modules an epimerization domain is located upstream of the thioester-forming domain. In addition to these domains, there is another putative elongation domain which is believed to be involved in peptide elongation, and may also facilitate module interactions.[91]

Individual domains associated with specific functions (like substrate-adenylation thiolation, epimerization, or N-methylation) were found to contain highly conserved specific core regions of three to eight amino acid residues. Their location and order within the different domains were found to be conserved, irrespective of the corresponding module. The order of known core sequences are core 1 KAGGA, core 2 SGTTGXPKG, core 3 GELCIGGXGXARGYL, core 4 TGD, core 5 KIRGXRIEL and core 6 LGGXS. In some sequence analyses (see Figure 6), the sequence NGK also seems to be a highly conserved core sequence and located between core 5 (KIRGXRIEL) and core 6 (LGGXS).

The physiological role of some conserved residues was analyzed by site-directed mutagenesis and functions were assigned for core sequences 2 to 5 in ATP-binding and hydrolysis within the adenylation domain, whereas the serine residue of core 6, located within the thiolation domain, was defined to be the site of 4'-phosphopantetheine (cofactor) binding. However, the biological significance of other core sequences (see Figure 2) located within other domains are still a matter of speculation. Additional biochemical and structural studies are needed to define their roles in substrate recognition and modification.

4.08.12   MULTIPLE P-PANT REQUIRING PATHWAYS

Most of the nonribosomal peptides, polyketides, and fatty acids are synthesized utilizing acyl group activation and transfer reactions by multienzyme complexes.[128] These multienzyme systems contain small functional units of about 100 amino acids, which can either represent a separate subunit or an integrated domain, that functions as a carrier protein for the growing acyl chain. This acyl (or peptidyl) carrier protein (ACP or PCP) is converted from inactive apo-forms to active holo-forms by the condensation of the 4'-phosphopantetheinyl (P-pant) prosthetic group to the β-hydroxy side chain of the conserved serine residue[126] of the LGGHSL (Figure 3) motif (core 6). This post-translational modification takes place by the nucleophilic attack of the β-hydroxy group of the conserved serine residue[127] on the pyrophosphate linkage of coenzyme A (CoASH), resulting in transfer of the 4'-phosphopantetheinyl moiety of CoASH onto the attacking serine.[128] The newly introduced -SH group of the 4'-phosphopantetheine acts as a nucleophile for acylation on a specific substrate. This can be an acyl-CoA for the fatty acid and polyketide synthases (PKSs) or can be aminoacyl-AMPs for the peptide and depsipeptide synthetases.[128]

The acyl-ACP in PKSs is attacked by a carbanion nucleophile for carbon skeleton assembly during polyketide construction, whereas in peptide synthetases, the aminoacyl-S-PCPs are attacked by nitrogen or oxygen nucleophiles in amide and ester bond-forming steps, respectively.

Lambalot *et al.*[128] have reported that the post-translational phosphopantetheinylation of apo-ACP/PCP domains is clearly essential for the activity of multidomain enzyme synthases responsible for the synthesis of fatty acids and an array of natural products. The *Escherichia coli* holo-acyl carrier protein synthase (ACPS) was the first phosphopantetheinyl transferase to be cloned and characterized that activates the ACP of fatty acid synthase by converting it to its holo-form.[129] Using the conversion of *E. coli* apo-ACP to holo-ACP as an assay, the ACPS was purified and identified[130] as the product of a previously described essential *E. coli* gene *dpj* of unknown function.

The *E. coli* ACPS also modifies apo-forms of *Lactobacillus casei* D-alanyl carrier protein (DCP), involved in D-alanylation of lipoteicholic acid,[131] *Rhizobia* NodF, involved in the acylation of the oligosaccharide-based nodulation factors,[132] and *Streptomycetes* ACPs involved in frenolicin, granaticin, oxytetracycline, and tetracenomycin polyketide antibiotic biosynthesis.[128] Through refinement of low-level sequence homologies and identification of two consensus motifs within ACPS, a large family of proteins, which share 12 to 22% homology with ACPS, were defined as putative 4'-phosphopantetheine transferases.[128]

In 1996, Lambalot *et al.*[128] presented the first direct evidence that organisms containing multiple 4'-phosphopantetheine-requiring enzymes also have partner-specific 4'-phosphopantetheine trans-

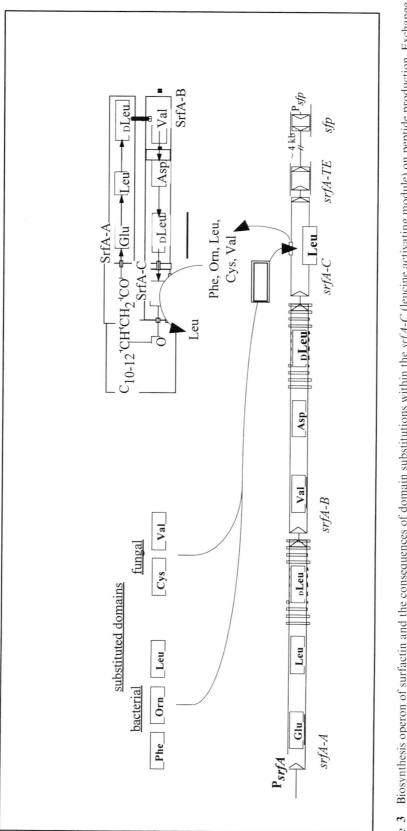

Figure 3 Biosynthesis operon of surfactin and the consequences of domain substitutions within the *srfA-C* (leucine activating module) on peptide production. Exchange of adenylation-domain-coding regions of bacterial and fungal origin led to the construction of hybrid genes that encoded peptide synthetases with altered amino acid specificities and the production of peptides with targeted modified amino acid sequences.

ferases. Identification of 4′-phosphopantetheine loading enzymes, that create the holo-ACP/PCP forms by post-translational modification, are therefore important to the mechanism of serine phosphopantetheinylation. This is essential to the design of strategies for heterologous production of functional polyketide or polypeptide antibiotics and to synthesis of inhibitors of specific 4′-phosphopantetheine loading reactions such as in fungal lysine biosynthesis.[128]

4.08.13 MODELS FOR NONRIBOSOMAL PEPTIDE BIOSYNTHESIS

Attempts using molecular biological approaches to investigate the nonribosomal biosynthesis of peptide antibiotics by microorganisms have given a wealth of knowledge on the different types of biosynthetic mechanisms. For example, surfactin and mycobacillin are synthesized nonribosomally like other peptides such as gramicidin S, tyrocidine, bacitracin, etc. However, surfactin biosynthesis differs in that the mechanism is apparently distinct from that of the multienzyme thiotemplate mechanism.[5] Therefore, it is interesting to examine the peptide biosynthetic mechanisms of different organisms. The following sections attempt to summarize some of these biosynthetic mechanistic models which have been studied in detail by different groups.

4.08.13.1 Surfactin Biosynthesis

The lipopeptide antibiotic surfactin was first isolated as a fibrin clotting inhibitor from *Bacillus subtilis* ATCC21332[133] is one of the most significant biosurfactants known today.[10] Although it is synthesized nonribosomally in part, by the multienzyme thiotemplate mechanism[24] at stationary phase,[134] this species also adopts a different mechanism to synthesize surfactin.[5] *B. subtilis* is highly amenable to genetic manipulation and therefore many studies were carried out towards understanding the molecular genetics of lipopeptide surfactin biosynthesis.[60,76] The surfactin is composed of seven amino acids and a β-hydroxy fatty acid moiety, predominantly 3-hydroxy-13-methyltetradecanoic acid.[135]

A set of three genetic loci, *sfp*, *srfA*, and *srfB* as shown in Figure 3, is necessary for the production of surfactin.[136] The *sfp* gene encodes the 4′-phosphopantetheinyl transferase, an essential component of the peptide synthesis system[137] needed for post-translational modification of all surfactin synthetases.[60] The *srfA* locus is an operon, composed of four open reading frames, *srfA-A*, *srfA-B*, *srfA-C*, and *srfA-D* (formerly *srfA-TE*). It encodes three multifunctional peptide synthetases as well as a thioesterase-like protein.[138]

The *SrfA-A* and *SrfA-B* code for modular peptide synthetases, each composed of three modules, which catalyze activation and incorporation of three amino acids, respectively. The seventh amino acid of surfactin (leucine) is activated by *SrfA-C*, a one-module enzyme. The *SrfA-D* gene encodes a protein with significant sequence similarity to thioesterases.[60] The *SrfB*, which is located apart from the *srfA*-operon, contains *comQ*, *comT*, *comP* and *comA* genes. The *comP* and *comA* belong to a signal-transduction, two-component system involved in the competence development pathway and are also required for the activation of *srfA* transcription.[139] This was proved when the *cis*-acting elements required for *comA*-dependent control were identified.[140,141]

4.08.13.2 Tyrocidine Biosynthesis

The cyclic decapeptide tyrocidine is produced by *Bacillus brevis* ATCC8185. The tyrocidine synthesizing system consists of three interacting multifunctional synthetases,[97] TycA, TycB, and TycC.[76] They are 123, 420, and 800 kDa in size, respectively, and activate one, three, and six amino acid constituents of tyrocidine. TycA, for example, consists of a single multifunctional polypeptide chain of 1098 amino acid residues that catalyzes the activation, thioesterification, and epimerization of phenylalanine.[142,143] This is a representative example of a single module containing all functions associated with adenylate formation, aminoacylation, and epimerization. TycA was the first peptide synthetase whose gene was cloned and expressed at high levels in *E. coli*, by Weckermann *et al.*[144]

TycB activates the proline at site 2 and the phenylalanine at site 3, as well as activating and racemizing another phenylalanine at site 4,[76] whereas TycC is responsible for the incorporation of six amino acids: Asn, Gln, Tyr, Val, Orn, and Leu.[5] Successive transpeptidation at each module is achieved with the help of several 4′-phosphopantetheine cofactors, which ultimately act as internal transport units for the growing peptide chain. The multifunctional components of the biosynthetic system also catalyze the cyclization of the linear decapeptide[97] yielding the cyclic decapeptide tyrocidine. Until 1994, the presence of 4′-phosphopantetheine as a cofactor had been demonstrated for TycB and TycC,[123] but not for TycA. However, in 1994 Gocht and Marahiel[145] demonstrated the incorporation of a 4′-phosphopantetheine by labeling of TycA with β-[³H]alanine.

In another study in 1996, Stachelhaus *et al.*[146] showed that, during nonribosomal peptide synthesis, a distinct thiolation domain — PCP — is responsible for the transfer of amino acyl residues by covalent binding. They also cloned and biochemically characterized the region encoding PCP of tyrocidine synthetase TycA. This study on phosphopantetheinylation and covalent thioester binding of phenylalanine by PCP, strongly supports both the multiple carrier model for nonribosomal peptide synthesis and the peptidyl carrier activity of the thiolation domain, PCP.

4.08.13.3 Gramicidin S Biosynthesis

The cyclic decapeptide gramicidin S is produced by *B. brevis* ATCC9999.[147] The synthesis of gramicidin S is catalyzed by two enzymes, designated GrsA and GrsB, and the genes responsible for these two enzymes have been identified.[148,149] The gramicidin S synthetase 1 (GrsA) has a molecular weight of 126 kDa. It activates, thioesterifies, and racemizes L-Phe to D-Phe. The gramicidin S synthetase 2 (GrsB), a multifunctional polypeptide with a molecular weight of 510 kDa, activates proline, valine, ornithine, and leucine.[76] Peptide elongation is initiated by the transfer of D-Phe from GrsA into the activated Pro residue on GrsB and the formation of the dipeptide. The thioester-bound amino acids are the immediate precursors in the subsequent oligomerization process via 4′-phosphopantetheine leading to the formation of the enzyme-bound pentapeptides. Two such pentapeptides cyclize to form the decapeptide gramicidin S via a head-to-tail condensation.[5,150]

The two genes *grsA* and *grsB* form an operon, as shown in Figure 4, with an additional open reading frame called *grsT*. The three genes, arranged in the order *grsT–grsA–grsB*, are transcribed from a single promoter located upstream of *grsT*. The *grsB* product, composed of four homologous modules, are responsible for activation of the four constituent amino acids Pro, Val, Orn, and Leu of gramicidin S. Additional studies support the view that the linear arrangement of the amino acid-activating modules of *grsB* determines the order in which the constituent amino acids are incorporated into the growing peptide chain.[5,151] In 1995,[152] evidence by mass spectrometric and amino acid analysis revealed that a cofactor 4′-phosphopantetheine is indeed attached to the L-Val module of GrsB and to the module of GrsA for D-phenylalanine. In 1996, a thorough study on the structures of all thiol template sites of GrsB has revealed for the first time[125] the modification of all four modules by the cofactor.

The gramicidin S synthesis shown in Figure 4 is considered to be one of the best-studied systems of nonribosomal peptide synthesis. A number of earlier studies, based on the analysis of primary structure and biochemical investigations mentioned above, had revealed the striking similarity between the mechanism of synthesis of gramicidin S and tyrocidine.[76] These mechanistic similarities open a new dimension in the comparative analysis of biosynthesis of peptides for novel drug design.

4.08.13.4 Enterobactin Biosynthesis

Enterobactin is a catechol-containing siderophore produced by *E. coli*. Under iron deprivation conditions, *E. coli* synthesizes these low molecular weight compounds, which bind ferric ions with high affinity and which are used to supply iron for metabolic pathways.[110]

In *E. coli*, the synthesis of the iron-chelating compound enterobactin requires a group of genes encoding the biosynthetic enzymes EntC, EntB, and EntA.[153] The *entC*, *entB*, and *entA* genes encode for enzymes[154] involved in the biosynthesis of 2,3-dihydroxybenzoic acid from chorismic acid, the last common intermediate in the biosynthetic pathway of the aromatic amino acids. Isochorismate synthetase, the produce of the *entC* gene, interconverts chorismate and isochorismate.

The EntB enzyme, isochorismatase, hydrolyzes the enol pyruvyl ether side-chain of isochorismate. The dehydrogenase product of the *entA* gene produces catechol.[154] In addition to *ent* genes, the *fep*

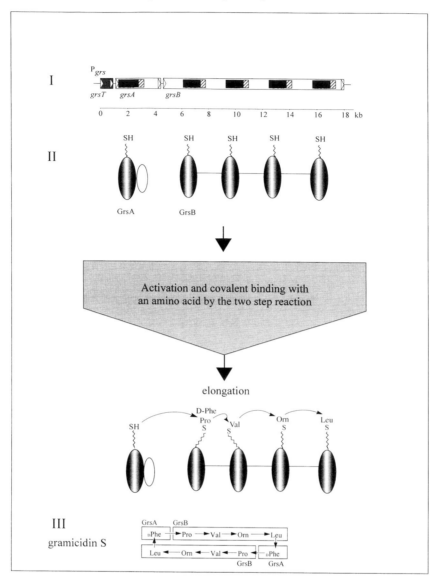

Figure 4 The schematic diagram shows the biosynthetic operon of gramicidin S from *Bacillus brevis*, as a model for the mechanism of nonribosomal peptide biosynthesis on the modular protein template (multicarrier model). I. The bacterial biosynthesis genes of the cyclic decapeptide gramicidin S are organized in an operon, which is transcribed from the P*grs* promoter. The *grsT* gene encodes a thioesterase-like protein, whose function remains unclear, whereas the structural genes *grsA* and *grsB* encode peptide synthetases with a highly conserved modular structure. Homologous type I modules are about 1000 aa in length and contain individual domains involved in amino acid adenylation (black boxes) and thioester formation (striped boxes). II. Peptide synthetases possess a highly conserved and ordered structure of semi-autonomous modules (type I shaded oval; racemase, unshaded oval). After activation and elongation, the termination of the enzyme-catalyzed peptide synthesis is induced by a head-to-tail condensation of two identical pentapeptides. III. Primary structure of gramicidin S and enzymes that catalyze peptide synthesis are shown. Here termination of the enzyme-catalyzed peptide synthesis is induced by a head-to-tail condensation synthesized by the multifunctional enzymes.

gene is required for the uptake of the ferric enterobactin complex, whereas the *fes* gene is required for iron release. These *ent*, *fep*, and *fes* genes are grouped in a 24 kb region as distinct transcriptional units. Subsequent condensation of 2,3-dihydroxybenzoate with serine followed by cyclization to yield enterobactin was shown to involve at least four more genes, *entD*, *entE*, *entF*, and *entG*.[112] EntF is a peptide synthetase-like enzyme that activates serine and EntD is the corresponding 4′-phosphopantetheinyl transferase that modifies EntF with the cofactor.

4.08.14 STRATEGIES FOR DESIGNING NOVEL PEPTIDE PRODUCTS

Many antibiotics in use today are derived from a handful of microorganisms, especially Actinomycetes and Gram-positive bacteria, the most prolific source of natural products. Most of these antibiotics are produced through secondary metabolic pathways. Understanding the basis of these processes has led to an enhanced use of recombinant microbes in the development of new antibiotics.

The age-old approach of identifying new antibiotic compounds had relied on a sequential screening method to identify bioactive compounds. This approach has also employed existing antibiotic compounds in order to generate novel analogues with better therapeutic properties. These sequential approaches are very time-consuming and expensive. For example, using a multidisciplinary approach, more than 500 000 entities were screened[155] against *Helicobacter pylori* over a four-year period. However, an agent that will completely eradicate this stomach bacterium is still only at the development stage.

Since the 1940s, natural sources such as *Streptomyces* and some other genera of Actinomycetes have been considered to be a vital source of antibiotic compounds for screening programs. Although many genera of Gram-positive bacteria, such as *Bacillus*, and Gram-negative bacteria, such as *Pseudomonas*, have emerged as new sources for antibiotic compounds, the rate of novel drug discoveries has markedly declined since the 1970s. The complex chemical structure of natural products and the lack of a sufficiently broad biologically active spectrum have been identified as possible reasons for the decline of natural products as antibiotics.[156]

Combinatorial libraries of peptides[157–160] can greatly reduce the time required for the discovery of antibiotic compounds for receptor-based screening programs.[161, 162] The structural and functional diversity of nonribosomally synthesized peptides could therefore be a suitable subject for drug engineering. Compounds with various modifications have the potential to increase the effectiveness and variability in the design of novel bioactive peptides. In general, studies have focused on the modification of natural secondary metabolites that are known to possess bioactivities.

4.08.14.1 Recombination Techniques

Production of some enzymes, such as proteases used in microbial processes and some proteins used for medical purposes, is from recombinant microorganisms. Although the compounds are diverse, for example vitamin C, indigo, benzoquinone, and also some antibiotics,[163] they are all commercially produced in large quantities this way. Due to the advancement of biotechnological tools, it has also become more convenient to use recombinant techniques in the search for novel drugs.

Biosynthesis of novel peptide antibiotics through genetic engineering is accomplished by reprogramming the peptide synthetase-encoding genes. For the realization of such domain substitutions using a recombination method, very detailed information regarding identification of peptide synthetase genes in the desired biosynthetic system is necessary.

Significant advances have been made towards the identification and cloning of genes encoding peptide synthetases.[147] These studies confirmed the identity of several novel enzymes identified by targeted domain replacement. Apart from this, the engineered enzymes restored a biological activity very similar to the wild-type protein. The approach developed is based on the enhanced knowledge of the modular structure of the enzymes involved and the mechanism by which they generate their corresponding peptide product. Because we can engineer genes responsible for peptide products, we should be able to engineer the modules to synthesize new peptide products.

The method of recombination comprises two successive steps. First is marking the chromosomal target site with a selectable gene, for example, through a drug-resistance gene by double crossing-over. The second step is the subsequent delivery of an engineered hybrid gene that encodes a peptide synthetase with an altered amino acid specificity using marker exchange.[46] This method has an advantage over an *in trans* approach, because, here, wild-type mechanisms are used for gene regulation and gene expression.

Based on the method shown schematically in Figure 5, two domain-flanking regions of target chromosome have been amplified by the polymerase chain reaction (PCR) method. These two regions, called A and C, were cloned in a vector, resulting in an integration vector that was used for the insertion of a selectable drug resistance marker called R. This was then inserted into the chromosomal target site through a double recombination event, as directed by the specific linker fragments (A and C). The gene disruption was accomplished by selection for drug resistance and screening for an antibiotic-deficient phenotype.

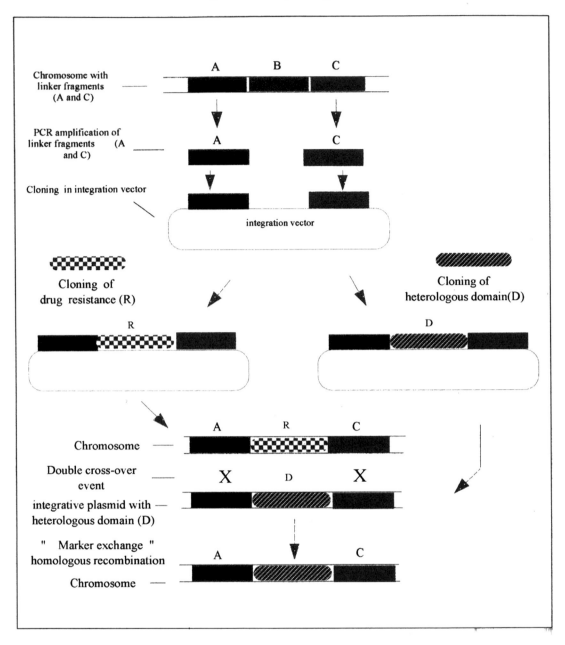

Figure 5 Domain replacement technique through homologous recombination. Construction of hybrid genes encoding peptide synthetases with altered amino acid specificities is schematically shown. The PCR-amplified linker fragments A (blue) and C (red) are cloned to construct an integration vector. This vector is used for the insertion of a selectable drug-resistance marker R (mini-squares). This is inserted into the chromosomal target site through a double recombination event, as directed by the specific linker fragments. The gene disruption is accomplished by selection for drug resistance. The heterologous domain, D (gray strip) is inserted *in frame* between the linkers fragments, resulting in different hybrid genes encoding peptide synthetases with the altered amino acid specificities of the substituted domains. This was used to substitute the selectable marker R and to reconstitute the interrupted biosynthesis gene(s) within the chromosome. The replacement plasmid delivers an engineered hybrid gene into the marked chromosome, by marker exchange reaction.

For the engineering of hybrid genes encoding peptide synthetases with the altered amino acid specificities, heterologous domains (D) from bacterial or fungal origin were used to insert *in frame* between the linker fragments of integration vector. These replacement plasmids could then be used to substitute the selectable marker (R) and to reconstitute the interrupted biosynthesis gene(s) within the chromosome. The replacement plasmid delivers an engineered hybrid gene into the marked chromosome, by marker exchange reaction. The lipopeptide surfactin, shown in Figure 3, was specifically modified by this method through the targeted replacement of amino acid activating modules originating from the filamentous fungi, ACV gene.[122]

4.08.14.2 Combinatorial Biosynthesis of Peptide Products

Many gene-encoding biosynthetic enzymes of peptides are combined into multiple positions to produce libraries of altered products. This approach, called combinatorial biosynthesis of peptide products,[38] is a novel method to manipulate natural product synthesis and to screen a number of compounds within a short time period.[164] This method yields new insight into the function and mechanisms of enzymes involved in natural product biosynthesis. It provides a path for novel drug discoveries and has been successfully used in polyketide products discovery. Since the mid-1990s, a revolution has been taking place in polyketide natural products through efficient design of novel polyketides by combinatorial shuffling of the genes coding for their biosynthetic enzymes. The nonribosomal synthesis of peptides by shuffling of modules between other systems, for example from enniatin to cyclosporin or gramicidin S to tyrocidine, should facilitate the combinatorial biosynthesis of novel peptide products[38] in the near future.

4.08.14.3 Gene Sequences

Targeted cloning of genes encoding peptide synthetases and the identification of their peptide products through improved genetic engineering, as well as screening studies and the subsequent application of bioinformatic technology with computational analysis methods, have improved and made it less time consuming to seek a novel product. Gene sequence recognition and comparison for conserved motifs present in peptide synthetases makes possible unprecedented opportunities to define putative genes encoding peptide synthetases from a wide range of organisms.

Genes for biosynthetic enzymes are assembled into a small number of units. The structure of these genes governing the utilization of a particular class of precursors is often highly conserved.[156] In some cases, the enzymes involved in the process of biosynthesis of two different peptide products are similar in nature, as in the case of gramicidin S and tyrocidine.[76] A number of significant peptide synthetase gene sequences from various natural sources have already been reported.[76] Table 2 is a list of sequenced peptide synthetases available commercially.

Table 2 Sequenced peptide synthetases.

Producer organisms	Antibiotics
Aspergillus nidulans	β-lactams
Bacillus brevis	tyrocidine
Bacillus brevis	gramicidin S
Bacillus licheniformis	bacitracin
Bacillus subtilis	fengycin
Bacillus subtilis	iturin
Bacillus subtilis	surfactin
Cochliobolus carbonum	HC-toxin
Escherichia coli	enterobactin
Flavobacterium SC12	β-lactams
Fusarium scripi	enniatin
Myxococcus xanthus	saframycin
Norcardia lactamdurans	β-lactams
Penicillium chrysogenum	β-lactams
Pseudomonas aeruginosa	pyoverdin
Pseudomonas syringae	coronatin
Pseudomonas syringae	syringomycin
Streptomyces clavuligerus	β-lactams
Streptomyces hygroscopicus	bialaphos
Streptomyces lavendulae	saframycin
Streptomyces pristinaespiralis	pristinamycin I
Streptomyces viridochromogenes	bialaphos
Tolypocladium niveum	cyclosporin A
Vibrio anguillarum	anguibactin

The systematic analysis of these genes or operons and uncovering the sequences of adjacent DNA provide the basis for the establishment of a large super-family of peptide synthetases from different microbial sources. This could allow effort invested in one family member to yield insights into another which has similar properties.[55] For example, such an alignment of derived amino acid sequences of various peptide synthetases compared with a newly sequenced putative peptide synthesizing region of *Pseudomonas fluorescens* is shown in Figure 6. Thus, by just comparing sequences among microorganisms,[155] it should be possible to define those encoding putative peptide synthetases.[35] The comparative analysis of gene sequences, which currently comprises over 70 to 80 peptide synthetase modules, is progressing more rapidly.[10]

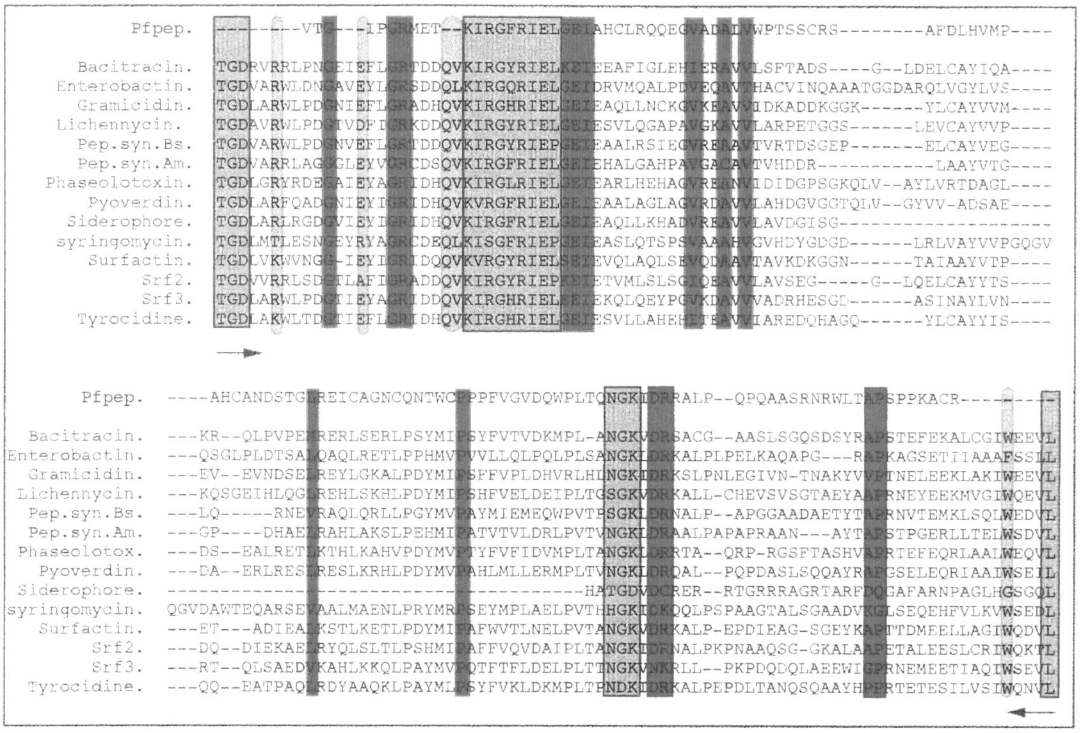

Figure 6 Alignment of the derived amino acid sequence of putative peptide synthetase fragments containing some core regions from *Pseudomonas fluorescens* with homologous fragments of different peptide synthetase genes obtained from various other microbial species. The sequence alignment was performed using computer data of CLUSTAL as implemented in HUSAR on GENIUS in the EMBLNET at DKFZ-Heidelberg (Germany). The red shades indicate highly conserved core regions, such as TGD, KIRGFRIEL, NGK, and LGG,[168, 170] the green shades indicate high homologous amino acid sequences of other peptides present in *P. fluorescens*, whereas yellow strips indicate the possible homologous amino acid sequences of *P. fluorescens*. The arrow in each direction shows the position of the used primers, the TGD and LGG.

In another approach, sequence analysis from genomic DNA of a pathogenic organism for target identification may point to another direction in drug discovery. The first such attempt involved picking out the aerobic and anaerobic genes for electron transport in *Helicobacter pylori*.[155] The entire genome sequences of *Haemophilus influenzae*[165] and *Mycoplasma genitalium*[166] have also been completed and those of *E. coli* and *B. subtilis* are expected to become available. Exploitation of the genome analysis technology is other pathogens such as *Streptomyces aureus* and *S. pneumoniae* are also under way. This approach may create expanded opportunities for the discovery of unknown pathogenic determinants.

4.08.14.4 Identification of Putative Peptide Synthetases

Identification of genes encoding putative peptide synthetases involved in biosynthesis of peptide antibiotics through specific amplification of DNA fragments by PCR is a useful and rapid detection method developed earlier by the authors' laboratory. Some strains, previously not identified as peptide producers, for instance *Planobispora rosea*[167] and *Pseudomonas fluorescens*,[168] are now identified as producers of nonribosomally synthesized peptides using this identification technique.

According to this method, two designed oligonucleotide cores, derived from two highly conserved regions of known peptide synthetases, were used to amplify chromosomal DNA. These cores, TGD and LGGXS regions of five to eight amino acid residues, shown in Figure 2, are highly conserved among peptide synthetases.[169] They were used to develop primers[170] for amplifying homologous putative peptide synthetase sequences in *Pseudomonas syringae* pv. *phaseolica*, by the PCR method.

For example, such a technique to identify the putative peptide synthetase genes was adopted in *P. fluorescens*, as shown in Figure 7. Southern hybridization of genomic DNAs from *P. fluorescens*, with PCR-amplified fragments, indicate the existence of discrete bands at 500 bp. The size of the observed band was well in excess of the expected size based on the structure of the oligopeptide(s) known to be produced by the producer strain. Sequence analysis of the PCR product obtained from *P. fluorescens* genomic DNA had confirmed their high relatedness to peptide synthetase.[168] Although the biological role of the products synthesized by these putative peptide synthetases are not presently known, results suggest the high biosynthetic versatility of these enzymes, which could greatly expand the number of amino acids that could be incorporated into antibiotics. Currently, this method has been widely accepted to identify peptide synthetases from many organisms.[170]

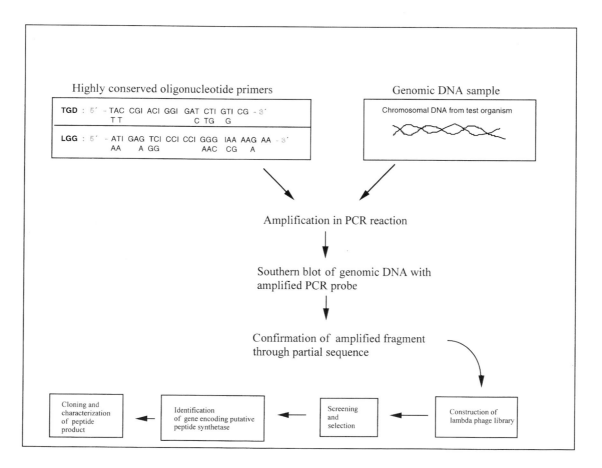

Figure 7 Identification of putative peptide synthetases. The scheme shows specific PCR amplification of DNA fragments belonging to genes encoding putative peptide synthetases and screening for the coding genes. Two designed oligonucleotides, TGD and LGG, of known peptide synthetases were used as primers to identify the peptide synthetase genes of *Pseudomonas fluorescens* using PCR. After amplification of putative peptide synthetase gene fragments from the genomic DNA, the amplified DNA was cloned. A lambda-phage library of the target strains was screened using the PCR fragment as probe in Southern hybridization. After identification of the clones containing genes encoding putative peptide synthetase, larger gene fragments were subcloned and characterized.

4.08.14.5   Transposon Mutagenesis

The extensive use of antibiotics has provided powerful pressure for the selection of microbes that either carry mutations conferring resistance or have enhanced ability to mutate.[35] Mutants have acquired constructed genes for producing novel machinery to overcome the action of antibiotics or acquired improved activities against many pathogens.[171]

Transposon mutagenesis, a versatile tool for the molecular analysis of various microorganisms that impair peptide antibiotic production, has been used to examine the mechanism and regulation of biosynthesis, as well as to assess the biological function of the peptide products.[172] According to this transposon mutagenesis technique, a plasmid construction containing Tn5 was used to obtain mutants from microorganisms. This commonly used Tn5 was constructed in a pUT plasmid.[173] The mobility of the recombinant transposon is determined by two insertion sequences (IS) flanking the DNA region determining the selectable phenotype and the essential elements such as the IS terminal sequences and transposage gene have been arranged such that the transposage gene is adjacent to, but outside of, the mobile DNA segment.[173] This series of TN5 constructions simplify substantially the generation of mutants from the target strain.

Using such TN5 mutagens, integration of transposon is possible through a triparental mating procedure[172] or electroporation[174] into specially treated competent cells. The former includes an additional third strain as helper plasmid during mating of donor and receiver strains, whereas the latter electrotransformation involves the application of a brief high-voltage pulse to a sample DNA. The competent cells of target strains can be obtained through either the $CaCl_2$ method[172] or using the electrocompetent cell procedure.[174]

Methods employed to identify antibiotic-negative mutants usually involve an initial screening of large numbers of transformants obtained through appropriate transposon-associated antibiotic resistance markers. If this is successful, the selection of specific mutants by bioassay technique using indicator strains commences. The confirmation of physical integration of transposon is achieved by a Southern blot assay and sequence analysis.[175,176]

These kinds of studies were made in many species, resulting in a number of valuable mutants. For example, mycobacillin nonproducing mutants of *B. subtilis* were isolated by this method, using *Aspergillus* as indicator strain. They were analyzed for the accumulation of intermediate peptides in mycobacillin-negative mutants.[177] Results obtained in *B. subtilis* indicated that the mycobacillin synthesizing system consists of at least three protein components.[178] Identification of mutations was thus an essential prerequisite for the isolation of the genes that function in biosynthesis.

4.08.14.6   Other Drug Design Strategies

Several techniques, such as structure-based strategies for drug design,[179] receptor-based *de novo* ligand design,[180] protoplast fusion,[181] and NMR techniques to determine the three-dimensional solution structures[182] were additionally demonstrated as part of other drug design strategies.[183] Some special computational tools have also indicated that the method of computer-aided drug design (CADD) could increase the efficiency of drug discovery. Basically, the clinical interest in neuropeptides and peptide hormones has stimulated such computational study and development of peptide drugs.[184]

The CADD technique is used to determine whether a molecule will bind to a specific target[185] or not. This is also used to evaluate the characteristics of the interaction between the molecule and the target.[186] This method describes a molecule exactly in terms of all interactions such as van der Waals and electrostatic interactions with the receptors, by utilizing crystallographic data and computer analysis.[184] Starting with a specific target site, the approach determines what types of molecules would have a desired effect on that target and subsequently constructs molecules that fit into these criteria. A complete engineering of a desired peptide biosynthetic system using the thiotemplate mechanism and the almost unlimited structural feasibilities of nonribosomally synthesized peptides on modular peptide synthetases may in the near future meet the demand to create drugs with improved activities.

4.08.15   CONCLUSIONS

Many microorganisms produce an array of bioactive peptide products through multifunctional peptide synthetases. Significant progress has been made in understanding the modular arrangement of the multifunctional enzymes and also the mechanisms by which they generate their corresponding peptide products. It has now been established that the synthesis of bioactive peptides and their specific sequences are brought about by a protein template that contains the appropriate number and the correct order of activating domains.

Progress made in the comprehension of structural aspects of the multienzyme systems that catalyze the polymerization, epimerization, and tailoring of peptides to a specific function have revealed many novel peptide products since the 1950s. However, a thirst for new peptide antibiotics and the increasing number of pathogenic microbes necessarily require innovative concepts to develop commercially viable and pharmaceutically useful drugs.

Nonribosomally synthesized peptide antibiotics have not yet had widespread clinical applications against many infectious pathogens. However, nonribosomally synthesized peptides like bacitracin, cyclosporin, and perhaps vancomycin have proved effective. Further research on tailored active compounds synthesized on protein templates will certainly enhance knowledge of both natural and synthetic products.

ACKNOWLEDGEMENTS

We wish to thank Dr Torsten Stachelhaus for critical reading of the manuscript. The work in the authors' laboratory in Germany was supported by the Deutsche Forschungsgemeinschaft and Fond der chemischen Industrie. Narayanan Rejendran wishes to thank the Alexander von Humboldt Foundation, Germany, for an AvH fellowship.

4.08.16 REFERENCES

1. H. Budzikiewicz, *FEMS Microbiol. Rev.*, 1993, **104**, 209.
2. L. C. Vining, *Adv. Appl. Microbiol.*, 1979, **25**, 147.
3. E. Cundliffe, *Annu. Rev. Microbiol.*, 1989, **43**, 207.
4. J. W. Bennett and R. Bentley, *Adv. Appl. Microbiol.*, 1989, **34**, 1.
5. M. M. Nakano and P. Zuber, *Crit. Rev. Biotechnol.*, 1990, **10**, 223.
6. V. Behal, *Trends Biol. Sci.*, 1986, **11**, 88.
7. S. Drew and A. L. Demain, *Annu. Rev. Microbiol.*, 1977, **31**, 343.
8. L. C. Vining, *Annu. Rev. Microbiol.*, 1990, **44**, 395.
9. P. Schaeffer, *Bacteriol. Rev.*, 1969, **33**, 48.
10. N. Rajendran, T. Stachelhaus, and M. A. Marahiel, in "Recent Research Developments in Antimicrobial Agents and Chemotherapy," ed. S. G. Pandalai, Research Signpost, Thiruvananthapuram, 1997, p. 341.
11. J. F. Martin and P. Liras, in "Biotechnology, Microbial Fundamentals," eds. H. J. Rehm and G. Reed, Verlag Chemie, Weinheim, 1981, vol. 1, p. 211.
12. P. Zuber and M. A. Marahiel, "Biotechnology of Industrial Antibiotics," Marcel Dekker, New York, 1997, submitted for publication.
13. E. Katz and A. L. Demain, *Bacteriol. Rev.*, 1977, **41**, 449.
14. A. L. Demain, in "Secondary Metabolites: Their Function and Evaluation," Wiley, New York, 1992, p. 2.
15. J. D. Bu'lock, *Adv. Appl. Microbiol.*, 1961, **3**, 293.
16. H. Kleinkauf and H. von Döhren, *Antonie van Leeuwenhoek*, 1995, **67**, 229.
17. J. W. Bennet and R. Bentley, *Adv. Appl. Microbiol.*, 1989, **34**, 1.
18. S. Horinouchi and T. Beppu, *Crit. Rev. Biotechnol.*, 1990, **10**, 191.
19. J. R. Willey, R. Santamaria, J. Guijarro, M. Geistlich, and R. Losick, *Cell*, 1991, **65**, 641.
20. K. Jayaraman and R. Kannan, *Biochem. Biophys. Res. Commun.*, 1972, **48**, 1235.
21. J. J. Lugtenberg and L. A. de Weger, in "*Pseudomonas* Molecular Biology and Biotechnology," ed. E. Galli, American Society for Microbiology, Washington DC, 1992, p. 13.
22. N. Rajendran and D. I. Arokiasamy, *Bionature*, 1989, **11**, 21.
23. M. A. Marahiel, M. Krause, and H. J. Skarpeid, *Mol. Gen. Genet.*, 1985, **201**, 231.
24. E. Katz and A. L. Demain, *Bacteriol. Rev.*, 1977, **41**, 449.
25. M. A. Marahiel and P. Zuber, in "Cell growth and Differentiation: Biology of the Prokaryotes," eds. H. G. Schlegel and J. Lengeler, Thieme Verlag, Weinheim, 1997, in press.
26. P. C. Loewen and R. H. Aronis, *Annu. Rev. Microbiol.*, 1994, **48**, 53.
27. J. F. Martin and A. L. Demain, *Microbiol. Rev.*, 1980, **44**, 230.
28. J. R. Tagg, A. S. Dajani, and L. W. Wannamaker, *Bacteriol. Rev.*, 1976, **40**, 722.
29. R. S. Pore, *Biosystems*, 1978, **10**, 189.
30. C. A. Schindler and V. T. Schuhardt, *Biochim. Biophys. Acta*, 1965, **97**, 242.
31. K. T. Venema, A. J. Abee, K. J. Haandrikman, J. Leenhouts, W. N. Kok, W. N. Konings, and G. Venema, *Appl. Environ. Microbiol.*, 1993, **59**, 1041.
32. M. S. Gilmore, G. M. Dunny, P. P. Cleary, and L. L. McKay, in "Genetics and Molecular Biology of Streptococci, Lactococci and Enterococci," American Society of Microbiology, Washington DC, 1991, p. 3.
33. H. Kleinkauf and H. von Döhren, *Eur. J. Biochem.*, 1990, **192**, 1.
34. J. J. Coque, F. J. Enguita, R. E. Cardoza, J. F. Martin, and P. Liras, *Appl. Microbiol. Biotechnol.*, 1996, **44**, 605.

35. T. W. Daniel, D. Chu, J. J. Plattner, and K. Katz, *J. Med. Chem.*, 1996, **39**, 3853.
36. E. Cundliffe and J. Thompson, *J. Gen. Microbiol.*, 1981, **126**, 185.
37. R. F. Service, *Science*, 1995, **270**, 724.
38. C. J. Tsoi and C. Khosla, *Chem. Biol.*, 1995, **2**, 355.
39. Z. Podlesek, A. Comino, B. H. Velikonja, D. Z. Bertok, R. Komel, and M. Grabnar, *Mol. Microbiol.*, 1995, **16**, 969.
40. D. C. Ellwood and D. W. Tempest, *Adv. Microb. Physiol.*, 1972, **7**, 83.
41. W. V. Shaw and D. A. Hopwood, *J. Gen. Microbiol.*, 1976, **94**, 159.
42. A. D. Argoudelis and J. H. Coats, *J. Antibiot.*, 1971, **24**, 206.
43. W. Piendl, A. Böck, and E. Cundliffe, *Mol. Gen. Genet.*, 1984, **197**, 24.
44. M. Zalacain and E. Cundliffe, *J. Bacteriol.*, 1989, **171**, 4254.
45. A. Kawaguchi, H. Tomoda, S. Okuda, J. Awaya, and S. Omura, *Arch. Biochem. Biophys.*, 1979, **197**, 30.
46. S. Fuma, Y. Fujishima, N. Corbell, C. D'Souza, M. M. Nakano, P. Zuber, and K. Yamane, *Nucleic Acids Res.*, 1993, **21**, 93.
47. P. Cosmina, F. Rodriguez, F. de Ferra, M. Guido Grandi Perego, G. Venema, and D. van Sinderen, *Mol. Microbiol.*, 1993, **8**, 821.
48. S. Kaneda, C. H. Young, K. Yazawa, K. Takahashi, Y. Mikami, and T. Arai, *J. Cancer Res.*, 1986, **77**, 1043.
49. G. Weber, K. Schörgendorfer, E. S. Scherzer, and E. Leitner, *Curr. Genet.*, 1994, **26**, 120.
50. S. S. Gill, E. A. Cowless, and P. V. Pretratonio, *Annu. Rev. Entomol.*, 1992, **37**, 615.
51. E. F. Gale, E. Cundliffe, P. E. Reynolds, M. H. Richmond, and M. J. Waring, in "The Biochemical Basis of Antibiotic Action," Wiley, London, 1972, p. 4.
52. S. Nandi, I. Lazaridis, and B. Seddon, *FEMS Microbiol. Lett.*, 1981, **10**, 71.
53. W. Pschorn, H. Paulus, J. Hansen, and H. Ristow, *Eur. J. Biochem.*, 1982, **129**, 403.
54. H. Federn and H. Ristow, *Eur. J. Biochem.*, 1987, **165**, 223.
55. E. Daher, E. Rosenburg, and A. L. Demain, *J. Bacteriol.*, 1985, **161**, 47.
56. D. R. Storm, *Ann. N. Y. Acad. Sci.*, 1974, **235**, 387.
57. D. R. Storm and J. L. Strominger, *J. Biol. Chem.*, 1974, **249**, 1823.
58. S. A. Waksman, *Antibiot. Chemother.*, 1954, **6**, 90.
59. J. Davies, *Mol. Microbiol.*, 1990, **4**, 1227.
60. M. A. Marahiel, M. M. Nakano, and P. Zuber, *Mol. Microbiol.*, 1993, **7**, 631.
61. L. Pasteur and J. E. Joubert, *C. R. Soc. Biol. Paris*, 1877, **85**, 101.
62. H. W. Florey, *Yale J. Biol. Med.*, 1946, **19**, 101.
63. A. Gratia, *C. R. Sea. Soc. Biol. Fil,*, 1925, **93**, 1040.
64. S. L. Dax, "Antibacterial Chemotherapeutic Agents," Blackie Academic & Professional, London, 1997, p. 1.
65. P. Zuber, M. M. Nakano, and M. A. Marahiel, in "*Bacillus subtilis* and Other Gram-positive Bacteria," eds. A. L. Sonenshein, J. A. Hoch, and R. Losick, American Society of Microbiology, Washington, DC, 1993.
66. M. A. Marahiel, *FEBS Lett.*, 1992, **307**, 40.
67. S. Borchert, T. Stachelhaus, and M. A. Marahiel, *J. Bacteriol.*, 1994, **176**, 2458.
68. B. Ray and M. A. Daeschel, in "Secondary Metabolites," CRC Press, Boca Raton, FL, 1992, p. 11.
69. M. E. Stiles and J. W. Hastings, *Trends Food Sci. Technol.*, 1991, **2**, 247.
70. C. E. Wolf and W. R. Gibbons, *J. Appl. Bacteriol.*, 1996, **80**, 453.
71. H. Fliri, G. Baumann, A. Enz, J. Kallen, M. Luyten, M. Vincent, R. Movva, V. Quesniaux, M. Schreier, *et al.*, *Ann. N. Y. Acad. Sci,*, 1993, **696**, 47.
72. T. Stachelhaus, A. Schneider, and M. A. Marahiel, *Biochem. Pharmacol.*, 1996, **52**, 177.
73. W. Gulewitsch and S. Amiradzibi, *Ber. Dt. Chem. Ges.*, 1902 (1990), 33.
74. P. C. Robert, *Science*, 1989, **246**, 883.
75. P. Zuber, *Curr. Opinion Cell Biol.*, 1991, **3**, 1046.
76. P. Zuber, M. M. Nakano, and M. A. Marahiel, in "*Bacillus subtilis* and Other Gram-positive Bacteria: Biochemistry, Physiology and Molecular Genetics," eds. A. L. Sonenshein, J. A. Hoch, and R. Losick, American Society of Microbiology, Washington, DC, 1993, p. 4.
77. K. Meissner, E. Dittman, and T. Börner, *FEMS Microbiol. Lett.*, 1996, **135**, 295.
78. A. Fleming, *Br. J. Exp. Pathol.*, 1929, **10**, 226.
79. S. A. Waksman, "Streptomycin," Williams and Wilkins, Baltimore, MD, 1949, p. 5.
80. E. B. P. Abraham and P. B. Loder, in "Cephalosporins and Penicillins," ed. E. H. Flyn, Academic Press, New York, 1972.
81. J. S. Glasby, "Encyclopedia of Antibiotics," 2nd edn., Wiley, Chichester, 1979, p. 8.
82. L. P. Garrod, H. P. Lambert and F. O'Grady, "Antibiotics and Chemotherapy," 4th edn., Churchill, London, 1973.
83. S. Umezawa, K. Tatsuta, and S. Koto, *J. Antibiot.*, 1968, **21**, 367.
84. J. Mann, *Nature*, 1995, **375**, 533.
85. M. J. Bibb, S. Biro, H. Motamedi, J. F. Collins, and C. R. Hutchinson, *EMBO J*, 1989, **8**, 2727.
86. J. Ye, M. L. Dickens, R. Plater, Y. Li, J. Lawrence, and W. R. Strohl, *J. Bacteriol.*, 1994, **176**, 6270.
87. R. McDaniel, E. S. Khosla, D. A. Hopwood, and C. Khosla, *Nature*, 1995, **375**, 549.
88. H. Kleinkauf and H. von Döhren, in "Biotechnology, Microbial Product," eds. H. Pape and H. J. Rehm, VCH, Weinheim, 1984, p. 283.
89. N. Schnell, K. D. Entian, U. Schneider, F. Gotz, H. Zahner, R. Kellner, and G. Jung, *Nature*, 1988, **333**, 276.
90. D. J. Smith, A. J. Earl, and G. Turner, *EMBO J.*, 1990, **9**, 2743.
91. K. Kurahashi, in "Antibiotics," ed. J. W. Corcoran, Springer, Berlin, 1981, p. 25.
92. T. Stachelhaus and M. A. Marahiel, *FEMS Microbiol. Lett.*, 1995, **125**, 3.
93. F. Lipmann, *Adv. Microbial Phys.*, 1980, **21**, 227.
94. C. Y. Shiau, M. F. Byford, J. E. Baldwin, and C. J. Schofield, *Biochem. Soc. Trans.*, 1995, **23**, 629S.
95. H. Kleinkauf and H. von Döhren, *Crit. Rev. Biotechnol.*, 1988, **8**, 1.
96. K. Kurahashi, *Annu. Rev. Biochem.*, 1974, **43**, 445.
97. E. Pfeifer, M. Pavela-Vrancic, H. von Döhren, and H. Kleinkauf, *Biochemistry*, 1995, **34**, 7450.
98. V. de Crecy-Lagard, P. Marliere, and W. Saurin, *C. R. Acad. Sci. Ser. 3*, 1995, **318**, 927.

99. H. Kleinkauf and H. von Döhren, *J. Antibiot.*, 1995, **48**, 563.

100. M. Pavela-Vrancic, H. V. Liempt, E. Pfeifer, W. Freist, and H. von Döhren, *Eur. J. Biochem.*, 1994, **220**, 535.

101. D. J. MacNeil, J. L. Occi, K. M. Gewain, and T. MacNeil, *Ann. N. Y. Acad. Sci.*, 1994, **721**, 123.

102. J. Heider, C. Baron, and A. Böck, *EMBO J.*, 1992, **11**, 3759.

103. M. J. Berry, J. W. Harney, T. Ohama, and D. L. Hatfield, *Nucleic Acids Res.*, 1994, **22**, 3753.

104. P. B. Rainey, C. L. Brodey, and K. Johnstone, *Mol. Microbiol.*, 1993, **8**, 643.

105. G. W. Xu and D. C. Gross, *J. Bacteriol.*, 1988, **170**, 5680.

106. M. K. Morgan and A. K. Chartterjee, *J. Bacteriol.*, 1988, **170**, 5689.

107. R. E. Moore, W. P. Niemczura, O. C. H. Kwok, and S. S. Patil, *Tetrahedron Lett.*, 1984, **25**, 3931.

108. F. Rusnak, M. Sakaitani, D. Drueckhammer, J. Reichert, and C. T. Walsh, *Biochemistry*, 1991, **30**, 2916.

109. J. F. Staab, M. Elkins, and C. F. Earhart, *FEMS Microbiol. Lett.*, 1989, **59**, 15.

110. C. D. Cox and P. Adams, *Infect. Immun.*, 1985, **48**, 130.

111. T. R. Merriman, M. E. Merriman, and I. L. Lamont, *J. Bacteriol.*, 1995, **177**, 252.

112. F. Rusnak, W. S. Faraci, and C. T. Walsh, *Biochemistry*, 1989, **28**, 6827.

113. W. W. Carmichael, *Sci. Am.*, 1994, **270**, 78.

114. R. N. Matsushima, T. Ohta, S. N. Waki, M. Suganama, K. Kohyama, T. Ishikawa, W. W. Carmichael, and H. Fujiki, *J. Cancer Res. Clin. Oncol.*, 1992, **118**, 420.

115. K. Sivonen, M. Namikoshi, W. R. Evans, W. W. Carmichael, F. Sun, L. Rouhiainen, R. Luukkainen, and K. L. Rinehart, *Appl. Environ. Microbiol.*, 1992, **58**, 2495.

116. M. Debono and R. S. Gordee, *Ann. Rev. Microbiol.*, 1994, **48**, 471.

117. T. Stachelhaus, A. Schneider, and M. A. Marahiel, *Science*, 1995, **269**, 69.

118. F. Lipmann, *Acc. Chem. Res.*, 1973, **6**, 361.

119. F. Lipman, in "The Mechanism of Enzyme Action," eds. W. D. M. Elory and B. Glass, Johns Hopkins University Press, Baltimore, MD, 1954, p. 599.

120. S. G. Laland and T. L. Zimmer, *Essays Biochem.*, 1973, **9**, 31.

121. S. G. Laland, Ø. Frøyshov, C. M. Gilhuus, and T. L. Zimmer, *Nature New Biol.*, 1974, **239**, 43.

122. T. Stachelhaus and M. A. Marahiel, *J. Biol. Chem.*, 1995, **270**, 6163.

123. C. J. Lee, S. D. Banks, and J. P. Li, *Crit. Rev. Microbiol.*, 1991, **18**, 89.

124. H. Kleinkauf and H. von Döhren, in "The Biochemistry of Peptide Antibiotics," Walter de Gruyter, New York, 1990, p. 3.

125. T. Stein, J. Vater, V. Kruft, A. Otto, B. W. Liebold, P. Franke, M. Panico, R. McDowell, and H. R. Morris, *J. Biol. Chem.*, 1996, **271**, 15 428.

126. W. Schlumbohm and B. Wittmann-Liebold, *J. Biol. Chem.*, 1991, **266**, 23 135.

127. J. Reichert, M. Sakaitani, and C. T. Walsh, *Prot. Sci.*, 1992, **1**, 549.

128. R. H. Lambalot, A. M. Gehring, R. S. Flugel, P. Zuber, M. La Celle, M. A. Marahiel, R. Reid, K. Khosla, and C. T. Walsh, *Chem. Biol.*, 1996, **3**, 923.

129. R. H. Lambalot and C. T. Walsh, *J. Biol. Chem.*, 1995, **270**, 24 658.

130. H. E. Takiff, T. Baker, T. Copeland, S. M. Chen, and D. L. Court, *J. Bacteriol.*, 1992, **174**, 1544.

131. D. V. Debabov, M. P. Heaton, Q. Zhang, K. D. Stewart, R. H. Lambalot, and F. C. Neuhaus, *J. Bacteriol.*, 1996, **178**, 3869.

132. T. Ritsema, O. Geiger, P. van Dillewijn, B. J. J. Lugtenberg, and H. P. Spaink, *J. Bacteriol.*, 1994, **176**, 7740.

133. K. Arima, A. Kakinuma, and G. Tamura, *Biochem. Biophys. Res. Commun.*, 1968, **31**, 488.

134. C. Ullrich, B. Kluge, Z. Palacz, and J. Vater, *Biochemistry*, 1991, **30**, 6503.

135. J. Vater, in "Biologically Active Molecules," ed. U. P. Schlunegger, Springer-Verlag, Berlin, 1989, p. 27.

136. M. M. Nakano and P. Zuber, in "Genetics and Biotechnology of Bacilli," eds. J. Hoch and A. T. Ganesan, Academic Press., New York, 1990, p. 397.

137. M. M. Nakano, N. Corbell, J. Besson, and P. Zuber, *Mol. Gen. Genet.*, 1992, **232**, 313.

138. M. M. Nakano, R. Magnuson, A. Myers, J. Curry, A. D. Grossman, and P. Zuber, *J. Bacteriol.*, 1991, **173**, 1770.

139. M. M. Nakano, L. Xia, and P. Zuber, *J. Bacteriol.*, 1991, **173**, 5487.

140. M. M. Nakano and P. Zuber, *J. Bacteriol.*, 1991, **173**, 7269.

141. M. M. Nakano and P. Zuber, *J. Bacteriol.*, 1989, **171**, 5347.

142. M. A. Marahiel, M. Krause, and H. J. Skarpeid, *Mol. Gen. Genet.*, 1985, **201**, 231.

143. M. A. Marahiel, P. Zuber, G. Czekay, and R. Losick, *J. Bacteriol.*, 1987, **196**, 2215.

144. R. Weckermann, R. Fuebass, and M. A. Marahiel, *Nucleic Acid Res.*, 1988, **16**, 11 841.

145. M. Gocht and M. A. Marahiel, *J. Bacteriol.*, 1994, **176**, 2654.

146. T. Stachelhaus, A. Hüser, and M. A. Marahiel, *Chem. Biol.*, 1996, **3**, 913.

147. H. J. Skarpeid, T. L. Zimmer, B. Shen, and H. von Döhren, *Eur. J. Biochem.*, 1990, **187**, 627.

148. J. Krätzschmar, M. Krause, and M. A. Marahiel, *J. Bacteriol.*, 1989, **171**, 5422.

149. K. Turgay, M. Krause, and M. A. Marahiel, *Mol. Microbiol.*, 1992, **6**, 529.

150. K. Tokita, K. Hori, T. Kurotsu, M. Kanda, and Y. Saito, *J. Biochem.*, 1993, **114**, 522.

151. K. Hori, M. Kanda, T. Kurotsu, S. Miura, Y. Yamada, and Y. Saito, *J. Biochem.*, 1981, **90**, 439.

152. T. Stein, B. Kluge, J. Vater, P. Franke, A. Otto, and B. W. Liebold, *Biochemistry*, 1995, **34**, 4633.

153. A. O. Bradley, J. B. Timothy, and A. M. Mark, *J. Bacteriol.*, 1989, **171**, 775.

154. T. Rusnak, M. Sakaitani, D. Drueckhammer, J. Reichert, and C. T. Walsh, *Biochemistry*, 1991, **30**, 2916.

155. Glaxo Wellcome's Research Report, *Nature*, 1996, **384** (suppl.), 1.

156. C. R. Hutchinson, *Bio/Technology*, 1994, **12**, 375.

157. M. R. Pavia, T. K. Sawyer, and W. H. Moos, *Bioorg. Med. Chem. Lett.*, 1993, **3**, 387.

158. I. Amato, *Science*, 1992, **257**, 330.

159. S. Brenner and R. A. Lerner, *Proc. Natl. Acad. Sci.*, 1992, **89**, 5381.

160. B. A. Bunin and J. A. Ellman, *J. Am. Chem. Soc.*, 1992, **114**, 10 997.

161. A. Borchardt and W. C. Still, *J. Am. Chem. Soc.*, 1994, **116**, 373.

162. J. K. Chen, W. S. Lane, A. W. Brauer, A. Tanaka, and S. L. Schreiber, *J. Am. Chem. Soc.*, 1993, **115**, 12 591.

163. P. L. Skatrudk, J. A. Tietz, T. D. Ingolia, C. A. Cantwell, D. L. Fisher, J. I. Chapman, and S. W. Queener, *Bio/Technology*, 1989, **7**, 477.

164. G. L. Verdine, *Nature*, 1996, **384** (Suppl.), 11.
165. N. Rajendran and M. A. Marahiel, in "Symposium of the Special Research Group Integrated Enzyme Systems," Technical University, Berlin, 1996, p. 44.
166. K. Turgay and M. A. Marahiel, *Peptide Res.*, 1994, **7**, 238.
167. R. D. Fleischmann, M. D. Adams, O. White, R. A. Clayton, E. F. Kirkness, A. R. Kerlavage, C. J. Bult, J. F. Tomb, B. A. Dougherty, J. M. Merrick *et al.*, *Science*, 1995, **269**, 496.
168. C. M. Fraser, J. D. Gocayre, O. White, M. D. Adams, R. A. Clayton, R. D. Fleischmann, C. J. Bult, A. R. Kerlavage, G. Sutton, J. M. Kelly *et al.*, *Science*, 1995, **270**, 397.
169. S. Donadio, M. Sosio, and E. Bossi, in "Beijerinck Centennial Symposium on Microbial Physiology and Gene Regulation," eds. W. A. Scheffers and J. P. van Dijkenp, Federation of European Microbiological Societies, Amsterdam, 1995, p. 353.
170. S. Borchert, S. S. Patil, and M. A. Marahiel, *FEMS Microbiol. Lett.*, 1992, **92**, 175.
171. S. S. Queener, O. K. Sebek, and C. Vezina, *Annu. Rev. Microbiol.*, 1978, **32**, 593.
172. N. Rajendran, D. Jahn, K. Jayaraman, and M. A. Marahiel, *FEMS Microbiol. Lett.*, 1994, **115**, 191.
173. V. de Lorenzo, M. Herrero, U. Jakubzik, and K. N. Timmis, *J. Bacteriol.*, 1990, **172**, 6568.
174. J. F. Miller, *Methods Enzymol.*, 1994, **235**, 375.
175. M. Iwaki, K. Shimura, M. Kanda, E. Kaji, and Y. Saito, *Biochem. Biophys. Res. Commun.*, 1972, **48**, 113.
176. M. Kambe, Y. Imae, and K. Kurahashi, *J. Biochem.*, 1974, **75**, 481.
177. S. Majumder, S. K. Ghosh, N. K. Mukhopadhyay, and S. K. Bose, *J. Gen. Microbiol.*, 1985, **131**, 119.
178. S. Majumder, N. K. Mukhopadhyay, S. K. Ghosh, and S. K. Bose, *J. Gen. Microbiol.*, 1988, **134**, 1147.
179. I. D. Kuntz, *Science*, 1992, **257**, 1078.
180. J. B. Moon and J. W. Howe, *Proteins*, 1991, **11**, 314.
181. N. Rajendran, E. Sivamani, and K. Jayaraman, *FEMS Microbiol. Lett.*, 1994, **122**, 103.
182. S. Freund, G. Jung, W. A. Gibbous, and H. G. Sahl, in "Nisin and Novel Lantibiotics," Escom Publishers, Leiden, 1991, p. 33.
183. B. Rosenwirth, A. Billich, R. Datema, P. Donatsch, F. Hammerschmid, R. Harrison, P. Hiestand, H. Jaksche, P. Mayer, and P. Peichl, *Antimicrob. Agents Chemother.*, 1994, **38**, 1763.
184. D. J. Ward, Y. Chen, E. Platt, and B. Robson, *J. Theor. Biol.*, 1991, **148**, 193.
185. G. R. Marshall, *Annu. Rev. Pharmacol. Toxicol.*, 1987, **27**, 193.
186. D. J. Ringe, *Nucl. Med.*, 1995, **36**, 28.

4.09
Catalysis of Amide and Ester Bond Formation by Peptide Synthetase Multienzymatic Complexes

VALÉRIE DE CRÉCY-LAGARD
Institut Pasteur, Paris, France

4.09.1 INTRODUCTION

Many peptide derivatives are synthesized in a nontranslational way by multienzyme complexes called peptide synthetases (PPS).[1] These enzymes are composed of modules each involved in the incorporation of one given amino acid.[2] Proteolysis and primary sequence data indicate that the modules are multifunctional and can be dissected into domains that colinearly catalyze the different reactions involved in the synthesis of the final molecule.[3,4] A mechanistic analogy between non-ribosomal peptide synthesis and fatty acid (or polyketide) synthesis was proposed by Lipmann[5] in the thiotemplate model and has been confirmed by the primary sequences of many PPS, fatty acid synthases (FAS), and polyketide synthase (PKS) enzymes.

The exact mechanism of peptide synthesis by PPS enzymes is not clear, however, the sequences of several complete systems involved in the synthesis of molecules of known structure (peptide, depsipeptide, peptidolactone, lipopeptide, or peptolide), have now been determined (Table 1, Figure 1, and Figure 2) and analysis of the gathered data may help to reveal the catalytic mechanisms involved.

Table 1 Peptide synthetase genes from bacteria or filamentous fungi, producing antibiotics or siderophores.

Organisms	Molecules synthesized	Genes	Enzymes	Number of activation modules	Accession numbers
Bacillus brevis ATCC 8185	Gramicidin S	*grsA*	GrsA	1	M29703
		grsB	GrsB	4	M29703
Bacillus subtilis F29-3	Fengycin	*fenB*	FenB	1	L42523
Bacillus subtilis ATCC 21332	Surfactin	*srfAA1*	SFRS2	3	X70356
		sfrAB1	SFRS3	3	X70356
		sfrAC	SFRS4	1	X70356
Streptomyces pristinaespiralis	Pristinamycin I	*snbA*	SnbA	1	X98515
		snbC	SnbC	2	Y11548
		snbDE	SnbD	4	Y11548
Escherichia coli	Enterocholin	*entE*	EntE	1	X15058
		entF	EntF	1	J05325
Aspergillus nidulans	β-Lactam precursor ACV	*acvA*	ACVS	3	X54853
Nocardia lactamdurans	β-Lactam precursor ACV	*pcbAB*	ACVS	3	X57310
Penicillium chrysogenum	β-Lactam precursor ACV	*pcbAB*	ACVS	3	M57425
Acremonium chrysogenum	β-Lactam precursor ACV	*pcbAB*	ACVS	3	P25464
Fusarium scirpi	Enniatin B	*esyn1*	Esyn	2	Z18755
Cochliobolus carbonum	HC-toxin	*hts1*	HTS1	4	A45086
Tolypocladium inflatum	Cyclosporin A	*simA*	Cysyn	11	Z28383
Streptomyces hygroscopicus	Rapamycin	*rapP*	RapP	1	X86780
		rapA	RAPS1	1	
Pseudomonas aeruginosa	Pyoverdin	*pvdE*	PvdE	1	U07359
		pvdD	PvdD	2	U07359

In 1971 Lipmann proposed two key concepts: activation of precursors as acyladenylates and transfer of the acyl intermediates as thioesters on a phosphopantetheinyl (Pan) cofactor attached to the enzyme. This model has been refined in the light of the sequence and biochemical data obtained during the 1990s. First, domains involved in amino acid activation,[2,6] which are members of the adenylate-forming enzymes superfamily, have been identified and shown to be structurally distinct from class I and class II aminoacyl-tRNA synthetase enzymes.[7] Second, an acyl carrier protein (ACP) domain has been identified in each incorporation module, with a Pan cofactor attached. This led to the proposal of a multiple carrier model.[8–10] However, the exact mechanisms of initiation, elongation, termination, and epimerization are still unclear. This review focuses on the mechanisms of acyl transfer catalyzed by PPS enzymes and elaborates on models which may be tested experimentally. The elucidation of the mechanism of peptide synthesis by these multienzymatic complexes is a prerequisite to exploring the synthesis of new metabolites.

4.09.2   ACYL TRANSFER REACTIONS CATALYZED BY PPS ENZYMES

4.09.2.1   Thioester Activation is the Driving Force of PPS Enzymes

The concept of the high-energy bond in metabolism was originally proposed by Lipmann,[11] and has facilitated the biological study of metabolism. The notion of the activation of biological molecules as adenylates was later derived from this concept.[12] The discovery of coenzyme A (CoA),[13] the central acyl carrier of the cell used as a cofactor by 4% of known enzymes[14] was also of key significance in the study of metabolism. Chemical reactions catalyzed by PPS enzymes are clearly linked with the chemistry of thioesters since all acyl intermediates are linked as thioesters to the Pan arm, attached to the enzyme. Sulfur acid esters act as electron-withdrawing groups and therefore activate the acyl group towards transfer by nucleophilic displacement as shown in Figure 3. The acidity of the α-C—H bond is also increased by the ester group and under certain conditions reacts with a base to form a carbanion (Figure 3(b)). These reactions are pertinent to understanding the

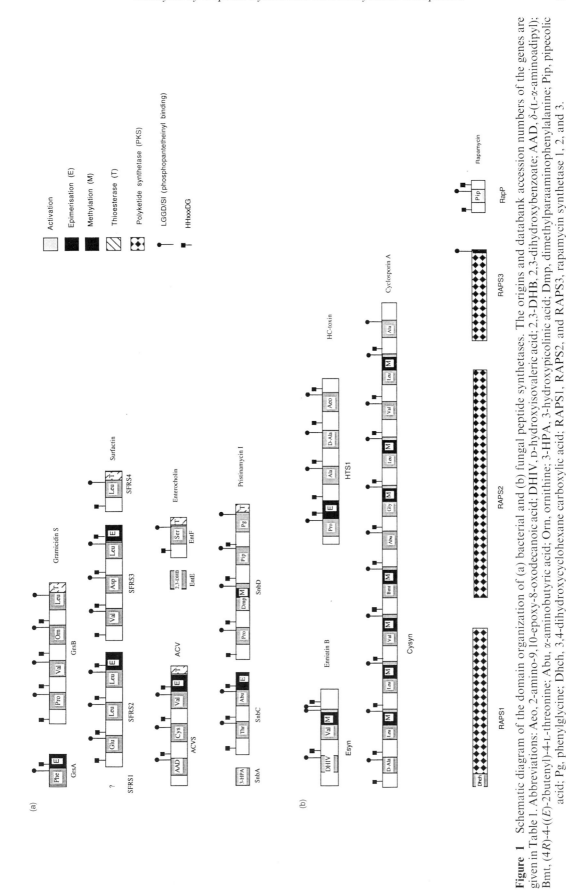

Figure 1 Schematic diagram of the domain organization of (a) bacterial and (b) fungal peptide synthetases. The origins and databank accession numbers of the genes are given in Table 1. Abbreviations: Aeo, 2-amino-9,10-epoxy-8-oxodecanoic acid; DHIV, D-hydroxyisovaleric acid; 2,3-DHB, 2,3-dihydroxybenzoate; AAD, δ-(L-α-aminoadipyl); Bmt, (4R)-4-((E)-2butenyl)-4-L-threonine; Abu, α-aminobutyric acid; Orn, ornithine; 3-HPA, 3-hydroxypicolinic acid; Dmp, dimethylparaaminophenylalanine; Pip, pipecolic acid; Pg, phenylglycine; Dhch, 3,4-dihydroxycyclohexane carboxylic acid; RAPS1, RAPS2, and RAPS3, rapamycin synthetase 1, 2, and 3.

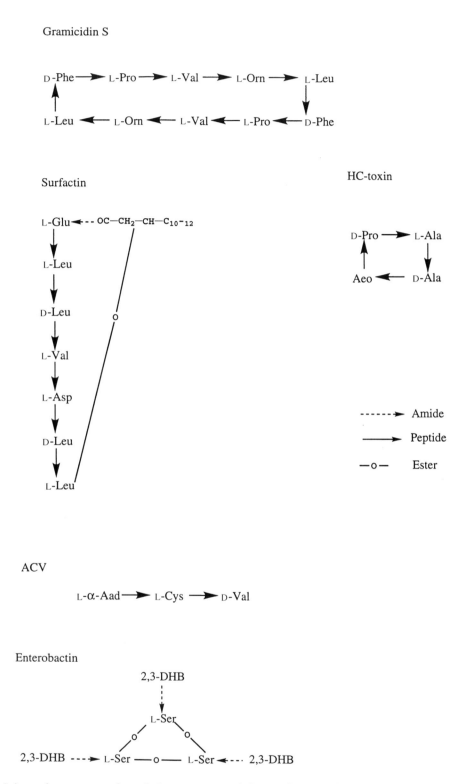

Figure 2 Schematic representation of the structures of the small molecules synthesized by PPS enzymes presented in Figure 1. The different bonds (amide, peptide, or ester) are indicated. Abbreviations are given in Figure 1 caption.

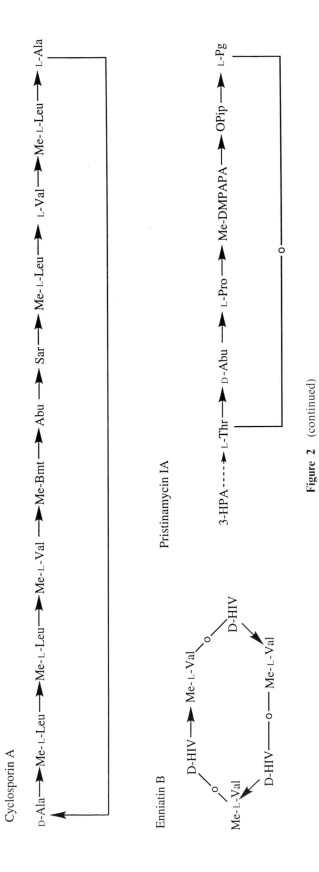

Figure 2 (continued)

mechanisms of acyl transfer catalyzed by PPS enzymes. A more precise understanding of how these enzymes promote the nucleophilic attack by a donor such as the amine of an amino acid or the hydroxy group of a side-chain (or water molecule) on the thioester carbonyl, or extract a proton from the α-carbon atom, is of crucial importance in PPS mechanisms and should help to explain how the initiation, elongation, epimerization, and termination steps are catalyzed.

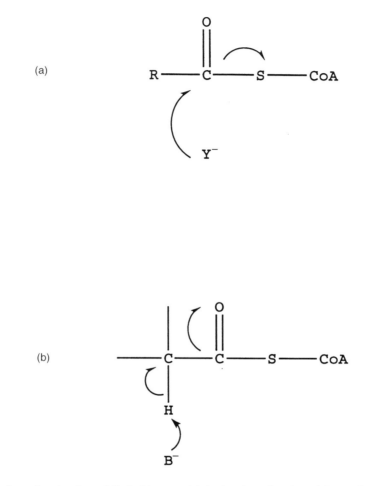

Figure 3   Mechanism of activation of CoA thioester. (a) Activation of carbonyl for nucleophilic substitution. (b) Activation of α-C hydrogen atom for formation of a carbanion.

4.09.2.2   PPSs Catalyze the Formation of Two Types of Bonds: Amide and Ester

A survey of the structures of the molecules synthesized by PPS enzymes[1] revealed that, unlike the ribosome, PPS enzymes catalyze the formation of amide and ester bonds. The amide bond does not have to be a peptide bond. In pristinamycin I, enterobactin, and surfactin, amide (nonpeptide) bonds are found between the starter unit (hydroxy acid or fatty acid) and the first incorporated amino acid as shown in Figure 2. Ester bonds are also present in these three molecules and in enniatin B. Although these molecules contain the two types of bonds, other molecules synthesized by PPS enzymes such as gramicidin B, ACV (δ-(L-α-aminoadipyl)-L-cysteinyl-D-valine)), HC-toxin and cyclosporin A, are more monotonous in structure having only peptide bonds. Acyltransferase domains of PPS must therefore catalyze two types of nucleophilic attack on the thioester carboxyl, one by an amine leading to the formation of amide bonds and the other by hydroxy leading to ester bond formation or hydrolysis. Analysis of the primary sequences of PPS enzymes suggest that two types of domains are involved in acyl transfer catalysis: the thioesterase domain and the elongation domain.

4.09.3 ACYLTRANSFERASE DOMAINS OF PPS ENZYMES

4.09.3.1 The Thioesterase Domains

The first PPS domains that were implicated in acyl transfer were the 250 amino acid C-terminal domains similar to the FAS and PKS thioesterase domains.[15] In type I fatty acid synthesis and type I polyketide synthesis, the thioesterase domains catalyze the liberation of the fatty acid or the cyclization of the polyketide,[16,17] suggesting that the thioesterase domains of PPS enzymes may be involved in termination of peptide synthesis by liberating the peptide from the Pan cofactor.[18] The position of the thioesterase domain is essential in polyketide synthesis: insertion of the DNA fragment encoding the thioesterase domain from the *eryAIII* (involved in the biosynthesis of the aglycone core of erythromycin A), at the end of the *eryAI* gene, led to the liberation of the expected triketide lactone.[19] Thioesterase domains have been identified in some, but not all peptide synthetase systems (Figure 1). The presence or absence of these domains does not appear to correlate with the type of termination catalyzed by the complex. A thioesterase domain is always present in the hydrolysis of a linear peptide (ACVS), however, in the cyclization of the final peptide by the formation of an amide or ester bond, thioesterase domains may be present (GrsB, SFRS4, SnbDE, EntF) or absent (HTS1, Cysyn, Esyn). To date, the sequences of 11 thioesterase domains of peptide synthetases are available in data banks (Figure 1 and Table 1). Thioesterases are members of the superfamily of serine dependent hydrolases including proteinases, lipases, and acetylcholine esterases.[20] This family of enzymes catalyze the hydrolysis of amide, ester, and thioester bonds with an acyl intermediate on the enzyme. The catalytic residues constitute a chymotrypsin-like triad where the nucleophile is usually serine (but may also be cysteine or aspartate). A histidine residue extracts a proton from the nucleophile during the reaction. Aspartic acid (or sometimes glutamic acid or tryptophan) constitutes the third member of the triad.[21] Comparison of the thioesterase domains of PPS enzymes with thioesterase domains of different origins led to the identification of three conserved residues,[15,22] a conserved serine in a GW/YSXG motif, followed by an aspartic acid residue 25 amino acids downstream from the conserved serine and histidine residues in a conserved C-terminal GXH motif.

In summary, the thioesterase domains of PPSs are members of the lipase/serine esterase super-family. They could therefore be expected to catalyze firstly the transfer of the acyl intermediate from the SH of Pan to the OH of the conserved serine residue and secondly, the attack of the acyl intermediate by the proton donor, which may be a water molecule, a terminal amine, or a hydroxy side-chain of the elongated molecule. These enzymes therefore appear to function as hydrolases or as acyltransferases, like the ACP-myristoyl acyltransferase.[20]

4.09.3.2 The Condensation Domains

The author has previously proposed[23] that each amino acid incorporating module contains not only the amino acid activating domain and the ACP domain, but also a 350 amino acid domain (Figure 4(a) and (b)) containing seven conserved motifs. The most highly conserved motif is called here the His motif (HHxxxDG), but has been called the M motif by other authors.[3,4] The second histidine and the aspartic acid residues are strictly conserved in all the domains sequenced so far (Figure 5(a)). The identification of a His motif in several families of acyltransferases and the correlation between the number of condensation domains and the number of acyl transfers necessary to synthesize the final molecule led to the proposal that it is the elongation domain, as discussed previously.[23]

4.09.3.2.1 An acyltransferase superfamily

The HHxxxDG motif is also found in other families of acyltransferases. It has been reported that an identical motif is present in type III chloramphenicol acetyltransferase (CAT)[24] and in the transferase enzymes (E2o, E2P, and E2b) of the 2-oxoacid dehydrogenase complexes (Figure 6).[23] Structural and site-directed mutagenesis data on these two families indicate that the second histidine residue is in the active site acting as a general base catalyst. It promotes the attack of the nucleophile, C-3 hydroxy of chloramphenicol or thiol of CoA to attack the carbonyl of acetylCoA, or the dihydrolipoamide acceptor, respectively.[25–27] This suggests that in PPS enzymes the corresponding histidine deprotonates the incoming amino acid to facilitate the attack on the carbonyl group of the

(a)   **Peptide synthetase incorporation module**

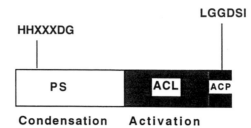

(b)

```
Motif I      [REQ]-L-x(1)-[GTA]-x(1)-[LIV]-x(3)-[ARL]-L-x(2)-A-x(3)-[LVI]
             -x(2)-R-[HYF]-[EP]-x(1)-L-R-T-x(1)-F-x(6)-G-x(3)-Q-x(1)-[VI]

                  -x(48-58)-

Motif II     L-x(4)-H-H-[ILA]-[ILA]-x(1)-D-G-[WVM]-S-x(3)-[LIF]-x(2)-[EDA]-[LV]-x(3)-Y

                  -x(4-19)-

Motif III    L-x(3)-Y-[RKA]-[DAE]-[FY]-x(2)-[WY]-[QLH]

                  -x(13-23)-

Motif IV     W-x(3)-[LRF]-x(1)-[DGR]-x(2)-P-x(3)-L-x(2)-[DR]-x(2)-R

                  -x(30-40)-

Motif V      T-x(3)-V-L-[LQA]-[ATF]-[AV]-x(4)-[LH]-x(3)-[TGS]-[GAH]-x(2)-[DHE]-
             x(3)-G-x(1)-[PTV]-x(1)-[AST]-[GN]-R-x(6)-[LIV]-[EDQ]-x(2)-[VIP]-G-
             x(1)-F-[VI]-N-[TVL]-[LQV]-[PCA]-[LMV]-[RVI]

                  -x(2-12)-

Motif VI     [FV]-x(2)-[LFA]-[VLI]-x(2)-[VIL]-x(6)-[AMS]-x(2)-[HNK]-[QER]-D-[VYL]
             -P-[FL]-E-x(1)-[IL]-x(3)-L-x(3)-[RS]-[RSA]-[DS]-x(1)-S-[RSK]-x(1)-P-L

                  -x(28-43)

Motif VII    [FYP]-D-[LIV]-x(1)-[LFV]-x(1)-[LVA]-x(1)-[EQP]
```

Figure 4 The PPS condensation domain. (a) Schematic representation of a peptide synthetase incorporation module. PS, peptide synthase (elongation domain containing the HHxxxDG motif); ACL, acyl CoA ligase (adenylate forming domain); ACP, acyl carrier protein. (b) The seven motifs conserved in the elongation domains. The number of residues separating the motifs are indicated as x(n-m). The syntax for the description of the motifs is identical to the syntax in the Prosite databank (release 12.2, February 1995). The HHxxxDG motif and the potential waiting positions are in bold type.

elongating chain. Research on conserved sequences of families of acyl-CoA dependent acyl-transferases has shown that several other acyltransferase families contain conserved histidine residues (data not shown). A catalytic role of the histidine has been reported for the choline/carnitine acyltransferases,[28,29] and for the *N*-myristyl acyltransferases[30] listed in Figure 6. Structural data may become available for the *O*-acetyltransferases GAT and NodL[31,32] and the UDP-*N*-acetyl-glucosamine *O*-acyltransferase,[33] also shown in Figure 6. Preliminary experiments suggest that the conserved His motifs have a catalytic function in elongation. The pH optimum for peptide synthesis suggests the involvement of a histidine residue.[4] Site-directed mutagenesis of the conserved Asp in a domain of SFRS4 abolished product formation.[34] If the involvement of the histidine extends to other acyltransferase families it would suggest a common mechanism of acyl transfer from thioester intermediates with histidine as a general base catalyst promoting nucleophilic attack of the thioester carbonyl (Figure 7).

(a)   Peptide synthetases interdomains

```
GrsB        142     WSFHHILMDGWCFN
            1182    MDMHHIISDGVSMQ
            2218    MDMHHIISDGVSMQ
            3264    FDMHHIISDGISSN
HTS1        1492    FQVNHLVFDGMSTD
            2668    ISAHHAIYDGWSLN
            3799    VGAHHSIYDAHVLS
            4921    WTLNHAAYDAWSLR
SFRS2       136     ANVHHVISDGMSMN
            1181    FDMHHLISDGVSIG
            2220    FDMHHIISDGASVG
SFRS3       140     WSNHHIMMDGWSMG
            1176    IDMHHIITDGSSTG
            2222    LDMHHIIADGVSRG
SFRS4       143     WSYHHIILDGWCFG
ACVS Ani    1077    FSCHHAILDHWSLP
            2151    LSFHHTCFDAWSLK
ACVS Pch    179     TIVTHENRDGWSVA
            1082    FSCHHAILDGWSLP
            2157    LAFHHTCFDAWSLK
ACVS Nlo    1011    FCSHHIILDGWSLP
            2076    VVVHHSCFDGWSWD
ACVS Ach    1026    FSCHHAILDGWSLP
            2108    IVCHHLAFDAMSWD
Cysyn       154     IVVHHIISDSRSLD
            1239    IVMHHIVSDGWSLD
            2738    TVMHHAISDGWSVD
            4224    VVMHHIISDGWSVD
            5717    IVMHHIISDGWSVE
            7214    IVMHHIIYDGWSVD
            8275    IVMHHIISDGWTVD
            9769    LVMHHMFSDGWSVD
            11265   IVMHHIISDGWSVD
            12336   IVMHHIISDGWSVD
            13836   IVMHHIISDGWSTE
EntF        134     QRYHHLLVDGFSFP
Esyn        1226    IVMHHIISDGWSID
SnbC        138     QHVHHLLLDGYGFR
            1165    LLMHHVAGDGWSLR
SnbD        143     LTTHHLLLDGWSMP
            1177    LLLHHIAGDGWSLA
            2608    LLMHHIAGDGWSWS
            3653    LLLHHIAGDGWSLA
RapP        164     LTVHHIAGDGWSLA
            1223    LMLHHIAGDGWSFD
```

(b)   Epimerization domains

```
HTS1        1000    VVIHHAVVDLVSWR
TycA        738     LAIHHLVVDGISWR
GrsA        749     MAIHHLVVDGISWR
SFRS2       3252    LAIHHLVVDGVSWR
SFRS3       3252    IAVHHLVVDGVSWR
ACVS Ani    3233    FSVHHIIIDIVSWQ
ACVS Nlo    3122    FALHHLVVDTVSWH
ACVS Pch    3242    FSVHHMAIDTVSWQ
ACVS Ach    3179    FACHHVMVDTVSWN
SnbC        2227    LTLHHLVVDGVSWR
```

Figure 5 Comparison of the HHxxxDG motifs of PPSs. (a) Peptide synthetases interdomains (elongation domains); (b) epimerization domains. The origins of the peptide synthetase genes are given in Table 1. Abbreviations: Ani, *Aspergillus nidulans*; *Nlo, Nocardia lactamdurans*; Pch, *Penicillium chrysogenum*; Ach, *Acremonium chrysogenum*.

(a)   Chloramphenicol acetyltransferases (EC 2.3.1.28)

```
P26825|CAT1.CLOPE     185       VQVHHAVCDGYHIS
P00485|CAT1.STAAU     184       LQVHHSVCDGYHAG
P26826|CAT2.CLOPE     181       IQVHHAVCDGFHIC
P22615|CAT2.ECOLI     184       VQVHHAVCDGFHAA
P22616|CAT2.HAEIN     184       VQVHHAVCDGFHAA
P00486|CAT2.STAAU     184       LQLHHAVCDGYHAS
P00484|CAT3.ECOLI     184       VQVHHAVCDGFHVA
P06135|CAT3.STAAU     184       LQLHHAVCDGYHAS
P36882|CAT4.STAAU     184       LQLHHAVCDGYHAS
P36883|CAT5.STAAU     184       LQVHHAVCDGYHVS
P00487|CAT.BACPU      182       IQVHHAVCDGYHAG
P22782|CAT.CAMCO      181       IQVHHAVCDGFHVC
Q02736|CAT.CLOBU      185       IQVHHAICDGYHAS
P11504|CAT.CLODI      181       IQVHHAVCDGFHIC
P00483|CAT.ECOLI      188       IQVHHAVCDGFHVG
P07641|CAT.PROMI      188       IQVHHAVCDGFHVG
P25309|CAT.STAIN      184       LQLHHAVCDGYHAS
P20074|CAT.STRAC      190       VQIHHAAADGFHTA
Q03058|CAT.STRAG      184       LQLHHSVCDGYHAS
```

Dihydrolipoamide acyltransferases (EC 2.3.1.12)

```
P37942|ODB2.BACSU     390       LSLDHRVLDGLVCG
P11181|ODB2.BOVIN     447       WSADHRIIDGATVS
P11182|ODB2.HUMAN     447       WSADHRVIDGATMS
P09062|ODB2.PSEPU     162       GRILHEDLDAFMSK
P20708|ODO2.AZOVI     364       LSYDHRLIDGKEAV
P16263|ODO2.BACSU     383       LSYDHRIVDGKEAV
P07016|ODO2.ECOLI     371       LSYDHRLIDGRESV
P36957|ODO2.HUMAN     419       LTYDHRLIDGREAV
Q01205|ODO2.RAT       408       LTYDHRLIDGREAV
P19262|ODO2.YEAST     358       VRLSHKARDGKLTL
P11961|ODP2.BACST     393       LSFDHRMIDGATAQ
P21883|ODP2.BACSU     407       LSFDHRMIDGATAQ
P36413|ODP2.DICDI     560       LSCDHRVIDGAVGA
P06959|ODP2.ECOLI     597       LSFDHRVIDGADGA
P10515|ODP2.HUMAN     583       LSCDHRVVDGAVGA
P20285|ODP2.NEUCR     426       ASFDHKVVDGAVGA
P08461|ODP2.RAT       523       LSCDHRVVDGAVGA
P12695|ODP2.YEAST     450       GTFDHRTIDGAKGA
P35489|OPD2.ACHLA     511       LAVDHRIIDGADGG
```

Figure 6 Comparison of the sequences surrounding catalytic histidine residues of acyltransferase enzymes. (a) Chloramphenicol acetyltransferases and dihydrolipamide acyltransferases; (b) other CoA dependent acyltransferases. The SwissProt accession numbers and sequence names are given.

4.09.3.2.2 *Role of condensation domains both in amide and ester bond formation*

Synthesis of a linear peptide *n* amino acids long requires *n* − 1 condensation reactions. Therefore, *n* − 1 elongation domains should be encoded by the genes specifying biosynthesis of the entire peptide. This has been confirmed in the synthesis of the linear tripeptide ACV, where ACV synthetase has two elongation domains. However, as most peptides are cyclic, and cyclization occurs by the formation of an ester or a peptide bond, an extra acyltransfer domain should be present in these cases. As discussed above, thioesterase domains are thought to catalyze the termination steps but not all peptide synthetase systems have internal thioesterase domains. This raises the question of how peptide synthesis is terminated in the biosynthesis of HC-toxin, enniatin B, and cyclosporin A. It has been proposed[4] that, in these cases, termination may be effected by yet undiscovered discrete

(b)

```
        Carnitine acyltransferases (EC 2.3.1.7)

        P80235|CACM_YEAST 173 ::    GEPSCHLQTDATSHHV
        P32796|CACP_YEAST 372 ::    GFLAEHSKMDGTPTLF
        P32198|CPT1_RAT   467 ::    GINAEHSWADAPIVGH
        P23786|CPT2_HUMAN 366 ::    AVHFEHSWGDGVAVLR
        P18886|CPT2_RAT   366 ::    AVHFEHSWGDGVAVLR



        Choline o-acetyltransferase (EC 2.3.1.6)

        P07668|CLAT_DROME 412 ::    GLCYEHSCSEGIAVVQ
        P28329|CLAT_HUMAN 436 ::    GVVCEHSPFDGIVLVQ
        Q03059|CLAT_MOUSE 328 ::    GVVCEHSPFDGIVLVQ
        P13222|CLAT_PIG   328 ::    GVVCEHSPFDGIVLVQ
        P32738|CLAT_RAT   327 ::    GVVCEHSPFDGIVLVQ



        Acyl-[acyl-carrier-protein]-UDP-N-acetylglucosamine
        o-acyltransferases (EC 2.3.1.129)

        P10440|LPXA_ECOLI  41 ::    TVLKSHVVVNGHTKIG
        P32199|LPXA_RICRI  39 ::    VELKSHVVIEGITEIG
        P32200|LPXA_SALTY  41 ::    TVLKSHVVVNGQTKIG
        P32201|LPXA_YEREN  41 ::    TELKSHVVVNGITKIG



        Thiogalactoside acyltransferases (EC 2.3.1.18)

        P09632|NODL_RHILV 113 ::    ADHPHDPEQR
        P28266|NODL_RHIME 113 ::    ADHPDDPEQR
        P07464|THGA_ECOLI 113 ::    TGHPVHHELR



        N- Myristoyl acyltransferases (EC 2.3.1.97)

        P34763|NMT_AJECA  300 ::    YYHRALDWLK
        P46548|NMT_CAEEL  249 ::    YYHRSLNPRK
        P30418|NMT_CANAL  225 ::    YQHRPINWSK
        P34809|NMT_CRYNE  230 ::    YFHRNLNPPK
        P30419|NMT_HUMAN  216 ::    YWHRSLNPRK
        P14743|NMT_YEAST  219 ::    YTHRPLNWKK
```

Figure 6 (continued)

thioesterase proteins similar to GrsT[35] or SfrT.[18] The functions of GrsT and SfrT are, however, still unclear: the *Bacillus subtilis sfrT* gene has been mutated with no effect on surfactin synthesis.[36] Both cyclosporin A and enniatin A have been synthesized *in vitro* using the appropriate precursors and the purified Cysyn or Esyn enzymes, respectively.[37,38] Although it is possible that contaminating thioesterase proteins were present in the pure Cysyn or Esyn enzyme preparations, thorough analysis of the sequence data suggests that the presence of additional thioesterase proteins is unnecessary. A catalytic role of the conserved His motif (HHxxxDG), present in the elongation domain in acyl transfer, has already been discussed.[23] An extra His motif is present, in the N-terminal domain of Cysyn and in the C-terminal domains of HC-toxin synthetase, and enniatin synthetase (Figure 5). Furthermore, in HC-toxin and cyclosporin A an extra peptide bond is present in the final peptide. In both cases the extra His motif could be involved in catalyzing the cyclization of the peptides, and the conserved histidine could catalyze the attack of the amine of the N-terminal amino acid on the carbonyl of the C terminus amino acid. Enniatin B cyclization is more complex as the final molecule

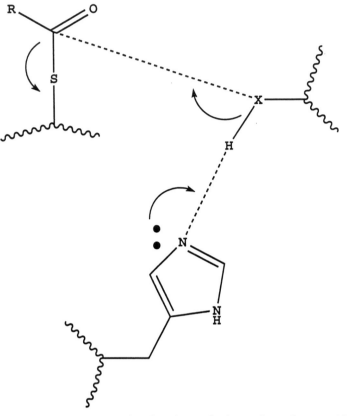

Figure 7   General mechanism for the catalysis of acyl transfer by acyltransferases with thioester activated acyl substrates.

is a hexadipeptidol. It is proposed that in this case the C-terminal His motif is involved in the formation of a lactone bond rather than a peptide bond. The sequence data on the rapamycin biosynthetic cluster suggests that a condensation domain of RapP may also be involved in a lactone bond formation in this case[39] (Figure 1).

The role of the discrete thioesterase proteins such as GrsT and SfrT is still unclear. Phylogenetic analysis of thioesterase proteins and domains of different origins (data not shown) revealed that they cluster with the mammalian thioesterase II proteins. These enzymes release medium length fatty acids in some specialized tissues,[40] whereas the thioesterase domain of the FAS enzyme liberates the acyl chains of 16 carbon atoms.[41] As nonribosomal peptide synthesis and fatty acid synthesis are mechanistically parallel in many ways, GrsT and SfrT may have an editing role releasing the peptides from the enzymes when an amino acid is missing or misincorporated or when synthesis cannot proceed to the final step, as has been previously suggested by Marahiel and co-workers.[35]

4.09.3.2.3   Mechanism of acyl transfer by condensation domains

As discussed above, it is suggested that condensation domains, like thioesterase domains, catalyze acyl transfers involved in ester and peptide bond formation. Thioesterase domains function as both hydrolases and acyltransferases, whereas condensation domains only function as acyltransferases. Further analysis of both the sequence data on enzymes involved in the synthesis of pristinamycin I and the biochemical data on the synthesis of the structural analogue actinomycin led to the proposal of a model for elongation by PPS enzymes.[42] This model mixes the initial and revised form of Lipmann's thiotemplate model, presented in Figure 8. The growing peptide chain is transferred from the Pan cofactor to a waiting position on the elongation domain before the actual condensation reaction occurs paralleling PKS synthesis. Analysis of the amino acid sequence of 55 condensation domains suggested several possible candidates for the waiting position (data not shown). Transfer onto a thiol group as in the FAS/PKS enzymes is unlikely since no cysteine residue is conserved. A good candidate is a serine residue two amino acids downstream from the His motif (HHxxxDGxxS).

A serine residue is present at this position in 85% of sequences, a cysteine residue in 9%, and a threonine residue in 2%. There are many precedents for acyl transfer from an acyl-CoA to a serine, for example the acyltransferase sites of FAS/PKS enzymes and the serine residue of thioesterase II.[43] An alternative position for ester intermediates is a conserved tyrosine 14 amino acids downstream from the His motif. The histidine residues of the His motif are alternative candidates for the waiting position. Prebiotic chemistry experiments have shown that histidinyl-histidine catalyzes the condensation of glycine residues in a clay environment.[44] The authors proposed that the glycyl residue is transferred onto the imidazole ring in an activated amide linkage which is then transferred to the amine of a glycine residue or a glycyl-peptide. The existence of a waiting position would strengthen the mechanistic analogy between the condensation and the thioesterase domain proposed in this chapter.

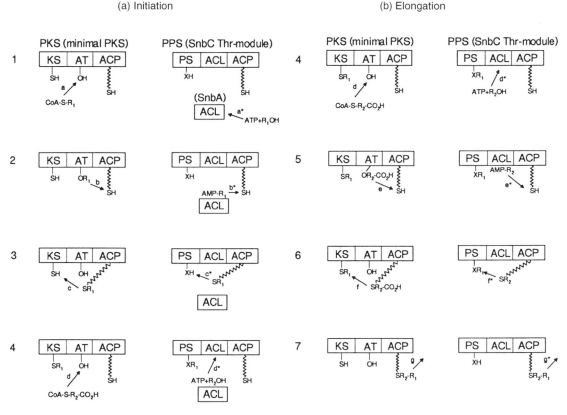

Figure 8 Model for peptide elongation by PPS indicating the parallel between polyketide and nonribosomal peptide synthesis. (a) Initiation; (b) elongation. Polyketide synthase, PKS: KS, β-ketoacylsynthase; AT, acyltransferase; ACP, acyl carrier protein. Peptide synthetase, PPS: PS, peptide synthase (elongation domain containing the HHxxxDG motif); ACL, acyl CoA ligase (adenylate forming domain); ACP, acyl carrier protein. Steps 1 to 3: initiation. a, a* = transfer of the starter chain to the loading site; b, b* = transfer to the 4'-phosphopantetheinyl (4-PP) arm; c, c* = transfer to the waiting/elongation site. Steps 4 to 7: elongation. d, d* = transfer of the elongation chain to the loading site; e, e* = transfer to the 4-PP arm; f, f* = condensation of the elongation chain on the 4-PP arm with the starter chain at the waiting/elongation site; g, g* = transfer of the elongated chain to the following waiting/elongation site.

4.09.3.2.4 *Evidence for the existence of the condensation domain as a discrete protein*

As the domains have similar activities, it would be interesting to determine whether the condensation domains exist as discrete polypeptide chains (like the thioesterase domains) and/or only as domains of multifunctional proteins. In searching proteins other than PPS enzymes containing the seven motifs of the condensation domain,[45] an open reading frame (*orf2*) was identified encoding a protein of unknown function located at the C terminus of the mycocerosic synthase gene *mas* of

Mycobacterium tuberculosis.[46] The *mas* gene encodes the specific fatty acid synthase that catalyzes the elongation of the tetramethyl-branched mycocerosic acids, which are then esterified to phenol-phtiocerol. The Mas enzyme has an ACP domain at its terminus but lacks a thioesterase domain. The similarity of the *orf2* gene product with the PPS condensation domain (for example 46% similarity with the second elongation domain of SnbC) and the possible role of these domains in esterification reactions discussed in this chapter, suggest that it may encode the acyltransferase involved in the transfer of the mycocerosic acyl chains to the hydroxy groups of phenolphtiocerol.[47]

4.09.3.3 Does the Structure of the Peptide Synthetase Complex Dictate the Final Length of the Peptide?

The proposal that condensation domains catalyze cyclization of the growing peptides, raises a question essential to understanding how peptide synthesis terminates. As each amino acid incorporating module contains an elongation domain, what determines elongation termination? In cyclosporin A synthesis *in vitro*, the reaction terminates if a precursor is omitted and cyclization of the linear peptide intermediates is not observed.[48] However, when the synthesis reaction proceeds to completion, linear intermediates containing 11 amino acids cannot be isolated, indicating that cyclization is immediate. It can be proposed that two criteria must be fulfilled in order to terminate the synthesis: (i) the structure of the peptide dictates the correct positioning of the terminal residues to favor cyclization substrate and (ii) the spatial organization of the peptide synthetase complex dictates the final length of the molecule. The simplest model for fulfilling the second criterion involves a channeling of the elongated peptide to correctly position the C-terminal thioester carbonyl function and the N-terminal amino or hydroxy group, in the terminating domain catalytic pocket (thioesterase or condensation). This reaction may be effected by the peptide synthetase complex (monomeric or multimeric) only when the peptide is of a given length, as shown in Figure 9.

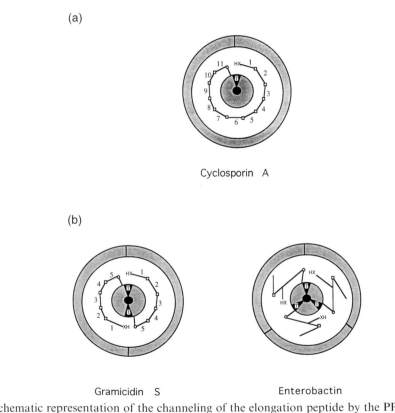

(a)

Cyclosporin A

(b)

Gramicidin S   Enterobactin

Figure 9 Schematic representation of the channeling of the elongation peptide by the PPS. T represents the terminating domain (condensation or thioesterase), XH can be the NH of the first amino acid (cyclosporin A or gramicidin S) or the OH of a side-chain (enterobactin), which attacks the thioester activated carbonyl symbolized by a circle. Peptide bonds are symbolized by squares. The number of inner circle lines symbolizes the number of subunits.

The synthesis of gramicidin S, enniatin, and enterobactin is more complex as the colinearity of the peptide structure and of modular arrangement of the PPS enzyme is not conserved. Gramicidin S is the product of head-to-tail condensation of the pentapeptide synthesized by the GrsAB complex. Enniatin and enterobactin are products of the trimerization of the dipeptidols synthesized by the Esyn and EntEF enzymes, respectively. Two distinct mechanisms may be proposed for these systems. The first requires a waiting position on the enzyme to accept the initial peptide synthesized prior to the synthesis of a second one. The second mechanism requires the formation of an enzyme dimer in gramicidin S synthesis or an enzyme trimer in enniatin B and enterobactin synthesis as shown in Figure 4(b), with each terminating domain of a given subunit (thioesterase or condensation) catalyzing the formation of an amide or lactone bond. In the case of gramicidin S and enterobactin, the second model seems more likely, as a waiting domain has not been identified in the sequence of GrsB or EntF. In the case of enniatin B however, an extra ACP domain that may be a potential waiting position has been identified in the Esyn enzyme sequence.[49]

Zocher and co-workers[50] proposed that the dipeptide, tetrapeptide, or hexapeptide may be transferred to the waiting position and cyclized in a cyclization pocket. However, the mechanisms of transthiolation and lactone formation are not easily explained in this model.[49,50] If the length of the peptide is dictated by the peptide synthetase enzyme channeling and not as in PKS synthesis solely by the position of the thioesterase domain, this additional constraint has to be kept in mind when recombining peptide synthetase modules in the purpose of synthesizing new peptides.

4.09.4 MECHANISM OF EPIMERIZATION BY PPS ENZYMES

Many peptides synthesized by PPS enzymes contain D-amino acids.[1] They are all epimerized by the PPS enzyme after aminoacylation on the Pan arm,[51-54] with the exception of alanine, which is epimerized by a racemase[55] in cyclosporin A synthesis and also possibly in HC-toxin synthesis. The epimerization domains are located distal to the ACP domain of every epimerized precursor (Figure 1). From the sequence data available, two types of epimerization can be deduced: epimerization of a starter amino acid (tyrocidin or gramicidin S) or an internal amino acid (surfactin, ACV, pristinamycin I). Stein *et al.*[51] have studied gramicidin S synthetase 1 (GrsA), a prototype of a new class of epimerases with 4′-phosphopantethein cofactor dependent epimerization activity. They postulated the existence of a conserved proton acceptor group in the epimerization domain. Strong evidence for epimerization in the peptide-bound state via a carbanion intermediate has been proposed for actinomycin and ACV and the presence of an unknown proton acceptor group on the peptide synthetase enzyme has also been postulated.[52,54] Sequence analysis of the epimerization domain revealed similarities with the elongation domain at least at the N terminus.[56] The His motif is also present in the epimerization domain of peptide synthetases (Figure 5(b)). It is possible that one of the histidine residues of the His motif present in the epimerization domain may be involved in the deprotonation of the carbon atom adjacent to the activated carbonyl of the Pan bound amino acid or peptide as proposed in Figure 10. The hydroxy group of the serine, or the indole group of the tryptophan, conserved in the extended epimerization His motif are potential candidates for proton donation.

The rupture in the polypeptide chain found after each epimerization domain (except with HC-toxin synthetase; Figure 1) can be explained by the elongation model proposed above: the incoming chain is aligned with the waiting position of the next elongation domain to enable elongation to proceed. When an epimerization domain is present the distance is not conserved and the Pan arm is out of reach. Rupture of the chain may allow the flexibility necessary to correctly reposition the acyl donor and the acyl acceptor.

4.09.5 CONCLUSIONS

Analysis of sequence and biochemical data on nonribosomal peptide synthesis suggests that two distinct enzyme domains are involved in the catalysis of acyl transfer. Thioesterase domains (from the superfamily of serine hydrolases, e.g., FAS and PKS), and condensation domains, both utilize NH or OH as proton donors to catalyze amide or ester bond formation and cyclization of the final peptide. The PPS condensation domains may therefore constitute a novel family of acyltransferases. Structural data is now essential to allow the comparison of the mechanisms of acyl transfer by these enzymes and the thioesterase family.

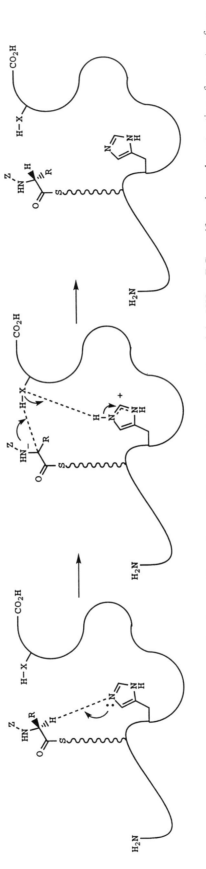

Figure 10 Proposed mechanism for epimerization by peptide synthetases. The second histidine residue of the HHxxDG motif catalyzes the extraction of a proton from the α-C of the peptide linked to the Pan, in the epimerization reaction. Abbreviations: Z, peptide (length from 0 to *n*); XH, acid moiety from a peptide synthetase residue, or water molecule held by the enzyme; wavy line, Pan cofactor.

ACKNOWLEDGMENTS

The studies which have led to the observations and conclusions discussed in this chapter were initially supported by Rhône-Poulenc-Rorer. I would like to thank William Saurin, Philippe Marlière, Denis Thibaut, Jean-Claude Barrière, Véronique Blanc, and Nathalie Bamas-Jacques for fruitful discussions and critical reading of the manuscript, and Philippe Mazodier and Georges Rapoport for their support.

4.09.6 REFERENCES

1. H. Kleinkauf and H. von Döhren, *Eur. J. Biochem*, 1990, **192**, 1.
2. K. Turgay, M. Krause, and M. A. Marahiel, *Mol. Microbiol.*, 1992, **6**, 529.
3. T. Stein and J. Vater, *Amino Acids*, 1996, **10**, 201.
4. H. Kleinhauf and H. von Döhren, *Eur. J. Biochem.*, 1996, **236**, 335.
5. F. Lipmann, *Science*, 1971, **173**, 875.
6. A. P. MacCabe, H. van Liempt, H. Palissa, S. E. Unkles, M. B. R. Riach, E. Pfeifer, H. von Döhren, and J. R. Kinghorn, *J. Biol. Chem.*, 1991, **266**, 12 646.
7. E. Conti, N. P. Franks, and P. Brick, *Structure*, 1996, **4**, 287.
8. W. Schlumbohm, T. Stein, C. Ullrich, J. Vater, M. Krause, M. A. Marahiel, V. Kruft, and B. Wittman-Liebold, *J. Biol. Chem.*, 1991, **266**, 23 135.
9. T. Stein, J. Vater, V. Kruft, B. Wittman-Liebold, P. Franke, M. Panico, R. McDowell, and H. R. Morris, *FEBS Lett.*, 1994, **340**, 39.
10. T. Stein, J. Vater, V. Kruft, A. Otto, B. Wittmannliebold, P. Franke, M. Panico, R. McDowell, and H. R. Morris, *J. Biol. Chem.*, 1996, **271**, 15 428.
11. F. Lipmann, *Adv. Enzymol.*, 1941, **1**, 99.
12. R. Wolfenden, D. Rammler, and F. Lipmann, *Biochemistry*, 1964, **3**, 329.
13. F. Lipmann, *Bacteriol. Rev.*, 1953, **17**, 1.
14. C.-H. Lee and A. F. Chen, in "The Pyridine Nucleotide Coenzymes," eds. J. Everse, B. Anderson, and K. You, Academic Press, New York, 1982, p. 189.
15. A. P. MacCabe, M. B. R. Riach, and J. R. Kinghorn, *J. Biotech.*, 1991, **17**, 97.
16. S. J. Wakil, *Biochemistry*, 1989, **28**, 4523.
17. S. Donadio, M. J. Staver, J. B. McAlpine, S. J. Swanson, and L. Katz, *Science*, 1991, **252**, 675.
18. P. Cosmina, F. Rodriguez, F. de Ferra, G. Grandi, M. Perego, G. Venema, and D. van Sinderen, *Mol. Microbiol.*, 1993, **8**, 821.
19. J. Cortes, K. E. H. Wiesmann, G. A. Roberts, M. J. B. Brown, J. Staunton, and P. F. Leadlay, *Science*, 1995, **268**, 1487.
20. D. M. Lawson, U. Derewenda, L. Serre, S. Ferri, R. Szittner, Y. Wei, E. A. Meighen, and Z. S. Derewenda, *Biochemistry*, 1994, **33**, 9382.
21. S. Brenner, *Nature*, 1988, **334**, 528.
22. V. de Crécy-Lagard, D. Thibaut, W. Saurin, P. Gil, L. Naudin, J. Crouzet, and V. Blanc, *Antimicrob. Agents Chemother.*, 1997, **41**, 1904.
23. V. de Crécy-Lagard, P. Marlière, and W. Saurin, *C. R. Acad. Sci. Paris, Life Sci.*, 1995, **318**, 927.
24. W. V. Shaw, *CRC Crit. Rev. Biochem.*, 1983, **14**, 1.
25. W. V. Shaw, *Sci. Prog. Oxford*, 1992, **76**, 565.
26. A. G. W. Leslie, *J. Mol. Biol.*, 1990, **213**, 167.
27. A. Mattevi, G. Obmolova, E. Schulze, K. H. Kalk, A. H. Westphal, A. De Kok, and W. G. J. Hol, *Science*, 1992, **255**, 1544.
28. N. F. Brown, R. C. Anderson, S. L. Caplan, D. W. Foster, and J. D. McGarry, *J. Biol. Chem.*, 1994, **269**, 19 157.
29. L. A. Carbini and L. B. Hersh, *J. Neurochem.*, 1993, **61**, 247.
30. S. Peseckis and M. D. Resh, *J. Biol. Chem.*, 1994, **269**, 30 888.
31. A. Lewendon, J. Ellis, and W. V. Shaw, *J. Biol. Chem.*, 1995, **270**, 26 326.
32. S. M. Dunn, P. C. E. Moody, A. Downie, and W. V. Shaw, *Protein Sci.*, 1996, **5**, 538.
33. C. R. H. Raetz and S. L. Roderick, *Science*, 1995, **270**, 997.
34. J. Vater, T. Stein, D. Vollenbroich, V. Kruft, B. Wittman-Liebold, P. Francke, L. Liu, and P. Zuber, *J. Protein Chem.*, 1997, **16**, 557.
35. J. Krätzschmar, M. Krause, and M. A. Marahiel, *J. Bacteriol.*, 1989, **171**, 5422.
36. F. de Ferra, F. Rodriguez, O. Tortora, C. Tosi, and G. Grandi, *J. Biol. Chem.*, 1997, **272**, 25 304.
37. A. Lawen and R. Zocher, *J. Biol. Chem.*, 1990, **265**, 11 355.
38. R. Zocher, U. Keller, and H. Kleinkauf, *Biochemistry*, 1982, **21**, 43.
39. T. Schwecke, J. F. Aparicio, I. Molnar, A. König, L. E. Khaw, S. F. Haydock, M. Oliynyk, P. Caffrey, J. Cortes, J. B. Lester, G. A. Bohm, J. Staunton, and P. F. Leadlay, *Proc. Natl. Acad. Sci. USA*, 1995, **92**, 7839.
40. L. J. Libertini and S. Smith, *J. Biol. Chem.*, 1978, **253**, 1393.
41. C. Y. Lin and S. Smith, *J. Biol. Chem.*, 1978, **253**, 1954.
42. V. de Crécy-Lagard, V. Blanc, P. Gil, L. Naudin, S. Lorenzon, A. Famechohn, and N. Bamas-Jacques, *J. Bacteriol.*, 1997, **179**, 705.
43. A. Witkowski, H. E. Witkowska, and S. Smith, *J. Biol. Chem.*, 1994, **269**, 379.
44. D. H. White and J. C. Erickson, *J. Mol. Evol.*, 1980, **16**, 279.
45. W. Saurin, unpublished results.
46. M. Mathur and P. Kolattukudy, *J. Biol. Chem.*, 1992, **267**, 19 388.
47. A. Vercellone and G. Puzo, *J. Biol. Chem.*, 1989, **264**, 7447.
48. J. Dittmann, R. M. Wenger, H. Kleinkauf, and A. Lawen, *J. Biol. Chem.*, 1994, **269**, 2841.

49. A. Haese, M. Schubert, M. Herrmann, and R. Zocher, *Mol. Microbiol.*, 1993, **7**, 905.
50. H. von Döhren, U. Keller, J. Vater, and R. Zocher, *Chem. Rev.*, 1997, **97**, 2675.
51. T. Stein, B. Kluge, J. Vater, P. Franke, A. Otto, and B. Wittmann-Liebold, *Biochemistry*, 1995, **34**, 4633.
52. A. Stindl and U. Keller, *Biochemistry*, 1994, **33**, 9358.
53. T. Stachelhaus and M. A. Marahiel, *J. Biol. Chem.*, 1995, **270**, 6163.
54. C.-Y. Shiau, J. E. Baldwin, M. F. Byford, W. J. Sobey, and C. J. Schofield, *FEBS Lett.*, 1995, **358**, 97.
55. K. Hoffman, E. Schneider-Scherzer, H. Kleinkauf, and R. Zocher, *J. Biol. Chem.*, 1994, **269**, 12 710.
56. S. Fuma, Y. Fujishima, N. Corbell, C. D'Souza, M. M. Nakano, P. Zuber, and K. Yamane, *Nucleic Acids Res.*, 1993, **21**, 93.

4.10
Enzymatic Synthesis of Penicillins

JOSÉ M. LUENGO
Universidad de León, Spain

4.10.1   GENERAL INTRODUCTION

Penicillins are natural products belonging to a special group of compounds, the β-lactam antibiotics, of great clinical, pharmacological, and economical importance.[1-7] This group includes different families of molecules (nocardicins, monobactams, clavulanic acid, carbapenems, cephalosporins, cephamycins, and penicillins) (Figure 1) characterized by two common properties: (i) the presence in their structure of a four-membered ring, the β-lactam ring, which gives its name to this type of compounds and (ii) the antibacterial activity shown by most of them against different microbes.[3,8-13] Most β-lactam antibiotics are produced by *Streptomyces* species as well as by other prokaryotes belonging to the genera *Nocardia*, *Lysobacter*, *Xanthomonas*, *Serratia*, *Erwinia*, *Pseudomonas*, *Agrobacter*, *Gluconobacter*, *Chromobacter*, *Acetobacter*, and *Flavobacterium*. Eukaryotic microbes also able to synthesize these compounds are *Emericellopsis*, *Epidermophyton*, *Trycophyton*, *Polypaecilum*, *Sartorya*, *Spiroidium*, *Scopulariopsis*, *Anixiopsis*, *Malbranchea*, *Pleurophomopsis*, *Arachnomyces*, *Diheterospora*, *Pacilomyces*, *Aspergillus*, *Acremonium*, and *Penicillium*, *Penicillium chrysogenum* and *Acremonium chrysogenum* being the most important producers.[6,7,14,15] Through the use of empirical mutation techniques, overproducer mutants of *A. chrysogenum* and *P. chrysogenum* have been isolated. In submerged cultures these reach high cephalosporin C (30 000 IU ml^{-1}) or benzylpenicillin (about 80 000 IU ml^{-1}) titres.[16,17] In spite of this considerable success, new approaches are currently required to improve the industrial production of these antibiotics. The limitation in the biosynthetic capacity of the strains, the engineering problems arising from the need to apply new fermentation technologies, and the appearance of microbial resistances has forced many scientists to change older research strategies, based on traditional mutation procedures, in order to achieve alternative and more imaginative solutions. Chemists, biochemists, microbiologists, and molecular biologists have chosen different industrial strains as models of study, opening new perspectives to the industry of antibiotic fermentation. Thus, advances in protein technology and the application of genetic engineering to overproducer strains have contributed to our understanding of the biosynthetic steps involved in the synthesis of different antibiotics as well as to establishing some of the regulatory mechanisms which control these pathways.[6,7,18-23] This has led to well-founded hopes that the biosynthetic capacity of many strains can be altered to overproduce new or modified molecules showing: (i) different physicochemical properties; (ii) broader antibacterial activity; (iii) higher resistance to enzymatic attacks; and (iv) less harmful secondary effects.

This chapter describes the most important features achieved with one of these new approaches: the enzymatic synthesis of penicillins. How proteins purified from different microbial origins could be used to obtain different penicillins *in vitro* is discussed.

4.10.2   PENICILLINS: STRUCTURE AND GENERAL PROPERTIES

All molecules of penicillins contain in their structure two fused rings, the β-lactam and the thiazolidine rings (Figure 2), and a variable side chain specific to each of these[3,24,25] (R in Figure 2). When R is a hydrogen atom, the compound obtained—6-aminopenicillanic acid (6-APA)—is considered the simplest penicillin. This molecule, usually considered as a biosynthetic intermediate, can also be accumulated as a final product when *P. chrysogenum* is cultured in fermentation media lacking penicillin side-chain precursors.[26,27] In spite of its low antibacterial activity, 6-APA is a very important product since it is employed for the chemical synthesis of different β-lactam antibiotics (semisynthetic penicillins and cephalosporins). The industrial production of 6-APA is achieved either by chemical synthesis or by enzymatic digestion of benzylpenicillin with *Escherichia coli* penicillin amidase.[28,29] In both processes, yields are near 90%.

Most penicillins are white or pale-yellow powders that are very stable in the absence of humidity, but are sensitive to mild acid and alkali treatments (see Scheme 1). Their sodium and potassium salts are very soluble in water and they are readily absorbed. However, the free acids are extremely unstable in aqueous solution and are therefore of little pharmacological interest. Procaine, benzathine, and other penicillin salts are slightly soluble in water. Their slow absorption results in low serum levels of antibiotics over prolonged periods of time.[5,30]

The antibacterial activity of penicillins is facilitated by the interference of these antibiotics in the synthesis of peptidoglycan.[31,32] Penicillin acylates the transpeptidase, inactivating the enzyme and making it unable to form the cross-linking bond between two linear units of polymer.[33,34] Under these conditions, bacteria synthesize abnormal cell walls that are unable to protect them from osmotic changes, meaning that they are dramatically affected by many different environmental variables.[35,36]

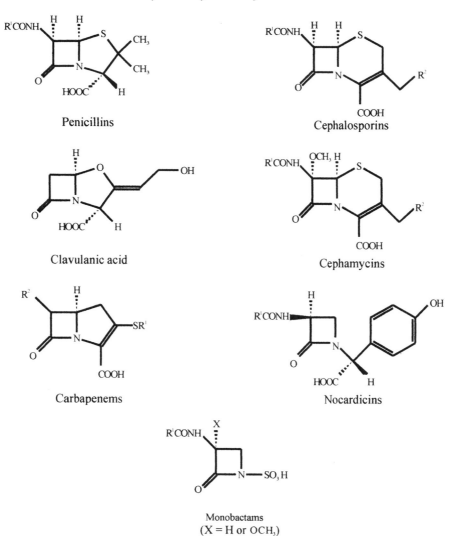

Figure 1 Structure of the different families of naturally occurring β-lactam antibiotics.

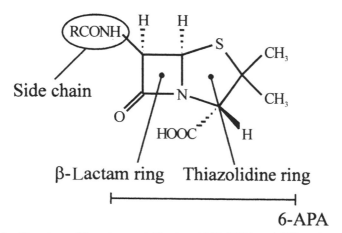

Figure 2 Structure of 6-aminopenicillanic acid (6-APA) and different penicillins.

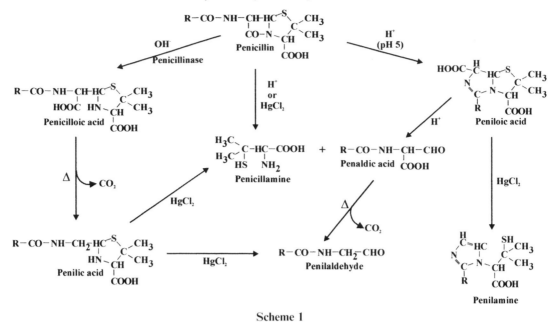

Scheme 1

4.10.3 CLASSIFICATION OF PENICILLINS

Depending on the nature of the side chain present on the molecule of penicillin, these antibiotics are divided in two classes: (i) hydrophobic penicillins (those containing a nonpolar side chain) and (ii) hydrophilic penicillins (with polar side chains)[37] (Figure 3). Hydrophilic penicillins are biosynthetic intermediates which are not generally accumulated in the fermentation broths, whereas hydrophobic penicillins are final biosynthetic products. It is interesting to note that whereas bacterial penicillins are hydrophilic, fungal penicillins may be hydrophobic or hydrophilic.[7,37] Both kinds of penicillin have another important microbiological difference. Hydrophobic penicillins have potent antibacterial activity and a broad antibacterial spectrum, whereas hydrophilic penicillins show lower antibiotic potency and a narrower spectrum.[5,7,14]

Further information about the characteristics of penicillins, the possible role of these secondary metabolites, as well as their ecological implications, can be found in reviews elsewhere.[6,7,38,39]

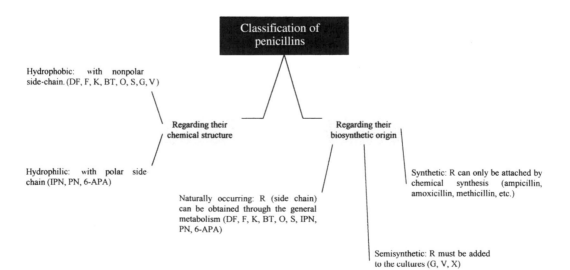

Figure 3 Classification of penicillins.

4.10.4 BENZYLPENICILLIN BIOSYNTHETIC PATHWAY

The biosynthetic pathway leading to benzylpenicillin and L-lysine in *P. chrysogenum* (Scheme 2; A, common steps; B, specific routes) is a common route which starts with the condensation of acetyl-CoA and α-ketoglutarate to afford homocitrate (HA), the first intermediate of the route.[40] The enzyme catalyzing this reaction, homocitrate synthase (HS), is inhibited and repressed by L-lysine, this being a critical regulatory point of the pathway.[41–46] HA is later transformed in L-α-aminoadipic acid in a series of reactions similar to those reported for the citric acid cycle.[47] In the specific branch of L-lysine, L-α-aminoadipic acid is activated to α-aminoadipyl-AMP, reduced, and then finally hydrolyzed to generate α-aminoadipic acid-δ-semialdehyde.[40,47] In a second step this molecule is condensed with L-glutamic acid to form saccharopine. Finally, the enzyme saccharopine dehydrogenase (SD) catalyzes the transformation of saccharopine into L-lysine by releasing a molecule of α-ketoglutarate. SD is another important regulatory point since its activity is modulated by L-lysine as in the case of HS.[47] Therefore, in the presence of an excess of L-lysine, the metabolic flux through this branch is blocked and all the L-α-aminoadipic acid synthesized is utilized for the production of penicillin.[43,46]

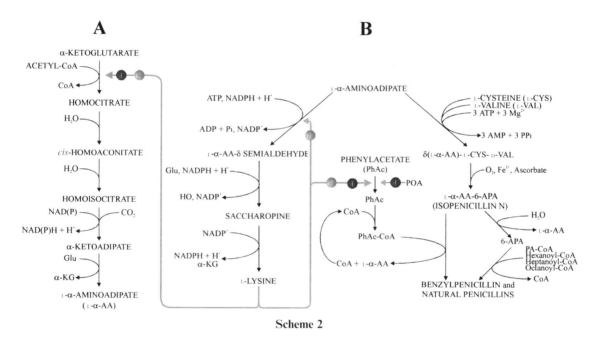

Scheme 2

In the first step of the specific pathway of benzylpenicillin (see Scheme 2), L-α-aminoadipic acid is nonribosomally condensed with two other amino acids, L-cysteine and L-valine, into a linear tripeptide—L-α-aminoadipyl-L-cysteinyl-D-valine (ACV)—which is the common precursor of penicillins, cephalosporins, and cephamycins.[2,14,48–50] This reaction is catalyzed by the enzyme δ-(L-α-aminoadipyl)-L-cysteinyl-D-valine synthetase (ACVS).[18,51–54]

In a second step, the L-cysteinyl-D-valine moieties present in the molecule of ACV are cyclized to generate the molecule of isopenicillin N (IPN).[19] This compound, which is the first antibiotic in the pathway, contains the common structure of all penicillins (the β-lactam and the thiazolidine rings), with the L-α-aminoadipic acid molecule remaining as a side chain. The enzyme that catalyzes this reaction is named isopenicillin N synthase (IPNS) and was the first protein of the pathway that was purified to homogeneity[55] and whose gene has been cloned and sequenced.[56] IPN is a hydrophilic penicillin having low antibacterial activity. It is therefore active against gram positive bacteria but not effective against gram negative bacteria.[2,24]

In *P. chrysogenum* and in *Aspergillus nidulans* IPN is finally transformed into hydrophobic penicillins by exchanging the α-aminoadipic moiety present in its molecule for other side chains, thus generating different products (see below) with stronger antibiotic potency and broader antibacterial spectra.[5,57] The enzyme that catalyzes this reaction is acyl-CoA:6-APA (IPN) acyl-transferase (AT).[58–61] This protein has only been identified in fungal strains producing hydrophobic penicillins, but has not been found either in bacteria or in eukaryotic microbes, which only produce hydrophilic penicillins.[15,62]

The biosynthesis of benzylpenicillin (penicillin G) requires two additional enzymes. First, a permease is needed to take up phenylacetic acid (PhAc) (the side-chain precursor of benzylpenicillin) since *P. chrysogenum* is unable to synthesize this aromatic acid. The PhAc transport system (PhAcTS) has been studied and characterized in two members of the class Plectomycetes (*P. chrysogenum* and *A. nidulans*) able to make β-lactam antibiotics.[63–66] In both cases it seems to be a critical regulatory point that effectively modulates the metabolic flux of the biosynthesis intermediates[65,66] (Figure 11).

A second enzyme, phenylacetyl-CoA ligase (PhAcCoAL), seems to be involved in the activation of PhAc to phenylacetyl-CoA (PhAc-CoA).[67,68] Currently, it is not clear whether a specific enzyme catalyzes this reaction or whether, by contrast, different acyl-CoA-activating enzymes could unspecifically catalyze the activation of PhAc to PhAc-CoA. In any case, the work carried out until now suggests that the formation of PhAc-CoA is a true limiting step in the biosynthetic pathway of benzylpenicillin (see below).

The different steps involved in the biosynthetic pathway of hydrophobic penicillins as well as the enzymes that catalyze them are summarized in Scheme 2.

4.10.5   LOCATION OF PENICILLIN BIOSYNTHETIC ENZYMES

Since the original discovery of penicillin,[1] many ultrastructural studies of *P. chrysogenum* have been carried out in attempts to pinpoint the exact location of the different penicillin biosynthetic enzymes in this fungus.[69–73] It is now generally accepted that penicillin biosynthesis is a complex process requiring the participation of different organelles. For example, the PhAc transport system seems to be a particulate enzyme located in the plasma membrane[63] whereas PhAcCoAL, as well as other acyl-CoA-activating enzymes involved in the activation of aromatic and medium chain fatty acids, are soluble cytosolic enzymes.[67,68,74,75] ACVS seems to be bound to membranes of Golgi vesicles in *P. chrysogenum*, although it is also found associated with membrane preparations of other related small organelles.[76–78] Immunological studies have also shown that a certain portion of ACV could also be located inside vacuoles.[79] However, it is still not clear whether this fraction corresponds to functional enzymes or to proteins that are being catabolized. Using immunoelectron fluorescence microscopy, Kurylowicz *et al.*[80] reported the existence of IPN in vesicular compartments belonging to the Golgi apparatus, although other authors have indicated that this enzyme has a wide cytosolic distribution.[76]

Immunoprecipitation analysis has also allowed the identification of AT in microbody-like organelles.[76] As reported for ACVS,[76–78] AT has also been found as a free soluble enzyme in the cytosol.[22] It is still not clear whether the different location of those enzymes corresponds to a complex subcellular organization of the pathway or whether, by contrast, the soluble enzymes are a consequence of the disruption of the different organelles during the fractionation process.[76] In any case, the distribution among different organelles of a single pathway suggests that many other unknown proteins (permeases, binding proteins, and so forth) may be required to facilitate the transference of substrates (or biosynthetic intermediates) from one to another. This complicated organization could be simplified if, as has been suggested, several penicillin precursors are synthesized in a single organelle.[14,15] The demonstration that β-oxidation occurs in certain fungal microbodies[81] suggests that some acyl-CoA derivatives, which can be used as precursors for different hydrophobic penicillins (G, V, F, DF, K), may be synthesized in these organelles. If this were so, the location of AT in microbodies would facilitate the exchange reaction between the L-α-aminoadipic acid present in IPNS and the acyl moiety of these acyl-CoA variants. In summary, although considerable advances have been made regarding the location of penicillin biosynthetic enzymes, many important points (particularly those related to the transference of biosynthetic intermediates) remain to be elucidated and will be addressed in forthcoming work.

4.10.6   PENICILLIN BIOSYNTHETIC GENES

Genetic studies on the biosynthetic pathway of penicillins and cephalosporins have received considerable impetus since the late 1980s and some important progress has been made.[6,15,82–84] The genes (*pcb*AB, *pcb*C and *pen*DE) encoding the three specific benzylpenicillin biosynthetic enzymes (ACVS, IPNS, and AT) in *P. chrysogenum* are linked in a single cluster[85–87] (Figure 4). The open reading frame (ORF) for the *pcb*AB gene (encoding ACVS) contains 11 376 nucleotides and starts 1 kb upstream from the IPNS gene.[85] The *pcb*C gene, which encodes IPNS, has been isolated from

A. chrysogenum and is the first gene of the pathway that has been cloned, sequenced, and expressed in different microbes.[56] It has been routinely used as a probe to identify DNA fragments encoding other IPNS in different β-lactam producer microorganisms.[88–94] In *P. chrysogenum* the *pcb*C gene contains 996 nucleotides.

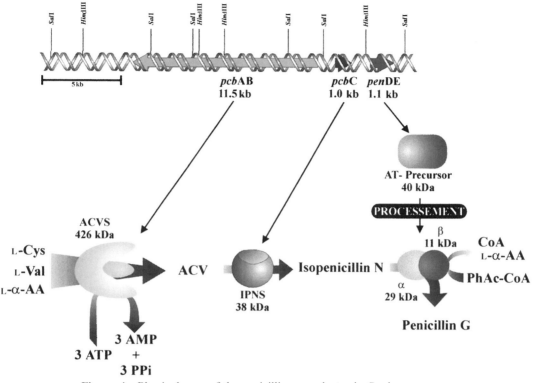

Figure 4 Physical map of the penicillin gene cluster in *P. chrysogenum*.

It is generally accepted that both genes (*pcb*AB and *pcb*C) are divergently transcribed from a short intergenic region.[95–97] However, some doubt persists as to the transcriptional starting point of the *pcb*C gene in *P. chrysogenum*.[93,96,97] Different nuclear proteins that interact with this intergenic promoter region have been identified and characterized.[98–100] The most abundant one (named nuclear factor A) binds at a single site of the promoter and recognizes the sequence: GCCAAGCC.[98] Another specific protein, which seems to be a transcriptional activator (appearing in *P. chrysogenum* crude extracts at the same time as *pcb*AB mRNA), interacts with a definite DNA region (a 7 bp motif TGCCAAG) located -387 to -242 relative to the *pcb*AB translational start codon.[99] Investigation into the role played by this intergenic region in the regulation of the gene expression has been carried out by Haas and Marzluf,[100] who have shown that the major nitrogen regulatory protein of *P. chrysogenum* (NR) binds to specific sequences present in this region.

The third gene, *pen*DE, encoding acyl-CoA:6-APA (IPN) AT, was cloned by Barredo *et al.*[101] This gene, with 1274 nucleotides, is located 3′ to the *pcb*C gene in *P. chrysogenum*.[85] The close linkage of these three genes (see Figure 4) suggests the existence of a position-dependent mechanism of co-regulation of gene expression, as has been reported in other fungal clusters.[102]

Furthermore, all three genes (*pcb*AB, *pcb*C, and *pen*DE) have different chromosomal locations. Thus, they map on chromosome I in *P. chrysogenum*, and on chromosomes II and VI in *Penicillium notatum* and in *A. nidulans*, respectively.[103] Conversely, the two additional genes required for benzylpenicillin biosynthesis, those encoding PhAcCoAL and PhAc permease, have not been found linked to the other biosynthetic genes. This result suggests that they either encode enzymes involved in general metabolism (which could be unspecifically used for the production of benzylpenicillin) or that, although specific, they require a different subcellular location.

From an evolutionary point of view, it seems obvious that the first three genes (*pcb*AB, *pcb*C, and *pen*DE) have different origins.[6,15,90] Thus, whereas *pcb*AB and *pcb*C are common to many β-lactam antibiotic-producing microbes and seem to have a prokaryotic origin (their appearance in eukaryotes could be caused by horizontal gene transference),[104,105] the third gene (*pen*DE) has a typical eukaryotic structure:[101] it contains a percentage of G + C lower than those of the other two and it has three introns in the 5′ region of the ORF. Moreover, *pen*DE seems to be specific for the biosynthesis of hydrophobic penicillins since it is lacking in prokaryotes and also in cephalosporin- or cephamycin-producing microbes.

Genetic analyses carried out in different industrial strains of *P. chrysogenum* have shown that the DNA fragment containing the *pcb*AB, *pcb*C, and *pen*DE genes is amplified in tandem repeats linked by conserved TTTACA sequences.[89,106] Penicillin production seems to be related to the number of copies of these genes in the genomes of the different overproducer mutants.[89,106–108]

4.10.7 BENZYLPENICILLIN BIOSYNTHETIC ENZYMES

In this section, the most important characteristics of the different enzymes involved in the synthesis of penicillin G in *P. chrysogenum* are described. In each particular case, the properties of the enzyme, the reaction catalyzed by it, the optimal physicochemical conditions of assay, and its enzymatic requirements are focused on. In those cases where the mechanism of catalysis has been established, this is indicated. Substrate specificity is also analyzed, placing special emphasis on such cases in which the reaction products have antibacterial activity. The advances achieved in penicillin synthesis by the use of enzymatic coupling systems, which have permitted the *in vitro* reproduction of the last steps involved in benzylpenicillin biosynthesis, are analyzed. Although as a general rule the results obtained from using enzymes isolated from β-lactam-producing microorganisms are discussed, in other cases, there is analysis of the data obtained when enzymes purified from other microbial sources (PhAcCoAL from *Pseudomonas putida* U and different acyl-CoA synthetases) are coupled *in vitro* with acyl-CoA:6-APA (IPN) AT from *P. chrysogenum*. Finally, as a new biotechnological approach, the effects that cloning of some of these genes in *P. chrysogenum* have had on benzyl-penicillin production are discussed.

4.10.7.1 δ-(L-α-Aminoadipyl)-L-Cysteinyl-D-Valine Synthetase (ACVS)

4.10.7.1.1 General characteristics

The first enzyme in the biosynthetic pathway of penicillins, cephalosporins, and cephamycins—ACVS—has been studied in different β-lactam-producing microbes.[18,20,51–54,94,95] Its molecular mass ranges between 424 kDa in *P. chrysogenum*[85] and 283 kDa in *Streptomyces clavuligerus*.[108,109] The *P. chrysogenum* enzyme has three different independent domains with conserved amino acid sequences separated by unrelated regions[87] (Figure 5). All three domains correspond to individual sites involved in the ATP-activation of each L-amino acid (L-α-aminoadipic acid, L-cysteine, and L-valine). This organization is similar to that reported for other adenylating enzymes as well as for many multi-enzyme peptide synthetases,[110–112] suggesting that the genes encoding these proteins have evolved from a common ancestor.[6,15,103,105,113]

4.10.7.1.2 ACVS reaction

ACVS catalyzes the nonribosomal condensation of L-α-aminoadipic acid, L-cysteine, and L-valine into a linear tripeptide molecule, ACV, according to the equation shown in Figure 6. van Liempt *et al.*[20] suggested that ACV synthesis might occur through a thiotemplate mechanism and that the binding of acyl intermediates could occur through pantethine residues[60] (Figure 6). Maximal rates of catalysis were achieved when the reaction mixtures were incubated in a temperature range between 24 °C and 26 °C in buffer solutions with pH values ranging between 7.5 and 8.2. Kadima *et al.*[114] have studied the immobilization of ACVS from *S. clavuligerus* in a reactor, reporting that this enzyme was stimulated by phosphoenol pyruvate at lower concentrations of ATP, whereas it was inhibited in the presence of an excess of AMP.

4.10.7.1.3 Substrate specificity and mechanistic studies

ACVS was first assayed by Fawcett and Abraham[51] using a partially purified preparation obtained from *A. chrysogenum*. These authors reported that the enzyme catalyzed the *in vitro* synthesis of

```
1     GACCAATGCAGCAGGCCCAGTATAAGGAATTCCCCTCGAGCTTGTCTGTGATTGCGTTTT
61    TTCTAACACTTGTTGTTGCATCCGATCCGTCCCTACCAATTATTGGTCATTGACAGACAT
121   GACTCAACTGAAGCCACCGAACGGAACCACGCCGATAGGCTTCTCGGCCACTACATCCCT
181   GAACGCCAGTGGGAGCTCGAGTGTGAAAAATGGGACCATCAAACCCAGCAATGGCATACT
                                            M  G  P  S  N  P  A  M  A  Y
241   TCAAGCCCAGCACTAGGGACACCATGGACCCTTGCAGTGGGAATGCGGCCGATGGCAGTA
      F  K  P  S  T  R  D  T  M  D  P  C  S  G  N  A  A  D  G  S
301   TCCGCGTCCGTTTCCGTGGAGGAATCGAACGGTGGAAGGAGTGCGTCAACCAGGTCCCCG
      I  R  V  R  F  R  G  G  I  E  R  W  K  E  C  V  N  Q  V  P
361   AGCGCTGCGACCTGAGTGGTCTGACAACCGACTCCACGCGATATCAGCTCGCATCGACTG
      E  R  C  D  L  S  G  L  T  T  D  S  T  R  Y  Q  L  A  S  T
421   GGTTCGGTGACGCGAGCGCTGCCTACCAGGAGCGCTTGATGACGGTCCCTGTTGACGTAC
      G  F  G  D  A  S  A  A  Y  Q  E  R  L  M  T  V  P  V  D  V
481   ATGCCGCGCTCCAAGAGCTGTGCCTAGAACGCCGTGTGAGCGTGGGATCCGTCATTAATT
      H  A  A  L  Q  E  L  C  L  E  R  R  V  S  V  G  S  V  I  N
541   TCTCCGTGCACCAGATGCTGAAAGGGTTTGGAAATGGCACACACACTATCACCGCCTCTC
      F  S  V  H  Q  M  L  K  G  F  G  N  G  T  H  T  I  T  A  S
601   TGCACCGTGAGCAGAATTTGCAGAATTCTTCGCCATCCTGGGTAGTCTCCCCCACAATCG
      L  H  R  E  Q  N  L  Q  N  S  S  P  S  W  V  V  S  P  T  I
661   TCACCCATGAGAACAGAGACGGATGGTCCGTCGCGCAGGCGGTCGAGAGTATCGAAGCGG
      V  T  H  E  N  R  D  G  W  S  V  A  Q  A  V  E  S  I  E  A
721   GGCGCGGTTCCGAGAAGGAGTCAGTGACTGCGATTGACTCCGGGTCAAGTCTCGTGAAAA
      G  R  G  S  E  K  E  S  V  T  A  I  D  S  G  S  S  L  V  K
781   TGGGGTTATTTGACTTACTCGTCAGCTTTGTCGATGCAGACGATGCTCGTATTCCATGTT
      M  G  L  F  D  L  L  V  S  F  V  D  A  D  D  A  R  I  P  C
841   TCGACTTTCCCCTCGCAGTGATAGTGCGTGAGTGTGATGCCAACCTCTCGCTGACTCTGC
      F  D  F  P  L  A  V  I  V  R  E  C  D  A  N  L  S  L  T  L
901   GTTTCTCCGACTGTCTCTTCAACGAGGAGACGATATGCAATTTTACCGATGCCCTAAACA
      R  F  S  D  C  L  F  N  E  E  T  I  C  N  F  T  D  A  L  N
961   TCTTGCTCGCCGAAGCAGTGATAGGAAGAGTGACCCCGGTTGCCGATATCGAACTACTAT
      I  L  L  A  E  A  V  I  G  R  V  T  P  V  A  D  I  E  L  L
1021  CCGCGGAGCAGAAGCAGCAGCTGGAAGAGTGGAACAACACGGATGGCGAGTACCCTTCAT
      S  A  E  Q  K  Q  Q  L  E  E  W  N  N  T  D  G  E  Y  P  S
1081  CAAAGCGACTGCACCATCTCATTGAAGAGGTGGTTGAACGGCATGAAGACAAAATAGCCG
      S  K  R  L  H  H  L  I  E  E  V  V  E  R  H  E  D  K  I  A
1141  TTGTCTGCGACGAGCGAGAGCTCACTTACGGCGAGCTCAATGCCCAAGGCAACAGCCTCG
      V  V  C  D  E  R  E  L  T  Y  G  E  L  N  A  Q  G  N  S  L
1201  CACGCTATCTCCGTTCCATTGGTATCCTGCCCGAGCAGCTAGTCGCATTGTTTCTAGATA
      A  R  Y  L  R  S  I  G  I  L  P  E  Q  L  V  A  L  F  L  D
1261  AGAGCGAGAAGCTCATTGTTACCATCCTCGGCGTGTGGAAATCCGGCGCCGCCTACGTGC
      K  S  E  K  L  I  V  T  I  L  G  V  W  K  S  G  A  A  Y  V
1321  CCATCGACCCGACTTATCCGGATGAGCGAGTGCGCTTCGTGCTGGATGACACCAAGGCAC
      P  I  D  P  T  Y  P  D  E  R  V  R  F  V  L  D  D  T  K  A
1381  GGGCCATCATCGCCAGTAATCAACATGTGGAGAGGCTCCAGCGAGAGGTCATCGGCGATA
      R  A  I  I  A  S  N  Q  H  V  E  R  L  Q  R  E  V  I  G  D
1441  GAAACCTATGCATTATCCGTCTGGAGCCCTTGTTGGCCTCCCTTGCTCAGGATTCCTCAA
      R  N  L  C  I  I  R  L  E  P  L  L  A  S  L  A  Q  D  S  S
1501  AATTCCCCGCGCATAACTTGGACGACCTACCCCTCACAAGCCAGCAGCTCGCCTATGTGA
      K  F  P  A  H  N  L  D  D  L  P  L  T  S  Q  Q  L  A  Y  V
1561  CTTACACCTCTGGGACCACTGGCTTCCCAAAGGGCATATTTAAACAACACACCAATGTGG
      T  Y  T  S  G  T  T  G  F  P  K  G  I  F  K  Q  H  T  N  V
1621  TGAACAGTATCACCGACCTGTCTGCAAGGTACGGGGTGGCCGGGCAGCACCACGAAGCCA
      V  N  S  I  T  D  L  S  A  R  Y  G  V  A  G  Q  H  H  E  A
1681  TTCTGCTTTTCTCGGCCTGCGTGTTCGAGCCGTTCGTTCGACAGACGCTCATGGCACTCG
      I  L  L  F  S  A  C  V  F  E  P  F  V  R  Q  T  L  M  A  L
1741  TGAATGGCCATCTCCTCGCAGTTATCAATGACGTGGAAAAATATGATGCCGATACGCTCC
      V  N  G  H  L  L  A  V  I  N  D  V  E  K  Y  D  A  D  T  L
1801  TGCCGTTCATACGCAGACACAGCATCACCTACCTCAATGGTACTGCCTCTGTCCTGCAAG
      L  P  F  I  R  R  H  S  I  T  Y  L  N  G  T  A  S  V  L  Q
1861  AGTACGACTTTTCCGACTGCCCATCACTGAATCGGATAATCCTGGTGGGTGAGAACCTGA
      E  Y  D  F  S  D  C  P  S  L  N  R  I  I  L  V  G  E  N  L
1921  CAGAAGCCCGGTATCTGGCGCTGCGCCAGCGGTTCAAGAATCGCATCCTCAACGAGTATG
      T  E  A  R  Y  L  A  L  R  Q  R  F  K  N  R  I  L  N  E  Y
1981  GTTTTACCGAGTCAGCCTTTGTAACGGCCCTCAAGATTTTCGACCCGGAGTCGACCCGTA
      G  F  T  E  S  A  F  V  T  A  L  K  I  F  D  P  E  S  T  R
2041  AGGACACGAGTCTGGGGAGACCGGTGCGCAACGTCAAGTGCTACATCCTCAATCCATCCC
      K  D  T  S  L  G  R  P  V  R  N  V  K  C  Y  I  L  N  P  S
2101  TTAAACGTGTCCCGATTGGAGCTACGGGTGAGTTGCATATCGGAGGGTTGGGCATTTCCA
      L  K  R  V  P  I  G  A  T  G  E  L  H  I  G  G  L  G  I  S
2161  AGGGATACCTCAACCGCCCCGAACTCACGCCGCACCGCTTCATTCCCAACCCCTTCCAAA
      K  G  Y  L  N  R  P  E  L  T  P  H  R  F  I  P  N  P  F  Q
2221  CGGATTGCGAGAAGCAGCTCGGGATCAACAGCTTGATGTACAAGACCGGTGACCTGGCCC
      T  D  C  E  K  Q  L  G  I  N  S  L  M  Y  K  T  G  D  L  A
2281  GCTGGCTTCCGAACGGCGAGGTTGAGTATCTCGGACGCGCAGATTTCCAGATCAAACTGC
      R  W  L  P  N  G  E  V  E  Y  L  G  R  A  D  F  Q  I  K  L
2341  GAGGTATTCGAATTGAACCTGGTGAAATTGAGACGATGCTGGCTATGTACCCTAGGGTCC
      R  G  I  R  I  E  P  G  E  I  E  T  M  L  A  M  Y  P  R  V
2401  GGACCAGTTTAGTGGTGTCCAAAAAGCTCCGCAACGGTCCAGAGGAAACTACCAACGAGC
      R  T  S  L  V  V  S  K  K  L  R  N  G  P  E  E  T  T  N  E
2461  ACCTCGTGGGTTATTATGTTTGTGATAGCGCCTCAGTGTCCGAGGCAGACCTGCTGTCAT
      H  L  V  G  Y  Y  V  C  D  S  A  S  V  S  E  A  D  L  L  S
2521  TTTTAGAGAAGAAACTGCCTCGATACATGATTCCCACGCGGTTGGTACAGCTGTCGCAGA
      F  L  E  K  K  L  P  R  Y  M  I  P  T  R  L  V  Q  L  S  Q
2581  TCCCAGTGAATGTGAACGGGAAGGCGGACCTACGCGCCTTGCCGGCCGTCGATATCTCCA
      I  P  V  N  V  N  G  K  A  D  L  R  A  L  P  A  V  D  I  S
2641  ATTCCACGGAGGTCCGTTCCGACCTTCGAGGCGATACGGAAATCGCCCTCGGGGAAATCT
      N  S  T  E  V  R  S  D  L  R  G  D  T  E  I  A  L  G  E  I
2701  GGGCCGACGTGTTGGGAGCCCGCCAGAGATCCGTCTCTCGCAACGACAACTTCTTCCGCC
      W  A  D  V  L  G  A  R  Q  R  S  V  S  R  N  D  N  F  F  R
2761  TAGGAGGGCACAGCATCACCTGCATCCAACTGATCGCTCGCATCCGACAACGACTCTCGG
      L  G  G  H  S  I  T  C  I  Q  L  I  A  R  I  R  Q  R  L  S
2821  TCAGCATCTCCGTCGAAGATGTTTTTGCAACAAGGACACTTGAGCGCATGGCAGACCTTC
      V  S  I  S  V  E  D  V  F  A  T  R  T  L  E  R  M  A  D  L
2881  TACAGAACAAGCAGCAGGAGAAATGCGACAAACCCCATGAGGCGCCGACAGAGCTGCTTG
      L  Q  N  K  Q  Q  E  K  C  D  K  P  H  E  A  P  T  E  L  L
2941  AGGAGAATGCAGCAACGGACAATATCTATCTGGCAAACAGTCTTCAGCAGGGCTTCGTCT
      E  E  N  A  A  T  D  N  I  Y  L  A  N  S  L  Q  Q  G  F  V
3001  ACCATTACCTCAAGAGCATGGAACAATCCGACGCCTATGTAATGCAGTCCGTTCTTCGGT
      Y  H  Y  L  K  S  M  E  Q  S  D  A  Y  V  M  Q  S  V  L  R
3061  ACAACACCACATTGTCTCCAGATCTGTTTCAGAGAGCCTGGAAGCATGCACAGCAGTCCT
      Y  N  T  T  L  S  P  D  L  F  Q  R  A  W  K  H  A  Q  Q  S
3121  TTCCAGCGCTGCGGCTGCGGTTCTCATGGGAAAAGGAGGTTTTCCAACTGCTCGATCAGG
      F  P  A  L  R  L  R  F  S  W  E  K  E  V  F  Q  L  L  D  Q
3181  ATCCACCATTGGACTGGCGTTTCCTCTACTTCACCGACGTTGCCGCGGGTGCTGTCGAGG
      D  P  P  L  D  W  R  F  L  Y  F  T  D  V  A  A  G  A  V  E
3241  ACCGGAAATTGGAAGACTTGCGGCGCCAAGACCTTACGGAGAGATTCAAGCTGGATGTTG
      D  R  K  L  E  D  L  R  R  Q  D  L  T  E  R  F  K  L  D  V
3301  GCAGACTGTTCCGCGTCTATCTGATTAAACACAGCGAGAATCGCTTCACGTGTCTTTTCA
      G  R  L  F  R  V  Y  L  I  K  H  S  E  N  R  F  T  C  L  F
3361  GCTGCCACCATGCAATCCTCGATGGTTGGAGTCTGCCACTCTTGTTCGAAAAGGTTCACG
      S  C  H  H  A  I  L  D  G  W  S  L  P  L  L  F  E  K  V  H
3421  AGACCTACCTGCAACTGCTGCATGGGGACAATCTCACTTCGTCCATGGATGACCCTTACA
      E  T  Y  L  Q  L  L  H  G  D  N  L  T  S  S  M  D  D  P  Y
3481  CTCGCACCCAGCGGTATCTCCACGCTCACCGTGAGGATCACCTCGACTTTTGGGCCGGTG
      T  R  T  Q  R  Y  L  H  A  H  R  E  D  H  L  D  F  W  A  G
3541  TGGTTCAAAAGATCAACGAACGGTGTGATATGAACGCCTTGTTGAACGAGCGCAGTCGTT
      V  V  Q  K  I  N  E  R  C  D  M  N  A  L  L  N  E  R  S  R
3601  ACAAAGTCCAGCTGGCAGACTATGACCAGGTGCAGGAGCAGCGACAGCTGACAATTGCTC
      Y  K  V  Q  L  A  D  Y  D  Q  V  Q  E  Q  R  Q  L  T  I  A
3661  TCTCTGGAGACGCATGGCTAGCAGACCTTCGTCAGACCTGCTCCGCCCAGGGTATTACCT
      L  S  G  D  A  W  L  A  D  L  R  Q  T  C  S  A  Q  G  I  T
3721  TACATTCGATTCTCCAATTTGTTTGGCACGCCGTGCTGCACGCTTATGGCGGTGGCACCC
      L  H  S  I  L  Q  F  V  W  H  A  V  L  H  A  Y  G  G  G  T
3781  ACACCATAACCGGCACGACCATTTCTGGAAGGAACCTGCCCATCTTGGGAATTGAACGAG
      H  T  I  T  G  T  T  I  S  G  R  N  L  P  I  L  G  I  E  R
3841  CAGTTGGTCCGTATATCAACACTCTACCGCTGGTACTCGATCATTCGACGTTCAAGGATA
      A  V  G  P  Y  I  N  T  L  P  L  V  L  D  H  S  T  F  K  D
3901  AGACAATCATGGAGGCCATCGAGGATGTGCAGGCCAAGGTAAACGTCATGAACAGCCGGG
      K  T  I  M  E  A  I  E  D  V  Q  A  K  V  N  V  M  N  S  R
3961  GCAATGTGGAACTGGGCCGTTTGCACAAAACCGACTTAAAGCACGGATTATTCGATTCTT
      G  N  V  E  L  G  R  L  H  K  T  D  L  K  H  G  L  F  D  S
4021  TATTCGTGCTTGAAAACTACCCGAATTTGGACAAATCGCGAACACTTGAGCACCAGACTG
      L  F  V  L  E  N  Y  P  N  L  D  K  S  R  T  L  E  H  Q  T
4081  AACTGGGGTATTCGATTGAAGGCGGCACTGAGAAGCTGAATTATCCACTGGCTGTCATCG
      E  L  G  Y  S  I  E  G  G  T  E  K  L  N  Y  P  L  A  V  I
4141  CGCGCGAAGTCGAGACGACTGGCGGATTCACAGTATCCATCTGCTACGCCAGTGAGCTAT
      A  R  E  V  E  T  T  G  G  F  T  V  S  I  C  Y  A  S  E  L
4201  TTGAGGAGGTTATGATCTCCGAGCTTCTTCATATGGTCCAGGACACACTGATGCAGGTTG
      F  E  E  V  M  I  S  E  L  L  H  M  V  Q  D  T  L  M  Q  V
4261  CCCGAGGTTTGAATGAACCCGTCGGCAGCCTGGAGTATCTCTCATCTATCCAATTGGAGC
      A  R  G  L  N  E  P  V  G  S  L  E  Y  L  S  S  I  Q  L  E
4321  AACTCGCCGCGTGGAATGCCACGGAGGCTGAGTTTCCCGATACCACGCTTCATGAGATGT
      Q  L  A  A  W  N  A  T  E  A  E  F  P  D  T  T  L  H  E  M
4381  TTGAAAACGAAGCGAGCCAGAAGCCGGACAAGATAGCAGTGGTCTATGAGGAGACGTCCT
      F  E  N  E  A  S  Q  K  P  D  K  I  A  V  V  Y  E  E  T  S
4441  TGACTTACCGCGAGTTGAATGAGCGGGCGAACCGTATGGCACATCAGCTAAGGTCCGACG
      L  T  Y  R  E  L  N  E  R  A  N  R  M  A  H  Q  L  R  S  D
4501  TCAGCCCCAACCCCAACGAGGTCATTGCGCTGGTGATGGACAAGAGCGAGCATATGATCG
      V  S  P  N  P  N  E  V  I  A  L  V  M  D  K  S  E  H  M  I
4561  TCAACATTCTGGCCGTATGGAAGAGCGGCGGTGCCTATGTCCCCATTGACCCTGGATATC
      V  N  I  L  A  V  W  K  S  G  G  A  Y  V  P  I  D  P  G  Y
4621  CTAACGACCGCATTCAATACATCCTAGAGGACACACAAGCCCTCGCAGTCATCGCGGACT
      P  N  D  R  I  Q  Y  I  L  E  D  T  Q  A  L  A  V  I  A  D
4681  CCTGCTATCTGCCTCGCATCAAGGGAATGGCTGCCTCCGGCACGCTTCTTTATCCCTCTG
      S  C  Y  L  P  R  I  K  G  M  A  A  S  G  T  L  L  Y  P  S
4741  TCTTGCCTGCCAATCCGGATTCCAAGTGGAGCGTATCGAACCCTTCACCGTTGAGTCGGA
      V  L  P  A  N  P  D  S  K  W  S  V  S  N  P  S  P  L  S  R
4801  GCACGGACTTAGCTTATATCATCTATACCTCTGGAACGACAGGTCGGCCCAAGGGCGTCA
      S  T  D  L  A  Y  I  I  Y  T  S  G  T  T  G  R  P  K  G  V
4861  CGGTAGAGCATCATGGAGTGGTCAACCTGCAGGTGTCGCTATCCAAAGTATTCGGACTAC
      T  V  E  H  H  G  V  V  N  L  Q  V  S  L  S  K  V  F  G  L
4921  GGGATACTGACGACGAGGTAATTCTCTCCTTTTCCAACTATGTGTTCGACCATTTCGTGG
      R  D  T  D  D  E  V  I  L  S  F  S  N  Y  V  F  D  H  F  V
4981  AGCAGATGACCGACGCCATTCTCAATGGCCAAACCCTCCTGGTCCTCAACGATGGAATGC
      E  Q  M  T  D  A  I  L  N  G  Q  T  L  L  V  L  N  D  G  M
5041  GCGGGGACAAAGAGCGACTCTACAGATACATTGAGAAGAACCGAGTGACCTACTTGTCTG
      R  G  D  K  E  R  L  Y  R  Y  I  E  K  N  R  V  T  Y  L  S
5101  GCACCCCATCCGTGGTCTCCATGTACGAATTTAGCCGGTTCAAGGACCATCTACGCCGTG
      G  T  P  S  V  V  S  M  Y  E  F  S  R  F  K  D  H  L  R  R
5161  TGGACTGCGTGGGGGAGGCGTTCAGCGAACCGGTCTTCGACAAGATCCGCGAAACGTTCC
      V  D  C  V  G  E  A  F  S  E  P  V  F  D  K  I  R  E  T  F
5221  ATGGCCTCGTTATCAACGGCTACGGCCCAACTGAAGTTTCCATCACCACCCACAAGCGGC
      H  G  L  V  I  N  G  Y  G  P  T  E  V  S  I  T  T  H  K  R
5281  TCTATCCATTCCCAGAGCGGCGAATGGACAAAAGTATTGGCCAACAGGTCCACAATAGCA
      L  Y  P  F  P  E  R  R  M  D  K  S  I  G  Q  Q  V  H  N  S
5341  CGAGCTATGTGCTGAACGAGGACATGAAGCGCACCCCCATAGGTTCTGTCGGCGAGCTCT
      T  S  Y  V  L  N  E  D  M  K  R  T  P  I  G  S  V  G  E  L
5401  ACCTGGGTGGTGAAGGAGTGGTACGGGGATATCACAATCGCGCAGATGTGACCGCGGAGC
      Y  L  G  G  E  G  V  V  R  G  Y  H  N  R  A  D  V  T  A  E
5461  GTTTTATTCCTAATCCATTCCAGTCGGAAGAAGATAAGCGAGAAGGTCGTAACTCCCGTT
      R  F  I  P  N  P  F  Q  S  E  E  D  K  R  E  G  R  N  S  R
5521  TGTACAAGACCGGTGACCTGGTACGCTGGATTCCTGGAAGCAGCGGGGAGGTCGAGTATC
      L  Y  K  T  G  D  L  V  R  W  I  P  G  S  S  G  E  V  E  Y
5581  TAGGTCGTAATGACTTCCAGGTCAAGATTCGCGGACTGCGCATCGAACTAGGCGAGATTG
      L  G  R  N  D  F  Q  V  K  I  R  G  L  R  I  E  L  G  E  I
5641  AGGCCATCCTATCGTCTTATCACGGAATCAAACAGTCTGTGGTGATTGCCAAGGATTGCA
      E  A  I  L  S  S  Y  H  G  I  K  Q  S  V  V  I  A  K  D  C
5701  GAGAAGGGGCCCAGAAATTCCTGGTTGGTTACTATGTCGCCGATGCAGCGCTGCCGTCCG
      R  E  G  A  Q  K  F  L  V  G  Y  Y  V  A  D  A  A  L  P  S
5761  CTGCCATTCGGCGCTTCATGCAGTCTCGGCTCCCTGGCTACATGGTGCCCTCTCGTCTCA
      A  A  I  R  R  F  M  Q  S  R  L  P  G  Y  M  V  P  S  R  L
5821  TTCTCGTCAGCAAGTTCCCCGTCACTCCTAGTGGAAAATTAGACACCAAGGCTTTGCCCC
      I  L  V  S  K  F  P  V  T  P  S  G  K  L  D  T  K  A  L  P
5881  CAGCCGAGGAAGAGAGCGAGATTGACGTGGTGCCGCCGCGTAGTGAAATCGAACGCTCCT
      P  A  E  E  E  S  E  I  D  V  V  P  P  R  S  E  I  E  R  S
5941  TGTGTGACATCTGGGCGGAACTACTCGAGATGCACCCAGAGGAGATCGGCATTTACAGCG
      L  C  D  I  W  A  E  L  L  E  M  H  P  E  E  I  G  I  Y  S
6001  ATTTCTTCAGCCTGGGAGGTGACAGCCTAAAGAGCACAAAGCTTTCCTTCATGATTCACG
      D  F  F  S  L  G  G  D  S  L  K  S  T  K  L  S  F  M  I  H
6061  AGTCCTTTAACCGCGCCGTCTCAGTCAGCGCCCTTTTCTGTCACCGGACAGTTGAAGCCC
      E  S  F  N  R  A  V  S  V  S  A  L  F  C  H  R  T  V  E  A
6121  AGACGCACTTGATCCTGAACGATGCTGCAGATGTGCACGAAATTACTCCCATAGATTGCA
      Q  T  H  L  I  L  N  D  A  A  D  V  H  E  I  T  P  I  D  C
6181  ATGATACGCAGATGATTCCCGTGTCCCGTGCCCAGGAGCGACTCCTCTTCATCCACGAAT
      N  D  T  Q  M  I  P  V  S  R  A  Q  E  R  L  L  F  I  H  E
6241  TTGAGAATGGCAGCAATGCATACAATATCGACGCTGCATTTGAACTGCCTGGCTCGGTTG
      F  E  N  G  S  N  A  Y  N  I  D  A  A  F  E  L  P  G  S  V
6301  ACGCGTCGCTTCTCGAGCAGGCGCTGCGTGGAAACCTTGCTCGACATGAGGCGTTGAGAA
      D  A  S  L  L  E  Q  A  L  R  G  N  L  A  R  H  E  A  L  R
6361  CTTTACTGGTCAAGGATCACGCAACCGGCATCTATCTTCAGAAGGTATTGAGTCCCGATG
      T  L  L  V  K  D  H  A  T  G  I  Y  L  Q  K  V  L  S  P  D
6421  AAGCCCAGGGCATGTTCTCCGTCAACGTGGACACAGCCAAGCAGGTGGAGCGGCTGGACC
      E  A  Q  G  M  F  S  V  N  V  D  T  A  K  Q  V  E  R  L  D
6481  AGGAGATAGCCAGTCTATCCCAGCATGTTTTCCGCCTCGATGATGAACTGCCTTGGGAGG
      Q  E  I  A  S  L  S  Q  H  V  F  R  L  D  D  E  L  P  W  E
6541  CCCGCATCCTTAAACTCGAATCCGGCGGCCTGTATCTCATTCTGGCGTTCCACCATACCT
      A  R  I  L  K  L  E  S  G  G  L  Y  L  I  L  A  F  H  H  T
6601  GCTTCGATGCATGGTCATTGAAAGTCTTCGAGCAAGAGCTTCGGGCCTTGTACGCAGCGC
      C  F  D  A  W  S  L  K  V  F  E  Q  E  L  R  A  L  Y  A  A
```

Figure 5 Nucleotide and amino acid sequences of the *pcb*AB gene and δ-(L-α-aminoadipyl)-L-cysteinyl-D-valine synthetase (ACVS) from *P. chrysogenum*. The three domains are indicated in bold, red, and blue letters, respectively.

```
6661  TCCAGAAAACCAAAAGTGCAGCGAACTTACCAGCCCTCAAAGCGCAGTACAAGGAATACG
      L  Q  K  T  K  S  A  A  N  L  P  A  L  K  A  Q  Y  K  E  Y
6721  CGCTCTACCATCGCCGGCAGCTGTCTGGCGATCGCATGCGCAACCTGTCAGACTTTTGGC
      A  L  Y  H  R  R  Q  L  S  G  D  R  M  R  N  L  S  D  F  W
6781  TGCGGAAACTCATTGGCTTGGAACCATTGCAGCTGATCACGGACCGCCCACGTCCTGTGC
      L  R  K  L  I  G  L  E  P  L  Q  L  I  T  D  R  P  R  P  V
6841  AATTCAAATACGACGGTGACGACCTCAGTATCGAACTGAGCAAGAAGGAAACGGAGAACC
      Q  F  K  Y  D  G  D  D  L  S  I  E  L  S  K  K  E  T  E  N
6901  TGAGGGGGGGTGGCCAAACGTTGCAAGTCGAGTCTGTACGTCGTGTTGGTTTCCGTTTATT
      L  R  G  V  A  K  R  C  K  S  S  L  Y  V  V  L  V  S  V  Y
6961  GCGTTATGCTAGCCTCGTACGCGAACCAGTCCGATGTTTCCGTGGGTATCCCAGTCAGCC
      C  V  M  L  A  S  Y  A  N  Q  S  D  V  S  V  G  I  P  V  S
7021  ACCGAACGCATCCTCAGTTCCAATCGGTCATTGGATTCTTCGTCAACCTTGTGGTGCTAA
      H  R  T  H  P  Q  F  Q  S  V  I  G  F  F  V  N  L  V  V  L
7081  GGGTGGATATTTCTCAGTCAGCCATTTGCGGGCTCATCAGAAGGGTAATGAAAGAGCTCG
      R  V  D  I  S  Q  S  A  I  C  G  L  I  R  R  V  M  K  E  L
7141  TGGACGCCCAACTGCACCAAGACATGCCGTTCCAGGAAGTGACGAAGCTGCTGCAGGTGG
      V  D  A  Q  L  H  Q  D  M  P  F  Q  E  V  T  K  L  L  Q  V
7201  ATAATGACCCCAGCCGGCATCCGCTGGTACAGAACGTGTTCAACTTCGAATCCCGTGCGA
      D  N  D  P  S  R  H  P  L  V  Q  N  V  F  N  F  E  S  R  A
7261  ACGGAGAACACGATGCCAGGTCGGAGGATGAAGGATCGCTTGCATTCAATCAATACCGGC
      N  G  E  H  D  A  R  S  E  D  E  G  S  L  A  F  N  Q  Y  R
7321  CGGTTCAGCCCGTGGATTCCGTTGCGAAGTTTGATCTGAACGCAACGGTCACGGAATTGG
      P  V  Q  P  V  D  S  V  A  K  F  D  L  N  A  T  V  T  E  L
7381  AGTCGGGATTGAGAGTCAACTTCAACTATGCGACCAGCCTATTCAACAAAAGCACGATCC
      E  S  G  L  R  V  N  F  N  Y  A  T  S  L  F  N  K  S  T  I
7441  AGGGTTTTTTGCATACCTATGAGTATCTCCTGCGCCAGCTGTCCGAACTGAGTGCAGAAG
      Q  G  F  L  H  T  Y  E  Y  L  L  R  Q  L  S  E  L  S  A  E
7501  GGATCAATGAGGATACGCAGCTGTCGTTAGTTCGCCCGACAGAGAATGGCGATCTGCACT
      G  I  N  E  D  T  Q  L  S  L  V  R  P  T  E  N  G  D  L  H
7561  TGCCATTGGCACAGTCCCCGTTGCGACGACTGCTGAGGAGCAGAAAGTAGCGTCGTTGA
      L  P  L  A  Q  S  P  L  A  T  T  A  E  E  Q  K  V  A  S  L
7621  ACCAGGCCTTTGAGCGCGAAGCTTTCCTTGCCGCAGAGAAGATTGCCGTCGTGCAGGGAG
      N  Q  A  F  E  R  E  A  F  L  A  A  **E  K  I  A  V  V  Q  G**
7681  ATAGAGCACTTAGTTATGCTGATCTTAACGGGCAGGCTAACCAGCTCGCCCGGTACATAC
      **D  R  A  L  S  Y  A  D  L  N  G  Q  A  N  Q  L  A  R  Y  I**
7741  AGTCCGTGTCCTGTATTGGGGCAGACGACGGAATAGCTTTGATGCTGGAAAAAGAGTATCG
      **Q  S  V  S  C  I  G  A  D  D  G  I  A  L  M  L  E  K  S  I**
7801  ACACGATTATTTGCATTCTCGCGATTTGGAAGGCTGGTGCAGCATACGTGCCCTTGGATC
      **D  T  I  I  C  I  L  A  I  W  K  A  A  A  Y  V  P  L  D**
7861  CGACTTACCCCACCCGGACGCGTCCAGCTGATTCTGGAGGAAGATTAAAGCGAAGGCTGTCC
      **P  T  Y  P  P  G  R  V  Q  L  I  L  E  E  I  K  A  K  A  V**
7921  TTGTGCACTCCAGTCATGCTTCGAAATGTGAACGCCATGGCGCGAAGGTGATTGCAGTCG
      **L  V  H  S  S  H  A  S  K  C  E  R  H  G  A  K  V  I  A  V**
7981  ACTCCGCCGCCATCGAGACGGCGGTCAGCCAACAGTCAGCTGCTGACCTGCCCACAATTG
      **D  S  P  A  I  E  T  A  V  S  Q  Q  S  A  A  D  L  P  T  I**
8041  CTAGCCTCGGCAATCTAGCGTATATAATCTTTACTTCAGGCACTTCCGGTAAGCCAAAGG
      **A  S  L  G  N  L  A  Y  I  I  F  T  S  G  T  S  G  K  P  K**
8101  GAGTCCTAGTTGAGCAAAAGGCAGTTCTTCTTCTACGCGATGCCCTCCGGGAGCGGTATT
      **G  V  L  V  E  Q  K  A  V  L  L  L  R  D  A  L  R  E  R  Y**
8161  TCGGTCGAGACTGTACCAAGCATCATGGCGTCCTGTTCCTGTCCAACTACGTCTTCGACT
      **F  G  R  D  C  T  K  H  H  G  V  L  F  L  S  N  Y  V  F  D**
8221  TCTCCGTCGAACAACTTGTGTTGTCGGTGCTCAGCGGACACAAGCTGATCGTTCCCCCAG
      **F  S  V  E  Q  L  V  L  S  V  L  S  G  H  K  L  I  V  P  P**
8281  CTGAGTTCGTCGCAGATGATGAATTTACAGAATGGCCAGCACGCACGGTCTCTCCTATC
      **A  E  F  V  A  D  D  E  F  Y  R  M  A  S  T  H  G  L  S  Y**
8341  TCAGCGGCACACCATCCTTACTGCAGAAGATCGATCTGGCACGACTGGACCATCTGCAGG
      **L  S  G  T  P  S  L  L  Q  K  I  D  L  A  R  L  D  H  L  Q**
8401  TTGTTACCGCCGCGGGCGAAGAGCTTCACGCCACCCAGTACGAGAAGATCGCGCCGCCGAT
      **V  V  T  A  A  G  E  E  L  H  A  T  Q  Y  E  K  M  R  R  R**
8461  TCAACGGTCCCATCTACAATGCTATGGTGTCACCGAGACCACGGTGTACAACATTATCG
      **F  N  G  P  I  Y  N  A  Y  G  V  T  E  T  T  V  Y  N  I  I**
8521  CGGAATTCACAACGAATTCGATATTTGAGAATGCTCTTCGGGAAGTGCTCCCTGGTACCC
      **A  E  F  T  T  N  S  I  F  E  N  A  L  R  E  V  L  P  G  T**
8581  GAGCGTATGTGCTGAACGCGGCACTTCAGCCCGTCCCCTTCGATGCTGTCGGAGAACTCT
      **R  A  Y  V  L  N  A  A  L  Q  P  V  P  F  D  A  V  G  E  L**
8641  ATCTTGCCGGCGACAGCGTTACGCGTGGTTATCTCAACCAACCTCTTCTAACGGATCAGC
      **Y  L  A  G  D  S  V  T  R  G  Y  L  N  Q  P  L  L  T  D  Q**
8701  GATTCATTCCCAACCCTTTCTGCAAAGAGGAGGACATCGCTATGGGGCGCTTCGCGCGGC
      **R  F  I  P  N  P  F  C  K  E  E  D  I  A  M  G  R  F  A  R**
8761  TCTACAAGACCGGCGACCTGGTTCGATCGCGTTTCAACCGTCAGCAGCAGCCGCAGCTGG
      **L  Y  K  T  G  D  L  V  R  S  R  F  N  R  Q  Q  Q  P  Q  L**
8821  AATACCTAGGAAGAGGCGATCTGCAGATCAAGATGAGGGGATACCGGATCGAGATTTCTG
      **E  Y  L  G  R  G  D  L  Q  I  K  M  R  G  Y  R  I  E  I  S**
8881  AAGTTCAGAACGTGCTCACTTCAAGTCCCGGTGTCGGGAGGGTGCAGTCGTTGCCAAGT
      **E  V  Q  N  V  L  T  S  S  P  G  V  R  E  G  A  V  V  A  K**
8941  ATGAGAACAACGATACCTATTCCCCGGACCGCTCACTCTCTGGTCGGTTACTATACCACGG
      **Y  E  N  N  D  T  Y  S  R  T  A  H  S  L  V  G  Y  Y  T  T**
9001  ACAATGAAACAGTATCGGAAGCCGATATTCTCACTTTCATGAAAGCAAGGCTTCCAACGT
      **D  N  E  T  V  S  E  A  D  I  L  T  F  M  K  A  R  L  P  T**
9061  ACATGGTGCCAAGCCACCTCTGCTGTCTGGAAGGCGCACTGCCTGTGACGATTAACGGAA
      **Y  M  V  P  S  H  L  C  C  L  E  G  A  L  P  V  T  I  N  G**
9121  AGCTCGACGTCCGGAGATTGCGCGGAGATTATCAACGACTCCGCGCAGTCCTCGTACAGCC
      **K  L  D  V  R  R  L  P  E  I  I  N  D  S  A  Q  S  S  Y  S**
9181  CACCAAGGAACATAATCGAGGCCAAGATGTGCAGACTGTGGGAATCCGCCTTGGGAATGG
      **P  P  R  N  I  I  E  A  K  M  C  R  L  W  E  S  A  L  G  M**
9241  AGCGATGCGGTATCGACGACGACCTGTTCAAACTGGGTGGCGACAGCATCACATCTTTGC
      **E  R  C  G  I  D  D  D  L  F  K  L  G  G  D  S  I  T  S  L**
9301  ATCTCGTGGCCCAGATTCACAACCAGGTGGGCTGCAAGATCACCGTTCGGGATATATTTG
      **H  L  V  A  Q  I  H  N  Q  V  G  C  K  I  T  V  R  D  I  F**
9361  AACATCGTACCGCCCCGAGCCCTCCATGATCACGTCTTCATGAAGGACTCCGACCGGAGTA
      **E  H  R**  T  A  R  A  L  H  D  H  V  F  M  K  D  S  D  R  S
9421  ATGTGACTCAGTTCCGAACCGAACAAGGGCCGGTCATCGGCGAGGCGCCCCTACTGCCGA
      N  V  T  Q  F  R  T  E  Q  G  P  V  I  G  E  A  P  L  L  P
9481  TTCAAGACTGGTTTTTGTCAAAGGCTCTGCAGCATCCGATGTATTGGAATCACACTTTCT
      I  Q  D  W  F  L  S  K  A  L  Q  H  P  M  Y  W  N  H  T  F
9541  ACGTCCGAACGCCAGAGCTGGATGTTGATTCCTTAAGCGCTGCTGTCAGGGACTTGCAAC
      Y  V  R  T  P  E  L  D  V  D  S  L  S  A  A  V  R  D  L  Q
9601  AGTATCACGATGTTTTCCGCATGCGACTCAAGCGCGAGGAAGTCGGATTCGTGCAGTCCT
      Q  Y  H  D  V  F  R  M  R  L  K  R  E  E  V  G  F  V  Q  S
9661  TTGCTGAGGACTTCTCTCCTGCCCAGCTTCGGGTGCTGAACGTAAAAGATGTTGACGGGT
      F  A  E  D  F  S  P  A  Q  L  R  V  L  N  V  K  D  V  D  G
9721  CCGCGGCCGTCAACGAGATATTGGATGGGTGCAGTCTGGCTTCAACCTTGAGAACCGGAC
      S  A  A  V  N  E  I  L  D  G  W  Q  S  G  F  N  L  E  N  G
9781  CCATTGGTTCCATTGGCTACCTACATGGGTATGAAGACCGATCCGCGCGAGTCTGGTTCT
      P  I  G  S  I  G  Y  L  H  G  Y  E  D  R  S  A  R  V  W  F
9841  CCGTTCACCATATGGCCATTGACACCGTCAGCTGGCAGATCCTTGTCCGTGACCTGCAGA
      S  V  H  H  M  A  I  D  T  V  S  W  Q  I  L  V  R  D  L  Q
9901  CGCTGTACCGAAATGGAGCCTCGGAAGCAAGGGCAGCAGTTTCCGGCAGTGGGCTGAAG
      T  L  Y  R  N  G  S  L  G  S  K  G  S  S  F  R  Q  W  A  E
```

```
9961   CCATCCAAAATTACAAGGCGTCAGACTCTGAGAGGAACCATTGGAATAAGCTCGTCATGG
       A  I  Q  N  Y  K  A  S  D  S  E  R  N  H  W  N  K  L  V  M
10021  AAACAGCTTCCAGCATATCCGCATTGCCTACGTCAACCGGTTCGCGCGTGCGCCTGAGCA
       E  T  A  S  S  I  S  A  L  P  T  S  T  G  S  R  V  R  L  S
10081  GAAGTTTGAGCCCTGAGAAGACAGCCTCACTGATCCAAGGAGGAATCGATCGACAGGATG
       R  S  L  S  P  E  K  T  A  S  L  I  Q  G  G  I  D  R  Q  D
10141  TCTCCGTGTACGACTCCCTCCTGACTTCAGTTGGATTGGCGCTCCAACATATCGCTCCAA
       V  S  V  Y  D  S  L  L  T  S  V  G  L  A  L  Q  H  I  A  P
10201  CCGGCCCAAGTATGGTTACGATCGAGGGACATGGCCGTGAAGAAGTGGATCAGACACTGG
       T  G  P  S  M  V  T  I  E  G  H  G  R  E  E  V  D  Q  T  L
10261  ATGTGAGCCGCACCATGGGTTGGTTCACCACCATGTATCCATTTGAAATTCCCCGTCTCA
       D  V  S  R  T  M  G  W  F  T  T  M  Y  P  F  E  I  P  R  L
10321  GCACCGAGAACATTGTTCAAGGAGTCGTCGCTGTGAGCGAACGGTTCAGACAGGTGCCTG
       S  T  E  N  I  V  Q  G  V  V  A  V  S  E  R  F  R  Q  V  P
10381  CCCGTGGCGTCGGGTATGGAACCTTGTACGGCTATACTCAACACCCGCTGCCCCAGGTGA
       A  R  G  V  G  Y  G  T  L  Y  G  Y  T  Q  H  P  L  P  Q  V
10441  CCGTCAACTACCTGGGCCAGCTCGCCCGCAAGCAATCGAAGCCAAAGGAATGGGTCCTCG
       T  V  N  Y  L  G  Q  L  A  R  K  Q  S  K  P  K  E  W  V  L
10501  CGGTGGGCGACAACGAATTTGAATACGGACTCATGACTAGCCCAGAGGACAAAGACCGGA
       A  V  G  D  N  E  F  E  Y  G  L  M  T  S  P  E  D  K  D  R
10561  GCTCTTCTGCCGTCGACGTCACGGCCGTGTGTATTGACGGCACTATGATCATCGATGTGG
       S  S  S  A  V  D  V  T  A  V  C  I  D  G  T  M  I  I  D  V
10621  ACAGTGCTTGGAGCCTTGAGGAGAGCGAGCAATTCATCTCGAGCATCGAGGAAGGACTGA
       D  S  A  W  S  L  E  E  S  E  Q  F  I  S  S  I  E  E  G  L
10681  ACAAGATCCTCGACGGCAGGGCAAGTCAGCAAACCTCGCGATTCCCGGATGTTCCTCAAC
       N  K  I  L  D  G  R  A  S  Q  S  T  R  F  P  D  V  P  Q
10741  CGGCGGAGACATATACGCCGTATTTCGAGTATCTGGAACCTCCACGACAGGGACCCGACGC
       P  A  E  T  Y  T  P  Y  F  E  Y  L  E  P  P  R  Q  G  P  T
10801  TGTTCCTGCTGCCGCCGGGCGAAGGAGGCGCCGAGAGTTACTTCAACAACATCGTCAAGC
       L  F  L  L  P  P  G  E  G  G  A  E  S  Y  F  N  N  I  V  K
10861  GCCTGCGTCAGACAAATATGGTGGTCTTCAACAACTACTACTTGCACAGCAAACGCCTGC
       R  L  R  Q  T  N  M  V  V  F  N  N  Y  Y  L  H  S  K  R  L
10921  GCACGTTCGAGGAGCTGGCGGAAATGTATCTCGACCAAGTACGCGGCATCCAACCACACG
       R  T  F  E  E  L  A  E  M  Y  L  D  Q  V  R  G  I  Q  P  H
10981  GACCGTACCACTTCATCGGATGGAGCTTCGGAGGAATTCTCGCAATGGAAATGTCGCGGC
       G  P  Y  H  F  I  G  W  S  F  G  G  I  L  A  M  E  M  S  R
11041  GACTGGTAGCCTCGGACGAGAAGATTGGCTTCCTCGGTATTATCGACACCTATTTCAACG
       R  L  V  A  S  D  E  K  I  G  F  L  G  I  I  D  T  Y  F  N
11101  TGCGGGGAGCGACACGCACCATTGGCTTGGGGGACACTGAGATTCTGGACCCGATCCATC
       V  R  G  A  T  R  T  I  G  L  G  D  T  E  I  L  D  P  I  H
11161  ACATCTACAATCCCGATCCGGCCAACTTCCAACGCCTGCCCTCTGCAACAGATCGCATTG
       H  I  Y  N  P  D  P  A  N  F  Q  R  L  P  S  A  T  D  R  I
11221  TGCTGTTCAAGGCCATGAGGCCGAACAACAAGTACGAATCCGAGAACCAGCGTCGCCTGT
       V  L  F  K  A  M  R  P  N  N  K  Y  E  S  E  N  Q  R  R  L
11281  ACAGATACTATGACGGCACTCGACTCAACGGACTGGACAGCTTGTTACCAAGCGATTCCG
       Y  E  Y  Y  D  G  T  R  L  N  G  L  D  S  L  L  P  S  D  S
11341  ACGTCCAGCTGGTCCCGCTTACGGACGATACACACTTTTCCTGGGTCGGAAATCCACAAC
       D  V  Q  L  V  P  L  T  D  D  T  H  F  S  W  V  G  N  P  Q
11401  AGGTGGAGCAGATGTGTGCGACTATCAAGGAACACCTCGCTCGCTATTGATCCGTCACTA
       Q  V  E  Q  M  C  A  T  I  K  E  H  L  A  R  Y  -
```

Figure 5   (continued)

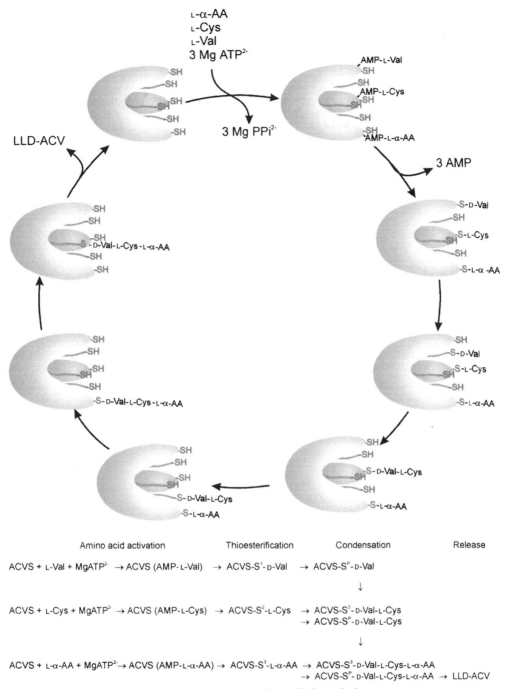

Figure 6 Mechanism for ACVS catalysis.

ACV from free L-isomers but failed to do so when δ-D-α-aminoadipyl-L-cysteine and D-valine or δ-(L-α-aminoadipyl)-L-cysteine and D-valine were tested as substrates. However, it was not clear whether a single enzyme (ACVS) or two different proteins (AC synthetase and ACVS) were needed. Banko and co-workers[52,53] further analyzed this enzymatic reaction and they concluded that a single protein, ACVS, catalyzes the synthesis of the tripeptide and that ACV formation is a complex process requiring the following steps: (i) recognition of the three L-enantiomers of its constituent amino acids; (ii) their adenylation; (iii) condensation of the three precursors; (iv) epimerization of L-valine to D-valine; and (v) release of the products. These authors further indicated that ACVS catalyzes the synthesis of ACV as well as that of other tripeptides in which the L-α-aminoadipic acid and the D-valine moieties have been replaced by carboxymethyl-cysteine and glutamic acid, respectively. They also suggested that the epimerization from L-valine to D-valine might occur during ACV formation since D-valine was not recognized by the enzyme as a substrate.[53] Shiau *et al.*[115] have reported that the replacement of cysteine by L-*O*-methylserine leads to the formation of both L-*O*-methylserinyl-L-valine and L-*O*-methylserinyl-D-valine dipeptides, suggesting that the order of peptide bond formation in ACV synthesis involves the formation of the cysteinyl–valine dipeptide bond before the epimerization of valine and condensation with L-α-aminoadipic acid takes place. These authors isolated the dipeptide L-cysteinyl-D-valine from an ACVS reaction,[116] lending strong support to the hypothesis[115] that the α-peptide bond between cysteine and valine is formed before the δ-peptide bond between α-aminoadipic acid and cysteine is established (Figure 6).

4.10.7.2   Isopenicillin N Synthase (IPNS)

4.10.7.2.1   General characteristics

The second step in the pathway is catalyzed by the enzyme IPNS, a so-called cyclase.[19] This protein catalyzes the oxidative cyclization of the L-cysteinyl and D-valine moieties of ACV, generating the β-lactam and thiazolidine rings (Scheme 2). Maximal rates of catalysis were observed at pH 7.5–8.2 at 25 °C. The enzyme was first assayed in crude cell-free extracts of *A. chrysogenum*[55,117,118] although it was later purified from different microbes: *P. chrysogenum*,[119,120] *Aspergillus*,[90,91] *S. clavuligerus*,[121] *Streptomyces jumonjinensis*,[122] *Streptomyces lipmanii*,[90,122] *Flavobacterium* sp.[123] and *Nocardia lactamdurans*.[94] All the IPNSs studied have a single polypeptide chain and their molecular mass ranges between 37–40 kDa with the exception of *N. lactamdurans*, whose molecular mass seems to be lower.[94] IPNS has been crystallized by Baldwin and co-workers,[124,125] who showed that this enzyme is the first member of a new structural family of enzymes.

The IPNS gene (*pcb*C) was cloned by Samson *et al.*[56] and was the first gene of the penicillin biosynthetic pathway whose sequence was established. The gene was mapped on chromosomes I (*P. chrysogenum*), II (*P. notatum*), and VI (*A. chrysogenum* and *A. nidulans*),[103,126] and is linked to the *pcb*AB in a single cluster.[85–87,95] The sequence of the *pcb*C gene and the IPNS protein isolated from *P. chrysogenum* are shown in Figure 7.

4.10.7.2.2   Mechanistic studies and substrate specificity

The following participate in the reaction catalyzed by IPNS: ACV, molecular oxygen, and ferrous iron.[19,116,117] In a single enzymatic step, this protein removes four hydrogen atoms from the tripeptide ACV in a desaturative ring closure, thereby generating the β-lactam and the thiazolidine rings typical of the structure of penicillins.[19] Unlike other nonheme iron-dependent oxygenases and oxidases, which need electron donors or oxidizable substrates,[127] IPNS uses its four-electron oxidizing power to catalyze this step.[19] It has been proposed that during the reaction, iron binds directly to ACV through its cysteinyl thiol and that a ferryl–oxo intermediate, required for thiazolidine ring closure, is formed.[128–130] All these findings, supporting a new enzymatic mechanism, are summarized in Figure 8.

The participation of the cysteine moiety of ACV in the cyclization reaction suggested the formation of a thiol disulfide bond between IPNS and ACV (see above). To clarify this point, the role of the two cysteines present in IPNS molecules was analyzed genetically. Site-directed mutagenesis of the IPNS gene from *A. chrysogenum* expressed in *E. coli* afforded different recombinant IPNS, in which one of the cysteine residues present at position 106 or 255 had been replaced by L-serine, as well as other mutagenized enzymes in which both cysteines had been replaced by serines.[131] It was observed that when Cys-255 was replaced by L-serine a 33% reduction in the V_{max} occurred, whereas substitution at position 106 or double replacement decreased the V_{max} calculated for the wild-type recombinant protein by 63%. It was also found that mutation in the codon of Cys-106 increased the K_m value, whereas this was not affected when mutagenesis occurred in Cys-255.[131] These findings suggest that although cysteines are important for catalysis they do not play an essential role in it. The participation of three residues of His and one Asp in the binding of ACV to IPNS has also been reported.[132,133]

Because the IPNS-catalyzed reaction leads to the formation of a product with antibacterial activity, the substrate specificity of the enzymes purified from different microbes (*P. chrysogenum*,

```
         10        20        30        40        50        60
ATGGCTTCCACCCCCAAGGCCAATGTCCCCAAGATCGACGTGTCGCCCCTGTTCGGCGAC
 M   A   S   T   P   K   A   N   V   P   K   I   D   V   S   P   L   F   G   D

         70        80        90       100       110       120
AATATGGAGGAGAAGATGAAGGTTGCCCGCGCGATTGACGCTGCCTCGCGCGACACCGGC
 N   M   E   E   K   M   K   V   A   R   A   I   D   A   A   S   R   D   T   G

        130       140       150       160       170       180
TTCTTCTACGCGGTCAACCACGGTGTGGATGTGAAGCGACTCTCGAACAAGACCAGGGAG
 F   F   Y   A   V   N   H   G   V   D   V   K   R   L   S   N   K   T   R   E

        190       200       210       220       230       240
TTCCACTTTTCTATCACAGACGAAGAGAAGTGGGACCTCGCGATTCGCGCCTACAACAAG
 F   H   F   S   I   T   D   E   E   K   W   D   L   A   I   R   A   Y   N   K

        250       260       270       280       290       300
GAGCACCAGGACCAGATCCGTGCCGGATACTACCTGTCCATTCCGGAGAAAAAGGCCGTG
 E   H   Q   D   Q   I   R   A   G   Y   Y   L   S   I   P   E   K   K   A   V

        310       320       330       340       350       360
GAATCCTTCTGCTACCTGAACCCCAACTTCAAGCCCGACCACCCTCTCATCCAGTCGAAG
 E   S   F   C   Y   L   N   P   N   F   K   P   D   H   P   L   I   Q   S   K

        370       380       390       400       410       420
ACTCCCACTCACGAGGTCAACGTGTGGCCGGACGAGAAGAAGCATCCGGGCTTCCGCGAG
 T   P   T   H   E   V   N   V   W   P   D   E   K   K   H   P   G   F   R   E

        430       440       450       460       470       480
TTCGCCGAGCAATACTACTGGGATGTGTTCGGGCTCTCGTCTGCCTTGCTGCGAGGCTAT
 F   A   E   Q   Y   Y   W   D   V   F   G   L   S   S   A   L   L   R   G   Y

        490       500       510       520       530       540
GCTCTGGCGCTGGGCAAGGAGGAGGACTTCTTTAGCCGCCACTTCAAGAAGGAAGACGCG
 A   L   A   L   G   K   E   E   D   F   F   S   R   H   F   K   K   E   D   A

        550       560       570       580       590       600
CTCTCCTCGGTTGTTCTGATTCGTTACCCGTACCTGAACCCCATCCCACCTGCCGCCATT
 L   S   S   V   V   L   I   R   Y   P   Y   L   N   P   I   P   P   A   A   I

        610       620       630       640       650       660
AAGACGGCGGAGGACGGCACCAAATTGAGTTTCGAATGGCATGAGGACGTGTCGCTCATT
 K   T   A   E   D   G   T   K   L   S   F   E   W   H   E   D   V   S   L   I

        670       680       690       700       710       720
ACCGTCCTGTACCAGTCAGACGTGGCGAACCTGCAGGTGGAGATGCCCCAGGGTTACCTC
 T   V   L   Y   Q   S   D   V   A   N   L   Q   V   E   M   P   Q   G   Y   L

        730       740       750       760       770       780
GATATCGAGGCGGACGACAACGCCTACCTGGTCAATTGCGGCAGCTACATGGCACACATC
 D   I   E   A   D   D   N   A   Y   L   V   N   C   G   S   Y   M   A   H   I

        790       800       810       820       830       840
ACCAACAACTACTACCCCGCTCCCATCCACCGGGTCAAGTGGGTGAACGAGGAGCGCCAA
 T   N   N   Y   Y   P   A   P   I   H   R   V   K   W   V   N   E   E   R   Q

        850       860       870       880       890       900
TCCCTCCCGTTCTTCGTCAATCTGGGATTTAATGATACCGTCCAGCCGTGGGATCCTAGC
 S   L   P   F   F   V   N   L   G   F   N   D   T   V   Q   P   W   D   P   S

        910       920       930       940       950       960
AAGGAAGACGGCAAGACCGATCAGCGGCCAATCTCGTACGGCGACTATCTGCAGAACGGA
 K   E   D   G   K   T   D   Q   R   P   I   S   Y   G   D   Y   L   Q   N   G

        970       980       990
TTAGTTAGTCTAATCAACAAGAACGGCCAGACATGA
 L   V   S   L   I   N   K   N   G   Q   T   -
```

Figure 7 Nucleotide and amino acid sequences of the *pcb*C gene and isopenicillin N synthase (IPNS) from *P. chrysogenum*.

A. chrysogenum, and *S. clavuligerus*) has been studied exhaustively.[19,129,130] The substrate specificity of IPNSs was analyzed using different tripeptide analogues of the natural substrate (ACV) in which the L-α-aminoadipic acid, the L-cysteine, or the D-valine moieties had been replaced by other molecules. Baldwin and Bradley[19] synthesized an impressive number of tripeptides, permitting study of their transformation into penicillins or into cephalosporin-like products as well as unraveling the mechanism of the enzymatic reaction[130] (Figure 8). Some of the most important findings reported on this topic will be described below.

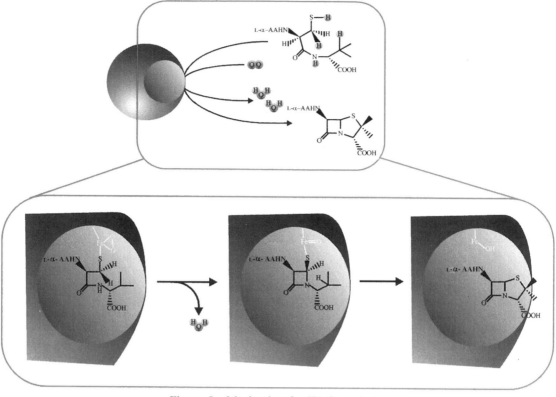

Figure 8 Mechanism for IPNS catalysis.

(i) Replacement of the L-α-aminoadipic acid moiety

Baldwin *et al.*[134] first reported that IPNSs can convert some tripeptide molecules that are analogues of ACV, in which the L-α-aminoadipic moiety had been replaced by D-α-aminoadipic acid, into β-lactam antibiotics. Other workers showed that IPNS also tolerates removal of the terminal amino group of L-α-aminoadipic acid present in the ACV molecule as well as the substitution of a methylene group by a sulfur atom (L-carboxymethyl cysteinyl-CV).[135,136] Some time later, the author and others reported that certain heteropeptides such as phenylacetyl-cysteinyl-D-valine (PCV),[137] its *meta*-carboxy derivative, and other aromatic variants of PCV can be directly converted into penicillin G, *meta*-carboxypenicillin G, and related penicillins, indicating that the *N*-acyl moiety must have a length of six carbon atoms (or an equivalent length if a rigid structure is present in the substrate molecule). Furthermore, the presence of a terminal carboxy group seems to be required since it helps the optimal orientation of the substrate to be reached for ring closure at the active site of IPNS.[137-141]

The structure of the tripeptides obtained by replacement of the L-α-aminoadipic acid moiety of ACV that can be converted into penicillins by different IPNSs are shown in Figure 9.

(ii) Substitution in the cysteinyl residue

In the IPNS reaction, the L-cysteinyl moiety of ACV plays an important role in binding to the iron atom[128-130] as well as in the formation of a hypothetical disulfur intermediate at the active site.[142] Thus, it appeared that the replacement of L-cysteine by other analogous compounds (without SH or with a larger molecular size) should lead to synthesis of tripeptides that cannot be converted into penicillins. This was the case when the L-cysteinyl moiety of ACV was replaced by L-amino-butyric acid, L-serine, and S-methylcysteine.[143] However, Baldwin *et al.*[144,145] showed that a tri-peptide containing 3-methylcysteine in the place of cysteine could be efficiently used as substrate by IPNS, indicating that this increment in the volume of the substrate molecule (presence of an extra methyl group) was also tolerated by IPNS. In summary, the use of ACV analogues in which the

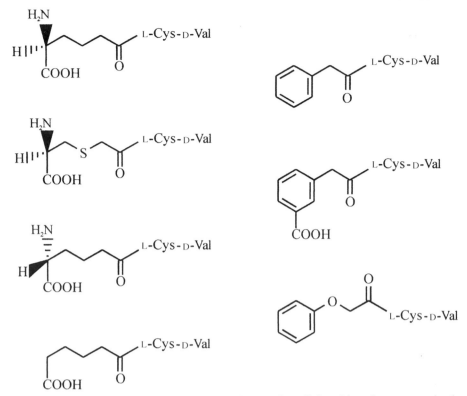

Figure 9 Structure of the molecules able to replace the α-aminoadipic acid moiety present in the tripeptide ACV.

L-cysteinyl moiety had been replaced by other close structural molecules has been crucial for determining the mechanism of ring closure as well as for the synthesis of new antibiotics containing the γ-lactam structure.[146,147]

(iii) Substitution in the D-valine residue

IPNS can also transform different tripeptides in which the D-valine moiety has been replaced by N-hydroxy-D-valine, D-alloisoleucine, D-aminobutyric acid, D-allylglycine, D(−)-isoleucine, D-norvaline, or O-methyl-allothreonine into β-lactam antibiotics, allowing the synthesis of penicillin and cephalosporin-type products (Figure 10).[148,149] It has also been reported that some of these unnatural tripeptides, as well as the noncyclizable AC-glycine, strongly inhibit the formation of IPN from ACV in standard reactions.[150] Taking into account the broader substrate specificity of this enzyme, it seems evident that many different penicillins could be produced *in vivo* by *P. chrysogenum* (or by other strains able to synthesize β-lactam antibiotics) as long as the corresponding precursor tripeptides are synthesized by these cells. It is therefore evident that genetic manipulation of the ACVS gene (*pcb*AB) will lead to mutated proteins (abnormal ACVS) with broader substrate specificity, enabling synthesis of a larger number of compounds which can be later used as substrates by IPNS.

4.10.7.3 Specific Enzymes of Hydrophobic Penicillins

As indicated above, the last step of benzylpenicillin (penicillin G) biosynthesis requires the participation of at least three different enzymes: a PhAcTS,[63–66] which is responsible for the intracellular incorporation of the side-chain precursor;[151] an activating enzyme, PhAcCoAL,[67,68] and a third enzyme, acyl-CoA:6-APA (IPN) AT.[7,22,58–62] The latter enzyme catalyzes either the replacement of the residue of α-aminoadipic acid present in the IPN molecule or N-acylation of the amino group

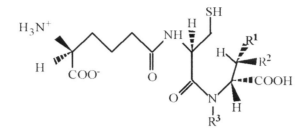

R¹ = R² = Me; R³ = H (Valine)
R¹ = Et; R² = Me; R³ = H (D-Alloisoleucin)
R¹ = H; R² = Me; R³ = H (D-α-Aminobutyric acid)
R¹ = R² = Me; R³ = OH (Hydroxy-D-valine)

R¹ = H; R² = CH = CH₂; R³ = H (D-Allylglycine) ⟵ Penam / Homoceph-3-em / Hydroxycepham } type product

R¹ = Et; R² = Me; R³ = H (D(−)-Isoleucine) } Cepham-type products
R¹ = H; R² = Et; R³ = H (D-Norvaline)
R¹ = Me; R² = Me; R³ = H (*O*-Methylallothreonine)

Figure 10   Structure of the molecules able to replace the D-valine moiety present in the tripeptide ACV.

of 6-APA with different aliphatic (hexanoic, *trans*-3-hexenoic, octanoic acids) or aromatic acids (phenylacetic, phenoxyacetic, 4-hydroxyphenylacetic acid), thus generating several hydrophobic penicillins (DF, F, K, G, V, X) with broader antibacterial spectra and higher antibiotic potency than their biosynthetic precursors (IPN or 6-APA).[152–159] In this section the most important characteristics of these three enzymes are analyzed, placing special emphasis on the substrate specificity of PhAcCoAL and AT.

4.10.7.3.1   Phenylacetic acid transport system

It is well known that several hydrophobic penicillins (F, DF, K) contain as a side chain some aliphatic compound (fatty acids) that can be synthesized endogenously by *P. chrysogenum*.[2,4,7,24] Others, however, such as benzylpenicillin (penicillin G) or phenoxymethylpenicillin (penicillin V), have PhAc and phenoxyacetic acid (PhOAc), respectively,[37] as side chains. These precursors must be added exogenously to the fermentation broths since the fungus lacks the biosynthetic pathway for these compounds.[160,161] Therefore, the synthesis of penicillin G and V requires an uptake system which will allow the cellular incorporation of these side-chain precursors. Initially, it was believed that PhAc could cross the plasma membrane by passive diffusion[162] and some results supported this hypothesis,[163] but a specific energy-dependent transport system responsible for the uptake of PhAc in *P. chrysogenum* and in *A. nidulans* was later characterized.[63–66,151,164] The PhAcTS is an important regulatory point involved in the control of benzylpenicillin biosynthesis in both fungi[14,65,66] (Figure 11).

(i)   General properties

The PhAcTS is an active system[63] that is induced by PhAc, by 2-hydroxyphenylacetic acid, and also by 4-phenylbutyric acid.[64] Other aromatic compounds fail to accomplish this, for example benzoic acid, phenoxyacetic acid, phenylpropionic acid, different PhAc analogues, as well as other molecules containing a different ring.[64] These data allow us to conclude that a benzene ring linked to an alkanoic acid moiety containing an even number of carbon atoms is required for induction.[64] However, 2-phenylbutyric acid does not induce PhAcTS either, suggesting that β-oxidation is needed to generate the true endogenous inducer (probably PhAc-CoA).[64,65]

Kinetic measurements carried out with labeled [¹⁴C]-PhAc show that maximal rates of uptake are achieved when the mycelia of *P. chrysogenum* are incubated at 25 °C in 0.006 M phosphate buffer

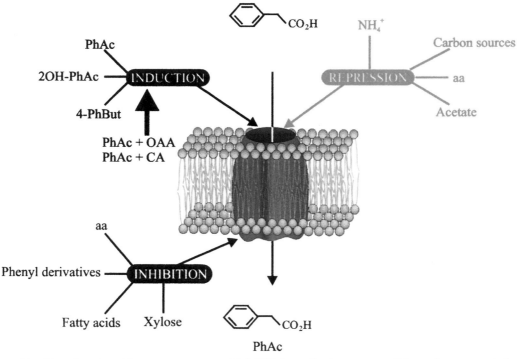

Figure 11 Phenylacetic acid transport system (PhAcTS) from *P. chrysogenum*. Schematic representation of its regulation. OAA, oxalacetic acid; CA, citric acid; PhBut, phenylbutyric acid; and aa, amino acid.

at pH 6.5. Under these conditions the K_m value was 5.2 µM. L-Tyrosine, L-phenylalanine, and several derivatives of these amino acids (other than PhAc) did not affect the uptake rate, suggesting that the PhAcTS is not related to the cellular uptake of those compounds.[63]

(ii) Regulation

The appearance of the PhAcTS in fungal membranes occurs some time before benzylpenicillin starts to be produced,[63] and is strictly modulated in the sense that it is repressed by several carbon sources (sugars, acetate, and glycerol) and by free amino acids (especially those whose catabolic pathways lead to acetyl-CoA).[65] Other metabolic intermediates, such as Krebs cycle intermediates, do not affect the rate of induction, with the exception of oxalacetic acid and citric acid, which strongly stimulate it.[65]

It is interesting to note that the addition of acetate to cultures of *P. chrysogenum* strongly delays the appearance of PhAcTS, this effect being more dramatic when acetate and methylsuccinic acid (an inhibitor of the glyoxylic acid cycle) are supplied together to the culture broths. These data suggest that the intracellular pool of acetyl-CoA, or the acetyl-CoA/CoA ratio, plays an important role in the biosynthesis of benzylpenicillin, by regulating the uptake of the PhAc.[65]

The PhAcTS has also been characterized in *A. nidulans*.[66] In this fungus, the transport system is also energy dependent, showing a maximal rate of uptake at 37 °C and pH 7.0. It is induced by PhAcA and to a lesser extent by PhOAc, as well as by other phenyl derivatives. As reported for the PhAcTS in *P. chrysogenum*,[65] the utilization of easily metabolizable carbon sources (glucose, glycerol, or acetate) results in a substantial reduction in uptake. This effect is independent of the *cre*A gene, which is the regulatory gene mediating carbon catabolite repression.[165] Furthermore, negative regulation of acetate is prevented in *A. nidulans* by a function-loss mutation in the gene encoding acetyl-CoA synthase, suggesting that this regulation is mediated by the intracellular pool of acetyl-CoA.[66]

Although different reports have been published about the uptake of PhAc in fungal and bacterial species,[63–66,151,162–164,166] certain important points remain to be elucidated. Thus, the number of proteins involved in uptake is unclear and the genes encoding this protein(s) have not been cloned.

Only in the bacterium *P. putida* U has a gene encoding a permease involved in the uptake of PhAc been identified and its nucleotide sequence established.[167]

A schematic representation depicting the substrate specificity and regulation of the PhAcTS in *P. chrysogenum* is shown in Figure 11.

4.10.7.4 Acyl-CoA Synthetases or Ligases

Once PhAc has been taken up by *P. chrysogenum*, it must be activated to its CoA thioester. The reaction seems to be catalyzed by a phenylacetyl-CoA ligase, according to Equation (1).

$$\text{PhAc} + \text{CoA} + \text{ATP} \xrightarrow[\text{Mg}^{2+}]{\text{AcCoAS}} \text{PhAc-CoA} + \text{AMP} + \text{PPi} \qquad (1)$$

The existence of this enzyme was first reported by Brunner and Rohr[67] in cell-free extracts of *P. chrysogenum*. Some time later, Kogekar and Deshpande[68] reported that this enzyme was able to activate different monosubstituted acetic acids (PhAc, PhOAc, and acetic acid) but failed to do so when phenylpropionic, α-aminoadipic and different fatty acids were tested as substrates. In spite of the importance of this enzyme, all the attempts to purify it from *P. chrysogenum* and *A. nidulans* have proven unsuccessful. The reasons for this remain unclear, though it could be that PhAcCoAL is a very unstable enzyme or that it is present at very low concentrations in the mycelia of these fungi.

It was reported in 1992 that the activation of PhAc to PhAc-CoA can also be accomplished by other unspecific enzymes that are involved in different metabolic pathways and that are required for the activation of acetate and several medium-chain fatty acids.[75] It was shown that the acetyl-CoA synthetase (AsCoAS) from *P. chrysogenum* could catalyze the formation of PhAc-CoA *in vitro*, suggesting that it may be indirectly involved in the biosynthesis of benzylpenicillin.[75] This enzyme, as well as the AcCoAS purified from *A. nidulans*, recognizes PhAc, PhOAc, acetic acid, and several fatty acids (*trans*-3-hexenoic acid, hexanoic acid, and octanoic acid) present as side chains in different hydrophobic penicillins (F, DF, K) as substrates. In both fungi, this enzyme has a M_r of 140 kDa and the native enzyme is composed of two apparently identical units (70 kDa) in an α_2 dimeric structure. AcCoAS is regulated *in vivo* by carbon catabolite inactivation and can be coupled *in vitro* to AT, thus allowing the reconstitution of a functional enzymatic system (see below), similar to the one that catalyzes the last step of hydrophobic penicillin biosynthesis.[75] The AcCoAS gene of *P. chrysogenum* has a typical eukaryotic organization, showing five introns located at the same position as those found in *A. nidulans*[168] (Figure 12). In both fungi this gene is not included in the penicillin cluster.[15,85–87]

4.10.7.4.1 *Acyl-CoA-activating enzyme from other origins*

The fact that the AcCoASs from *P. chrysogenum* and *A. nidulans* activate some penicillin side-chain precursors to their CoA thioesters[75] suggested that other similar enzymes may also play this role. Under this assumption, several microbes able to grow in chemically defined media containing PhAc as the sole carbon source were selected and the presence of enzymes able to activate PhAc to PhAc-CoA were tested in them. A PhAcCoAL enzyme that catalyzes the first step in a new catabolic pathway involved in the aerobic degradation of PhAc was identified in *P. putida* U.[74] This enzyme was purified to homogeneity and characterized in detail.[169–170] Study of its substrate specificity revealed that PhAcCoAL could activate a large number of compounds to CoA thioesters and that these reaction products could later be used as substrates by AT.[171–173] Moreover, the optimal physicochemical conditions of assay (25–30 °C and pH values ranging between 7.5–8.2) were similar to penicillin biosynthetic enzymes, suggesting that PhAcCoAL could be coupled *in vitro* with IPNS and AT, thus reproducing in the test tube the last two steps of benzylpenicillin biosynthesis.[171,174] The gene encoding PhAcCoAL in *P. putida* has been cloned, sequenced, and expressed in different microbes[175] (Figure 13).

Comparative study of PhAcCoAL from *P. putida* and AcCoAS from *P. chrysogenum* showed that the two enzymes are quite different and although both can activate PhAc to PhAc-CoA, their substrate specificity is quite different.[74,75] Furthermore, the genes encoding PhAcCoAL and AcCoAS

```
AAGCTTACCCCGGAGCAACGGAAAGAACCCCCGCATGGCCGAACCCAAACTCGTATGGGA          60
CAAGGCAATTTACTGAAATTTACTGAAATTTACTGAATTGGACCGTATTCGGAATGTATC         120
TTATTCCTGATTCGGAGATGAGAGTGGATCGTCCGAATGTCCAATGCACAATGTACTTTC         180
TCTAGGCCGTCTGCGGCTAGCGAGACAGCCGGAGTTGGGTAGTTTGAAGTGGTATTGTAA         240
CTTATTGTAATTTATTGTAAGGGGCACGGACCACTGATGAAAAGGGAAGTGGCACATCCT         300
CCCGGGACAGCTGGACTACTAATATTGTCGCGAGTCCCCCCTCCTTGAGTTTCTTTTTCT         360
CTTTCTCTTCTCTAATATCTTCTCTAATTGCTATACATACCCTGTTTGATCATTACTCTT         420
AGTATATTATATAGTTCATCCCCCACATTTATTATTCCCATTGGACTACCGCAATCATGT         480
                                                           M
CGGACGGCCCAATTCAGCCTCCCAAGCCCGGCAGTGGTAAGAATCACCGACCTCCAGACCG         540
S  D  G  P  I  Q  P  P  K  P  A  V
AGATGACCAGACCCGTGTCGCACTGGTGACCGAAGTATCATGGGCTAACTGGTGATATAG         600
GTGCATGAGGCACACGAGGTCGACACTTTCCACGTCCCCAAGGCGTTCCACGATAAGCAC         660
V  H  E  A  H  E  V  D  T  F  H  V  P  K  A  F  H  D  K  H
CCCTCCGGCACTCACATCAAGGACATTGAGGAGTACAAGAAGCTTTACGAAGAATCAATC         720
P  S  G  T  H  I  K  D  I  E  E  Y  K  K  L  Y  E  E  S  I
AAGAGCCCCGACACCTTCTGGGCACGCATGGCCCGCGAGCTCCTCACATTTGACAAGGAC         780
K  S  P  D  T  F  W  A  R  M  A  R  E  L  L  T  F  D  K  D
TTTGAAACCACACATCACGGCTCGTTTGAGAACGGCGACAATGCCTGGTTCGTCGAGGGT         840
F  E  T  T  H  H  G  S  F  E  N  G  D  N  A  W  F  V  E  G
CGGTTGAACGCATCGTTCAACTGTGTCGATCGCCATGCCCTCAAGAACCCAGATAAGGTC         900
R  L  N  A  S  F  N  C  V  D  R  H  A  L  K  N  P  D  K  V
GCCATTATTTATGAGGCCGACGAGCCCAACGAGGGCCGTAAGATCACCTACGGAGAGCTG         960
A  I  I  Y  E  A  D  E  P  N  E  G  R  K  I  T  Y  G  E  L
ATGCGCGAGGTGTCCCGGGTTGCCTGGACTCTGAAGGAGCGGTGGCGTCAAGAAGGGCGAC        1020
M  R  E  V  S  R  V  A  W  T  L  K  E  R  G  V  K  K  G  D
ACGGTCGGTATCTACCTGCCCATGATTCCCGAGGCCGTAATCGCTTTCCTGGCTTGCTCG        1080
T  V  G  I  Y  L  P  M  I  P  E  A  V  I  A  F  L  A  C  S
CGTATTGGTGCCGTGCACTCCGTTGTCTTCGCTGGTTTCTCTTCCGACTCCCTCCGGGAC        1140
R  I  G  A  V  H  S  V  V  F  A  G  F  S  S  D  S  L  R  D
CGTGTCCTGGACGCCTCCTCCAAGGTCATCATTACCTCCGACGAGGGCAAGCGCGGTGGC        1200
R  V  L  D  A  S  S  K  V  I  I  T  S  D  E  G  K  R  G  G
AAGATCATTGGCACTAAGAAGATTGTGGACGAGGCCATGAAGCAGTGCCCCGATGTGCAC        1260
K  I  I  G  T  K  K  I  V  D  E  A  M  K  Q  C  P  D  V  H
ACCGTGCTGGTGTACAAGCGCACCGGTGCCGAGGTGCCCTGGACCGCTGGCCGTGACATT        1320
T  V  L  V  Y  K  R  T  G  A  E  V  P  W  T  A  G  R  D  I
TGGTGGCACGAGGAGGTCGAGAAGTACCCCAACTACCTCGCCCCTGAGTCGGTCAGCTCC        1380
W  W  H  E  E  V  E  K  Y  P  N  Y  L  A  P  E  S  V  S  S
GAGGATCCTCTCTTCCTGTTGTACACCTCCGGTTCCACCGGTAAGCCCAAGGGTGTTATG        1440
E  D  P  L  F  L  L  Y  T  S  G  S  T  G  K  P  K  G  V  M
CACACCACTGCCGGTTACCTGCTCGGTGCGGCCATGACTGGAAAGTACGTGTTTGATATC        1500
H  T  T  A  G  Y  L  L  G  A  A  M  T  G  K  Y  V  F  D  I
CACGACGATGATCGCTACTTCTGCGGTGGTGATGTCGGTTGGATTACAGGTCACACCTAT        1560
H  D  D  D  R  Y  F  C  G  G  D  V  G  W  I  T  G  H  T  Y
GTCGTGTACGCCCCTCTATTGCTTGGCTGCGCCACCGTCGTGTTCGAGAGTACCCCCGCC        1620
V  V  Y  A  P  L  L  L  G  C  A  T  V  V  F  E  S  T  P  A
TACCCTAACTTCTCGCGCTACTGGGATGTCATTGACAAGCACGACGTCACACAATTCTAC        1680
Y  P  N  F  S  R  Y  W  D  V  I  D  K  H  D  V  T  Q  F  Y
GTTGCACCCACCGCTCTGCGTCTGCTGAAGCGCGCTGGAGATGAGCACATTCACCACAAG        1740
V  A  P  T  A  L  R  L  L  K  R  A  G  D  E  H  I  H  H  K
ATGCACAGTCTGCGTATTCTTGGCTCCGTCGGAGAGCCCATTGCCGCGGAAGTCTGGAAG        1800
M  H  S  L  R  I  L  G  S  V  G  E  P  I  A  A  E  V  W  K
TGGTACTTCGAGTGTGTTGGCAAGGAGGAAGCTCACATCTGCGACGTTCGTTCCCCCTTA        1860
W  Y  F  E  C  V  G  K  E  E  A  H  I  C  D
CCCCTTGGACCTTTTGGAATAACTTCTAATTTTTGGATCTGTAGACATACTGGCAAACCGA        1920
                                           T  Y  W  Q  T  E
GACCGGCTCACATGTCATCACCCCTCTCGGCGGTATCACCCCCCACCAAGCCCGGCAGTGC        1980
T  G  S  H  V  I  T  P  L  G  G  I  T  P  T  K  P  G  S  A
CTCCCTACCCTTCTTCGGTATCGAGCCTGCCATTATCGACCCCGTCTCCGGAGAGGAGAT        2040
S  L  P  F  F  G  I  E  P  A  I  I  D  P  V  S  G  E  E  I
TGTCGGCAATGATGTCGAGGGTGTTTTGGCCTTCAAGCAGCCGTGGCCCAGCATGGCCCG        2100
V  G  N  D  V  E  G  V  L  A  F  K  Q  P  W  P  S  M  A  R
CACCGTGTGGGGTGCCCACAAGCGTTACATGGACACTTACTTGAACGTGTACAAGGGTTA        2160
T  V  W  G  A  H  K  R  Y  M  D  T  Y  L  N  V  Y  K  G  Y
CTACGTAAGACGCTTCGCAGCCTGCCTTGCAGGGTTGATACTAACTCATATATAGTTCAC        2220
Y                                               F  T
CGGAGATGGTGCTGGCCGTGACCACGACGGCTATTACTGGATCCGCGGTCGTGTTGACGA        2280
G  D  G  A  G  R  D  H  D  G  Y  Y  W  I  R  G  R  V  D  D
TGTCGTCAACGTTTCTGGACACCGTCTGTCCACCGCTGAGATCGAGGCCGCTCTTCTCGA        2340
V  V  N  V  S  G  H  R  L  S  T  A  E  I  E  A  L  L  E
GCACCGTAAGTCCAACCACAGTATCTGCCAAAAATTGCAACTGAGCCCAAACTAACTATG        2400
H
AACAGCTTCCGTTGCCGAGGCTGCTGTCGTTGGTATTGCCGACGAGCTGACCGGTCAGGC        2460
A  S  V  A  E  A  A  V  V  G  I  A  D  E  L  T  G  Q  A
TGTCAATGCCTTTGTCTCTCTCAAGGAGGGCAAGCCCACAGAACAGATCAGCAAGGACCT        2520
V  N  A  F  V  S  L  K  E  G  K  P  T  E  Q  I  S  K  D  L
TGCCAATGCAAGTTCGCAAGTCCATTGGTCCCTTCGCCGCCCCCAAGGCTGTCTTCGTCGT        2580
A  M  Q  V  R  K  S  I  G  P  F  A  A  P  K  A  V  F  V  V
GGATGACCTCCCCAAGACCCGCAGTGGCAAGATCATGCGCCGAATCCTCCGGAAGATTCT        2640
D  D  L  P  K  T  R  S  G  K  I  M  R  R  I  L  R  K  I  L
CAGTGGCGAGGAGGACAGCCTCGGTGATACATCAACGGTAAGCATCATCTCTCAGCAAGA        2700
S  G  E  E  D  S  L  G  D  T  S  T
TAGTACCCGCAATCGTATCGTCCGAACAATAGCTAACGAAATATTCTTCACAGCTCTCCG        2760
                                                    L  S
ACCCCAGTGTCGTGGACAAGATCATAGAAACCGTCCACAGTGCTCGCCAGAAGTAAAGTG        2820
D  P  S  V  V  D  K  I  I  E  T  V  H  S  A  R  Q  K  -
```

Figure 12 Nucleotide and amino acid sequences of the acyl-CoA synthetase (AcCoAS) from *P. chrysogenum*. Introns are indicated with red letters.

have important differences. The fungal gene has five introns[168] while the bacterial gene has a typical prokaryotic structure (Figure 13). Moreover, the G + C content differs, being much lower in the AcCoAS gene. Both enzymes (PcAcCoAL and AcCoAS) are also different in their amino acid sequences, the similarities being restricted to an AMP-binding site consensus sequence (see Figure 13).

```
   1   CATGACACTCACCGCGTGGCTTGCAACCGCTGGCGCGCGGCGTACAAGAACAATTCGAGT

  61   GAAGCCATGAACATGTACCATGATGCCGACCGTGCCCTGTTGGACCCGATGGAAACCGCC
             M  N  M  Y  H  D  A  D  R  A  L  L  D  P  M  E  T  A

 121   AGTGTCGACGCCCTGCGCCAGCACCAGCTGGAGCGCCTGCGCTGGAGCCTGAAGCACGCC
         S  V  D  A  L  R  Q  H  Q  L  E  R  L  R  W  S  L  K  H  A

 181   TACGACAATGTGCCGCTGTACCGCCAGCGCTTTGCCGAATGCGGCGCCCACCCCGACGAC
         Y  D  N  V  P  L  Y  R  Q  R  F  A  E  C  G  A  H  P  D  D

 241   CTCACGTGCCTGGAAGACCTGGCGAAGTTCCCCTTCACCGGCAAGAACGACCTGCGCGAC
         L  T  C  L  E  D  L  A  K  F  P  F  T  G  K  N  D  L  R  D

 301   AACTACCCCTACGGGATGTTCGCCGTCCCCCAGGAAGAGGTGGTGCGCCTGCATGCTTCC
         N  Y  P  Y  G  M  F  A  V  P  Q  E  E  V  V  R  L  H  A  S

 361   AGCGGCACCACCGGCAAGCCGACGGTGGTCGGTTACACCCAGAATGACATCAACACCTGG
         S  G  T  T  G  K  P  T  V  V  G  Y  T  Q  N  D  I  N  T  W

 421   GCCAATGTCGTGGCGCGCTCGATCCGTGCGGCCGGCGGGCGCAAGGGTGACAAAGTGCAT
         A  N  V  V  A  R  S  I  R  A  A  G  G  R  K  G  D  K  V  H

 481   GTTTCCTACGGCTATGGGCTTTTCACTGGCGGGCTTGGTCGGCACTACGGCGCCGAGCGC
         V  S  Y  G  Y  G  L  F  T  G  G  L  G  R  H  Y  G  A  E  R

 541   CTGGGCTGTACGGTAATCCCGATGTCGGGTGGCCAGACCGAGAAGCAGGTGCAGCTGATC
         L  G  C  T  V  I  P  M  S  G  G  Q  T  E  K  Q  V  Q  L  I

 601   CGCGACTTTCAGCCCGACATCATCATGGTCACACCGTCCTACATGCTCAACCTGGCCGAC
         R  D  F  Q  P  D  I  I  M  V  T  P  S  Y  M  L  N  L  A  D

 661   GAGATCGAGCGCCAGGGCATCGACCCGCATGACCTCAAGCTACGCCTGGGCATTTTCGGT
         E  I  E  R  Q  G  I  D  P  H  D  L  K  L  R  L  G  I  F  G

 721   GCCGAACCTTGGACCGATGAACTACGTCGCTGATCGAGCAGCGCCTGGGCATCAATGCC
         A  E  P  W  T  D  E  L  R  R  S  I  E  Q  R  L  G  I  N  A

 781   CTCGACATCTATGGTTTGTCGGAAATCATGGGCCCCGGGGTGGCCATGGAATGCATCGAA
         L  D  I  Y  G  L  S  E  I  M  G  P  G  V  A  M  E  C  I  E

 841   ACCAAGGACGGCCCGACCATATGGGAAGACCACTTCTACCCCGAAATCATCGACCCGGTC
         T  K  D  G  P  T  I  W  E  D  H  F  Y  P  E  I  I  D  P  V

 901   ACCGGCGAAGTATTGCCAGACGGTCAGCTGGGCGAACTGGTGTTCACCTCGCTAAGCAAA
         T  G  E  V  L  P  D  G  Q  L  G  E  L  V  F  T  S  L  S  K

 961   GAGGCGCTTCCGATGGTGCGCTACCGCACCCGTGACCTCACCCGCCTGCTGCCCGGCACC
         E  A  L  P  M  V  R  Y  R  T  R  D  L  T  R  L  L  P  G  T

1021   GCCAGGCCGATGCGGCGGATCGGCAAGATTACCGGGCGCAGTGACGACATGCTGATCATT
         A  R  P  M  R  R  I  G  K  I  T  G  R  S  D  D  M  L  I  I

1081   CGCGGCGTCAACGTGTTCCCGACCCAGATCGAGGAACAGGTATTAAAAATAAAACAGCTT
         R  G  V  N  V  F  P  T  Q  I  E  E  Q  V  L  K  I  K  Q  L

1141   TCCGAGATGTATGAGATTCATTTGTATCGCAATGGCAACCTGGACAGCGTAGAGGTGCAT
         S  E  M  Y  E  I  H  L  Y  R  N  G  N  L  D  S  V  E  V  H

1201   GTAGAGTTGCGTGCGGAGTGCCAGCACCTCGATGAAGGCCAGCGCAAGCTGGTTATCGGG
         V  E  L  R  A  E  C  Q  H  L  D  E  G  Q  R  K  L  V  I  G

1261   GAGCTGAGCAAACAGATCAAGACCTACATCGGCATCAGCACCCAGGTGCACCTGCAGGCT
         E  L  S  K  Q  I  K  T  Y  I  G  I  S  T  Q  V  H  L  Q  A

1321   TGCGGCACGCTCAAGCGTTCCGAGGGCAAGGCGTGCCACGTGTACGACAAACGGTTGGCC
         C  G  T  L  K  R  S  E  G  K  A  C  H  V  Y  D  K  R  L  A

1381   AGCTGATTCATTCGGCTGCCT
         S  -
```

Figure 13 Nucleotide and amino acid sequences of the phenylacetyl-CoA ligase (PhAcCoAL) from *P. putida* U.

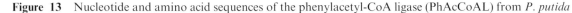

The above findings suggest that the use of enzymes that catalyze the same or a similar reaction in different microorganisms (or in other living systems) could offer a good biotechnological approach to qualitatively and quantitatively modify the final products obtained in certain enzymatic reactions. Accordingly, several acyl-CoA synthetases obtained from different biological sources were employed to synthesize R-CoAs that could later be used by the AT from *P. chrysogenum*, leading to the enzymatic production of new, modified, or well-known penicillins.[176,177] The general characteristics of some of these enzymes as well as their optimal physicochemical assay conditions are summarized in Table 1.

Table 1 Comparative study of different microbial acyl-CoA activating enzymes.

Enzyme	MW (kDa)	Optimal pH	Optimal T^c (°C)	K$_m$
Acyl-CoA synthetase (*Escherichia coli*)	130 (dimer or trimer)	7.5	35	11.2 µM (hexanoate)
Acyl-CoA synthetase (*Candida lipolytica*)	84	7.4	25	2.5 mM (palmitate)
Acyl-CoA synthetase (*Pseudomonas* sp.)	ND^a	8.1	25	ND
Acyl-CoA synthetase (*Pseudomonas fragi*)	ND	ND	37	ND
Acyl-CoA synthetase (*Pseudomonas putida* U)	67	7.0	40	4 mM (acetate)
Acyl-CoA synthetase (*Rhodospirillum rubrum*)	ND	7.5	32	41 mM (acetate)
Acetyl-CoA synthetase (*Methanotrix soehngenii*)	148 (dimer)	8.5	35	0.86 mM (acetate)
Acetyl-CoA synthetase (*Saccharomyces cerevisiae*)	151 (dimer)	7.6	25	0.28 mM (acetate)
Acetyl-CoA synthetase (*Penicillium chrysogenum*)	139 (dimer)	8.0	37	6.8 mM (acetate)
Phenylacetyl-CoA ligase (*P. putida* U)	48	82	30	16.6 mM (phenylacetate)
Benzoyl-CoA synthetase (*Pseudomonas* sp.)	120	8.5–9.2	30	11 µM (benzoate)
Benzoyl-CoA synthetase (*Rhodopseudomonas palustris*)	60	8.4–8.9	30	0.6–2 µM (benzoate)
2-Aminobenzoyl-CoA synthetase (*Pseudomonas* sp.)	60	9.3	30	13 µM (2-aminobenzoate)
2-Aminobenzoyl-CoA synthetase (*Pseudomonas* sp.)	65	8.5	30	70 µM (2-aminobenzoate)
Furoyl-CoA synthetase (*P. putida* FU$_1$)	85	8.5–9.5	30	0.75 mM (furoate)

^{a}ND, not determined.

4.10.7.5 Acyl-CoA:6-Aminopenicillanic Acid (Isopenicillin N) Acyltransferase

4.10.7.5.1 General properties

The last step in the biosynthetic pathway of benzylpenicillin in *P. chrysogenum* comprises the *N*-acylation of the amino group of 6-APA, or the replacement of the L-α-aminoadipic acid residue present in the IPN molecule by the acyl moiety of different acyl-CoA derivatives (Scheme 2). The enzyme catalyzing this reaction is an acyl-CoA:6-APA (IPN) AT, so-called transacylase or acyltransferase.[7,15,22,58–62,178] This protein, purified from *P. chrysogenum*[22] and *A. nidulans*,[62] was first assayed by Spencer and Maung[179] in cell-free extracts of *P. chrysogenum*. Those authors reported the existence of four activities attributable to a single protein: (i) isopenicillin *N*-acyltransferase activity; (ii) amidolyase activity (6-APA is generated from IPN); (iii) 6-APA acyltransferase activity; and (iv) hydrolytic activity on PhAc-CoA and other thioesters.[157] The molecular mass of the native AT from *P. chrysogenum* is 40 kDa and the enzyme has two different subunits of 10.5 kDa and 20 kDa, respectively.[14,22,62] Although different roles for the two subunits in the reaction of transacylation have been suggested,[180–182] their physiological functions have not yet been clarified[7] (see below). Maximal catalysis rates for AT were observed when reactions were carried out at 25 °C in buffer solutions with pH values ranging between 7.8–8.4.[22,152,153] These values are quite similar to those reported for other penicillin biosynthetic enzymes as well as for certain acyl-CoA ligases.[19,20,74,75,119,120,176,177] The gene encoding AT (*pen*DE) has been identified in *P. chrysogenum*[101] and in *A. nidulans*[183,184] and, in both cases, three introns were found. Barredo *et al.*[101] showed that in *P. chrysogenum* this gene encodes a protein of about 40 kDa, suggesting that posttranslational cleavage must occur. The sequences of the gene (*pen*DE) and protein (AT) from *P. chrysogenum* are shown in Figure 14.

4.10.7.5.2 Mechanistic studies and substrate specificity

Spencer and Maung[179] reported that the reaction catalyzed by AT can be explained by a Ping-Pong Bi–Bi mechanism. Baldwin and co-workers[181] have studied the modification of AT during the acyltransferase reaction using electrospray mass spectrometry, establishing the posttranslational changes. They concluded that the α-subunit (10.5 kDa) is acylated with the penicillin side-chain precursor as long as *P. chrysogenum* is grown in medium containing this compound. However, if the fungus is cultured in the absence of the precursor, the α-subunit remains unacylated. Surprisingly, when AT (α–β dimer form) was incubated *in vitro* with several acyl-CoA variants, only the β-subunit (29 kDa) was acylated, the acyl group being released when 6-APA was added to the reaction mixture.[179] The difference between the *in vivo* and *in vitro* acylating mechanisms has not yet been

```
          10                              50
TCGAGGTGCCCCAGTTGATGTCCCATCAGTGTCATGCTATGGTCCCAGATTGGTGGCTAC
                    100
GGCCAATATAAATCTCAGCATGCAGTTCCGCCTGCATGATCATCCCCAGGACGCGTTTGT
          150
CATCTCCGTCAGCCAGGTCTCAGTTGTTTACCCATCTTCCGACCCGCAGCAGAAATGCTT
                                                           M  L
              200
CACATCCTCTGTCAAGGCACTCCCTTTGAAGTAAGTGCTGCACTGAATACCAGATTTTTT
 H  I  L  C  Q  G  T  P  F  E
            250                                        300
CCTTCTGAATCTTCCGAGTTCTGACCTGATCCAGATCGGCTACGAACATGGCTCTGCTGC
                                   I  G  Y  E  H  G  S  A  A
CAAAGCCGTGATAGCCAGAAGCATTGACTTCGCCGTCGATCTCATCCGAGGGAAAACGAA
 K  A  V  I  A  R  S  I  D  F  A  V  D  L  I  R  G  K  T  K
                                     400
GAAGACGGACGAAGAGCTTAAACAGGTACTCTCGCAACTGGGGCGCGTGATCGAGGAAAG
 K  T  D  E  E  L  K  Q  V  L  S  Q  L  G  R  V  I  E  E  R
                              450
ATGGCCCAAATACTACGAGGAGATTCGCGGTGAGTGCCACTTCGGTCTTTCCTACATTTT
 W  P  K  Y  Y  E  E  I  R
                      500
CTGCACCAATGCTGACCGATGACCCCCGAAAAACCAGGTATTGCAAAGGGCGCTGAACGC
                                   G  I  A  K  G  A  E  R
      550                          600
GATGTCTCCGAGATTGTCATGCTTAATACCCGCACGGAATTTGCATACGGGCTCAAGGCA
 D  V  S  E  I  V  M  L  N  T  R  T  E  F  A  Y  G  L  K  A
                                     650
GCCCGTGATGGCTGCACCACTGCCTATTGTCAACTTCCAAATGGAGCCCTCCAGGGCCAA
 A  R  D  G  C  T  T  A  Y  C  Q  L  P  N  G  A  L  Q  G  Q
      970    980    990    700    1010   1020
AACTGGGATGTACGTTAAGAGATTTTACCTCCTCATTTTATTCAATCGAATTTGCGCCGA

      1030   1040          750    1060   1070   1080
CTAATTTGGTTGTTCAAGTTCTTTTCTGCCACCAAAGAGAACCTGATCCGGTTAACGATC
                    F  F  S  A  T  K  E  N  L  I  R  L  T  I
      1090   800    1110   1120   1130   1140
CGTCAGGCCGGACTTCCCACCATCAAATTCATAACCGAGGCTGGAATCATCGGGAAGGTT
 R  Q  A  G  L  P  T  I  K  F  I  T  E  A  G  I  I  G  K  V
      850    1160   1170   1180   1190   900
GGATTTAACAGTGCGGGCGTCGCCGTCAGTTACAACGCCCTTCACCTTCAGGGTCTTCGA
 G  F  N  S  A  G  V  A  V  N  Y  N  A  L  H  L  Q  G  L  R
                                     950
CCCACCGGAGTTCCTTCGCATATTGCCCTCCGCATAGCGCTCGAAAGCACTTCTCCTTCC
 P  T  G  V  P  S  H  I  A  L  R  I  A  L  E  S  T  S  P  S
                              1000
CAGGCCTATGACCGGATCGTGGAGCAAGGCGGAATGGCCGCCAGCGCTTTTATCATGGTG
 Q  A  Y  D  R  I  V  E  Q  G  G  M  A  A  S  A  F  I  M  V
                       1050
GGCAATGGGCACGAGGCATTTGGTTTGGAATTCTCCCCCACCAGCATCCGAAAGCAGGTG
 G  N  G  H  E  A  F  G  L  E  F  S  P  T  S  I  R  K  Q  V
                  1100
CTCGACGCGAATGGTAGGATGGTGCACACCAACCACTGCTTGCTTCAGCACGGCAAAAAT
 L  D  A  N  G  R  M  V  H  T  N  H  C  L  L  Q  H  G  K  N
      1150                                 1200
GAGAAAGAGCTCGATCCCTTACCGGACTCATGGAATCGCCACCAGCGTATGGAGTTCCTC
 E  K  E  L  D  P  L  P  D  S  W  N  R  H  Q  R  M  E  F  L
      1510   1520   1530   1540   1250   1560
CTCGACGGGTTCGACGGCACCAAACAGGCATTTGCCCAGCTCTGGGCCGACGAAGACAAT
 L  D  G  F  D  G  T  K  Q  A  F  A  Q  L  W  A  D  E  D  N
      1570   1580   1590   1300   1610   1620
TATCCCTTTAGCATCTGCCGCGCTTACGAGGAGGGCAAGAGCAGAGGCGCGACTCTGTTC
 Y  P  F  S  I  C  R  A  Y  E  E  G  K  S  R  G  A  T  L  F
      1630   1640   1350   1660   1670   1680
AATATCATCTACGACCATGCCCGTAGAGAGGCAACGGTGCGGCTTGGCCGGCCGACCAAC
 N  I  I  Y  D  H  A  R  R  E  A  T  V  R  L  G  R  P  T  N
      1690   1700   1710   1720   1730   1740
CCTGATGAGATGTTTGTCATGCGGTTTGACGAGGAGGACGAGAGGTCTGCGCTCAACGCC
 P  D  E  M  F  V  M  R  F  D  E  E  D  E  R  S  A  L  N  A
      1750   1760   1770   1780   1790   1800
AGGCTTTGAAGGCTCTTCATGACGAGCCAATGCATCTTTTGTATGTAGCTTCAACCGACT
 R  L  -
```

Figure 14 Nucleotide and amino acid sequence of the acyl-CoA:6-aminopenicillanic acid (isopenicillin N)-acyltransferase (AT) from *P. chrysogenum*. Introns are indicated with red letters.

clarified, but it could be due to certain variations in the structural organization of the functional enzymes when cell-free extracts are used.

AT is able to catalyze the conversion of 6-APA and phenylacetyl-CoA into penicillin G. Although the partially purified enzyme also recognizes IPN, this author's group have never been able to synthesize benzylpenicillin *in vitro* when AT is incubated with IPN and PhAc-CoA.[152] It is therefore suggested that the amidolyase activity is attributable to the α-subunit or to the complex α–β-subunits.[7] Conversely, other authors have reported that purified AT can use IPN as substrate[62,182–185] and hence further work is required to clarify this aspect. Below, the substrate specificity of AT is analyzed in depth. For this kind of experiment, three different groups of molecules were selected: (i) analogues of 6-APA; (ii) acyl-CoA variants containing an aliphatic molecule in the acyl moieties; and (iii) certain PhAc-CoA variants.

(i) Analogues of 6-APA

In these cases, the substrate specificity of AT was analyzed by replacing 6-APA by other β-lactam antibiotics (Figure 15) in the reaction shown in Equation (2).

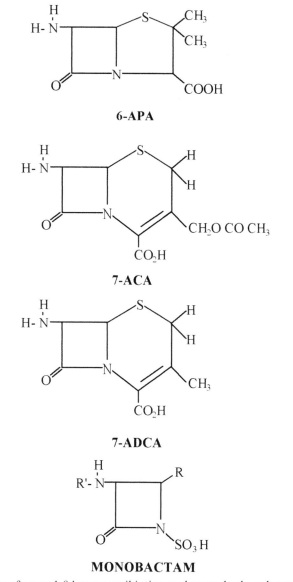

Figure 15 Structure of several β-lactam antibiotics used to study the substrate specificity of AT.

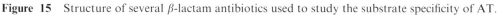

$$\textbf{PhAc-CoA} + \textbf{6-APA} \xrightarrow[\textbf{DTT}]{\textbf{AT}} \textbf{Penicillin G} + \textbf{CoA} \qquad (2)$$

When 6-APA was replaced by molecules containing the dihydrothiazine ring (7-amino-cephalosporanic acid, 7-ACA; or 7-aminodeacetoxycephalosporanic acid, 7-ADCA) the synthesis of cephalosporin-like products did not take place.[14,152] 3-Aminomonobactamic acid was also resistant to acylation, suggesting that the presence of a thiazolidine ring fused to the β-lactam is a basic structural requirement of the substrate molecule.[7] Other modifications of 6-APA, such as the presence of 6-methyl and 6-methoxy groups, prevents these compounds from being used as substrates. Modifications on the methyl groups of 6-APA as well as esterification of the carboxy group, or its replacement by a different function are currently under study. However, all the variants assayed suggest that none can replace 6-APA in the transacylation reaction.[186]

(ii) R-CoA in which R is an aliphatic molecule

In these cases, the substrate specificity of AT was tested by incubating AT in the presence of 6-APA and different acyl-CoA derivatives whose carbon length ranged between two (acetyl-CoA) and 18 (stearoyl-CoA). Penicillins were only obtained in sufficient quantity when hexanoyl-CoA, heptanoyl-CoA, and octanoyl-CoA were used as substrates. With valeryl-CoA and nonaoyl-CoA only traces of antibiotic were detected and the other compounds were not recognized by AT.[153] These data indicate that AT was the enzyme responsible for the synthesis of penicillins DF and K (which contains hexanoic acid and octanoic acid as side chains, respectively) as well as a different one, called penicillin H, which contains heptanoic acid linked to the amino group of 6-APA. These results were later confirmed by Applin *et al.*,[180,181] who demonstrated that the β-subunit of AT was acylated with the same penicillin side-chain precursors, with the exception of octanoyl-CoA, as those reported by this author's group.

Furthermore, *P. chrysogenum* produces an aliphatic penicillin (F)[2,4,24,37] (the first penicillin isolated from fermentation broths) that contains a double bond in its side chain (*trans*-3-hexenoic acid). To establish whether a single enzyme was able to synthesize this natural penicillin, different CoA derivatives of aliphatic molecules (C_6 to C_8) containing double or triple bonds were tested as substrates (Figure 16). Penicillins were only produced when *trans*-3-hexenoyl-CoA and *trans*-3-octenoyl-CoA were used as substrates.

In conclusion, AT can synthesize different penicillins containing either saturated or unsaturated linear aliphatic molecules as side chains as long as their carbon lengths range between C_6 to C_8 and free rotation about the C-2—C-3 bond exists.[155] When branched or cyclic aliphatic moieties were linked to CoA, these substrates were poorly used by AT and longer incubation times were usually required for penicillin to be synthesized. Other modifications in the aliphatic structure of the acyl-CoA derivatives (substitution of a methyl group by a carboxy function or replacement of a methylene by a keto group) consistently led to reduction in the synthetic efficiency of the corresponding penicillins with respect to controls in which no modifications had been induced.[173]

(iii) Phenylacetyl-CoA variants

To further characterize the substrate specificity of AT, PhAc-CoA, the precursor of benzyl-penicillin, was modified according to the following criteria. The substrate molecule was divided into three different parts (aromatic ring, acyl moiety, and CoA, see Figure 17) and each of these was modified, keeping two constant in a given experiment. All the PhAc-CoA analogues synthesized were included in three different groups:

(a) Analogues of PhAc-CoA, in which either different hydrogen atoms of the aromatic ring had been replaced by other atoms (or groups of atoms), or in which the benzene ring had been replaced by a different one (aromatic or not). Below, the substrates in this group are referred to as substrates with substitution on the ring.

(b) Analogues of PhAc-CoA with variations in the length of the acyl chain or with substitutions in the α-carbon atom (acyl variants).

(c) Analogues of PhAc-CoA in which the CoA moiety had been replaced by other sulfur-containing molecules (CoA variants).

(a) *With substitutions on the ring.* Substitution of one or more hydrogen atoms of the benzene ring by fluorine atoms led to different compounds that were efficiently used as substrates by AT and were transformed into penicillins with similar antibacterial activity to that reported for benzyl-

Good substrates

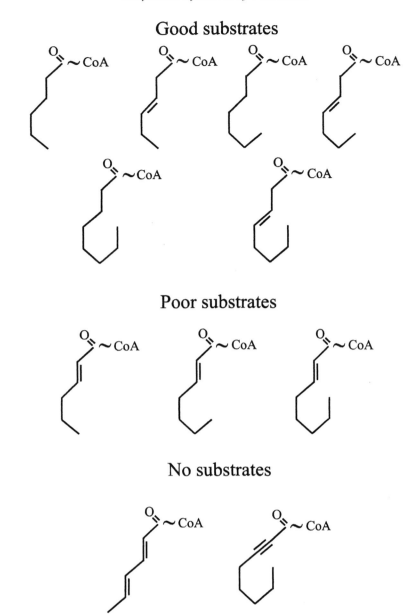

Poor substrates

No substrates

Figure 16 Hypothetical conformations of aliphatic-CoA molecules and their influence on recognition as substrates by AT.

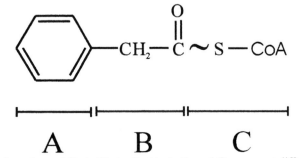

Figure 17 Structure of phenylacetyl-CoA (PhAc-CoA). A, B, and C represent different parts of the molecule that are modified to synthesize several derivatives.

penicillin.[173] However, their substitution by other halogen atoms (Cl, Br, I) afforded compounds which strongly inhibited AT.[187]

The presence of a hydroxy group in one of the three different positions (*o*-, *m*-, or *p*-) of the ring was accepted by AT and hence three different penicillins were obtained.[7] Enzymatic synthesis of penicillin X (which contains a molecule of 4-hydroxyphenylacetic acid as side chain) again showed that AT was the only enzyme involved in the synthesis of all the natural or semisynthetic penicillins accumulated in the fermentation broths.[7,154] When the size of the substituent was increased (methyl, methoxy, or amino group), only those compounds in which the replacements had been carried out at the *m*- or *p*-positions were transformed into penicillins, whereas their presence in the *o*-position prevented these molecules from being used as substrates.[154,156] This effect was also observed when a hydrogen atom was replaced by a NO_2 group. In these cases, penicillins were not produced, regardless of the position at which the substituent had been introduced.[173]

The foregoing data suggest that the molecular size of the aromatic ring linked to CoA is critical in the formation of the enzyme–substrate complex or, at least, that the interaction between the enzyme and the substrate at the catalytic site (active site) imposes certain structural limitations. It is therefore proposed that all the compounds included in this group used as substrates by AT are in fact used because they mimic rigid structural variants of several natural substrates of AT (hexanoyl-CoA, 3-hexenoyl-CoA, octanoyl-CoA)[7] (Figure 18). The results obtained when molecules containing methyl or methoxy groups were tested as substrates (see above) reinforce this hypothesis.[154,156] Figure 18 shows that *p*- and *m*-methylphenylacetyl-CoA as well as *p*- and *m*-methoxyphenylacetyl-CoA could represent rigid structures incorporating six to eight carbon atoms. However, in *o*-methyl-or *o*-methoxyphenylacetyl-CoA, the length of the rigid chains lies outside these limits in either orientation (a or b, equivalent to five and nine carbon atoms, respectively).

Another group of substrates includes molecules in which the benzene ring of PhAc-CoA had been replaced by a different ring system. In these reactions penicillins were efficiently produced when the natural substrate was substituted by another aromatic ring, whereas in other cases either penicillins were not produced, or only traces of antibiotic were synthesized after long incubation times (3 h or more).[173] It may therefore by concluded that the existence of a rigid planar substructure of the substrate molecule as well as the appropriate size are critical for recognition by the enzyme.[7]

(*b*) *With substitutions in the acyl chain*. This group includes molecules in which the acyl-CoA moiety was modified, keeping the benzene ring constant.[22,173] It has been shown that all compounds containing linear acyl chains shorter or longer than an acetyl-CoA residue (benzoyl-CoA, phenyl-propionyl-CoA, phenylbutyryl-CoA, etc.) are not recognized by AT. Other PhAc-CoA analogues in which a hydrogen atom linked to the α-carbon atom had been replaced by a different atom (or group of atoms) displayed different behavior, depending on the substituent. Whereas compounds in which the substituent was a hydroxy, methyl, or ethyl group led to the corresponding penicillins, other substitutions at this position (amino, nitro, keto group, or the presence of halogen atoms) prevented these compounds from being used as substrates.[173] The presence of an oxygen atom between the benzene ring and the acetyl-CoA moiety (phenoxyacetyl-CoA) was also accepted by AT and, therefore, penicillin V (phenoxymethylpenicillin) was also synthesized.[22] When the oxygen atom was replaced by a sulfur atom (thiophenoxyacetyl-CoA) it was not used by AT, pointing to the restricted ability of the enzyme to accept single variations on this part of the molecule.[173]

(*c*) *With replacement of the CoA moiety*. Although it has been shown that the synthesis of penicillin G and many other penicillins is carried out by direct acylation of 6-APA with different acyl-CoA variants (see above), the possibility has not yet been discussed of whether AT is able to use thiol esters other than R-CO ∼ S-CoA as substrates. Ferrero *et al.*[158] showed that penicillins G and V were synthesized when AT was incubated with 6-APA and (*S*-phenylacetyl)glutathione or (*S*-phenoxyacetyl)glutathione, respectively. These two antibiotics were also synthesized when certain *S*-glutathione-*S*-derivatives, (*S*-phenylacetyl)cysteinylglycine or (*S*-phenoxyacetyl)cysteinylglycine, were used as substrates, suggesting that some intermediates of the γ-glutamyl cycle would be directly involved in the biosynthesis of penicillin. Conversely, although (*S*-octanoyl)glutathione was not accepted as a substrate by AT, penicillin K was synthesized as long as CoA was supplied to the enzymatic assays.[158] These results suggested that either CoA stimulates the *N*-acylation of 6-APA from the glutathione-*S*-derivative or that formation of octanoyl-CoA (one of the natural substrates of AT) is generated in the assay. Different experiments revealed that in the presence of (*S*-octanoyl)glutathione (or other *S*-alkylglutathione derivatives) and CoA, a rapid nonenzymatic *S*→*S* intermolecular acyl migration between the octanoyl moiety of (*S*-octanoyl)glutathione or (*S*-octanoyl)cysteinylglycine and CoA occurs.[158] Taking into account these observations, it is proposed that *P. chrysogenum* may take up some penicillin side-chain precursors (those which, like PhAc, might be exogenously added to the cultures due to the incapacity of the fungus to synthesize

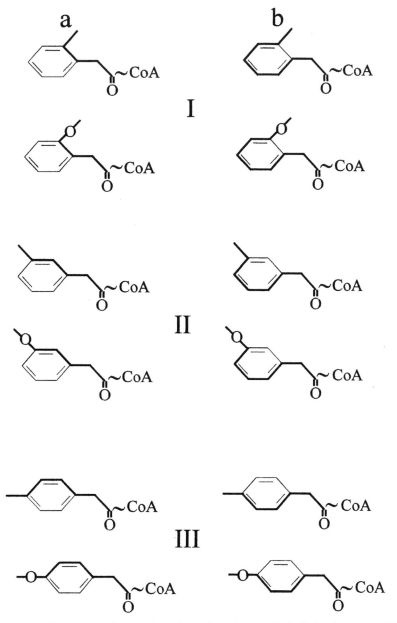

Figure 18 Structure of different methyl- and methoxyphenylacetyl-CoA derivatives. a and b correspond to different rigid conformations of these molecules.

them) by binding them to glutathione (GSH) in a reaction similar to the one that is catalyzed by glutathione-*S*-transferase.[158] In this reaction, GSH provides electrons through its sulfur atom for nucleophilic attack on, or reduction of, a second electrophilic substrate, resulting in the formation of a conjugate[158,188,189] (see Scheme 3). The action of a different enzyme, γ-glutamyltranspeptidase,[190,191] affords as a product (*S*-phenylacetyl)cysteinylglycine or similar (*S*-aryl)cysteinylglycine derivatives, which can then be used as substrates by AT.[158] The fact that greater quantities of benzylpenicillin and penicillin V were accumulated *in vitro* when AT was incubated with 6-APA and the two (*S*-aryl)cysteinylglycine derivatives suggests that the γ-glutamyl cycle could be directly involved in the production of the side-chain precursors of penicillin G and V. Although (*S*-aryl)glutathione or (*S*-aryl)cysteinylglycine derivatives are good substrates for AT, it is also possible that when a high intracellular concentration of CoA exists (conditions typical of secondary metabolism) an $S \to S$ intermolecular acyl migration between (*S*-phenylacetyl)- or (*S*-phenoxyacetyl)-glutathione or (*S*-phenylacetyl)- or (*S*-phenoxyacetyl)-cysteinylglycine and CoA, could occur resulting in the production of PhAc-CoA (or phenoxyacetyl-CoA) and GSH (or γ-glutamyl intermediates). Thus, once the exogenously added penicillin side-chain precursors have been taken up, they could be transformed into acyl-CoA derivatives via GSH or γ-glutamyl cycle intermediates,[191] whereas precursors that can be synthesized by the fungus (octanoyl-CoA, hexanoyl-CoA, and others) would be synthesized directly through β-oxidation or by fatty acid synthesis.

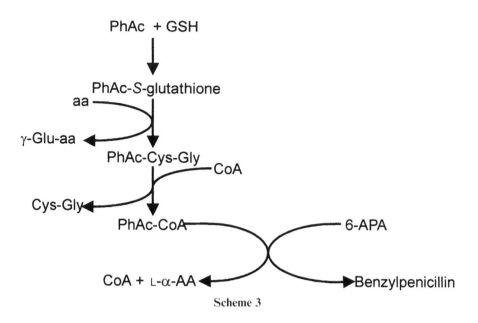

Scheme 3

In view of all the above-reported data concerning the substrate specificity of AT, several conclusions can be drawn.

(i) The carbon length of the acyl moiety in the substrate molecule (R in R-CoA) must range between six and eight carbon atoms (or equivalent rigid chain);

(ii) mobility at C-2 is required;

(iii) free rotation at C-2 is needed;

(iv) an increase in the volume of the side-chain precursor restricts the utilization of some molecules as substrates; and

(v) a thiol ester linked to the acyl moiety is required in the substrate molecule.

In summary, the use of very different substrates (PhAc-CoA, 3-thiopheneacetyl-CoA, 3-furoylacetyl-CoA, octanoyl-CoA) by a single enzyme demonstrates that these acyl-CoA derivatives could acquire very similar steric conformations during catalysis. The lack of conversion observed when rigid chains containing double or triple bonds at C-2 (which prevents free rotation of this linkage, thus avoiding the steric adaptability of such molecules) were tested as substrates[155] further supports the above conclusions.

4.10.8   ENZYMATIC COUPLED SYSTEMS

As shown above, the enzymatic synthesis of different penicillins requires the use of many acyl-CoA derivatives. These compounds can be obtained either commercially or through chemical synthesis.[22,192–195] In some cases, chemical synthesis of these molecules requires the protection and deprotection of certain reactive groups present in the precursor molecule, thus hindering the process of synthesis or lowering final yields. Accordingly, enzymatic synthesis of acyl-CoA derivatives seemed to be a good strategy to improve the quality as well as the number of reaction products (acyl-CoA thioesters). With this objective, different AcCoASs were selected: (i) AcCoAS from *Pseudomonas fragi*; (ii) AcCoAS from *P. putida* U; (iii) AcCoASs from *P. chrysogenum* and *A. nidulans*; and (iv) phenylacetyl-CoA ligase from *P. putida* U (see Table 1). These enzymes showed optimal physicochemical parameters very similar to those reported for the IPNS[119,120] and AT[7] purified from *P. chrysogenum*, suggesting that they could be coupled *in vitro*, thus reproducing the last step in the biosynthetic pathway of hydrophobic penicillins.[14,15,171,172] The sequences of the reactions leading to the production of different penicillins by incubating some AcCoAS with AT (a), or with IPNS and AT (b) are indicated in Scheme 4. The results obtained with the different coupled systems are summarized in the following sections.

(a) AcCoAS + AT

$$\text{R-COOH} + \text{ATP} + \text{6-APA} \xrightarrow[\text{Mg}^{2+}, \text{ DTT, CoA}]{\text{AcCoAS / AT}} \text{Penicillin (R-6-APA)} + \text{AMP} + \text{PPi}$$

(b) AcCoAS + IPNS + AT

$$\text{ACV} + \text{R-COOH} + \text{ATP} \xrightarrow[\substack{\text{Mg}^{2+}, \text{ Fe}^{2+}, \text{ Ascorbate,} \\ \text{DTT, CoA}}]{\text{AcCoAS / IPNS / AT}} \text{Penicillin} + \text{L-}\alpha\text{-AA} + \text{AMP} + \text{PPi}$$

Scheme 4

4.10.8.1 Enzymatic Synthesis of Penicillins Using AcCoAS from *P. fragi* and AT from *P. chrysogenum*

AcCoAS from *P. fragi* was purified from a commercial sample. With this coupled system penicillins were produced when AcCoAS and AT were incubated with ATP, Mg^{2+}, DTT, 6-APA, CoA, and several penicillin side-chain precursors (hexanoic, *trans*-3-hexenoic, heptanoic, octanoic, and *trans*-3-octenoic acids). The greatest amount of penicillin was obtained when octanoic and *trans*-3-octenoic acid were used as substrates. However, when PhAc-CoA or other aromatic analogues were tested, no penicillins were synthesized, indicating that the ligase does not recognize those molecules as substrates.[176]

4.10.8.2 Enzymatic Synthesis of Penicillins Using AcCoAS from *P. putida* U and AT from *P. chrysogenum*

AcCoAS was purified from cell-free extracts of *P. putida* U grown in a chemically defined medium containing octanoic acid as the sole carbon source.[177] This enzyme has broad substrate specificity and is able to activate many different compounds to their CoA derivatives. Thus, when it was coupled with AT (AcCoAS-AT), penicillins were synthesized in those reactions in which PhAc, phenoxyacetic acid, and several aliphatic molecules whose carbon length ranged between six and eight carbon atoms, were used as substrates.[177]

4.10.8.3 Enzymatic Synthesis of Penicillins when AcCoASs from *P. chrysogenum* and *A. nidulans* were coupled *in vitro* with AT from *P. chrysogenum*

AcCoASs were purified from *P. chrysogenum* and *A. nidulans* when these fungi were grown in a chemically defined medium containing acetate and other molecules as carbon sources.[75] According to the sequence of reactions indicated in Scheme 4, this coupled system leads to the production of different penicillins when R-COOH corresponds to PhAc, phenoxyacetic, 3-thiopheneacetic acids as well as several aliphatic compounds (hexanoic, *trans*-3-hexenoic, heptanoic, *trans*-3-octenoic, and octanoic acids).[75] The best substrate was found to be *trans*-3-hexenoic acid, the side-chain precursor of penicillin F.[24] However, when acetate was supplied to these reaction mixtures no antibiotic was formed in any case, since mainly acetyl-CoA (which is not a substrate of AT) was synthesized. Moreover, it was shown that penicillins began to be synthesized when the concentration of acetate in the assays was reduced.[75] These data, together with other observations about the effect of acetate on PhAc uptake,[65] allow one to speculate that acetate concentrations *in vivo*, or those of its derivative acetyl-CoA, might modulate the biosynthesis of several penicillins.[75] Therefore, it could be suggested that the transition from primary metabolism, or the trophophase (conditions that are unsuitable for penicillin production), to secondary metabolism, or the idiophase (a stage of growth in which penicillin is efficiently synthesized), could be regulated by the acetyl-CoA/CoA ratio.[7,66]

4.10.8.4 Enzymatic Synthesis of Penicillins Using PhAcCoAL from *P. putida* U and AT from *P. chrysogenum* in a Coupled System

PhAcCoAL was purified from *P. putida* U grown in a chemically defined medium containing PhAc as the sole carbon source.[74] The coupled system PhAcCoAL-AT is the most efficient one so far reported.[171,173] Using these two enzymes, penicillin synthesis can be achieved when the penicillin side-chain precursors are:

  (i) aliphatic acids (linear, branched, saturated, or unsaturated);
 (ii) structural analogues of PhAc with substitution on the aromatic ring;
(iii) PhAc analogues with different substitution on the α-carbon atom;
 (iv) PhAc analogues containing different acyl chains joined to the aromatic molecules; and
  (v) compounds in which the benzene ring of PhAc has been replaced by a different one.

The limitations imposed on all substrates whose structure corresponds to one of these groups for use as penicillin side-chain precursors are those derived from the requirements described for the substrate specificity of AT (see above).[7] Using this enzymatic synthesis, more that 60 different penicillins have been obtained *in vitro*. Among them are natural penicillins (F, DF, K, and others), some semisynthetic (G, V, X, and others), and also other penicillins of considerable clinical importance that, like ticarcillin, until this approach emerged could only be obtained by chemical synthesis.[173]

Additionally, the PhAcCoAL isolated from *P. putida* U activated PhAc to PhAc-CoA at a higher rate (500–600 fold higher) than the other acyl-CoA-activating enzymes. The PhAcCoAL-AT coupled system was the most efficient in all the assays, since more than 95% of the 6-APA present in the assay was converted into benzylpenicillin.[171] Taking into account these considerations, it seemed obvious that cloning and the expression of the *pcl* gene (encoding PhAcCoAL in *P. putida*) in a benzylpenicillin-producing strain such as *P. chrysogenum* or *A. nidulans* might offer a good approach for increasing the biosynthetic capacity of these fungi, for selecting genetically manipulated mutants able to synthesize unusual penicillins, or for modifying the structure and properties of the penicillins accumulated in the broth.

4.10.9 BIOTECHNOLOGICAL APPROACH: EXPRESSION OF PhAcCoAL FROM *P. PUTIDA* U IN *P. CHRYSOGENUM*

Since PhAc-CoA ligase was first purified from *P. putida* U,[74] it seemed that the cloning and expression of the gene encoding PhAcCoAL in *P. chrysogenum* could contribute to increasing the titres of penicillin G reached by this fungus if, as expected, PhAcCoAL is indeed a limiting step in the benzylpenicillin biosynthetic pathway.[196,197]

Once the PhAcCoAL gene had been characterized (Figure 13) its expression in *P. chrysogenum* was addressed.[198] *P. chrysogenum* Wis 54–1255 was transformed with a plasmid containing the following genetic information: (i) a gene that confers the fungus resistance to the antibiotic fleomycin (*ble*[r])[85] and that is under the control of the promoter of *P. chrysogenum* glutamate dehydrogenase (*Pgdh*); (ii) the promoter of the *pcb*AB gene, which encodes ACVs in *P. chrysogenum* (*Ppcb*AB); (iii) the *pcl* gene isolated from *P. putida* U; and (iv) the terminator of the *trp*C gene (*Ttrp*C) of *P. chrysogenum*. The construction designated *pALPs9* is represented schematically in Figure 19. It may be seen that *pcl* is under the control of the promoter of the *pcb*AB gene of *P. chrysogenum*.

Fungal transformants expressing the *ble*[r] gene were selected[198] and analyzed. PCR amplification of an internal *pcl* sequence showed that all of them contained the *pcl* gene,[175] whereas in the controls (nontransformed strains of *P. chrysogenum*) or in the transformants containing the same construction without the *pcl* gene no amplification was observed.[175] These results suggested that: (i) the *pcl* gene does not exist in the genome of *P. chrysogenum*; and (ii) similar sequences which may be amplified are not present either.

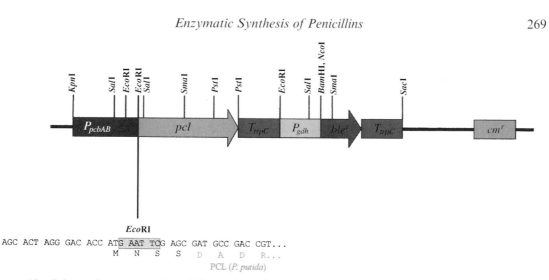

Figure 19 Schematic representation of the construction *pALPs*9 (9.0 kb) used to transform *P. chrysogenum*. P*pcb*AB, promoter of the *pcb*AB gene that encodes ACVS in *P. chrysogenum*; *pcl*, *P. putida* phenylacetyl-CoA ligase gene; T*trp*C, terminator of the *trp*C gene of *P. chrysogenum*; P*gdh*, promoter of *P. chrysogenum* glutamate dehydrogenase; *ble*^{r}, fleomycin resistance gene; and *cm*^{r}, chloramphenicol resistance gene.

However, these data did not reveal whether or not PCL was expressed in *P. chrysogenum*. Accordingly, different transformants, controls, and fungal strains transformed with a construction similar to *pALPs*9 but not containing the *pcl* gene were selected and, in them, the presence of PCL activity in fungal cell-free extracts was analyzed. It was observed that in all the transformants containing the *pcl* gene good PCL activity was found, whereas under the assay conditions used such activity was not detected either in the controls or in the transformants lacking this gene.[175] These data show that the *pcl* gene from *P. putida* was being efficiently expressed in *P. chrysogenum*.

HPLC determinations of penicillin G accumulated in the culture broths by these strains[199] reveal that the transformants containing *pcl* from *P. putida* U produced between 84% and 121% more benzylpenicillin than the controls, showing that the expression of this gene in *P. chrysogenum* is responsible for a strong increase in the rate of synthesis of penicillin G in this fungus.[175] Furthermore, this effect specifically concerns benzylpenicillin biosynthesis, since the rates of synthesis of 6-APA and natural penicillins were not affected when *P. chrysogenum* transformants were cultured in fermentation broths lacking PhAc.[200]

4.10.10 CONCLUDING REMARKS AND FUTURE OUTLOOK

The biochemical knowledge of the different enzymes (ACVS, IPNS, AT, PhAcTS, and acyl-CoA-activating enzymes) involved in the biosynthetic pathway of penicillins has allowed the collection *in vitro* of many different antibiotics that cannot be produced *in vivo*. It would therefore be expected that chemical synthesis of analogues of the natural substrates using single enzymatic reactions, as reported above, would enhance the production of new, modified penicillins or other β-lactam antibiotics. In several cases, such studies have also facilitated the establishment of structure–function relationships and could therefore contribute in the near future to the design of new molecules with broader or more specific antibacterial spectra.

Moreover, the isolation, cloning and expression of the genes encoding the penicillin biosynthetic enzymes could help: (i) to modify, through genetic engineering, the substrate specificity of these enzymes; (ii) to alter the regulatory mechanisms controlling the expression of these genes; (iii) to clone the biosynthetic genes in different microbes possibly more amenable to industrial fermentation (with better utilization of raw materials, requiring less shaking or airing, facilitating extraction etc.); and (iv) to obtain mutants containing altered genes, encoding abnormal proteins, that will allow the synthesis of a particular type of penicillin or mutants lacking some enzyme (peptidases or amidases) that destroy the penicillin accumulated in the broths. Several papers have been published on different aspects of genetic engineering and in them the reader will find new strategies or more detailed discussion of certain points not covered here.[6,15,105,201–206]

New discoveries and advances in industrial engineering are currently contributing to the development of a still incipient technology which should lead to the immobilization of these enzymes in reactors, thus enabling the enzymatic synthesis of these compounds from their single precursors at laboratory scale (or at pilot-plant scale).

ACKNOWLEDGMENTS

I am gratefully indebted to E. R. Olivera for his important contributions, friendly support, and helpful collaboration in the preparation of this manuscript. Thanks are also given to my students B. Miñambres, B. García, C. Muñiz, M. Fernández-Valverde, D. Carnicero, and A. Fernández-Medarde. I would like to thank R. Sánchez-Barbero for typing the manuscript. The investigation was supported by grants from the Comisión Interministerial de Ciencia y Tecnología (CICYT) Madrid, Spain (Grant No. BIO96-0403), Antibióticos, S.A. (León, Spain), and the Junta de Castilla y León (Consejería de Cultura y Turismo) Valladolid, Spain. Special thanks are given to the *Journal of Antibiotics* and to the *Journal of Biological Chemistry* for authorization to reproduce part of the results published by us in these journals. This chapter is dedicated to Dr Clemente Luengo, my dear father, who with his daily work aroused my own interest in science.

4.10.11   REFERENCES

1. A. Fleming, *Br. J. Exp. Pathol.*, 1929, **10**, 226.
2. A. L. Demain, in "Antibiotics Containing the β-Lactam Structure," eds. A. L. Demain and N. A. Solomon, Springer-Verlag, Berlin, 1983, vol. 1, p. 189.
3. R. Southgate and S. Elson, in "Progress in the Chemistry of Organic Products," eds. W. Herz, H. Grisebach, G. W. Kirby, and C. Tamm, Springer-Verlag, Wien, 1985, vol. 47, p. 1.
4. E. P. Abraham, in "50 Years of Penicillin Application," eds. H. Kleinhauf and H. von Döhren, Public, Czech Republic, 1991, p. 7.
5. G. M. Eliopoulos, in "β-Lactam Antibiotics for Clinical Use," eds. S. F. Queener, J. A. Webber, and S. W. Queener, Marcel Dekker, New York, 1986, vol. 14, p. 227.
6. Y. Aharonowitz, G. Cohen, and J. F. Martín, *Annu. Rev. Microbiol.*, 1992, **46**, 461.
7. J. M. Luengo, *J. Antibiotics*, 1995, **48**, 1195.
8. H. Aoki, H. Sakai, M. Koshaka, T. Konami, J. Hosada, Y. Kubochi, E. Iguchi, and H. Imanaka, *J. Antibiotics*, 1976, **29**, 492.
9. A. Imada, K. Kitano, K. Kintaka, M. Muroi, and M. Asai, *Nature*, 1981, **289**, 590.
10. F. P. Doyle and J. H. C. Nayler, in "Advances in Drug Research," eds. N. J. Harper and A. B. Simmonds, Academic Press, New York, 1964, vol. 1, p. 1.
11. A. G. Brown, D. Butterworth, M. Cole, G. Hanscomb, J. D. Hood, C. Reading, and G. N. Rolinson, *J. Antibiotics*, 1976, **29**, 668.
12. G. G. F. Newton and E. P. Abraham, *Nature*, 1955, **175**, 548.
13. R. Nagarajan, L. D. Boeck, M. Gorman, R. L. Hamill, C. E. Higgens, M. M. Hoehn, W. M. Stark, and J. G. Whitney, *J. Am. Chem. Soc.*, 1971, **93**, 2308.
14. J. M. Luengo, in "Progress in Industrial Microbiology," eds. M. E. Bushell and U. Gräfe, Elsevier, Amsterdam, 1989, vol. 27, p. 315.
15. J. M. Luengo and M. A. Peñalva, in "Progress in Industrial Microbiology," eds. D. Martinelli and J. R. Kinghorn, Elsevier, Amsterdam, 1994, vol. 29, p. 603.
16. R. P. Elander, in "Antibiotics Containing the β-Lactam Structure," eds. A. L. Demain and N. A. Solomon, Springer-Verlag, Berlin, 1983, vol. 1, p. 97.
17. A. L. Demain, *Adv. Appl. Microbiol.*, 1973, **16**, 177.
18. Y. Aharonowitz, J. Bergmeyer, J. M. Cantoral, G. Cohen, A. L. Demain, U. Fink, J. Kinghorn, H. Kleinkauf, A. MacCabe, H. Palissa, E. Pfeifer, T. Schweck, H. van Liempt, H. von Döhren, S. Wolfe, and J. Zang, *Bio/Technology*, 1993, **11**, 807.
19. J. E. Baldwin and M. Bradley, *Chem. Rev.*, 1990, **90**, 1079.
20. H. van Liempt, H. von Döhren, and H. Kleinkauf, *J. Biol. Chem.*, 1989, **264**, 3680.
21. J. E. Baldwin, in "50 Years of Penicillin Application," eds. H. Kleinkauf and H. von Döhren, Public, Czech Republic, 1991, p. 151.
22. M. J. Alonso, F. Bermejo, A. Reglero, J. M. Fernández-Cañón, J. M. González de Buitrago, and J. M. Luenge, *J. Antibiotics*, 1988, **41**, 1074.
23. F. Gómez-Pardo and M. A. Peñalva, *Gene*, 1990, **89**, 109.
24. E. P. Abraham, in "The Chemistry of Penicillins," eds. H. W. Florey, E. Chain, N. G. Heatley, M. A. Jennings, A. G. Sanders, E. P. Abraham, and M. E. Florey, Academic Press, London, 1949, p. 768.
25. R. B. Woodward, in "The Chemistry of Penicillins," eds. H. T. Clarke, J. R. Johnson, and R. Robinson, Princeton University Press, Princeton, NJ, 1949, vol. 49, p. 473.
26. F. R. Batchelor, F. D. Doyle, J. H. C. Nayler, and G. N. Rolinson, *Nature*, 1959, **183**, 257.
27. A. Ballio, E. B. Chain, F. Dentice di Accadia, M. Mauri, K. Rauer, M. J. Schlesinger, and S. Schlesinger, *Nature*, 1961, **191**, 909.
28. Y. B. Tewaari and R. N. Goldberg, *Biophys. Chem.*, 1988, **29**, 245.

29. N. Burteau, S. Burton, and R. R. Crichton, *FEBS Lett.*, 1989, **258**, 185.
30. R. L. Green, J. E. Lewis, S. J. Kraus, and E. L. Frederikson, *N. Engl. J. Med.*, 1974, **291**, 223.
31. U. Roze and J. L. Strominger, *Mol. Pharmacol.*, 1964, **2**, 92.
32. D. J. Waxman and J. L. Strominger, *Annu. Rev. Biochem.*, 1983, **52**, 825.
33. D. J. Tipper and J. L. Strominger, *Proc. Natl. Acad. Sci. USA*, 1965, **54**, 1133.
34. K. Izaki, M. Matsuhashi, and J. L. Strominger, *J. Biol. Chem.*, 1965, **243**, 3180.
35. D. A. Wilson and G. N. Rolinson, *Chemotherapy*, 1979, **25**, 14.
36. P. Giesbreckt, M. Franz, D. Krüger, H. Labischinski, and J. Wecke, in "50 Years of Penicillin Application," eds. H. Kleinkauf and H. von Döhren, Public, Czech Republic, 1991, p. 353.
37. B. B. Mukherjee and B. L. Lee, in "Journal Chromatography Library," eds. M. J. Weinstein and G. H. Wagman, Springer-Verlag, Berlin, 1978, vol. 15, p. 387.
38. A. L. Demain, *Ann. N.Y. Acad. Sci.*, 1974, **235**, 601.
39. D. Gottlieb, *J. Antibiotics*, 1976, **29**, 987.
40. D. Bonner, *Arch. Biochem. Biophys.*, 1947, **13**, 1.
41. A. L. Demain, *Arch. Biochem. Biophys.*, 1957, **67**, 244.
42. P. S. Masurekar and A. L. Demain, *Can. J. Microbiol.*, 1972, **18**, 1045.
43. A. L. Demain and P. S. Masurekar, *J. Gen. Microbiol.*, 1974, **28**, 143.
44. P. S. Masurekar and A. L. Demain, *Appl. Microbiol.*, 1974, **28**, 265.
45. J. M. Luengo, G. Revilla, J. R. Villanueva, and J. F. Martín, *J. Gen. Microbiol.*, 1979, **115**, 207.
46. J. M. Luengo, G. Revilla, M. J. López-Nieto, J. R. Villanueva, and J. F. Martín, *J. Bacteriol.*, 1980, **144**, 869.
47. A. L. Demain, *Lloydia*, 1974, **37**, 147.
48. H. R. Arnstein and D. Morris, *Biochem. J.*, 1960, **76**, 357.
49. P. B. Loder and E. P. Abraham, *Biochem. J.*, 1971, **123**, 477.
50. P. Adriaens, B. Messchaert, W. Wuyts, H. Vanderhaeghe, and H. Eyssen, *Antimicrob. Agents Chemother.*, 1975, **8**, 638.
51. P. Fawcett and E. P. Abraham, in "Methods in Enzymology," ed. J. H. Hash, Academic Press, New York, 1975, vol. 43, p. 471.
52. G. Banko, S. Wolfe, and A. L. Demain, *Biochem. Biophys. Res. Commun.*, 1986, **137**, 528.
53. G. Banko, A. L. Demain, and S. Wolfe, *J. Am. Chem. Soc.*, 1987, **109**, 2858.
54. J. Zhang and A. L. Demain, *Crit. Rev. Biotechnol.*, 1992, **12**, 245.
55. I. H. Hollander, Y.-Q. Shen, J. Heim, A. L. Demain, and S. Wolfe, *Science*, 1984, **224**, 610.
56. S. M. Samson, R. Belagaje, D. T. Blaukenship, J. L. Chapman, D. Perry, P. L. Skatrud, R. M. VanFrank, E. P. Abraham, J. E. Baldwin, S. W. Queener, and T. D. Ingolia, *Nature*, 1985, **318**, 191.
57. K. E. Price, in "Structure–Activity Relationships of Semisynthetic Penicillins," ed. D. Perlman, Academic Press, New York, 1977, p. 1.
58. R. C. Erickson and L. D. Dean, *Appl. Microbiol.*, 1966, **14**, 1047.
59. B. Spencer, *Biochem. Biophys. Res. Commun.*, 1968, **31**, 170.
60. S. Gatenbeck and H. Brunsberg, *Acta Chem. Scand.*, 1968, **22**, 1059.
61. S. Gatenbeck, in "Methods in Enzymology," ed. J. H. Hash, Academic Press, New York, 1975, vol. 43, p. 474.
62. P. A. Whiteman, E. P. Abraham, J. E. Baldwin, M. D. Fleming, C. J. Schofield, J. D. Sutherland, and A. C. Willis, *FEBS Lett.*, 1990, **262**, 342.
63. J. M. Fernández-Cañón, A. Reglero, H. Martínez-Blanco, and J. M. Luengo, *J. Antibiotics*, 1989, **42**, 1398.
64. J. M. Fernández-Cañón, A. Reglero, H. Martínez-Blanco, M. A. Ferrero, and J. M. Luengo, *J. Antibiotics*, 1989, **42**, 1410.
65. H. Martínez-Blanco, A. Reglero, M. A. Ferrero, J. M. Fernández-Cañón, and J. M. Luengo, *J. Antibiotics*, 1989, **42**, 1416.
66. J. M. Fernández-Cañón and J. M. Luengo, *J. Antibiotics*, 1997, **50**, 45.
67. R. Brunner and M. Rohr, in "Methods in Enzymology," ed. J. H. Hash, Academic Press, New York, 1975, vol. 43, p. 476.
68. R. Kogekar and V. D. Deshpande, *Indian J. Biochem. Biophys.*, 1982, **19**, 257.
69. A. P. J. Trinci and R. C. Righelato, *J. Gen. Microbiol.*, 1970, **60**, 239.
70. W. Kurylowicz, W. Kurzatkowski, W. Woznicka, H. Polowniak-Pracka, A. Paskiewicz, J. Luba, and J. Piorunowski (eds.) "Atlas of Ultrastructure of *Penicillium chrysogenum* in the Course of Biosynthesis of Penicillin," Chemia, Warsaw, 1980.
71. W. Kurzatkowski, *Med. Doswkladi Mikrobiol.*, 1981, **33**, 15.
72. W. Kurzatkowski, W. Kurylowicz, and A. Paskiewicz, *Eur. J. Appl. Microbiol. Biotechnol.*, 1982, **15**, 211.
73. J. M. Luengo, A. Domínguez, J. M. Cantoral, and J. F. Martín, *Curr. Microbiol.*, 1986, **13**, 203.
74. H. Martínez-Blanco, A. Reglero, L. B. Rodríguez-Aparicio, and J. M. Luengo, *J. Biol. Chem.*, 1990, **265**, 7084.
75. H. Martínez-Blanco, A. Reglero, M. Fernández-Valverde, M. A. Ferrero, M. A. Moreno, M. A. Peñalva, and J. M. Luengo, *J. Biol. Chem.*, 1992, **267**, 5474.
76. W. H. Muller, T. P. van der Krift, A. J. J. Krouwer, H. A. B. Wösten, L. H. M. van der Voort, E. B. S. Smaal, and A. J. Verkleig, *EMBO J.*, 1991, **10**, 489.
77. W. H. Muller, R. A. L. Bovenberg, M. H. Groothuis, F. Kattevilder, E. B. Smaal, L. H. M. van der Voort, and A. J. Verkleig, *Biochim. Biophys. Acta*, 1992, **1116**, 210.
78. W. Kurzatkowski, H. Palissa, H. van Liempt, H. von Döhren, H. Kleinkauf, W. P. Wolfe, and W. Kurylowicz, *Appl. Microbiol. Biotechnol.*, 1991, **35**, 517.
79. T. Lendenfeld, D. Ghali, M. Wolschek, E. M. Kubicek-Prank, and C. P. Kubicek, *J. Biol. Chem.*, 1993, **268**, 665.
80. W. Kurylowicz, W. Kurzatkowski, and J. Kurzatkowski, *Arch. Immunol. Ther. Exp.*, 1987, **35**, 699.
81. R. Thieringer and W.-H. Kunau, *J. Biol. Chem.*, 1991, **266**, 13110.
82. M. A. Peñalva, A. Vian, C. Patiño, A. Perez-Aranda, and D. Ramón, in "Genetics and Molecular Biology of Industrial Microorganisms," eds. C. L. Hershberger, S. W. Queener, and G. Hegeman, American Society for Microbiology, Washington, DC, 1989, p. 256.
83. G. Cohen, D. Schiffman, M. Mevarech, and Y. Aharonowitz, *Trends in Biotechnol.*, 1990, **8**, 105.
84. J. R. Miller and T. D. Ingolia, *Mol. Microbiol.*, 1989, **3**, 689.

85. B. Díez, G. Gutierrez, J. L. Barredo, P. van Solingen, L. H. M. van der Voort, and J. F. Martín, *J. Biol. Chem.*, 1990, **265**, 16 358.
86. D. J. Smith, M. K. R. Burham, J. H. Bull, J. E. Hodgson, J. M. Ward, P. Browne, B. Barton, A. J. Earl, and G. Turner, *EMBO J.*, 1990, **9**, 741.
87. D. J. Smith, A. J. Earl, and G. Turner, *EMBO J.*, 1990, **9**, 2743.
88. L. G. Carr, P. L. Skatrud, M. E. Scheetz III, S. W. Queener, and T. D. Ingolia, *Gene*, 1986, **48**, 257.
89. J. L. Barredo, B. Díez, E. Alvarez, and J. R. Martín, *Curr. Genet.*, 1989, **16**, 453.
90. B. J. Weigel, S. G. Burgett, V. J. Chen, P. L. Skatrud, C. A. Frolik, S. W. Queener, and T. D. Ingolia, *J. Bacteriol.*, 1988, **170**, 3817.
91. D. Ramón, L. Carramolino, C. Patiño, F. Sánchez, and M. A. Peñalva, *Gene*, 1987, **57**, 171.
92. B. K. Leskiw, Y. Aharonowitz, M. Mevarech, S. Wolfe, L. C. Vining, D. W. S. Westlake, and S. E. Jensen, *Gene*, 1988, **62**, 187.
93. J. L. Barredo, J. M. Cantoral, E. Alvarez, B. Díez, and J. F. Martín, *Mol. Gen. Genet.*, 1989, **216**, 91.
94. J. J. R. Coque, J. F. Martín, J. G. Calzada, and P. Liras, *Mol. Microbiol.*, 1991, **5**, 1125.
95. A. MacCabe, M. van Liempt, H. Palissa, S. E. Unkles, M. B. R. Riach, E. Pfeifer, H. von Döhren, and J. R. Kinghorn, *J. Biol. Chem.*, 1991, **266**, 12 646.
96. B. Pérez-Esteban, M. Orejas, E. Gómez-Pardo, and M. A. Peñalva, *Mol. Microbiol.*, 1993, **9**, 881.
97. M. Kolar, K. Holzmann, G. Weber, E. Leitner, and H. Schwab, *J. Biotechnol.*, 1991, **17**, 67.
98. B. Feng, E. Friedlin, and G. A. Marzluf, *Curr. Genet.*, 1995, **27**, 351.
99. Y.-W. Chu, D. Renno, and G. Daunders, *Curr. Genet.*, 1995, **28**, 184.
100. H. Haas and G. A. Marzluf, *Curr. Genet.*, 1995, **28**, 177.
101. J. L. Barredo, P. van Solingen, B. Díez, E. Alvarez, J. M. Cantoral, A. Kattevilder, E. B. Smaal, M. A. M. Groenen, A. E. Veenstra, and J. F. Martín, *Gene*, 1989, **83**, 291.
102. B. L. Miller, K. Y. Miller, K. A. Robert, and W. E. Timberlake, *Mol. Cell Biol.*, 1987, **7**, 427.
103. J. R. Martin and S. Gutierrez, *Anton. Leeuw. Int. J.G.*, 1995, **67**, 181.
104. M. A. Peñalva, B. Pérez-Esteban, E. Gómez-Pardo, M. Orejas, and E. Espeso, in "Molecular Biology of Filamentous Fungi," eds. U. Stahl and P. Tudzinsky, VCN, Weinheim, 1992, p. 217.
105. G. Landan, G. Cohen, Y. Aharonowitz, Y. Shuali, G. Graur, and D. Schiffman, *Mol. Biol. Evol.*, 1990, **7**, 399.
106. F. Fierro, J. L. Barredo, B. Diez, S. Gutierrez, F. J. Fernández, and J. F. Martín, *Proc. Natl. Acad. Sci. USA*, 1995, **92**, 13, 6200.
107. D. J. Smith, J. H. Bull, J. Edwards, and G. Turner, *Mol. Gen. Genet.*, 1989, **216**, 492.
108. S. E. Jensen, A. Wong, M. J. Rollins, and D. W. S. Westlake, *J. Bacteriol.*, 1990, **172**, 7269.
109. J. Y. Zhang and A. L. Demain, *Biotechnol. Lett.*, 1990, **12**, 649.
110. M. Krause and M. A. Marahiel, *J. Bacteriol.*, 1988, **170**, 4669.
111. G. Mittenhuber, R. Weckermann, and M. A. Marahiel, *J. Bacteriol.*, 1989, **171**, 4881.
112. H. Kleinkauf and H. von Döhren, *Annu. Rev. Microbiol.*, 1987, **41**, 259.
113. S. Gutiérrez, B. Díez, E. Montenegro, and J. F. Martín, *J. Bacteriol.*, 1991, **173**, 2354.
114. T. A. Kadima, S. E. Jensen, and M. A. Pickard, *J. Ind. Microbiol.*, 1995, **14**, 35.
115. C. Y. Shiau, J. E. Baldwin, M. F. Byford, W. J. Sobey, and C. J. Schofield, *FEBS Lett.*, 1995, **358**, 97.
116. C. J. Shiau, J. E. Baldwin, M. F. Byford, and C. J. Schofield, *FEBS Lett.*, 1991, **37**, 303.
117. J. O'Sullivan, R. C. Bleaney, J. A. Huddleston, and E. P. Abraham, *Biochem. J.*, 1979, **184**, 421.
118. T. Konomi, S. Herchen, J. E. Baldwin, B. Yoshida, N. A. Hunt, and A. L. Demain, *Biochem. J.*, 1979, **184**, 427.
119. C.-C. Pang, B. Chakravarti, R. M. Adlington, H.-H. Ting, R. L. White, G. S. Jayatilake, J. E. Baldwin, and E. P. Abraham, *Biochem. J.*, 1984, **222**, 789.
120. F. R. Ramos, M. J. López-Nieto, and J. F. Martín, *Antimicrob. Agents Chemother.*, 1985, **27**, 380.
121. S. E. Jensen, D. W. S. Westlake, and S. Wolfe, *J. Antibiotics*, 1982, **35**, 483.
122. D. Schiffman, M. Mevarech, S. E. Jensen, G. Cohen, and J. Aharonowitz, *Mol. Gen. Genet.*, 1988, **214**, 562.
123. H. Palissa, H. von Döhren, H. Kleinkauf, H.-H. Ting, and J. E. Baldwin, *J. Bacteriol.*, 1989, **171**, 5720.
124. P. L. Roach, C. J. Schofield, J. E. Baldwin, I. J. Clifton, and J. Hajdu, *Protein Sci.*, 1995, **4**, 1007.
125. P. L. Roach, I. J. Clifton, V. Fülöp, K. Harlos, G. J. Barton, J. Hajdu, I. Andersson, C. J. Schofield, and J. E. Baldwin, *Nature*, 1995, **375**, 700.
126. F. Fierro, S. Gutiérrez, B. Diez, and J. F. Martín, *Mol. Gen. Genet.*, 1993, **241**, 573.
127. A. L. Feig and S. J. Lippard, *Chem. Rev.*, 1994, **94**, 759.
128. J. E. Baldwin and E. P. Abraham, *Nat. Prod. Rep.*, 1988, **5**, 129.
129. J. E. Baldwin, G. P. Lynch, and C. J. Schofield, *Tetrahedron Lett.*, 1992, **48**, 9085.
130. J. M. Blackburn, J. D. Sutherland, and J. E. Baldwin, *Biochemistry*, 1995, **34**, 7548.
131. A. Krianciunas, C. A. Frolik, T. C. Hassell, P. L. Skatrud, M. G. Johnson, N. L. Holbrook, and V. C. Chen, *J. Biol. Chem.*, 1991, **266**, 11 779.
132. F. Jiang, J. Peisach, L.-J. Ming, L. Que, Jr., and V. Chen, *Biochemistry*, 1991, **30**, 11 437.
133. L.-J. Ming, L. Que, Jr, A. Kriaucinas, C. A. Frolik, and V. J. Chen, *Biochemistry*, 1991, **30**, 11 653.
134. J. E. Baldwin, E. P. Abraham, R. M. Adlington, G. A. Bahadur, B. Chakravarti, B. P. Domayne-Hayman, L. D. Field, S. L. Flitsch, G. S. Jayatilake, A. Spakovskis, H. H. Ting, N. J. Turner, R. L. White, and J. J. Usher, *J. Chem. Soc. Chem. Commun.*, 1984, 1225.
135. S. Wolfe, I. H. Hollander, and A. L. Demain, *Bio/Technology*, 1984, **2**, 635.
136. S. E. Jensen, D. W. S. Westlake, R. Bowers, C. F. Ingold, M. Jouany, L. Lyubechansky, and S. Wolfe, *Can. J. Chem.*, 1984, **62**, 2712.
137. J. M. Luengo, M. T. Alemany, F. Salto, F. Ramos, M. J. López-Nieto, and J. F. Martín, *Bio/Technology*, 1986, **4**, 44.
138. J. E. Baldwin, E. P. Abraham, G. L. Burge, and H. H. Ting, *J. Chem. Soc. Chem. Commun.*, 1985, 1808.
139. S. E. Jensen, D. W. S. Westlake, R. Y. Bowers, L. Lyubechansky, and S. Wolfe, *J. Antibiotics*, 1986, **39**, 822.
140. J. M. Luengo, M. J. López-Nieto, and F. Salto, *J. Antibiotics*, 1986, **39**, 1144.
141. J. E. Baldwin, R. M. Addlington, M. James, C. Crabbe, C. G. Knight, T. Nomoto, and C. J. Schofield, *J. Chem. Soc. Chem. Commun.*, 1987, 806.
142. S. M. Samson, J. L. Chapman, R. Belegaje, S. W. Queener, and T. D. Ingolia, *Proc. Natl. Acad. Sci. USA*, 1987, **84**, 5705.

143. E. P. Abraham, J. A. Huddleston, G. S. Jayatilake, J. O'Sullivan, and R. L. White, in "Proceedings of the 2nd International Symposium, The Royal Society of Chemistry, London," ed. G. I. Gregory, 1981, p. 25.

144. J. E. Baldwin, R. M. Addlington, N. Moss, and N. Robinson, *J. Chem. Soc. Chem. Commun.*, 1987, 1664.

145. J. E. Baldwin, R. M. Addlington, and N. Moss, *Tetrahedron*, 1989, **45**, 2841.

146. J. E. Baldwin, W. J. Norris, R. T. Freeman, M. Bradley, R. M. Addlington, S. Long-Fox, and C. J. Schofield, *J. Chem. Soc. Chem. Commun.*, 1988, 1128.

147. J. E. Baldwin, G. P. Lynch, and J. Pitlik, *J. Antibiotics*, 1991, **44**, 1.

148. G. A. Bahadur, J. E. Baldwin, J. J. Usher, E. P. Abraham, G. S. Jayatilake, and R. L. White, *J. Chem. Soc. Chem. Commun.*, 1981, **103**, 7650.

149. R. L. Baxter, G. A. Thompson, and A. I. Scott, *J. Chem. Soc. Chem. Commun.*, 1984, 32.

150. J. E. Baldwin, E. P. Abraham, C. G. Lovel, and H. H. Ting, *J. Chem. Soc. Chem. Commun.*, 1984, 902.

151. I. K. P. Tan, J. M. Fernández-Cañón, A. Reglero, and J. M. Luengo, *Appl. Microbiol. Biotechnol.*, 1993, **40**, 113.

152. J. M. Luengo, J. L. Iriso, and M. J. López-Nieto, *J. Antibiotics*, 1986, **39**, 1565.

153. J. M. Luengo, J. L. Iriso, and M. J. López-Nieto, *J. Antibiotics*, 1986, **39**, 1754.

154. J. Martín-Villacorta, A. Reglero, and J. M. Luengo, *J. Antibiotics*, 1989, **42**, 1502.

155. J. Martín-Villacorta, A. Reglero, and J. M. Luengo, *J. Antibiotics*, 1990, **43**, 1559.

156. J. Martín-Villacorta, A. Reglero, and J. M. Luengo, *Biotech. Forum Eur.* (*BFE*). *Int. J. Biotech.*, 1991, **8**, 60.

157. J. Martín-Villacorta, A. Reglero, and J. M. Luengo, *J. Antibiotics*, 1991, **44**, 108.

158. M. A. Ferrero, A. Reglero, J. Martín-Villacorta, J. M. Fernández-Cañón, and J. M. Luengo, *J. Antibiotics*, 1990, **43**, 684.

159. M. A. Ferrero, A. Reglero, H. Martínez-Blanco, M. Fernández-Valverde, and J. M. Luengo, *Antimicrob. Agents Chemother.*, 1991, **35**, 1931.

160. K. Higuchi, F. G. Harvis, W. H. Peterson, and M. J. Johnson, *J. Am. Chem. Soc.*, 1946, **68**, 1669.

161. A. Y. Moyer and R. D. Coghill, *J. Bacteriol.*, 1947, **53**, 329.

162. G. J. M. Hersbach, C. P. van der Beeck, and P. W. M. Dijck, *Biotechnol. Ind. Antibiot.*, 1984, **22**, 45.

163. D. J. Hillenga, H. J. M. Versantvoort, S. van der Molen, A. J. M. Driessen, and W. N. Konings, *Appl. Environ. Microbiol.*, 1995, **61**, 2589.

164. S. H. Eriksen, B. Jensen, I. Schneider, S. Kaasgaard, and J. Olsen, *Appl. Microbiol. Biotechnol.*, 1995, **42**, 945.

165. C. Bailey and H. N. Arst, Jr, *Eur. J. Biochem.*, 1975, **51**, 573.

166. C. Schleissner, E. R. Olivera, M. Fernández-Valverde, and J. M. Luengo, *J. Bacteriol.*, 1994, **176**, 7667.

167. E. R. Olivera, B. Miñambres, B. García, C. Muñiz, M. A. Moreno, A. Ferrández, E. Diaz, J. L. Garcia, and J. M. Luengo, *Proc. Natl. Acad. Sci. USA*, 1998, in press.

168. H. Martínez-Blanco, M. Orejas, A. Reglero, J. M. Luengo, and M. Peñalva, *Gene*, 1993, **130**, 265.

169. H. Martínez-Blanco, A. Reglero, and J. M. Luengo, *Biochem. Biophys. Res. Commun.*, 1990, **167**, 891.

170. L. B. Rodríguez-Aparicio, A. Reglero, H. Martínez-Blanco, and J. M. Luengo, *Biochim. Biophys. Acta*, 1991, **1073**, 431.

171. H. Martínez-Blanco, A. Reglero, J. Martín-Villacorta, and J. M. Luengo, *FEMS Microbiol. Lett.*, 1990, **72**, 113.

172. O. Ferrero, A. Reglero, J. Martín-Villacorta, H. Martínez-Blanco, and J. M. Luengo, *FEMS Microbiol. Lett.*, 1991, **83**, 1.

173. H. Martínez-Blanco, A. Reglero, and J. M. Luengo, *J. Antibiotics*, 1991, **44**, 1252.

174. J. M. Luengo, in "50 Years of Penicillin Application," eds. H. Kleinkauf and H. von Döhren, Public, Czech Republic, 1991, 287.

175. B. Miñambres, H. Martínez-Blanco, E. R. Olivera, B. García-Alonso, J. L. Barredo, B. Díez, M. A. Moreno, C. Schleissner, F. Salto, and J. M. Luengo, *J. Biol. Chem.*, 1996, **271**, 33 531.

176. M. Fernández-Valverde, A. Reglero, and J. M. Luengo, *FEMS Microbiol. Lett.*, 1992, **96**, 111.

177. M. Fernández-Valverde, A. Reglero, H. Martínez-Blanco, and J. M. Luengo, *Appl. Environ. Microbiol.*, 1993, **59**, 1149.

178. P. Fawcett, J. J. Usher, and E. P. Abraham, *Biochem. J.*, 1975, **151**, 741.

179. B. Spencer and C. Maung, *Biochem. J.*, 1970, **118**, 29p.

180. R. T. Applin, J. E. Baldwin, S. C. J. Cole, J. D. Sutherland, and M. T. Tobin, *FEBS Lett.*, 1993, **319**, 166.

181. R. T. Applin, J. E. Baldwin, P. L. Roach, C. V. Robinson, and C. J. Schofield, *Biochem. J.*, 1993, **294**, 357.

182. M. B. Tobin, S. C. J. Cole, S. Kovacevic, J. R. Miller, J. E. Baldwin, and J. D. Sutherland, *FEMS Microbiol. Lett.*, 1994, **121**, 39.

183. M. B. Tobin, M. D. Fleming, P. L. Skatrud, and J. R. Miller, *J. Bacteriol.*, 1990, **172**, 5908.

184. E. Montenegro, J. L. Barredo, S. Gutiérrez, B. Díez, E. Alvarez, and J. F. Martín, *Mol. Gen. Genet.*, 1990, **221**, 322.

185. E. Alvarez, J. M. Cantoral, J. L. Barredo, B. Diez, and J. F. Martín, *Antimicrob. Agents Chemother.*, 1987, **31**, 1675.

186. O. Ferrero and J. M. Luengo, unpublished results.

187. H. Martínez-Blanco, A. Reglero, and J. M. Luengo, *J. Ind. Microbiol.*, 1994, **13**, 144.

188. C. B. Pickett and Y. H. Lu, *Annu. Rev. Biochem.*, 1989, **58**, 743.

189. P. Arca, M. Rico, A. F. Braña, C. J. Villar, C. Hardisson, and J. E. Saurez, *Antimicrob. Agents Chemother.*, 1988, **32**, 1552.

190. A. Meister, S. S. Tate, and O. W. Griffith, in "Methods in Enzymology," ed. W. B. Jakoby, Academic Press, New York, 1981, vol. 77, p. 237.

191. A. Meister and M. E. Anderson, *Annu. Rev. Biochem.*, 1983, **52**, 711.

192. P. J. Bebby, *Tetrahedron Lett.*, 1977, 3379.

193. G. W. J. Fleet and P. J. C. Harding, *Tetrahedron Lett.*, 1979, 975.

194. M. H. Zenk, B. Ulbrich, J. Busse, and J. Stöckigt, *Anal. Biochem.*, 1979, **101**, 182.

195. R. Padmakumar, S. Gantla, and R. Banerjee, *Anal. Biochem.*, 1993, **214**, 318.

196. J. J. Usher, D. W. Hughes, M. A. Lewis, and S.-J. D. Chiang, *J. Ind. Microbiol.*, 1992, **10**, 157.

197. S. J. Pirt, *Trends Biotechnol.*, 1987, **5**, 69.

198. F. Sánchez, M. Lozano, V. Rubio, and M. A. Peñalva, *Gene*, 1987, **51**, 97.

199. M. A. Moreno and J. M. Luengo, *Anal. Biochem.*, 1987, **164**, 559.

200. B. Miñambres, B. García, and J. M. Luengo, 1996, unpublished results.

201. J. R. S. Fincham, *Microbiol. Rev.*, 1989, **53**, 148.

202. W. Timberlake and M. Marshall, *Science*, 1989, **244**, 1313.
203. J. L. Barredo and J. M. Martín, *Microbiologia SEM*, 1991, **7**, 1.
204. G. F. Sprague, Jr, *Curr. Opin. Genet. Develop.*, 1991, **1**, 530.
205. A. E. Brakhage, P. Browne, and G. Turner, *J. Bacteriol.*, 1992, **174**, 3789.
206. T. D. Petes and C. W. Hill, *Annu. Rev. Genet.*, 1988, **22**, 147.

4.11
Genetics of Lantibiotic Biosynthesis

GABRIELE BIERBAUM
Universität Bonn, Germany

4.11.1 INTRODUCTION TO A GROUP OF UNIQUE PEPTIDES

The term "lantibiotic" is derived from "*lan*thionine-containing peptide an*tibiotic*."[1] Lantibiotics are a group of antibacterial peptides that contain the thioether amino acids lanthionine (Lan) and/or methyllanthionine (MeLan) (Figure 1). The presence of Lan and MeLan residues in these peptides leads to the formation of characteristic, intrachain ring structures. Additionally, several other nonproteinogenic amino acids are found in lantibiotics. For example, the dehydroamino acids 2,3-dehydroalanine (Dha) and (Z)-2,3-dehydrobutyrine (Dhb) (Figure 1) are present in most lantibiotics discovered so far. The rare amino acids that occur only in individual peptides, for example S-aminovinyl-D-cysteine or D-alanine, have been listed in Table 1. The unusual amino acids found in lantibiotics and the pathways of their synthesis have attracted a great deal of attention, a fact which is documented by the number of reviews and books on this subject.[15–31]

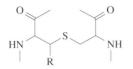

R = H; (2S,6R)-lanthionine
R = Me; (2S,3S,6R)-3-methyllanthionine

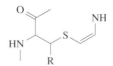

R = H; (S)-[(Z)-2-aminovinyl]-D-cysteine
R = Me; (S)-[(Z)-2-aminovinyl]-3-methyl-D-cysteine

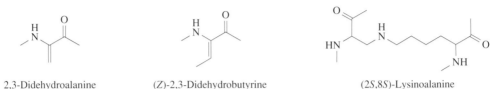

2,3-Didehydroalanine   (Z)-2,3-Didehydrobutyrine   (2S,8S)-Lysinoalanine

Figure 1   Structures of dehydroamino acids and bridge-forming amino acids that occur in lantibiotics. Isomeric configurations are indicated where appropriate.

Table 1   Nonprotein amino acid residues found in lantibiotics with the exception of lanthionine, methyl-lanthionine, dehydroalanine, and dehydrobutyrine residues.

Nonproteinogenic amino acid	Lantibiotic	Ref.
D-Alanine	lactocin S	2
S-(Z)-Aminovinyl-D-cysteine	epidermin, gallidermin, cypemycin	3–5
S-Aminovinyl-D-3-methylcysteine	mersacidin	6
Alloisoleucine	cypemycin	5
Dimethylalanine	cypemycin	5
erythro-3-Hydroxy-L-aspartic acid	cinnamycin, duramycin, duramycin B and C	7–9
Hydroxypropionyl	epilancin K7	10
(2S, 8S)-Lysinoalanine	cinnamycin, duramycin, duramycin B and C	7–9
2-Oxobutyryl	Pep5	11
2-Oxopropionyl	lactocin S, epilancin K7 (3–31)	2, 12
Nα-Succinyltryptophan	subtilin variant	13
Valine amide	nisin-(1–32)amide	14

Lantibiotics are produced by Gram-positive bacteria and were originally considered to be a group of heat-stable, ribosomally synthesized, bacteriocin-like peptides with a pore-forming mode of action and a potential application in food preservation. This conception was deduced from the oldest and most prominent member of the group, nisin, a lantibiotic that is produced by *Lactococcus lactis* ssp. *lactis* and is currently applied as a safe food preservative in many countries.[32] This view of lantibiotics was shaken when the enterococcal hemolysin/bacteriocin, which is now called cytolysin, was characterized as a lantibiotic.[33,34] This peptide has been known since 1934[35] and was shown to be a toxin that contributes to the virulence of *Enterococcus faecalis* in eye and heart infections.[36,37] In contrast to other lantibiotics, which cannot form pores in membranes that have the low membrane potential characteristic of eukaryotic cells, the cytolysin peptides are able to damage both eukaryotic and prokaryotic cells.

On the other hand, mersacidin and actagardine could be classified and employed as antibiotics. Both peptides interfere with bacterial peptidoglycan biosynthesis and are potential therapeutic agents against infections caused by, for example, methicillin-resistant staphylococci.[38,39] Therefore, the idea that all lantibiotics are bacteriocins which have a potential application in food preservation does not hold. Lantibiotics are simply a group of peptides that contain the thioether amino acids Lan and MeLan conferring ring structures that are more stable under extreme conditions than, for example, the disulfide bridges formed by Cys residues. Obviously, the presence of Lan residues is no indication for the mode of action of these peptides, and every peptide has to be examined individually as to its applicability, safety, and classification.[25]

4.11.1.1   Structures of Lantibiotics

According to Jung,[40,41] lantibiotics can be divided into two distinct groups by their structure and mode of action. The elongated, amphipathic, screw-shaped peptides of the type A lantibiotics

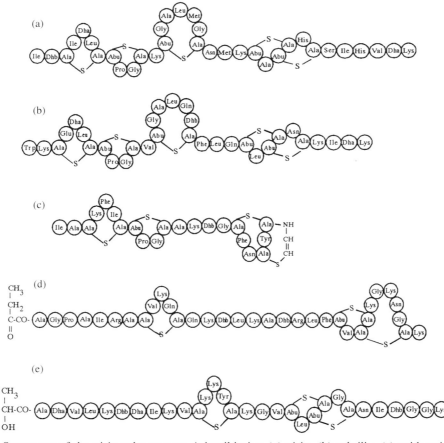

Figure 2 Structures of the nisin subtype type A lantibiotics: (a) nisin, (b) subtilin, (c) epidermin, (d) Pep5, and (e) epilancin K7 (Ala-S-Ala, lanthionine; Abu-S-Ala, methyllanthione).

(Figures 2 and 3) act primarily by pore formation in the bacterial membrane. Their molecular masses range from 2164 Da (epidermin) to 3764 Da (lactocin S). The small, globular peptides of the type B lantibiotics (Figure 4) (molecular mass 1825–2042 Da) appear to inhibit enzymatic activities by binding to lipids in the bacterial membrane (see Section 4.11.1.2). The type A lantibiotics can be further subdivided into the nisin subtype (Figure 2), the streptococcin A-FF22 subtype, and the lactocin S subtype (Figure 3), which are distinguished by structure, prepeptide sequence, and modification pathway.[25] Some of the new lantibiotics cannot yet be classified because structural and/or genetic information, for example the thioether bridging pattern or the sequence of the

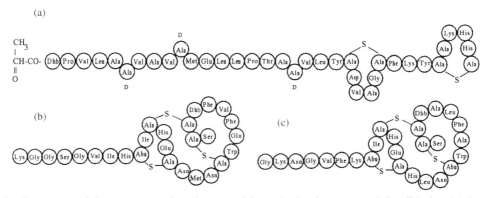

Figure 3 Structures of the streptococcin subtype and lactosin S subtype type A lantibiotics: (a) lactocin S, (b) lacticin 481, and (c) streptococcin A-FF22. The residues denoted "ᴅ" in lactocin S are in the ᴅ-configuration.

structural gene, is still missing. These lantibiotics, for example carnocin, cypemycin, and mutacin II, are included in Table 2.

Table 2   Overview of lantibiotic peptides.

Lantibiotic	No. of intrachain bridges	Mass (Da)	Producer strain	Ref.
Type A lantibiotics				
Nisin subtype				
Nisin A	5	3353	*Lactococcus lactis*	42
Nisin Z	5	3330	*Lactococcus lactis*	43
Subtilin	5	3317	*Bacillus subtilis*	44
Epidermin	4	2164	*Staphylococcus epidermidis*	3
Gallidermin	4	2164	*Staphylococcus epidermidis*	4
[Val¹/Leu⁶]epidermin	4	2151	*Staphylococcus epidermidis*	45
Pep5	3	3488	*Staphylococcus epidermidis*	11
Epilancin K7	3	3032	*Staphylococcus epidermidis*	10, 12
Streptococcin A-FF22 subtype				
Streptococcin A-FF22	3	2795	*Streptococcus pyogenes*	46
Lacticin 481	3	2901	*Lactococcus lactis*	47
Salivaricin A	3?	2315	*Streptococcus salivarius*	48
Variacin	3?	2658	*Micrococcus varians*	49
Lactocin S subtype				
Cytolysin A1	?	3438	*Enterococcus faecalis*	34
Cytolysin A2	?	2032	*Enterococcus faecalis*	34
Lactocin S	2	3764	*Lactobacillus sake*	2
Not yet classified				
Carnocin UI 49	?	4635	*Carnobacterium piscicola*	50
Cypemycin	1	2094	*Streptomyces* sp.	5
Mutacin II	?	3245	*Streptococcus mutans*	51
Type B lantibiotics				
Mersacidin subtype				
Mersacidin	4	1825	*Bacillus* sp.	6
Actagardine	4	1890	*Actinoplanes* sp.	52
Cinnamycin subtype				
Cinnamycin	4	2042	*Streptomyces cinnamoneus*	9
Duramycin	4	2014	*Streptomyces cinnamoneus*	9
Duramycin B	4	1951	*Streptoverticillium* sp.	9
Duramycin C	4	2008	*Streptomyces griseoluteus*	9
Ancovenin	3	1959	*Streptomyces* sp.	53

The difficulties that are encountered during the structure elucidation of a lantibiotic are caused by the presence of the dehydroamino acids. As soon as a dehydroamino acid becomes located at the N-terminus of the peptide during Edman degradation, a spontaneous deamination occurs, which results in an oxopropionyl or oxobutyryl group that blocks further sequencing reactions.[40,41] Therefore, a derivatization procedure was developed which allows direct sequencing of dehydroamino acids and thioethers.[54] Since lantibiotics are gene-encoded, ribosomally synthesized peptides, the rare amino acids are formed by the modification of precursor peptides. The hydroxyl amino acids Ser and Thr are dehydrated to give the dehydroamino acids Dha and Dhb. Cys residues then react with the dehydro residues to form the thioether bridges (see Section 4.11.2). Therefore, sequencing of the structural genes is also a powerful tool for the elucidation of the primary structure of lantibiotics. Today the bridging pattern is most frequently established by NMR-based methods.

For all type A lantibiotics that have been studied so far, the thioether bridges are always formed from a dehydro residue that is located toward the N-terminus of the peptide and from a Cys residue that is located in the C-terminal part of the peptide. The lantibiotics of the nisin subtype comprise elongated, flexible peptides that carry a net positive charge and act by pore formation in the bacterial membrane. Nisin (also designated as nisin A) was first described in 1928 as a Lancefield group N inhibitory substance and, therefore, is the oldest and most prominent lantibiotic of this subtype.[55] In 1952, Berridge *et al.* showed that Lan is present in nisin, but the complete structure of nisin (Figure 2) was not fully elucidated until 1971.[42,56] Nisin is produced by certain *L. lactis* ssp. *lactis* strains, and is a 34 amino acid peptide that contains five ring structures. It has a very broad antibacterial spectrum and it is active against *Micrococcus, Clostridium, Bacillus, Streptococcus, Staphylococcus, Neisseria, Listeria,* and even *Salmonella* and *Helicobacter* spp. in the presence of EDTA.[32,57–59] In addition, nisin effectively inhibits the germination of clostridial spores and is widely used as a preservative. A natural variant of nisin A, called nisin Z ([Asn[26]]nisin A), has also been reported and is more soluble at neutral pH than nisin A.[43]

The lantibiotic subtilin,[60] which is excreted by *Bacillus subtilis* 6633, is similar to nisin in its bridging pattern.[44] In addition to its pore-forming activity, subtilin is also active against bacterial spores.[61] Subtilin is not employed for food preservation because it is rather unstable at room temperature, a feature which is caused by the instability of the Dha residue at position 5.[62]

Epidermin and its natural variant gallidermin ([Leu[6]]epidermin) are produced by *Staphylococcus epidermidis* Tü3298 and *S. gallinarum* DSM4616, respectively.[3,4] Epidermin and gallidermin are tetracyclic peptides of 22 amino acids with a mass of 2164 Da. The arrangement of rings A and B in these peptides is similar to that of rings A and B in nisin and subtilin. The C-terminal ring structure is formed by the rare amino acid *S*-aminovinyl-D-cysteine. Both peptides show activity against staphylococci, streptococci, and *Propionibacterium acnes*, which might make them suitable for the topical treatment of juvenile acne. The production of epidermin and its natural variants, such as [Val[1]/Leu[6]]epidermin, is widespread among coagulase-negative staphylococci,[45] and staphylococcin 1580 was also shown to be [Leu[6]]epidermin.[63]

Another lantibiotic that is also excreted by *S. epidermidis* 5 is Pep5,[64] a tricyclic, 34 amino acid peptide with a molecular mass of 3488 Da.[11] Pep5 carries a net charge of +7, and its N-terminus is formed by a 2-oxobutyryl residue which arises from spontaneous deamination of the Dhb residue at position 1. Pep5 has a narrow spectrum of activity, with *Staphylococcus* and *Micrococcus* spp. being the most sensitive organisms.

The last member of the nisin subtype lantibiotics, epilancin K7, is also produced by *S. epidermidis*. Like Pep5, epilancin K7 is a basic peptide with a net charge of +6 and three ring structures. Whereas ring A in the N terminus of epilancin K7 resembles Pep5, the C-terminal double ring system is reminiscent of the C-terminal rings of nisin and subtilin. The N-terminus of epilancin K7 is formed by an unusual 2-hydroxypropionyl group. This group probably derives from a Ser residue, which is dehydrated and deaminates upon cleavage of the leader peptide. This reaction results in an oxopropionyl residue that must then be reduced, possibly by an enzymatic modification process.[10,12]

The lantibiotics of the streptococcin A-FF22 subtype form a homogeneous group (Figure 3). The bridging patterns of lacticin 481[65] and streptococcin A-FF22[66] have been elucidated, and were shown to be identical. It has been suggested that the other two members of the group, which contain ring-forming amino acids in identical positions, might have the same or at least similar bridging patterns. The presence of three overlapping rings makes these peptides less elongated than the lantibiotics of the nisin subtype, and they carry only a single or no net charge. In the case of streptococcin A-FF22, these structural properties are reflected by the fact that the peptide forms smaller and less stable pores than the nisin subtype lantibiotics.[67] The first lantibiotic of this group, streptococcin A-FF22, was detected in 1971 and is produced by a clinical isolate of *Streptococcus pyogenes*.[68] It is a tricyclic peptide of 2795 Da and 26 amino acids.[46,69] Lacticin 481,[70] which is identical to lactococcin DR,[71] is produced by some *L. lactis* ssp. *lactis* strains, and shows a broad spectrum of activity against a number of Gram-positive bacteria. Lacticin 481 is composed of 27 amino acids and has a molecular mass of 2901 Da. Variacin, a lantibiotic that is excreted by *Micrococcus varians* MCV8 and MCV12, is very similar to, although at only 25 amino acids slightly shorter than, lacticin 481. The characterization of the structural gene showed that there is 84% identity between lacticin 481 and variacin.[49] The fourth lantibiotic of the streptococcin A-FF22 subtype is salivaricin A, a 22 amino acid peptide which was isolated from *Streptococcus salivarius* 20P3 and is active against *S. pyogenes*.[48]

The third subgroup of type A lantibiotics, the lactocin S subtype, contains lactocin S and the cytolysins.[25] The bridging pattern of cytolysin peptides has not yet been established, but both lantibiotics are clearly distinct from the above groups. Lactocin S was purified from *Lactobacillus sake* L45, and is active against *Lactobacillus, Pediococcus,* and *Leuconostoc* spp.[72,73] It is a 37 amino acid peptide (3764 Da) which contains three D-Ala residues and two thioether rings.[2] In the N-terminus of the propeptide a Ser residue is located, and from the mass spectra it was concluded that this residue is dehydrated to Dha and then undergoes oxidative deamination, which gives a 2-oxopropionyl group.

In 1934, Todd reported that some strains of *E. faecalis* express a hemolysin.[35] Subsequently it was shown that this hemolysin, which is now called cytolysin, also acts as a bacteriocin.[74] Active cytolysin is a mixture of two different peptides: a long product CylA1 (formerly CylL$_L$, 3438 Da) and a short product CylA2 (formerly CylL$_S$, 2032 Da). Both peptides contain mainly hydrophobic residues. The presence of Lan and/or MeLan residues has been shown in hydrolysates of both peptides, but the bridging patterns have not yet been elucidated.[34,75]

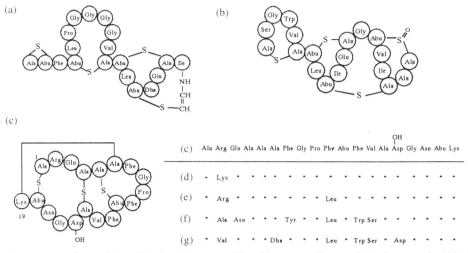

Figure 4   Structures of type B lantibiotics: (a) mersacidin, (b) actagardine, and (c) cinnamycin (Ala-NH-Lys, lysinoalanine; Asp-OH, hydroxyaspartic acid). The exchanges within the lantibiotics of the cinnamycin type are shown in the table: (c) cinnamycin, (d) duramycin, (e) duramycin B, (f) duramycin C, and (g) ancovenin.

The type B lantibiotics (Figure 4) can also be divided into two subgroups. The first group is formed by the lantibiotics of the cinnamycin subtype, which includes cinnamycin (also described as lanthiopeptin and Ro09-0198), duramycin (also called leucopeptin), duramycin B and duramycin C, and ancovenin.[7-9,53,76-82] All lantibiotics of the cinnamycin subtype are produced by species of *Streptomyces* or *Streptoverticillium*. The structural arrangement of the thioether ring structures (one Lan and two MeLan residues) is conserved within the group, and the overall globular structure of the peptides is caused by a head-to-tail connection between the first and the last-but-one amino acid. In contrast to the type A lantibiotics, the thioether bridges are built from an N-terminal Cys residue and a C-terminally located dehydroamino acid. Within the 19 amino acids of these peptides, there are only seven variable positions.[40,41] Lysinoalanine, which forms a fourth ring, and *erythro*-3-hydroxyaspartic acid are found only in this group. The lysinoalanine residue is presumably formed by addition of the amino group of Lys19 to the Dha residue at position 6. The type B lantibiotics of the cinnamycin subgroup show only a weak antibacterial action but have elicited interest as inhibitors of phospholipase A_2, and cinnamycin was shown to interfere with the replication of the *Herpes simplex* virus-1.[81]

Mersacidin and actagardine form the second subtype of type B lantibiotics. Both peptides have a globular structure,[6,52] but their bridging pattern is clearly different from that of the cinnamycin subgroup. Actagardine, which was formerly called gardimycin, is produced by *Actinoplanes* spp.[83] An erroneous structure elucidated by NMR was published in 1990,[84] and only since this was corrected[52] has the similarity between the second ring of actagardine and the third ring of mersacidin been obvious. It is not yet clear whether or not the sulfoxide group found on the fourth ring of actagardine is an artifact formed during the purification procedure. Actagardine is most active against streptococci.

Mersacidin is the smallest lantibiotic with a molecular mass of 1825 Da, and is produced by a *Bacillus* sp. (*Bacillus* HIL Y-85,54728).[6] Mersacidin is a relatively hydrophobic peptide that does not carry a net charge. It is the only lantibiotic that contains an *S*-aminovinyl-3-methyl-D-cysteine residue and a ring structure that bridges two neighboring amino acids. The activity of mersacidin is directed against staphylococci.[85] Mersacidin and actagardine are of considerable interest because they inhibit cell wall biosynthesis of bacteria via a novel mechanism[38] and might be of use in the treatment of infections caused by methicillin-resistant staphylococci.

4.11.1.2   Modes of Action of Lantibiotics

So far, most interest has focused on the type A lantibiotics of the nisin subtype, which kill the cells by formation of short-lived pores in the cytoplasmic membrane. A more detailed report of the progress in this field can be found in several reviews.[22,25,86-88] The first indications of an interaction between lantibiotics and the cytoplasmic membrane were obtained as early as 1960, when the lysis of growing clostridial cells by nisin was reported.[89] Addition of a type A lantibiotic, for example Pep5, to sensitive cells immediately inhibits all cellular biosyntheses of DNA, RNA, protein, and polysaccharides.[90] This indicates de-energetization of the cells, and is the result of membrane depolarization, as shown by Pep5[91] and nisin,[92] for example. Small metabolites, such as amino acids or ATP, leak out of the cells after treatment with Pep5,[93] nisin,[92] subtilin,[94] and epidermin/gallidermin,[87] and the uptake of labeled amino acids by pretreated cells is inhibited. De-energized cells are not susceptible to the action of the lantibiotics; a membrane potential of at least 50–100 mV is necessary for the activity of the peptides.[91,92] In black lipid bilayer experiments, membrane conductance in the presence of lantibiotics only increased when membrane potentials in excess of −50 mV to −80 mV were applied. The data obtained indicated that the lantibiotics form pores with diameters of approximately 1 nm and lifetimes in the millisecond range for Pep5,[95] nisin,[96] and subtilin,[94] but up to 30 s for epidermin and gallidermin.[87]

Based on these results as well as conformational data derived from NMR experiments employing nisin in the presence of micelles[97,98] a model for pore formation by lantibiotics has been postulated[88,98] (Figure 5). In aqueous solution the peptides are far from rigid, and defined conformations are mostly found only in the region of the ring structures.[25] Experiments with nonenergized membranes demonstrated that the cationic lantibiotics bind to the membrane by ionic interaction with the anionic phospholipids.[99,100] Upon binding to the membrane, the peptides adopt an amphiphilic rod-like conformation, with the hydrophilic groups interacting with the headgroups of the phospholipids and the hydrophobic groups in contact with the fatty acid chains. For example, the hydrophobic N-terminus of nisin was shown to be embedded in the membrane.[97,98,100] Lantibiotics are too small to form a pore by transversing the membrane several times and, therefore, several peptides have to aggregate to form a pore. In the presence of a trans-negative membrane potential the peptides are pulled into the membrane. From NMR data of nisin bound to micelles, it was postulated that the peptides reorient themselves perpendicularly to the membrane without losing their contact with the phospholipid headgroups and, thus, fuse the outer and inner leaflets of the membrane. This results in formation of a pore bordered by the charged and hydrophilic residues of the peptides whereas the hydrophobic residues are still within the lipid membrane. In this position, the peptide is bound on a curved surface and, therefore, adopts a bent conformation. The bilayer structure of the phospholipids is disturbed, a conformation which is thermodynamically unfavorable. Thus, the pore is closed again when the proton flux lowers $\Delta\psi$ below the threshold required for pore formation.

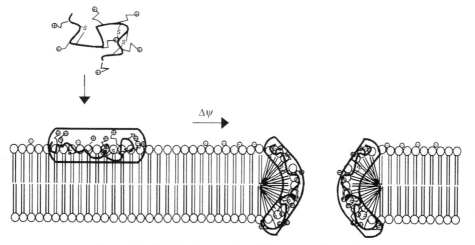

Figure 5   Model of type A lantibiotic pore formation.

In addition to forming pores, nisin and subtilin inhibit the germination of bacterial spores, and this mode of action seems to depend on the presence of the Dha residue at position 5.[61] Furthermore, inhibition of bacterial cell wall biosynthesis by nisin in an *in vitro* assay was reported[101] and, as a secondary effect, Pep5 and nisin induce lysis of susceptible staphylococcal cells in media of low ionic strength. This is caused by a ionic interaction of the cationic lantibiotics with the teichoic and lipoteichoic acids in the cell wall. These anionic polymers are noncompetitive inhibitors of the lytic enzymes, which are cationic proteins themselves and are displaced by the cationic lantibiotics.[102,103]

The only detailed mode of action studies recorded for the streptococcin A-FF22 subtype lantibiotics have been performed with streptococcin A-FF22.[67] Streptococcin forms smaller (0.5–0.6 nm in diameter) and less stable pores than the nisin subtype lantibiotics. The pores are relatively short lived. A threshold potential of -100 mV was required for their formation. Since streptococcin A-FF22 and the other lantibiotics of the streptococcin subtype are structurally closely related, a similar mode of action may be assumed for the other members of this group. For lacticin 481, a higher affinity to zwitterionic than to anionic lipids was recorded. This correlates with the fact that the peptide does not carry a net charge and with its antibacterial spectrum.[100]

A lantibiotic that was isolated only in 1994 and is still only poorly characterized as to its structural features is mutacin II from *Streptococcus mutans*.[51] An investigation of its mode of action showed that although a transient depolarization of the membrane was recorded, a depletion of the intracellular ATP pool seemed to be the main bactericidal effect. Therefore, it was concluded that mutacin II might act by inhibiting the generation of ATP in the cells, which, if proven, would be a novel mode of action for lantibiotics.[104]

Early experiments, which were performed in order to investigate the mode of action of the type B lantibiotics of the cinnamycin subtype, indicated an interaction between the lantibiotics and biological membranes. For example, an increase in membrane permeability in *Bacillus*,[105] inhibition of the proton pump and transport by clathrin-coated vesicles,[106] inhibition of the sodium- and potassium-dependent ATPases,[107] and an inhibition of the ATP-dependent calcium uptake[108] were described. Black lipid bilayer and patch clamp experiments with duramycin even gave evidence that these lantibiotics might be able to form ion channels or pores, since increases in conductance, sometimes step-like, were observed.[109] A first indication for the primary mode of action was obtained when resistant mutants of *Bacillus*[108] were studied. The membrane of a resistant *Bacillus firmus* strain did not contain any phosphatidylethanolamine, but plasmenylethanolamine was found instead.[110] Further experiments showed that cinnamycin interacts only with liposomes if phosphatidylethanolamine is present,[111] that duramycin affects the mitochondrial membrane whose major component is phosphatidylethanolamine,[112] and that the activity of cinnamycin can be inhibited by a preincubation in the presence of phosphatidylethanolamine. In conclusion, cinnamycin was found to bind tightly to the amino group and glycerol backbone of phosphatidylethanolamine.[113] The formation of this complex also results in an inhibition of phospholipase A_2 by direct competition in the form of substrate binding.[114]

In the meantime, the amphiphilic conformations of cinnamycin[80] and duramycins B and C[115] as well as the three-dimensional structure of cinnamycin in the presence of phosphatidylethanolamine[116] have been elucidated by NMR. The complex between cinnamycin and lysophosphatidylethanolamine has a cylindrical shape which is 11 Å in diameter and 26 Å long. The headgroup of phosphatidylethanolamine is bound in a hydrophobic pocket which is formed by the ring encompassing amino acids 4–14 and is surrounded by the residues Phe7 to Ala-S14. Intermolecular contacts were recorded for Gly8, Pro9, and Val13 and the carboxyl group of the hydroxyaspartic acid at position 15 interacting with the ammonium group of phosphatidylethanolamine. The high flexibility which was recorded for single residues in free cinnamycin[80] is not seen within the complex; here the structure is rather rigid, which indicates a tight fit of the two ligands and explains the strict substrate specificity that was observed in the earlier experiments.[116]

Mersacidin and actagardine both interfere with bacterial cell wall biosynthesis. In 1977, actagardine, then called gardimycin, was reported to induce the accumulation of the cytoplasmic cell wall precursor (UDP-*N*-acetylmuramylpentapeptide) of *B. subtilis* and inhibit the biosynthesis of polymeric peptidoglycan in an *in vitro* assay. The product that accumulated in these reactions was characterized as lipid II or the disaccharide form of the cell wall precursor still bound to the undecaprenol carrier.[39] From these results it was concluded that actagardine inhibits the transglycosylation reaction, that is, the synthesis of the polysaccharide chains of the bacterial cell wall. The first erroneous structure that had been obtained by NMR experiments for actagardine[84] did not reveal its similarity to mersacidin. Only when the revised structure was published[52] did it become obvious that one ring structure has been conserved between mersacidin and actagardine. Detailed studies with mersacidin showed that this lantibiotic shares the mode of action of actagardine. Interestingly, the target site of mersacidin and actagardine on lipid II is different from the binding site of vancomycin; therefore, these lantibiotics obviously recognize a novel binding site on lipid II.[38,117]

Since the mid-1980s a plethora of nonlantibiotic bacteriocins have been described,[28] and, therefore, the most intriguing question of the function of the modified residues in lantibiotics arises. Some aspects of this problem have been addressed by site-directed mutagenesis experiments or by characterization of natural by-products and degradation products. The ring structures, which are the distinct feature of this class of peptides, were proposed to provide thermostability, tolerance against acids, and to stabilize the antibacterial conformation of nisin.[118,119] In the case of Pep5, the ring structures also protect the peptide against proteolytic degradation by the producer strain. In addition, the antibacterial activity of lantibiotics seems to depend on the presence of the ring structures. For example, the antibacterial activity of Pep5 mutant peptides strongly decreased when formation of the central ring structure had been inhibited by site-directed mutagenesis.[120] A degradation product of nisin, (des-Dha⁵)-nisin, has lost the Dha residue at position 5, which results in opening of ring A; this peptide is 31-fold less active than wild-type nisin.[121] The engineered peptide T13C nisin carries a disulfide bridge in ring C instead of the thioether. Reduction of the disulfide bond, as well as tryptic digestion of M17K nisin, which cleaves the bond between Leu16 and Lys17, results in the opening of ring C of nisin, and the peptides lose their antibacterial activity.[122]

The results obtained after exchange or loss of the dehydroamino acid residues do not provide a uniform theory as to their function. Some dehydroamino acids, for example Dha5 or Dha33 in nisin, can be removed without dramatic effects on pore formation, although the double mutant Dha5A/Dha33A nisin is less active than the wild-type peptide.[122,123] For subtilin a loss of its sporicidal effects in the absence of Dha5 was described, whereas its bactericidal activity against vegetative cells remains intact.[61] In contrast, NMR experiments have shown that the dehydroamino acids of gallidermin and Pep5, which are not located in the ring structures, stabilize β-turn structures in the flexible central hinge region of the peptides,[124-126] and Dhb14A gallidermin, Dhb16A Pep5, and Dhb20A Pep5 are significantly less active than the wild-type peptides.[120,127]

4.11.2 BIOSYNTHETIC PATHWAY OF LANTIBIOTICS

As early as 1969, Ingram demonstrated that after incubation of a culture of *Lactococcus lactis* in the presence of radioactively labeled Thr and Cys, the labels appear in the MeLan residues of mature nisin.[128,129] Moreover, subtilin was not synthesized in the presence of inhibitors of ribosomal protein biosynthesis.[130] From these results it was concluded that the biosynthesis pathway of subtilin and nisin differs from that of peptide antibiotics, such as surfactin and gramicidin, which do not possess structural genes and are formed by multienzyme complexes via the thiotemplate mechanism.[131] In contrast, the results demonstrated that lantibiotics must possess structural genes which encode precursor peptides that are made by the ribosomal biosynthesis pathway. The Ser, Thr, and Cys residues of these peptides are post-translationally modified to form dehydroamino acids, Lan and MeLan residues. This hypothesis was confirmed when the first lantibiotic structural genes, *epiA*, *spaA*, *nisA*, and *pepA*, encoding the prepeptides of epidermin, subtilin, nisin, and Pep5, respectively, were detected.[1,132-134] As predicted, these genes encode precursor peptides, called prepeptides, consisting of an N-terminal extension (the leader peptide) and a C-terminal part, which is the actual precursor of the mature lantibiotic and is called the propeptide domain. In the propeptide domain, Ser and Thr residues are encoded at the Dha and Dhb residue positions. In type A lantibiotics, the N-terminal half of the Lan or MeLan residues is always occupied by Ser or Thr residues, respectively, whereas in the C-terminal half a Cys residue is present. In a first step which is considered to be the key reaction of lantibiotic biosynthesis,[135] the hydroxyl amino acids are dehydrated by an as yet unknown mechanism. The products of this reaction are α,β-unsaturated amino acids, thus the Ser residues are transformed into Dha residues, and the Thr residues are transformed into Dhb residues. Some of these dehydroamino acids are intermediates in the formation of the lanthionine residues, others are not modified further and represent the dehydroamino acids that are present in nearly all lantibiotics. This hypothesis was confirmed when the dehydrated prepeptides of the lantibiotic Pep5 were isolated and characterized.[136] The dehydroamino acids were found solely in the propeptide part and were not present in the leader peptide. The thioethers had not yet been formed in these peptides.[136] The dehydroamino acids contain a reactive double bond. In the second biosynthetic step this double bond reacts with the sulfhydryl group of a Cys residue to form the thioether of the Lan or MeLan residue. This reaction is stereospecific, because the Lan and MeLan residues are always found in the *meso* configuration; the part that derives from the dehydroamino acid is present in the D configuration, and the other half formed by the Cys residue remains in the L configuration. The enzymes involved in both modification steps have been identified, the larger B enzyme catalyzes the dehydration step whereas the smaller C enzyme is involved in the ring formation.[137] The designations B and C derive from the open reading frames of the subtilin and epidermin biosynthetic gene cluster.[138,139] The catalytic mechanisms of the modification reactions, especially that of the first step, have not yet been elucidated, and any coenzymes that might take part in this process, etc., are still unknown. Ser dehydratases cannot serve as a model. These enzymes dehydrate the free amino acids in a pyridoxal phosphate-dependent mechanism that involves the free amino groups and could not be active on residues bound within a peptide chain.[135] The stereoselective formation of a (2S,6R)-*meso*-lanthionine between 2,3-Dhb and a Cys residue is energetically possible at room temperature, and was also shown to take place spontaneously with a fragment of epidermin.[140] Therefore, it is presumed that the LanC enzyme acts in a chaperone-like fashion, ensuring the correct folding of the peptide.[137]

In the last maturation step, which may take place before or after export from the cell, the leader peptide is removed by a peptidase which releases the mature lantibiotic. If a dehydroamino acid is present at position +1, then spontaneous deamination takes place and results in an oxobutyryl residue or an oxopropionyl residue from Thr or Ser, respectively. The hydroxypropionyl residue that was found in epilancin K7 must be formed by reduction of an oxopropionyl residue. The enzyme that is involved in this reaction is still unknown. The S-aminovinyl-D-cysteine residue of epidermin is formed by an oxidative decarboxylation of the C-terminal Cys or MeLan residue. The enzyme responsible for this step, EpiD, has been well characterized (see Section 4.11.2.1.3). In lactocin S, D-Ala residues have been discovered in three positions,[2] which are formed from Ser residues that are present in the propeptide. The reaction pathway that has been suggested for this unique modification involves dehydration of the Ser residues to Dha and tautomerization to an enamine which is then stereospecifically hydrogenized to D-Ala. The enzyme(s) involved in these reactions have not yet been identified.[141] The biosynthesis of all other rare residues that are found in lantibiotics (see Table 1) remains to be elucidated.

4.11.2.1 Biosynthetic Gene Clusters of Nisin Subtype Lantibiotics

The genes for the accessory proteins that are necessary for the modification procedures, export, and proteolytic processing of the lantibiotics, as well as self-protection of the producer strains and biosynthetic regulation, are organized in biosynthetic gene clusters together with the structural genes. The databank accession numbers and references for the nucleotide sequences are given in Table 3. The biosynthetic gene clusters have been the subject of reviews,[25–27,29,168] and a common nomenclature has been proposed for all genes.[168,169] The locus symbol *lan* is used collectively, if all lantibiotic genes are referred to, and the identical capital letters represent proteins with an identical function. Thus, the capital letter "A" represents the structural gene, although *spaA* is still often designated *spaS*. *lanB*, *lanC*, and *lanM* encode the modification enzymes that introduce the dehydro and thioether residues, and LanP is the processing protease that cleaves off the leader peptide. LanT is the exporter involved in export of the lantibiotic from the producer cell, and LanH is an accessory protein to this transporter. LanI is the "immunity" protein or peptide that confers self-protection against the lantibiotic to the producer strain. LanE, LanF, and LanG constitute a second transporter that is most probably also involved in immunity. The regulation of lantibiotic biosynthesis is often mediated by a two-component regulatory system, comprising a histidine kinase, LanK, and the response regulator, LanR. In a part of the early, original literature, especially in those papers published earlier than 1991 or before a particular peptide had been identified as a lantibiotic, a different nomenclature was used by the authors. This comment refers especially to cytolysin,[33] lacticin 481, which was also described as lactococcin DR,[163] and subtilin,[152,153] but also to nisin; for example, the structural gene of nisin, *nisA*, was originally called *spaN*[132] in analogy to the original designation of the structural gene of subtilin *spaS*.[133]

The gene clusters may be found on chromosomal DNA or on mobile elements, such as plasmids or transposons. Structural features of the mature peptides and of the prepeptides as well as the types of modification enzymes present in the gene clusters distinguish two subclasses of type A lantibiotics:[25] the biosynthetic gene clusters of the nisin subtype lantibiotics are characterized by the presence of two separate modification enzymes collectively designated LanB and LanC, a serine protease (LanP), and an ABC transporter (LanT); in contrast, the streptococcin A-FF22 and lactocin S subtype lantibiotics possess a single modification enzyme (LanM), and a protein that combines protease and transporter activity is also found in most of these gene clusters. Therefore, this section will concentrate on the nisin subtype lantibiotics; the other subtypes will be addressed in a separate section (Section 4.11.2.2). The biosynthetic gene clusters of the nisin subtype lantibiotics usually consist of several operons, and are shown in Figure 6. With the exception of the subtilin gene cluster, the genes of the modification enzymes are located downstream of the structural gene. Frequently, a weak terminator structure is located between the structural gene and the biosynthetic

Table 3 Accession codes for nucleotide sequences of lantibiotic gene clusters in the EMBL DataBank and GenBank.

Lantibiotic	Genes (as originally referred to)[a]	Consensus nomenclature	Accession code	Ref.
Nisin A	spaN	nisA	J04057	132
	nisA		M24527	142
	nisA		M79445	143
	nisA		M27277	144
	nisAB		M65089	145
	nisABTC		X68307 S52234	146
	nisABTCI		L16226	147
	nisPR		L11061	148
	nisIPRK		X76884	149
	nisFEG, sacR		U17255	150
Nisin Z	nisZ		X61144 S61034	43
	nisZBTCI		Z18947	151
	nisP		Z22725	151
	nisRK		Z22813	151
Subtilin	spaS	spaA	J03767	133
	spaDBCS	3′spaB, spaTCA	M83944	152
	spaEDBCS	spaBTCA	M99263	153
	spaTCS	spaTCA	M86869	139
	spaRK		L07785	154
	spaB		L24075	155
	spaSIFG	spaAIFG	U09819	156
Pep5	pepIA		L23967	157
	pepTIAPBC		Z49865	137
Epidermin	epiY′Y″ABCDQP	epiT′T″ABCDQP	X62386	1, 138
	epiFEG		U29130	158
	epiH		U29130	159
Gallidermin	gdmA		A61072	160
	gdmTH		U61158	159
Epilancin K7	elkA		U20348	10
Streptococcin A	scnA		L11653	161
	scnA′scnA″		L36235	162
Lacticin 481	lctA		X71410	47
	lcnDR1DR2DR3	lctAMT	U04057	163
Variacin	varA		X93303	49, 164
Salivaricin	salA		L07740	48
Lactosin S	lasAMTP[b]		Z54312	141
Cytolysin	cylB	cylT	M38052	165
	cylL_LL_SMBA	cylA1A2MTP	L37110	33
Mersacidin	mrsA		Z47559	166
Cinnamycin	cinA		X58545	167

[a] Only genes which were completely sequenced are included.   [b] Includes five additional open reading frames with unknown functions.

genes. This terminator allows a moderate readthrough and ensures expression of an excess of the lantibiotic prepeptide in comparison to the enzymes.

The nisin A biosynthetic gene cluster is located on the 70 kb conjugative transposon Tn5276 in *L. lactis* NIZO R5[170] or Tn5301 in *L. lactis* NCFB894.[143,171] The production of nisin Z was found on Tn5278; this gene cluster shows between 97% and 100% identity to the nisin A biosynthetic gene clusters.[151] All these transposons also carry the information for the utilization of sucrose encoded in *sacA*, *sacB*, and *sacR*, as well as a copy of the insertion sequence IS904. The integrated transposon is flanked on both sides by the nucleotide sequence 5′-TTTTTG-3′;[170,171] 750 bp upstream of *nisA*, a transposase is encoded.[132] However, in a few strains the nisin biosynthetic gene cluster seems to be located on plasmid DNA.[142,172] The gene cluster is composed of *nisABTCIPRKFEG*,

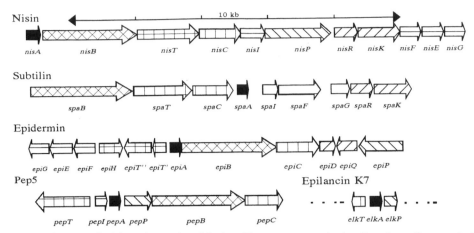

Figure 6 Gene clusters of nisin subtype lantibiotics. The arrows mark the direction of transcription. The structural genes are shown in black, and all genes with similar functions are shown with identical patterns.

and encompasses about 13 kb of sequence (see Figure 6). Early studies were controversial with respect to the promoters that initiate the transcription of *nisA*,[145–147] but a later report that used primer extension experiments and transcription of a reporter gene showed that the genes are organized in three operons in *L. lactis* NIZO R5: *nisABTCIP*, *nisRK*, and *nisFEG*. The promoters in front of *nisA* and *nisF* are inducible by the addition of nisin, whereas the promoter in front of *nisR* is independent of the presence of nisin.[173] This latter promoter was not found in the nisin Z biosynthetic gene cluster in *L. lactis* N8, which consists of two operons (*nisZBTCIPRK* and *nisFEG*) with two nisin-dependent promoters.[174] Downstream of *nisA* lies an inverted repeat which may function as a weak ρ-dependent terminator[132] and allow a partial readthrough.[144,145]

The subtilin biosynthetic gene cluster is located on the chromosome of *B. subtilis* 6633, and could be transferred by competence transformation into *B. subtilis* 168.[175] So far, the genes *spaBT-CAIFGRK* have been identified. In the early literature, the nomenclature of the biosynthetic genes does not correspond to the nomenclature agreed on for lantibiotics (see also Table 3): The group of Hansen identified two open reading frames, called *spaE* and *spaD*,[153] which were later shown to constitute *spaB*,[155] and the structural gene *spaA* was called *spaS*. A protease gene was not found in this gene cluster, but the subtilin producer contains and excretes various proteases that could function in the processing of subtilin. Two promoter structures in front of *spaA*[133] and *spaB*[153] have been described. The primary transcript of *spaA* has been reported to have an unusually long half-life of 45 min. Downstream of *spaA*, a ρ-independent terminator was identified.[133]

The epidermin biosynthetic gene cluster is located on the 54 kb plasmid pTü32 in *Staphylococcus epidermidis* Tü3298/DSM 3095. So far, about 13 kb of sequence have been published. The gene cluster consists of at least five operons: *epiABCD*, *epiPQ*, *epiT'T''*,[138] *epiH*,[159] and *epiFEG*.[158] For gallidermin (or [Leu⁶]epidermin), a natural variant of epidermin, only *gdmA*,[160] *gdmT*, and *gdmH*[159] have been described. EpiT'T'' is probably inactive since *epiT'T''* contains a frameshift and two deletions in comparison to the homologous transporter in the gallidermin biosynthetic gene cluster, and expression of *gdmT* in the epidermin producer leads to an increase in the level of epidermin production.[159] EpiD is an oxidative decarboxylase that is involved in the biosynthesis of the C-terminal *S*-aminovinyl-D-cysteine residue of epidermin.[176] The promoters in front of *epiA*, *gdmT*, *gdmH*, and *epiF* are all under the control of the transcriptional activator *epiQ*.[158,159,177] Again, there is a putative terminator downstream of *epiA* which may allow partial readthrough. A short 300 bp *epiA* transcript and a long 5 kb mRNA that hybridized with an *epiB* probe have been detected.[138] Complementation experiments with mutant strains, which had been inactivated by ethyl methanesulfonate mutagenesis, indicated that *epiB* is co-transcribed with *epiA*. In contrast, *epiC* and *epiD* could be complemented *in trans* with the intact genes and, thus, functional promoters should be present in front of *epiC* and *epiD*.[178] A second terminator is located downstream of *epiQ*. *epiABCD* and *epiQ* are essential for production of epidermin in *Staphylococcus carnosus*,[177,178] but reconstitution of *epiFEG*,[158] *gdmT*, and *gdmH*[159] increases production in this heterologous host. The proteolytic activity of EpiP does not seem to be essential, but the promoter of *epiP* which confers expression of *epiQ* must be present.[177]

The Pep5 biosynthetic gene cluster is encoded on the 18 kb plasmid pED503 in the producer strain *S. epidermidis* 5. So far, the genes *pepTIAPBC* have been described,[137] and a DNA fragment harboring *pepIAPBC* is sufficient to reconstitute Pep5 production in the absence of pED503.[179] *pepT* is located on the opposite strand and therefore arranged in the same position as *epiT'T''* in relation to *epiA*. There are possible promoter structures upstream of *pepI* and *pepA*, and a weak ρ-independent terminator was found downstream of *pepA*[157] which should allow a limited readthrough, as there are no promoters present in front of *pepP*, *pepB*, or *pepC*. A strong terminator is located downstream of *pepIAPBC*. *pepI* and *pepA* are most probably co-transcribed.[180]

Only a short fragment of the epilancin K7 gene cluster has been cloned and sequenced from chromosomal DNA of *S. epidermidis* K7, but the data obtained indicate that the arrangement of *elkT*, *elkA*, and *elkP* is similar to that found in the Pep5 biosynthetic gene cluster, with the exception that the immunity gene is missing.[10]

4.11.2.1.1 Prepeptides

The *lanA* genes code for the lantibiotic precursor peptides which are designated prepeptides. These prepeptides consist of an N-terminal extension, called the leader peptide, and the C-terminal propeptide domain, into which the unusual amino acids are introduced by post-translational modification reactions.[169] Finally, the fully modified propeptide domain is released from the leader peptide and, thus, the mature lantibiotic is set free. The leader peptides of the nisin subtype lantibiotics contain about one-third charged amino acids and carry no or a net negative charge, whereas the propeptide part contains a net positive charge of between +2 (subtilin) and +7 (Pep5). Cys residues are not present in the leader peptides, but Ser and Thr residues occur. Prepeptides in varying stages of dehydration have been isolated from the Pep5-producing strain *S. epidermidis* 5, and were analyzed by mass spectroscopy and Edman degradation; here, the dehydrated amino acids were detected only in the C-terminal propeptide part, demonstrating that the hydroxyl amino acids of the leader peptide had not been dehydrated by the modification apparatus. Unmodified PepA was only isolated from a clone that did not contain the modifying system.[136] The leader peptides of the nisin subtype lantibiotics (Figure 7) are distinguished by two conserved motifs from the leader peptides of the other lantibiotic subtypes. A Pro residue at position −2 is characteristic of the leader peptide cleavage site[40,41] (see Section 4.11.2.1.5 for a detailed discussion) and a F-N/D-L-D/E/N-X motif (with X being one of the hydrophobic amino acids I, L, or V) is conserved around position −20 to −15.[181] However, the lantibiotic leader peptides do not contain the characteristic membrane-spanning helix and peptidase cleavage site that direct proteins to the *sec*-dependent export machinery, and the role of the leader peptides during biosynthesis is still unclear. Several possible functions have been suggested.[25]

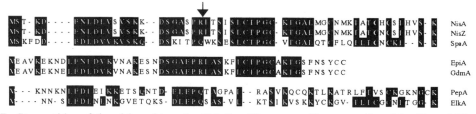

Figure 7 Prepeptides of the nisin subtype lantibiotics. The prepeptides were aligned by Clustal W software, and identical residues that occur at the same position in peptides that are not natural variants (e.g., nisin A and nisin Z or epidermin and gallidermin) are indicated by black boxes. The protease cleavage site between leader and propeptide is shown by an arrow.

In analogy to proteases or hormones, the leader peptides may function to keep the lantibiotic inactive and, thus, protect the producer cells before the peptides are exported from the cells. For example, fully modified nisin prepeptides have been isolated from the supernatant of clones that did not contain an active leader peptidase. These peptides were inactive as long as the leader peptides were not cleaved off.[148] This mechanism may protect all those lantibiotic-producing cells that process their products by extracellular proteases, that is, the producer strains of epidermin and gallidermin, nisin, and subtilin. In contrast, Pep5[137] and, presumably, epilancin K7 possess intracellular proteases and are processed within the producer cell.

Another function of the leader peptide could be the stabilization of a conformation of the propeptide that is recognized by the modification enzymes and guarantees the modification of only those (and sometimes not even all of those, e.g., Ser29 in nisin) hydroxyl amino acids that are located in the propeptide domain. An α-helical conformation with a strong amphiphilic nature has been predicted by computational analyses for the lantibiotic leader peptides. For the cleavage site, a β-turn was assumed, and the propeptide domains of most lantibiotics, with the exception of Pep5, probably adopt a β-sheet conformation.[40,41] In order to examine the conformation of the prepeptides and verify a possible interaction between the leader peptide and propeptide domains, prenisin,[182] pregallidermin, and the leader peptide and propeptide of Pep5 have been synthesized.[183] Circular dichroism spectroscopy of the peptides confirmed the helical structure of the leader peptides in a lipophilic solvent. The helical structure of the isolated Pep5 prepeptide was even more marked than those of the isolated leader and propeptide domains. The same effects were recorded with pregallidermin. It was therefore suggested that the propeptide domain might indeed interact with the leader peptide domain.[183] For prenisin a more flexible structure, which does not prefer a defined conformation, was observed.[126] Surprisingly, the nisin prepeptide is able to bind Zn^{2+} ions stoichiometrically, which might be important for the modification reactions.[184]

A third possible function was suggested for the FNLDL motif in the leader peptide of nisin, which might be a recognition signal for the modification enzymes NisB and NisC or the transporter NisT. Site-directed mutagenesis in the leader peptide sequence of nisin showed that any mutation of Phe(-18), Leu(-16) or Asp(-15) completely abolished production of nisin.[181] In analogous experiments with the lantibiotic Pep5 a lower production was observed after exchange of Phe(-19) for Ser and Asp(-16) for Lys.[185]

On the other hand, experiments with gene fusions of nisin and subtilin showed that the presence of a lantibiotic leader alone is not sufficient for correct modification. For example, two fusions between the propeptide regions of subtilin and nisin were assayed for expression in *B. subtilis*: whereas a fully modified product was isolated after expression of the subtilin leader fused to a chimeric propeptide nisin(1–11)–subtilin(12–32), the reverse construct (subtilin(1–11)–nisin(12–34)) was expressed but not correctly modified. Hence, the presence of the C-terminal part of the subtilin propeptide was necessary for correct modification.[186] The same point is stressed by the observation that a fusion of the subtilin leader to the nisin propeptide did not give rise to an active product in a subtilin-producing strain.[187] In contrast, when a similar construct, employing the subtilin leader and the nisin Z propeptide domain, was expressed in a nisin-producing *Lactococcus* sp., the fully and correctly modified prepeptide accumulated in the culture supernatant. Only the N-terminal Met residue had been lost, and the protease NisP was inhibited by the presence of the Gln residue at position -1.[188] However, when the prepeptide contained only the seven N-terminal amino acids of the subtilin leader fused to the 17 C-terminal amino acids of the nisin leader and nisin propeptide domains, an active product was excreted by the *Bacillus* producer strain that was eluted by HPLC at the same time as nisin.[187] In conclusion, these results indicate that there are structural requirements for modification which concern both domains, leader and propeptide, and differ between the various lantibiotic biosynthetic systems.

4.11.2.1.2 *Modifying enzymes LanB and LanC*

Two open reading frames that do not show any homologies to proteins in the databases are present in each gene cluster of a nisin subtype lantibiotic and were designated *lanB* and *lanC*.[137–139,146,147,152] Replacement of *spaB* and *spaC* by resistance genes[139] or ethyl methanesulfonate mutagenesis in *epiB* and *epiC*[178] resulted in the isolation of mutants that were no longer able to produce the lantibiotic. The production was restored upon reconstitution of *epiB* and *epiC in trans*.[178] Therefore, it was suggested that these proteins might be novel enzymes responsible for the modification of the lantibiotic propeptide which results in biosynthesis of the lanthionine residues: dehydration of the hydroxyl amino acids and formation of the thioether bridges.[138]

The *lanB* genes code for large proteins of about 1000 amino acids (EpiB 990, NisB 993, SpaB 1030, and PepB 967 amino acids). They possess only 16–29% identity but seven stretches of stronger homology can be identified, among them a conserved YX_2EX_2RYGG motif near the C terminus.[168] The B enzymes are mostly hydrophilic proteins with a short hydrophobic stretch in their C-terminal domain. EpiB has been expressed in *S. carnosus*, and was found in both the membrane and cytoplasmic fraction of the cells.[189] In Western blots an association with the vesicle fraction has also been demonstrated for NisB[146] and SpaB[155] from which both enzymes were only released by treatment with sodium dodecyl sulfate. For NisB, some evidence has been presented that the enzyme is located in a membrane-bound complex together with NisC and NisT.[190]

The open reading frames of the C enzymes code for smaller proteins consisting of more than 400 amino acids (EpiC 455, NisC 418, SpaC 441, and PepC 398 amino acids). NisC and SpaC show the highest degree of similarity (33% sequence identity), otherwise 24–32% identity is found.[168] Hydrophilic and hydrophobic domains alternate in the proteins, and seven conserved motifs have been found in the hydrophobic regions. Here, several Gly residues with a potential structural function and two His, two Cys, and one Trp residue, which could be involved in catalysis, are conserved.[168] Sequence analysis of three *epiC* genes that had been inactivated by ethyl methanesulfonate mutagenesis, showed that all three strains harbored point mutations in the two conserved motifs NXGXAHGX$_2$G and SX$_3$GX$_2$G. The inactivated proteins contained Glu residues instead of the conserved Gly residues in the positions that have been underlined.[191] NisC and EpiC have been expressed in *Escherichia coli*.[147,191] EpiC is a hydrophobic but soluble protein, was isolated from the cytoplasmic fraction and seems to possess an intramolecular disulfide bond. From the N-terminal sequence of the isolated protein MININNI it was deduced that, at least in *E. coli*, the GTG at position 118 is used as a start codon, giving a 416 amino acid protein instead of the 455 amino acid protein that had been postulated from the nucleotide sequence of *Staphylococcus* with a TTG start codon.[191]

Successful *in vitro* enzyme assays have been reported neither for the B nor the C proteins.[189,191] Therefore, the only indications for the role of LanB and LanC in lantibiotic biosynthesis derive from inactivation experiments in the Pep5 biosynthetic gene cluster. *pepC* was inactivated by a 693 bp deletion and then tested in the Pep5 expression system.[137] Analysis of the peptides excreted by this clone showed that Pep5 was no longer produced, but shorter, proteolytically degraded peptides were present in the culture supernatant. These peptides contained dehydroamino acids, but Lan or MeLan residues could not be detected. Furthermore, the proteolytic processing had taken place at those positions which are normally adjacent to a ring (between amino acids 8 and 9) or within the C-terminal double-ring system (between amino acids 29 and 30). Normally Pep5 is protected from processing at these positions by the ring structures, and truncated peptides arise only when ring formation is inhibited by, for example, site-directed mutagenesis.[120] There was one exception: a peptide species was found that was not processed at its C-terminus and contained a single ring which was attributed to spontaneous formation of the C-terminal Lan as described by Toogood.[140] In conclusion, the formation of thioethers was assigned to PepC, leaving the dehydrating function to PepB.[137]

4.11.2.1.3 *Modifying enzyme LanD*

Only one LanD enzyme, EpiD, an oxidative decarboxylase required for biosynthesis of the C-terminal *S*-aminovinyl-D-cysteine residue of epidermin, has so far been characterized; extensive *in vitro* research has been carried out and, therefore, EpiD is the best known lantibiotic modification enzyme.[176,192–194]

Ethyl methanesulfonate mutagenesis and complementation studies in the epidermin-producing strain *S. epidermidis* Tü3298 showed that *epiD*, a further open reading frame downstream of *epiC*, is involved in epidermin biosynthesis.[178] *epiD* codes for a protein of 181 amino acids, and homologous proteins could not be found in the databanks.[138] The protein is present only in low amounts in *S. epidermidis*, and was, therefore, purified after overexpression in *E. coli*.[192] EpiD is a soluble, hydrophobic protein and contains four intramolecular Cys residues. Only under reducing conditions was the protein retarded on an EpiA affinity column. The purified protein appeared yellow, displayed the typical adsorption spectrum of a flavoprotein, and after heat treatment the coenzyme flavin mononucleotide was released. The mass of the purified protein after release of the coenzyme (20 827 Da) corresponded closely to the value calculated from the gene sequence, indicating that the protein is not modified. The necessity of FMN for catalysis was also shown when the mutant gene *epiD**, which had been inactivated by ethyl methanesulfonate mutagenesis, was sequenced and overexpressed in *E. coli*. This inactive protein, which harbors a point mutation at position 93, was not able to bind FMN. The presence of a flavin coenzyme identified EpiD as an oxidoreductase, and it was concluded that it might be involved in the synthesis of the C-terminal *S*-aminovinyl-D-cysteine residue of epidermin.[192] An open reading frame coding for a homologous protein was found in the mersacidin biosynthetic gene cluster,[195] and is probably involved in biosynthesis of the C-terminal aminovinyl-3-methyl-D-cysteine. The enzymatic activity of EpiD was analyzed *in vitro*, employing the unmodified EpiA prepeptide or the epidermin propeptide as substrates. The products of the reaction were analyzed by mass and UV spectroscopy, and found to have masses that had decreased by 46 Da. This corresponded to an oxidized and decarboxylated peptide, and the increased absorbance at 260 nm indicated the presence of a thioenol group. The reaction products were unstable, and a peptide that had lost the C-terminal Cys residue was also formed *in vitro*. Accordingly, the model shown in Scheme 1 was proposed for the action of EpiD with the EpiA prepeptide:[176] EpiD catalyzes the oxidation of the C-terminal Cys residue in EpiA. The removal of two hydrogen atoms reduces the FMN coenzyme, and a double bond is formed. In a second step, the thioenol compound is decarboxylated, a reaction which may occur spontaneously. Only afterwards is the *S*-aminovinyl-D-cysteine residue formed by addition of the thioenol group to the dehydroalanine. The intermediate thioenol compound has been isolated, and was characterized by mass spectroscopy and NMR.[194]

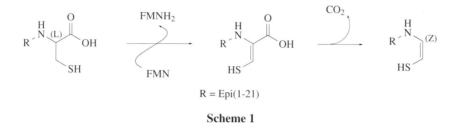

R = Epi(1-21)

Scheme 1

It was further shown that EpiD will not catalyze any reaction in the absence of the C-terminal Cys residue at position 22, thus it will not catalyze dehydration of Cys21. The complete propeptide does not have to be present for catalysis, since the C-terminal heptapeptide SFNSYCC of EpiA is sufficient for oxidation. Employing both a library of SFNSYCC heptapeptides with one variable position and tandem mass spectroscopy to record the oxidative decarboxylation, the substrate specificity of EpiD was characterized: in the first four positions any one amino acid exchange was possible; even for the peptides of a XXXXYCC library, a shift to lower masses was observed, indicating that the substrate specificity of EpiD resides in the last three residues of EpiA. At position 5, Val, Ile or Leu, Phe, or Trp could be substituted for the Tyr residue, while in contrast a Met residue gave a very low activity. At position 6, Ala, Ser, Thr, or Val could replace the Cys residue, while Ile and Leu strongly decreased the reaction rates. A study of the kinetics of the SFNS(F/W/Y)(C/S/T)C and SFNSYXC libraries showed that the original substrate gave the highest reaction rate and that a Val residue at position 6 resulted in the same activity of EpiD as the original Cys residue.[193]

4.11.2.1.4 Transporter LanT

All lantibiotic gene clusters contain genes for putative transporters, *lanT*, whose function is the export of the lantibiotic or lantibiotic precursor from the producer cell following biosynthesis. Replacement of *spaT* by a resistance gene led to cessation of biosynthesis in the case of subtilin.[139] In contrast, for epidermin and Pep5 biosynthesis, the transport proteins are not essential; their function can be substituted by other transporters of the cell. For example, EpiT′T″ (formerly designated EpiY′Y″) is not active in the natural epidermin producer or the heterologous expression system in *S. carnosus*, as it contains two deletions and a frame shift mutation.[138,159,178] PepT is also not present in the Pep5 expression system; as a consequence the yields of this system are only 10–30% of the production levels reached with the natural producer strain.[179] The LanT proteins contain between 500 and 600 residues, and belong to the ATP-binding cassette (ABC) transport protein superfamily.[137–139,146,152,159] One subunit of such a transporter usually consists of two domains, a membrane domain with six membrane-spanning helices and an intracellular ATP-binding domain with a highly conserved consensus motif, called the Walker motif (GXGKST).[196] These domains may be present as a single (group A transporters) or two different polypeptides (group B transporters). Characteristically, both domains of the LanT proteins are encoded by a single open reading frame. The membrane domain is found at the N terminus, and the ATP-binding domain forms the C-terminus. Two of these subunits are thought to comprise a functional membrane protein.[196] In the case of GdmT and EpiT′T″, a second hydrophobic 330 amino acid protein, GdmH and EpiH, respectively, was found to be encoded upstream of the transporter gene. The presence of *gdmT* and *gdmH* led to a seven- to 10-fold increase in production of epidermin in *S. carnosus*, and, likewise, the production of epidermin could be increased twofold in the natural producer *S. epidermidis* Tü3298, which contains *epiH* and the inactive *epiT′T″*, by introduction of *gdmT*. The H proteins were absolutely essential for this effect, indicating that they are involved in epidermin export.[159]

For the biosynthesis of nisin, evidence has been presented that the modification enzymes and the transporter interact with each other. Employing the yeast two-hybrid system to detect interactions of the proteins involved in biosynthesis of nisin, an interaction between NisC and NisB, as well as NisC and NisT, has been suggested. This led to the proposal of a so-called lanthionine synthetase, consisting of NisB, two molecules of NisC, and two molecules of NisT. This complex might be anchored at the membrane via NisT.[190]

4.11.2.1.5 *Protease LanP*

The nisin,[148] epidermin,[138] and Pep5[137] biosynthetic gene clusters contain genes coding for proteases; the N-terminal sequence of a similar gene from the epilancin K7 gene cluster has also been published.[10] These proteases remove the leader peptide from the inactive, fully modified lantibiotic precursor peptide and, thus, release the active form of the mature lantibiotic.[181,188] In the subtilin biosynthetic gene cluster, no protease gene seems to be present; possibly this function is taken over by one of the various serine proteases that have been identified in *Bacillus*,[197] for example subtilisin.

The *lanP* genes code for proteins ranging from 285 (PepP) to 682 amino acids (NisP), which all display homology to the subtilisin-like serine proteases. The conserved residues of the active site (Asp, His, Ser) and the Asn residue of the oxyanion hole are present in every case but the proteases vary with respect to their localization and possession of a prepropeptide and a C-terminal extension, which explains the differences in size of the *lanP* genes.[168] NisP and EpiP are extracellular proteins. EpiP was isolated from the culture supernatant after overexpression in *S. carnosus*, and the N-terminal amino acid sequence of the purified protein suggested that EpiP is synthesized as a preproprotein with a 99 amino acid preprosequence. Thus, the mature protein consists of only 362 amino acids and is synthesized with a 25 amino acid signal peptide which is responsible for export and a 74 amino acid prosequence that acts as an intramolecular chaperone.[198] A preprosequence of about 220 amino acids is also present in NisP, which is a 54 kDa protein when expressed in *E. coli* and, in addition, a consensus membrane anchor sequence is found at the C terminus. This suggests that NisP is bound to the cell membrane of *L. lactis* after export.[148] In the absence of NisP, the fully modified but unprocessed NisA precursor peptide accumulates in the culture supernatant. This confirms that the substrate of NisP, the fully modified lantibiotic prepeptide, is excreted from the cells and is subsequently activated by proteolytic processing.

In contrast, PepP, which does not contain a prepro sequence, is presumably an intracellular enzyme.[137] After inactivation of PepP by site-directed mutagenesis, no PepA prepeptides could be detected in the culture supernatant. However, fragments of Pep5, for example Pep5(6–34), Pep5(15–34), and Pep5(17–34), were isolated. These fragments had probably been formed by inappropriate proteolytic processing by another host cell protease. This demonstrated that PepP is essential for the processing of Pep5.[137] The fragment of ElkP that has been sequenced shows a high degree of homology to PepP (47% identity) and also encodes an intracellular enzyme without a prepro sequence.[10]

The substrate of the proteases, that is, the cleavage sites of the lantibiotic leader peptides of the nisin subtype, can be characterized as follows: at position −4 lies a hydrophobic amino acid (Ala or Leu), followed by a negatively charged or polar residue at position −3 (Glu or Ser), position −2 is always occupied by Pro, and a positively charged or polar amino acid is located at position −1 (Arg or Gln). In spite of these similarities, the enzymes are specific. For example, the Arg residue at position −1 in EpiA and NisA cannot be exchanged for the Gln residue that is found in PepA and SpaA in this position, without inhibiting EpiP and NisP.[181,198] Furthermore, a substitution of the Ala residue at position −4 by Asp, but not an exchange of the highly conserved Pro residue at position −2, abolished the activity of NisP.[181] In contrast, mutations of the first amino acids in the propeptide sequence did not inhibit NisP,[181] and EpiP cleaved the unmodified EpiA precursor peptide.[198] On the basis of these data, the active sites of NisP and EpiP have been modeled in analogy to subtilisin BPN and thermitase.[199] As the experiments with the prepeptide mutations had already indicated, the most important interactions between enzyme and substrate were found to reside in the leader peptide, at position −1 and −4 for nisin and −1 and −3 for epidermin. The model also predicted a surface patch which is extremely rich in aromatic residues and is located next to the catalytic domain in NisP. It could be involved in the binding of the propeptide domain of the precursor or the association of NisP with the cell membrane or other hydrophobic proteins.[199]

4.11.2.2   Biosynthetic Gene Clusters of Streptococcin A-FF22 Subtype Lantibiotics and Lactocin S Subtype Lantibiotics

The biosynthetic gene clusters of the streptococcin A-FF22 and lactocin S subtype lantibiotics are shown in Figure 8. In each case, the structural gene *lanA* is followed downstream by the modifying enzyme LanM that substitutes the LanB and LanC enzymes of the nisin subtype lantibiotic biosynthetic gene clusters in this group.

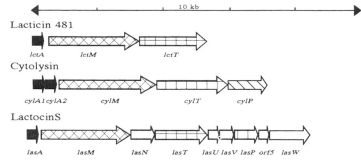

Figure 8   Lantibiotic gene clusters of streptococcin subtype lantibiotics (lacticin 481) and lactocin S subtype lantibiotics. The structural genes are shown in black, genes with similar functions are shown with identical patterns, and the arrows indicate the direction of transcription.

The gene cluster of lacticin 481, which was also referred to as lactococcin DR, is the only gene cluster of the streptococcin A-FF22 subtype that has been published so far.[163] It is found on a 70 kb plasmid in *L. lactis* ssp. *lactis*, and a 4.1 kb fragment carrying the structural genes *lctA*, *lctM*, and the 5′ end of *lctT* was sufficient to promote production of lacticin 481 in a nonproducing *Lactococcus* strain.[163] Upstream of *lctA*, an open reading frame that shows sequence similarity to the transposase of the *E. coli* IS4 insertion element is encoded.[47]

Only the structural genes of the other lantibiotics of the streptococcin A-FF22 subtype have been sequenced so far. *SalA* of salivaricin A was subcloned from chromosomal DNA of *Streptococcus salivarius* 20P3;[48] most interestingly, this structural gene or very closely related genes have been detected in a number of *S. salivarius* and *S. pyogenes* strains.[200] The structural gene for streptococcin A-FF22, *scnA*, was cloned from chromosomal DNA of *S. pyogenes* FF22; this structural gene is also present in duplicate in M type 49 group A *S. pyogenes*.[162] The structural gene of variacin, *varA*,[164] was found on genomic DNA of *Micrococcus varians*.[49]

The cytolysin biosynthetic gene cluster is located on the large conjugative plasmid pAD1 in *Enterococcus faecalis*.[36] Two structural genes coding for two prepeptides are present, and both gene products are necessary for the bacteriolytic and hemolytic activity.[34] Downstream of the structural genes, three open reading frames code for the modifying enzyme, CylM, a transporter-associated protease, CylT, which was originally referred to as CylB, and a subtilisin-like Ser protease, CylP, which was first described as CylA[201] as the protein had been identified as activating component A as early as 1969.[75] Linker and transposon insertion mutagenesis and complementation *in trans* have been used to demonstrate that the three enzymes are involved in the biosynthesis of cytolysin. From polar effects that were detected in the insertion experiments, it was concluded that *cylP* is not transcribed from the same promoter as *cylA1A2MT*.[33,34,201]

The lactocin S operon is located on the 50 kb plasmid pCIM1 in *Lactobacillus sake*.[72] Downstream of *lasA*, which encodes prelactocin S,[2] eight open reading frames are located. Three of the proteins encoded by these open reading frames were identified by their homology to other lantibiotic proteins: LasM, a putative modifying enzyme, LasT, an ABC transport protein, and LasP, a leader peptidase. The significance of the other five open reading frames remains to be elucidated; some of these proteins could be involved in the biosynthesis of the unique D-Ala residues that are present in lactocin S as well as in immunity. Transcriptional analyses showed the presence of a long transcript that started in front of the structural gene and ended only downstream of the last open reading frame; additionally, two small transcripts containing only *lasA* were also detected. These transcripts terminated with a palindrome sequence downstream of *lasA*. It was therefore suggested that this terminator structure might limit expression of the biosynthetic genes compared to the prepeptide.[141]

4.11.2.2.1 Prepeptides

The prepeptides of the streptococcin and lactocin S subtype lantibiotics are also formed by an N-terminal extension, the leader peptide, and the C-terminal propeptide domain, which is processed to yield the mature lantibiotic. However, the prepeptides of the streptococcin A-FF22 subtype lantibiotics and cytolysin are characterized by a cleavage site that differs from that of the nisin subtype lantibiotics and always contains Gly at position -2 and Gly, Ala, or Ser at position -1 (Figure 9).[25] This cleavage site is also found in many nonlantibiotic bacteriocins that are produced by lactic acid bacteria.[28,47,48] Here, Gly occurs most frequently at position -1, and because of this the designation "double Gly cleavage site" was chosen. The lantibiotics and the nonlantibiotic bacteriocins that possess a double Gly cleavage site are often processed and transported by a single protein, an ABC transporter that is associated with a leader peptidase (see Section 4.11.2.2.3).

Figure 9 Prepeptides of the streptococcin A-FF22 subtype lantibiotics, the cytolysins, and lactocin S.

Another difference between the leader peptides of the nisin subtype and streptococcin subtype lantibiotics are the conserved motifs of the leader sequence; with the exception of lacticin 481, the FNLDL motif of the nisin subtype lantibiotics is not found in these peptides. In contrast, the central part of the leader peptide sequence is conserved, carries a net negative charge, and contains several Glu residues.[25] As in the leader peptides of the nisin subtype lantibiotics, there are no Cys residues present. The leader peptides of lacticin and streptococcin AFF-22 have been synthesized and show considerable helical character;[202] a β-turn at the cleavage site and a β-sheet structure for the propeptide domain have been predicted for SalA.[48] Lacticin 481 and variacin are very similar in their propeptide sequence: there are only three exchanges within the sequence, and the N-terminal Lys-Gly of lacticin 481 is missing from variacin. In the leader peptide sequence, nine exchanges and two additional residues are found in lacticin 481.[49]

The cytolysin biosynthetic gene cluster contains two structural genes, *cylA1* and *cylA2*, coding for a larger prepeptide of 68 amino acids (CylA1) and a smaller prepeptide of 63 amino acids (CylA2), respectively.[33] Cytolysin is the only lantibiotic with precursor peptides that are cleaved twice during the maturation process. The first cleavage is performed by a transporter-associated leader peptidase (CylT, originally referred to as CylB) at the double Gly cleavage site GS↓GD. Afterwards, a dedicated Ser protease, which is encoded by *cylP* (originally called *cylA*) and active in the culture supernatant, releases the mature lantibiotics at the VQAE↓TTP site. This second processing is necessary for the activation of the peptides.[34] The nucleotide sequence of both genes is different with the exception of a large 78 bp repeat that codes for a PSFEELSV/LEEN-MEAIQGS↓GDVQAE↓TTP motif. This sequence encompasses the C-terminal part of the leader peptide, both the protease cleavage sites (marked by arrows), and the first three amino acids of the N-terminus of both peptides.[34]

When the lactocin S structural gene *lasA* was sequenced, three Ser residues were found to be encoded by the nucleotide sequence whereas Ala residues were detected in the mature peptide.[2] These Ala residues were present in the unusual D configuration in the lantibiotic, making lactocin S the first ribosomally synthesized peptide that contains D-Ala. LasA is processed by an intracellular Ser protease, and its cleavage site does not show any similarity to other lantibiotic cleavage sites.[141]

4.11.2.2.2 Modifying enzyme LanM

LanB and LanC proteins have not been identified in the streptococcin A-FF22 and lactocin S subtype gene clusters that have been sequenced so far. Instead, large open reading frames are located directly downstream of the structural genes. Since these proteins seemed to be involved in lantibiotic modification, they were called LanM. LctM (originally named LcnDR2) was predicted to be a 922 amino acid protein, and the transfer of *lctAM* into another *L. lactis* strain, IL1403, resulted in production of lacticin 481.[163] *lasM* from the lactocin S biosynthetic gene cluster codes for a putative protein of 925 amino acids.[141]

Lactocin S production is unstable, and loss of production is often caused by the insertion of a transposon (IS*1163*) into *lasM* or *lasT*. RNA analyses showed that the insertion of IS*1163* did not produce any polar effects, which proves the importance of LasM for lactocin S biosynthesis.[141] CylM is a putative 993 amino acid protein.[33] The necessity of CylM for biosynthesis of cytolysin was shown by transposon or linker insertion mutagenesis and complementation analysis.

The C-terminal 350–450 amino acid domains of all three proteins show sequence similarity to the LanC proteins (20–27% identity), and contain the same pattern of seven conserved amino acid motifs with several Gly, two His, two Cys, and one Trp residues.[168] It was therefore suggested that in analogy to the C proteins, the C-terminal domain might be involved in the correct folding of the peptides and thioether formation. In contrast, the N-terminal 500–650 amino acid domains display no similarity to the B enzymes, so that the LanM enzymes probably do not originate from a gene fusion of LanB and LanC enzymes. The N-terminal domains show only a weak mutual similarity (23–26% identity), but a conserved stretch of 40 amino acids within the N-terminus displaying about 50% identity was identified.[168] A search of the databases did not show any homologous proteins. The M proteins have not yet been overexpressed or assayed *in vitro* and, thus, the question whether the N-terminal domain might be also responsible for dehydration of the precursor peptides still remains open.

4.11.2.2.3 *Transport and peptidase functions*

The lacticin 481 and cytolysin biosynthetic gene clusters contain proteins called LctT[163] and CylT (original nomenclature CylB)[165] that catalyze both proteolytic processing and export of the lantibiotics. These transporters, with an associated protease activity, are also common in many nonlantibiotic bacteriocins, for example lactococcin G[203] and pediocin PA-1,[204] which contain the same conserved cleavage site with Gly at position −2 and Gly, Ala, or Ser at position −1. The proteins are composed of about 700 amino acids and, upon comparison with the LanT proteins that are found in the gene clusters of the nisin subtype lantibiotics and which are smaller by 100–200 amino acids, an N-terminal extension is found as part of these proteins. This domain contains two conserved amino acid sequence motifs, $QX_4D/ECX_2AX_3MX_4Y/FGX_4I/L$ and $HY/FY/VVX_{10}I/LXDP$. The experiments that revealed the role of these unusual proteins within the biosynthesis of bacteriocins were performed with the transporters that are involved in export of the nonlantibiotic bacteriocins lactococcin G and pediocin PA-1. The N-terminal domain of the lactococcin G transporter (LagD) was expressed without the transporter domain and shown to cleave the precursor peptide of lactococcin G *in vitro*.[203] The N-terminal domain of the pediocin transporter was also sufficient for cleavage of prepediocin *in vivo*.[204] Upon exchange of the conserved Cys residue at position 13 of LagD, the enzymatic activity of the enzyme was completely lost, indicating that these domains could be Cys proteases. The transporter domains of these proteins show homology to ABC transport proteins, with six helical membrane-spanning stretches and a C-terminal ATP-binding domain. A computational analysis of the topology of the LagD protein indicated that most probably the N-terminal protease domain and the C-terminal ATP-binding domain are located in the cytoplasm. It was therefore suggested that processing and transport might function simultaneously for these proteins.[203]

The cytolysin gene cluster and the lactocin S gene cluster contain additional genes for Ser proteases of the subtilisin family. As already mentioned, the cytolysin precursor peptides, CylA1 and CylA2, are processed twice: Once upon export by the protease transporter CylT (originally called CylB), which cleaves off the leader sequence at a double Gly cleavage site, and then in the culture supernatant by the Ser protease CylP (original nomenclature CylA), which removes a further GDVQAE fragment. N-terminal sequencing showed that the enzyme is expressed as a prepro-enzyme. After removal of the 95 N-terminal amino acids, the mature protein has a mass of 34 kDa, as determined by mass spectroscopy.[34,75]

LasP also codes for a subtilisin-like Ser protease with a calculated molecular mass of 29.2 kDa. In contrast to CylP, this protease does not possess a prepro sequence, and, therefore, should be located in the cytoplasm of the lactocin S producer strain as described previously for PepP and ElkP.[141] In addition, about 30 amino acids seem to be missing from the C-terminus of LasP when it is aligned with other Ser proteases. The substrate specificity of both CylP (VQAE↓TT) and LasP

(MNAD↓ST) is different from that of the subtilisin-like Ser proteases of the nisin subtype lanti-biotics, and in both cases dehydrated residues are present at positions +1 and +2 of the cleavage site.[34,75]

Lactocin S is exported from the producer cells by a typical ABC transporter, LasT, that comprises 535 amino acids. The protein was identified by its sequence similarity to ABC transport proteins, and seems essential for lactocin S biosynthesis because insertion of IS*1163* into *lasT* effectively blocks production of lactocin S.[141]

4.11.2.3 Prepeptides and Biosynthetic Gene Clusters of Type B Lantibiotics

Only the structural genes of cinnamycin, *cinA*, and mersacidin, *mrsA*, have been cloned from chromosomal DNA of the producer strains and sequenced.[166,167] Both prepeptides are extraordinarily long (Figure 10), since CinA and MrsA contain leader peptides of 59 and 48 amino acids, respectively. In contrast to the type A lantibiotics, a lower percentage of charged residues is present in these leader peptides. For example, the leader peptide of the type A lantibiotic Pep5 contains 10 charged amino acids in a total of 26 residues, whereas in CinA 11 charged residues occur in a total of 59 amino acids, and in MrsA 12 charged residues are found in a total of 48 amino acids. The conserved amino acid motifs of the type A lantibiotics are not present, and the protease cleavage sites (MEAA↓C and EAFA↓C) do not show similarity to the cleavage sites of the type A lantibiotics. Upon comparison of MrsA and CinA, an NPA and an EAQ motif can be found, but the importance of these motifs is not yet clear.

Figure 10 Prepeptides of the type B lantibiotics cinnamycin CinA and mersacidin MrsA.

The biosynthetic gene cluster of mersacidin is currently being sequenced, and preliminary results indicate the presence of genes for two-component regulatory systems, a type B ABC transporter, a LanD, and the 5′ terminus of a putative LanM protein.[195]

4.11.2.4 Regulation of Lantibiotic Biosynthesis

Lantibiotics are not produced constitutively; the expression of subtilin, for example, starts in the mid-exponential growth phase and reaches a maximum in the early stationary phase,[154] and the nisin prepeptide was first detected in the cells after 4 h of incubation.[149] Both biosynthetic gene clusters contain open reading frames which show sequence similarity to two-component regulatory systems.[148,149,154] These systems are frequently found in bacteria and consist of two separate proteins, a membrane-bound His kinase (LanK) and an intracellular response regulator protein (LanR). The His kinase functions as a sensor protein. It contains membrane-spanning sequences and an extracellular domain that is thought to bind a specific extracellular signal molecule. Upon contact with the signal molecule, a conserved His residue is autophosphorylated in the intracellular domain of the kinase. The phosphate residue is then transferred to a conserved Asp residue in the N-terminal domain of the second protein, the response regulator. This protein adopts an active conformation upon phosphorylation, which enables it to bind to DNA with its C-terminal domain and activate or repress transcription. As disruption of these genes abolishes production of the lantibiotic, they are presumed to regulate biosynthesis of the lantibiotics.[154,205,206]

NisR and SpaR are proteins of about 220 amino acids and have six conserved segments when compared to *E. coli* and *Bacillus* response regulators. The essential residues of the N-terminal domain (in NisR Asp10, Asp53 (phosphate acceptor), and Lys102) are present as well as the C-terminal GXGY motif.[148,154]

NisK is a 447 amino acid protein which contains two membrane-spanning sequences in its N-terminal domain. All conserved residues that are characteristic for His kinases, the phosphate acceptor His, Asn, and two Gly-rich motifs, are found in its C-terminal domain.[149] In contrast, SpaK consists of only 387 amino acids,[154] and seems to be missing the C-terminal Gly-rich motifs, which are thought to be ATP-binding sites. This might be caused by a frame shift mutation, as at least 27 following residues with one Gly-rich motif were detected in another reading frame.[168]

In order to confirm the putative regulatory role of the two component systems, the expression of NisB and SpaB has been monitored by Western blots, and was found to start in the mid-exponential and early stationary phases, respectively.[149,155] During the construction of the nisin expression systems for site-directed mutagenesis it became apparent that in the absence of a functional nisin structural gene, the immunity gene *nisI* and the inactivated nisin structural gene, Δ*nisA*, carrying a 4 bp deletion, were not expressed. The expression of *nisI* could be restored by complementation with *nisA* *in trans*. From these results it was deduced that the expression of NisA together with NisR was a requirement for transcription of *nisI*.[147] The transcription of Δ*nisA* or a reporter gene that was fused to the promoter of *nisA* could also be activated by addition of very low and subinhibitory concentrations of mature nisin to the culture supernatant. This induction was inhibited after disruption of *nisK*, demonstrating that the presence of mature nisin in the culture supernatant is the signal that is sensed by the two-component regulatory system of the nisin biosynthetic gene cluster.[205] Deletion of *nisR* also abolished transcription.[206] An analysis of all promoters present in the nisin biosynthetic gene cluster revealed that in the nisin A producer strain *L. lactis* MG1363 the operons *nisABTCIP* and *nisFEG* are controlled by nisin in the supernatant, whereas *nisRK* is expressed constitutively.[173] In contrast, the nisin Z producer *L. lactis* N8 contains only two operons, *nisZBTCIPRK* and *nisFEG*, which are both induced by nisin.[174,207] Experiments with fragments of nisin indicated that the N terminus (amino acid residues 1–11),[205] with an intact ring A, is essential for induction.[206] The nisin mutant peptides also vary with respect to their induction capacity, ranging from T2S nisin Z, which is the most effective peptide with an induction capacity more than 10 times higher than that of wild-type nisin, to I1W nisin Z, which is more than 100-fold less effective than wild-type nisin.[205]

The epidermin biosynthetic gene cluster contains a single response regulator protein called EpiQ. A gene for the corresponding His kinase was not found, and the N-terminal domain of EpiQ does not show homologies to characteristic response regulator proteins. For example, the Asp residue that usually functions as the phosphate acceptor is not found in the conserved position. In contrast, the C-terminal domain of EpiQ shows homology to response regulator proteins, for example PhoB and OmpR.[138] In the absence of EpiQ or after inhibition of transcription of *epiQ*, no epidermin is produced by *S. carnosus*, demonstrating the necessity of EpiQ for epidermin biosynthesis. On the other hand, overexpression of EpiQ in the wild-type producer strain increases production of epidermin twofold. In further experiments it was demonstrated that EpiQ stimulates transcription of the promoter of *epiA*,[177] *epiFEG*,[158] *epiT*, and *epiH*.[159] In gel mobility shift experiments, an inverted repeat immediately upstream of the −35 region of the *epiA* promoter was identified as an operator site for EpiQ.[177] The *epiQ* gene product was detected in Western blots after overexpression in *E. coli* and chromatography on heparin sepharose. It is not yet clear whether EpiQ has to be activated by phosphorylation and, if so, which enzyme fulfills this function in the wild-type producer, *S. epidermidis*, and the heterologous expression host, *S. carnosus*.

The only regulatory gene that was described for the lantibiotics of the streptococcin A-FF22 subtype is *scnR*, which codes for a putative response regulator protein. This gene and the structural gene *scnA* are preceded by a putative phosphate box sequence which might be the binding site for the response regulator.[208]

4.11.2.5 Producer Self-protection (Immunity)

Lantibiotics are bacteriocins, most often active against organisms that are closely related to the producer strains, and which are potentially dangerous for the producing organism. Therefore, the production of a bacteriocin is always associated with the so-called "immunity" phenomenon, which means that the producer strain displays only a low sensitivity to its own product. Normally, immunity is genetically linked to lantibiotic production, as shown, for example, for Pep5,[209] nisin,[210] epidermin,[158] lactocin S,[72] and subtilin.[175] So far, the gene products involved in immunity have only been characterized for Pep5, subtilin, nisin, and epidermin: the *lanI* genes of the Pep5, nisin, and subtilin biosynthetic gene clusters code for so-called "immunity" peptides or proteins that are localized on the outside of the cytoplasmic membrane, and protect the cells by an, as yet, unknown mechanism. The LanEFG proteins of the subtilin, nisin, and epidermin gene clusters represent ABC transport proteins that are thought to expel the pore-forming peptides from the membrane of the producer cells.

The first lantibiotic immunity gene to be described, *pepI*, is encoded upstream of the structural gene *pepA* on the plasmid pED503 in *S. epidermidis* 5, and codes for a peptide of 69 amino acids and 8065 Da.[157] Upon loss of pED503, the resulting strain, *S. epidermidis* 5 Pep5⁻, is cured of Pep5 production and sensitive to exogenous Pep5. The immunity can be restored by transformation of the strain with a fragment carrying *pepIA*[157] or *pepI* and the terminator structure that is located downstream of *pepA*.[180] Within the N-terminal domain of PepI, a stretch of 20 apolar residues is flanked by two Lys residues at positions 6 and 27, whereas the C-terminal domain carries a positive net charge and is hydrophilic. In Western blots, PepI was detected in about equal amounts in both cytoplasmic and membrane fractions of the wild-type producer strain or *S. epidermidis* 5 Pep5⁻ that habored *pepIA*. Mild treatment of the cells with lysostaphin (a cell wall lytic enzyme) did not release PepI, but 90% of the peptide could be removed from the membrane fraction by washing the cells with EDTA, and no more PepI was detected after washing with Triton X-100 or treatment of protoplasts with proteases. In conclusion, this points to a loose association of PepI with the outside of the cytoplasmic membrane. Pore formation by Pep5 was inhibited in cells that harbor PepI. No significant efflux of labeled amino acids was observed with cells expressing PepI after addition of Pep5. Similarly, the accumulation of labeled amino acids was not inhibited in the presence of Pep5. In contrast, in the absence of PepI the cells were fully sensitive.[157] Apparently, the immunity peptide is able to antagonize pore formation by Pep5 by an interaction with the lantibiotic at the membrane.

Nisin and subtilin producer strains each contain a gene called *nisI* or *spaI* that codes for a protein with a 22 amino acid signal sequence and an N-terminal hydrophobic and a C-terminal hydrophilic domain. NisI[147] and SpaI[156] are composed of 245 and 165 amino acids, respectively. The signal sequence of these proteins is characteristic for a lipoprotein signal sequence, and is followed by a Cys residue at position +1; in conclusion, it was assumed that both proteins might be coupled to a lipid after export and, therefore, remain anchored on the outside of the cytoplasmic membrane. This has been proven by *in vivo* labeling of NisI with [³H]palmitic acid.[211] Interestingly, there is neither any similarity in the amino acid sequence of the mature NisI and SpaI nor could any similarity of NisI or SpaI with protein sequences in the databases be found. NisI was expressed in *E. coli*, and the molecular mass of the protein was estimated to be about 32 kDa. The *E. coli* cells expressing NisI showed a higher level of immunity against nisin than control cells in the presence of nisin and EDTA. Expression of NisI in a *L. lactis* strain that carried all genes necessary for nisin production, including *nisFEG* but with the exception of *nisA*, restored about one-tenth of the immunity level of the wild-type strain. Full immunity could be achieved by introduction of an expression vector carrying the structural gene *nisA*.[147] It was shown later that this is a regulatory phenomenon and that the transcription of the nisin biosynthetic gene cluster is activated by nisin via NisR and NisK.[205,206] The role of NisI in immunity against nisin has been controversially discussed, because expression of *nisI* in the absence of *nisFEG* by the *L. lactis* P45 promoter or on an expression vector restored only 5% of the wild-type level of immunity to nisin.[147,211,212]

SpaI has been disrupted by insertion of a resistance gene, and the strains harboring this mutation were still able to produce subtilin but were less immune to exogenous subtilin.[156] Most interestingly, in addition to disruption of *spaFG* (see below), the immunity against subtilin was also decreased by inactivation of the structural gene *spaA* and the regulatory genes *spaR* and *spaK*.[156]

Three lantibiotic biosynthetic gene clusters carry genes for an ATP-binding cassette transport protein that seems to be involved in immunity. These genes were first described for subtilin[156] and subsequently found in the nisin[150] and epidermin[158] biosynthetic gene clusters. In contrast to the LanT proteins (see Section 4.11.2.1.4) that are involved in export of the lantibiotics from the producer cells and are encoded by a single open reading frame, the LanEFG transporters always consist of two or three separately encoded proteins. SpaF in the subtilin gene cluster is composed of 456 amino acids. It contains an ATP-binding N-terminal domain and a hydrophobic C-terminal domain with six membrane-spanning sequences. SpaF shows similarity to the microcin B17 transporter McbF, which is also involved in immunity against this nonlantibiotic bacteriocin. The second open reading frame, *spaG*, encodes a hydrophobic protein of 187 or 203 amino acids, depending on the ATG that functions as the start codon in this protein. SpaG might be localized in the membrane, and shows homology to McbE, which forms the active transporter with McbF.[213] Presumably, two molecules each of SpaF and SpaG represent the active protein.

nisE (247 amino acids) and *nisG* (214 amino acids) encode the membrane-spanning domains of the homologous proteins of the nisin gene cluster, and hydrophobicity profiles indicate six membrane-spanning domains in each of the proteins. NisF (225 amino acids) represents the ATP-binding domain, and is a cytoplasmic protein.[150] The sequence of *nisFEG* is preceded by a promoter structure and forms an operon. NisF shows 45.5% similarity to SpaF, whereas the similarity of the membrane domains is lower. The active transporter is probably composed of two NisF proteins and one NisG and one NisE subunit. NisF, NisE, and NisG have been inactivated by insertion of selection markers. The resulting strains were all able to produce nisin, a fact which indicated that this transporter is not involved in biosynthesis of nisin, but showed increased sensitivity to exogenous nisin. This effect was least obvious for the NisG mutants.[150] The involvement of NisFEG in immunity has also been discussed controversially in a review.[212]

A similar division into three separate proteins was also found for the corresponding transporter in the epidermin gene cluster.[158] Three open reading frames, *epiF*, *epiE*, and *epiG*, which probably form an operon, are located upstream of *epiH*. *epiE* and *epiG* encode two membrane proteins of 255 amino acids and 230 amino acids, respectively. The ATP-binding domain EpiF consists of 231 residues and shows 49% sequence identity with SpaF, whereas the homologies of the membrane-spanning domains are lower. Expression of *epiFEG* in the heterologous host *S. carnosus* decreased the sensitivity of this indicator strain threefold. A sevenfold higher immunity was detected in the presence of the positive regulatory protein EpiQ that was shown to activate transcription of the *epiF* promoter.[158]

The role of the ABC transporters in immunity is not yet clear. These proteins could either mediate import with subsequent proteolytic degradation of the peptides or remove those molecules that have associated with the outside of the membrane or even flipped to the inside of the membrane when the pore closed. The only experiment that indicates that the transporters can export lantibiotics from the cells was performed with the epidermin transporter. Epidermin production in *S. carnosus*, which does not contain a functional epidermin exporter (since *epiT' T''* is inactivated by a frameshift and two deletions), was increased fivefold by expression of *epiFEG*. This result points to the conclusion that EpiFEG is able to export epidermin from the cell and, therefore, it was suggested that the transporter may remove the peptides that have associated with the inside or outside of the membrane *in vivo*.[158]

4.11.3 GENETIC ENGINEERING OF LANTIBIOTICS

4.11.3.1 Construction of Expression Systems

Since lantibiotics possess structural genes, the modification of the amino acid sequence by a comparatively technically simple genetic method, site-directed mutagenesis, is possible. Variant peptides have been constructed for several lantibiotics of the nisin subtype, and the results have been reviewed.[122] On the other hand, site-directed mutagenesis of lantibiotics is more complicated than mutagenesis of other peptides or proteins because, for correct biosynthesis of the Lan and MeLan residues, the expression of the biosynthetic apparatus is necessary, that is, all of the enzymes and accessory factors that introduce the rare amino acids into the precursor peptide have to be present and active when the modified gene is expressed. If such an "expression system" can be constructed, mutagenesis of lantibiotics is a very powerful tool that even allows the introduction of nonproteinogenic amino acids into the peptides as in, for example, the construction of novel ring structures or dehydroamino acids.[119,120] Heterologous expression of a lantibiotic has only been achieved for epidermin, which is normally produced by *S. epidermidis*, in *S. carnosus*[178] and, generally, the modified genes are expressed in a variant of the original producer strain which carries an inactivated or a closely related structural gene. Several of these expression systems have been constructed so far for nisin, epidermin/gallidermin, subtilin, and Pep5.

In the simplest systems, the variant genes are introduced on a plasmid into a strain that carries a closely related lantibiotic gene. For example, the first nisin Z mutant peptides were expressed in a nisin A producer strain.[119] In the case of nisin, which autoregulates its own biosynthesis, this system has the advantage that nisin A will induce biosynthesis of the accessory proteins, if the mutant peptide has only a low induction capacity. A disadvantage of this system is that nisin A and the variant nisin Z peptides have to be separated by HPLC runs and the antimicrobial activity of the new constructs cannot be judged by measuring the inhibition zones that are formed by the strain expressing these peptides.

In contrast, this assay is possible with systems in which the structural genes have been inactivated. In the case of epidermin, this was achieved by ethyl methanesulfonate mutagenesis.[178] The expression system in *S. epidermidis* harbors the plasmid pTü32-6 that codes for G10E epidermin. This mutation inhibits production of the epidermin prepeptide as well as mature epidermin.[127] The structural gene of nisin, *nisA*, has been inactivated by a restriction digest, and removal of the overhanging ends followed by religation, resulting in a four-base deletion and frame shift mutation. This Δ*nisA* gene was introduced into the gene cluster by homologous recombination.[147] Another group has inactivated *nisA* by insertion of a resistance gene. As this insertion exerted a polar effect and inhibited transcription of the downstream biosynthetic genes which include the immunity gene *nisI*, a strain was chosen that had integrated an insertion sequence (IS905) providing a promoter downstream of *nisA* and which therefore displayed immunity.[144]

For some lantibiotics, expression of the structural genes *in trans*, that is, on a separate small plasmid, is not possible. It has been suggested that the copy number of the structural genes on these vectors, which is higher than that of the structural genes in the wild-type strains, leads to an overexpression of the lantibiotic to levels which are potentially lethal to the host bacterium. This is the case for subtilin and Pep5.[62,179] Therefore, in these systems the mutated structural gene has to replace the original *lanA* gene. In the case of subtilin, the biosynthetic gene cluster was transformed into *B. subtilis* 168 by competence transformation.[175] The structural gene was then replaced by an erythromycin resistance gene. The modified *spaA* gene is cloned into a cassette that contains an additional chloramphenicol resistance gene that serves as a marker for integration of the mutant gene into the chromosome of *B. subtilis* 168 by homologous recombination.[62]

The Pep5 biosynthetic gene cluster is located on the plasmid pED503 in the producer strain *S. epidermidis* 5.[209] The mutagenesis is performed with a fragment that harbors *pepI*, *pepA*, and the 5′ terminus of *pepP*. Then this fragment, which carries the mutated gene, is cloned into a vector that carries the 3′ terminus of *pepP* and *pepBC* and, thus, *pepIAPBC* is reconstituted. This plasmid is expressed in a variant of the producer strain that has lost pED503. Since *pepT* is not present in this system, the production rate is only 10–30% of that of the wild-type producer strain, *S. epidermidis* 5.[179]

For the expression of mutated nisin peptides *in cis*, *nisA* was inactivated by deletion of 300 bp, which includes most of the promoter of nisin. The mutated *nisA* genes are cloned into a vector that contains both deleted and overlapping sequences. A double homologous recombination then reconstitutes *nisA* and the nisin promoter to the gene cluster. Only after reconstitution of the promoter of *nisA* can the immunity gene *nisI* be expressed, and recombinant strains are selected by their resistance to exogenously added nisin. This system has been reported to give higher levels of production of nisin mutant peptides than expression on a vector *in trans*.[123]

One drawback to site-directed mutagenesis of lantibiotics is that not all mutant genes are expressed by the strains. In the nisin and epidermin systems, for example, mutant peptides that affect the formation of rings (S23A nisin A, S19A epidermin, ΔC22 epidermin) were not produced.[122,127] In the Pep5 system, F23A Pep5 also seemed to be incompatible with the modification system.[179]

On the other hand, these systems have been used to introduce nonproteinogenic amino acids into nisin and Pep5: Dha5Dhb nisin Z which contains an exchange of Dhb for Dha at position 5 is more stable than nisin Z but shows an activity that is 2–10-fold lower than that of the original peptide.[214] A new dehydroamino acid was successfully introduced into K18Dha Pep5.[120] The double mutation M17Q/G18T in the *nisZ* gene gives rise to two products, M17Q/G18T nisin Z and M17Q/G18Dhb nisin Z.[119] A similar incomplete modification was also seen with Dhb14S gallidermin, where equimolar amounts of both analogues (Dhb14S gallidermin and Dhb14Dha gallidermin) were purified. These experiments indicate that the affinity of the dehydrating enzymes for hydroxyl amino acids in unusual positions is lower than for the original template.[127]

The construction of a ring structure in the central linear fragment of Pep5 was possible by the exchange of the Ala residue at position 19 for Cys. A novel four-membered ring, which spans the chymotryptic cleavage site of Pep5 situated between Leu17 and Lys18, was formed with the Dhb residue at position 16. The resulting peptide is very resistant to proteolytic digestion but considerably less active. This loss of activity is presumably due to the fact that the fourth ring is located in the former flexible hinge region of Pep5, a region which has to bend for pore formation and whose mobility is most probably restricted by the ring structure.[120]

4.11.3.2 Optimization of Lantibiotics by Site-directed Mutagenesis

Site-directed mutagenesis is a powerful tool for the optimization of the chemical and physical characteristics of lantibiotics on a rational basis. The first experiments of this kind were performed with subtilin and nisin. Nisin is the only lantibiotic that is used commercially, currently in about 50 countries. It was introduced into the UK in the 1960s as a preservative (E234) for processed cheese products, pasteurized dairy products, and vegetable conserves. These, and novel applications of nisin, have been reviewed.[32] A disadvantage of nisin is its low solubility in water at neutral pH. This can be improved by introduction of a charged residue, and the mutant peptides N27K nisin Z and H31K nisin Z show a four- to sevenfold higher solubility at pH 7. The antibacterial activity was not affected by these exchanges.[214] M17K nisin also displays an increased solubility but is easily inactivated by trypsin.[122] Dha5Dhb nisin Z is more stable than wild-type nisin but less active.[214]

In contrast, subtilin, which is structurally related to nisin and very effective against spores, cannot be employed commercially because, with a half-life of only 0.8 d, the peptide is unstable at room temperature. This loss of activity is accompanied by a loss of the Dha residue at position 5. This reaction is caused by the neighboring Glu residue at position 4, which can act as a nucleophile and catalyze the addition of water to the double bond of Dha5. As shown for nisin, where this reaction can be stimulated by lyophilization in 0.1 M HCl, this leads to formation of an α-OH-Ala residue which is not stable and ultimately converts to an amide and keto acid, thereby opening ring A.[78,214] Therefore, the Glu residue of subtilin was exchanged for Ile, which occupies the homologous position in the nisin molecule. The resulting subtilin mutant, E4I subtilin, displayed a greatly enhanced half-life of 48 d and a threefold higher sporicidal activity.[62] Another mutant subtilin, E4I/Dha5A subtilin, still showed the same bacteriolytic activity but the sporicidal activity was completely lost.[61] This effect has also been recorded for the sporicidal mode of action of Dha5A nisin.[215] Site-directed mutagenesis of gallidermin has been used to generate variant peptides that are less sensitive to trypsin, for example A12L gallidermin and Dhb14P gallidermin.[127]

The above examples show that the optimization of the chemical and physical properties of a peptide is possible and that the effects of single exchanges can be more or less predicted. However, such a rational procedure is not yet possible for the optimization of the antibacterial activity of a lantibiotic. T2S nisin, M17Q/G18T nisin Z, and L6V gallidermin possess an increased antibacterial activity, at least with some indicator strains.[122] L6V gallidermin, for example, displays a higher antibacterial activity against *Micrococcus luteus* and *Clostridium glutamicum* and a lower activity against *Staphylococcus aureus*.[127] The reason for this phenomenon is most probably that the target of the type A lantibiotics, the cytoplasmic membrane, varies between the different species with respect to composition, membrane potential, etc. The phospholipid composition of a membrane, for example, influences the efficacy of a peptide, as was shown for nisin.[216–218] Additionally, secondary effects, such as the induction of autolysis by some lantibiotics[102] or the existence of slime capsules, modulate the activity of the peptides against a specific organism. Even the production of a lantibiotic influences the sensitivity of the strain against other lantibiotics; nisin producers were reported to be hypersensitive to carnocin, and the data suggested that NisP could act as a receptor for carnocin;[219] on the other hand, strains that produce very similar lantibiotics may or may not display cross-immunity.[49] This lack of knowledge about the changes that are necessary in order to optimize the antibacterial activity of a peptide against a given strain has led to the establishment of a method for random mutagenesis of nisin: the mutant gene is synthesized by polymerase chain reaction in the presence of dITP and a limiting amount of one of the other dNTPs. These conditions favor the incorporation of dITP into the nucleotide sequence and, in the following cycle, the inosine will pair with any one of the other bases.[220] Employing this method, L16P, M17V, M21L, T25A, and I30T nisin have been produced, but the peptides excreted by these strains still have to be characterized.[122]

In conclusion, detailed insight into the interaction of lantibiotics with their target, the bacterial membrane, is most urgently needed in order to enable rational optimization of the antibacterial activity of type A lantibiotics. Presumably this problem can be solved more easily for the type B lantibiotics, which bind to a single target molecule such as phosphatidylethanolamine. For the conformation of these complexes, computer models are or will soon be available. Another problem that limits site-directed mutagenesis of lantibiotics is the immunity of the producer strains, since very active peptides might be lethal for the producing cells. The most elegant solution to this problem would be the establishment of *in vitro* modification systems. Alternatively, the protease cleavage site of the precursor peptides could be modified in such a way that the fully matured precursor is excreted by the cells and the active lantibiotic is only released by *in vitro* proteolytic processing.[122]

4.11.4 OUTLOOK

Lantibiotics are gene-encoded peptides that contain temperature-stable thioether bonds and act on gram-positive bacteria. Most lantibiotics are not toxic to humans or animals and show promising features for application as biopreservatives, bactericidal agents in cosmetics, or antibiotics. The above data demonstrate that many questions remain to be resolved even though interest in lantibiotics, especially their biosynthesis and mode of action, has increased since the 1980s. For example, the catalytic mechanisms and the substrate specificity of the LanB, LanC, and LanM enzymes remain to be elucidated. Similarly, there are only limited data available on the biosynthesis of type B lantibiotics, the immunity mechanisms against the lantibiotics of this subtype, or the factors that determine the efficacy of a type A lantibiotic against a specific strain. Fortunately, the key features of lantibiotics, their content of unusual amino acids and their ribosomal biosynthesis mechanism, provide exciting perspectives for this work. More knowledge on the mode of action of lantibiotics will enable the rational adaptation of the peptides to target bacteria, and further progress in lantibiotic research may result in the establishment of systems that exploit the post-translational modification system of lantibiotics for the incorporation of thioether amino acids into nonlantibiotic peptides and proteins, resulting in completely novel structures.

4.11.5 REFERENCES

1. N. Schnell, K.-D. Entian, U. Schneider, F. Götz, H. Zähner, R. Kellner, and G. Jung, *Nature*, 1988, **333**, 276.
2. M. Skaugen, J. Nissen-Meyer, G. Jung, S. Stevanović, K. Sletten, C. I. Mørtvedt Abildgaard, and I. F. Nes, *J. Biol. Chem.*, 1994, **269**, 27183.
3. H. Allgaier, G. Jung, R. G. Werner, U. Schneider, and H. Zähner, *Eur. J. Biochem.*, 1986, **160**, 9.
4. R. Kellner, G. Jung, T. Hörner, H. Zähner, N. Schnell, K.-D. Entian, and F. Götz, *Eur. J. Biochem.*, 1988, **177**, 53.
5. Y. Minami, K.-i. Yoshida, R. Azuma, A. Urakawa, T. Kawauchi, T. Otani, K. Komiyama, and S. Ōmura, *Tetrahedron Lett.*, 1994, **35**, 80001.
6. S. Chatterjee, S. Chatterjee, S. J. Lad, M. S. Phansalkar, R. H. Rupp, B. N. Ganguli, H.-W. Fehlhaber, and H. Kogler, *J. Antibiot.*, 1992, **45**, 832.
7. H. Kessler, S. Steuernagel, M. Will, G. Jung, R. Kellner, D. Gillessen, and T. Kamiyama, *Helv. Chim. Acta*, 1988, **71**, 1924.
8. T. Wakamiya, K. Fukase, N. Naruse, M. Konishi, and T. Shiba, *Tetrahedon Lett.*, 1988, **29**, 4771.
9. A. Fredenhagen, F. Märki, G. Fendrich, W. Märki, J. Gruner, J. van Oostrum, F. Raschdorf, and H. H. Peter, in "Nisin and Novel Lantibiotics," eds. G. Jung and H.-G. Sahl, ESCOM, Leiden, 1991, p. 131.
10. M. van de Kamp, H. W. van den Hooven, R. N. H. Konings, G. Bierbaum, H.-G. Sahl, O. P. Kuipers, R. J. Siezen, W. M. de Vos, C. W. Hilbers, and F. J. M. van de Ven, *Eur. J. Biochem.*, 1995, **230**, 587.
11. R. Kellner, G. Jung, M. Josten, C. Kaletta, K.-D. Entian, and H.-G. Sahl, *Angew. Chem., Int. Ed. Engl.*, 1989, **28**, 618.
12. M. van de Kamp, L. M. Horstink, H. W. van den Hooven, R. N. H. Konings, C. W. Hilbers, A. Frey, H.-G. Sahl, J. W. Metzger, and F. J. M. van de Ven, *Eur. J. Biochem.*, 1995, **227**, 757.
13. W. C. Chan, B. W. Bycroft, M. L. Leyland, L.-Y. Lian, and G. C. K. Roberts, *Biochem. J.*, 1993, **291**, 23.
14. W. C. Chan, B. W. Bycroft, L.-Y. Lian, and G. C. K. Roberts, *FEBS Lett.*, 1989, **252**, 29.
15. J. Delves-Broughton, *Food Technol.*, 1990, **44**, 100.
16. G. Jung and H.-G. Sahl (eds.), "Nisin and Novel Lantibiotics," ESCOM, Leiden, 1991.
17. R. James, C. Lazdunski, and F. Pattus (eds.) "Bacteriocins, Microcins and Lantibiotics," NATO ASI Series H: Cell Biology, Springer-Verlag, Berlin, 1992, vol. 65.
18. B. Ray and M. Daeschel (eds.), "Food Biopreservatives of Microbial Origin," CRC Press, Boca Raton, FL, 1992.
19. R. Kolter and F. Moreno, *Annu. Rev. Microbiol.*, 1992, **46**, 141.
20. T. R. Klaenhammer, *FEMS Microbiol. Rev.*, 1993, **12**, 39.
21. J. N. Hansen, *Annu. Rev. Microbiol.*, 1993, **47**, 535.
22. G. Bierbaum and H.-G. Sahl, *Zentralbl. Bakteriol.*, 1993, **278**, 1.
23. D. G. Hoover and L. R. Steenson (eds.) "Bacteriocins of Lactic Acid Bacteria," Academic Press, San Diego, CA, 1993.
24. L. de Vuyst and E. J. Vandamme (eds.) "Bacteriocins of Lactic Acid Bacteria," Chapman and Hall, London, 1994.
25. H.-G. Sahl, R. W. Jack, and G. Bierbaum, *Eur. J. Biochem.*, 1995, **230**, 827.
26. R. W. Jack, G. Bierbaum, C. Heidrich, and H.-G. Sahl, *BioEssays*, 1995, **17**, 793.
27. R. W. Jack and H.-G. Sahl, *Tibtech*, 1995, **13**, 269.
28. R. W. Jack, J. R. Tagg, and B. Ray, *Microbiol. Rev.*, 1995, **59**, 171.
29. W. M. de Vos, O. P. Kuipers, J. R. van der Meer, and R. J. Siezen, *Mol. Microbiol.*, 1995, **17**, 427.
30. R. N. H. Konings and C. W. Hilbers (eds.), *Antonie van Leeuwenhoek*, 1996, **69**.
31. R. W. Jack, G. Bierbaum, and H.-G. Sahl, "Lantibiotics and Related Peptides," Springer-Verlag, Berlin and Landes Bioscience, Georgetown, TX, 1998.
32. J. Delves-Broughton, P. Blackburn, R. J. Evans, and J. Hugenholtz, *Antonie van Leeuwenhoek*, 1996, **69**, 193.
33. M. S. Gilmore, R. A. Segarra, M. C. Booth, C. P. Bogie, L. R. Hall, and D. B. Clewell, *J. Bacteriol.*, 1994, **176**, 7335.
34. M. C. Booth, C. P. Bogie, H.-G. Sahl, R. J. Siezen, K. L. Hatter, and M. S. Gilmore, *Mol. Microbiol.*, 1996, **21**, 1175.
35. E. W. Todd, *J. Pathol. Bacteriol.*, 1934, **39**, 299.
36. Y. Ike, D. B. Clewell, R. A. Segarra, and M. S. Gilmore, *J. Bacteriol.*, 1990, **172**, 155.
37. J. W. Chow, L. A. Thai, M. B. Perri, J. A. Vazquez, S. M. Donabedian, D. B. Clewell, and M. J. Zervos, *Antimicrob. Agents Chemother.*, 1993, **37**, 2474.
38. H. Brötz, G. Bierbaum, A. Markus, E. Molitor, and H.-G. Sahl, *Antimicrob. Agents Chemother.*, 1995, **39**, 714.
39. S. Somma, W. Merati, and F. Parenti, *Antimicrob. Agents Chemother.*, 1977, **11**, 396.
40. G. Jung, *Angew. Chem., Int. Ed. Engl.*, 1991, **30**, 1051.
41. G. Jung, in "Nisin and Novel Lantibiotics," eds. G. Jung and H.-G. Sahl, ESCOM, Leiden, 1991, p. 1.
42. E. Gross and J. L. Morell, *J. Am. Chem. Soc.*, 1971, **93**, 4634.
43. J. W. M. Mulders, I. J. Boerrigter, H. S. Rollema, R. J. Siezen, and W. M. de Vos, *Eur. J. Biochem.*, 1991, **201**, 581.
44. E. Gross, H. H. Kiltz, and E. Nebelin, *Hoppe-Seyler's Z. Physiol. Chem.*, 1973, **354**, 810.
45. A. M. Israel, R. W. Jack, G. Jung, and H.-G. Sahl, *Zentralbl. Bakteriol.*, 1996, **284**, 285.
46. R. W. Jack, A. Carne, J. Metzger, S. Stevanović, H.-G. Sahl, G. Jung, and J. Tagg, *Eur. J. Biochem.*, 1994, **220**, 455.
47. J.-C. Piard, O. P. Kuipers, H. S. Rollema, M. J. Desmazeaud, and W. M. de Vos, *J. Biol. Chem.*, 1993, **268**, 16361.

48. K. F. Ross, C. W. Ronson, and J. R. Tagg, *Appl. Environ. Microbiol.*, 1993, **59**, 2014.
49. D. Pridmore, N. Rekhif, A.-C. Pittet, B. Suri, and B. Mollet, *Appl. Environ. Microbiol.*, 1996, **62**, 1799.
50. G. Stoffels, J. Nissen-Meyer, A. Gudmundsdottir, K. Sletten, H. Holo, and I. F. Nes, *Appl. Environ. Microbiol.*, 1992, **58**, 1417.
51. J. Novák, P. W. Caufield, and E. J. Miller, *J. Bacteriol.*, 1994, **176**, 4316.
52. N. Zimmermann, J. W. Metzger, and G. Jung, *Eur. J. Biochem.*, 1995, **228**, 786.
53. T. Wakamiya, Y. Ueki, T. Shiba, Y. Kido, and Y. Motoki, *Tetrahedron Lett.*, 1985, **26**, 665.
54. H. E. Meyer, M. Heber, B. Eisermann, H. Korte, J. W. Metzger and G. Jung, *Anal. Biochem.*, 1994, **223**, 185.
55. L. A. Rogers and E. O. Whittier, *J. Bacteriol.*, 1928, **16**, 211.
56. N. J. Berridge, G. G. F. Newton, and E. P. Abraham, *Biochem. J.*, 1952, **52**, 529.
57. A. T. R. Mattick and A. Hirsch, *Nature*, 1944, **154**, 551.
58. N. Benkerroum and W. E. Sandine, *J. Dairy Sci.*, 1988, **71**, 3237.
59. K. A. Stevens, B. W. Sheldon, N. A. Klapes, and T. R. Klaenhammer, *Appl. Environ. Microbiol.*, 1991, **57**, 3613.
60. E. F. Jansen and D. J. Hirschmann, *Arch. Biochem.*, 1944, **4**, 297.
61. W. Liu and J. N. Hansen, *Appl. Environ. Microbiol.*, 1993, **59**, 648.
62. W. Liu and J. N. Hansen, *J. Biol. Chem.*, 1992, **267**, 25078.
63. H.-G. Sahl, *Appl. Environ. Microbiol.*, 1994, **60**, 752.
64. H.-G. Sahl and H. Brandis, *J. Gen. Microbiol.*, 1981, **127**, 377.
65. H. W. van den Hooven, F. M. Lagerwerf, W. Heerma, J. Haverkamp, J.-C. Piard, C. W. Hilbers, R. J. Siezen, O. P. Kuipers, and H. S. Rollema, *FEBS Lett.*, 1996, **391**, 317.
66. D. Kaiser and R. W. Jack, personal communication.
67. R. Jack, R. Benz, J. Tagg, and H.-G. Sahl, *Eur. J. Biochem.*, 1994, **219**, 699.
68. J. R. Tagg, R. S. D. Read, and A. R. McGiven, *Pathology*, 1971, **3**, 277.
69. R. W. Jack and J. R. Tagg, *J. Med. Microbiol.*, 1992, **36**, 132.
70. J.-C. Piard, P. M. Muriana, M. J. Desmazeaud, and T. R. Klaenhammer, *Appl. Environ. Microbiol.*, 1992, **58**, 279.
71. A. Dufour, D. Thuault, A. Boulliou, C. M. Bourgeois, and J. P. Le Pennec, *J. Gen. Microbiol.*, 1991, **137**, 2423.
72. C. I. Mørtvedt and I. F. Nes, *J. Gen. Microbiol.*, 1990, **136**, 1601.
73. C. I. Mørtvedt, J. Nissen-Meyer, K. Sletten, and I. F. Nes, *Appl. Environ. Microbiol.*, 1991, **57**, 1829.
74. T. D. Brock and J. M. Davie, *J. Bacteriol.*, 1963, **86**, 708.
75. M. S. Gilmore, M. Skaugen and I. Nes, *Antonie van Leeuwenhoek*, 1996, **69**, 129.
76. R. G. Benedict, W. Dvonch, O. L. Shotwell, T. G. Pridham, and L. A. Lindenfelser, *Antibiot. Chemother.*, 1952, **2**, 591.
77. A. Fredenhagen, G. Fendrich, F. Märki, W. Märki, J. Gruner, F. Raschdorf, and H. H. Peter, *J. Antibiot.*, 1990, **43**, 1403.
78. E. Gross, *Adv. Exp. Med. Biol. B*, 1977, **86**, 131.
79. H. Kessler, S. Steuernagel, D. Gillessen, and T. Kamiyama, *Helv. Chim. Acta*, 1987, **70**, 726.
80. H. Kessler, S. Seip, T. Wein, S. Steuernagel, and M. Will, in "Nisin and Novel Lantibiotics," eds. G. Jung and H.-G. Sahl, ESCOM, Leiden, 1991, p. 76.
81. N. Naruse, O. Tenmyo, K. Tomita, M. Konishi, T. Miyaki, K. Kawaguchi, K. Fukase, T. Wakamiya, and T. Shiba, *J. Antibiot.*, 1989, **42**, 837.
82. O. L. Shotwell, F. H. Stodola, W. R. Michael, L. A. Lindenfelser, G. Dworschak, and T. G. Pridham, *J. Am. Chem. Soc.*, 1958, **80**, 3912.
83. F. Parenti, H. Pagani, and G. Beretta, *J. Antibiot.*, 1976, **29**, 501.
84. J. K. Kettenring, A. Malabarba, K. Vekey, and B. Cavalleri, *J. Antibiot.*, 1990, **43**, 1082.
85. S. Chatterjee, D. K. Chatterjee, R. H. Jani, J. Blumbach, B. N. Ganguli, N. Klesel, M. Limbert, and G. Seibert, *J. Antibiot.*, 1992, **45**, 839.
86. H.-G. Sahl, in "Nisin and Novel Lantibiotics," eds. G. Jung and H.-G. Sahl, ESCOM, Leiden, 1991, p. 347.
87. R. Benz, G. Jung, and H.-G. Sahl, in "Nisin and novel lantibiotics," eds. G. Jung and H.-G. Sahl, ESCOM, Leiden, 1991, p. 359.
88. G. N. Moll, G. C. K. Roberts, W. N. Konings and A. J. M. Driessen, *Antonie van Leeuwenhoek*, 1996, **69**, 185.
89. H. R. Ramseier, *Arch. Microbiol.*, 1960, **37**, 57.
90. H.-G. Sahl and H. Brandis, *Zentralbl. Bakteriol., Mikrobiol. Hyg., Abt. I, Orig. A*, 1982, **252**, 166.
91. H.-G. Sahl, *J. Bacteriol.*, 1985, **162**, 833.
92. E. Ruhr and H.-G. Sahl, *Antimicrob. Agents Chemother.*, 1985, **27**, 841.
93. H.-G. Sahl and H. Brandis, *FEMS Microbiol. Lett.*, 1983, **16**, 75.
94. F. Schüller, R. Benz, and H.-G. Sahl, *Eur. J. Biochem.*, 1989, **182**, 181.
95. M. Kordel, R. Benz, and H.-G. Sahl, *J. Bacteriol.*, 1988, **170**, 84.
96. H.-G. Sahl, M. Kordel, and R. Benz, *Arch. Microbiol.*, 1987, **149**, 120.
97. H. W. van den Hooven, C. C. M. Doeland, M. van de Kamp, R. N. H. Konings, C. W. Hilbers, and F. J. M. van de Ven, *Eur. J. Biochem.*, 1996, **235**, 382.
98. H. W. van den Hooven, C. A. E. M. Spronk, M. van de Kamp, R. N. H. Konings, C. W. Hilbers, and F. J. M. van de Ven, *Eur. J. Biochem.*, 1996, **235**, 394.
99. M. Kordel, F. Schüller, and H.-G. Sahl, *FEBS Lett.*, 1989, **244**, 99.
100. R. A. Demel, T. Peelen, R. J. Siezen, B. de Kruijff, and O. P. Kuipers, *Eur. J. Biochem.*, 1996, **235**, 267.
101. P. Reisinger, H. Seidel, H. Tschesche, and W. P. Hammes, *Arch. Microbiol.*, 1980, **127**, 187.
102. G. Bierbaum and H.-G. Sahl, *J. Bacteriol.*, 1987, **169**, 5452.
103. G. Bierbaum and H.-G. Sahl, *FEMS Microbiol. Lett.*, 1989, **58**, 223.
104. M. L. Chikindas, J. Novák, A. J. M. Driessen, W. N. Konings, K. M. Schilling, and P. W. Caufield, *Antimicrob. Agents Chemother.*, 1995, **39**, 2656.
105. E. Racker, C. Riegler, and M. Abdel-Ghany, *Cancer Res.*, 1984, **44**, 1364.
106. D. K. Stone, X. S. Xie, and E. Racker, *J. Biol. Chem.*, 1984, **259**, 2701.
107. S. Nakamura and E. Racker, *Biochemistry*, 1984, **23**, 385.
108. J. Navarro, J. Chabot, K. Sherrill, R. Aneja, S. A. Zahler, and E. Racker, *Biochemistry*, 1985, **24**, 4645.
109. T. R. Sheth, R. M. Henderson, S. B. Hladky, and A. W. Cuthbert, *Biochim. Biophys. Acta*, 1992, **1107**, 179.

110. S. Clejan, A. A. Guffanti, M. A. Cohen, and T. A. Krulwich, *J. Bacteriol.*, 1989, **171**, 1744.
111. S.-Y. Choung, T. Kobayashi, J.-I. Inoue, K. Takemoto, H. Ishitsuka, K. Inoue, *Biochim. Biophys. Acta*, 1988, **940**, 171.
112. P. M. Sokolove, P. A. Westphal, M. B. Kester, R. Wierwille, and K. Sikora-VanMeter, *Biochim. Biophys. Acta*, 1989, **983**, 15.
113. S.-Y. Choung, T. Kobayashi, K. Takemoto, H. Ishitsuka, and K. Inoue, *Biochim. Biophys. Acta*, 1988, **940**, 180.
114. F. Märki, E. Hänni, A. Fredenhagen, and J. van Oostrum, *Biochem. Pharmacol.*, 1991, **42**, 2027.
115. N. Zimmermann, S. Freund, A. Fredenhagen, and G. Jung, *Eur. J. Biochem.*, 1993, **216**, 419.
116. K. Hosoda, M. Ohya, T. Kohno, T. Maeda, S. Endo, and K. Wakamatsu, *J. Biochem.*, 1996, **119**, 226.
117. H. Brötz, G. Bierbaum, P. E. Reynolds, and H.-G. Sahl, *Eur. J. Biochem.*, 1997, **246**, 193.
118. L. de Vuyst and E. J. Vandamme, in "Bacteriocins of Lactic Acid Bacteria," eds. L. de Vuyst and E. J. Vandamme, Chapman and Hall, London, 1994, p. 151.
119. O. P. Kuipers, H. S. Rollema, W. M. G. J. Yap, H. J. Boot, R. J. Siezen, and W. M. de Vos, *J. Biol. Chem.*, 1992, **267**, 24340.
120. G. Bierbaum, C. Szekat, M. Josten, C. Heidrich, C. Kempter, G. Jung, and H.-G. Sahl, *Appl. Environ. Microbiol.*, 1996, **62**, 385.
121. H. S. Rollema, P. Both, and R. J. Siezen, in "Nisin and Novel Lantibiotics," eds. G. Jung and H.-G. Sahl, ESCOM, Leiden, 1991, p. 123.
122. O. P. Kuipers, G. Bierbaum, B. Ottenwälder, H. M. Dodd, N. Horn, J. Metzger, T. Kupke, V. Gnau, R. Bongers, P. van den Bogaard, H. Kosters, H. S. Rollema, W. M. de Vos, R. J. Siezen, G. Jung, F. Götz, H.-G. Sahl, and M. J. Gasson, *Antonie van Leeuwenhoek*, 1996, **69**, 161.
123. H. M. Dodd, N. Horn, C. J. Giffard, and M. J. Gasson, *Microbiology*, 1996, **142**, 47.
124. S. Freund, G. Jung, W. A. Gibbons, and H.-G. Sahl, in "Nisin and Novel Lantibiotics," eds. G. Jung and H.-G. Sahl, ESCOM, Leiden, 1991, p. 103.
125. S. Freund, G. Jung, O. Gutbrod, G. Folkers, W. A. Gibbons, H. Allgaier, and R. Werner, *Biopolymers*, 1991, **31**, 803.
126. S. Freund and G. Jung, in "Bacteriocins, Microcins and Lantibiotics," eds. R. James, C. Lazdunski, and F. Pattus, Springer-Verlag, Berlin, 1992, p. 75.
127. B. Ottenwälder, T. Kupke, S. Brecht, V. Gnau, J. Metzger, G. Jung, and F. Götz, *Appl. Environ. Microbiol.*, 1995, **61**, 3894.
128. L. C. Ingram, *Biochim. Biophys. Acta*, 1970, **224**, 263.
129. L. C. Ingram, *Biochim. Biophys. Acta*, 1969, **184**, 216.
130. A. Hurst, *Adv. Appl. Microbiol.*, 1981, **27**, 85.
131. H. Kleinkauf and H. von Döhren, *Eur. J. Biochem.*, 1996, **236**, 335.
132. G. W. Buchmann, S. Banerjee, and J. N. Hansen, *J. Biol. Chem.*, 1988, **263**, 16260.
133. S. Banerjee and J. N. Hansen, *J. Biol. Chem.*, 1988, **263**, 9508.
134. C. Kaletta, K.-D. Entian, R. Kellner, G. Jung, M. Reis, and H.-G. Sahl, *Arch. Microbiol.*, 1989, **152**, 16.
135. T. Kupke and F. Götz, *Antonie van Leeuwenhoek*, 1996, **69**, 139.
136. H.-P. Weil, A. G. Beck-Sickinger, J. Metzger, S. Stevanović, G. Jung, M. Josten, and H.-G. Sahl, *Eur. J. Biochem.*, 1990, **194**, 217.
137. C. Meyer, G. Bierbaum, C. Heidrich, M. Reis, J. Süling, M. I. Iglesias-Wind, C. Kempter, E. Molitor, and H.-G. Sahl, *Eur. J. Biochem.*, 1995, **232**, 478.
138. N. Schnell, G. Engelke, J. Augustin, R. Rosenstein, V. Ungermann, F. Götz, and K.-D. Entian, *Eur. J. Biochem.*, 1992, **204**, 57.
139. C. Klein, C. Kaletta, N. Schnell, and K.-D. Entian, *Appl. Environ. Microbiol.*, 1992, **58**, 132.
140. P. L. Toogood, *Tetrahedron Lett.*, 1993, 7833.
141. M. Skaugen, C. I. M. Abildgaard, and I. F. Nes, *Mol. Gen. Genet.*, 1997, **253**, 674.
142. C. Kaletta and K.-D Entian, *J. Bacteriol.*, 1989, **171**, 1597.
143. H. M. Dodd, N. Horn, and M. J. Gasson, *J. Gen. Microbiol.*, 1990, **136**, 555.
144. H. M. Dodd, N. Horn, Z. Hao, and M. J. Gasson, *Appl. Environ. Microbiol.*, 1992, **58**, 3683.
145. M. T. Steen, Y. J. Chung, and J. N. Hansen, *Appl. Environ. Microbiol.*, 1991, **57**, 1181.
146. G. Engelke, Z. Gutowski-Eckel, M. Hammelmann, and K.-D. Entian, *Appl. Environ. Microbiol.*, 1992, **58**, 3730.
147. O. P. Kuipers, M. M. Beerthuyzen, R. J. Siezen, and W. M. de Vos, *Eur. J. Biochem.*, 1993, **216**, 281.
148. J. R. van der Meer, J. Polman, M. M. Beerthuyzen, R. J. Siezen, O. P. Kuipers, and W. M. de Vos, *J. Bacteriol.*, 1993, **175**, 2578.
149. G. Engelke, Z. Gutowski-Eckel, P. Kiesau, K. Siegers, M. Hammelmann, and K.-D Entian, *Appl. Environ. Microbiol.*, 1994, **60**, 814.
150. K. Siegers and K.-D. Entian, *Appl. Environ. Microbiol.*, 1995, **61**, 1082.
151. T. Immonen, S. Ye, R. Ra, M. Qiao, L. Paulin and P. E. J. Saris, *DNA Sequence*, 1995, **5**, 203.
152. Y. J. Chung, M. T. Steen, and J. N. Hansen, *J. Bacteriol.*, 1992, **174**, 1417.
153. Y. J. Chung and J. N. Hansen, *J. Bacteriol.*, 1992, **174**, 6699.
154. C. Klein, C. Kaletta, K.-D. Entian, *Appl. Environ. Microbiol.*, 1993, **59**, 296.
155. Z. Gutowski-Eckel, C. Klein, K. Siegers, K. Bohm, M. Hammelmann, and K.-D. Entian, *Appl. Environ. Microbiol.*, 1994, **60**, 1.
156. C. Klein and K.-D. Entian, *Appl. Environ. Microbiol.*, 1994, **60**, 2793.
157. M. Reis, M. Eschbach-Bludau, M. I. Iglesias-Wind, T. Kupke, and H.-G. Sahl, *Appl. Environ. Microbiol.*, 1994, **60**, 2876.
158. A. Peschel and F. Götz, *J. Bacteriol.*, 1996, **178**, 531.
159. A. Peschel, N. Schnell, M. Hille, K.-D. Entian, and F. Götz, *Mol. Microbiol.*, 1997, **254**, 312.
160. N. Schnell, K.-D. Entian, F. Götz, T. Hörner, R. Kellner, and G. Jung, *FEMS Microbiol. Lett.*, 1989, **58**, 263.
161. W. L. Hynes, J. J. Ferretti, and J. R. Tagg, *Appl. Environ. Microbiol.*, 1993, **59**, 1969.
162. W. L. Hynes, V. L. Friend, and J. J. Ferretti, *Appl. Environ. Microbiol.*, 1994, **60**, 4207.
163. A. Rince, A. Dufour, S. Le Pogam, D. Thuault, C. M. Bourgeois, and J. P. Le Pennec, *Appl. Environ. Microbiol.*, 1994, **60**, 1652.

164. D. Pridmore, N. Rekhif, A.-C. Pittet, B. Suri, and B. Mollet, personal communication.
165. M. S. Gilmore, R. A. Segarra, and M. C. Booth, *Infect. Immun.*, 1990, **58**, 3914.
166. G. Bierbaum, H. Brötz, H.-P. Koller, and H.-G. Sahl, *FEMS Microbiol. Lett.*, 1995, **127**, 121.
167. C. Kaletta, K.-D. Entian, and G. Jung, *Eur. J. Biochem.*, 1991, **199**, 411.
168. R. J. Siezen, O. P. Kuipers, and W. M. de Vos, *Antonie van Leeuwenhoek*, 1996, **69**, 171.
169. W. M. de Vos, G. Jung, and H.-G. Sahl, in "Nisin and Novel Lantibiotics," eds. G. Jung and H.-G. Sahl, ESCOM, Leiden, 1991, p. 457.
170. P. J. G. Rauch and W. M. de Vos, *J. Bacteriol.*, 1992, **174**, 1280.
171. N. Horn, S. Swindell, H. Dodd, and M. Gasson, *Mol. Gen. Genet.*, 1991, **228**, 129.
172. H.-J. Tsai and W. E. Sandine, *Appl. Environ. Microbiol.*, 1987, **53**, 352.
173. P. G. G. A. de Ruyter, O. P. Kuipers, M. M. Beerthuyzen, I. van Alen-Boerrigter, and W. M. de Vos, *J. Bacteriol.*, 1996, **178**, 3434.
174. S. R. Ra, M. Qiao, T. Immonen, I. Pujana, and P. E. J. Saris, *Microbiology*, 1996, **142**, 1281.
175. W. Liu and J. N. Hansen, *J. Bacteriol.*, 1991, **173**, 7387.
176. T. Kupke, C. Kempter, V. Gnau, G. Jung, and F. Götz, *J. Biol. Chem.*, 1994, **269**, 5653.
177. A. Peschel, J. Augustin, T. Kupke, S. Stevanović, and F. Götz, *Mol. Microbiol.*, 1993, **9**, 31.
178. J. Augustin, R. Rosenstein, B. Wieland, U. Schneider, N. Schnell, G. Engelke, K.-D Entian, and F. Götz, *Eur. J. Biochem.*, 1992, **204**, 1149.
179. G. Bierbaum, M. Reis, C. Szekat, and H.-G. Sahl, *Appl. Environ. Microbiol.*, 1994, **60**, 4332.
180. U. Pag, M. Reis, and H.-G. Sahl, unpublished results.
181. J. R. van der Meer, H. S. Rollema, R. J. Siezen, M. M. Beerthuyzen, O. P. Kuipers, and W. M. de Vos, *J. Biol. Chem.*, 1994, **269**, 3555.
182. B. W. Bycroft, W. C. Chan, and G. C. K. Roberts, in "Nisin and Novel Lantibiotics," eds. G. Jung and H.-G. Sahl, ESCOM, Leiden, 1991, p. 204.
183. A. G. Beck-Sickinger and G. Jung, in "Nisin and Novel Lantibiotics," eds. G. Jung and H.-G. Sahl, ESCOM, Leiden, 1991, p. 218.
184. A. Surovoy, D. Waidelich, and G. Jung, in "Proceedings of the 22nd European peptide symposium, Peptides 1992," eds. C. H. Schneider and A. N. Eberle, ESCOM, Leiden, 1993, p. 563.
185. S. Neis, G. Bierbaum, M. Josten, U. Pag, C. Kempter, G. Jung and H.-G. Sahl, *FEMS Microbiol. Lett.*, 1997, **149**, 249.
186. A. Chakicherla and J. N. Hansen, *J. Biol. Chem.*, 1995, **270**, 23533.
187. H. Rintala, T. Graeffe, L. Paulin, N. Kalkkinen, and P. E. J. Saris, *Biotechnol. Lett.*, 1993, **15**, 991.
188. O. P. Kuipers, H. S. Rollema, W. M. de Vos, and R. J. Siezen, *FEBS Letters*, 1993, **330**, 23.
189. A. Peschel, B. Ottenwälder, and F. Götz, *FEMS Microbiol. Lett.*, 1996, **137**, 279.
190. K. Siegers, S. Heinzmann, and K.-D. Entian, *J. Biol. Chem.*, 1996, **271**, 12294.
191. T. Kupke and F. Götz, *J. Bacteriol.*, 1996, **178**, 1335.
192. T. Kupke, S. Stevanović, H.-G. Sahl, and F. Götz, *J. Bacteriol.*, 1992, **174**, 5354.
193. T. Kupke, C. Kempter, G. Jung, and F. Götz, *J. Biol. Chem.*, 1995, **270**, 11282.
194. C. Kempter, T. Kupke, D. Kaiser, J. W. Metzger, and G. Jung, *Angew. Chem., Int. Ed. Engl.*, 1996, **35**, 2104.
195. K. Altena, G. Bierbaum, and H.-G. Sahl, unpublished results.
196. M. J. Fath and R. Kolter, *Microbiol. Rev.*, 1993, **57**, 995.
197. R. J. Siezen, W. M. de Vos, J. A. M. Leunissen, and B. W. Dijkstra, *Protein Eng.*, 1991, **4**, 719.
198. S. Geissler, F. Götz, and T. Kupke, *J. Bacteriol.*, 1996, **178**, 284.
199. R. J. Siezen, H. S. Rollema, O. P. Kuipers, and W. M. de Vos, *Protein Eng.*, 1995, **8**, 117.
200. W. J. Simpson, N. L. Ragland, C. W. Ronson, and J. R. Tagg, in "Genetics of Streptococci, Enterococci and Lactococci," Developments in Biological Standardization, vol. 85, eds. J. J. Ferretti, M. S. Gilmore, T. R. Klaenhammer, and F. Brown, Karger, Basel, 1995, p. 639.
201. R. A. Segarra, M. C. Booth, D. A. Morales, M. M. Huycke, and M. S. Gilmore, *Infect. Immun.*, 1991, **59**, 1239.
202. R. W. Jack, personal communication.
203. L. S. Håvarstein, D. B. Diep, and I. F. Nes, *Mol. Microbiol.*, 1995, **16**, 229.
204. K. Venema, J. Kok, J. D. Marugg, M. Y. Toonen, A. M. Ledeboer, G. Venema, and M. L. Chikindas, *Mol. Microbiol.*, 1995, **17**, 515.
205. O. P. Kuipers, M. M. Beerthuyzen, P. G. G. A. de Ruyter, E. J. Luesink, and W. M. de Vos, *J. Biol. Chem.*, 1995, **270**, 27299.
206. H. M. Dodd, N. Horn, W. C. Chan, C. J. Giffard, B. W. Bycroft, G. C. K. Roberts, and M. J. Gasson, *Microbiology*, 1996, **142**, 2385.
207. M. Qiao, S. Ye, O. Koponen, R. Ra, M. Usabiaga, T. Immonen, and P. E. J. Saris, *J. Appl. Bacteriol.*, 1996, **80**, 626.
208. W. L. Hynes and J. J. Ferretti, in "Genetics of Streptococci, Enterococci and Lactococci," Developments in Biological Standardization, vol. 85, eds. J. J. Ferretti, M. S. Gilmore, T. R. Klaenhammer, and F. Brown, Karger, Basel, 1995, p. 635.
209. H. Ersfeld-Dressen, H.-G. Sahl, and H. Brandis, *J. Gen. Microbiol.*, 1984, **130**, 3029.
210. M. J. Gasson, *FEMS Microbiol. Lett.*, 1984, **21**, 7.
211. M. Qiao, T. Immonen, O. Koponen, and P. E. J. Saris, *FEMS Microbiol. Lett.*, 1995, **131**, 75.
212. P. E. J. Saris, T. Immonen, M. Reis and H.-G. Sahl, *Antonie van Leeuwenhoek*, 1996, **69**, 151.
213. M. C. Garrido, M. Herrero, R. Kolter, and F. Moreno, *EMBO J.*, 1988, **7**, 1853.
214. H. S. Rollema, O. P. Kuipers, P. Both, W. M. de Vos, and R. J. Siezen, *Appl. Environ. Microbiol.*, 1995, **61**, 2873.
215. W. C. Chan, H. M. Dodd, N. Horn, K. Maclean, L.-Y. Lian, B. W. Bycroft, M. J. Gasson, and G. C. K. Roberts, *Appl. Environ. Microbiol.*, 1996, **62**, 2966.
216. F. H. Gao, T. Abee, and W. N. Konings, *Appl. Environ. Microbiol.*, 1991, **57**, 2164.
217. M. J. G. Garcerá, M. G. L. Elferink, A. J. M. Driessen, and W. N. Konings, *Eur. J. Biochem.*, 1993, **212**, 417.
218. A. J. M. Driessen, H. W. van den Hooven, W. Kuiper, M. van de Kamp, H.-G. Sahl, R. N. H. Konings, and W. N. Konings, *Biochemistry*, 1995, **34**, 1606.
219. G. Stoffels, A. Gudmundsdottir, and T. Abee, *Microbiology*, 1994, **140**, 1443.
220. J. H. Spee, W. M. de Vos, and O. P. Kuipers, *Nucleic Acids Res.*, 1993, **21**, 777.

4.12
Glycosylphosphatidylinositol (GPI)-anchor Biosynthesis

JUNJI TAKEDA, NORIMITSU INOUE, and TAROH KINOSHITA
Osaka University, Japan

4.12.1 INTRODUCTION

Many proteins are anchored to the plasma membrane by a glycolipid termed glycosylphosphatidylinositol (GPI) rather than by hydrophobic regions of the proteins themselves.[1,2] The core of GPI consists of phosphatidylinositol (PI), glucosamine (GlcN), three mannoses (Man), and ethanolaminephosphate (EtNP), and is conserved in eukaryotes from yeast to mammals (Figure 1).[3] Proteins that are anchored to the cell membrane with GPI are termed GPI-anchored proteins. The polypeptide sequences in the GPI-anchored proteins differ, in contrast to the conserved GPI. The precursor peptides that are destined to be GPI-anchored bear carboxy-terminal signal sequences that are cleaved off and replaced by a preassembled GPI-anchor in the endoplasmic reticulum (ER).

If GPI-anchor biosynthesis is completely blocked, no GPI-anchored proteins are expressed on the cell surface, due to either their degradation in the ER or secretion.

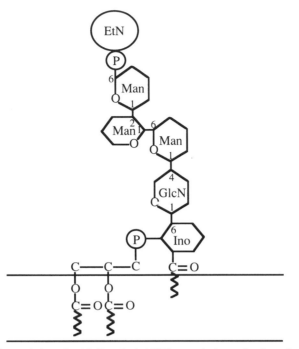

Figure 1   Structure of glycosylphosphatidylinositol (GPI) consisting of ethanolaminephosphate (EtN-P), three mannoses (Mans), glucosamine (GlcN), and phosphatidylinositol (PI).

The human hematopoietic disease, paroxysmal nocturnal hemoglobinuria (PNH), occurs via this mechanism. Somatic mutation in a hematopoietic stem cell causes deficiency of GPI-anchored proteins in all lineages of hematopoietic cells derived from it. The gene responsible for PNH is *PIG-A*, which is involved in the first step of GPI-anchor biosynthesis[4] (see below). The clinical manifestation of PNH is an anemia caused by high sensitivity of affected erythrocytes to the autologous complement, due to the absence of GPI-anchored complement regulatory proteins, such as CD55 and CD59, on the cell surface. The abnormality of GPI-anchor biosynthesis in PNH resides only in hematopoietic cells. An inherited human disease caused by a global GPI-anchor deficiency has not been reported, presumably because it would be lethal. Knockout experiments using *Pig-a* (a mouse homologue of *PIG-A*)-deficient ES cells revealed that complete GPI-anchor biosynthesis deficiency caused embryonic lethality.[5] It is not surprising that GPI-anchored proteins have important roles in embryonic development. GPI-anchored proteins are minor but important components in mammals. Interestingly GPI-anchored proteins are very abundant in unicellular organisms such as protozoas and yeasts. GPI-anchoring has been shown to be essential to the viability of these unicellular organisms as well.

4.12.2   INTRACELLULAR SITE AND ORIENTATION OF GPI-ANCHOR BIOSYNTHESIS

The GPI-anchor is assembled on the ER. The assembled GPI is then transferred to polypeptides bearing a carboxy-terminal signal sequence on the lumenal side of the ER. The GPI-anchor biosynthesis starts from a binding of *N*-acetylglucosamine (GlcNAc) to PI to form GlcNAc-PI. The proteins involved in this biosynthetic step and GlcNAc-PI itself are oriented toward the cytoplasmic side of the ER. A GPI intermediate or mature precursor of the GPI-anchor should, therefore, move into the lumen.

Dolicholphosphatemannose (Dol-P-Man) is a donor of Man in both GPI-anchoring and *N*-glycanation, because mutants lacking synthesis of Dol-P-Man are defective in GPI-anchoring and also are abnormal in *N*-glycanation.

Dol-P-Man is used on the lumenal side of the ER to add Man to *N*-glycan intermediates. The PIG-B protein involved in the third mannose transfer to the GPI intermediate from Dol-P-Man has been shown to be localized in the lumenal side of the ER.[6] Moreover, a mutant Lec35 cell is defective in Dol-P-Man usage for both GPI and *N*-glycan.[7] It is, therefore, speculated that GPI-anchor intermediates move into the lumen of the ER either before or during the transfer of Man from Dol-P-Man.

4.12.3 STRUCTURE OF GPI AND GPI-ANCHORED PROTEIN

Although the core structure of GPI, ethanolaminephosphate-Manα1-2Manα1-6Manα1-4GlcNα1-6inositol phospholipid, is conserved,[8–10] many modifications of GPI have been reported.[3] In higher eukaryotes, additional ethanolaminephosphates are detected in GPI. The presence of additional Man residues as seen in yeast, *Trypanosoma cruzi*, slime mold, and mammals appears common. Other sugars such as galactose, *N*-acetylglucosamine, and sialic acid are also found in the variant surface glycoprotein (VSG) and the procyclic acidic repetitive protein (PARP or procyclin) from *T. brucei*.[3] The lipid moieties in the GPI-anchor also vary. Alkylacylglycerols and diacylglycerols are commonly found in mammals and fish whereas ceramide is common in yeast and slime mold.

Most mammalian GPI-anchored proteins were originally identified by their release from the cell surfaces by bacterial phosphatidylinositol-specific phospholipase C (PI-PLC).[11,12] PI-PLC reacts with mammalian cells, generating alkylacyl-glycerol in the membrane and releasing the soluble part of GPI-anchored protein. Additional acylation (palmitoylation) of the inositol ring in GPI imparts resistance to PI-PLC treatment.[13,14]

4.12.4 GENES INVOLVED IN GPI-ANCHOR BIOSYNTHESIS

Several mammalian mutant cell lines deficient in the biosynthesis of the GPI-anchor have been established.[15–17] Thy-1, a murine GPI-anchored protein, is expressed in thymoma cell lines. Many thymoma cell lines deficient in Thy-1 expression on the plasma membrane have been classified into complementation groups by somatic cell hybridizations.[18] Defects in most of them, classes A–C, E, F, and H, resided in the GPI-anchor biosynthesis and not the Thy-1 gene itself.[19–22] Human GPI-anchored protein negative cell lines have been reported and classified into complementation groups. Two other complementation classes, classes J and K, which were not identified in murine thymoma mutants, were established in the human erythroleukemia cell line K562.[23,24]

Yeast mutants deficient in GPI-anchor biosynthesis have been independently established and also classified into complementation groups (Gpi1–8).[25,26] Some of the genes for these yeast complementation groups correspond to those for mammalian mutants.[27–29] Yeast dolicholphosphatemannose synthase gene (*yDPM1*) was able to restore mammalian class E deficiency.[30] However, no homologue of the class H gene (*PIG-H*) was found in the whole yeast genome,[29] suggesting that either the homology between the genes is very low or the reaction step involving *PIG-H* is different in yeast.

4.12.5 GPI-ANCHOR BIOSYNTHESIS

4.12.5.1 Synthesis of the First GPI Intermediate, *N*-acetylglucosaminylphosphatidylinositol (GlcNAc-PI)

The initial step of GPI-anchor biosynthesis is the transfer of *N*-acetylglucosamine (GlcNAc) to PI from UDP-GlcNAc to form GlcNAc-PI (**1**) (Figure 2). Three genes (*PIG-A*, *PIG-C*, and *PIG-H*) have been identified and expression-cloned[29,31,32] using mammalian mutant cells, which have an abnormality in the first step.[33,34]

GlcNAc-PI (**1**)

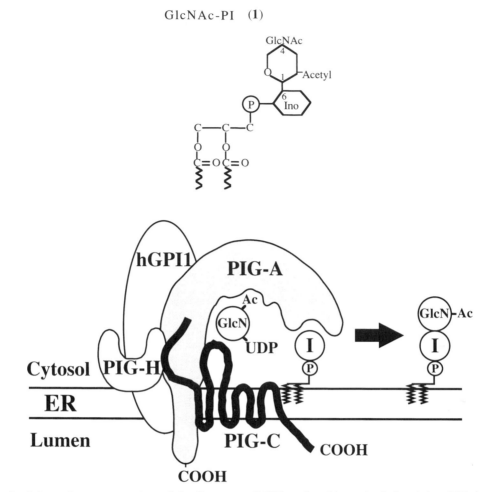

Figure 2 Schematic representation of the first step of GPI-anchor biosynthesis involving PIG-A, PIG-C, PIG-H, and hGPI1 proteins.

4.12.5.1.1 *Cloning and characterization of the class A gene,* **PIG-A,** *and its homologue*

Human lymphoblastoid cell line JY5, which belongs to complementation class A, was used as the recipient for expression cloning of class A gene.[31] A cDNA which restored surface expression of GPI-anchored proteins was cloned and termed *PIG-A* for *p*hosphatidyl*i*nositol*g*lycan class *A*.[31] *PIG-A* encodes a putative protein containing 484 amino acids. There is no hydrophobic region at the N-terminus of PIG-A protein.[31] A hydrophobic region is located close to the C-terminus (Figure 2). A region spanning the N-terminus to the transmembrane domain shows homology to the bacterial GlcNAc transferase, Rfak, which is involved in the synthesis of lipopolysaccharide, implying that PIG-A protein contains a catalytic site for *N*-acetylglucosamine transferase.[35]

Chromosomal localization of *PIG-A* was determined by fluorescence *in situ* hybridization (FISH) and mapped at Xp22.1.[36] This localization may provide an important insight into the pathogenesis of human hematopoietic disease, paroxysmal nocturnal hemoglobinuria (PNH). PNH is an acquired hematopoietic stem cell disorder in which all the hematopoietic cells have an abnormality, but not the other somatic cells.[4] Erythrocytes from patients with PNH are abnormally sensitive to autologous complement.[4] The hypersensitivity of erythrocytes to complement is due to a lack of complement regulatory proteins on erythrocytes, such as CD55 and CD59, which are GPI-anchored proteins. Mutations of *PIG-A* are detected in hematopoietic cells but not non-hematopoietic cells of patients with PNH. These facts suggest that mutations on the *PIG-A* gene are responsible for PNH.[4] Because of the X-chromosomal localization of *PIG-A*, a single somatic mutation would cause a loss of function. There is only one active X chromosome even in female somatic cells, due to random X-chromosome inactivation. Genes involved in GPI-anchor biosynthesis other than *PIG-A* have been cloned and mapped on autosomes. The different chromosomal localizations of *PIG-A* and the others would account for the uniformity of the gene responsible for PNH.[36] The mouse homologue of *PIG-A* was cloned by cross-hybridization using *PIG-A* cDNA as a probe and termed *Pig-a*.[35] The size of the predicted Pig-a protein (485 amino acids) is very similar to that of PIG-A protein and the two have 88% amino acid identity. Transfection of *Pig-a* complemented not only *Pig-a* deficient murine cell lines but also *PIG-A* deficient human cell lines, demonstrating that functions of PIG-A and Pig-a are conserved.[35]

An essential yeast gene, *SPT14*, that encodes a protein with 47% amino acid identity to PIG-A protein was originally isolated as a suppressor of histidine auxotroph caused by Ty element insertion in his4 promoter.[37] The phenotype of the *SPT14* mutant has a long doubling time, transcriptional abnormalities, cell separation defects, and a lower rate of sphingolipid synthesis than wild-type cells.[37] A subsequent study showed that the *spt14* mutant is severely defective in GPI-anchor biosynthesis, particularly in the synthesis of GlcNAc-PI, the first step in GPI-anchor biosynthesis.[28] It is concluded that *SPT14* is a homologue of *PIG-A* and that the many defects seen in *SPT14* mutants are indirect effects of the loss of the GPI-anchor. Yeast GPI-anchor mutants in which colonies were screened for defects in the incorporation of (^{3}H)-inositol into proteins were independently isolated and one such mutant, *gpi3*, was found to have a defective *SPT14* gene.[38] Human or mouse class A cell lines bearing mutations in *PIG-A* show no apparent abnormality of cell growth or morphology. These facts, taken together, suggest that dependencies of cell functions on GPI-anchor are different between yeast and mammals.

4.12.5.1.2   *Cloning and characterization of* PIG-C *and its homologue*

Yeast *GPI2* encodes a predicted hydrophobic protein of 269 amino acids.[38] *GPI2* gene is essential for growth of yeast *Saccharomyces cerevisiae*.[38] Overexpression of the *GPI2* gene partially rescued *gpi1* mutants, indicating that GPI1p and GPI2p interact *in vivo*, probably forming a part of the GlcNAc transferase.[38] A homology search in the expressed sequence *tag* (EST) database found a human GPI2 homologue.[29] It had 297 amino acids and 20% amino acid identity to GPI2. It also complemented the defect of class C mutants, so it was termed PIG-C.[29]

4.12.5.1.3   *Cloning and characterization of* PIG-H

Human class H gene, *PIG-H*, was cloned by expression cloning using murine Ltk-cells as recipients.[32] The deduced protein of *PIG-H* has 188 amino acids.[32] PIG-H has no significant homology to any glycosyltransferase.[32] It is, therefore, not clear how PIG-H protein participates in the first reaction. There is no structural homologue of *PIG-H* in the genome of the yeast *Saccharomyces cerevisiae*.[29]

4.12.5.1.4   *Cloning and characterization of* GPI1 *and its homologue*

The *GPI1* gene, which encodes a 609 amino acid protein, was cloned by complementation of the temperature sensitivity of yeast *gpi1* mutant.[39] Although *GPI1* is a non-essential gene, the phenotype of *GPI1*-disrupted cells demonstrated that the GPI1 protein is clearly involved in GPI-anchor biosynthesis. *GPI1* disruptant cells are temperature-sensitive for growth and can incorporate (^{3}H)-inositol into proteins at permissive temperature.[39] Since it has been speculated that protein involved in GlcNAc-PI synthesis forms a complex (see below), a putative role of non-essential GPI1 protein may be to stabilize such a complex or modulate enzymatic activity[39] (Figure 2).

A human *GPI1* homologue has been cloned from an EST database using the tblastn program. This cDNA encodes a 581 amino acid protein that has 24% amino acid identity with yeast GPI1 protein.[40]

4.12.5.1.5   *Protein complex involved in the first step of GPI-anchor biosynthesis*

There are three complementation groups of mutants in the first step in both mammals and the yeast *S. cerevisiae*. As described above, *PIG-A* and *PIG-C* are homologous to *SPT14/GPI3* and

GPI2, respectively. There is no structural homologue of PIG-H in the genome of the yeast *S. cerevisiae*. In contrast, a human homologue of *GPI1* has been identified.[40] Thus, four gene products are involved in the first step of mammalian GPI-anchor biosynthesis as shown in Figure 2.

To study the interaction between these gene products, various tagged constructs were produced and expressed in cells. Both GST-tagged *PIG-A* and *PIG-H* were found in the ER membrane,[41] and their orientations were determined by proteinase K protection assay. A large amino-terminal portion of PIG-A protein was sensitive to proteinase K, demonstrating that it faces the cytoplasm.[41] The small lumenal domain is important for ER localization because deletion of the domain caused a mislocation of PIG-A proteins. Proteinase K completely digested GST-tagged PIG-H expressed in microsomal vesicles. Thus most of the PIG-H is oriented to the cytoplasm.[41] When differentially tagged *PIG-A* and *PIG-H* were expressed in the same cells, both proteins were precipitated together. Thus, PIG-A and PIG-H form a protein complex.[41] The involvement of *PIG-C* and *hGPI1* in the formation of the initial enzyme complex was also tested by tagged constructs.[40] The proteins of *PIG-A*, *PIG-C*, *PIG-H*, and *hGPI1* form a stable complex which retains *N*-acetylglucosaminyl transferase activity.[40] This indicates that at least four proteins participate in the *N*-acetylglucosaminyl transferase reaction.

4.12.5.2 Synthesis of the Second GPI Intermediate, Glucosaminylphosphatidylinositol (GlcN-PI)

The second step is the deacetylation of GlcNAc to form GlcN-PI (**2**) (Figure 3). Since the deacetylation of GlcNAc is a unique reaction, the presence of GlcN is a usable marker to distinguish GPI-anchor from other glycoconjugates. Class J mutants of K562 cells synthesize GlcNAc-PI but not GlcN-PI.[24] Similar mutants have been established in Chinese hamster ovarian cells (CHO) and termed class L.[42] The gene responsible for class J has not been cloned yet. It is not known whether class J and L mutants are the same or not. A class L gene, *PIG-L*, has been cloned by expression cloning using class L cells as recipient cells. *PIG-L* encodes a 252 amino acid protein that resides in the ER.[42] Since the enzymatic activity of the PIG-L protein has not been demonstrated, it is not clear whether PIG-L is a deacetylase. The intact microsome fraction bearing PIG-L proteins was

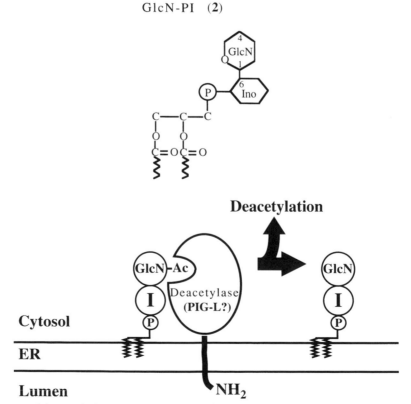

GlcN-PI (**2**)

Figure 3   The second deacetylation step of GPI-anchor biosynthesis with PIG-L proteins occurs in the cytosol.

prepared and treated with proteinase K. PIG-L proteins were not protected from the treatment whereas ER luminal proteins were, suggesting that most of the PIG-L protein faces the cytoplasm.[42]

The orientation of GlcNAc-PI and GlcN-PI was determined by treatment of the intact ER membrane with PI-PLC. Both GlcNAc-PI and GlcN-PI were sensitive to PI-PLC suggesting that they face the cytoplasm[43] (Figure 3). The orientations of both the first and second GPI intermediates and the *PIG-L* gene product indicate that the second step of GPI-anchor biosynthesis also occurs on the cytoplasmic side of the ER membrane.[42]

This step was activated by GTP *in vitro* when the microsome fraction was used.[44] The effect of GTP on the second reaction was not coupled to the first reaction because the deacetylation of exogenously added GlcNAc-PI in microsomes from cells defective in the first reaction was enhanced by GTP.[44] A non-hydrolyzable GTP analogue completely inhibited GTP-mediated enhancement of GlcNAc-PI deacetylation, suggesting that the hydrolysis of GTP was required for this phenomenon.[44]

4.12.5.3   Synthesis of the Third GPI Intermediate, GlcN-acyl PI

The 2-position of the inositol in GlcN-PI is acylated (palmitoylation) in this step[45,46] (3) (Figure 4). This modification occurs in the yeast *S. cerevisiae* and mammalian cells but not in trypanosoma. The significance of this modification has been therefore discussed by many investigators. When the mannosylation was blocked in mammalian cells and yeast, GlcN-acyl PI was markedly accumulated. This indicates that acyl-modification is a prerequisite step before mannosylation of GlcN-PI. If this concept is correct, GPI-anchor biosynthesis and/or the enzymes involved in the biosynthesis are different between trypanosoma and mammals/yeast *S. cerevisiae*. This issue has been addressed by the use of a synthetic dioctanoyl GlcN-PI analogue, GlcN-PI(C8).[47] The efficiencies of mannose addition to GlcN-PI(C8) or GlcN-acyl PI(C8) were compared in lysates from Chinese hamster ovarian cells (CHO).[47] The mannosylation was examined in various concentrations of GlcN-PI(C8) in the absence or presence of palmitoyl-CoA. The addition of mannose occurred at a much lower concentration of GlcN-PI(C8) when palmitoyl-CoA was present, suggesting that GlcN-acyl PI is a better substrate for mannosylation in the mammalian system.[47] In trypanosomes, the efficiencies of mannosylation to GlcN-PI and GlcN-acyl PI were similar indicating that the acylation of inositol is not a prerequisite.[47] The different specificities of acylation between trypanosome and mammals/ yeast *S. cerevisiae* open the possibility for chemotherapy targeted to this step. A donor molecule for palmitoylation is thought to be palmitoyl-CoA because the addition of palmitoyl-CoA enhanced the palmitoylation of inositol.[48] In the mouse system, transfer of palmitate from labeled palmitoyl-CoA was unsuccessful, so free CoA seems to be an activator for the formation of GlcN-acyl PI.[49]

GlcN-acylPI   (3)

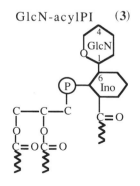

4.12.5.4   Synthesis of the Fourth GPI Intermediate, Mannose(Man)-GlcN-acyl PI

Man is bound to GlcN by an α1–4 linkage (Figure 5). The fourth step is catalyzed by α1-4 mannosyltransferase, although its gene has not been identified. As described above, the specificity of this mannosyltransferase differs in mammals/yeast and trypanosoma. All three Mans in the GPI-anchor are provided from dolicholphosphatemannose (Dol-P-Man).[50] Deficient Dol-P-Man synthesis causes accumulations of GlcN-acyl PI and GlcN-PI. Dol-P-Man synthase (DPM1) was originally cloned from the yeast *S. cerevisiae*.[51] A human homologue of DPM1 has also been cloned.[52] The synthesis of Dol-P-Man in mammalian cells is thought to be complicated because two

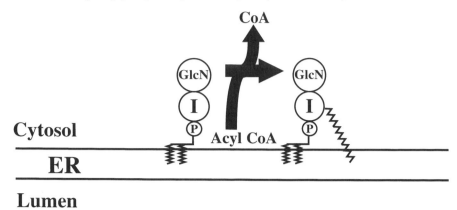

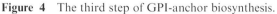

Figure 4   The third step of GPI-anchor biosynthesis.

genetically different mutants, Lec15[53] and class E[54] mutants, cannot produce Dol-P-Man. A novel rat gene termed *DPM2* has been cloned by expression cloning using Lec15 cells as a recipient cell.[55] Lec15 and class E mutants have been shown to be defective in *DPM2* and mouse *DPM1* genes.[55] The existence of another mutant, Lec35, further complicates our understanding of Dol-P-Man synthesis.[53] In Lec35, mannosyl transfer to GlcN-acyl PI was not detected, but Dol-P-Man was synthesized.[53] Mannosylation of *N*-glycan precursors was also impaired in Lec35, suggesting that utilization of Dol-P-Man was perturbed.[53] Consistent with this idea, mannosyl transfer to GlcN-acyl PI was observed with the lysate of Lec35. Biosynthesis of *N*-glycan precursors occurs on the cytoplasmic side of the ER until formation of Man$_5$GlcNAc$_2$-P-P-dolichol. The subsequent step, a transfer of Man from Dol-P-Man, occurs on the luminal side. Since Dol-P-Man is generated from dolichol and GDP-Man, which is localized in cytoplasm (Figure 5), Dol-P-Man should move from the cytoplasmic side to the lumen either after or during the formation (Figure 5). Interestingly, *SL15* cDNA complemented the defect of Lec35.[56] This suggests that SL15 is a flippase itself or a part of it.

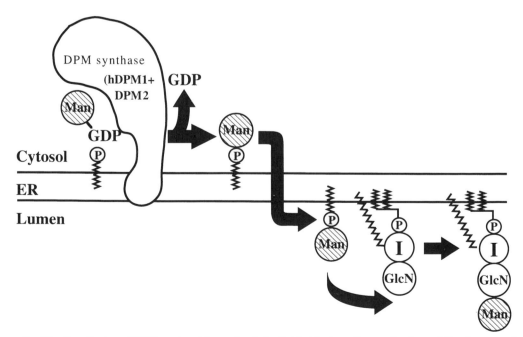

Figure 5   The fourth step of GPI-anchor biosynthesis is divided in two, the production of the donor molecule dolicholphosphatemannose (Dol-P-Man) involving hDPM1 and DPM2 proteins and the transfer of Dol-P-Man to GlcN-acyl PI to form Man-GlcN-acyl PI.

Man-GlcN-acylPI   (**4**)

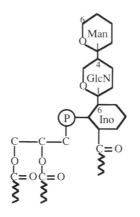

Conzelmann and his colleagues have produced several yeast mutants of GPI-anchor biosynthesis and extensively analyzed them.[26] They identified four different mutants (*gpi4–6* and *gpi9*) in which GlcN-acyl PI was accumulated.[26] It has been speculated that the defect resides in the α1,4-mannosyltransferase, which adds the first mannose of the GPI core, or in the step of generation of Dol-P-Man required for this transferase reaction. The addition of N-linked sugar was also impaired in three of four mutants, implying that these mutants have an abnormality in either the synthesis or usage of Dol-P-Man because Dol-P-Man is needed for synthesis of both GPI-anchor and N-linked sugar.[26] No abnormality in the production of N-linked sugar was observed in the gpi5 mutant, suggesting the activity of the α1,4-mannosyltransferase is perturbed in this mutant.[26]

4.12.5.5   Synthesis of the Fifth GPI Intermediate, Man2-GlcN-acyl PI

This step involves the second Man transfer to Manα1–4GlcNα1–6inositol to form Manα1–6Manα1–4GlcNα1–6inositol (Figure 6). Although Dol-P-Man is known to be a donor,[50] this step is not extensively characterized because of a lack of mutants deficient in putative α1–6 transferase.

Man2-GlcN-acylPI   (**5**)

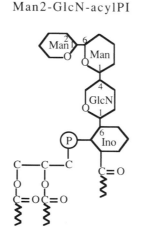

4.12.5.6   Synthesis of the Sixth GPI Intermediate, Man3-GlcN-acyl PI

The third mannose of the GPI-anchor binds to the second mannose by an α1,2 linkage as shown in Figure 6. The class B mutant is defective in the third mannosylation and therefore accumulates

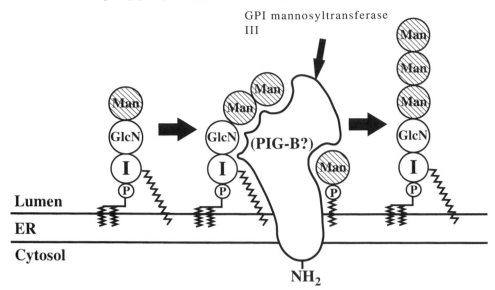

Figure 6   The fifth and sixth steps of GPI-anchor biosynthesis occurring in the lumen of the ER. PIG-B protein is involved in the sixth step.

an intermediate having two Mans.[57] A cDNA termed *PIG-B* complemented the deficiency of class B mutant.[6] *PIG-B* encodes an ER transmembrane protein of 554 amino acids.[6] PIG-B has homology to ALG9,[58] a putative mannosyltransferase for the seventh Man of N-glycan, suggesting that PIG-B is GPI mannosyltransferase III itself. There is no typical amino-terminal hydrophobic sequence for the signal peptide.[6] Analysis of PIG-B proteins expressing glutathione-*S*-transferase (GST)-*tag* demonstrated that the amino-terminal hydrophilic portion (60 amino acids) faces the cytoplasm and the relatively hydrophobic and large carboxy-terminal portion (470 amino acids) faces lumen of the ER[6] (Figure 6). The amino-terminal hydrophilic portion was not required for the function of PIG-B because a deletion mutant lacking the amino-terminal portion showed full activity.[6] These facts suggest that the functional domain resides in the lumen and the third GPI mannosylation would proceed in the lumen.[6] This notion, however, is inconsistent with the cytoplasmic orientation of the late GPI intermediates of *T. brucei*[59] and *Leishmania major*.[60] The potential differences in the topology of GPI mannosylation between mammals and protozoa need to be clarified.

Man3-GlcN-acylPI   (**6**)

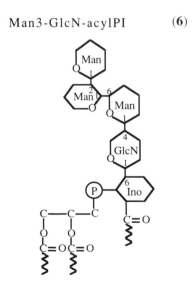

4.12.5.7 Synthesis of the Seventh GPI Intermediate, Ethanolaminephosphate (EtNP)-Man3-GlcN-PI

The seventh step of GPI-anchor biosynthesis involves the transfer of ethanolamine phosphate (EtN-P) from its donor phosphatidylethanolamine to the third mannose (Figure 7). The class F mutant derived from the murine thymoma cell line is defective in the EtN-P transfer and accumulates an intermediate of GPI-anchor precursor having three Mans. A human cDNA termed *PIG-F* was cloned by expression cloning using the class F mutant.[61] The *PIG-F* encodes a very hydrophobic protein of 219 amino acid residues, most of which is probably embedded in the membrane[61] (Figure 7). The cellular localization of PIG-F has not been determined.

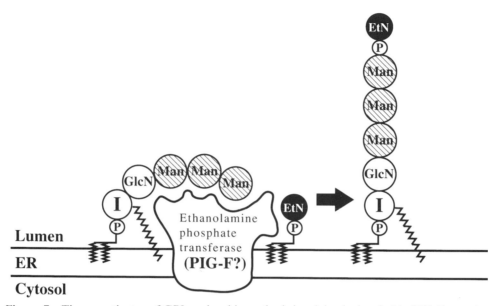

Figure 7   The seventh step of GPI-anchor biosynthesis involving hydrophobic PIG-F proteins.

4.12.5.8 Attachment of GPI to Proteins, Peptide-EtNP-Man3-GlcN-acyl PI

This is the final step in the GPI-anchor biosynthesis that provides the complete GPI core for proteins, as shown in Figure 8. The reaction consists of a replacement of a carboxy-terminal signal sequence in the nascent protein with a complete GPI core to form a GPI-anchored protein. EtNP of the GPI-anchor binds to the newly formed carboxy-terminus of the polypeptides by an amide-linkage. Udenfriend's group extensively analyzed this step *in vitro* using molecularly engineered placental alkaline phosphatase (miniPLAP) and microsomal fractions.[62] The miniPLAP is a model GPI-anchored protein that contains both amino- and carboxy-terminal signal sequences which are processed. Processing involves cleavage of both signal sequences and a replacement of the latter by GPI, which can be assessed by SDS–PAGE.[62] ATP is required in this step[63] because the immuno-globulin heavy chain binding proteins (Bip; GRP 78) participate in the processing of GPI-anchored proteins.[64] The cleavage and GPI-anchor attachment site of polypeptides is referred to as the ω residue. The ω site permits only small amino acids (Gly, Ala, Cys, Ser, Asp, and Asn),[65] while the $\omega + 2$ site, which is two residues downstream, allows only four small amino acids (Ala, Cys, Gly, and Ser). At the $\omega + 1$ site, there is less stringency, so most amino acids except Pro are allowed. There are 15 to 20 hydrophobic amino acids composing the carboxy-terminus of the signal sequence. In addition, there is a hydrophilic region (5–10 amino acids) between the ω site and the hydrophobic domain. Although a precise consensus sequence is not found within the carboxy-terminal signal sequence, its length and hydropathy profile are well conserved.

The putative GPI : protein transamidase is required in the final step. To investigate the nature of GPI : protein transamidase, mutant cells that make complete GPI-anchor but do not express

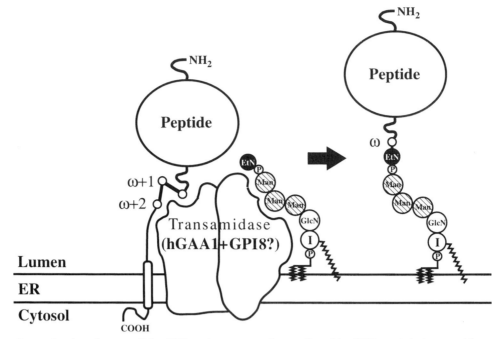

Figure 8   Attachment of the GPI-anchor to proteins mediated by GPI:protein transamidase.

GPI-anchored proteins on the cell surface should be very useful. A human K562 mutant cell line designated class K produced the fully assembled GPI-anchor donor, but failed to express GPI-anchored proteins.[24] The characterization of the mutant fits nicely with the investigation of the final step of GPI-anchor biosynthesis. Small nucleophilic chemicals such as hydrazine and hydroxylamine have been shown to be alternate substrates for GPI.[66] The addition of these nucleophiles to microsomal membranes from wild or mutant cells (class B or F) produced the hydrazide or hydroxamate of proteins whereas the addition of these nucleophiles to microsomal membranes from mutant K cells did not yield these products, suggesting that mutant K cells lack GPI:protein transamidase itself or its cofactor.[67] In yeast there are at least two mutants, *gaa1*[68] and *gpi8*,[26] in which complete GPI-anchor is produced. Since gaa1 and gpi8 mutants have complementary defects, *GAA1* and *GPI8* genes are different. It has been reported that the mutant K cells are defective in the human homologue of *GPI8*.[69] These findings suggest that GPI:protein transamidase is also the protein complex as seen in the first step of GPI-anchor biosynthesis (Figure 8).

4.12.6   MODIFICATION OF GPI AFTER THE GPI:PROTEIN TRANSAMIDASE REACTION

In trypanosomes, newly synthesized GPI which is not myristoylated undergoes fatty acid remodeling to replace its fatty acids with myristate after the transamidase reaction.[1] The myristate analogue 10-propoxydecanoic acid is incorporated into the GPI and is toxic to the trypanosome, suggesting that fatty acid remodeling with myristate is essential for the organism's survival.[70] In yeast, some GPIs contain ceramide.[71] The addition of ceramide to GPI is believed to occur by a remodeling after the GPI:transamidase reaction takes place, as in trypanosome.[71]

4.12.7   REFERENCES

1. P. T. Englund, *Annu. Rev. Biochem.*, 1993, **62**, 121.
2. J. Takeda and T. Kinoshita, *Trends Biochem. Sci.*, 1995, **20**, 367.

3. M. J. McConville and M. A. J. Ferguson, *Biochem. J.*, 1993, **294**, 305.
4. T. Kinoshita, N. Inoue, and J. Takeda, *Adv. Immunol.*, 1995, **60**, 57.
5. K. Kawagoe, D. Kitamura, M. Okabe, I. Taniuchi, M. Ikawa, T. Watanabe, T. Kinoshita, and J. Takeda, *Blood*, 1996, **87**, 3600.
6. M. Takahashi, N. Inoue, K. Ohishi, Y. Maeda, N. Nakamura, Y. Endo, T. Fujita, J. Takeda, and T. Kinoshita, *EMBO J.*, 1996, 4264.
7. Y. Zeng and M. A. Lehrman, *J. Biol. Chem.*, 1990, **265**, 2296.
8. M. A. Ferguson and A. F. Williams, *Annu. Rev. Biochem.*, 1988, **57**, 285.
9. S. W. Homans, M. A. Ferguson, R. A. Dwek, T. W. Rademacher, R. Anand, and A. F. Williams, *Nature*, 1988, **333**, 269.
10. W. L. Roberts, S. Santikarn, V. N. Reinhold, and T. L. Rosenberry, *J. Biol. Chem.*, 1988, **263**, 18 776.
11. M. G. Low, M. A. Ferguson, A. H. Futerman, and I. Silman, *Trends Biochem. Sci.*, 1986, **11**, 212.
12. M. G. Low, *Biochem. J.*, 1987, **244**, 1.
13. W. L. Roberts, J. J. Myher, A. Kuksis, M. G. Low, and T. L. Rosenberry, *J. Biol. Chem.*, 1988, **263**, 18 766.
14. E. I. Walter, W. L. Roberts, T. L. Rosenberry, W. D. Ratnoff, and M. E. Medof, *J. Immunol.*, 1990, **144**, 1030.
15. R. Hyman and V. Stallings, *J. Natl. Cancer Inst.*, 1974, **52**, 429.
16. I. S. Trowbridge, R. Hyman, and C. Mazauskas, *Cell*, 1978, **14**, 21.
17. R. Hyman, K. Cunningham, and V. Stallings, *Immunogenetics*, 1980, **10**, 261.
18. R. Hyman, *Trends Genet.*, 1988, **4**, 5.
19. A. Conzelmann, A. Spiazzi, R. Hyman, and C. Bron, *EMBO J.*, 1986, **5**, 3291.
20. A. Conzelmann, A. Spiazzi, C. Bron, and R. Hyman, *Mol. Cell. Biol.*, 1988, **8**, 674.
21. S. H. Fatemi and A. M. Tartakoff, *Cell*, 1986, **46**, 653.
22. S. H. Fatemi and A. M. Tartakoff, *J. Biol. Chem.*, 1988, **263**, 1288.
23. S. Hirose, R. P. Mohney, S. C. Mutka, L. Ravi, D. R. Singleton, G. Perry, A. M. Tartakoff, and M. E. Medof, *J. Biol. Chem.*, 1992, **267**, 5272.
24. R. P. Mohney, J. J. Knez, L. Ravi, D. Sevlever, T. L. Rosenberry, S. Hirose, and M. E. Medof, *J. Biol. Chem.*, 1994, **269**, 6536.
25. S. D. Leidich, D. A. Drapp, and P. Orlean, *J. Biol. Chem.*, 1994, **269**, 10 193.
26. M. Benghezal, P. N. Lipke, and A. Conzelmann, *J. Cell Biol.*, 1995, **130**(6), 1333.
27. J. H. Vossen, A. F. J. Ram, and F. M. Klis, *Biochim. Biophys. Acta*, 1995, **1243**, 549.
28. M. Schonbachler, A. Horvath, J. Fassler, and H. Riesman, *EMBO J.*, 1995, **14**, 1637.
29. N. Inoue, R. Watanabe, J. Takeda, and T. Kinoshita, *Biochem. Biophys. Res. Commun.*, 1996, **226**, 193.
30. R. DeGasperi, L. J. Thomas, E. Sugiyama, H. M. Chang, P. J. Beck, P. Orlean, C. Albright, G. Waneck, J. F. Sambrook, C. D. Warren, and E. T. H. Yeh, *Science*, 1990, **250**, 988.
31. T. Miyata, J. Takeda, Y. Iida, N. Yamada, N. Inoue, M. Takahashi, K. Maeda, T. Kitani, and T. Kinoshita, *Science*, 1993, **259**, 1318.
32. T. Kamitani, H. M. Chang, C. Rollins, G. L. Waneck, and E. T. H. Yeh, *J. Biol. Chem.*, 1993, **268**, 20 733.
33. V. L. Stevens and C. R. Raetz, *J. Biol. Chem.*, 1991, **266**, 10 039.
34. E. Sugiyama, R. DeGasperi, M. Urakaze, H. M. Chang, L. J. Thomas, R. Hyman, C. D. Warren, and E. T. H. Yeh, *J. Biol. Chem.*, 1991, **266**, 12 119.
35. K. Kawagoe, J. Takeda, Y. Endo, and T. Kinoshita, *Genomics*, 1994, **23**, 566.
36. J. Takeda, T. Miyata, K. Kawagoe, Y. Iida, Y. Endo, T. Fujita, M. Takahashi, T. Kitani, and T. Kinoshita, *Cell*, 1993, **73**, 703.
37. J. S. Fassler, W. Gray, J. P. Lee, G. Yu, and G. Gingerich, *Mol. Gen. Genet.*, 1991, **230**, 310.
38. S. D. Leidich, Z. Kostova, R. R. Latek, L. C. Costello, D. A. Drapp, W. Gray, J. S. Fassler, and P. Orlean, *J. Biol. Chem.*, 1995, **270**, 13 029.
39. S. D. Leidich and P. Orlean, *J. Biol. Chem.*, 1996, **271**, 27 829.
40. R. Watanabe, N. Inoue, B. Westfall, C. H. Taron, P. Orlean, J. Takeda, and T. Kinoshita, *EMBO J.*, 1998, **17**, 877.
41. R. Watanabe, T. Kinoshita, R. Masaki, A. Yamamoto, J. Takeda, and N. Inoue, *J. Biol. Chem.*, 1996, **271**, 26 868.
42. N. Nakamura, N. Inoue, R. Watanabe, M. Takahashi, J. Takeda, V. L. Stevens, and T. Kinoshita, *J. Biol. Chem.*, 1997, **272**, 15 834.
43. J. Vidugiriene and A. K. Menon, *J. Cell Biol.*, 1993, **121**, 987.
44. V. L. Stevens, *J. Biol. Chem.*, 1993, **268**, 9718.
45. S. Mayor, A. K. Menon, and G. A. Cross, *J. Biol. Chem.*, 1990, **265**, 6174.
46. M. Urakaze, T. Kamitani, R. DeGasperi, E. Sugiyama, H. M. Chag, C. D. Warren, and E. T. H. Yeh, *J. Biol. Chem.*, 1992, **267**, 6459.
47. W. T. Doerrler, J. Ye, J. R. Falck, and M. A. Lehrman, *J. Biol. Chem.*, 1996, **271**, 27 031.
48. L. C. Costello and P. Orlean, *J. Biol. Chem.*, 1992, **267**, 8599.
49. V. L. Stevens and H. Zhang, *J. Biol. Chem.*, 1994, **269**, 31 397.
50. A. K. Menon, S. Mayor, and R. T. Schwarz, *EMBO J.*, 1990, **9**, 4249.
51. P. Orlean, C. Albright, and P. W. Robbins, *J. Biol. Chem.*, 1988, **263**, 17 499.
52. S. Tomita, N. Inoue, Y. Maeda, K. Ohishi, J. Takeda, and T. Kinoshita, *J. Biol. Chem.*, 1998, **273**, 9249.
53. L. A. Camp, P. Chauhan, J. D. Farrar, and M. A. Lehrman, *J. Biol. Chem.*, 1993, **268**, 6721.
54. A. Chapman, K. Fujimoto, and S. Kornfeld, *J. Biol. Chem.*, 1980, **255**, 4441.
55. Y. Maeda and T. Kinoshita, unpublished results.
56. F. E. Ware and M. A. Lehrman, *J. Biol. Chem.*, 1996, **271**, 13 935.
57. A. Puoti, C. Desponds, C. Fankhauser, and A. Conzelmann, *J. Biol. Chem.*, 1991, **266**, 21 051.
58. P. Burda, S. te Heesen, A. Brachat, A. Wach, A. Dusterhoft, and M. Aebi, *Proc. Natl. Acad. Sci. USA*, 1996, **93**, 7160.
59. J. Vidugiriene and A. K. Menon, *J. Cell Biol.*, 1994, **127**, 333.
60. K. Mensa Wilmot, J. H. LeBowitz, K. P. Chang, A. al Qahtani, B. S. McGwire, S. Tucker, and J. C. Morris, *J. Cell Biol.*, 1994, **124**, 935.
61. N. Inoue, T. Kinoshita, T. Orii, and J. Takeda, *J. Biol. Chem.*, 1993, **268**, 6882.
62. K. Kodukula, R. Micanovic, L. Gerber, M. Tamburrini, L. Brink, and S. Udenfriend, *J. Biol. Chem.*, 1991, **266**, 4464.

63. R. Amthauer, K. Kodukula, L. Brink, and S. Udenfriend, *Proc. Natl. Acad. Sci. USA*, 1992, **89**, 6124.
64. R. Amthauer, K. Kodukula, L. Gerber, and S. Udenfriend, *Proc. Natl. Acad. Sci. USA*, 1993, **90**, 3973.
65. R. Micanovic, L. D. Gerber, J. Berger, K. Kodukula, and S. Udenfriend, *Proc. Natl. Acad. Sci. USA*, 1990, **87**, 157.
66. S. Ramalingam, S. E. Maxwell, M. E. Medof, R. Chen, L. D. Gerber, and S. Udenfriend, *Proc. Natl. Acad. Sci. USA*, 1996, **93**, 7528.
67. R. Chen, S. Udenfriend, G. M. Prince, S. E. Maxwell, R. Ramalingam, L. D. Gerber, J. Knez, and M. E. Medof, *Proc. Natl. Acad. Sci. USA*, 1996, **93**, 2280.
68. D. Hamburger, M. Egerton, and H. Riezman, *J. Cell Biol.*, 1995, **129**, 629.
69. J. Yu, S. Nagarajan, J. J. Knez, S. Udenfriend, R. Chen, and M. E. Medof, *Proc. Natl. Acad. Sci. USA*, 1997, **94**, 12 580.
70. T. L. Doering, J. Raper, L. U. Buxbaum, S. P. Adams, J. I. Gordon, G. W. Hart, and P. T. Englund, *Science*, 1991, **252**, 1851.
71. A. Conzelmann, A. Puoti, R. L. Lester, and C. Desponds, *EMBO J.*, 1992, **11**, 457.

4.13
Structure, Function, and Biosynthesis of Gramicidin S Synthetase

JOACHIM VATER and TORSTEN HELGE STEIN
Technische Universität Berlin, Germany

4.13.1 INTRODUCTION

The finding of Sir Alexander Fleming in 1928, that a low molecular weight metabolite of the filamentous fungus *Penicillium notatum*—penicillin—was able to inhibit the growth of microbes,[1] initiated the modern area of antibiotic research. The fundamental work of Fleming, Chain,[2] and Abraham[3] in the late 1930s and early 1940s—the isolation, structural characterization, as well as the successful application of penicillin against inflammatory diseases—introduced a worldwide revolution in the therapy of microbial infections. An intensive screening of natural resources for their bioactive potential by Dubos[4] led to the discovery of a peptide fraction, tyrothricin, produced by the soil bacterium *Bacillus brevis*, which exhibited antibiotic activity against Gram-positive bacteria. This fraction was subfractionated into two classes of biologically active peptides, the gramicidins and the tyrocidines.[5,6] The isolation of a bioactive agent from the spore-forming soil organism *B. brevis* ATCC 9999 (now *Bacillus migulanus*, according to the new nomenclature of the American Type Culture Collection (ATCC)) was first reported by Gause and Brazhnikova in 1944.[7,8] The so-called "*Soviet gramicidin*" (gramicidin S) has been shown to be identical to gramicidin J, a peptide agent isolated by Otani and Saito[9] in Japan from the Nagano strain of *B. brevis* a decade later. In addition, by the comprehensive investigation of the bioactive peptides produced by *B. brevis* the linear gramicidins were discovered, which are structurally different from gramicidin S and tyrocidine.[10] The structures of all of these bioactive peptides are summarized in Figure 1. Fundamental knowledge concerning the biosynthesis of these compounds was established in the 1960s. A key observation in deciphering the mechanism was that *B. brevis* cells treated with specific inhibitors of protein and RNA synthesis, such as chloramphenicol or actinomycin D, did not interfere with the production of peptide antibiotics, such as the gramicidins[11] and tyrocidines.[12,13] This feature implied that they were synthesized by the microbial cell nonribosomally, independent from the ribosomal machinery. Further experimental evidence for such an enzymatic pathway of peptide biosynthesis was obtained from the finding that cell- and RNA-free extracts of *B. brevis* were able to catalyze the formation of these peptides.[14,15]

Gramicidin S

```
      1       2       3       4       5
   D-Phe→ L-Pro→ L-Val→ L-Orn→ L-Leu
    ↑                             ↓
   L-Leu← L-Orn ← L-Val ← L-Pro ← D-Phe
    5'      4'      3'      2'      1'
```

Tyrocidine

```
      1       2       3       4       5
   D-Phe→ L-Pro→ L-Phe→ D-Phe→ L-Asn
    ↑                             ↓
   L-Leu← L-Orn ← L-Val ← L-Tyr ← L-Gln
    10      9       8       7       6
```

Linear gramicidin

```
          1       2      3       4       5
   (HCO)-L-Val→ Gly→ L-Ala→ D-Leu→ L-Ala
       10      9      8      7        ↓
   D-Leu ← L-Trp← D-Val← L-Val← D-Val  6
       ↓
   L-Trp→ D-Leu→ L-Trp→ D-Leu→ L-Trp-(NHCH₂CH₂OH)
    11      12      13      14      15
```

Figure 1 Primary structure of the bioactive peptides produced by *B. brevis* sp.

In the early history of protein synthesis research (1954), Lipmann[16] suggested polyenzymes as catalysts for both protein and peptide production. According to this hypothesis a linear sequence of amino acid-activating regions would determine the sequence of the (poly)peptide products. With the discovery of the ribosomal machinery this view became untenable for protein synthesis. However, in the late 1960s the polyenzyme model was taken as the basis for a mechanistic concept for nonribosomal peptide biosynthesis. At that time, several laboratories succeeded in the isolation, purification, and characterization of the multienzyme systems which catalyze the production of gramicidin S and the closely related tyrocidine in *B. brevis*. These peptide synthetases were studied as model systems for the investigation of this pathway. The knowledge accumulated by the groups of Lipmann,[17] Laland,[18,19] and Kurahashi[20] culminated in the presentation of a "multienzymatic thiotemplate mechanism" at the beginning of the 1970s. In the original version of this model, in analogy to the activating regions in the polyenzyme hypothesis of Lipmann,[16] it was assumed that peptide synthetases are composed of amino acid-activating domains which catalyze the activation of the amino acid substrates in two steps involving (i) aminoacyl adenylation and (ii) thioester formation at specific thiotemplate sites. It was postulated that the formation of the peptide product occurs in a step-by-step condensation of the thioesterified substrates in a series of transpeptidation and transthiolation reactions involving a central 4'-phosphopantetheine (4'PPan) cofactor managing the internal transport of the growing peptide chain. This mechanistic concept resembled the early model of Lynen[21,22] for the biosynthesis of fatty acids by fatty acid synthetase multienzyme systems, which also contain 4'PPan as the prosthetic groups of an acyl carrier protein/domain (ACP) catalyzing the transfer of the intermediates between the reaction centers.

The isolation of the peptide synthetases from *B. brevis* was the starting point of the enzymatic era of nonribosomal peptide biosynthesis which developed rapidly and intensively during the following decade.[23–26] By the middle of the 1980s, detailed knowledge had been obtained of the structural and functional organization of numerous other peptide-forming multienzymes and extensive studies had been performed to understand the biosynthesis of pharmaceutically and biotechnologically important bioactive peptides such as the lipopeptide biosurfactant surfactin,[27–31] δ-L-α-aminoadipylcysteinylvaline (ACV) as the tripeptide precursor in β-lactam biosynthesis,[32–34] the immunomodulator cyclosporin,[35–37] or the depsipeptide enniatin and related compounds.[38–40] A major part of this work has been contributed by Kleinkauf and von Döhren.[41,42]

At the beginning of the 1980s the genetics and molecular biology of peptide biosynthesis were approached. Pioneering work in this field was contributed by Marahiel and co-workers.[43] A milestone in the genetic analysis of peptide synthetases was the cloning and sequencing of tyrocidine synthetase 1 from *B. brevis*, which marked the starting point of a new, exciting period in nonribosomal peptide biosynthesis research, the genetic era. During the 1990s the nucleotide sequences of numerous gene clusters encoding peptide-forming systems have been obtained. The availability of the primary sequences of these multienzymes allowed intensive sequence comparisons which revealed the modular structure of peptide synthetases. In this way, highly conserved consensus motifs were detected. By active site mutagenesis and protein engineering it was possible to detect the catalytically active amino acid residues of functional domains and to probe the modular structure of peptide synthetases in detail. This development manifested a first breakthrough in the structural analysis of these very large multienzymatic systems. Important progress has now been made in the protein chemical analysis of peptide synthetases, mainly by the immense progress in the field of mass spectrometry, but also by the development of novel electrophoretic, chromatographic, and microsequencing methodologies, as well as the availability of microseparation systems. These achievements mark the onset of the protein chemical era as an important milestone towards the understanding of nonribosomal peptide biosynthesis at the molecular level.[44–47] The crystal structure of an amino acid-activating fragment of gramicidin S synthetase 1 cocrystallized with its substrates determined at 1.9 Å by Brick *et al.*[48] initiated a new era of nonribosomal peptide biosynthesis research towards the determination of the three-dimensional structure of complete peptide-forming systems.

4.13.2 PEPTIDE ANTIBIOTICS FROM *BACILLUS BREVIS* SP.

4.13.2.1 Gramicidin S

As shown in Figure 1, gramicidin S is a cyclic decapeptide containing two identical pentapeptide moieties (D-Phe-L-Pro-L-Val-L-Orn-L-Leu; Orn, ornithine), which are condensed in a head-to-tail manner.[49,50] The first crystal structure analysis of gramicidin S was performed by Hodgkin and Oughton.[51] The three-dimensional structure of gramicidin S obtained by X-ray analysis of gramicidin S cocrystallized with urea is shown in Figure 2. The two tripeptide units L-Val-L-Orn-L-Leu of the cyclodecapeptide gramicidin S form a double-stranded antiparallel β-sheet structure, which is stabilized by four hydrogen bonds between both strands. These hydrogen bridges are formed by each of the amide protons and the carbonyl groups of the Leu and Val residues, respectively. In addition, two hydrogen bonds were detected between the δ-amino groups of both Orn residues and the carbonyl groups of the D-Phe moieties of the gramicidin S molecule.[52] Due to its basic Orn residues, gramicidin S is able to form complexes with metal ions. Modern mass spectrometric methods have been used to probe the kinetics of complex formation between gramicidin S and various alkali metal salts. The apparent gas phase basicities have been determined for the protonated species $[GS+H]^+$ ($219 \, kcal \, mol^{-1}$) as well as for the alkali adducts $[GS+X]^+$ ($223 \, kcal \, mol^{-1}$; $X = Li^+$, Na^+, K^+). From this difference a charge separation distance of $11.5 \, Å$ was determined between the δ-amino groups of both Orn residues of gramicidin S.[53]

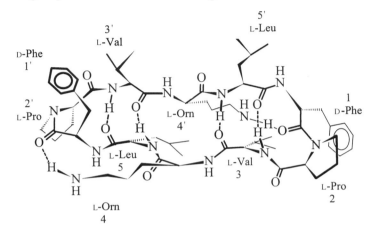

Figure 2   The three-dimensional structure of the cyclodecapeptide gramicidin S produced by *B. brevis* ATCC 9999.

Gramicidin S exhibits a broad antibiotic spectrum against Gram-positive bacteria. However, the utility of gramicidin S as a clinically useful antibiotic agent is restricted by its relatively low bioactivity and appreciable cytotoxicity. In order to improve the antimicrobial activity of gramicidin S, to suppress its property to lyze eukaryotic cells, as well as for a better definition of its structure–activity relationships, a huge number of structural analogues have been synthesized. From studies on more than 200 synthetic analogues, a number of structural requirements important for the antimicrobial activity have been determined. Most important for the bioactivity of gramicidin S is its amphiphilic character governed by combination of the ornithine residues as the hydrophilic part[54] with the high overall hydrophobicity of the rest of the molecule.[55] The second important feature is the antiparallel β-sheet structure in solution or the ability to form such a structure in the presence of lipid bilayers. Studies have demonstrated that these features lead to a significant bioactivity only in combination with a ring size of 10 or more residues.[56] Very important in the context of the development of new antibacterial agents was the observation that gramicidin S is not only active against Gram-positive microorganisms but also against Gram-negative ones when tested using liquid-based antibacterial assays, in contradiction to the agar-based assays of former studies.[57,58]

The bioactivity of gramicidin S is mainly based upon its ability to interact with cell membranes. For example, the binding of gramicidin S to the plasma membranes of microorganisms induces a sharp increase in membrane permeability for cellular low molecular weight compounds which are released into the medium. Due to our current understanding that the biological activity of gramicidin S depends on its interaction with cell membranes, the peptide antibiotic was shown to disrupt and solubilize lecithin liposomes.

The interaction of gramicidin S with dimyristoyllecithin bilayers has been studied by ^{31}P and ^2H NMR as well as by DSC.[59,60] At low concentrations, gramicidin S exhibited strong interactions with the bilayer, disordering both the polar phospholipid headgroups[60] and the hydrophobic acyl side chains.[59] Increasing the concentration of gramicidin S led to a disintegration of the bilayer membrane structure.

4.13.2.2   Tyrocidine and Linear Gramicidins

The tyrocidines produced by *B. brevis* ATCC 8185 (the Dubos strain; now *Bacillus parabrevis* instead of *B. brevis* ATCC 8185, according to the new nomenclature of the ATCC) and ATCC

10068 are cyclic decapeptides containing two phenylalanines in the D configuration (positions 1 and 4) as indicated in Figure 1.[61] The Val-Orn-Leu-D-Phe–Pro half of the tyrocidine molecule is identical to the respective amino acid sequence in gramicidin S. Tyrocidines B–D are isoforms of tyrocidine A with specific replacements of one (tyrocidine B, L-Trp versus L-Phe at position 3), two (tyrocidine C, L-Trp→L-Phe and D-Trp→D-Phe at positions 3 and 4), or three amino acid residues (tyrocidine D, the changes observed in tyrocidine C plus an L-Trp→L-Tyr exchange at position 7). The incorporation of amino acids in the D configuration as well as microheterogeneity are characteristic features of microbial low molecular weight peptides.[42,62] As is discussed in Section 4.13.3.2 for gramicidin S synthetase, the replacement of amino acid residues by molecules with a similar structure may be the result of a relatively low specificity of the biosynthetic machinery responsible for the enzymatic synthesis of peptide antibiotics.

In addition to the tyrocidines, the *B. brevis* strains ATCC 8185 and 10068 also produce the linear gramicidin A–C isoforms of a pentadecapeptide either with the aromatic amino acid residues L-Trp, L-Phe, or L-Tyr at position 11, respectively. The Ile gramicidins A–C contain specifically an L-Ile residue at position 1 instead of L-Val (see Figure 1). Because these compounds have modified N- and C-terminal ends, they were first thought to have cyclic structures as in the case of tyrocidine or gramicidin S. However, the complete structure elucidation gave evidence for linear structures in which the N terminus is formylated and the C-terminal end is amidated with ethanolamine.[11,63]

Linear gramicidins are membrane-active agents.[64–66] They form channels in biological membranes, planar bilayers, and lipid vesicles which are selective for monovalent cations. Because of these properties they have been used as prototypes of channel-forming molecules to investigate the mechanism of ion transport and lipid–protein interactions. Linear gramicidins can adopt various conformations in different lipid environments. A variety of models has been suggested for the structure of these agents in organic solvents and lipid membranes from spectroscopic and diffraction studies as well as from theoretical considerations. There is now convincing evidence that gramicidin channels are head-to-head (formyl-NH to formyl-NH) single-stranded β-helical dimers. In addition to their channel-forming properties, linear gramicidins also show strong modulating effects on lipid structure. For example, in bilayer-forming lipids, such as diacylphosphatidyl choline, they are able to induce phase transitions. For these effects their tryptophan residues seem to be essential, but the complete three-dimensional structure is also important.

Sarkar and Paulus[67] isolated a mutant of *B. brevis* which was unable to synthesize gramicidin D and which produced defective spores. The defects could be cured by the addition of exogenous gramicidin. Ristow *et al.*[68] demonstrated that sporulation of early vegetatively growing *B. brevis* cells can be induced by tyrothricin, a mixture of tyrocidine and gramicidin D, when the cells were exposed to a culture medium lacking the nitrogen source. The induction of sporulation occurred concomitantly with a significant increase in RNA synthesis. From these results the authors concluded that linear gramicidin and tyrocidine interact with the transcriptional process during sporogenesis and proposed that these compounds may play a positive regulatory role in cell differentiation. Research has been initiated to clarify the effects of both peptides at the molecular level. Tyrocidine inhibited RNA synthesis *in vivo* and *in vitro*, forming a complex with DNA and inhibiting the initiation of transcription. Gramicidin D inhibited RNA synthesis by destabilizing the initiation complex. Apparently by interacting with the DNA-binding site, gramicidin D can selectively modulate the affinity of RNA polymerase for different promoters. When gramicidin D was added to a tyrocidine-inhibited transcription system a partial reactivation of transcription was observed which was specifically dependent on the RNA polymerase from *B. brevis*. Studies on the molecular mechanism of the reactivation process suggested that gramicidin D interacts with the DNA–tyrocidine complex. Thus it is weakened, allowing the RNA polymerase to transcribe. Obviously, tyrocidine could act as a nonspecific repressor, turning off transcription, whereas gramicidin D would derepress a part of the genome allowing transcription to those genes which are involved in the spore-forming process.

4.13.3 BIOSYNTHESIS OF GRAMICIDIN S AND TYROCIDINE

4.13.3.1 Intermediate Steps and Product Pattern

In *B. brevis* cultures, the production of the peptide antibiotics gramicidin S as well as the tyrocidines appears in the late logarithmic growth phase (tropophase) and ceases at the beginning of the stationary growth phase (idiophase) before the onset of sporulation. The biosynthesis of the cyclodecapeptide gramicidin S is catalyzed by two large multienzymes, gramicidin S synthetase 1 (GS1) and 2 (GS2), and comprises at least 15 individual reaction steps, which are summarized in Scheme 1(a). Gramicidin S synthetase activates its substrate amino acids in a two-step process which is characteristic for nonribosomal peptide biosynthesis. The first step involves the nucleophilic attack of the α-carboxyl group of the amino acid substrate at the α-phosphate linkage of a Mg–ATP complex (Mg–ATP^{2-}), forming a highly reactive adenylate (reactions (1)–(5) in Scheme 1(a)). This mixed anhydride between the carboxylic substrate and AMP is stabilized by the enzyme in a noncovalent manner to prevent side reactions such as hydrolysis. As indicated in Scheme 2, mechanistically this reaction resembles the amino acid activation process catalyzed by aminoacyl tRNA ligases in the ribosomal system. However, the two classes of enzymes, though catalyzing the same amino acid activation reactions, do not share structural homologies. The second activation step distinguishes the nonribosomal pathway (step 2a) from ribosomal protein/peptide biosynthesis (step 2b). Peptide synthetases catalyze the incorporation of their amino acid substrates as reactive aminoacyl thioesters at internal sulfhydryl groups (the "thiotemplate sites"; reactions (6)–(10) in

(a) Biosynthesis of gramicidin S

(b) Biosynthesis of tyrocidine

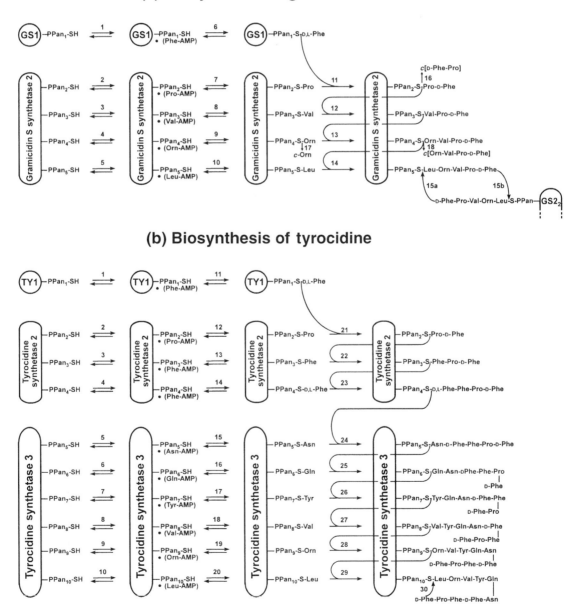

Scheme 1

Scheme 1(a), as well as step 2a in Scheme 2). In contrast, aminoacyl tRNA ligases do not contain thiolation centers. Instead of an intramolecular reaction, the adenylated substrates are linked to the 2′- or 3′-hydroxy groups at the 3′-terminal end of the cognate tRNA molecules as oxygen esters (Scheme 2, step 2b).

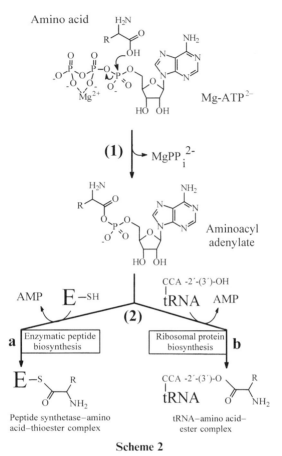

Scheme 2

The thioester-binding sites of the peptide-forming multienzymes are represented by enzyme-bound 4′PPan cofactors (numbered 1–5), one at each amino acid activation unit (biosynthetic module) of gramicidin S synthetase,[69] as outlined in detail in Section 4.13.6.2. Following the multiple carrier concept,[69] which is discussed in detail in Section 4.13.7, the polymerization of the activated amino acid substrates occurs by specific interaction of adjacent 4′PPan cofactors in a series of transpeptidation steps (Scheme 1(a), reactions (11)–(14)). The sequence of amino acid-activating modules on the multienzymatic polypeptide chains defines the sequence of the amino acids within the peptide product. GS1 represents a single biosynthetic unit with a molecular mass of 127 kDa. It activates and racemizes Phe, and has been classified as Phe racemase, EC 5.1.1.11.[70–72] The elongation cycle starts with the transfer of the D-Phe enantiomer to the condensing enzyme GS2. This enzyme comprises four biosynthetic modules activating the other four amino acid constituents of gramicidin S, Pro, Val, Orn, and Leu (if not indicated, always the L configuration),[73–75] and catalyzes the synthesis of a pentapeptide.[76,77] Two of these pentapeptide moieties are finally cyclized by two head-to-tail condensations (reactions (15a) and (15b) in Scheme 1(a)), forming the cyclo-decapeptide antibiotic. The mechanism of the termination reaction still remains to be clarified. The intermolecular cyclization, which implies the interaction of two GS2 molecules,[78] is shown in Scheme 1(a). Also, an intramolecular mechanism[79] has been discussed which requires the postulation of a specific waiting position for the first pentapeptide moiety within GS2.

The biosynthesis of tyrocidine is accomplished by three multienzymes, tyrocidine synthetases 1, 2, and 3 (TY1, TY2, and TY3), which catalyze at least 30 individual reactions (see Scheme 1(b)).[80–82] TY1 and GS1 are isoenzymes, both performing the activation and racemization of Phe (reactions (1) and (11) in Scheme 1(b) for the biosynthesis of tyrocidine A). TY2 activates three amino acids (one Pro and two Phe residues, reactions (2)–(4) and (12)–(14) in Scheme 1(b)) and elongates the peptide chain, yielding the tetrapeptide D,L-Phe-Phe-Pro-D-Phe (Scheme 1(b), reactions

(21)–(23)), which is subsequently transferred to TY3. This large multienzyme activates the remaining amino acid constituents of tyrocidine (Scheme 1(b), reactions (5)–(10) and (15)–(20)), polymerizes them to a linear decapeptide (Scheme 1(b), reactions (24)–(29)), and catalyzes the cyclization to the decapeptide product tyrocidine A (reaction (30)).[83]

Gramicidin S synthetase shows a complex product pattern. In Scheme 3 the side reactions of this multienzyme (reactions (16)–(18) in Scheme 1(a)), sterically favored cyclization processes involving intramolecular aminolysis, which lead to the dissociation of biosynthetic intermediates, are summarized. D-Phenylalanyl-L-prolyl-piperazinedione (cyclo-D-Phe-Pro; see Scheme 3(a)), formed by an internal cyclization by nucleophilic attack of the amino group of D-Phe on the carboxyl group of the thioester bound L-Pro, had already been observed at an early stage of investigation of the gramicidin S biosynthetic process.[84]

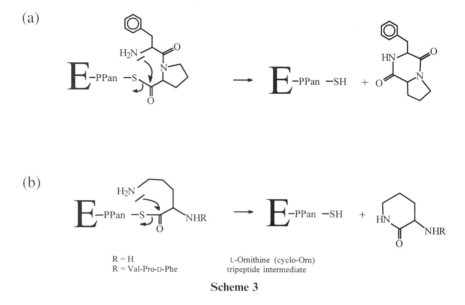

(a)

(b)

R = H
R = Val-Pro-D-Phe

L-Ornithine (cyclo-Orn)
tripeptide intermediate

Scheme 3

The thioester-bound Orn as well as the tetrapeptide intermediate D-Phe-Pro-Val-Orn can be released from GS2 by an internal nucleophilic addition of the δ-amino group of Orn to its thioesterified carboxyl group.[85,86] As indicated in Scheme 3(b), in this way Orn can be converted into 3-amino-2-piperidone (cyclo-L-Orn) in a rapid cyclization process. Similar side reactions can be expected for tyrocidine synthetase. This multienzymatic system also experiences the dissociation of biosynthetic intermediates by intramolecular aminolysis as well as formation of D-phenylalanyl-L-prolyl-piperazinedione (cyclo-D-Phe-Pro), cyclo-Orn, and the cyclo-Orn-Val-Tyr-Glu-Asn-D-Phe-Phe-Pro-D-Phe nonapeptide, as possible side reactions.

4.13.3.2   Substrate Specificity of Gramicidin S Synthetase

The available experimental data imply that both amino acid activation reactions, adenylation and thioester fixation, are reversible processes. The reverse reaction of the amino acid-dependent cleavage of ATP, the incorporation of ^{32}PP$_i$ into ATP, is the most widely used detection technique for testing the reactivity of peptide synthetases. However, in many cases peptide synthetases and aminoacyl tRNA ligases catalyze the adenylation of the same amino acids. For the detection of peptide synthetases in crude extracts or in the course of the purification of a specific multienzyme of this class it would be helpful to have ATP analogues available which are specific for either system. For this purpose, different nucleotide triphosphates were probed for their ability to replace ATP in the synthesis of gramicidin S.[87] Compared with 100% gramicidin S formation in the case of the natural substrate ATP, 2′-deoxy-ATP (80%), 7-deaza-ATP (77%), 2-chloro-ATP (29%), 3-deoxy-ATP (26%), and, surprisingly, ADP (13%) show a reduced biosynthetic rate. In particular, the affinity of peptide synthetases for deoxy-ATP as well as 7-deaza-ATP, which are not accepted by aminoacyl tRNA ligases,[88] can be used to discriminate between both classes of amino acid-activating enzymes.

Peptide-forming multienzymes show a relatively broad specificity regarding the recognition and binding of their amino acid substrates. Usually they are less specific in their acceptance of substrate analogues than aminoacyl tRNA ligases in the ribosomal system, which contains a specific proof-reading mechanism to avoid the incorporation of incorrect amino acids into the protein products.

As indicated in Table 1, gramicidin S synthetase has been tested for its ability to activate a large number of substrate analogues as adenylates or thioesters as well as for the incorporation of these compounds into a gramicidin S analogue peptide product. In general, the activation of an amino acid substrate or one of its analogues coincides with its incorporation into a gramicidin S-like peptide product. Some compounds, which are readily activated but not used for product formation, are the Phe analogues *threo*-phenylserine and *β*-2-thionylserine as well as some D isomers of the naturally occurring substrates or substrate analogues (D-Leu, D-Ile). The structurally similar amino acids L-Leu and L-Val can be bound to either thiotemplate site. The thioester-binding site of L-Leu is able to bind L-Val more tightly than L-Leu is bound to the thiotemplate of L-Val. Substrate analogues with branched side chains, such as L-Ile and, especially L-allo-Ile, show a higher affinity for the thiotemplate sites of both L-Val and L-Leu than the linear derivatives L-norvaline or L-norleucine. Substrate analogues which are structurally related to the natural substrates by replacement of a hydrophobic structural element by a polar group, such as L-Thr, are also accepted by the thiotemplate sites.

The formation of the D-Phe-Pro-diketopiperazinedione by-product catalyzed by GS1 and GS2 has been used for the production of several 2,5-piperazinediones by *in vitro* assays.[62] As summarized in Table 2, the nature of the amino acid activated by GS1 instead of Phe and by GS2 instead of L-Pro determines the tendency for the intramolecular cyclization of the intermediary dipeptide intermediate. In particular, if *γ*-thioproline as well as 3,4-dehydroproline and azetidine-2-carboxylic acid are applied in combination with the natural substrate Phe, high rates and yields of 2,5-piperazinedione formation are obtained.

4.13.4 PURIFICATION OF GRAMICIDIN S SYNTHETASE

Various purification techniques for the gramicidin S synthetase multienzyme system have been developed in several laboratories.[89–92] In general, combinations of gel filtration, sucrose gradient ultracentrifugation, and ion exchange and affinity chromatography procedures were used (for a review, see Zimmer and Laland[93]). The observation of Kurahashi and co-workers[94] that a cell-free crude extract of *B. brevis*, which catalyzed the biosynthesis of gramicidin S can be divided into two complementary fractions by gel filtration on Sephadex G200 was of considerable importance for the further purification progress. The light enzyme fraction which activates and racemizes Phe corresponds to GS1. The residual amino acid constituents of gramicidin S were activated by a heavy enzyme fraction containing GS2, which in addition catalyzed the condensation reactions in gramicidin S formation. Several groups applied affinity chromatography as an efficient purification step for gramicidin S synthetase. In combination with gel filtration, both components of this multienzyme were successfully purified by chromatography on 3,3′-diaminodipropylamine-substituted Sepharose as well as Orn- or Phe-Sepharose 4B.[95–99] However, the results of these experiments suggest that the interaction of gramicidin S synthetase with the respective affinity matrix may occur not only by affinity interactions. Adsorptive effects of the spacer molecules of the matrix as well as ion exchange effects have also been discussed in this context.[99]

Gramicidin S synthetase contains specific thiol groups within the catalytic centers for substrate activation which are inhibited by certain disulfides. This property has been used for coupling of GS1 to cystamine-Sepharose 6B. The immobilized enzyme could be cleaved from the matrix either by disulfide reduction with thiol reagents or by a salt step/gradient.[100]

Important progress in the purification of gramicidin S synthetase has been achieved by high-resolution anion exchange fast liquid chromatography (FPLC) on Mono Q[86] and Fractogel TMAE-650.[87] By these procedures, highly purified enzymes have been obtained which show specific activities in the range of 30–40 nmol of gramicidin S produced per minute and per milligram of enzyme.

4.13.5 GRAMICIDIN S AND TYROCIDINE BIOSYNTHETIC GENES

The analysis of the complex gene structures for the biosynthesis of antibiotics was initiated in the 1980s.[101–105] Isolation of the genes encoding the multienzymes involved in the biosynthesis of gramicidin S and tyrocidine was approached by the groups of Marahiel and Saito. Sources for the chromosomal DNA for isolation of the genes coding for the biosynthesis of gramicidin S were the

Table 1   Amino acid activation pattern of the biosynthetic modules of gramicidin S synthetase.

Enzyme	Module	Substitution[a]	Activation	Cyclodecapeptide synthesis[b]
GS1	Phenylalanine	D-Phenylalanine	+ +	+ +
		DL-*p*(*o*,*m*)-Fluorophenylalanine	+ +(+ +,+ +)	+ +(+ +,+ +)
		Tyrosine	+ +	+ +
		D-Tyrosine	+ +	+ +
		DL-*o*-Tyrosine	+ +	+
		DL-*m*-Tyrosine	+ +	+ +
		o-Methyltyrosine	+	+
		Tryptophan	+ +	+
		5-Methyltryptophan	+	+
		6(7)-Methyltryptophan	+(+)	−(−)
		β-Phenyl-β-alanine	+	+
		β-2-Thienyl-DL-alanine	+ +	+ +
		DL-*p*-Aminophenylalanine	+ +	+
		threo-Phenylserine	+ +	−
		β-2-Thienylserine	+	−
GS2	Proline	3,4-Dehydroproline	+ +	+ +
		Azetidin-2-carboxylic acid	+ +	+ +
		Hydroxyproline	+	+ +
		γ-Thioproline	+	+ +
		DL-β-Thioproline	+	+ +
		Allo-4-hydroxyproline	+	+ +
		Sarcosine	+	+
		Pipecolic acid	−	−
	Valine	Isoleucine	+ +	+ +
		Norleucine	+ +	+ +
		Norvaline	+ +	+ +
		Alloisoleucine	+ +	+ +
		Leucine	+	+ +
		Threonine	+	+
		D-Valine	−	−
		D-Isoleucine	+	−
		D-Leucine	+	−
	Ornithine	Lysine	+	+ +
		Arginine	+	+ +
		D-Ornithine	+	−
		α-*N*-Acetylornithine	+	−
		2,4-Diaminobutyric acid	+	−
		Diaminopropionic acid	−	−
	Leucine	Isoleucine	+ +	+ +
		Norleucine	+ +	+ +
		Alloleucine	+ +	+ +
		D-Leucine	+ +	−
		D-Isoleucine	+	−

[a]If not indicated, always the L configuration.   [b]Rate of incorporation into gramicidin S: + +, >10%; +, >1%; −, not (<1%).

ATCC 9999 and the Nagano strains of *B. brevis*, respectively. A breakthrough was obtained by Marahiel *et al.*[106] and Krause *et al.*[107] by the cloning of the entire structural gene of TY1 from *B. brevis* ATCC 8185 and of an Orn-activating fragment of the GS2 gene from *B. brevis* ATCC 9999 in *Escherichia coli*. At the beginning of the 1990s the complete gene structure coding for the biosynthesis of gramicidin S and a part of the tyrocidine biosynthetic genes had been cloned and sequenced.[43,108–112] From these studies it became evident that the genes encoding these peptide synthetases are clustered and organized as operons similar to the gene structures for the biosynthesis of numerous other polyketide and peptide antibiotics.[101–105]

As illustrated in Figure 3, in the *grs* operon encoding the enzymes for the biosynthesis of gramicidin S, three open reading frames, *grsT*, *grsA*, and *grsB*, have been identified.[109] They are located on a DNA fragment of the *B. brevis* genome with a total length of 19 000 bp. *grsA* and *grsB* represent the structural genes for the two components of gramicidin S synthetase, GS1 (the GrsA protein) and GS2 (the GrsB protein). The *grs* operon is expressed postexponentially from a σ^B-like promoter which is located 81 bp upstream from *grsT*, the first open reading frame at the 5′ end of this gene cluster with a size of 768 bp. It encodes a protein of 256 amino acid residues with a

Table 2 Production of 2,5-piperazinediones by gramicidin S synthetase.

Activation site (amino acids used)		Rate of 2,5-piperazinedione formation[a]	Rate of cyclodecapeptide synthesis
Phenylalanine (GS1)	Proline (GS2)		
Phenylalanine	Proline	20	100
Phenylalanine	3,4-Dehydroproline	91	85
Phenylalanine	Azetidin-2-carboxylic acid	73	58
Phenylalanine	Hydroxyproline	11	15
Phenylalanine	γ-Thioproline	179	12.5
Phenylalanine	Allo-4-hydroxyproline	8.4	10.5
Phenylalanine	Sarcosine	22	0.7
β-Phenyl-β-alanine	Proline	2	2
5-Methyltryptophan	Proline	2	1
6-Methyltryptophan	Proline	2	<1
7-Methyltryptophan	Proline	1	<1
Tryptophan	Proline	16	4
D-Tyrosine	Proline	7	35
Tyrosine	Proline	7	29
DL-o-Tyrosine	Proline	4	3.5
DL-m-Tyrosine	Proline	5	20
o-Methyltyrosine	Proline	32	1
β-2-Thienyl-DL-alanine	Proline	21	40

[a]All rates are compared to the rate of gramicidin S formation (= 100%).

molecular mass of 29 191 Da of still unknown function which shows significant homology to eukaryotic thioesterases of type II.[113,114] After an untranslated region of 26 nucleotides, the *grsA* gene follows. It comprises 3294 nucleotides coding for GS1 composed of 1098 amino acid residues with a molecular mass of 126 663 Da. The *grsB* gene is located 71 bp downstream of the *grsA* stop codon. It contains an open reading frame of 13 359 (13 350) bp encoding GS2 composed of 4453 (4450) amino acids with a mass of 510 287 (508 658) Da. About 20 bp downstream of the *grsB* stop codon a sequence resembling a ρ-independent transcription termination site was detected. The data cited first were obtained by Marahiel and his colleagues,[109,111] while the data of the Japanese group[110,112] are given in parentheses. Similar results have been obtained by both groups. Saito's group, which isolated the *grs* operon from the *B. brevis* Nagano strain, found 94 bp mismatches in their *grsB* gene compared with the *grsB* sequence obtained from the ATCC 9999 strain, causing a variation of 132 amino acids in their GS2 protein and a difference of three amino acids in its total length.[112]

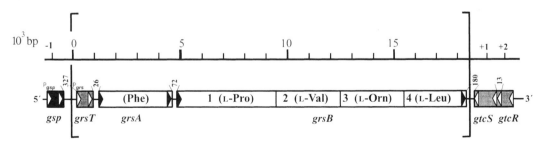

Figure 3 Structure of the *grs* operon of *B. brevis* responsible for gramicidin S biosynthesis and adjacent gene structures.

Near the 5' end of the *grs* operon the *gsp* gene was found to be closely linked with the gramicidin S biosynthetic genes.[115] It is located 326 bp upstream from the initiation codon of *grsT*, and codes for a protein of 237 amino acids with a predicted molecular mass of 27 856 Da. A putative ribosome-binding site precedes the *gsp* open reading frame, and is located about 57 bp downstream of the transcription initiation site. The Gsp protein shows 34% identity and 52% similarity to the Sfp gene product which is involved in the biosynthesis of the lipoheptapeptide surfactin by *Bacillus subtilis*. Both Gsp and Sfp have been demonstrated to be members of a new enzyme superfamily, the phosphopantetheine transferases.[116] Such enzymes play an essential role in nonribosomal peptide biosynthesis by transformation of the nascent inactive apo forms of peptide synthetases into their holo forms. This posttranslational event involves the transfer of 4′PPan moieties from coenzyme A molecules to each biosynthetic module of the peptide synthetases.

Adjacent to the 5′-terminal end of the *grs* operon is located the *gtcRS* operon (*grs*-associated two-component system), which codes for a two-component signal transduction system. The first gene product, GtcR, represents the response regulator protein; the second gene, *gtcS*, encodes the corresponding sensor protein of the two-component system. However, a possible linkage to the genes responsible for gramicidin S biosynthesis is still under investigation.[117]

The gene products of the *grsA* and *grsB* genes were identified by functional characterization and cross-reaction with specific antibodies raised against both components of gramicidin S synthetase.[108,110,112,117] The GrsA protein was expressed in *E. coli* and purified to an essentially homogeneous state. It catalyzed the D-Phe-dependent ATP–PP$_i$ exchange with nearly the same specific activity as the purified GS1 from *B. brevis*, but showed only a weak Phe racemase activity with less than 3% of the wild-type enzyme.[108,110] The GrsB gene product, GS2, has a pronounced modular structure. It can be divided into four conserved and repeated segments of high homology comprising about 1000–1200 amino acids.[111,112] The first domain of this large multienzyme was identical to a Pro-activating fragment which had been cloned and sequenced previously by Hori *et al.*[118] Translation products of subcloned fragments containing the third and fourth modules of GrsB catalyzed Orn- and Leu-dependent ATP–PP$_i$ exchange reactions, respectively.[112] The second and fourth domains of GS2 show a high similarity in their entire sequence. In addition, the amino acid sequences of the proteins derived from the *grsA* and *grsB* genes are consistent with the N-terminal amino acid sequences obtained for the purified GS1 and GS2 proteins.[119,120] From these features it was demonstrated that the sequence of the amino acid-activating modules are arranged along the multifunctional peptide chains of gramicidin S synthetase colinearly with the sequence of the amino acid components in the peptide product. This colinearity rule has been confirmed for all peptide-forming multienzyme systems that have so far been characterized in detail.

Genes similar to *grsT* of *B. brevis* are *srfAD* found in the *srfA* operon encoding the biosynthesis of surfactin in *B. subtilis*[28,29] and the two open reading frames (ORFs 1 and 2) in the bialaphos biosynthetic gene cluster of *Streptomyces hygroscopicus*.[121] Information on the nature of the corresponding proteins has been obtained from sequence comparisons. The proteins show a significant degree of homology of more than 30% to medium-chain *S*-acyl fatty acid synthetase thioester hydrolases (thioesterases II) from rat mammary gland (29 471 Da)[113] and mallard duck uropygial gland (28 800 Da),[114] and both enzymes catalyze the release of the thioester-bound fatty acid products linked to the 4′PPan prosthetic group of an acyl carrier domain. In this process an active Ser is involved as part of the consensus motif GHSFG, which is also highly conserved in the GrsT protein (GHSMG extending from position 99 to 103).[109]

For the detection of the tyrocidine biosynthetic genes, ~4500 recombinant phages containing inserts of the *B. brevis* ATCC 8185 DNA were screened by *in situ* immunoassays using antibodies raised against both components of gramicidin S synthetase.[106,122] These antibodies show cross-reactions with the structurally related TY1 and TY2 enzymes (the TycA and TycB proteins). From a genomic library of this organism, a recombinant phage was detected which contained the structural genes coding for both proteins. Immunoreactive proteins were detected in cell lysates by anti-GrsB antibodies which exhibited cross-reactions with two proteins of ~120 and 190 kDa, respectively. The 120 kDa protein was identified enzymatically by the specific D-Phe-dependent ATP–PP$_i$ exchange reaction as TY1. This multienzyme shows a 56% homology to GS1, which increases to almost 70% if similar amino acids are considered in the sequence comparison. The TycA protein is 11 amino acids shorter than GS1. The 190 kDa protein contains the Pro- and Phe-activating domains of TycB. The noncoding region of 94 bp which separates the *tycA* and *tycB* genes revealed no consensus sequences of RNA polymerase-binding sites.

The *tyc* operon is expressed from a σ^{43}-like promoter. Its transcription depends on the SpoOA, SpoOB, and SpoOR gene products. The complete *tyc* operon has also been cloned and sequenced.[123] The structural genes *tycA*, *tycB*, and *tycC* which code for the three enzyme components of tyrocidine synthetase comprise 37 000 bp of DNA in total. Approximately 4000 bp downstream from the *tycC* gene, the *tycTE* gene was found, corresponding to GrsT and SrfAD in the gramicidin S and surfactin systems, which are homologous to thioesterases from vertebrates of type II.

From gene sequencing, precise information on the molecular masses of peptide synthetases was obtained. The data derived from the *grsA* and *grsB* genes revealed that the molecular masses of GS1 (100 kDa) and GS2 (280 kDa), which were originally obtained by sucrose density gradient centrifugation[76] and sodium dodecyl sulfate–polyacrylamide gel electrophoresis (SDS–PAGE)[124] have by far been underestimated, presumably because of the lack of suitable high molecular marker proteins in the mass range of several hundred kilodaltons. The mass numbers for both components of gramicidin S synthetase deduced from the gene sequences correspond to the apo forms of these proteins. Accounting for one 4′PPan cofactor for GS1 and four 4′PPan carriers for GS2, which are attached posttranslationally to these multienzymes, molecular masses of 127 003 Da and 511 648 Da were calculated for the active holo forms of GS1 and GS2, respectively. Similar data have been obtained for both proteins by a careful revised study by SDS–PAGE.[119,125] A molecular mass of 509.2 ± 0.5 kDa has been measured for GS2 by matrix-assisted laser desorption mass spectrometry (MALDI-MS) using an erbium–yttrium aluminum garnet (Er–YAG) infrared laser (Figure 4),[126] which matches the gene-derived mass number with high precision.[127,128] GS2 is the largest protein that has so far been desorbed into the gas phase without fragmentation, demonstrating the high potential of mass spectrometry in protein chemistry attained since the 1980s.

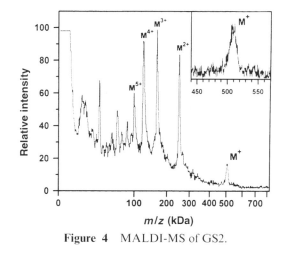

Figure 4 MALDI-MS of GS2.

4.13.6 THE MODULAR STRUCTURE OF GRAMICIDIN S SYNTHETASE

The genetic data so far available for peptide-forming multienzyme systems have revealed that peptide synthetases are distinguished by a characteristic modular organization,[42,44–47,129,130] as demonstrated in Figure 5 for some representative peptide synthetases. Multienzymes, such as the gramicidin S and tyrocidine synthetases, consist of homologous biosynthetic units comprising 1000–1500 amino acid residues. Gramicidin S synthetase is composed of five of these biosynthetic modules. GS1 is a one module enzyme, while GS2 can be divided into four repeated, homologous segments. Tyrocidine synthetase is organized into 10 of such biosynthetic units, one in TY1, three in TY2, and six in TY3. The modules of both multienzyme systems are arranged colinearly with the amino acid components in the peptide products. By limited proteolysis of GS2, it has been demonstrated that its biosynthetic units function as independent enzymes which recognize, bind, and activate their cognate amino acids.[120,131,132] Each of the modules is distinguished by a linear array of highly conserved consensus motifs that are part of the functional domains involved in the biosynthetic process. These reactive structural elements are arranged in series along the multifunctional polypeptide chain like assembly points on a production line. Peptide synthetases, therefore, can be regarded as factories at cellular dimensions. At least 18 of such motifs (named in alphabetical order A–R) have been identified by sequence comparisons of the so far known multienzyme structures.[28,29,32,37,39,133–137] Some of them have been functionally characterized in detail by affinity labeling[45,69,125,128,138–141] and site-directed mutagenesis[45,129,142–146] as well as specific dissection of the gene structures and expression of the truncated proteins in heterologous host systems.[146–148] A summary of these results is given in Figure 6.

The biosynthetic modules of peptide synthetases can be divided into three main sections, responsible for: (i) substrate recognition and binding as well as amino acid adenylation (dark gray in Figure 6); (ii) thiolation of the amino acid substrates (black); and (iii) peptide elongation and, in some cases, also amino acid epimerization (light gray). Each thiolation domain contains a 4′PPan cofactor attached to a strictly conserved Ser, and is involved in an acyl/peptidyl carrier protein (PCP) in the elongation process. Dissection studies[146–148] and the specific interaction of separated modules obtained by genetic engineering[148] gave experimental evidence that the functional domains a–c can be regarded as independent enzymes. Both the adenylation and thiolation domains

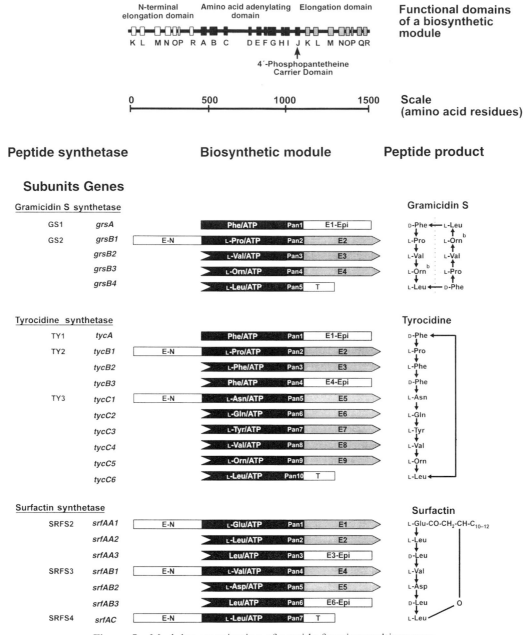

Figure 5 Modular organization of peptide-forming multienzymes.

of peptide synthetases belong to specific enzyme superfamilies. The adenylation domains are members of a class of adenylate-forming proteins which include firefly luciferase[149,150] and acyl-CoA ligases[151–154] (for a review, see Fulda *et al.*[155]). The PCP domains show a high homology and structural similarities to ACPs involved in fatty acid and polyketide biosynthesis.[45,116,147,148,156–161]

The elongation/epimerizing domains which comprise about 350 amino acid residues seem to be unique functional units that have been evolved specifically in nonribosomal peptide biosynthesis. However, De Crécy-Lagard *et al.*[162] have outlined that motif M (HHIxxDGW), the signature sequence of these species, is involved in acyl transfer reactions of chloramphenicol acyltransferases[163,164] and dehydrolipoamide acyltransferase components of oxo-acid dehydrogenase multienzyme complexes[165] too. Obviously, also the elongation domains of peptide synthetases are related to a superfamily of acyltransferases utilizing the acid–base properties of the conserved histidines of motif M as proton donor/acceptor systems in the catalysis of acyl transfer processes.

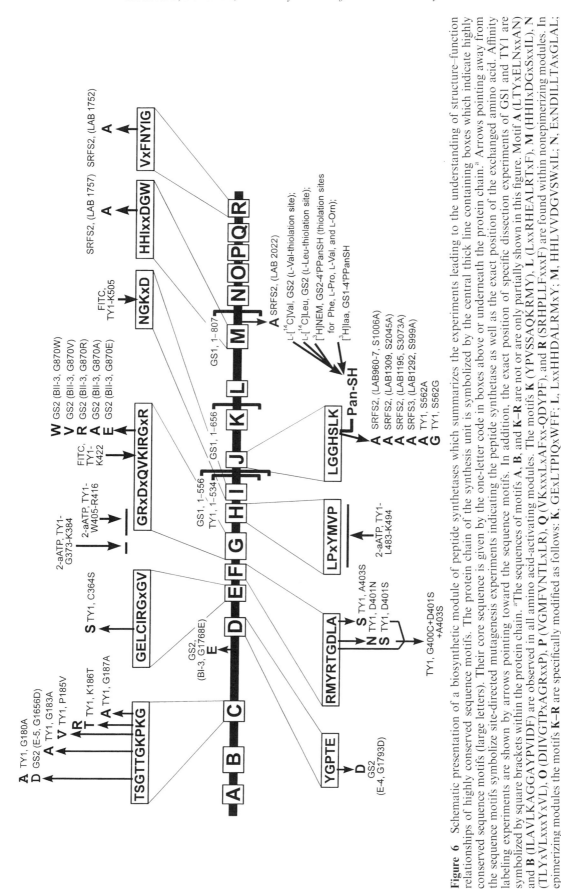

Figure 6 Schematic presentation of a biosynthetic module of peptide synthetases which summarizes the experiments leading to the understanding of structure–function relationships of highly conserved sequence motifs. The protein chain of the synthesis unit is symbolized by the central thick line containing boxes which indicate highly conserved sequence motifs (large letters). Their core sequence is given by the one-letter code in boxes above or underneath the protein chain.[a] Arrows pointing away from the sequence motifs symbolize site-directed mutagenesis experiments indicating the exchanged amino acid. Affinity labeling experiments are shown by arrows pointing toward the sequence motifs. In addition, the exact position of specific dissection experiments of GS1 and TY1 are symbolized by square brackets within the protein chain. [a]The sequences of motifs **A**, **B**, and **K–R** are not or are only partially shown in this figure. Motif **A** (LTYxELNxxAN) and **B** (ILAVLKAGGAYPVIDF) are observed in all amino acid-activating modules. The motifs **K** (YPVSSAQKRMY), **L** (LxxRHEALRTxF), **M** (HHIIxDGxSxxIL), **N** (TLYxVLxxxYxVL), **O** (DIIVGTPxAGRxxP), **P** (VGMFVNTLxLR), **Q** (VKxxxLxAFxx-QDYPF), and **R** (SRHPLLFxxxF) are found within nonepimerizing modules. In epimerizing modules the motifs **K–R** are specifically modified as follows: **K**, GExLTPIQxWFF; **L**, LxxHHDALRMxY; **M**, HHLVVDGVSWxiL; **N**, ExNDILLTAxGLAL; **O**, LEGHGREII; **P**, SRTVGWFTSMYP; **Q**, VPxKGVGYGILxY; and **R**, PxxxFNYLGQF.

4.13.6.1   The Aminoacyl Adenylation Domain

In particular, the first 600 amino acid residues of each module of gramicidin S synthetase show 45–50% identity and a high degree of similarity (20–70%) to numerous other bacterial and fungal peptide synthetases, as well as to the superfamily of adenylate-forming enzymes of diverse origin. At least nine consensus motifs (A–I) have been identified in the aminoacyl adenylation domains by sequence alignment of numerous multifunctional peptide synthetases. Motifs A and B have not been attributed to specific functions so far. The highly conserved motif C (SGTTGxPKG) shows similarities to the phosphate-binding loop (P loop or Walker A motif), which is found in all guanosine- and some adenosine-binding proteins.[166–168] The three-dimensional structure analysis of a 60 kDa fragment of GS1 revealed that sequence motif C is located within a flexible region formed by a distorted loop connecting two antiparallel strands of a β-sheet structure of this protein, whereas most of motif C is disordered.[48] The amino acid composition of the Gly-rich motif C suggests high flexibility and a susceptibility of this functional domain to conformational changes which may control access to the substrate-binding site, modify binding affinities of substrates, or relocate the catalytically active groups at the aminoacyl adenylate activation centers. Substitution of K186 in TY1 by Arg reduced the original activity of this enzyme to 25%, while replacement by the neutral amino acid L-Thr led to an almost complete loss of the aminoacyl adenylation of phenylalanine.[145,169] Though in both cases the catalytic activity of this protein is strongly affected, no significant effect on the binding affinity of these mutants for Phe was observed. Obviously K186 of motif C plays an important role in either stabilization of the bound substrate by charge–charge interactions or fixing of the loop for maintenance of the active site conformation or both. When the strictly conserved G180 of TY1 was exchanged by the structurally related Ala residue the Phe-dependent ATP–PP$_i$ exchange activity was not significantly affected (82% of the wild-type enzyme).[145] However, the replacement of this Gly residue by the negatively charged Asp residue (G1656D) in the Val-activating domain of GS2 from the *B. brevis* Nagano mutant strain E-5 completely abolished adenylate formation.[170]

The Gly→Glu exchange G1768E in the region between motifs C and D as well as the Gly→Asp substitution in motif D (G1793D) of the Val-activating module of GS2 from the *B. brevis* Nagano BI-3 mutant strain gave rise to the complete loss of the adenylation activity.[170] Secondary structure analysis showed that as a consequence of the G1768E mutation a β turn is lost, which is a highly conserved structural element among adenylate-forming enzymes, while in the case of the G1793D replacement a structural change was not observed. Nevertheless, G1793 is an indispensable functional element in aminoacyl adenylate formation. These results imply that G1768 between motifs C and D is important for maintenance of the three-dimensional structure in this domain and that motif D is essential for aminoacyl adenylate formation.[170]

Motif E contains a Cys residue that is highly conserved in bacterial peptide synthetases. When this residue in TY1 was replaced by Ser the Phe-dependent ATP–PP$_i$ exchange activity was not significantly affected.[145] The function of motif E still has to be clarified in more detail. The central part of motif F resembles the highly conserved TGD motif, which represents an essential structural element of cation-pumping ATPases forming a phosphorylated intermediate. Affinity labeling of these enzymes with 2-azido-ATP[171] in combination with site-specific mutagenesis of the yeast plasma ATPase within the TGD core[172] demonstrated that this motif is involved in ATP binding. Analogous experiments have been performed with peptide synthetases.

Mutation of the Asp residue in the TGD motif in TY1 by Asn and Ser (D401N and D401S) decreased the adenylation activity to 78% and 12%, respectively. A triple mutation (G400C, D401S, and A403S) led to the complete loss of the activity of the enzyme, presumably because of a change of the tertiary structure at the reaction center.[145] After affinity labeling of TY1 with the fluorescent ATP-derivative 2-azido-ATP and subsequent tryptic digestion of the resulting complex, a fluorescent peptide fragment was isolated which is located near consensus motif F (G373–K348),[139] supporting the view that the TGD motif is part of an ATPase reaction center in peptide synthetases.

Five mutants with substitutions of the Gly residue of motif G in position 870 of the Pro-activating module of GS2 (*B. brevis* Nagano, mutant BII-3) showed a significant or total loss of Pro adenylation compared with the wild-type strain.[144] Secondary structure analysis of this region of the enzyme indicated that the mutated Gly residue is an essential component of a β loop which is interspersed between two α helices.

All point mutations at this position led to a loss of the loop structure. Affinity labeling and analysis of active site peptides support the view that motifs G, H, and I are involved in nucleotide binding. With 2-azido-ATP as a probe, the active site peptide of TY1 was identified as W405–R416, which is a part of motif G.[140] In addition, another fragment (L483–K494) with labeling in motif H was found. Using FITC the highly conserved Lys residues K422 (motif G) and K505 (motif I) were modified.[139] In the luciferase structure the reaction center is located at the interface between the two structural subdomains of the enzyme, which are connected by a distorted loop comprising residues 436–440.[150] This linker region accommodates R437, which is one of the highly conserved residues of this protein corresponding to R416 in TY1.[169] An R416T mutation decreased the catalytic efficiency of the mutant enzyme. The linker loop is the preferential target for limited proteolysis of TY1 by proteases, such as proteinase K and trypsin.[169] In the presence of the substrates ATP and Phe, the cleaving rate was significantly reduced. Obviously, aminoacyl adenylate formation restricted the access of proteases to their cleaving site. A still higher rate of proteolysis was observed in the case of the R416T mutant without change of the fragment pattern.[169] However, in this case the presence of substrates did not protect the enzyme against degradation.

By specific dissection of the *grsA* and *tycA* genes between motifs I and J and heterologous expression of the related proteins (for GS1, residues 1–556; for TY1, residues 1–534) catalytically active fragments were obtained. The truncated proteins are similar in size and composition to other members of the adenylate-forming superfamily including consensus motifs (A–I). These fragments were deficient in thioester formation, but still showed aminoacyl adenylation of phenylalanine.[145–147]

Certain bioactive peptides of eukaryotic origin, for example the depsipeptide enniatin and the lipopeptide cyclosporin A, contain several *N*-methylated amino acid residues. The related peptide synthetases catalyze the *N*-methylation reaction of these residues using *S*-adenosylmethionine as the methyl donor at the thioesterified amino acid stage.[35,36]

Modules containing the *N*-methylation activity can be considered as hybrids between normal modules and *N*-methyltransferases.[39] They have been classified as type II modules.[44] Here a segment of ∼450 amino acid residues is inserted between motif G and H of the adenylation domain.[37,39] This segment, especially the Gly-rich motif (E/D)x(G/F)xG, shows similarities to SAM-dependent methyltransferases specific for proteins, rRNAs, and DNA.[39] The *N*-methylation domain is integrated into the last fourth of the adenylation domain between motif G and H, both motifs being involved in nucleotide binding.[139,140,144] It has yet to be elucidated whether the *N*-methylation of the substrates has already occurred at the adenylated substrate stage.

4.13.6.2 The Aminoacyl Thiolation Domain

4.13.6.2.1 *4′-Phosphopantetheine cofactors represent the thiotemplate sites*

The key to detailed mechanistic insight into the chain elongation process of nonribosomal peptide biosynthesis was the structure elucidation of the thiotemplate sites by Stein *et al.*[69,128] These catalytic centers, obligate structural elements of the nonribosomal system, were identified by analyzing peptide fragments of the thiolation centers of gramicidin S synthetase.

On the basis of the results of thioester-binding studies of gramicidin S synthetase and its characteristic inhibition patterns, two techniques for specific labeling of the thiolation sites of gramicidin S synthetase were developed, as indicated in Table 3. In the first protocol (Table 3(a)) the radiolabeled substrate amino acid is attached directly to the thioester-binding site. In the second procedure, one of the thiotemplate sites of gramicidin S synthetase was alkylated by a radiolabeled SH modifier (e.g., *N*-ethylmaleimide (NEM)) in a nucleophilic addition reaction of the 1,4 Michael type. To achieve selective labeling of only one thiolation site of GS2, a multistep procedure was necessary (Table 3(b)). After protection of one thiolation site with its amino acid substrate, the reactive sulfhydryl groups in the remainder of the molecule were alkylated applying millimolar concentrations of the nonradioactive SH modifier. The reactive multienzyme–thioester complex was hydrolyzed by dithioerythritol. Thereafter, the deblocked thiolation site could be specifically labeled with lower concentrations of the SH modifier in the micromolar range.

These radiolabeling procedures formed the basis for the structural characterization of the thiotemplate sites of gramicidin S synthetase on the molecular level. After modification with the tracer, the multienzyme–substrate/inhibitor complexes were subsequently cut into small pieces and the thiotemplate site peptides were identified by their radioactive modification. Several techniques were developed to isolate the thiolation site peptides of gramicidin S synthetase in pure form.[69] Using the fast direct labeling technique, the protein/peptide fragmentation and the separation techniques were restricted to acidic conditions, because of the instability of the gramicidin S synthetase–thioester complexes in neutral and alkaline media. Following this method the structures of the thiolation

Table 3   Specific radiolabeling of gramicidin S synthetase.

(a) Direct affinity labeling.

---

1.   *Affinity radiolabeling of gramicidin S synthetase*
     • Formation of a thioester complex between gramicidin S synthetase and a ^{14}C-labeled amino acid substrate (2 mM ATP, 10 mM MgCl$_2$, 15 μM substrate amino acid; 37 °C, 10 min)

2.   *NEM saturation*
     • Modification of the bulk of gramicidin S synthetase with *N*-ethylmaleimide (2 mM NEM; 37 °C, 30 min)

3.   *Isolation of the radiolabeled gramicidin S synthetase complex*
     • Gel filtration (Sephadex G25, 3 °C)
     • Lyophylization

(b) Substrate protection.

---

1.   *Substrate protection of gramicidin S synthetase*
     • Formation of a thioester complex between gramicidin S synthetase and a substrate amino acid (2 mM ATP, 10 mM MgCl$_2$, 15 μM substrate amino acid; 37 °C, 10 min)

2.   *NEM saturation*
     • Modification of the bulk of gramicidin S synthetase with high concentrations of *N*-ethylmaleimide (2 mM NEM; 37 °C, 30 min)
     • Isolation of the modified gramicidin S synthetase[a]

3.   *Cleavage of the gramicidin S synthetase substrate amino acid complex*
     • 2 mM dithioerythritol (37 °C, 60 min)
     • Isolation of the modified gramicidin S synthetase[a]

4.   *Specific radiolabeling of a thioester-binding site of gramicidin S synthetase*
     • 20–40 μM *N*-[^{3}H]ethylmaleimide (37 °C, 30 min)
     • Isolation of the modified gramicidin S synthetase[a]

---

[a]The modified enzymes were isolated by gel filtration (Sephadex G25, 3 °C) and concentrated by ultrafiltration (Amicon ultrafilter XM 50, 3 °C).

sites of L-Val and L-Leu of GS2 have been determined, demonstrating for the first time that reactive Ser residues are involved in the thioester-binding process of peptide synthetases.[125] The second, more time-consuming labeling method normally led to lower yields than the direct procedure. However, the advantage of the latter technique is the stability of the tracer, for example the *S*-(*N*-[^{3}H]ethylsuccinimido) (NES) moiety in the case of the [^{3}H]NEM-alkylated sulfhydryl group.

To obtain labeled peptide fragments of the thiolation sites of gramicidin S synthetase, the radioactively modified enzymes were cleaved with cyanogen bromide (L-[^{14}C]Leu- or L-[^{14}C]Val-labeled GS2) or digested with trypsin in the cases of the [^{3}H]NES-alkylated Phe, Pro, Val, and Orn sites of GS1 and GS2. These procedures resulted in complex fragment patterns. By tryptic proteolysis of GS2, a very complex mixture of 391 peptides was generated assuming quantitative fragmentation of the Lys–Xaa or Arg–Xaa peptide bonds. As summarized in Figure 7, radioactively labeled active site peptide fragments were purified from these mixtures to homogeneity by multistep reversed-phase HPLC methodology using combinations of analytical and preparative C$_{18}$ columns and different eluent systems. If necessary this procedure was repeated after additional cleavage(s) with other proteases.

Following this strategy, peptide fragments of all five thiotemplate sites of gramicidin S synthetase were obtained in pure form. Their structures were elucidated by the combination of three techniques: (i) N-terminal sequencing by automated Edman degradation; (ii) mass spectrometry; and (iii) amino acid analysis. The obtained sequences of the peptide fragments of the thiotemplate sites of GS1 for L-Phe (D564–K575) and of GS2 for L-Pro (I564–K1008), L-Val (I2029–R2044), L-Orn (V3075–K3090), and L-Leu (F4120–L4132) are in accordance with the corresponding sequences derived from the *grsA* and *grsB*1–4 gene segments.[109,111] They all contained the highly conserved motif LGG(H/D)**S**(L/I), which has been identified as the thioester-binding region of each biosynthetic unit of peptide synthetases in general.[125,141] In neither case was a Cys residue found in the structure of the active site peptide fragments,[69,128,138,141] as would have been expected from the original thiotemplate model.[17–20] However, the chemical nature of the thioester-binding site could not be elucidated by peptide sequencing, because the radiolabel was always lost in the first Edman degradation step. Instead of the invariant Ser residue of the thiolation motif claimed by the gene-derived sequence, a dehydroalanine was always found at this position.

To identify the chemical nature of the thioester-binding sites of gramicidin S synthetase, the isolated active site peptide fragments were investigated by modern mass spectrometric techniques,

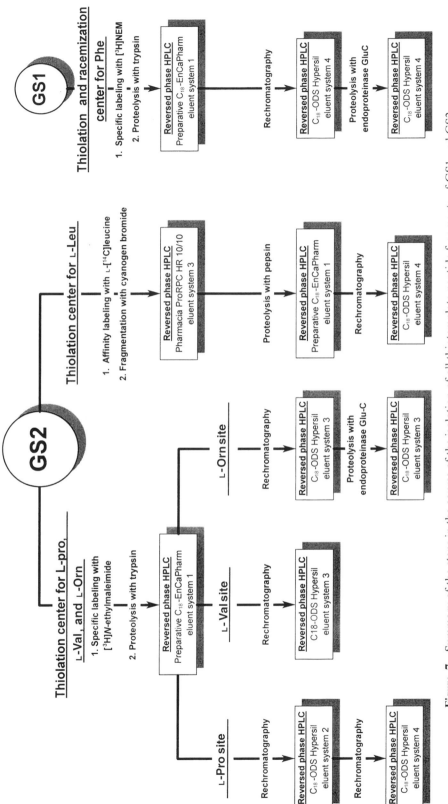

Figure 7 Summary of the steps in the course of the isolation of all thiotemplate peptide fragments of GS1 and GS2.

fast atom bombardment, and electrospray ionization mass spectrometry (FAB-MS and ESI-MS). As summarized in Table 4 the molecular masses of all radioactively labeled thiolation site peptide fragments are appreciably higher than the values calculated from the amino acid sequence of GS1 and GS2.[109,111] In each case the observed mass difference is consistent with a covalent substitution of the peptide moiety with a 4′PPan cofactor which is modified with the radioactive tracer molecule. The structure, covalent linkage, and site of location of the 4′PPan substituent attached to the Ser residue of the thiolation motif was proven by interpretation of the fragmentation data obtained from FAB-MS as well as collision-induced dissociation ESI-MS.[138] This is demonstrated for the active site heptapeptide of GS2 for the thioesterification of L-Val in Figure 8 as a representative example. For this peptide fragment the sequence of the heptapeptide LGGHΔSLR was obtained by Edman degradation. By FAB-MS a quasimolecular signal for the protonated heptapeptide species $[M+H]^+$ at $m/z = 1204.5$ u was detected. Obviously this mass is appreciably higher than 738.4 u, the mass calculated for the free heptapeptide derived from the *grsB2* gene sequence,[111] assuming proteolysis at Glu2037 and Arg2044 by GluC protease in combination with trypsin. A covalent substitution of the heptapeptide with an S-(N-ethylsuccinimido)-4′-phosphopantetheinyl adduct (NES-4′PPan; for nomenclature guidelines as well as the chemical structure of the substituent see Figure 9) is consistent with the mass difference of 465 Da. The weaker quasimolecular ion signal at $m/z = 1079$ u is attributed to the 4′PPan heptapeptide itself. Apparently, a portion of the molecules does not carry the [^{3}H]NES label. The complete structure of the active site peptide fragment was determined by interpretation of the fragmentation data (Figure 9) according to established rules for the fragmentation of peptides by mass spectrometric techniques[173,174] (for a review, see Biemann[175]). The fragment pattern observed for the active site heptapeptide is a series of N- and C-terminal sequence ions (b, c, y, and z) determining the structure of the peptide as well as locating the modification with a 4′PPan-NES substituent at position 5, which is the reactive Ser residue.

Table 4 Summary of the structure elucidation of peptide fragments from the thioester-binding sites of gramicidin S synthetase.

| | | | Molecular mass | | |
| | | | Calculated | | Measured |
Enzyme	Thiotemplate	Structure of the radiolabeled fragments from the thiolation sites	Gene	4′PPan adduct	(ESI-MS)
GS1	Phe	DNFYALGGDSIK └ 4′PPan-[^{3}H]NES	1299.4	1764.9	1764.4
GS2	L-Pro	IWEEVLGISQIGIQDNFFSLGGHSLR └ 4′PPan-[^{3}H]NES	2888.4	3353.8	3352.4
	L-Val	LGGHSLR └ 4′PPan-[^{3}H]NES	739.0	1204.3	1204.0
	L-Orn	VGIHDDFFTIGGHSLK └ 4′PPan-[^{3}H]NES	1742.9	2208.3	2208.0
	L-Leu	FELGGHSLKATLL └ 4′PPan-L[^{14}C]Leu	1385.6	1848.9	1849.1

This interpretation was verified by collision-induced tandem mass spectrometry (CID-ESI) of the active site heptapeptide. Applying this technique, signals were detected that correspond to the phosphorylated heptapeptide (M_p, $m/z = 819.3$ u) and the dehydrated species (M_Δ, 721.5 u) bearing a dehydroalanine at position 5, as well as the N-terminal and C-terminal fragments of these species (a, b, c and x, y, z series). In addition, fragments were obtained that represent the eliminated carrier ion with ($m/z = 484.0$ u) and without phosphate ($m/z = 386.2$ u), respectively, as well as characteristic fragments of the 4′PPan moiety (see Figure 9). The electrospray data demonstrate that the phosphoryl group is located on the Ser residue at position 5 of the heptapeptide, corroborating the hypothesis of the FAB-MS analysis that a 4′PPan carrier is attached to the active site Ser residue by a phosphodiester linkage. From the FAB-MS and ESI-MS data the structure of the heptapeptide fragment of the thiotemplate site of GS2 for L-Val shown in Figure 9 was derived, which is consistent with the amino acid sequence determined for this peptide by Edman degradation. The nature of the substituent covalently bound to the reactive Ser residue of the peptide moiety is supported by amino acid analysis. All the proteinogenic amino acid components of the active site peptides were found

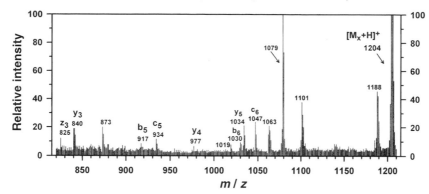

Figure 8   FAB-MS of the purified thiotemplate site peptide fragment of GS2 for L-Val.

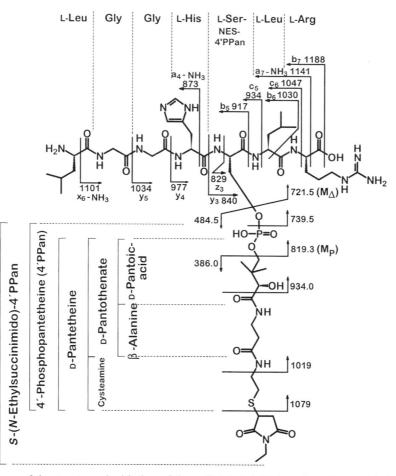

Figure 9   Summary of the structure elucidation of the radiolabeled thiotemplate site peptide fragment of GS2 for L-Val by mass spectrometry.

in the expected relationships. In addition, 1 mol of each active site peptide fragment contained 1–1.3 mol of β-alanine (3-aminopropionic acid), which normally is not found in proteins. This molecule, a constituent of 4′PPan, originated from the hydrolysis of the cofactor molecules.[69,127,128]

This research was of fundamental importance for the elucidation of the mechanism of non-ribosomal peptide biosynthesis, providing experimental evidence for the first time that all five amino acid-activating modules of gramicidin S synthetase are equipped with a separate 4′PPan prosthetic group connected by a phosphodiester linkage to the active Ser residue in the (H/D)**S**(L/I) core of their thiolation motifs.[69] The cysteamine thiol groups of the cofactors represent the thioester-binding sites for the substrate amino acids, instead of cysteines as proposed in the original version of the thiotemplate hypothesis.

4.13.6.2.2   *The cofactor binding sites are part of PCPs*

Sequence alignments of a region comprising 80–100 amino acid residues around the 4′PPan binding motif J (LGG(H/D)**S**(L/I)) of gramicidin S synthetase and other representative peptide forming multienzymes with ACPs of fatty acid and polyketide synthases as well as NodF gene products revealed an appreciable homology, as demonstrated in Table 5.

Secondary structure prediction of the PCP domain of peptide synthetase modules was compared with the structure analysis of the ACP of *E. coli* in solution determined by NMR spectroscopy.[45,147,176,177] The N-terminal part of the ACP of *E. coli* is distinguished by a short left-handed β turn changing its direction in the core region of the 4′PPan-binding motif, followed by an α-helical segment (α helix 2, residues 37–51) extending 15 amino acid residues from the active site Ser residue, which forms the 4′PPan cofactor attachment site. This α-helical region as well as two other α-helical segments (helix 1, residues 3–15 and helix 4, residues 65–75) can be related to α helices predicted within the 4′PPan-binding domains of peptide synthetases.[45,147] Obviously each amino acid-activating module of the peptide-forming multienzymes is equipped with a separate 4′PPan carrier domain, of 8–10 kDa. It represents the thiotemplate domain equipped with a covalently attached cofactor which binds the amino acid substrate as thioester and is involved in the elongation of the growing peptide chain. This structural element, essential for nonribosomal peptide biosynthesis, is integrated between the adenylation and elongation domains of a peptide synthetase module. The analysis of its secondary structure implies that the three-dimensional structure of the 4′PPan domain of peptide synthetases should be similar to acyl carrier proteins of fatty acid and polyketide synthases.

Stachelhaus *et al.*[148] have cloned and overexpressed the 4′PPan-binding domain of TY1 in *E. coli*. By coexpression of the separated PCP domain with Gsp, the phosphopantetheinyl transferase which is associated with the gramicidin S biosynthesis operon (see also Figure 4), an increased production of the holo form of PCP was achieved. This domain of TY1 was fused to a hexahistidine tag, expressed in *E. coli*, and purified from a crude extract using a Ni chelate affinity column. The recombinant 4′PPan-binding domain of TY1 has been shown to be functionally active. Together with the aminoacyl adenylation domain of GS1, the recombinant PCP of TY1 was able to bind phenylalanine as thioester.[148] These results demonstrated for the first time that the peptidyl carrier domains of peptide synthetases can work as independent enzymes. The PCP of TY1 is composed of 115 amino acid residues including the His$_6$ tag corresponding to a molecular mass of 13.12 kDa. Remarkably, the isoelectric point and the net charge of this domain at pH 7.0 significantly deviates from the data obtained for type II ACPs involved in fatty acid biosynthesis. For example, the corresponding ACP–His$_6$ fusion protein from *E. coli* (84 residues; $m = 9.82$ kDa) is a small acidic protein with a pI of 4.43 and a net charge of -13.89 at this pH.[156] In contrast, the thiolation domain of TY1 is almost neutral, showing a pI of 7.09 and a net charge of 0.2 under these conditions.[148] These striking differences in the properties of ACP and PCP proteins have to be clarified in the context of their integration into the multifunctional polypeptide chains of fatty acid and peptide synthetases, as well as their cooperation with adjacent functional domains of these multienzymes.

4.13.6.3   The Elongation Domain

The C-terminal segment of an amino acid-activating module of peptide synthetases with a size of 300–400 amino acid residues is less conserved than the adenylation and the 4′PPan carrier domain. However, within this region a series of eight conserved sequences has been identified (motifs K–R), including the so-called spacer motif M.[28] Their site of location behind the 4′PPan carrier domain and their appearance only in modules catalyzing condensation processes led to the proposal that some of these sequence motifs are involved in the elongation of the growing peptide chain as well as in specific interactions between adjacent modules.[42,45,162,178] The position of these consensus motifs is strictly conserved within amino acid activation modules, but information is still lacking about their specific functions. The invariant distance between motif M (HHxxxDG) and the 4′PPan attachment site of the adjacent PCP is remarkable. This signature sequence of the elongation

Table 5 Sequence comparison of PCPs and ACPs of peptide synthetases, polyketide, and fatty acid synthases, as well as NodF gene products.

	Motif I — Pan-binding motif J
grsA	..LPEPDLTFGMRVDYEAPRNEIEETLVTIWQDVLG----IEKIGI--KDNFYALGGDSIKAIQVAARLHSY-QLKLETKDLLKYPTIDQLVH-YIKDSKRRSQEGIVEG..
grsB1	..LPNLEGIVNTNAKYVVPTNELEEKLAKIWEEVLG----ISQIGI--QDNFFSLGGHSLKAITLISRMNKECNVDIPLRLLFEAPTIQEISN-YINGAKKESYVAIQPV..
grsB2	..LPKPDGEFGTATEYVAPSSDIEMKLAEIWHNVLG---VNKIGV--LDNFFELGGHSLRAMTMISQVHKEFDVELPLKVLFETPTISALAQ-YIADGQKGMYLAIQPV..
grsB3	..LPEPDGSISIGTEYDRPRTMLEGKLEEIWKDVLG---LQRVGI--HDDFFTIGGHSLKAMAVISQVHKECQTEVPLRVLFETPTIQGLAK-YIEETDTEQYMAIQPV..
grsB4	..LPEPQTIGLMAREYVAPRNEIEAQLVLIWQEVLG---IELIGI--TDNFFELGGHSLKATLLVAKIYEYMQIEMPLNVVFKHSTIMKIAE-YITHQESENNVH-QPI..
srfAA1	..LFALEVKAVSGTAYTAPRNETEKAIAAIWQDVLN----VEKAGI--FDNFFETGGHSLKAMTLLTKIHKETGIEIPLQFLFEHPTITALAE-EADHRSKAFAVIEPA..
srfAA2	..LPIPDANVSRGVSYVAPRNGTEQKVADIWAQVLQ----AEQVGA--YDHFFDIGGHSLAGMKMLALVHQELGVELSLKDLFQSPTVEGLAQ-VIASAEKGTAASISPA..
srfAA3	..LPEPDIEAG-SGEYKAPTTDMEELLAGIWQDVLG----MSEVGV--TDNFFSLGGDSIKGIQMASRLNQH-GWKLEMKDLFQHPTIEELTQ-YVERAEGKQAD-QGPV..
srfAB1	..LPAPQSE-AVQPEYAAPKTESEKKLAEIWEGILG----VKAGV--TDNFFMIGGHSLKAMMMTAKIQEHFHKEVPIKVLFEKPTIQELAL-YLEEIESKEEQTFEPI..
srfAB2	..LPKPNAAQSGGKALAAPETALESLCRIWQKTLG----IEAIGI--DDNFFDLGGHSLKGMMLIANIQAELEKSVPLKALFEQPTVCQLAV-YMEASAVSGGH-QVLK..
srfAB3	..LPEPDIEAG-SGEYKAPTTDMEELLAGIWQDVLG---MSEVGV--TDNFFSLGGDSIKGIQMASRLNQH-GWKLEMKDLFQHPTIEELTQ-YVERAEGKQAD-QGPP..
srfAC	..LPKPDQDQ-LAEEWIGPRNEMEETIAQIWSEVLG---RKQIGI--HDDFFALGGHSLKAMTAVPHQQEL-GIDLPVKLLFEAPTIAGISA-YLKNGGSDGL--QDVT..
acvA	..LPSVDLIQPKVSSCEL-TDEVEIALGKIWADVLG----AHHLSI-SRKDNFFRLGGHSITCIQLIARIRQQLGVIISEDVFSSRTLERMAE-LLRSKESNGTPDERAR..
acvB	..LPTAEEKGAMN--VLAPRNEISILCGISAGLLD----ISAQTI--GSDFFTLGGDSLKSTKLSFKIHEVFGRTISVSALFRHRTIESL-AHLIMNNVGD-IQEITPV..
acvC	..LPDIGNPQHQIS-YNPPRDVLEADLCRLWASALG---TERCGI--DDDLFRLGGDSITALHLAAQIHHQIGRKVTVRDIFDHPTIRGIHDNVMVKLVP-HVPQFQA..
	XXXXXXXXX XXXXXXXXXX XXXXXXXXX ////// XXXXXXXXXX
	STIEERVKKIIGEQLG----VKQEEVTNNASFVEDLGADSLDTVELVMALEEEFDTEIPDEEAEEKITTVQAAIDYINGHQA
ACP *E. coli*	MDRKEIFE-RIEQVLAEQLG----IPAEQITEEADLREDLGMDSLDLVELVSALEDEVGMRVEQSQLEGIETVGHVME-LTLDLVARLATASAADK..
ACP yeast	AAKQETVE-KVSEIVKKQLS---LKDDQQVV-AETKFVDLGADSLDTVEIVMGLEEEFGIQMAEEKAQKIATVEQAAE-LIEELMQAKK
ACP rape	AKKETID-KVSDIVKEKIALG---ADVVVT-ADSEFSKLGADSLDTVEIVMGLEEEFGINVDEDKAQDISTIQQAAD-VIE-LLEKKA
ACP spinach	SEGGSQR-DLVEAVAHILG---VRDVSSLNAESSLADLGLDSLMGVEVRQTLERDYDIVMTMREIRLL-TINKLRE-LSSKTGTAELKPSQVL
ACP chicken	EKKAVAHGDGEAQR-DLVKAVAHILG---IRDLAGINL-DSSLADLGLDSLMGVEVRQLEREHDLVLPIREVRQL-TLRKLQE-MSSKAGSDTEL
ACP rat	EQQENLLE-LVANAVAEVLG---HES-AAEIN-VRAFSELGLDSLNAMALRKRLSASTGLRLPASLVFDHPTVTALAQHLRARLVGDAD--QAA..
ACP DEBS2N	REREHLAH-LIRAEVAAVLG---HGDDAAIDR-DRAFRDLGFDSMTAVDLRNRLAAVTGVREAATVVFDHPTITRLADHYLERLVGAAEAEQAP..
ACP DEBS3N	MATLLTTD-DLRRALVECAGETDGTDLSGDFL-DLRFEDIGYDSLALMETAARLESRYGVSIPDDVAGRVDTPRELLD-LINGALAEAA..
ACP ARS	MARLTLDGLRTILVA-CAGEDDGVDLSGDIL-DITFEELGYDSLALMESASRIERELGVALADGDINEELTPRVLLD-LVNGAQAEAA..
ACP GRANS	MPQIGLPR--LVEIIRECAGDPDERDLDGDIL-DVTYQDLGYDSIALLEISAKLEQDLGVSIPGEELK----TPRHTLH-LVNTETAGEVA..
ACP TCMS	MVDQLESEIIGIIKNRVESEGGDGETALIVGDLTAATELTALGVDSLGLADIIWDVEQAYGIR-------IEMNTAEAWSD-LQNVGDIVGAIRGLLTKGA
NodF *R. meliloti*	MADQLTVEIIAAIKNLAQSENG-GRIPAAIGDTADRQLTSLGLSALADVLMDLEQAYGIR-------IEMNTADAWSN-LKNIGDVVEAVRGLIAKEA
NodF *R. trifolii*	MADQLTLEIISAINKLVKAENG-ERTSVALGEITTDTELTSLGIDSLGLADVLWDLEQLYGIK------IEMNTADAWSN-LNNIGDVVEAVRGLLTKEV

domains resembles a sequence motif detected as an indispensable functional element in chloramphenicol acetyltransferase[163,164] and dihydrolipoamide acyltransferase[165] that is involved in the acyl transfer processes catalyzed by these enzymes. Structural studies on the former enzymes imply that the two conserved His residues of this motif play a crucial role in the catalysis of the acyl transfer reactions. In particular, the first of them shows an unusual conformation allowing a proton abstraction from the C-3 hydroxy group of chloramphenicol to promote a nucleophilic attack on the carbonyl carbon of acetyl-CoA.[179] Also, from the atomic structure of the E2p subunit of the pyruvate dehydrogenase multienzyme complex of *Azotobacter vinelandii*[180] it is obvious that this His residue functions as a general base catalyst in the deprotonation of the thiol of coenzyme A to attack the reactive carbonyl of the dihydrolipoamide acceptor. On the basis of these results, a model for the acyl transfer reactions involved in the peptide elongation and epimerization reactions in nonribosomal peptide biosynthesis was proposed.[162] According to this concept, the second His residue of motif M would catalyze two types of reactions: a nucleophilic attack of the incoming amino acid on the activated carbonyl of the preceding peptide phosphopantetheinyl thioester in the elongation process and an abstraction of the proton at C-α of the peptide linked to the 4′PPan carrier in the case of a following epimerization step.

The hypothesis that motif M of the peptide synthetases is putatively involved in acyl and peptidyl transfer reactions has been probed by site-directed mutagenesis experiments in surfactin synthetase.[130] The exchange of both His residues and the Asp residue of the spacer motif HHIMMDGWS in the N-terminal region of surfactin synthetase 3 by Ala residues abolished surfactin formation. These results provide the first evidence that the replaced amino acid residues of motif M are essential for the elongation process and/or the interaction between surfactin synthetases 2 and 3.[129]

In the case of the interaction of individual proteins of peptide-forming multienzymes, such as GS1 and GS2 or the components of surfactin synthetase, a specific domain is observed at the N terminus of the acceptor enzyme, showing homology to the elongation domains.[28,42,45,129] Presumably this structural element plays an important role in the protein–protein interaction of the consecutive enzymes.

4.13.6.4   The Racemization/Epimerization Domain of Gramicidin S Synthetase 1

GS1 and the closely related enzyme TY1 catalyze the racemization of Phe. At the beginning of the 1970s, experimental efforts were made to determine the 4′PPan content of gramicidin[181] and tyrocidine synthetase.[82] The results obtained for both systems demonstrated that only GS2 and TY3 contained significant amounts of 4′PPan. Therefore, it was assumed for more than 20 years that one-module synthetases, such as GS1 and TY1, do not contain a prosthetic group.

With knowledge of the primary sequences of GS1[109] and TY1[111] it became clear that both enzymes contain the characteristic LGG(H/D)S(L/I) thiolation motif revealing a 4′PPan-binding site.[125] By careful reinvestigation of the pantothenic acid content of GS1 by a microbiological assay using *Lactobacillus plantarum* as the test organism, in combination with the structural characterization of the thiotemplate site of the multienzyme, 4′PPan could be detected as the prosthetic group. The experiments to determine the pantothenic acid content of Mono Q fractions of GS1 are shown in Figure 10. Pantothenic acid liberated from GS1 after mild hydrolysis by potassium hydroxide in combination with the action of alkaline phosphatase exactly corresponded with the enzymatic activities of gramicidin S synthetase, that is, racemization of Phe, and adenylation and thioesterification of D-Phe, as well as biosynthesis of gramicidin S.[141] The crucial factor for the detection of the 4′PPan cofactor was the analysis of freshly prepared GS1. The highly purified enzyme showed a rapid decrease of its activities upon aging. Most probably this is the reason that 4′PPan was not found as a constituent of GS1 in previous studies. The new data are in full agreement with the structural characterization of radiolabeled fragments of the thiotemplate site of GS1. Also in the case of the one-module synthetase GS1 a 4′PPan cofactor was attached to the reactive Ser residue of the active site peptide fragment DNFYALGGDSIK. In this case the radiolabeled thiol agent iodoacetic acid was used for labeling of the thiolation center (see also Table 4).[128,141]

GS1 exhibits the characteristics of a racemase enzyme using a single base as the specific proton acceptor/donor system at the racemization center.[182] GS1 does not show any sequence homologies to other amino acid racemases that follow the one-base pathway, such as the pyridoxal-dependent alanine racemases (for a review, see Walsh[183]). The identification of 4′PPan also at the thiotemplate site of GS1 led to the classification of this enzyme as a prototype of amino acid racemases/epimerases

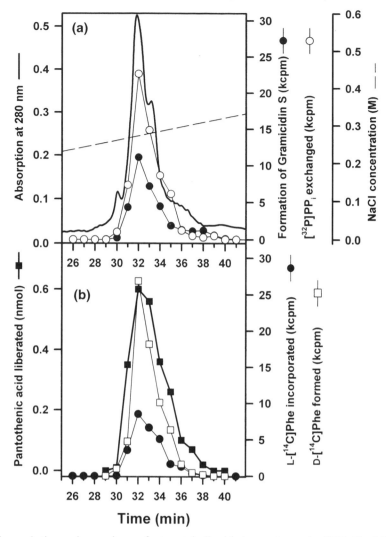

Figure 10 High-resolution anion exchange fast protein liquid chromatography (FPLC) of GS1, its enzymatic activities, and content of a 4′PPan prosthetic group.

containing 4′PPan as a cofactor by Stein *et al.*[128,141] This class of enzymes uses either a single amino acid (GS1[141] or TY1[147]) or a specific peptide intermediate (ACV, surfactin, or actinomycin synthetases)[184,185] as substrates in thioester-bound form. In order to identify the catalytic centers which are involved in such epimerization processes, intensive amino acid sequence alignments have been performed. In the C-terminal segment of certain peptide synthetase modules in particular, significant differences in the consensus sequence N–R between epimerizing and nonepimerizing modules have been identified.[28,42,45,141,146,178] A segment of the *grsA* gene containing these motifs has been cut off. Overexpression of the truncated enzyme in *E. coli* resulted in a GS1 fragment (Figure 6; GS1, 1–807) which is catalytically active in adenylate and thioester formation of Phe, but deficient in the epimerization of this amino acid, demonstrating that the C-terminal part of the enzyme containing the consensus motifs N–R is essential for the racemization/epimerization process.[146] One of the six to seven highly conserved basic amino acid residues within motifs N–R has been hypothesized either to function as the proton acceptor/donor system which is involved in the epimerization process[141] or as a constituent of an acid/base charge transfer system which stabilizes the negative charge of the enolic intermediate in a similar manner as that observed for mandelate racemase.[186–189] In first approaches to identify these catalytically active centers, the highly conserved Tyr residues in the conserved sequences AYDTEF between motifs M and N and in motif Q (FNYLGQ) of the third module of surfactin synthetase 3 were replaced by Ala residues. Analysis of the reactivity of the mutant enzymes and their capability for product formation showed that the mutated Tyr residues are not essential for the epimerization process. The Y→A replacements affected neither the composition and reactivity of the surfactin synthetase components isolated from the mutant strains nor the surfactin formation process. Further genetic and biochemical analysis is necessary to identify the catalytically active basic residue within the epimerization domain.[129]

4.13.6.5 The Thioesterase Domain/Protein

At the C-terminal end of bacterial peptide-forming systems, such as gramicidin S or surfactin synthetase, as well as in the case of the ACV synthetase, a segment is included replacing the elongation domain. This segment shows similarities to thioesterases[28,33,190] and contains a GxSxG core sequence in its reaction center. In addition, at the 5′ end of the *grs* operon,[109] the 3′ end of the *srfA* operon,[28,30] and associated with the gene cluster encoding the proteins involved in bialaphos biosynthesis, structural genes are located which encode proteins with molecular masses of 29 kDa, 25 kDa, and 27 kDa, respectively, showing about 30% homology among themselves and to vertebrate thioesterase II enzymes involved in fatty acid biosynthesis.[121] These proteins also contain the GxSxG core, the "lipase/esterase" consensus sequence,[191] which has been observed in different classes of hydrolytic enzymes involved in many different pathways, as in the biosynthesis of fatty acids,[192] various polyketides,[193,194] and penicillin.[189] The biological function of the thioesterase-like enzymes associated with the peptide synthetase genes in the same operons is not known. Further biochemical and genetic analysis is necessary to attribute specific functions to these proteins in nonribosomal peptide biosynthesis. De Ferra *et al.* have demonstrated for surfactin synthetase that the thioesterase domain within the last amino acid-activating module of this peptide-forming system is involved in the termination reaction, the cyclization of the thioesterified peptide product.[195]

4.13.7 THE MECHANISM OF NONRIBOSOMAL PEPTIDE BIOSYNTHESIS—THE MULTIPLE CARRIER THIOTEMPLATE MODEL

The present knowledge of structure–function relationships of peptide synthetases implies that the prosthetic group of each biosynthetic module can exist in three different functional states: (i) the free thiol group, or thioesterified either with (ii) an amino acid substrate or (iii) a peptide intermediate. Therefore, each module of gramicidin S synthetase should contain three specific positions for a 4′PPan cofactor which in each case remains covalently bound to the reactive Ser residue of module *i* (Figure 11). Within the first position (charging position; 1 in Figure 11) the thioester formation of the cofactor with the substrate amino acid AA_i occurs. Most probably the charging site is located within the aminoacyl adenylation domain. Important for the elongation process are the peptidyl acceptor and the peptidyl donor positions (2 and 3 in Figure 11), specific positions for interaction of the 4′PPan carrier of module *i* with the 4′PPan carriers of the former module *i* − 1 as the donor of the growing peptide chain and of the subsequent module *i* + 1 which functions as the acceptor for the peptide intermediate elongated by one amino acid in module *i*. These interactions take place at presently unidentified transpeptidation centers of peptide synthetase modules, where the formation of a new peptide bonds occurs. Most probably these functional elements are located within the elongation domains of peptide synthetases. The postulated mechanism of the transpeptidation reaction resembles the peptidyl transfer process between aminoacyl-tRNA and peptidyl-tRNA within the ribosomal A and P sites during ribosomal protein synthesis.

The new concept of nonribosomal peptide biosynthesis, the "multiple carrier thiotemplate model" hypothesized by Stein *et al.*,[69,128] allows a much simpler and more straightforward description of the biosynthetic process than the old thiotemplate version.[17–20] Within the former model one single central 4′PPan carrier was postulated to interact with reactive Cys residues as the peripheral thiols at the thiotemplate sites. In analogy to the early Lynen model of fatty acid biosynthesis,[21,22] the old thiotemplate hypothesis needed the assumption of transthiolation reactions to explain the transport of peptide intermediates in the process according to the head growth principle. The charging of a multienzyme with all intermediates of the growing peptide chain, mechanistically inexplicable by the former hypothesis, can be easily explained by the multiple 4′PPan carrier mechanism.[69,128] The biosynthesis of gramicidin S as a representative example for the multiple carrier thiotemplate model is illustrated in Figure 12. The biosynthetic units of this multienzyme catalyze the aminoacyl adenylation and subsequently the thioesterification of their cognate amino acids to their 4′PPan carriers (Figure 12(a)), most probably at the charging position within their adenylation domain $(C_{amino\ acid})$. The energy of the activation reaction is provided by the hydrolysis of an ATP α–β

Module *n*–1   # Module *n*   # Module *n*+1

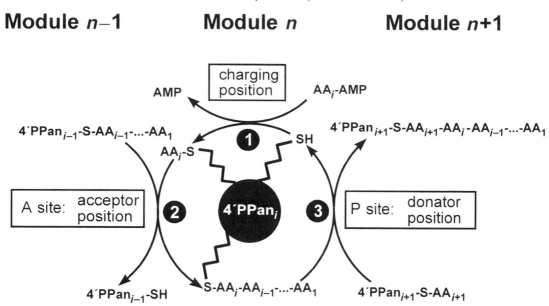

Figure 11 Schematic presentation of the three different functional states and positions of a 4′PPan cofactor within a biosynthetic module of peptide synthetases.

phosphate linkage, resulting in the release of AMP and pyrophosphate (PP_i). As illustrated in Figure 12(b), the elongation cycle of gramicidin S synthesis starts with the interaction of both multienzymes of gramicidin S synthetase, GS1 and GS2. The 4′PPan carrier of GS1 transports Phe to the elongation site of GS1 (E_1-Epi). As discussed in Section 4.13.6.4, the stereoinversion of L-Phe to D-Phe in the thioester-bound stage occurs within this domain.[146] The initial peptide bond is formed by an interaction of GS1 and GS2. In the first elongation domain E_1 a transpeptidation reaction is induced by nucleophilic attack of the imino group of L-Pro at the carboxyl C atom of D-Phe, resulting in the dipeptide D-Phe-L-Pro (F-P). During this process the 4′PPan$_{Pro}$ carrier of GS2 within its acceptor position interacts with the 4′PPan$_{Phe}$ carrier of GS1, which is located within its donator position. Most probably the domain E_N of GS2 is involved in the interaction of the multienzymes as well as the transpeptidation reaction. A free thiol group of the 4′PPan$_{Phe}$ carrier is recovered in this reaction, which can be charged by Phe again. This GS1 molecule is able to start a new elongation cycle. In the next step the 4′PPan carrier of the L-Pro module of GS2, thioesterified with the dipeptidyl intermediate (F-P), is translocated to the donator position of the Pro module, as demonstrated in Figure 12(c). By interaction with the 4′PPan carrier of the Val module, thioesterified with L-Val, in its acceptor position the tripeptide D-Phe-L-Pro-L-Val is formed in the second transpeptidation reaction. In a similar manner the tetra- and pentapeptide intermediates D-Phe-L-Pro-L-Val-L-Orn and D-Phe-L-Pro-L-Val-L-Orn-L-Leu (Figures 12(d) and 12(e)) are assembled involving the individual 4′PPan carriers of the Orn- and Leu-activating modules of GS2. Finally, the decapeptide gramicidin S is formed by cyclization of two pentapeptide moieties (Figure 12(f)).

4.13.8 APPROACHING THE THREE-DIMENSIONAL STRUCTURE OF GRAMICIDIN S SYNTHETASE

A challenge for future progress in the field of nonribosomal peptide biosynthesis is the elucidation of the three-dimensional structure of peptide synthetases by physicochemical techniques. For this difficult task, gramicidin S synthetase is well qualified as a model system. It is one of the best characterized peptide-forming multienzymes which can be prepared in high purity and in sufficient amounts for crystallization studies. A milestone in this regard is the electron microscopic structure analysis of this multienzyme. The first electron micrographs of GS2 were obtained by Wecke *et al.*,[196] demonstrating ring-shaped particles with outer and inner diameters of 12 nm and 6 nm, respectively. The quaternary structure of both enzymes of gramicidin S synthetase was investigated in more detail by Tesche and by Vater and his co-workers.[26] Electron micrographs of GS1 revealed particles of oblate ellipsoidal conformation with a diameter of approximately 7.5 nm. For GS2 a ring structure was also found with similar dimensions as reported by Wecke *et al.* (see Kleinkauf

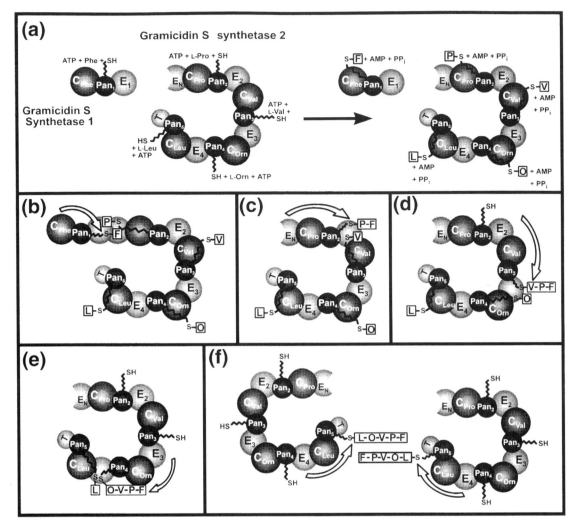

Figure 12   Formation of gramicidin S illustrating the "multiple carrier thiotemplate model" of nonribosomal peptide biosynthesis at multienzymatic templates.

and Koischwitz[196]). The height of the GS2 particles was determined from freeze-dried preparations shadowed with tungsten either unidirectionally or by rotational evaporation. From the observed shadow, a height of 6 nm was estimated. Important progress in the electron microscopic analysis of GS2 was made by cooperation between the groups of van Heel and Vater (see the thesis of T. H. Stein[128]) using a cryotechnique and negative staining with phosphotungstate in combination with modern computer image analysis. Cryoelectron micrographs of GS2 in vitrified ice showed ring-shaped particles with an outer diameter of 125 Å, as shown in Figure 13. Unfortunately, under these mild conditions only top and bottom views were visible. To obtain side views, a prerequisite for structure analysis, GS2 was negatively stained with phosphotungstate (Figure 14). A total of 7200 images were selected, and divided into 250 classes of particle views. A three-dimensional structure reconstruction of GS2 was obtained by the angular reconstitution method.[197] As shown in Figure 15, this refined computational analysis revealed doughnut-shaped particles organized into at least eight structural domains which are arranged below an angle of 45° in relation to the vertical axis of the cylinders. These cylindrical particles show outer and inner diameters of 125 Å and 35 Å, respectively, and a height of 105 Å.[128]

An important milestone concerning the structure elucidation of a peptide synthetase was achieved by determining the crystal structure of the N-terminal fragment of GS1 (residues 1–557) at 1.9 Å resolution by Brick *et al.*[48] The crystallized protein consists of the Phe adenylation domain of GS1. This enzyme was expressed in *E. coli*, purified, and cocrystallized with the substrates Mg-ATP^{2-} and both stereoisomers of Phe. Its three-dimensional structure is folded into two compact domains

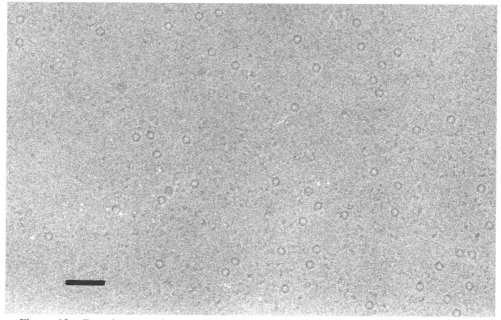

Figure 13 Cryoelectron micrographs of GS2 in vitrified ice (bar length represents 50 nm).

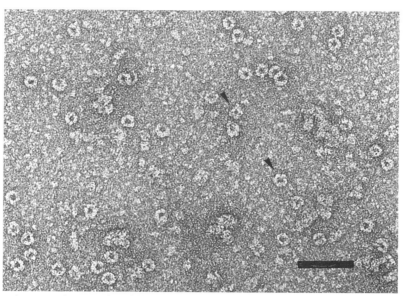

Figure 14 Cryoelectron micrographs of GS2 negatively stained with phosphotungstate (side views indicated by arrows; bar length represents 50 nm).

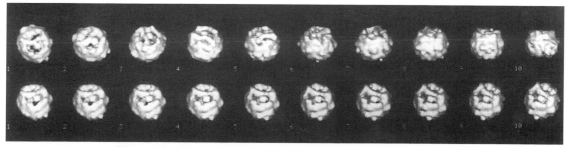

Figure 15 Three-dimensional reconstruction of a GS2 molecule.

formed by residues 17–428 (N-terminal domain) and residues 429–530 (C-terminal domain) with the active site localized at the interface between both domains. However, the electron density is not consistent with a Phe-adenylate bound at the reaction center. Obviously, this highly reactive mixed anhydride is hydrolyzed under the crystallization conditions, resulting in Phe and AMP.

The analysis of the crystal structure of the adenylation domain of GS1 revealed that recognition of both substrates, the nucleotide as well as Phe, is accomplished by an extensive network of hydrogen bonds, as well as several hydrophobic and van der Waals interactions contributed by the large N-terminal domain. A flexible region of poor electron density has been detected which is exposed to the solvent in the large cleft separating the N- and C-terminal domains. The invariant Lys517 residue of the C-terminal domain plays a key role in two polar interactions with the α-carboxylic group of the amino acid substrate and the adenosine of the nucleotide, fixing the positions of these substrates within the active site.

The adenylate substrate is bound in a cleft on the surface of the large N-terminal domain. For example, the adenine moiety is sandwiched between the hydrophobic side chains of Tyr323, Ile348, and Tyr425 from one side and the main carbon atoms of residues 302–304 on the other side. Hydrogen bonds between the N-6 amino group of the adenine base and the enzyme structure (Ala322 and Asn321) are the major specific determinants by which GS1 discriminates between ATP (substrate) and GTP (no substrate). All these residues are situated within the highly conserved motifs D, E, F, and G of the adenylation domain of peptide synthetases, demonstrating the importance of these structural elements for ATP binding. Two oxygen molecules of the ribose moiety (O-4′ and O-5′) are hydrogen bonded to the obligate Lys517 residue of the small C-terminal domain.

The binding site for the side chain of the amino acid substrate is a pocket on the surface of the large N-terminal domain formed by amino acid residues between motifs C and E, as hypothesized by Cosmina *et al.*[29] In the case of the adenylation domain of GS1, Trp239 is localized on the bottom of the pocket while both sites are formed by the residues Ala236, Ile330, and Cys331, as well as Ala322, Ala301, and Thr278. Obviously, the three-dimensional structure of this pocket determines the specificity of the biosynthetic module for its amino acid substrate. The knowledge of the amino acid residues forming the substrate-binding pocket should allow manipulation of the amino acid specificity of a biosynthetic module in the course of the production of novel nonribosomal synthesized peptides with modified biological activities.

The tertiary structure of GS1 has been compared with the structure of firefly luciferase from *Photinus pyralis*, another protein of the adenylate-forming superfamily, determined by Conti *et al.*[150] The overall structure of both enzymes is similar. However, the most striking difference between both crystal structures is the orientation between the large N-terminal domain and the smaller C-terminal domain. Compared with the luciferase tertiary structure, the smaller C-terminal domain of GS1 is rotated by more than 90° relative to the larger N-terminal domain, and both domains are 5 Å closer together, due to the cocrystallization of this protein with its substrates forming an active conformation.[48] Evidence for such an induced-fit mechanism has also been obtained by Dieckmann *et al.*[169] from functional studies, active site mutagenesis experiments, and limited proteolysis of TY1. Obviously the binding of the nucleotide and amino acid substrate induces a conformational change of the adenylation domain from an open to a closed conformation. Efforts have been initiated in several laboratories to crystallize other isolated functional domains of a biosynthetic module, complete bioactive modules of peptide synthetases, and native peptide synthetases consisting of several modules. From these approaches more detailed information on the three-dimensional structure of the peptide-forming cellular factories can be expected in the future.

4.13.9 PROSPECTS

During the 1990s immense progress has been made in the field of nonribosomal peptide biosynthesis. This burst of knowledge was mainly induced by deciphering the gene structures coding for the biosynthesis of numerous bioactive peptides of microbial origin. The availability of the gene sequences led to the elucidation of the modular structure of peptide synthetases and evolutionary relationships. Based on the genetic information, extensive sequence comparisons, and computational structure analysis of such multienzymes, functional domains were detected that are highly conserved in all known peptide forming cellular factories. In addition, new detailed knowledge concerning the thiotemplate mechanism of nonribosomal peptide biosynthesis has been obtained. Most important was the identification of a 4′PPan cofactor as the thiotemplate site at each reaction center of the peptide synthetases. The multiple carrier concept shed new light on the elongation reactions which involve the interaction of multiple 4'PPan carriers in a series of consecutive transpeptidation steps. Progress in the functional characterization of peptide synthetases has been stimulated by the availability of novel protein chemical methodologies with greatly improved sensitivities and efficiencies, enabling the analysis of such large multienzyme systems. A major part of this new information has been accumulated by the investigation of gramicidin S synthetase and the closely related tyrocidine synthetase at both the gene and protein levels.

Priority tasks for the near future are:

(i) The identification of the structural elements determining substrate recognition and selection by site-specific mutagenesis and genetic engineering.

(ii) The analysis of the elongation and termination reactions in nonribosomal peptide biosynthesis. Insights into this function will come from the elucidation of the structural and functional significance of the conserved motifs K–R in the elongation and epimerizing domains of peptide synthetases as well as the location and characterization of the three postulated positions of the 4'PPan carrier in each module in the course of an elongation step by site-specific mutagenesis in combination with artificial peptide substrates.

(iii) Closely related to task (ii) is the investigation of module and subunit interactions in the assembly of the growing peptide chain. Here, clarification of the function of the N-terminal elongation domains in subunit recognition and acyl transfer will be of special relevance.

(iv) The structural and functional characterization of the thioesterase/acyltransferase enzymes detected by sequencing of biosynthetic genes.

(v) The elucidation of the three-dimensional structure of native peptide synthetases by electron microscopy and X-ray crystallography as well as the investigation of the geometry and topography of the reaction centers by NMR and ESR methodologies. This is the most important task.

A challenge for the future will be the construction of recombinant peptide synthetases for a rational design of novel bioactive peptides by module and domain rearrangement and/or engineering of the recognition structures for the amino acid substrates.[195,198–201] Such combinatorial biosynthesis has the potential for the generation of a great diversity of new peptide agents (peptide design) which can be screened for pharmaceutical and biotechnological applications. A breakthrough has been achieved in establishing the fundamental routes of this attractive technology in the fields of both peptide and polyketide biosynthesis. For the successful resolution of all these important tasks, gramicidin S and tyrocidine synthetases qualify as excellent model systems.

4.13.10   REFERENCES

1. A. Fleming, *Br. J. Exp. Pathol.*, 1929, **10**, 226.
2. E. B. Chain, H. W. Florey, A. D. Gardner, N. G. Heatley, M. A. Jennings, J. Orr-Ewing, and A. G. Sanders, *Lancet*, 1940, **239**, 226.
3. E. P. Abraham, E. Chain, C. M. Fletcher, H. W. Florey, A. D. Gardner, N. G. Heatley, and M. A. Jennings, *Lancet*, 1941, **241**, 177.
4. R. J. Dubos, *J. Exp. Med.*, 1939, **70**, 1.
5. R. D. Hotchkiss and R. J. Dubos, *J. Biol. Chem.*, 1940, **132**, 791.
6. R. D. Hotchkiss and R. J. Dubos, *J. Biol. Chem.*, 1941, **141**, 155.
7. G. F. Gause and M. G. Brazhnikova, *Lancet*, 1944, **247**, 715.
8. G. F. Gause and M. G. Brazhnikova, *Nature*, 1944, **154**, 703.
9. S. Otani and Y. Saito, *J. Proc. Jpn. Acad.*, 1954, **30**, 991.
10. R. L. M. Synge, *Biochem. J.*, 1945, **39**, 355.
11. T. S. Eikhom, J. Jonsen, S. G. Laland, and T. Refsvik, *Biochim. Biophys. Acta*, 1963, **76**, 465.
12. B. Mach, E. Reich, and E. L. Tatum, *Proc. Natl. Acad. Sci. USA*, 1963, **50**, 175.
13. T. S. Eikhom and S. Laland, *Biochim. Biophys. Acta*, 1965, **100**, 451.
14. I. Uemura, K. Okuda, and T. Winnick, *Biochemistry*, 1963, **2**, 719.
15. M. Yukioka, Y. Tsukamoto, Y. Saito, T. Tsuji, S. Otani, and S. Otani, *Biochem. Biophys. Res. Commun.*, 1965, **19**, 204.
16. F. Lipmann, in "The Mechanism of Enzyme Action," eds. W. D. McElroy and B. Glass, John Hopkins University Press, Baltimore, MD, 1954, p. 599.
17. F. Lipmann, *Accts. Chem. Res.*, 1973, **6**, 361.
18. S. G. Laland, Ø. Frøshov, C. Gilhuus-Moe, and T.-L. Zimmer, *Nature New Biol.*, 1972, **239**, 43.
19. S. G. Laland and T.-L. Zimmer, *Essays Biochem.*, 1973, **9**, 31.
20. K. Kurahashi, *Annu. Rev. Biochem.*, 1974, **43**, 445.
21. F. Lynen, *Biochem. J.*, 1967, **102**, 381.
22. F. Lynen, *Eur. J. Biochem.*, 1980, **112**, 431.
23. F. Lipmann, *Adv. Microbiol. Physiol.*, 1980, **21**, 227.
24. H. Kleinkauf and H. von Döhren, in "Peptide Antibiotics—Biosynthesis and Functions," eds. H. Kleinkauf and H. von Döhren, Walter de Gruyter, Berlin, 1982, p. 3.

25. K. Kurahashi, *Antibiotics (NY)*, 1981, **4**, 325.
26. J. Vater, in "Biochemistry of Peptide Antibiotics," eds. H. Kleinkauf and H. von Döhren, Walter de Gruyter, Berlin, 1990, p. 33.
27. C. Ullrich, B. Kluge, Z. Palacz, and J. Vater, *Biochemistry*, 1991, **30**, 6503.
28. S. Fuma, Y. Fujishima, N. Corbell, C. D'Souza, M. M. Nakano, P. Zuber, and K. Yamane, *Nucleic Acids Res.*, 1993, **21**, 93.
29. P. Cosmina, F. Rodriguez, F. de Ferra, G. Grandi, M. Perego, G. Venema, and D. van Sinderen, *Mol. Microbiol.*, 1993, **8**, 821.
30. M. Menkhaus, C. Ullrich, B. Kluge, J. Vater, D. Vollenbroich, and R. M. Kamp, *J. Biol. Chem.*, 1993, **268**, 7678.
31. G. Galli, F. Rodriguez, P. Cosmina, C. Pratesi, R. Nogarotto, F. de Ferra, and G. Grandi, *Biochim. Biophys. Acta*, 1994, **1205**, 19.
32. A. P. MacCabe, H. van Liempt, H. Palissa, S. E. Unkles, M. B. R. Riach, E. Pfeifer, H. von Döhren, and J. Kinghorn, *J. Biol. Chem.*, 1991, **266**, 12646.
33. H. van Liempt, H. von Döhren, and H. Kleinkauf, *J. Biol. Chem.*, 1989, **264**, 3680.
34. Y. Aharonowitz, J. Bergmeyer, J. M. Cantoral, G. Cohen, A. L. Demain, U. Fink, J. Kinghorn, H. Kleinkauf, A. MacCabe, H. Palissa, E. Pfeifer, T. Schwecke, H. van Liempt, H. von Döhren, S. Wolfe, and J. Zhang, *Biotechnology*, 1993, **11**, 807.
35. A. Billich and R. Zocher, *Biochemistry*, 1987, **26**, 8417.
36. A. Lawen and R. Zocher, *J. Biol. Chem.*, 1990, **265**, 11355.
37. G. Weber, K. Schörgendorfer, E. Schneider-Scherzer, and E. Leitner, *Curr. Genet.*, 1994, **26**, 120.
38. R. Zocher, U. Keller, and H. Kleinkauf, *Biochemistry*, 1982, **21**, 43.
39. A. Haese, M. Schubert, M. Herrmann, and R. Zocher, *Mol. Microbiol.*, 1993, **7**, 905.
40. R. Pieper, A. Haese, W. Schröder, and R. Zocher, *Eur. J. Biochem.*, 1995, **230**, 119.
41. H. Kleinkauf and H. von Döhren, *Eur. J. Biochem.*, 1990, **192**, 1.
42. H. Kleinkauf and H. von Döhren, *Eur. J. Biochem.*, 1996, **236**, 335.
43. R. Weckermann, R. Fürbass, and M. A. Marahiel, *Nucleic Acids Res.*, 1988, **16**, 11841.
44. T. Stachelhaus and M. A. Marahiel, *FEMS Microbiol. Lett.*, 1995, **125**, 3.
45. T. Stein and J. Vater, *Amino Acids*, 1996, **10**, 201.
46. R. Zocher and U. Keller, *Adv. Microb. Physiol.*, 1997, **38**, 85.
47. H. Kleinkauf and H. von Döhren, in "Biotechnology," eds. H. Kleinkauf and H. von Döhren, VCH, Weinheim, 1997, vol. 7, p. 277.
48. E. Conti, T. Stachelhaus, M. A. Marahiel, and P. Brick, *EMBO J.*, 1997, **16**, 4174.
49. R. Consden, A. H. Gordon, J. P. Martin, and R. L. M. Synge, *Biochem. J.*, 1946, **40**, 43.
50. R. Consden, A. H. Gordon, J. P. Martin, and R. L. M. Synge, *Biochem. J.*, 1947, **41**, 596.
51. D. C. Hodgkin and B. M. Oughton, *Biochem. J.*, 1957, **65**, 752.
52. G. Némethy and H. A. Scheraga, *Biochem. Biophys. Res. Commun.*, 1984, **118**, 643.
53. D. S. Gross and E. R. Williams, *J. Am. Chem. Soc.*, 1996, **118**, 202.
54. Y. Yagi, S. Kimura, and Y. Imanishi, *Int. J. Pept. Protein Res.*, 1990, **36**, 18.
55. M. Tamaki, M. Takimoto, S. Nozaki, and L. Muramatsu, *J. Chromatogr. Biomed. Appl.*, 1987, **413**, 287.
56. T. Katayama, K. Nakao, M. Akamatsu, T. Ueno, and T. Fujita, *J. Pharm. Sci.*, 1994, **83**, 1357.
57. L. H. Kondejewski, S. W. Farmer, D. S. Wishart, R. E. Hancock, and R. S. Hodges, *Int. J. Pept. Protein Res.*, 1996, **47**, 460.
58. L. H. Kondejewski, S. W. Farmer, D. S. Wishart, C. M. Kay, R. E. W. Hancock, and R. S. Hodges, *J. Biol. Chem.*, 1996, **271**, 25261.
59. R. Zidovetzki, U. Banerjee, D. W. Harrington, and S. I. Chan, *Biochemistry*, 1988, **27**, 5686.
60. K. P. Datema, K. P. Pauls, and M. Bloom, *Biochemistry*, 1986, **25**, 3796.
61. A. Paladini and L. C. Craig, *J. Am. Chem. Soc.*, 1954, **76**, 688.
62. H. Kleinkauf and H. von Döhren, *Progr. Drug Res.*, 1990, **34**, 287.
63. R. Sarges and B. Witkop, *J. Am. Chem. Soc.*, 1964, **86**, 1862.
64. O. S. Andersen, *Annu. Rev. Physiol.*, 1984, **46**, 531.
65. D. W. Urry, in "The Enzymes of Biological Membranes," ed. A. N. Martonosi, Plenum Press, New York, 1985, vol. 1, p. 227.
66. A. B. Wallace, *Annu. Rev. Biophys. Biophys. Chem.*, 1990, **19**, 127.
67. N. Sarkar and H. Paulus, *Nature New Biol.*, 1972, **239**, 228.
68. H. Ristow, W. Pschorn, J. Hansen, and U. Winkel, *Nature (London)*, 1979, **280**, 165.
69. T. Stein, J. Vater, V. Kruft, A. Otto, B. Wittmann-Liebold, P. Franke, M. Panico, R. M. Dowell, and H. R. Morris, *J. Biol. Chem.*, 1996, **271**, 15428.
70. M. Yamada and K. Kurahashi, *J. Biochem. (Tokyo)*, 1969, **66**, 529.
71. M. Yamada and K. Kurahashi, *J. Biochem. (Tokyo)*, 1968, **63**, 59.
72. W. Gevers, H. Kleinkauf, and F. Lipmann, *Proc. Natl. Acad. Sci. USA*, 1968, **60**, 269.
73. W. Gevers, H. Kleinkauf, and F. Lipmann, *Proc. Natl. Acad. Sci. USA*, 1969, **63**, 1335.
74. S. Otani, T. Yamanoi, and Y. Saito, *Biochem. Biophys. Res. Commun.*, 1968, **33**, 620.
75. S. Otani, Jr., T. Yamanoi, and Y. Saito, *Biochim. Biophys. Acta*, 1970, **208**, 496.
76. H. Kleinkauf, W. Gevers, and F. Lipmann, *Proc. Natl. Acad. Sci. USA*, 1969, **62**, 226.
77. H. Kleinkauf, R. Roskoski, Jr., and F. Lipmann, *Proc. Natl. Acad. Sci. USA*, 1971, **68**, 2069.
78. R. Roskoski, Jr., W. Gevers, H. Kleinkauf, and F. Lipmann, *Biochemistry*, 1970, **9**, 4839.
79. E. Stoll, Ø. Frøyshov, H. Holm, T.-L. Zimmer, and S. G. Laland, *FEBS Lett.*, 1970, **11**, 348.
80. R. Roskoski, Jr., G. Ryan, H. Kleinkauf, W. Gevers, and F. Lipmann, *Arch. Biochem. Biophys.*, 1971, **143**, 485.
81. R. Roskoski, Jr., H. Kleinkauf, W. Gevers, and F. Lipmann, *Biochemistry*, 1970, **9**, 4846.
82. H. Kleinkauf, W. Gevers, R. Roskoski, Jr, and F. Lipmann, *Biochem. Biophys. Res. Commun.*, 1970, **41**, 1218.
83. S. G. Lee and F. Lipmann, *Proc. Natl. Acad. Sci. USA*, 1974, **71**, 607.
84. S. Otani, T. Yamanoi, Y. Saito, and S. Otani, *Biochem. Biophys. Res. Commun.*, 1966, **25**, 590.
85. A. Gadow, J. Vater, W. Schlumbohm, Z. Palacz, J. Salnikow, and H. Kleinkauf, *Eur. J. Biochem.*, 1983, **132**, 229.

86. J. Vater, W. Schlumbohm, Z. Palacz, J. Salnikow, A. Gadow, and H. Kleinkauf, *Eur. J. Biochem.*, 1987, **163**, 297.
87. M. Pavela-Vrancic, H. van Liempt, E. Pfeifer, W. Freist, and H. von Döhren, *Eur. J. Biochem.*, 1994, **220**, 535.
88. T. W. Traut, *Eur. J. Biochem.*, 1994, **222**, 9.
89. M. Yamada and K. Kurahashi, *J. Biochem.* (*Tokyo*), 1969, **66**, 529.
90. J. Vater and H. Kleinkauf, *Biochim. Biophys. Acta*, 1976, **429**, 1062.
91. H. Koischwitz and H. Kleinkauf, *Biochim. Biophys. Acta*, 1976, **429**, 1041.
92. C. Christiansen, K. Aarstad, T.-L. Zimmer, and S. G. Laland, *FEBS Lett.*, 1977, **81**, 121.
93. T.-L. Zimmer and S. G. Laland, *Methods Enzymol.*, 1975, **43**, 567.
94. S. Tomino, M. Yamada, H. Itoh, and K. Kurahashi, *Biochemistry*, 1967, **8**, 2552.
95. L. Pass, T.-L. Zimmer, and S. G. Laland, *Eur. J. Biochem.*, 1973, **40**, 43.
96. L. Pass, T.-L. Zimmer, and S. G. Laland, *Eur. J. Biochem.*, 1974, **47**, 607.
97. K. Hori, T. Kurotsu, M. Kanda, S. Miura, A. Nozoe, and Y. Saito, *J. Biochem.* (*Tokyo*), 1978, **84**, 425.
98. M. Kanda, K. Hori, T. Kurotsu, S. Miura, A. Nozoe, and Y. Saito, *J. Biochem.* (*Tokyo*), 1978, **84**, 435.
99. C. Schröter, W. Rönspeck, M. Altmann, H. von Döhren, and H. Kleinkauf, in "Peptide Antibiotics—Biosynthesis and Functions," eds. H. Kleinkauf and H. von Döhren, 1982, Walter de Gruyter, Berlin, p. 259.
100. C. Schröter-Kermani, H. von Döhren, and H. Kleinkauf, *Biochim. Biophys. Acta*, 1986, **883**, 345.
101. F. Malpartida and D. A. Hopwood, *Nature* (*London*), 1984, **309**, 462.
102. K. F. Chater and C. J. Bruton, *EMBO J.*, 1985, **4**, 1893.
103. F. Malpartida and D. A. Hopwood, *Mol. Gen. Genet.*, 1986, **205**, 66.
104. S. E. Fishman, K. Cox, J. L. Larson, P. A. Reynolds, E. T. Seno, W. K. Yen, R. Van Frank, and C. L. Hershberger, *Proc. Natl. Acad. Sci. USA*, 1987, **84**, 8248.
105. J. F. Martin and P. Liras, *Annu. Rev. Microbiol.*, 1989, **43**, 173.
106. M. A. Marahiel, M. Krause, and H.-J. Skarpeid, *Mol. Gen. Genet.*, 1985, **201**, 231.
107. M. Krause, M. A. Marahiel, H. von Döhren, and H. Kleinkauf, *J. Bacteriol.*, 1985, **162**, 1120.
108. M. Krause and M. A. Marahiel, *J. Bacteriol.*, 1988, **170**, 4669.
109. J. Krätzschmar, M. Krause, and M. A. Marahiel, *J. Bacteriol.*, 1989, **171**, 5422.
110. K. Hori, Y. Yamamoto, T. Minetoki, T. Kurotsu, M. Kanda, S. Miura, K. Okamura, J. Furuyama, and Y. Saito, *J. Biochem.* (*Tokyo*), 1989, **106**, 639.
111. K. Turgay, M. Krause, and M. A. Marahiel, *Mol. Microbiol.*, 1992, **6**, 529.
112. F. Saito, K. Hori, M. Kanda, T. Kurotsu, and Y. Saito, *J. Biochem.* (*Tokyo*), 1994, **116**, 357.
113. A. J. Polouse, L. Rogers, T. M. Cheesbrough, and P. E. Kolattukudy, *J. Biol. Chem.*, 1985, **260**, 15 953.
114. Z. I. Randhawa and S. Smith, *Biochemistry*, 1987, **26**, 1365.
115. S. Borchert, T. Stachelhaus, and M. A. Marahiel, *J. Bacteriol.*, 1994, **176**, 2458.
116. R. Lambalot, A. M. Gehring, R. S. Flugel, P. Zuber, M. LaCelle, M. A. Marahiel, R. Reid, C. Khosla, and C. T. Walsh, *Chem. Biol.*, 1996, **3**, 923.
117. K. Turgay and M. A. Marahiel, *DNA Sequ.-J. Sequ. Mapp.*, 1995, **5**, 283.
118. K. Hori, Y. Yamamoto, K. Tokita, F. Saito, T. Kurotsu, M. Kanda, K. Okamura, J. Furuyama, and Y. Saito, *J. Biochem.* (*Tokyo*), 1991, **110**, 111.
119. J. Vater, W. Schlumbohm, J. Salnikow, K.-D. Irrgang, M. Miklus, T. Choli, and H. Kleinkauf, *Biol. Chem. Hoppe-Seyler*, 1989, **370**, 1013.
120. T. Kurotsu, K. Hori, M. Kanda, and Y. Saito, *J. Biochem.* (*Tokyo*), 1991, **109**, 763.
121. A. Raibaud, M. Zalacain, T. G. Holt, R. Tizard, and C. J. Thompson, *J. Bacteriol.*, 1991, **173**, 4454.
122. G. Mittenhuber, R. Weckermann, and M. A. Marahiel, *J. Bacteriol.*, 1989, **171**, 4881.
123. H. D. Moutz and M. A. Marahiel, *J. Bacteriol.*, 1997, **179**, 6843.
124. H. Koischwitz and H. Kleinkauf, *Biochim. Biophys. Acta*, 1976, **429**, 1052.
125. W. Schlumbohm, T. Stein, C. Ullrich, J. Vater, M. Krause, M. A. Marahiel, V. Kruft, and B. Wittmann-Liebold, *J. Biol. Chem.*, 1991, **266**, 23 135.
126. S. Berkenkamp, M. Karas, and F. Hillenkamp, *Proc. Natl. Acad. Sci. USA*, 1996, **93**, 7003.
127. T. Stein, J. Vater, P. Franke, M. Panico, R. McDowell, H. Morris, S. Berkenkamp, and F. Hillenkamp, in "Proceedings of the XIth International Conference on Methods in Protein Structure Analysis (MPSA)," 1996, P 111, p. 68.
128. T. H. Stein, Ph.D. Thesis, Technical University of Berlin, 1996.
129. J. Vater, T. Stein, D. Vollenbroich, V. Kruft, B. Wittmann-Liebold, P. Franke, L. Liu, and P. Zuber, *J. Prot. Chem.*, 1997, **16**, 557.
130. H. von Döhren, U. Keller, R. Zocher, and J. Vater, *Chem. Rev.*, 1997, **97**, 2675.
131. H.-J. Skarpeid, T.-L. Zimmer, and H. von Döhren, *Eur. J. Biochem.*, 1990, **189**, 517.
132. H.-J. Skarpeid, T.-L. Zimmer, B. Shen, and H. von Döhren, *Eur. J. Biochem.*, 1990, **187**, 627.
133. J. S. Scott-Craig, D. G. Panaccione, J.-A. Pocard, and J. D. Walton, *J. Biol. Chem.*, 1992, **267**, 26 044.
134. D. J. Smith, A. J. Earl, and G. Turner, *EMBO J.*, 1990, **9**, 2743.
135. B. Díez, S. Gutiérrez, J. L. Barredo, P. van Solingen, L. H. M. van der Voort, and J. F. Martín, *J. Biol. Chem.*, 1990, **265**, 16 358.
136. S. Gutiérrez, B. Díez, E. Montenegro, and J. F. Martín, *J. Bacteriol.*, 1991, **173**, 2354.
137. J. J. R. Coque, J. F. Martín, J. G. Calzada, and P. Liras, *Mol. Microbiol.*, 1991, **5**, 1125.
138. T. Stein, J. Vater, V. Kruft, B. Wittmann-Liebold, P. Franke, M. Panico, R. M. Dowell, and H. R. Morris, *FEBS Lett.*, 1994, **340**, 39.
139. M. Pavela-Vrancic, E. Pfeifer, H. van Liempt, H.-J. Schäfer, H. von Döhren, and H. Kleinkauf, *Biochemistry*, 1994, **33**, 6276.
140. M. Pavela-Vrancic, E. Pfeifer, W. Schröder, H. von Döhren, and H. Kleinkauf, *J. Biol. Chem.*, 1994, **269**, 14 962.
141. T. Stein, B. Kluge, J. Vater, P. Franke, A. Otto, and B. Wittmann-Liebold, *Biochemistry*, 1995, **34**, 4633.
142. D. Vollenbroich, B. Kluge, C. D'Souza, P. Zuber, and J. Vater, *FEBS Lett.*, 1993, **325**, 220.
143. D. Vollenbroich, N. Mehta, P. Zuber, J. Vater, and R. M. Kamp, *J. Bacteriol.*, 1994, **176**, 395.
144. K. Tokita, K. Hori, T. Kurotsu, M. Kanda, and Y. Saito, *J. Biochem.* (*Tokyo*), 1993, **114**, 522.
145. M. Gocht and M. A. Marahiel, *J. Bacteriol.*, 1994, **176**, 2654.
146. T. Stachelhaus and M. A. Marahiel, *J. Biol. Chem.*, 1995, **270**, 6163.

147. R. Dieckmann, Y.-O. Lee, H. van Liempt, H. von Döhren, and H. Kleinkauf, *FEBS Lett.*, 1995, **357**, 212.
148. T. Stachelhaus, A. Hüser, and M. A. Marahiel, *Chem. Biol.*, 1996, **3**, 913.
149. J. R. de Wet, K. V. Wood, M. DeLuca, D. R. Helsinki, and S. Subramani, *Mol. Cell. Biol.*, 1987, **7**, 725.
150. E. Conti, N. P. Franks, and P. Brick, *Structure*, 1996, **4**, 287.
151. E. Lozoya, H. Hoffman, C. Douglas, W. Schulz, D. Scheel, and K. Halbrock, *Eur. J. Biochem.*, 1988, **176**, 661.
152. F. Rusnak, M. Sakaitani, D. Drueckhammer, J. Reichert, and C. T. Walsh, *Biochemistry*, 1991, **30**, 2916.
153. I. F. Connerton, J. R. S. Fincham, R. A. Sandeman, and M. J. Hynes, *Mol. Microbiol.*, 1990, **4**, 451.
154. C. F. Higgins, I. D. Hiles, G. P. C. Salmond, D. R. Gill, J. A. Downie, I. J. Evans, I. B. Holland, L. Gray, S. D. Buckel, A. W. Bell, and M. A. Hermodson, *Nature (London)*, 1986, **323**, 448.
155. M. Fulda, E. Heinz, and F. P. Wolter, *Mol. Gen. Genet.*, 1994, **242**, 241.
156. T. C. Vanaman, S. J. Wakil, and R. L. Hill, *J. Biol. Chem.*, 1968, **243**, 6420.
157. D. H. Farrell, P. Mikesell, L. A. Actis, and J. H. Crosa, *Gene*, 1990, **86**, 45.
158. R. S. Hale, K. N. Jordan, and P. F. Leadlay, *FEBS Lett.*, 1987, **224**, 133.
159. A. Witkowski, J. Naggert, J. Mikkelsen, and S. Smith, *Eur. J. Biochem.*, 1987, **165**, 601.
160. D. H. Sherman, F. Malpartida, M. J. Bibb, H. M. Kieser, M. J. Bibb, and D. A. Hopwood, *EMBO J.*, 1989, **8**, 2717.
161. C. Khosla, S. Ebert-Khosla, and D. A. Hopwood, *Mol. Microbiol.*, 1992, **6**, 3237.
162. V. De Crécy-Lagard, P. Marlière, and W. Saurin, *CR Acad. Sci. Paris, Life Sci.*, 1995, **318**, 927.
163. W. V. Shaw, *CRC Crit. Rev. Biochem.*, 1983, **14**, 1.
164. A. Lewendon and W. V. Shaw, *J. Biol. Chem.*, 1993, **268**, 20 997.
165. G. C. Russel, R. S. Machado, and J. R. Guest, *Biochem. J.*, 1992, **287**, 611.
166. W. Taylor and N. M. Green, *Eur. J. Biochem.*, 1989, **179**, 241.
167. M. Saraste, P. R. Sibbald, and A. Wittinghofer, *TIBS*, 1990, **15**, 430.
168. A. Wittinghofer and E. Pai, *TIBS*, 1991, **16**, 382.
169. R. Dieckmann, M. Pavela-Vrancic, E. Pfeifer, H. von Döhren, and H. Kleinkauf, *Eur. J. Biochem.*, 1997, **247**, 1074.
170. M. Saito, K. Hori, T. Kurotsu, M. Kanda, and Y. Saito, *J. Biochem. (Tokyo)*, 1995, **117**, 276.
171. C. B. Davis, K. E. Smith, B. N. Campbell, Jr., and G. G. Hammes, *J. Biol. Chem.*, 1990, **265**, 1300.
172. F. Portillo and R. Serrano, *EMBO J.*, 1988, **7**, 1793.
173. H. R. Morris, M. Panico, M. Barber, R. S. Bordoli, R. D. Sedgwick, and A. N. Tyler, *Biochem. Biophys. Res. Commun.*, 1981, **101**, 623.
174. H. R. Morris, M. Panico, A. Karplus, P. E. Lloyd, and B. Riniker, *Nature (London)*, 1982, **300**, 643.
175. K. Biemann, *Annu. Rev. Biochem.*, 1992, **61**, 977.
176. T. A. Holak, S. K. Kearsley, Y. Kim, and J. H. Prestegard, *Biochemistry*, 1988, **27**, 6135.
177. T. A. Holak, M. Nilges, and H. Oschkinat, *FEBS Lett.*, 1989, **242**, 218.
178. E. Pfeifer, M. Pavela-Vrancic, H. von Döhren, and H. Kleinkauf, *Biochemistry*, 1995, **34**, 7450.
179. A. G. W. Leslie, *J. Mol. Biol.*, 1990, **213**, 167.
180. A. Mattevi, G. Obmolova, E. Schulze, K. H. Kalk, A. H. Westphal, A. de Kok, and W. G. J. Hol, *Science*, 1992, **255**, 1544.
181. C. C. Gilhuus-Moe, T. Kristensen, J. E. Bredesen, T.-L. Zimmer, and S. G. Laland, *FEBS Lett.*, 1970, **7**, 287.
182. M. Kanda, K. Hori, T. Kurotsu, S. Miura, and Y. Saito, *J. Biochem. (Tokyo)*, 1989, **105**, 653.
183. C. T. Walsh, *J. Biol. Chem.*, 1989, **264**, 2393.
184. A. Stindl and U. Keller, *Biochemistry*, 1994, **33**, 9358.
185. C.-Y. Shiau, J. A. Baldwin, M. F. Byford, W. J. Sobey, and C. J. Schofield, *FEBS Lett.*, 1995, **358**, 97.
186. D. J. Neidhart, P. L. Howell, G. A. Petsko, V. M. Powers, R. Li, G. L. Kenyon, and J. A. Gerlt, *Biochemistry*, 1991, **30**, 9264.
187. J. A. Gerlt and P. G. Gassman, *Biochemistry*, 1993, **32**, 11 943.
188. J. A. Landro, J. A. Gerlt, K. W. Kozarich, C. W. Koo, V. J. Shah, G. L. Kenyon, D. J. Neidhart, S. Fujita, and G. A. Petsko, *Biochemistry*, 1994, **33**, 635.
189. P. C. Babbitt, G. T. Mrachko, M. S. Hasson, G. W. Huisman, R. Kolter, D. Ringe, G. A. Petsko, G. L. Kenyon, and J. A. Gerlt, *Science*, 1995, **267**, 1159.
190. E. Alvarez, B. Meesschaert, E. Montenegro, S. Gutiérrez, B. Díez, J. L. Barredo, and J. F. Martín, *Eur. J. Biochem.*, 1993, **215**, 323.
191. S. Brenner, *Nature (London)*, 1988, **334**, 528.
192. P. C. Babbitt, G. L. Kenyon, B. M. Martin, H. Charest, M. Sylvestre, J. D. Scholten, K.-H. Chang, P.-H. Liang, and D. Dunaway-Mariano, *Biochemistry*, 1992, **31**, 5594.
193. S. Donadio, M. J. Staver, J. B. McAlpine, S. J. Swanson, and L. Katz, *Science*, 1991, **252**, 675.
194. J. Cortes, S. F. Haydock, G. A. Roberts, D. J. Bevitt, and P. F. Leadley, *Nature (London)*, 1990, **348**, 176.
195. F. de Ferra, F. Rodriguez, O. Tortora, C. Tosi, and G. Grandi, *J. Biol. Chem.*, 1997, **272**, 25 304.
196. H. Kleinkauf and H. Koischwitz, *Progr. Mol. Subcell. Biol.*, 1978, **6**, 59.
197. E. V. Orlova, P. Dube, J. R. Harris, E. Beckmann, F. Zemlin, J. Markl, and M. van Heel, *J. Mol. Biol.*, 1997, **271**, 417.
198. T. Stachelhaus, A. Schneider, and M. A. Marahiel, *Science*, 1995, **269**, 69.
199. R. McDaniel, S. Ebert-Khosla, D. A. Hopwood, and C. Khosla, *Nature (London)*, 1995, **375**, 549.
200. J. Cortes, K. E. H. Wiesmann, G. A. Roberts, M. J. B. Brown, J. Staunton, and P. F. Leadley, *Science*, 1995, **268**, 1487.
201. A. Elsner, H. Engert, W. Saenger, L. Hamoen, G. Venema, and F. Bernhard, *J. Biol. Chem.*, 1997, **272**, 4814.

4.14
Biosynthesis of Selenocysteine and its Incorporation into Proteins as the 21st Amino Acid

DOLPH L. HATFIELD, VADIM N. GLADYSHEV,
SANG ICK PARK, HAROLD S. CHITTUM,
BRADLEY A. CARLSON, and MOHAMED E. MOUSTAFA
National Institute of Health, Bethesda, MD, USA

and

JIN MO PARK, JUN RYUL HUH, MIJIN KIM, and
BYEONG JAE LEE
Seoul National University, Republic of Korea

4.14.1   INTRODUCTION

Selenocysteine (Sec) is the 21st naturally occurring amino acid in proteins.[1-11] It has its own tRNA and its own codeword within the genetic code which is UGA. A feature of Sec is that it is biosynthesized on tRNA. Sec tRNA is aminoacylated with serine, which initiates Sec biosynthesis. The identity elements in Sec tRNA which specify its aminoacylation with serine correspond to those encoded in seryl-tRNA synthetase. Free Sec (i.e., that which is not attached to tRNA) is also biosynthesized in some cells, but selenocysteyl-tRNA synthetase has not been described. Thus, for specific site incorporation of Sec into protein, its biosynthesis must occur on Sec tRNA[1-11] and this phenomenon appears to be almost universal in nature.[5,6,8] In this chapter, the biosynthesis of Sec and its incorporation into proteins will be discussed. The chapter will emphasize primarily current understanding of Sec metabolism in higher eukaryotes, as excellent reviews on this subject in prokaryotes have been presented elsewhere.[1,2,4,7,9] However, Sec metabolism in prokaryotes will be discussed in detail whenever the corresponding biochemical events are not fully understood in higher eukaryotes. It should be noted that reviews on the occurrence of selenoproteins,[2,9,10-12] on the incorporation of selenium into proteins in prokaryotes[1,2,4,7,9,11] and eukaryotes,[6,8,11] and on the role of selenium in human health[8,13,14] have also been published.

4.14.1.1   Structure of Sec and Other Related Amino Acids

The structure of Sec is similar to that of cysteine with the exception that Sec contains selenium in place of sulfur (**1**) and (**2**). As serine has a role in the biosynthesis of Sec, and phosphoserine may also have a role, the structures of serine (**3**) and phosphoserine (**4**) are also shown. The correct chemical name for Sec is 2-amino-3-selenopropionic acid, but since this amino acid was initially named Sec,[15] this nomenclature will be retained.

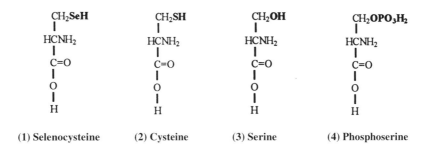

| (1) Selenocysteine | (2) Cysteine | (3) Serine | (4) Phosphoserine |

4.14.2   BIOSYNTHESIS OF Sec

As noted in the introduction, the biosynthesis of Sec may occur by one of two pathways. Sec may be synthesized on tRNA or it may be synthesized as free Sec where selenium replaces sulfur in sulfur metabolism. The replacement of sulfur with selenium may occur in either the pathway for cysteine or methionine biosynthesis. The resulting Sec and selenomethionine may then be attached to tRNA^{Cys} or tRNA^{Met} by the corresponding aminoacyl-tRNA synthetase and incorporated nonspecifically (i.e., in response to cysteine or methionine codons) into protein. Both the specific and nonspecific pathways of Sec biosynthesis and the incorporation of this amino acid into protein are discussed below.

4.14.2.1   Biosynthesis of Sec on tRNA (Specific Site Incorporation of Sec into Protein)

One of the unique characteristics of Sec is that its biosynthesis occurs on tRNA.[1-11] Only one other example of the biosynthesis of an amino acid occurring on tRNA is known. The biosynthesis of glutamine by amination of glutamic acid occurs after tRNA[Gln] is aminoacylated with glutamic acid in the presence of glutamyl-tRNA synthetase.[16,17] Whereas the biosynthesis of glutamine on tRNA has been reported in certain bacteria and chloroplasts of plants,[16,17] the biosynthesis of Sec on tRNA appears to be almost universal in nature (see below).

The first step in Sec biosynthesis is the aminoacylation of Sec tRNA with serine.[1-11] Sec tRNA has therefore been designated tRNA[Ser]Sec.[3] Aminoacylation of Sec tRNA[Ser]Sec with serine has been characterized in eukaryotes where the identity elements within tRNA[Ser]Sec correspond to those for serine and not those for Sec.[18-20] The sequences of the canonical serine tRNAs and the Sec tRNA[Ser]Sec isoacceptors have little homology to each other. Site-specific mutagenesis of synthetic genes encoding either Sec tRNA[Ser]Sec or one of the serine isoacceptors were used to generate mutant tRNAs for analysis of their ability to be aminoacylated by seryl-tRNA synthetase.[18,19] In this manner, it was determined that the discriminator base (G73) is essential for aminoacylation of both Sec and serine tRNAs and, in addition, the long extra arm plays an important role in the identity process. In 1996, an analysis of the structural features in tRNA[Ser]Sec that are required for aminoacylation indicated that the 13 paired bases in the acceptor and TψC stems, as well as the D stem, are also important.[20]

After tRNA[Ser]Sec is aminoacylated with serine, the serine moiety then becomes the backbone for Sec biosynthesis in both prokaryotes[21] and in eukaryotes.[22,23] The biosynthetic pathway in *Escherichia coli* has been established using genetic and biochemical techniques to identify four genes (*SelA, B, C, D*) and the corresponding gene products (SELA, B, C, D).[1,2,4,7,9-11] The serine moiety is modified in prokaryotes to an intermediate, aminoacrylyl-tRNA[Ser]Sec, by the 2,3-elimination of a water molecule. This reaction is catalyzed by selenocysteine synthase which is designated SelA[1,2] (Figure 1(a)). SelA is a large protein consisting of 50 kDa subunits that remain stably bound to the aminoacrylyl intermediate. The activated form of selenium is selenophosphate[24] that is synthesized by selenophosphate synthetase which is designated SelD.[1,2,7] Selenophosphate is donated to the aminoacrylyl intermediate which yields Sec attached to tRNA. Sec-tRNA[Ser]Sec is then ready to be utilized in protein synthesis (see below).

Details of the biosynthesis of Sec have not yet been characterized in mammals. Possible biosynthetic pathways in mammals are shown in Figure 1(b). As the serine moiety on tRNA[Ser]Sec was initially found to be phosphorylated by a kinase in higher vertebrate systems,[25,26] it was subsequently speculated that phosphoseryl-tRNA[Ser]Sec was an intermediate in the biosynthesis of Sec.[23] In fact, Mizutani and Hitaka initially reported that phosphoseryl-tRNA was the intermediate in the biosynthesis of Sec in eukaryotes[27,28] and also in prokaryotes,[29] but later reported that phosphoseryl-tRNA is not the intermediate in eukaryotes.[30,31] It was also subsequently found that the true intermediate in prokaryotic systems was aminoacrylyl-tRNA[Ser]Sec.[32] Mizutani and co-workers[20] have proposed that phosphoseryl-tRNA[Ser]Sec is "an inactive storage form" that upon dephosphorylation "regenerates seryl-tRNA for Sec synthesis." The important point to emphasize in each of these studies is that the pathway of Sec biosynthesis in mammals and the role of phosphoseryl-tRNA[Ser]Sec have not been clearly characterized. It should also be noted that an earlier study provided evidence that addition of labeled selenium to rat liver extracts and rat liver slices resulted in Sec attachment to tRNA.[33] The pathway of Sec biosynthesis, however, was not characterized in the latter studies.

Three forms of selenophosphate synthetase (SPS) have been described in mammals[34-36] and the genes for these proteins[34,36] have been designated *Sps*1 and *Sps*2.[36] Although the structure of the product generated by SPS (which is monoselenophosphate) has been carefully characterized in prokaryotes,[24] it has not been characterized in eukaryotes.[34-36] The mammalian *Sps*1 gene was introduced into an *E. coli* mutant lacking the corresponding host gene, and the mammalian gene product promoted Sec biosynthesis in *E. coli*, albeit less efficiently than the homologous *E. coli SelD* gene product.[34] This observation and the homology of SELD to SPS1 and SPS2 suggest that the selenium donor in mammals is similar, if not identical, to that generated in *E. coli*. It should also be noted that tRNA[Ser]Sec from *Xenopus* can replace the corresponding tRNA[Ser]Sec from *E. coli* when a vector encoding the *Xenopus* gene is introduced into an *E. coli* mutant lacking the host Sec tRNA[Ser]Sec gene.[37] Although Sec and selenoprotein syntheses are much less efficient when *Xenopus* tRNA[Ser]Sec is utilized in *E. coli*, this study demonstrates that the overall structural features of Sec tRNAs[Ser]Sec from higher vertebrates and prokaryotes are conserved. Interestingly, *Sps*2 contains a TGA codon that corresponds to the proposed active site of the gene product and these investigators suggest that the incorporation of Sec into SPS2 might be involved in the autoregulation of its synthesis.[36]

The identity elements for selenocysteylation of tRNA[Ser]Sec have been examined. The length of the acceptor arm is critical in the synthesis of Sec from serine[20,38] and the D-stem may also have a role[20] in this process. Previously, the identity elements for phosphorylation of seryl-tRNA[Ser]Sec were shown to be restricted to the D-stem.[39]

(a) *E. coli*

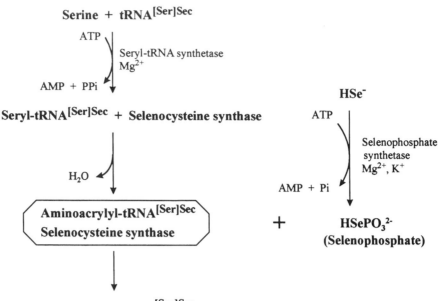

(b) Mammals

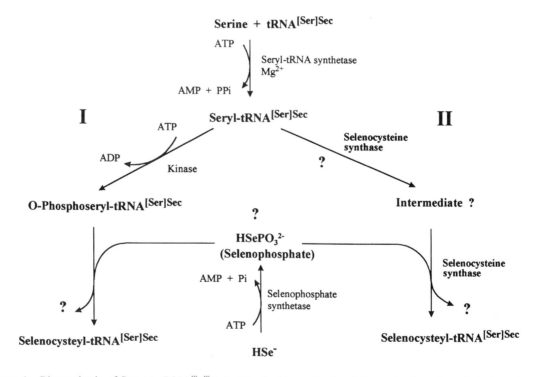

Figure 1 Biosynthesis of Sec on tRNA[Ser]Sec. In (a), the biosynthesis of Sec and selenophosphate have been established in *E. coli* as discussed in the text. In (b), the biosynthesis of Sec through phosphoseryl-tRNA[Ser]Sec as an intermediate or through an unknown intermediate involving the SELA homologue selenocysteine synthase, is shown. Both pathways have been implicated in Sec biosynthesis in mammals, but since neither pathway nor the precise active selenium donor have been completely established, each of these steps are shown with a question mark.

4.14.2.2   Biosynthesis of Free Sec (Nonspecific Incorporation of Sec into Protein)

The biosynthesis of free Sec occurs in sulfur metabolic pathways of certain bacteria and plants.[9,40] For example, in the genus, *Astragalus*, enzymes involved in sulfur amino acid metabolism were unable to distinguish between sulfur-containing substrates and their selenium analogues.[41–43] Furthermore, *E. coli* and *Salmonella typhimurium* mutants defective in their ability to incorporate selenium specifically into Sec-containing proteins synthesize selenomethionine and free Sec and incorporate these amino acids nonspecifically into protein.[44] These observations are in contrast to the biosynthesis of Sec on tRNA[Ser]Sec, where specific proteins are required and none of the steps involves free Sec (see Section 4.14.2.1).

Sec has been shown to be aminoacylated directly to tRNA^Cys by bacterial cysteyl-tRNA synthetase *in vitro*.[45,46] Moreover, a system was developed that allowed quantitative incorporation of Sec in place of cysteine in *E. coli* thioredoxin.[47] In this case, synthesis of selenothioredoxin was induced in a cysteine-deficient medium in the presence of a moderately toxic concentration of Sec. As expected, the nonspecific incorporation of Sec in place of cysteine has a profound effect on the properties of the resulting protein. For example, the diselenide bond of selenothioredoxin could not be reduced by bacterial thioredoxin reductase.[47] In other studies, chemical substitution of the active center serine with Sec in subtilisin (which is a protease) produced an enzyme with peroxidase activity.[48] In contrast, methionine auxotroph-based substitution of selenomethionine for methionine has little effect on the properties of the mutant proteins in most, but not all, cases.[49] Replacement of methionine with selenomethionine in protein has been used extensively in X-ray crystallography, resolving the phase problem for protein crystal structure determination.[50]

Several members within the genus *Astragalus* can grow at a high selenium concentration and accumulate up to several mg of selenium per g of dry weight.[51] In these plants, selenium is accumulated in the form of low molecular weight organic compounds, usually Se-methyl-selenocysteine. At the same time, the nonspecific incorporation of Sec into proteins in place of cysteine is significantly reduced in this species. Sec methyltransferase was purified and characterized from cultured *Astragalus* cells.[51] This enzyme does not utilize cysteine as a substrate and most likely plays a crucial role in conferring selenium tolerance by selenium-accumulating plants due to its ability to methylate Sec. In this manner, Sec is prevented from entering the cysteine pathway of incorporation into protein. In nonselenium-accumulating plants, the accumulation of Se-methyl-Sec was not observed, suggesting that Sec will intrude into the pathway involving cysteyl-tRNA synthetase and tRNA^Cys and be incorporated into protein.[52] Replacement of cysteine in protein with Sec can be considered as one of the causes for selenium toxicity.[53] Possible biosynthetic pathways showing the means by which selenium can enter protein through methionine and cysteine synthesis are shown in Figure 2.

Some enzymes that degrade sulfur-containing amino acids can also degrade selenium-containing amino acids.[54] However, selenocysteine-β-lyase was identified as an enzyme that can specifically degrade Sec.[55]

4.14.3   THE CODON FOR Sec IS UGA

Sec was first identified as an amino acid residue in clostridial glycine reductase,[15] which followed earlier observations of the presence of selenium in this enzyme[56] and glutathione peroxidase.[57,58] Zinoni *et al.*[59] showed that the formate dehydrogenase H gene in *E. coli* and Chambers *et al.*,[60] Sukenaga *et al.*,[61] and Mullenbach *et al.*[62] showed that the glutathione peroxidase gene in mammals contained a TGA codon in the open reading frame and that this TGA codon corresponded to the location of Sec in the corresponding gene product. Prior to the work of Chambers *et al.*,[60] the amino acid sequence of bovine glutathione peroxidase was determined by Günzler *et al.*,[63] which permitted Chambers *et al.*[60] to align the TGA codon in the mouse glutathione peroxidase gene with the Sec moiety in the glutathione peroxidase gene product. The precise position of the Sec moiety in formate dehydrogenase was not determined until Stadtman *et al.*[64] sequenced the peptide containing this amino acid and showed that it aligned perfectly with the UGA in the corresponding mRNA. The above studies show that the codon for Sec is UGA. It should be emphasized that these studies did not demonstrate whether the Sec moiety present in the corresponding protein arose as a result of direct incorporation or as a result of posttranslational modification.

It is also of interest to note that of all the codewords in the genetic code, UGA has undergone the most dramatic changes in evolution as evidenced by its multiple uses in genetic language.[5,65,66] UGA serves as both a termination codon[67–69] and as a Sec codon in the (almost) universal genetic code (see Section 4.14.4), as a cysteine codon in *Euplotes octocarinatus*[70] and as a tryptophan codon

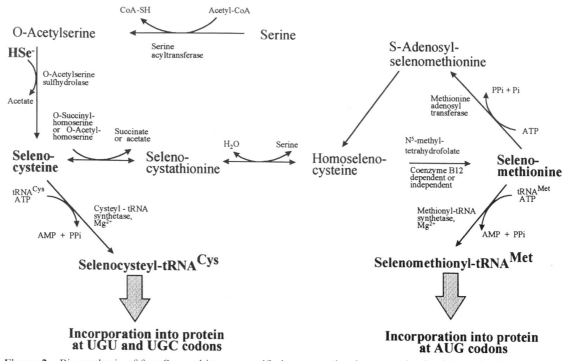

Figure 2   Biosynthesis of free Sec and its nonspecific incorporation into protein. Sec biosynthesis may occur by replacement of sulfur with selenium in methionine and cysteine synthetic pathways. The pathways shown are not intended to imply that all organisms synthesize Sec and selenomethionine in this manner, but the figure illustrates how Sec may enter protein nonspecifically.

in *Mycoplasma* and mitochondria.[65,66] It also serves as an inefficiently read codon, which is decoded by tryptophanyl-tRNA in *Bacillus subtilis*[71] and as an inefficiently read codon in *E. coli*, which is presumably decoded by tryptophanyl-tRNA.[72]

4.14.4   UNIVERSALITY OF UGA AS A CODON FOR Sec

The genetic code was previously thought to be used in the same way by all organisms and therefore was considered to be universal. However, it is now known that many changes have occurred in the genetic code during evolution[65,66] and thus, it has been described as the "almost universal genetic code".[73] In the present discussion of the assignment of Sec to the universal genetic code, the code will be referred to as the almost universal genetic code.[73]

Sec tRNAs that decode UGA in protein synthesis are widespread in nature.[5,6,8] Initially, a Sec-tRNA that decodes UGA in protein synthesis was identified intracellularly in *E. coli*[21] and in mammals.[23] These studies clearly established the existence of selenocysteyl-tRNA[[Ser]Sec] and provided strong evidence that the Sec moiety in selenoproteins must arise by direct incorporation of Sec and not by posttranslational modification. The gene encoding Sec tRNA[[Ser]Sec] was subsequently found to be ubiquitous in the subkingdom Eubacteria[74] and tRNA[[Ser]Sec] or its gene was found to be ubiquitous in the animal kingdom.[75] Sec-tRNAs that decode UGA were also found in two very diverse protists, *Tetrahymena borealis* and *Thalassiosira pseudoonana*,[76] in a higher plant, *Beta vulgaris*, and in a filamentous fungus, *Gliocladium virens*.[77] Several potential Sec-containing protein genes (i.e., genes that contained TGA in an open reading frame) and a Sec tRNA gene were found in the genome sequence of the archaeon, *Methanococcus jannaschii*.[78] Each of these studies shows that UGA as a codon for Sec occurs in representative organisms from all five life kingdoms, Monera (with its two subkingdoms, Eubacteria and Archaebacteria), Protists, Plants, Animals, and Fungi (see Figure 3 for the delineation of organisms into five life kingdoms).[79] Therefore, Sec should be assigned to UGA in the almost universal genetic code[5] as shown in Figure 4.

In addition to UGA, AUG also has a dual function in the almost universal genetic code (see Figure 4).[67,69] AUG serves both as a codon that initiates protein synthesis and a codon for methionine at internal positions of protein. The dual role of AUG has been known since the code was first

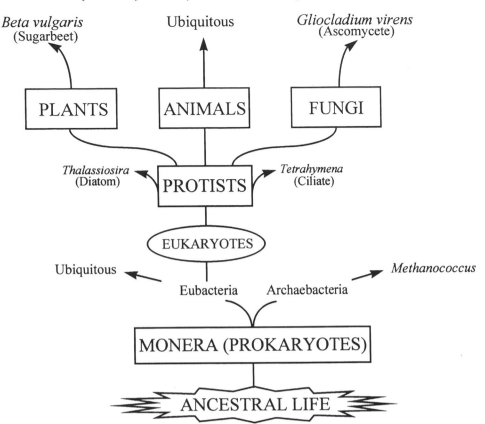

Figure 3   Evolutionary tree showing the distribution of Sec in nature.

Middle base / 5' Base	U	C	A	G	Middle base / 3' Base
U	Phenylalanine	Serine	Tyrosine	Cysteine	U
	Phenylalanine	Serine	Tyrosine	Cysteine	C
	Leucine	Serine	Terminator	Selenocysteine } Terminator }	A
	Leucine	Serine	Terminator	Tryptophan	G
C	Leucine	Proline	Histidine	Arginine	U
	Leucine	Proline	Histidine	Arginine	C
	Leucine	Proline	Glutamine	Arginine	A
	Leucine	Proline	Glutamine	Arginine	G
A	Isoleucine	Threonine	Asparagine	Serine	U
	Isoleucine	Threonine	Asparagine	Serine	C
	Isoleucine	Threonine	Lysine	Arginine	A
	Methionine } Initiator }	Threonine	Lysine	Arginine	G
G	Valine	Alanine	Aspartic Acid	Glycine	U
	Valine	Alanine	Aspartic Acid	Glycine	C
	Valine	Alanine	Glutamic Acid	Glycine	A
	Valine	Alanine	Glutamic Acid	Glycine	G

Figure 4   The almost universal genetic code showing the inclusion of Sec as the 21st amino acid.

deciphered[67,68] and thus it is not surprising that a second codon, UGA, also has a dual function. Furthermore, the fact that two codons have now been identified in the almost universal genetic code with multiple functions raises the possibility that other codewords may also exist with multiple roles.

The genome of *Saccharomyces cerevisiae* has been sequenced.[80] This organism does not appear to encode a Sec tRNA[SerSec] gene or any potential selenoprotein genes.[8] A *S. cerevisiae* homologue of the glutathione peroxidase gene in mammals was found to contain a cysteine codon (TGT) at the position where the mammalian gene encodes Sec (codon TGA). Thus, this organism appears to lack the biosynthetic pathways for specific site incorporation of Sec into protein found in other life forms.[8] This observation reflects genetic diversity and should not affect our proposal that Sec belongs in the almost universal genetic code. Other yeast forms, such as *Candida albicans*, are known to encode variations in the almost universal genetic code in their genomes.[65,66,81]

The fact that *S. cerevisiae* does not appear to have the system for incorporating Sec into specific sites of protein demonstrates that this means of utilizing selenium is not essential to life. Furthermore, *E. coli* mutants lacking the ability to incorporate Sec into specific sites of protein can grow normally under certain conditions.[82] Do these findings imply that the incorporation of Sec into specific sites of protein is not important in nature? In mammalian systems, the synthesis of specific selenoproteins is essential to sustain life as removal of the Sec tRNA[SerSec] gene from the mouse genome by gene replacement or "gene knockout" is embryonically lethal.[83] In *E. coli*, the selenoprotein, formate dehydrogenase, is required to detoxify formate under aerobic growth conditions.[82] The ability to incorporate Sec into specific selenoproteins is, therefore, not essential to sustain life in *E. coli*, but provides these organisms with a selective advantage. Thus, the ability to synthesize specific selenoproteins is essential to some life forms, while it appears to provide only a selective advantage to others. In addition, since this process is widespread in nature, it must then be important as a requirement and/or as a selective advantage to virtually all organisms.

4.14.5   Sec tRNA[SerSec]

Sec tRNA[SerSec] has been described as the key molecule[2] and the central component[6] in the biosynthesis of selenoproteins. It has a dual function of serving as a carrier molecule for the biosynthesis of Sec and as the donor of Sec to the growing polypeptide chain in response to UGA Sec codons. Sec tRNA[SerSec] was originally identified as a minor seryl-tRNA in higher vertebrates that formed phosphoseryl-tRNA[25] and recognized UGA in a ribosomal binding assay.[84] This "seryl-tRNA" was resolved into two discrete species[85] which were subsequently shown to form phosphoseryl-tRNA[86,87] and selenocysteyl-tRNA.[23]

4.14.5.1   Primary Sequences of Sec tRNAs

The sequences of Sec tRNAs[SerSec] from rat liver,[88] bovine liver,[89,90] mouse and HeLa cells,[91] and *E. coli*[92] have been determined. The presumed secondary structures of rat liver and *E. coli* tRNA[SerSec] are shown in Figure 5. Sec tRNAs from mammals contain 90 bases making them the longest eukaryotic tRNAs sequenced to date. Sec tRNA[SerSec] from *E. coli* contains 95 bases and represents the longest tRNA sequenced to date. It should be noted that several prokaryotic genes have been sequenced that encode potentially even longer Sec tRNAs.[74] Eukaryotic and prokaryotic tRNA[SerSec]s have little sequence homology to each other, but share several features in their secondary structures that are different from other tRNAs. Both animal[93,94] and prokaryotic[2] Sec tRNA[SerSec]s have: (i) longer base pairings in their acceptor stems (8 in *E. coli* and 9 in animals) than the normal 7 found in most other tRNAs; (ii) 13 base pairs in the acceptor- and TψC-stem helices instead of the normal 12 found in other tRNAs; and (iii) 6 possible base pairs in the dihydrouracil stem instead of the normal three or four found in other tRNAs. Both higher vertebrate and *E. coli* Sec tRNA[SerSec]s have few modified bases (4 each) compared to most other tRNAs which can have as many as 15–17 modified bases and both Sec tRNA[SerSec]s have a UCA anticodon that is responsible for translating UGA (Sec) codons. As shown in Figure 5, the two tRNA[SerSec] species in mammalian tissues differ from each other by a single methylated 2'-O-ribose in the wobble position of one of the tRNAs.[88,90] The 5-methylcarboxymethyluridine-2'-O-methylribose (mcmUm) nucleoside has been found only in mammalian Sec tRNA[SerSec],[88,90] while 5-methylcarboxymethyluridine (mcmU) has been found in a yeast arginine tRNA.[95] The other modified bases in mammalian Sec tRNA[[Ser]Sec]s are N^6-isopentenyladenosine (i^{6}A) at position 37, pseudouridine (ψ) at position 55, and 1-methyladenosine (m^{1}A) at position 58.

(a) Rat liver   (b) *E. coli*

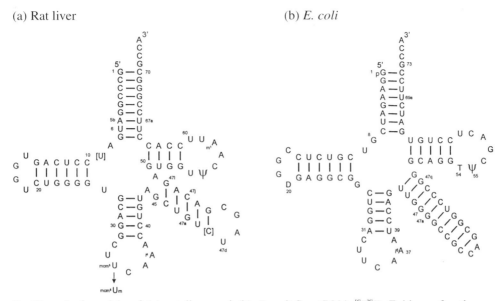

Figure 5 Cloverleaf models of (a) rat liver and (b) *E. coli* Sec tRNAs[[Ser]Sec]. Evidence for the secondary structures of these tRNAs is discussed in the text.

The model depicting the secondary structure of the mammalian Sec tRNA[[Ser]Sec]s shown in Figure 5 was originally proposed by Böck *et al.*[2] Evidence that this structure is indeed the correct one has been provided by Krol and co-workers.[93,94]

4.14.5.2 The Sec tRNA[[Ser]Sec] Gene and Possible Editing of the Gene Product

The gene for tRNA[[Ser]Sec] occurs in single copy in the genome of humans,[96] rabbits,[97] chickens,[98] cattle,[99] rats,[100] mice,[101,102] *Xenopus*,[75,103] *Drosophila*,[75] and *Caenorhabditis elegans*.[75] A pseudogene for tRNA[[Ser]Sec] was also found in human[96] and rabbit[97] genomes. The tRNA[[Ser]Sec] gene was mapped to human chromosome 19.[104] It was localized to bands q13.2–q13.3 and ordered with respect to other genes in this region.[105] The mouse gene has also been mapped and its chromosomal position determined.[101,102] The mouse gene occupies a similar position in its genome as that found for the tRNA[[Ser]Sec] gene in humans.

Since the Sec tRNA[[Ser]Sec] gene is present in only a single copy in the genomes of higher animals, all species of Sec tRNA[106] must arise by modification of the primary transcript. Of all the multiple, chromatographically distinct species of Sec tRNA[[Ser]Sec] described in the literature,[85,86,88,90,106] only the two major forms have been characterized and shown to differ from each other by a single 2′-O-methylribose of the nucleoside at position 34.[88,90] What then accounts for the differences in other species of Sec tRNA[[Ser]Sec] that have been observed? A Sec tRNA[[Ser]Sec] from bovine liver was sequenced[89] and shown not to be colinear with the Sec tRNA[[Ser]Sec] gene from cattle, suggesting that maturation of this tRNA includes posttranscriptional editing,[99] as has been shown to occur for other RNA species.[107] The bovine Sec tRNA[[Ser]Sec] population has been studied by Amberg *et al.*[90] and they were unable to detect any evidence for edited isoacceptors. However, the edited tRNA[[Ser]Sec] that was isolated and sequenced earlier from bovine liver was shown to be the most hydrophobic Sec tRNA[[Ser]Sec] isoacceptor,[89] and the Amberg *et al.* study reports the detection of a similarly hydrophobic isoacceptor which was not characterized.[90] If the existence of multiple Sec tRNA[[Ser]Sec] isoacceptors implies that these molecules may have distinct functions, then the structural resolution of all such species[106] will be important in understanding the regulation of selenoprotein biosynthesis. It should be noted that RNA editing is an extremely important phenomenon in that gene structures are altered posttranscriptionally and thus the function of the final RNA product may be changed from that encoded in the gene.[107]

4.14.5.3   Occurrence and Biosynthesis of Sec Isoacceptors in Mammalian Cells and Tissues

As noted above (Section 4.14.5.2), of the multiple isoacceptors within the Sec tRNA$^{[Ser]Sec}$ population in mammals,[106] the structures of only two, mcmU and mcmUm, have been characterized.[88,90] The relationship of these two isoacceptors to each other and their biosynthesis have been elucidated. Their relative amounts and distributions have been shown to vary in mammalian cells grown in culture[108] and in different mammalian tissues.[88] Selenium levels in the media or diet influence the steady state levels of both isoacceptors. The amount of the Sec tRNA$^{[Ser]Sec}$ population increased as much as 2.5-fold in some tissues[88] and the distribution of mcmU and mcmUm manifested a dramatic shift from the former to the latter form in response to selenium.[88,108] Extremely selenium-deficient rats were replenished with selenium and the amounts and distributions of the mcmU and mcmUm isoacceptors were followed over a 72 h period in liver, heart, kidney, and muscle.[109] The increase in the amounts of the Sec tRNA$^{[Ser]Sec}$ population was consistent with that found earlier[89] and mcmU and mcmUm isoacceptors shifted from the former to the latter isoacceptor in response to selenium. The rate of change in distribution varied, however, in the different tissues studied.

The pathway of biosynthesis of the mcmU and mcmUm Sec isoacceptors and the role of selenium in enhancing the levels of the Sec tRNA$^{[Ser]Sec}$ population have been reconstituted in *Xenopus* oocytes.[110–112] The modified bases, ψ at position 55 and m^{1}A at position 58, are biosynthesized in the nucleus and i^{6}A, mcmU[110–112] and mcmUm[110,112] in the cytoplasm. The biosynthetic pathway resulting in the fully modified isoacceptors is shown in Figure 6. Addition of mcmU and mcmUm, but without i^{6}A, occurs in the cytoplasm.[110,112] The mcmU species is converted to the mcmU-i^{6}A form which is then methylated to the mcmUm-i^{6}A form.[110,112]

Supplementation of the *Xenopus* oocyte culture media with exogenous selenium results in an enrichment of the Sec tRNA$^{[Ser]Sec}$ population.[110,112] Microinjection of the Sec tRNA$^{[Ser]Sec}$ gene or the Sec tRNA$^{[Ser]Sec}$ transcript (prepared *in vitro* from the synthetic gene) and subsequent recovery of the products resulting from maturation demonstrated that selenium stabilizes the Sec tRNA$^{[Ser]Sec}$ population, but does not increase gene transcription. Therefore, biosynthesis of all modified nucleosides in the two major Sec isoacceptors and the mechanism by which selenium enhances the levels of the Sec tRNA$^{[Ser]Sec}$ population in mammalian cells and tissues (see above) have been elucidated in *Xenopus* oocytes.

The levels and distributions of the two isoacceptors within the Sec tRNA$^{[Ser]Sec}$ population were measured in mouse embryonic stem cells heterozygous for the tRNA$^{[Ser]Sec}$ gene.[113] The level of the Sec tRNA$^{[Ser]Sec}$ population was about 60% of that found in wild type cells in two mutant cell lines. The distribution of the mcmU and mcmUm isoacceptors was about 2:1 for mcmU and about 1:1 for the mcmUm in wild type and mutant cells. Thus, the amount of mcmUm, which is the biosynthetic product of mcmU, is approximately the same in both wild type and mutant cells. The level of glutathione peroxidase in wild type and mutant cells was the same, suggesting that the Sec isoacceptors are not limiting in selenoprotein biosynthesis.[113]

4.14.5.4   Transcription of the Sec tRNA$^{[Ser]Sec}$ Gene

As Sec tRNA$^{[Ser]Sec}$ is considered to be the key molecule and the central component in the biosynthesis of Sec and its incorporation into protein (see Section 4.14.5), it is essential to understand the regulatory mechanisms involved in the expression of this unique tRNA. In this subsection, the transcription of tRNA$^{[Ser]Sec}$ and the regulatory factors involved in RNA polymerase III (Pol III) transcribed genes will be discussed.

4.14.5.4.1   Basal promoter elements for Sec tRNA transcription

Eukaryotic tRNA genes are transcribed by Pol III as are other small cellular and viral RNAs. Pol III-transcribed genes (designated class III genes) can be divided into two groups according to the organization of their promoter elements.[114] The principal controlling elements for transcription of most tRNAs, 5S RNA, and the adenovirus VA gene are located inside the gene and designated A box and B box in tRNA/VA genes and A box and C box in 5S RNA genes (Figure 7). Basal promoter elements for the U6 snRNA, 7SK, the Epstein Barr virus EBER, and the Sec tRNA$^{[Ser]Sec}$ genes reside outside and upstream of the gene. Most elements include a TATA box and a proximal sequence element (PSE) which are located in the 5-flanking region near the gene. As discussed in

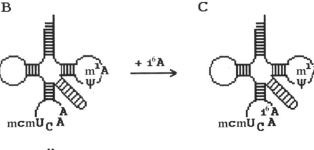

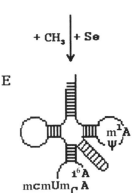

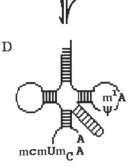

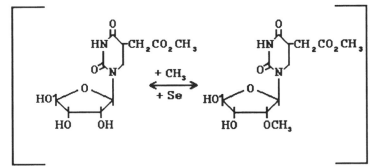

Figure 6 Biosynthesis of the Sec tRNAs[Ser]Sec in *Xenopus* oocytes. The cloverleaf structures show the synthesis of the modified nucleosides and the details of each reaction are discussed in the text. The structures of mcmU and mcmUm and their interconversion are also shown in the lower portion of the figure.

Section 4.14.5, Sec tRNA genes have been isolated and sequenced from various animals including nematodes,[75] fruit flies,[75] frogs,[75,103] chickens,[98] and numerous mammals.[96,97,99–102] Using the *Xenopus* Sec tRNA[Ser]Sec gene as a model, it was shown that Sec tRNA[Ser]Sec is synthesized by a pathway unlike that of any known eukaryotic tRNA[115] in that the primary transcript of the gene does not contain a 5′-leader sequence and transcription of this tRNA is initiated at the first nucleotide within the coding sequence. The 5′-terminal triphosphate remains intact through 3′-terminal maturation as well as through subsequent transport of the tRNA to the cytoplasm.[115]

Comparative analysis of both the coding and noncoding sequences of all the sequenced animal tRNA[Ser]Sec genes revealed several interesting features relevant to their expression. First, these genes contain intragenic promoter elements consisting of an A box and B box as found in other eukaryotic tRNA genes. However, the A box in each tRNA[Ser]Sec gene has two additional nucleotides (a T at position 36 and a C at position 77), and thus, the sequence of each Sec tRNA[Ser]Sec A box deviates from the consensus sequence found in other eukaryotic tRNA genes. This observation suggests that the role of the A box in transcription of the tRNA[Ser]Sec gene might be different from those of other tRNA genes. Second, a TATA-like sequence resides at or near −30 in each of the sequenced Sec tRNA[Ser]Sec genes which suggested that transcription of the gene might be modulated by this element.

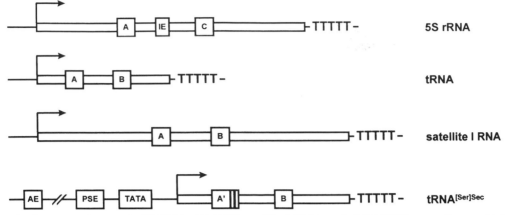

Figure 7 Promoter elements in 5S RNA, tRNA[Met], satellite I RNA, and tRNA[Ser]Sec genes. Boxes with letters represent promoter elements as follows: internal A, IE, and C boxes in 5S RNA genes, internal A and B boxes in tRNA genes and in satellite I RNA genes and upstream AE, PSE, and TATA box elements and internal A (designated in the figure as A′) and B boxes in the tRNA[Ser]Sec genes. AE designates the activator element and PSE the proximal sequence element as discussed in the text. The two extra lines in the A box of the tRNA[Ser]Sec gene indicate the two extra bases found in all such boxes of this gene and the poly T stretch at the 3′ end of each gene shows the termination signal for Pol III transcribed genes.

In fact, the upstream region of the tRNA[Ser]Sec genes including the TATA element was shown to determine efficiency of transcription in *Xenopus* oocytes[103,116] and HeLa cell extracts.[116] Mutation analyses of the upstream TATA box demonstrates that this element is essential for Sec transcription both *in vivo* and *in vitro*. These observations were similar to those involving the U6 snRNA and 7SK genes which also depend on an upstream TATA box for transcription; and the role of the TATA box in transcription of U6 genes was to recruit the TATA-binding protein (TBP).[117–119] TBP is the universal factor required for all three RNA polymerases (Pol I–III) in eukaryotes. Transcription of the U6 genes required the combined action of a TATA box and a PSE. The PSE was originally identified as a basal promoter element in snRNA genes that were transcribed by Pol II.[20] It was subsequently found also to be indispensable for Pol III-transcribed U6[121] and 7SK genes.[122]

PSE sequences are quite divergent among different species.[123] Substantial variation within the nucleotide sequences of PSEs within a given species also occur, indicating that factor(s) binding to the PSE may have a relaxed sequence specificity. The Sec tRNA[Ser]Sec gene PSE was not initially recognized[116] because of significant differences in the sequences of the PSEs among Sec tRNA genes cloned from different species. Table 1 shows a comparison of PSE sequences from U6, 7SK, and Sec tRNA[Ser]Sec genes of various organisms as well as those for the TATA elements and the sequences around the transcription start points (tsp) from various organisms. Most notable among the PSEs are two successive cytidine nucleotides conserved in each of the PSEs shown in Table 1. It should be noted that the relative efficiencies of transcription, both *in vitro* and *in vivo*, of human, rabbit, chicken, and *Xenopus* Sec tRNA genes[116] follow the order of conformity of these PSEs and TATA elements to the corresponding consensus sequence.

Dependence of Sec tRNA[Ser]Sec gene transcription on the PSE was confirmed by microinjection of mutant genes into *Xenopus* oocytes that harbor multiple point mutations at consensus positions of the PSE.[103,124] In one study, it was proposed that the G-33/G-32 base pairs immediately 5′ to the *Xenopus* Sec tRNA[Ser]Sec TATA box should be regarded as an integral component of the TATA box.[124] This proposal was based on the observation that a double point mutation, converting G-32/G-31 to T-32/T-31 (thus changing the overall sequence from -34GGGTATAAAAGG-23 to -34GTTTATAAAAGG-23), caused reduction of Sec tRNA[Ser]Sec transcription both *in vivo* and *in vitro*.[125] Such results might reflect a novel means by which a TATA box gains promoter specificity through positional context. However, this observation should be interpreted cautiously as the sequence change may influence the level of Sec tRNA[Ser]Sec transcription indirectly. Clearly, the above described dinucleotide substitution generates a reverse TATA box that can potentially interfere with formation of the TBP-TATA complex in a normal topology. In addition, the plasmid encoding the substituted sequence encodes a short A-tract in the complementary strand which may bend DNA toward the minor groove. A TATA element which is constrained such that it is present towards the minor groove, loses its affinity for TBP.[126,127] Nonetheless, it became evident that Sec tRNA[Ser]Sec gene transcription is governed by an upstream TATA box located at about −30 and a

Table 1 Upstream promoter elements in TATA-containing class III genes.

Organism	Gene	PSE[a]		TATA		tsp[b]		Genbank accession number
Human	U6	−65 TtA--CCgTAAcT-TgaAa--gT	−48	−34 ggcttTATATAtctt	−20	−10 acgaaacaCC Gtgct	+5	X07425
Human	7SK	−65 TgA-CC-TAAgTgTAaA---gT	−49	−35 aggttTATATAgctt	−21	−10 tgggtaccTC Ggatg	+5	X05490
Human	Sec tRNA	−66 cgA--CCaTAAcTcTAaAag-gT	−47	−35 atcctTATATAgctg	−21	−10 aggtgtcCT Gcccg	+5	K02923
Cow	Sec tRNA	−67 TcA--CCaTAAcTgTAaAaaagT	−47	−35 gctttTATATAgctg	−21	−10 ttttatgTT Gcccg	+5	X54551
Rabbit	Sec tRNA	−66 ccA--CCacAAcctTAaAaa-gT	−47	−35 agcctTATATAgcgc	−21	−10 ggggtattTC Gcccg	+5	X02667
Rat	Sec tRNA	−65 TcA--CCaTAAgTgTAaAaattT	−45	−34 cccttTATATAgctg	−20	−10 agttgtctCT Gcccg	+5	L34835
Mouse	U6	−65 TcA--CCCTAAcTgTaAA----gT	−48	−36 gagacTATAAAtatc	−22	−10 agccttgtTT Gtgct	+5	X06980
Mouse	7SK	−66 TcA--CCacAtgTgTAaA---gT	−49	−35 gccttTATATAccta	−21	−10 gagtactcTC Ggatg	+5	M63671
Mouse	Sec tRNA	−65 TgA--CCaTAAgTgTAaAaa-gT	−46	−34 gccttTATATAaacta	−20	−10 agctccttTC Gcccg	+5	L22019
Chicken	Sec tRNA	−64 cgc--CC-TgAcctgAcAacggc	−45	−34 agcttcATATAccttg	−20	−10 ggttataaTT Gcccg	+5	K01941
Xenopus	U6	−65 Tct--CCtTAAgT-TAcAggggT	−46	−34 gcgctTATAAggcgg	−20	−10 ctggagttTC Gtgct	+5	M31687
Xenopus	Sec tRNA	−65 TcA--CCccAA-TaTatAat-aT	−47	−36 gggggTATAAAagga	−22	−10 tgagtatTT Gcccg	+5	M34507
Drosophila	U6	−64 TgAttCC-TAAgTacAtAtt-cT	−44	−36 tacagTATATAtagg	−22	−10 ggtgaactTC Gttct	+5	M24605
Drosophila	Sec tRNA	−63 aaAttCC-cAAgTgcttAtt-ca	−43	−37 acgtgTATATAatta	−23	−10 cacactaaaT Gcccg	+5	M34509
C. elegans	Sec tRNA	−70 cgAaaCCgTgAgcgTcggttgcT	−48	−35 gtgttTATAAgcgac	−21	−10 agtttgaTC Gcccg	+5	M34508
		G		A				
Consensus		TNA--CCNTAA TNTANANN-NT		TATA A		YY G		
		C		T				

[a] PSE, proximal sequence element. [b] tsp, transcription start points. [c] Y, pyrimidine.

PSE located at about −60[103,116] in much the same way that this occurs in U6 and 7SK gene transcription in *Xenopus* oocytes.

The function of a basal promoter in transcription is always twofold: it interacts with a set of factors that can support proper initiation of transcription and simultaneously serves to dictate the tsp. The location of the tsp in the Sec tRNA[Ser]Sec gene was determined using various spacer mutants in which distances between the upstream TATA box and the tsp or between the tsp and the internal A box were increased by integrals of three DNA helix turns.[124] Transcription initiation of the Sec tRNA[Ser]Sec gene occurred 30 base pairs downstream of the TATA box in which a fixed spacing between the TATA box and tsp was maintained.[124] These results showed unequivocally that the TATA box and PSE fulfill all the requirements to constitute a basal promoter of the Sec tRNA gene.

The role of internal A and B boxes in Sec tRNA[Ser]Sec gene transcription has not been resolved and is still controversial. Mutations in these two internal regions resulted in varying levels of tRNA[Ser]Sec gene transcription. For example, insertion of Sec tRNA[Ser]Sec B box into the coding region of a *Xenopus* U6 gene compensated for the lack of an upstream PSE[103] and certain mutations in the internal A and B boxes partially reduced transcription while others had no effect.[103,124] However, complete removal of the internal elements by their replacement with heterologous sequences supported gene transcription.[124] The latter results demonstrate that these elements are not essential for basal level transcription. Several possibilities would explain these varying results. It is plausible that transcripts harboring mutations within the internal elements are unstable compared to transcripts with intact internal elements which would alter the steady-state levels of the respective transcripts. Another possibility is that the internal elements could play an indirect role in stimulating Sec tRNA transcription by, for instance, antagonizing nucleosomal repression of transcription. Alleviation of nucleosomal repression by internal promoter elements of class III genes is not unprecedented. Binding of transcription factor (TF) IIIC to the A and B boxes in tRNA genes and binding of TFIIIA to the internal element in 5S genes prevented nucleosomal repression of the yeast U6 gene[128,129] and the human 5S gene,[130] respectively. Unlike the metazoan U6 genes examined to date, the yeast U6 gene has natural internal elements within the coding sequence that can bind TFIIIC, but this gene lacks an upstream PSE. When the Sec tRNA[Ser]Sec B box was inserted at an intragenic position of the *Xenopus* U6 gene and at the same time the upstream PSE was removed,[103] promoter organization of the engineered gene looked quite similar to that of the yeast U6 gene. In fact, its transcription efficiency was similar to that of the wild type gene. In this regard, it is interesting to note that the transcription levels of the Sec tRNA[Ser]Sec gene harboring mutations in the A and B boxes were unaffected in *in vitro* assays using *Xenopus* oocyte extracts and naked plasmid templates (see below).[131] In summary, these studies show unequivocally that the Sec tRNA[Ser]Sec genes depend on upstream elements consisting of a TATA box and a PSE for basal level transcription in a manner similar to metazoan U6 and 7SK genes. Several other class III genes were also found to contain an upstream TATA box and a PSE. These include the human H1,[132] human MRP/7-2,[133] rat vault,[134] and rat BC1 genes.[135] It is likely that additional TATA-dependent Pol III-transcribed genes will be discovered. The human H1, human MRP/7-2, rat vault, and rat BC1 genes have provided, and new discoveries of such genes will most certainly provide, this unique group of class III genes with much more diversity than first expected.

4.14.5.4.2   *Transcription factors for TATA-dependent Pol III transcription*

As discussed above, the essential function of the upstream TATA box and PSE in transcription initiation has been demonstrated in a unique group of class III genes encoding U6 snRNAs, 7SK RNAs, and Sec tRNAs[Ser]Sec. It would seem likely that all of these genes are transcribed by a similar or identical set of transcription factors. Although the factors required for transcription of U6 genes and 7SK genes have been most extensively investigated, relatively little information is available for transcription of these genes compared with that found for Pol III transcription of genes lacking a TATA element. To understand transcription of the class III genes encoding a TATA box and the factors involved, it is imperative to examine what is known about transcription of class III genes as a whole. Since the initial studies on elucidating the mechanics involved in Pol III transcription began with the isolation and characterization of Pol III itself, examination of the factors involved in class III gene transcription begins with a discussion of Pol III.

(i) Pol III

Pol III represents one of the three principal classes of nuclear RNA polymerases found in eukaryotic cells. Pol III was originally distinguished from Pol I and Pol II on the basis of its distinct catalytic properties and its elution from DEAE-Sephadex at high salt concentrations.[136] Pol III from HeLa cells was reported to consist of 15 polypeptides with sizes ranging from 10 to more than 150 kDa.[137] Most of the genes for the yeast Pol III subunits have been cloned and sequenced.[138] The two largest subunits, C160 and C128, contain several regions with sequence homology to the β' and β subunits of bacterial RNA polymerase, while the two subunits shared with Pol I, AC40 and A19, show homology to the α subunit of the bacterial enzyme. In addition, yeast Pol III contains five common subunits, ABC27, ABC23, ABC14.5, ABC10d, and ABC10β, shared by all three RNA polymerases and seven Pol III-specific subunits, C82, C53, C37, C34, C31, C25, and C11. Genetic and biochemical studies, and the presence of analogous enzymes in bacteria, indicated that C160 and C128 subunits are most important for polymerase activity. Pol III-specific subunits are plausible targets for interaction with Pol III transcription factors assembled on a basal promoter of class III genes. Studies have shown that the C34 subunit of Pol III interacts with a subunit of TFIIIB and that this subunit contains the TATA box binding protein- (TBP-)Pol III initiation factor.[139,140]

(ii) TFIIIC

Pol III cannot recognize class III gene promoters nor initiate specific transcription without the association of a number of other factors for its function. TFIIIC interacts with the internal promoter elements of certain class III genes. This binding provides the initial step toward recruitment of Pol III to the tsp in TATA-less class III genes. TFIIIC interacts with both the A box and B box and these internal control sites can be separated from each other by a wide range of distances. Thus, TFIIIC shows remarkable flexibility in its binding to DNA. Once TFIIIC is positioned at the internal promoter of a class III gene, subsequent association of TFIIIB with the TFIIIC–DNA complex occurs. Yeast TFIIIC, which is also called τ, consists of six polypeptides designated τ 138, τ 131, τ 95, τ 91, τ 60, and τ 50.[141-143] τ 95 and τ 138 are thought to make direct contact with the A box and B box, respectively.[144] τ 131 interacts with the subunit of TFIIIB that also interacts with the Pol III C34 subunit.[145] Human TFIIIC has been chromatographically separated into two components called TFIIIC1 and TFIIIC2.[146] Human TFIIIC2 is composed of five polypeptides of 220, 110, 102, 90, and 63 kDa and it binds to the B box.[147,148] cDNAs encoding the two largest subunits of the TFIIIC2 were cloned and their sequences analyzed.[149-151] The 220 kDa TFIIIC2 subunit has only weak sequence homology to yeast τ 138. Little is known about the subunit composition or the precise role of TFIIIC1, although it is essential for transcription of tRNA genes and the adenovirus VA1 gene and, in addition, it stabilizes TFIIIC2 binding to Pol III promoters.[137] Unlike U6/7SK genes, the Sec tRNA[Ser]Sec gene contains a canonical B box which can possibly bind TFIIIC. However, even if TFIIIC binds to the Sec tRNA[Ser]Sec internal promoter sequence, this reaction is not essential to the basal level transcription of the gene as its upstream region is capable of supporting transcription of a heterologous sequence lacking internal control regions.[124]

(iii) Different TFIIIB complexes

TFIIIB reconstitutes Pol III transcription *in vitro* together with Pol III and TFIIIC. TFIIIB is positioned at an upstream region of class III, TATA-less genes by interacting with TFIIIC that is already complexed with the internal promoter.[152] Identification of each component of yeast TFIIIB coincided with the discovery that TBP is an essential factor for transcription of all three eukaryotic RNA polymerases.[153-155] TBP was originally identified as the TATA-binding core subunit of TFIID, a Pol II initiation factor. Binding of TFIID to the TATA box of a class II gene initiated the assembly of a Pol II preinitiation complex. During the formation of this complex, TFIIB interacted with, and thereby recruited, Pol II to the tsp. Genes encoding other components of yeast TFIIIB were identified either by genetic or biochemical methods. A suppressor gene (TDS4/BRF1/PCF4) was independently isolated by three groups that either rescued some TBP mutations that occurred in high copy number[156,157] or a tRNA gene A box mutation that existed as a dominant mutant.[158] The protein encoded by this gene showed remarkable homology to TFIIB and was thus called B-related factor (BRF). TBP and BRF are integral components of yeast TFIIIB.[159] TBP was also reported to be a subunit of human TFIIIB.[160-162] A 90 kDa polypeptide, designated "B", was the remaining component of yeast TFIIIB and the gene encoding "B" was cloned.[163–165] The findings that both TBP and TFIIB-like protein are involved in Pol III transcription suggest that a common initiation mechanism in transcription exists between the two RNA polymerases. Actually the TFIIB-like BRF1 binds to the C34 subunit of yeast Pol III[139,140] as well as the τ 131 subunit of yeast TFIIIC.[145] These interactions are thought to be essential for Pol III recruitment to a class III gene promoter. Transcription of the yeast U6 snRNA gene also depends on an upstream TATA box as is found in metazoan U6 genes. However, the yeast U6 gene lacks an upstream PSE and instead possesses internal A box and B box elements which are essential to transcription initiation. Although the yeast U6 TATA box binds TBP in a sequence-specific manner, the yeast TFIIIB component that is identical to those involved in TATA-less gene transcription participates in U6 transcription.[166] Both position and orientation of TFIIIB binding to the upstream region of the yeast U6 gene were found to be governed by TFIIIC bound to the internal promoter elements.[167,168] TFIIIC bound to the U6 internal promoter, which established Pol III specificity of yeast U6 transcription.[169] The gene encoding the human homologue of BRF was isolated and the recombinant human BRF reconstituted TFIIIB for TATA-less class III genes only with TBP alone.[137,170] It is still not clear whether a human homologue of yeast B″ exists and plays a role in Pol III transcription. Several studies have reported that different TFIIIB complexes are involved in Pol III transcription of TATA-less and TATA-containing class III genes in HeLa cells. In TBP-depleted, HeLa cell nuclear extracts, TATA-dependent U6 transcription was restored by addition of recombinant TBP alone, whereas TATA-independent VAI transcription was restored only by addition of the partially purified TFIIIB.[160] The TFIIIB activity required for TATA-dependent Pol III transcription was chromatographically separated from that required for Pol III transcription of TATA-less genes.[171] Transcription from the VAI promoter was inhibited in HeLa extracts depleted of BRF by immunoprecipitation, but transcription from the TATA-containing U6 promoter was not affected.[170] It thus seems that BRF is not involved in transcription of TATA-containing class III genes. The Sec tRNA[Ser]Sec gene also appears to utilize the TFIIIB complex distinct from the form used by TATA-less genes. A TBP mutant defective in Pol II transcription did not support Sec tRNA[Ser]Sec transcription, while a TBP mutant defective in Pol III transcription of TATA-less genes could restore transcription.[172] The latter mutant has a defect in interacting with the BRF protein, also indicating that TFIIIB containing BRF is not the form required for Pol III transcription of TATA-containing genes. Determination of the components in TFIIIB involved in TATA-containing class III genes and elucidation of their mode of action is a matter of utmost importance in elucidating the factors required in tRNA[Ser]Sec gene transcription.

(iv) *The Pol III transcription factor designated PTF, SNAP, or PBP*

The factor binding to the PSE of U6 or 7SK genes (designated PTF, SNAPc, or PBP) has been purified and found to be a multisubunit complex consisting of four polypeptides of 180, 55, 45, and 44 kDa.[173,174] Genes for the four PTF subunits were isolated and their proteins characterized.[174–177] Nuclear extracts immunodepleted with an antibody against any one of the cloned PTF subunits lost their ability to support TATA-dependent Pol III transcription without affecting transcription of TATA-less class III genes. The 45 kDa and 44 kDa subunits of PTF directly interact with TBP.[174–176] Only the largest subunit of PTF binds to DNA in a PSE-specific manner (see references 174 and 176 and references therein). Since the PSE sequence varies considerably among different species, it seems likely that the PTF subunits, or at least the DNA-binding subunit of PTF, will show species-specificity.

(v) *Distal elements in Pol III transcription and their binding factors*

Most of TATA-less class III genes contain a constitutively active basal promoter. TATA-dependent class III genes, however, depend on a distal element for maximal transcription. The distal elements for vertebrate U6 and 7SK genes contain an upstream octamer motif usually located between −100 to −250. The octamer motif binds factor Oct-1 or Oct-2 which stimulates U6 or 7SK transcription both *in vivo* and *in vitro*.[178,179] Binding of Oct-1 or Oct-2 to the distal element stimulates or stabilizes PTF binding to the basal promoter.[180,181] Unlike other TATA-containing class III genes, the *Xenopus* Sec tRNA[Ser]Sec gene does not contain an upstream octamer motif at a distal position. Instead, an SPH motif-containing 15 bp sequence located at about −200 supports optimal expression of tRNA[Ser]Sec gene.[182] Interestingly, this sequence, referred to as activator element (AE), stimulated Pol II transcription *in vivo*. The cDNA encoding the AE-binding protein (designated Staf) was isolated and found to encode a zinc finger protein.[183] Staf can activate tRNA[Ser]Sec transcription *in vivo*. It is not clear whether this protein can also stabilize PTF binding to the tRNA[Ser]Sec gene PSE and thereby activate tRNA[Ser]Sec transcription *in vitro*. An oligonucleotide containing the AE sequence did not inhibit tRNA[Ser]Sec transcription in *Xenopus* oocyte extracts while oligonucleotides containing the TATA or PSE sequences did inhibit tRNA[Ser]Sec transcription.[131] Thus, an alternative mechanism may exist in which AE stimulates tRNA[Ser]Sec transcription. For example, the AE-binding factor could be a protein which relieves the inhibitory effects of nucleosome assembly on transcription *in vivo*. This type of antirepression process may be bypassed in the cell-free system, since naked templates were employed in the study. Complete identification and isolation of the transcription factors for Sec tRNA gene transcription will allow more refined analyses of the activation by AE. Current understanding of the required transcription factors and their role in activating Sec tRNA[Ser]Sec gene transcription are summarized in Figure 8.

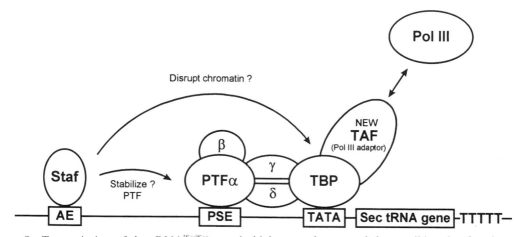

Figure 8 Transcription of the tRNA[Ser]Sec gene in higher vertebrates and the possible role of various transcription factors. Staf, PTFα, TBP, TAF and Pol III designate the Sec tRNA gene transcription activating factor, the proximal sequence element transcription factor α, the TATA box binding protein, TBP-associated factor and RNA polymerase III, respectively, and the DNA regulatory elements are defined in the text and the legend to Figure 7. Staf binds to the AE and possibly participates in stabilizing PTF binding and/or chromatin disruption. PTF consists of four subunits (α, β, γ, and δ). PTFα contacts directly with the PSE and its binding may be aided by the β subunit, and PTF may be linked to TBP that is bound to the TATA box through the γ and δ subunits. The TAF is a BRF equivalent of TATA-dependent Pol III transcription and serves as an adaptor for Pol III recruitment.

4.14.6 INCORPORATION OF Sec INTO PROTEIN

As UGA can dictate either the termination of protein synthesis or the insertion of Sec into protein, what then determines the function of a given UGA codon? Can the same UGA codon serve multiple roles? These questions were addressed and discussed by Low and Berry.[11] Clearly, the same UGA codon can serve multiple functions. It can function to (i) signal the insertion of Sec into protein, (ii) dictate the cessation of protein synthesis, and (iii) serve as a suppressor codon.[184,185] Furthermore, selenoprotein P (SelP) contains 10 UGA Sec codons[12,186] and at least one of these codons can serve a dual role.[187] The UGA codon nearest the amino terminus occurs roughly in a position similar to that of UGA codons in mRNAs of glutathione peroxidases in which Sec exists at the protein active center. The other 9 UGA codons correspond to amino acids in the carboxy terminal half of SelP and several of these codons exist in a nucleotide context that is favorable to termination.[11] Interestingly, the second UGA codon also serves as a termination codon since the truncated protein has been observed *in vivo*.[187] The role of this codon in Sec incorporation versus termination and the possibility that the truncated protein may have a specialized function has been discussed elsewhere.[11,187]

The presence of a stem-loop structure or structures in the 3'-untranslated region of selenoprotein mRNAs is essential for insertion of Sec into protein in eukaryotes.[11] These structures are called Sec insertion sequence elements or SECIS elements.[11] Prokaryotes also require a SECIS element for insertion of Sec into protein, but in contrast to eukaryotes, the SECIS element in prokaryotes occurs immediately downstream of the UGA Sec codon.[1,2,4,7,9,11] Another critical element for Sec insertion in prokaryotes is the presence of a specific elongation factor designated SELB.[1,2,4,7,9,11] SELB recognizes selenocystyl-tRNA$^{[Ser]Sec}$ and the Sec-tRNA$^{[Ser]Sec}$–SELB complex may anchor itself on the SECIS element for donation of Sec to the growing polypeptide chain in response to a UGA Sec codon (see Figure 9).

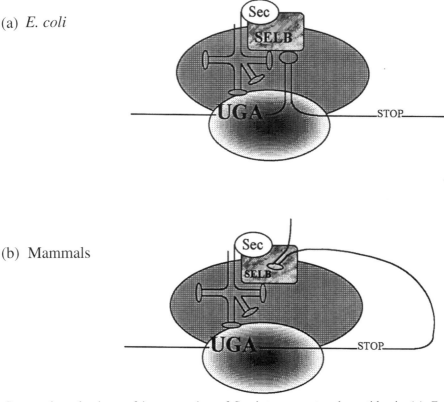

Figure 9 Proposed mechanisms of incorporation of Sec into nascent polypeptides in (a) *E. coli* and (b) mammals. SelB designates the specific elongation factor for Sec-tRNA$^{[Ser]Sec}$ and it recognizes Sec-tRNA$^{[Ser]Sec}$ and the SECIS element for insertion of Sec in response to the UGA Sec codon in a selenoprotein mRNA bound to the ribosome. The SECIS element occurs immediately downstream of the UGA Sec codon in *E. coli* mRNA and in the 3′-untranslated region in mammalian mRNA (after Low and Berry[11]).

Although a SELB homologue for Sec-tRNA$^{[Ser]Sec}$ has not been identified in eukaryotes, several lines of evidence suggest that this factor also exists in mammals. For example, EF-1α, the elongation factor for aminoacyl-tRNAs in mammalian cells, does not recognize tRNA$^{[Ser]Sec}$ with Sec attached, but does recognize tRNA$^{[Ser]Sec}$ with serine attached.[185] Ser-tRNA$^{[Ser]Sec}$ can suppress UGA codons and thus this form of the tRNA can serve as a suppressor.[86,185] Sec-tRNA,$^{[Ser]Sec}$ however, cannot suppress UGA codons *in vitro*.[185] Forster *et al.*[188] had previously shown that the bacterial elongation factor, EF-Tu, did not recognize the bacterial Sec-tRNA$^{[Ser]Sec}$ which provided evidence that this tRNA only participated in protein synthesis through SELB. Microinjection of mammalian Sec-tRNA$^{[Ser]Sec}$ and mRNA encoding the selenoprotein, glutathione peroxidase, into *Xenopus* oocytes shows that this tRNA can donate Sec to the nascent polypeptide chains of glutathione peroxidase, suggesting that *Xenopus* oocytes may provide an assay system and a source for the SELB homologue.[185] Furthermore, Gelpi *et al.*[189] have reported the presence of antibodies that react with a protein–Sec-tRNA complex in patients with an autoimmune chronic hepatitis leading the authors to speculate that the protein component may be a specific elongation factor for Sec-tRNA$^{[Ser]Sec}$. Further studies have largely examined gel mobility shifts and UV cross-linking assays to identify proteins that interact with the SECIS elements,[190,191] although the binding of Sec-tRNA$^{[Ser]Sec}$ to ribosomes in response to a UGA encoding oligonucleotide has been observed.[192] It is important to note that SELB homologue in mammals should (i) interact with Sec-tRNA$^{[Ser]Sec}$, (ii) interact with the SECIS element in selenoprotein mRNA, and (iii) promote Sec insertion into nascent polypeptides in response to UGA Sec codons. Thus far, such a factor that meets these criteria has not been reported.

Berry and co-workers[193,194] have measured the maximal and minimum distances between the SECIS element and the Sec UGA codon that can still support Sec insertion into protein. The maximal distance can be as much as 2.7 kb.[193] The minimum distance is between 51 and 111 bases, where 111 bases support Sec insertion, but 51 bases do not.[194] It is of interest to note that the termination codons for two selenoproteins, selenoprotein W[195] and glutathione peroxidase in *Schistosoma mansoni*,[196] are UGA and these codons occur within the minimum distances determined by Martin *et al.*[194] that do not support Sec insertion into protein.

Several studies have further elucidated the structural features and the role of the SECIS elements. For example, mutation of specific regions of the SECIS elements have identified certain conserved features that are essential for function.[191,193] These include an AUGA sequence in the 5′ arms, a GA sequence in the 3′ arms of the stems and an AAA sequence in the terminal loops. The role of the SECIS element has been further probed by experiments designed to read through a UGA Sec codon into the luciferase reporter gene.[197] This study suggested that the precise sequences of the SECIS elements are relatively unimportant as long as the length and thermostability of the base-paired structures are retained. Walczak *et al.*[198] examined the structures of SECIS elements in the 3′-untranslated region of 20 different animal selenoprotein mRNAs to determine if these elements fit into a model or a consensus sequence. SECIS elements from animals appeared to fit into a single model that consists of two helices with an internal loop and an apical loop that is contiguous with one of the helices. Most elements were also found to contain UGAN/YGAN at the proximal helix that is predicted to form four non-Watson–Crick base pairs. Low and Berry[11] previously proposed a similar structure that is consistent with the reported sequences.

4.14.7 SELENOPROTEINS

Sec-containing proteins that have been identified in prokaryotic and eukaryotic sources are shown in Table 2. As most of these proteins have been reviewed elsewhere,[1-13] the following sections will focus on: (i) the identification of human thioredoxin reductase as a selenoprotein[199,200] which marked an important finding in the field of selenium-containing enzymes in view of the importance of this protein for a variety of essential metabolic processes; (ii) the observation that a second form of mammalian selenophosphate synthetase is a selenoprotein;[36,201] and (iii) the discovery that a novel 15 kDa selenoprotein exists in mammalian cells.[202]

It should also be noted that glutathione peroxidase (GPX1) has been used as a model selenoprotein for elucidating many of the control mechanisms governing selenoprotein expression.[203,204] In rat liver, Sunde and co-workers[205] have shown that altering selenium status can have an order-of-magnitude effect on the level of GPX1 mRNA, but has little or no effect on mRNA levels for phospholipid hydroperoxide glutathione peroxidase (GPX3),[206] which shares approximately 40% nucleotide and amino acid sequence identity with GPX1. GPX1 mRNA reaches a maximum level at half the dietary selenium concentration necessary for maximal GPX1 enzyme activity,[207] and GPX4 enzyme activity reaches its maximum in rat liver at approximately the same selenium status as necessary for maximum GPX1 mRNA levels.[208] The regulation of GPX1 mRNA by selenium status occurs posttranscriptionally, presumably via mRNA stability.[12,209,210] Initial studies with recombinant GPX1 constructs in cultured cells indicate that the GPX1 SECIS element is necessary for the regulation of GPX1 mRNA levels by selenium status.[211-213] These studies have led Sunde to suggest that the role of GPX1 in normal metabolism may be as a "biological selenium buffer" that elongates the range of permissible total selenium concentrations by up-regulation of GPX1 mRNA leading to storage of excess selenium in GPX1.[203,204] Presumably the other selenium-dependent peroxidases (GPX2, GPX3, and GPX4, see Table 2) provide the essential hydroperoxide scavenging activities. The complete viability of the GPX1 knockout mouse[214] fed standard laboratory diets provides indirect support for this hypothesis.

4.14.7.1 Thioredoxin Reductase

Mammalian thioredoxin reductase is a FAD-containing NADPH-dependent oxidoreductase that was purified from bovine and rat liver, human placenta, and several mammalian cell

Table 2 Sec-containing proteins.

Protein	Organism	Comment	Ref.[a]
Eukaryotes			
Cytosolic glutathione peroxidase (GPX1)	mammals	4×23 kDa antioxidant protein with Sec at the active site	
Glutathione peroxidase-GI (GPX2)	mammals	4×23 kDa antioxidant protein with Sec at the active site	
Plasma glutathione peroxidase (GPX3)	mammals (plasma)	4×25 kDa antioxidant glycoprotein with Sec at the active site	
Phospholipid hydroperoxide glutathione peroxidase (GPX4)	mammals, *Schistosoma mansoni*	20 kDa antioxidant protein with Sec at the active site	249
Thioredoxin reductase	mammals	2×57 kDa, contains 2 FAD, reduces thioredoxin and peroxides with NADPH	199, 200
	C. elegans[b] (see text)	Gene sequence contains in-frame TGA	
Selenoprotein P	mammals (plasma)	57 kDa glycoprotein with 10–12 Sec residues, function is unknown (possibly storage or antioxidant)	
Thyroid hormone deiodinase (D1)	mammals	2×29 kDa, involved in thyroid hormone metabolism (activation of T_3)	
Thyroid hormone deiodinase (D2)	mammals	30 kDa, activation of T_3	250
Thyroid hormone deiodinase (D3)	mammals	31 kDa, inactivation of T_3	
Selenoprotein W	mammals	10 kDa, function is unknown	
15 kDa Selenoprotein	mammals	15 kDa protein subunit, function is unknown	202
Selenophosphate synthetase (SPS2)	mammals	50 kDa Sec-containing protein with homology to *E. coli* selenophosphate synthetase	36
Bacteria			
Formate dehydrogenase H	*E. coli*	80 kDa, contains Mo, 2 molybdopterin guanine dinucleotides, Fe4S4 cluster, Sec is involved in coordination to Mo and coupled electron–proton transfer	235
Formate dehydrogenase O	*E. coli*	110 kDa subunit, contains Mo, 2 molybdopterin guanine dinucleotides, FAD, FeS clusters	
Formate dehydrogenase N	*E. coli*	110 kDa subunit, contains Mo, 2 molybdopterin guanine dinucleotides, FAD, FeS clusters	
Formate dehydrogenase	anaerobic bacteria	contains Mo (or W), molybdopterin, FeS clusters, FAD	
Hydrogenase	sulfate-reducing bacteria	contains Ni, FeS clusters, Sec is coordinated to Ni	
Glycine reductase Selenoprotein A	anaerobic bacteria	18 kDa, active center Sec residue	
Glycine reductase Selenoprotein B	anaerobic bacteria	47 kDa, in-frame TGA in a gene sequence	251
Selenophosphate synthetase	*H. influenzae*	gene sequence is identified	36
Archaea			
Formate dehydrogenase	methanogenic archaea	contains Mo, molybdopterin, FeS clusters, FAD	
Hydrogenase	methanogenic archaea	contains Ni, FeS clusters, Ni–Se coordination	
Selenophosphate synthetase	*M. jannaschii*	gene sequence is identified	36
Viruses			
Glutathione peroxidase	human poxvirus[b]	homologous to mammalian GPX1	252, 254

[a]The original studies of the selenoproteins shown in Table 2 are found in references 2 and 7–13, with the exception of later studies, which are indicated by reference numbers.   [b]Gene sequence was identified.

lines.[109,200,215–219] It is a homodimer consisting of 55–60 kDa subunits and differs from bacterial and yeast thioredoxin reductase in its biochemical properties and in its amino acid sequence.

The initial studies that showed an association between selenium and thioredoxin reductase reported a Sec-containing protein purified from human lung adenocarcinoma cells which had thioredoxin reductase activity.[199] However, the gene sequence for human placental thioredoxin reductase showed the presence of a classical Cys-Val-Asn-Val-Gly-Cys motif in the N-terminal part of the gene product, which might represent a part of the active center of the protein, but no TGA codon that might be associated with a Sec residue.[220] To resolve this contradiction, selenium-75 labeled thioredoxin reductase was purified from a human T-cell line and the resulting peptides obtained by tryptic digestion were analyzed.[200] The sequence of the selenium labeled peptide was identical with the C-terminal deduced sequence of the placental enzyme except that a TGA codon, previously considered to represent termination, corresponded to a Sec residue. The unusual location of the Sec residue in a C-terminal conserved Gly-Cys-Sec-Gly motif provided an explanation, if this motif represents the enzyme active center, for the reported low substrate specificity and peroxidase activity of the mammalian thioredoxin reductase.[200]

Computer search comparison of the novel Gly-Cys-Sec-Gly motif to those observed in other organisms revealed the presence of thioredoxin reductase homologues in two other species as shown in Table 3. Thioredoxin reductase may exist in two forms, designated I and II, in *C. elegans* (see Table 3) and only the latter form contains a TGA codon in the corresponding gene. The TGA codon in this gene occurs at the same position as that in human thioredoxin reductase (Gladyshev and Hatfield, unpublished observations),[221] suggesting that *C. elegans* thioredoxin reductase II contains a Sec residue and is therefore a selenoprotein. This is the first indication of a Sec-containing protein in *C. elegans*. Consistent with this conclusion, a Sec tRNA gene was identified previously in this organism.[75] It has been found that thioredoxin reductase activity in selenium-deficient liver of rats is decreased compared to that observed in selenium sufficient (control) liver.[222] The overall sequence conservation of the C-terminal regions in thioredoxin reductases from different organisms (see Table 3) suggests that this region may be essential for catalytic activity and substrate recognition.

Table 3 Alignment of C-terminal regions of thioredoxin reductases.

Organism		*Sequence*[a]	
		* *** * * * * * *#*	
Human	468	TIGIHPVCAEVFTTLSVTKRSGASILQAG-CUG	499
C. elegans II	493	LIGIHPTVAENFTTLTLEKKEGDEELQAGSCUG	525
C. elegans I	468	TIAIHPCSSEEFVKLHITKRSGQDPRTQC-CCG	499
P. falciparum	505	CIGIHPTDAESFMNLFVTISSGLSYAAKGGCGG	537

[a]Numbers at left and right of sequences indicate the position of the N-terminal and C-terminal residue within the respective protein. The * indicates conserved residues and # shows the position of the Sec residue. Accession numbers for human, *Caenorhabditis elegans* I, *C. elegans* II, and *Plasmodium falciparum* sequences are gb|S79851, sp|P30635, gb|U61947, and pir‖S57658, respectively. Computer searches were performed with the BLAST program.[253] U represents Sec.

The finding of a Sec residue in mammalian thioredoxin reductase is especially significant from the standpoint of selenium biochemistry because the thioredoxin reductase–thioredoxin system is involved in a number of pathways including regulation of transcription factors, removal of hydroperoxides, reduction of disulfide bonds in various redox-active proteins, DNA synthesis, and cellular redox balancing.[223,224] Thus, if selenium is indeed essential for the catalytic activity of thioredoxin reductase, this micronutrient is involved in the control of various redox-related pathways, in addition to pathways controlled by glutathione peroxidase system.

4.14.7.2 Selenophosphate Synthetase, a Selenoprotein

As noted in Section 4.14.2.1, there are two forms of selenophosphate synthetase in mammals.[34–36] One of these forms, SPS2, represents the first example of a selenoprotein that is involved in Sec biosynthesis.[36] Hence, this enzyme may be involved in autoregulating its own synthesis. The gene for this second form of selenophosphate synthetase has been cloned[36] and the Sec TGA codon mutated to a cysteine TGC codon.[201] In an insect translation system in which the mutant selenophosphate synthetase gene product was overexpressed, selenophosphate synthetase activity was measurable.[201] Homologous Sec-containing proteins were also identified by computer analysis in mice, *Haemophilus influenzae* and *M. jannaschii*[36] and the form of the enzyme with cysteine in place of Sec may exist in *Brugia malayi* and that with glycine in *Dictyostelium discoideum* (see Table 4). The previously identified form of this protein in mammals, SPS1, contains threonine (predicted from the gene sequence) at the position corresponding to Sec in SPS2,[34] while *E. coli* selenophosphate synthetase contains cysteine at this site which is essential for its catalytic activity.[225]

Table 4   Alignment of sequences in the proposed active center of selenophosphate synthetases.[a]

Organism	Sequence	Accession number
	`*** *# *** * **`	
Human II	RLTGFSGMKGUGCKVPQKALLKLL	gb|AA099814
Human I	RLTRFTELKGTGCKVPQDVLQKLL	gb|U34044
Mouse	RLTSFSGMKGUGCKVPQETLLKLL	gb|AA117029
D. discoideum	RLTDFTKLKGGGCKVPQAELLSLL	gb|U68248
E. coli	RLTQYSHGAGCGCKISPKVLETIL	gb|M30184
H. influenzae	RLTQYSGGAGUGCKISPKVLGTIL	gb|U32705
M. jannaschii	KLTELVKLHGUACKLPSTELEFLV	gb|U67599
B. malayi	LLTKLTQMKGCGCKVPRAVLLEFL	gb|AA057958

[a]Computer searches were performed with the BLAST program.[253] The * indicates conserved residues and # shows the position of Sec, which is represented by U.

4.14.7.3   Mammalian 15 kDa Selenoprotein

A novel selenoprotein was identified in human T cells by labeling with selenium-75 and isolating and characterizing the resulting labeled proteins.[202] Sequences of two peptides from one of the labeled proteins corresponds to a gene sequence containing an inframe TGA codon, suggesting the resulting protein contains a Sec residue. The gene encodes a 15 kDa protein of unknown function that lacks homology to known proteins. Under native conditions, the characterized 15 kDa protein exists as a part of a large complex. A computer survey revealed the presence of genes in mice and rats with a high degree of homology to the novel selenoprotein (see corresponding peptide sequences in Table 5). Homologous sequences were also found in *C. elegans* and rice, but a cysteine codon, TGC, occurs in place of the TGA Sec codon. Furthermore, additional genes in mice and humans were found with lower homology to the 15 kDa selenoprotein than those observed above, suggesting that the human T-cell 15 kDa selenoprotein represents a family of novel selenoproteins.

Table 5   Sequences flanking selenocysteine residue in a novel 15 kDa selenoprotein.[a]

Organism	Sequence	Accession number
	`****** # * ****`	
Human	AILEVCGUKLGRFPQ	gb|AF051894
Mouse	AILEVCGUKLGRFPQ	gb|AA110476
Rat	AILEVCGUKLGRFPQ	dbj|C07061
C. elegans	AILEVCECNLARFPQ	dbj|C10051

[a]Computer searches were performed with the BLAST program.[253] The * indicates conserved residues and # shows the position of Sec, which is represented by U.

4.14.8   WHY IS Sec USED IN NATURE?

In Sec-containing enzymes characterized thus far, the Sec residue is located at the protein active center. Furthermore, in all cases where the essentiality of this Sec residue has been studied, it was found that Sec was important for catalytic activity of the enzyme. For example, changing the Sec codon to a cysteine codon in the *E. coli* formate dehydrogenase H gene led to a substantial (300-fold) decrease in the initial rate of formate oxidation in the resulting gene product,[226] while the corresponding serine mutant was inactive.[227] Similarly, replacement of Sec with cysteine residues led to a significant decrease in catalytic activities of Type 1[228] and Type 3 deiodinases.[229] No catalytic activity was detected in clostridial glycine reductase selenoprotein A, when the active center Sec residue was substituted with cysteine.[230]

The reason that Sec is essential at the active site of proteins rather than cysteine is that under identical environmental conditions a selenol is ionized at a much lower pH than thiol.[7] It is especially important that at physiological pH an average Sec residue is fully ionized, while an average cysteine residue is not. These considerations are supported by mechanistic studies with several selenoenzymes in which an ionized Sec is implicated in the direct interaction with the substrate. For example, clostridial glycine reductase Selenoprotein A provides an ionized selenol for attacking the intermediate formed between glycine reductase protein B and glycine and this interaction results in a carboxymethylated Sec intermediate (R—Se—CH_2COO^-).[231-233] In the glutathione peroxidase enzyme family, an ionized selenol at the active site is thought to interact directly with hydrogen peroxide resulting in a selenenic acid intermediate, R—Se—OH.[234] It has been suggested that an ionized selenol at the active center of Type 1 deiodinase reacts with thyroxine (T_4) yielding an iodinated Sec intermediate (R—Se—I).[7]

A new role for the Sec residue at the protein active center was established through crystallographic, spectroscopic and mechanistic studies with *E. coli* formate dehydrogenase H.[235-240] These studies resulted in a proposal of the most detailed structure-based catalytic mechanism for selenoenzyme action.[235] In formate dehydrogenase, the Sec residue is coordinated to a molybdenum (Mo) atom[237] and the latter changes its redox states in a Mo(VI)–Mo(IV)–Mo(V)–Mo(VI) cycle.[235-238] It was suggested that selenium–molybdenum coordination tunes the coordination of another Mo ligand influencing substrate binding.[235] During the catalytic reaction, 2 electrons are transferred to the Se–Mo center and the substrate, which is formate, is degraded into carbon dioxide and a proton.[238] The proton is further transferred to a nearby histidine residue through protonation of Sec.[238]

The proposed proton transfer and metal coordination roles of selenium provide additional insights into mechanisms of other metal-containing selenoenzymes. One example is nicotinic acid hydroxylase from *Clostridium barkeri*.[241-243] In this enzyme, the selenium atom is also coordinated to molybdenum and is essential for catalytic activity.[244] However, selenium is present in the active center as an unidentified labile cofactor.[242,243,245,246] Another example is NiSeFe hydrogenase. This enzyme contains a Sec residue that is coordinated to nickel[247,248] suggesting that a coupled electron–proton transfer reaction involving coordinated selenium occurs.

Each of the studies discussed in this section suggest that selenium is used in nature because of its unique protonation, coordination, and redox properties. Although utilization of Sec is restricted to a limited number of proteins, the complex machinery for inserting Sec into protein has been maintained in nature apparently to take advantage of the unique chemical properties of selenium that cannot be duplicated by other elements.

4.14.9 CONCLUSIONS

Sec is the most infrequently used of all the amino acids that are incorporated into proteins. However, it is one of the most fascinating amino acids in that its insertion into proteins at specific sites depends on its biosynthesis occurring on tRNA. The machinery for its biosynthesis and for ensuring its donation to nascent polypeptide chains at specific sites is complex compared to that for other amino acids. In spite of the infrequent use of Sec in polypeptides and of the complexity involved in its insertion into proteins, Sec's role as the 21st amino acid is widespread, if not virtually universal, in nature. Nature maintains Sec and its complex incorporation system to take advantage of the unique chemical properties of selenium.

The hazards of having too much selenium in the diet of mammals have been recognized since the 1930s and the risks of having too little in the diet since the 1950s.[8,13,14] However, the means by which this element makes its way into proteins as Sec have only been elucidated during the last 10 years. Much has been accomplished in this time period in unraveling the molecular biology and biochemistry of Sec as this chapter and other reviews demonstrate,[1-12] but there is still much to be done, particularly in mammals and other eukaryotes. Elucidating the mechanistic details of the biosynthesis of Sec and its insertion into proteins, determining the function of numerous selenoproteins, identifying new selenoproteins and their roles, and determining the role that many known selenoproteins may have in human disease[8] represent the challenges facing future investigations.

ACKNOWLEDGMENTS

The authors express their sincere appreciation to Drs Marla Berry, Alan Diamond, Kristina Hill, Thressa Stadtman, and Roger Sunde for their review of the manuscript and for their many helpful suggestions. The Genetic Engineering Research Fund from the Ministry of Education of Korea (1996) to BJL is also acknowledged for support.

4.14.10 REFERENCES

1. A. Böck, K. Forchhammer, J. Heider, W. Leinfelder, G. Sawers, B. Veprek, and F. Zinoni, *Mol. Microbiol.*, 1991, **5**, 515.
2. A. Böck, K. Forchhammer, J. Heider, and C. Baron, *Trends Biochem. Sci.*, 1991, **16**, 463.
3. D. L. Hatfield, I. S. Choi, B. J. Lee, and J.-E. Jung, in "Transfer RNA and Protein Synthesis," eds. D. L. Hatfield, B. J. Lee, and R. Pirtle, CRC Press, Boca Raton, FL, 1992, p. 269.
4. A. Böck, in "Selenium in Biology and Human Health," ed. R. F. Burk, Springer-Verlag, New York, 1994, p. 9.
5. D. L. Hatfield and A. M. Diamond, *Trends Genet.*, 1993, **9**, 69.
6. D. L. Hatfield, I. S. Choi, T. Ohama, J.-E. Jung, and A. M. Diamond, in "Selenium in Biology and Human Health," ed. R. F. Burk, Springer-Verlag, New York, 1994, p. 25.
7. T. C. Stadtman, *Annu. Rev. Biochem.*, 1996, **65**, 83.
8. B. J. Lee, S. I. Park, J. M. Park, H. S. Chittum, and D. L. Hatfield, *Mol. Cells*, 1996, **6**, 509.
9. J. Heider and A. Böck, *Adv. Microb. Physiol.*, 1993, **35**, 71.
10. T. C. Stadtman, *J. Biol. Chem.*, 1991, **266**, 16257.
11. S. C. Low and M. J. Berry, *Trends Biochem. Sci.*, 1996, **21**, 203.
12. R. F. Burk and K. E. Hill, *J. Nutr.*, 1994, **124**, 1891.
13. R. F. Burk, in "Selenium in Biology and Human Health," ed. R. F. Burk, Springer-Verlag, New York, 1994, p. 1.
14. G. F. Combs and S. B. Combs, "The Role of Selenium in Nutrition," Academic Press, New York, 1986, chap. 8, p. 532.
15. J. E. Cone, R. M. Del Rio, J. N. Davis, and T. C. Stadtman, *Proc. Natl. Acad. Sci. USA*, 1976, **73**, 2659.
16. M. Wilcox and M. Nirenberg, *Proc. Natl. Acad. Sci. USA*, 1968, **61**, 229.
17. A. Schön, C. Karrangara, S. Gough, and D. Söll, *Nature*, 1988, **331**, 187.
18. X.-G. Wu and H. J. Gross, *Nucleic Acids Res.*, 1993, **21**, 5589.
19. T. Ohama, D. Yang, and D. L. Hatfield, *Arch. Biochem. Biophys.*, 1994, **315**, 293.
20. R. Amberg, T. Mizutani, X.-Q. Wu, and H. J. Gross, *J. Mol. Biol.*, 1996, **263**, 8.
21. W. Leinfelder, T. C. Stadtman, and A. Böck, *J. Biol. Chem.*, 1989, **264**, 9720.
22. R. A. Sunde and J. K. Evenson, *J. Biol. Chem.*, 1987, **262**, 933.
23. B. J. Lee, P. J. Worland, J. N. Davis, T. C. Stadtman, and D. L. Hatfield, *J. Biol. Chem.*, 1989, **264**, 9724.
24. R. S. Glass, W. P. Singh, W. Jung, Z. Veres, T. D. Scholz, and T. C. Stadtman, *Biochemistry*, 1993, **32**, 12555.
25. P. H. Maenpaa and M. R. Bernfield, *Proc. Natl. Acad. Sci. USA*, 1970, **67**, 688.
26. S. J. Sharp and T. S. Stuart, *Nucleic Acids Res.*, 1977, **4**, 2123.
27. T. Mizutani and T. Hitaka, *FEBS Lett.*, 1988, **226**, 227.
28. T. Mizutani, *FEBS Lett.*, 1989, **250**, 142.
29. T. Mizutani, N. Maruyama, T. Hitaka, and Y. Sukenaga, *FEBS Lett.*, 1989, **247**, 345.
30. T. Mizutani, H. Kurata, and K. Yamada, *FEBS Lett.*, 1991, **289**, 59.
31. T. Mizutani, H. Kurata, K. Yamada, and T. Totsuka, *Biochem. J.*, 1992, **284**, 827.
32. K. Forchhammer and A. Böck, *J. Biol. Chem.*, 1991, **266**, 6324.
33. W. C. Hawkes, D. E. Lyons, and A. L. Tappel, *Biochim. Biophys. Acta*, 1982, **699**, 183.
34. S. C. Low, J. W. Harney, and M. J. Berry, *J. Biol. Chem.*, 1995, **270**, 21659.
35. I. Y. Kim and T. C. Stadtman, *Proc. Natl. Acad. Sci. USA*, 1995, **92**, 7710.
36. M. J. Guimaraes, D. Peterson, A. Vicari, B. G. Cocks, N. G. Copeland, D. J. Gilbert, N. A. Jenkins, D. A. Ferrick, R. A. Kastelein, J. F. Bazan, and A. Zlotnik, *Proc. Natl. Acad. Sci. USA*, 1996, **93**, 15086.
37. C. Baron, C. Sturchler, X.-Q. Wu, H. J. Gross, A. Krol, and A. Böck, *Nucleic Acids Res.*, 1994, **22**, 2228.
38. C. Sturchler-Pierrat, N. Hubert, T. Totsuka, T. Mizutani, P. Carbon, and A. Krol, *J. Biol. Chem.*, 1995, **270**, 18570.
39. X.-Q. Wu and H. J. Gross, *EMBO J.*, 1994, **13**, 241.
40. T. A. Brown and A. Shrift, *Biol. Rev.*, 1982, **57**, 59.
41. B. H. Ng and J. W. Anderson, *Phytochemistry*, 1978, **17**, 2069.
42. J. N. Burnell and A. Shrift, *Plant Physiol.*, 1979, **63**, 1095.
43. J. C. Dawson and J. W. Anderrson, *Phytochemistry*, 1988, **28**, 51.
44. T. C. Stadtman, J. N. Davis, E. Zehelin, and A. Böck, *BioFactors*, 1988, **2**, 35.
45. J. L. Hoffman, K. P. McConell, and D. R. Carpenter, *Biochim. Biophys. Acta*, 1970, **199**, 531.
46. P. A. Young and I. I. Kaiser, *Arch. Biochem. Biophys.*, 1975, **171**, 483.
47. S. Muller, H. Senn, B. Gsell, W. Vetter, C. Baron, and A. Böck, *Biochemistry*, 1994, **33**, 3404.
48. I. M. Bell and D. Hilvert, *Biochemistry*, 1993, **32**, 13969.
49. A. R. Bernard, T. N. C. Wells, A. Cleasby, F. Borlat, M. A. Payton, and A. E. I. Proudfoot, *Eur. J. Biochem.*, 1995, **230**, 111.
50. W. A. Hendrickson, J. R. Horton, and D. M. LeMaster, *EMBO J.*, 1990, **9**, 1665.
51. B. Neuhierl and A. Böck, *Eur. J. Biochem.*, 1996, **239**, 235.
52. J. N. Burnell, *Plant Physiol.*, 1981, **67**, 316.
53. E. A. Berger and L. A. Heppel, *J. Biol. Chem.*, 1972, **247**, 7684.
54. K. Soda, *Methods Enzymol.*, 1987, **143**, 453.
55. N. Esaki and K. Soda, *Methods Enzymol.*, 1987, **143**, 493.
56. D. C. Turner and T. C. Stadtman, *Arch. Biochem. Biophys.*, 1973, **154**, 366.
57. L. Flohé, W. A. Günzler, and H. H. Schock, *FEBS Lett.*, 1973, **32**, 132.
58. J. T. Rotruck, A. L. Pope, H. E. Ganther, A. B. Swanson, D. G. Hafeman, and W. G. Hoekstra, *Science*, 1973, **179**, 588.
59. F. Zinoni, A. Birkman, T. C. Stadtman, and A. Böck, *Proc. Natl. Acad. Sci. USA*, 1986, **83**, 4650.
60. I. Chambers, J. Frampton, P. Goldfarb, N. Affara, W. McBain, and P. R. Harrison, *EMBO J.*, 1986, **5**, 1221.
61. Y. Sukenaga, K. Ishida, T. Takeda, and K. Takagi, *Nucleic Acids Res.*, 1987, **15**, 7178.
62. G. T. Mullenbach, A. Tabrizi, B. D. Irvine, G. I. Bell, and R. A. Hallewell, *Prot. Engineer*, 1988, **2**, 239.
63. W. A. Günzler, G. J. Steffens, A. Grossmann, S.-M. A. Kim, F. Otting, A. Wendel, and L. Flohé, *Hoppe-Seyler's Z. Physiol. Chem.*, 1984, **365**, 195.

64. T. C. Stadtman, J. N. Davis, W.-M. Ching, F. Zinoni, and A. Böck, *BioFactors*, 1991, **3**, 21.
65. T. H. Jukes, *Experientia*, 1990, **46**, 1149.
66. S. Osawa, T. H. Jukes, K. Watanabe, and A. Muto, *Microbiol. Rev.*, 1992, **56**, 229.
67. M. Nirenberg, T. Caskey, R. Marshall, R. Brimacombe, D. Kellog, B. Doctor, D. Hatfield, J. Levin, F. Rottman, S. Pestka, M. Wilcox, and F. Anderson, *Cold Spring Harbor Symp. Quant. Biol.*, 1966, **31**, 11.
68. G. H. Khorana, H. Büchi, H. Ghosh, N. Gupta, T. M. Jacob, H. Kössel, R. Morgan, S. A. Narang, E. Ohtuska, and R. D. Wells, *Cold Spring Harbor Symp. Quant. Biol.*, 1966, **31**, 39.
69. R. E. Marshall, C. T. Caskey, and M. Nirenberg, *Science*, 1967, **155**, 820.
70. F. Meyer, H. J. Schmidt, E. Plumper, A. Hasilik, G. Mersmann, H. E. Meyer, A. Engstrom, and K. Heckmann, *Proc. Natl. Acad. Sci. USA*, 1991, **88**, 3758.
71. P. S. Lovett, *J. Bacteriol.*, 1991, **173**, 1810.
72. A. M. Weiner and K. Weber, *J. Mol. Biol.*, 1973, **80**, 837.
73. R. F. Gesteland, R. B. Weiss, and J. F. Atkins, *Science*, 1992, **257**, 1640.
74. J. Heider, W. Leinfelder, and A. Böck, *Nucleic Acids Res.*, 1989, **17**, 2529.
75. B. J. Lee, M. Rajagopalan, Y. S. Kim, K.-H. You, K. B. Jacobson, and D. L. Hatfield, *Mol. Cell Biol.*, 1990, **10**, 1940.
76. D. L. Hatfield, B. J. Lee, N. M. Price, and T. C. Stadtman, *Mol. Microbiol.*, 1991, **5**, 183.
77. D. L. Hatfield, I. S. Choi, S. Mischke, and L. D. Owens, *Biochem. Biophys. Res. Commun.*, 1992, **184**, 254.
78. C. J. Bult, O. White, G. L. Olsen, L. Zhou, R. D. Fleischmann, G. G. Sutton, J. A. Blake, L. M. FitzGerald, R. A. Clayton, and J. D. Gocayne *et al.*, *Science*, 1996, **273**, 1058.
79. L. Margulis and K. V. Schwartz, "Five Kingdoms, an Illustrated Guide to the Phyla of Life on Earth," 2nd edn., W. H. Freeman, San Francisco, CA, 1988, p. 376.
80. A. Goffeau, B. G. Barrell, H. Bussey, R. W. Davis, B. Dujon, H. Feldman, F. Galibert, J. D. Hoheisel, C. Jacq, and M. Johnston, *Science*, 1996, **274**, 546.
81. S. Osawa and T. H. Jukes, *J. Mol. Evol.*, 1995, **41**, 247.
82. R. Rossman, G. Sawers, and A. Böck, *Mol. Microbiol.*, 1991, **5**, 2807.
83. M. R. Bösl, K. Takaku, M. Oshima, S. Nishimura, and M. M. Taketo, *Proc. Natl. Acad. Sci. USA*, 1997, **94**, 5531.
84. D. L. Hatfield and F. H. Portugal, *Proc. Natl. Acad. Sci. USA*, 1970, **67**, 1200.
85. D. Hatfield, *Proc. Natl. Acad. Sci. USA*, 1972, **69**, 3014.
86. D. Hatfield, A. M. Diamond, and B. Dudock, *Proc. Natl. Acad. Sci. USA*, 1982, **79**, 6215.
87. T. Mizutani and A. Hashimoto, *FEBS Lett.*, 1984, **169**, 319.
88. A. M. Diamond, I. S. Choi, P. F. Crain, T. Hashizume, S. C. Pomerantz, R. Cruz, C. J. Steer, K. E. Hill, R. F. Burk, J. A. McCloskey, and D. L. Hatfield, *J. Biol. Chem.*, 1993, **268**, 14 215.
89. A. M. Diamond, B. Dudock, and D. Hatfield, *Cell*, 1981, **25**, 497.
90. R. Amberg, C. Urban, B. Reuner, P. Scharff, S. C. Pomerantz, J. A. McCloskey, and H. J. Gross, *Nucleic Acids Res.*, 1993, **21**, 5583.
91. N. Kato, H. Hoshino, and F. Harada, *Biochem. Int.*, 1983, **7**, 635.
92. A. Schon, A. Böck, G. Ott, M. Sprinzl, and D. Söll, *Nucleic Acids Res.*, 1989, **17**, 7159.
93. C. Sturchler, E. Westhof, P. Carbon, and A. Krol, *Nucleic Acids Res.*, 1993, **21**, 1073.
94. J. Gabryszuk, A. Przykorska, M. Monko, E. Kuligowska, C. Sturchler, A. Krol, G. Kirheimer, J. W. Szarkowski, and G. Keith, *Gene*, 1995, **161**, 259.
95. B. Kuntzel, J. Weissenbach, R. E. Wolff, T. D. Tumatitis-Kennedy, B. G. Lane, and G. Dirheimer, *Biochimie*, 1975, **57**, 61.
96. V. A. O'Neill, F. C. Eden, K. Pratt, and D. L. Hatfield, *J. Biol. Chem.*, 1985, **260**, 2501.
97. K. Pratt, F. C. Eden, K. H. You, V. A. O'Neill, and D. Hatfield, *Nucleic Acids Res.*, 1985, **13**, 4765.
98. D. L. Hatfield, B. S. Dudock, and F. C. Eden, *Proc. Natl. Acad. Sci. USA*, 1983, **80**, 4940.
99. A. M. Diamond, Y. Montero-Puerner, B. J. Lee, and D. Hatfield, *Nucleic Acids Res.*, 1990, **18**, 6727.
100. J. D. Kolker, J. Sharma, R. Cruz, and A. M. Diamond, *Gene*, 1995, **164**, 375.
101. T. Ohama, I. S. Choi, D. L. Hatfield, and K. R. Johnson, *Genomics*, 1994, **19**, 595.
102. M. R. Bösl, M. F. Seldin, S. Nishimura, and M. Taketo, *Mol. Gen. Genet.*, 1995, **248**, 247.
103. P. Carbon and A. Krol, *EMBO J.*, 1991, **10**, 599.
104. O. W. McBride, M. Rajagopalan, and D. Hatfield, *J. Biol. Chem.*, 1987, **262**, 11 163.
105. A. Mitchell, A. E. Bale, B. J. Lee, D. Hatfield, H. Harley, S. Rudle, Y. S. Fan, Y. Fukushima, T. B. Shows, and O. W. McBride, *Cytogenet. Cell Genet.*, 1992, **61**, 117.
106. T. Ohama, J.-E. Jung, S. I. Park, K. A. Clouse, B. J. Lee, and D. Hatfield, *Biochem. Mol. Biol. Int.*, 1995, **36**, 421.
107. B. Rauthor, *Curr. Opin. Genet. Dev.*, 1996, **6**, 221.
108. D. Hatfield, B. J. Lee, L. Hampton, and A. M. Diamond, *Nucleic Acids Res.*, 1991, **19**, 939.
109. H. S. Chittum, K. E. Hill, B. A. Carlson, B. J. Lee, R. F. Burk, and D. L. Hatfield, *Biochim. Biophys. Acta*, 1997, **1359**, 25.
110. I. S. Choi, A. M. Diamond, P. F. Crain, J. D. Kolker, J. A. McCloskey, and D. L. Hatfield, *Biochemistry*, 1994, **33**, 601.
111. C. Sturchler, A. Lescure, G. Keith, P. Carbon, and A. Krol, *Nucleic Acids Res.*, 1994, **22**, 1354.
112. E. S. Yang, D. L. Hatfield, and B. J. Lee, unpublished results.
113. H. S. Chittum, H. J. Baek, A. M. Diamond, P. Fernandez-Salguero, F. Gonzalez, T. Ohama, D. Hatfield, M. Kuehn, and B. J. Lee, *Biochemistry*, 1997, **36**, 8634.
114. E. P. Geiduschek and G. A. Kassavetis, in "Transcriptional Regulation," eds. S. L. McKnight and K. R. Yamamoto, Cold Spring Harbor Laboratory Press, New York, 1992, p. 247.
115. B. J. Lee, P. de la Pena, J. A. Tobian, M. Zasloff, and D. Hatfield, *Proc. Natl. Acad. Sci. USA*, 1987, **84**, 6384.
116. B. J. Lee, S. G. Kang, and D. Hatfield, *J. Biol. Chem.*, 1989, **264**, 9696.
117. S. M. Lobo, J. Lister, M. L. Sullivan, and N. Hernandez, *Genes Dev.*, 1991, **5**, 1477.
118. F. Margottin, G. Dujardin, M. Gerard, J.-M. Egly, J. Huet, and A. Sentenac, *Nature*, 1991, **251**, 424.
119. K. A. Simmen, J. Bernues, H. D. Parry, H. G. Stunnenberg, A. Berkenstam, B. Cavallini, J.-M. Egly, and I. W. Mattaj, *EMBO J.*, 1991, **10**, 1853.
120. J. M. Skuzeski, E. Lund, J. T. Murphy, R. R. Steinberg, R. R. Burgess, and J. E. Dahlberg, *J. Biol. Chem.*, 1984, **259**, 8345.

121. G. R. Kunkel, R. L. Maser, J. P. Calvet, and T. Pederson, *Proc. Natl. Acad. Sci. USA*, 1986, **83**, 8575.

122. S. Murphy, M. Tripodi, and M. Melli, *Nucleic Acids Res.*, 1986, **14**, 9243.

123. N. Hernandez, in "Transcriptional Regulation," eds. S. L. McKnight and K. R. Yamamoto, Cold Spring Harbor Laboratory Press, New York, 1992, p. 281.

124. J. M. Park, I. S. Choi, S. G. Kang, J. Y. Lee, D. L. Hatfield, and B. J. Lee, *Gene*, 1995, **162**, 13.

125. E. Myslinski, C. Schuster, J. Huet, A. Sentenac, A. Krol, and P. Carbon, *Nucleic Acids Res.*, 1993, **21**, 5852.

126. J. D. Parvin, R. J. McCormick, P. A. Sharp, and D. E. Fisher, *Nature*, 1995, **373**, 724.

127. A. TenHarmsel and M. D. Biggin, *Mol. Cell. Biol.*, 1995, **15**, 5492.

128. A. F. Burnol, F. Margottin, P. Schultz, M. C. Marsolier, P. Oudet, and A. Sentenac, *Nature*, 1993, **362**, 475.

129. M.-C. Marsolier, S. Tanaka, M. Livingstone-Zatachej, M. Grunstein, F. Thoma, and A. Sentenac, *Genes Dev.*, 1995, **9**, 410.

130. W. Stunkel, I. Kober, M. Kauer, G. Taimor, and K. H. Seifart, *Nucleic Acids Res.*, 1995, **23**, 109.

131. J. M. Park, E. S. Yang, D. L. Hatfield, and B. J. Lee, *Biochem. Biophys. Res. Commun.*, 1996, **226**, 231.

132. G. J. Hannon, A. Chubb, P. A. Maroney, G. Hannon, S. Altman, and T. W. Nilsen, *J. Biol. Chem.*, 1991, **266**, 22 796.

133. Y. Yuan and R. Reddy, *Biochim. Biophys. Acta*, 1991, **1089**, 33.

134. A. Vilalta, V. A. Kickhoefer, L. H. Rome, and D. L. Johnson, *J. Biol. Chem.*, 1994, **269**, 29 752.

135. J. A. Martignetti and J. Brosius, *Mol. Cell. Biol.*, 1995, **15**, 1642.

136. R. G. Roeder and W. J. Tutter, *Nature*, 1969, **224**, 234.

137. Z. Wang and R. G. Roeder, *Mol. Cell. Biol.*, 1996, **16**, 6841.

138. A. Sentenac, M. Riva, P. Thuriaux, J.-M. Buhler, I. Treich, C. Carles, M. Werner, A. Ruet, J. Huet, C. Mann, N. Chiannilkulchai, S. Stettler, and S. Mariotte, in "Transcriptional Regulation," eds. S. L. McKnight and K. R. Yamamoto, Cold Spring Harbor Laboratory Press, New York, 1992, p. 27.

139. M. Werner, N. Chaussivert, I. M. Willis, and A. Sentenac, *J. Biol. Chem.*, 1993, **268**, 20 721.

140. B. Khoo, B. Brophy, and S. P. Jackson, *Genes Dev.*, 1994, **8**, 2879.

141. O. S. Gabrielsen, N. Marzouki, A. Ruet, A. Sentenac, and P. Fromageot, *J. Biol. Chem.*, 1989, **264**, 7505.

142. B. Bartholomew, G. A. Kassavetis, B. R. Braun, and E. P. Geiduschek, *EMBO J.*, 1990, **9**, 2197.

143. M. C. Parsons and P. A. Weil, *J. Biol. Chem.*, 1990, **265**, 5095.

144. B. Bartholomew, G. A. Kassavetis, and E. P. Geiduschek, *Mol. Cell. Biol.*, 1991, **11**, 5181.

145. N. Chaussivert, C. Conesa, S. Shaaban, and A. Sentenac, *J. Biol. Chem.*, 1995, **270**, 15 353.

146. S. K. Yoshinaga, P. A. Boulanger, and A. J. Berk, *Proc. Natl. Acad. Sci. USA*, 1987, **84**, 3585.

147. S. K. Yoshinaga, N. D. L'Etoile, and A. J. Berk, *J. Biol. Chem.*, 1989, **264**, 10 726.

148. R. Kovelman and R. G. Roeder, *J. Biol. Chem.*, 1992, **267**, 24 446.

149. N. D. L'Etoile, M. L. Fahnestock, Y. Shen, R. Aebersold, and A. J. Berk, *Proc. Natl. Acad. Sci. USA*, 1994, **91**, 1652.

150. G. Lagna, R. Kovelman, J. Sukegawa, and R. G. Roeder, *Mol. Cell. Biol.*, 1994, **14**, 3053.

151. E. Sinn, Z. Wang, R. Kovelman, and R. G. Roeder, *Genes Dev.*, 1995, **9**, 675.

152. G. A. Kassavetis, B. R. Braun, L. H. Nguyen, and E. P. Gaiduschek, *Cell*, 1990, **60**, 235.

153. B. P. Cormack and K. Struhl, *Cell*, 1992, **69**, 685.

154. M. C. Schultz, R. H. Reeder, and S. Hahn, *Cell*, 1992, **69**, 697.

155. R. J. White, S. P. Jackson, and P. W. J. Rigby, *Proc. Natl. Acad. Sci. USA*, 1992, **89**, 1949.

156. S. Buratowski and H. Zhou, *Cell*, 1992, **71**, 221.

157. T. Colbert and S. Hahn, *Genes Dev.*, 1992, **6**, 1940.

158. A. Lopez-De-Leon, M. Librizzi, K. Puglia, and I. M. Willis, *Cell*, 1992, **71**, 211.

159. G. A. Kassavetis, C. A. P. Joazeiro, M. Pisano, E. P. Geiduschek, T. Colbert, S. Hahn, and J. A. Blanko, *Cell*, 1992, **71**, 1055.

160. S. M. Lobo, M. Tanaka, M. L. Sullivan, and N. Hernandez, *Cell*, 1992, **71**, 1029.

161. A. K. P. Taggart, T. S. Fisher, and B. F. Pugh, *Cell*, 1992, **71**, 1015.

162. R. J. White and S. P. Jackson, *Cell*, 1992, **71**, 1041.

163. G. A. Kassavetis, S. Nguyen, R. Kobayashi, A. Kumar, E. P. Geiduschek, and M. Pisano, *Proc. Natl. Acad. Sci. USA*, 1995, **92**, 9786.

164. J. Ruth, C. Conesa, G. Dieci, O. Lefebvre, A. Dusterhoft, S. Ottonello, and A. Sentenac, *EMBO J.*, 1996, **15**, 1941.

165. S. Robert, S. I. Miller, W. S. Lane, S. Lee, and S. Hahn, *J. Biol. Chem.*, 1996, **271**, 14 903.

166. C. A. P. Joazeiro, G. A. Kassavetis, and E. P. Geiduschek, *Mol. Cell. Biol.*, 1994, **14**, 2798.

167. V. L. Gerlach, S. K. Whitehall, E. P. Geiduschek, and D. A. Brow, *Mol. Cell. Biol.*, 1995, **15**, 1455.

168. S. K. Whitehall, G. A. Kassavetis, and E. P. Geiduschek, *Genes Dev.*, 1995, **9**, 2974.

169. S. Robert, T. Colbert, and S. Hahn, *Genes Dev.*, 1995, **9**, 832.

170. R. Mital, R. Kobayashi, and N. Hernandez, *Mol. Cell. Biol.*, 1996, **16**, 7031.

171. M. Teichmann and K. H. Seifart, *EMBO J.*, 1995, **14**, 5974.

172. J. M. Park, J. Y. Lee, D. L. Hatfield, and B. J. Lee, *Gene*, 1997, **196**, 99.

173. J.-B. Yoon, S. Murphy, L. Bai, Z. Wang, and R. G. Roeder, *Mol. Cell. Biol.*, 1995, **15**, 2019.

174. R. W. Henry, C. L. Sadowski, R. Kobayashi, and N. Hernandez, *Nature*, 1995, **374**, 653.

175. J.-B. Yoon and R. G. Roeder, *Mol. Cell. Biol.*, 1996, **16**, 1.

176. C. L. Sadowski, R. W. Henry, R. Kobayashi, and N. Hernandez, *Proc. Natl. Acad. Sci. USA*, 1996, **93**, 4289.

177. L. Bai, Z. Wang, J.-B. Yoon, and R. G. Roeder, *Mol. Cell. Biol.*, 1996, **16**, 5419.

178. P. Carbon, S. Murgo, J. P. Ebel, A. Krol, G. Tebb, and I. W. Mattaj, *Cell*, 1987, **51**, 71.

179. S. Murphy, A. Pierani, C. Scheidereit, M. Melli, and R. G. Roeder, *Cell*, 1989, **59**, 1071.

180. S. Murphy, J.-B. Yoon, T. Gerster, and R. G. Roeder, *Mol. Cell. Biol.*, 1992, **12**, 3247.

181. R. Mital, M. A. Cleary, W. Herr, and N. Hernandez, *Mol. Cell. Biol.*, 1996, **16**, 1955.

182. E. Myslinski, A. Krol, and P. Carbon, *Nucleic Acids Res.*, 1992, **20**, 203.

183. C. Schuster, E. Myslinski, A. Krol, and P. Carbon, *EMBO J.*, 1995, **14**, 3777.

184. M. J. Berry, J. W. Harney, T. Ohama, and D. L. Hatfield, *Nucleic Acids Res.*, 1994, **22**, 3753.

185. J.-E. Jung, V. Karoor, M. G. Sandbacken, B. J. Lee, T. Ohama, R. F. Gesteland, J. F. Atkins, G. T. Mullenbach, K. E. Hill, A. J. Wahba, and D. L. Hatfield, *J. Biol. Chem.*, 1994, **269**, 29 739.

186. K. E. Hill and R. F. Burk, in "Selenium in Biology and Human Health," ed. R. F. Burk, Springer-Verlag, New York, 1994, p. 118.

187. S. Himeno, H. S. Chittum, and R. F. Burk, *J. Biol. Chem.*, 1996, **271**, 15 769.
188. C. Forster, G. Ott, K. Forchhammer, and M. Sprinzil, *Nucleic Acids Res.*, 1990, **18**, 487.
189. C. Gelpi, E. J. Sontheimer, and J. L. Rodruigez-Sanchez, *Proc. Natl. Acad. Sci. USA*, 1992, **89**, 9739.
190. N. Hubert, R. Walczak, P. Carbon, and A. Krol, *Nucleic Acids Res.*, 1996, **24**, 464.
191. Q. Shen, J. L. Leonard, and P. E. Newburger, *RNA*, 1995, **1**, 519.
192. K. Yamada, *FEBS Lett.*, 1995, **377**, 313.
193. M. J. Berry, L. Banu, J. W. Harney, and P. R. Larsen, *EMBO J.*, 1993, **12**, 3315.
194. G. W. Martin, J. W. Harney, and M. J. Berry, *RNA*, 1996, **2**, 171.
195. S. C. Vendeland, M. A. Beilstein, J.-Y. Yeh, W. Ream, and P. D. Whanger, *Proc. Natl. Acad. Sci. USA*, 1995, **92**, 8749.
196. D. L. Williams, R. Pierce, E. Cookson, and A. Capron, *Mol. Biochem. Parasitol.*, 1992, **52**, 127.
197. H. Kollmus, L. Flohé, and J. E. McCarthy, *Nucleic Acids Res.*, 1996, **24**, 1195.
198. R. Walczak, E. Westhof, P. Carbon, and A. Krol, *RNA*, 1996, **2**, 367.
199. T. Tamura and T. C. Stadtman, *Proc. Natl. Acad. Sci. USA*, 1996, **93**, 1006.
200. G. N. Gladyshev, K.-T. Jeang, and T. C. Stadtman, *Proc. Natl. Acad. Sci. USA*, 1996, **93**, 6146.
201. V. Y. Kim, M. J. Guimaraes, A. Zlotnik, J. F. Bazan, and T. C. Stadtman, *Proc. Natl. Acad. Sci. USA*, 1997, **94**, 418.
202. V. N. Gladyshev, K.-T. Jeang, J. C. Woolton, and D. L. Hatfield, *J. Biol. Chem.*, 1998, **273**, 8910.
203. R. A. Sunde, in "Selenium in Biology and Human Health," ed. R. F. Burk, Springer-Verlag, New York, 1994, p. 45.
204. R. A. Sunde, in "Handbook of Nutritionally Essential Minerals," eds. B. L. O'Dell and R. A. Sunde, Marcel Dekker, New York, pp. 493–556.
205. M. S. Saedi, C. G. Smith, J. Frampton, I. Chambers, P. R. Harrison, and R. A. Sunde, *Biochem. Biophys. Res. Commun.*, 1988, **153**, 855.
206. R. A. Sunde, J. A. Dyer, T. V. Moran, J. K. Everson, and M. Sugimoto, *Biochem. Biophys. Res. Commun.*, 1993, **193**, 905.
207. S. L. Weiss, J. K. Evenson, K. M. Thompson, and R. A. Sunde, *J. Nutr.*, 1996, **126**, 2260.
208. X. G. Lei, J. K. Everson, K. M. Thompson, and R. A. Sunde, *J. Nutr.*, 1995, **125**, 1438.
209. H. Toyoda, S. Himeno, and N. Imura, *Biochim. Biophys. Acta*, 1990, **1049**, 213.
210. M. J. Christensen and K. W. Burgener, *J. Nutr.*, 1992, **122**, 1620.
211. S. L. Weiss and R. A. Sunde, *FASEB J.*, 1996, **10**, A557.
212. S. L. Weiss and R. A. Sunde, in "Trace Element Metabolism in Man and Animals," eds. P. W. F. Fischer, M. R. L'Abbe, K. A. Cockell, and R. S. Gibson, NRC Research Press, Ottawa, Canada, pp. 57–58.
213. S. L. Weiss and R. A. Sunde, *J. Nutr.*, 1997, **127**, 1304.
214. A. Spector, Y. Yang, Y. S. Ho, J. L. Magnenat, R. R. Wang, W. Ma, and W. C. Li, *Exp. Eye Res.*, 1996, **62**, 521.
215. E. Martinez-Galisteo, C. A. Padilla, C. Garcia-Alfonso, J. Lopez-Barea, and J. A. Barcena, *Biochimie*, 1993, **75**, 803.
216. M. Luthman and A. Holmgren, *Biochemistry*, 1982, **21**, 6628.
217. J. E. Oblong, P. Y. Gasdaska, K. Sherrill, and G. Powis, *Biochemistry*, 1993, **32**, 1006.
218. C.-C. Chen, B. L. B. McCall, and E. C. More, *Prep. Biochem.*, 1977, **7**, 165.
219. M. L.-S. Tsang and J. A. Weatherbee, *Proc. Natl. Acad. Sci. USA*, 1981, **78**, 7478.
220. P. Y. Gasdaska, J. R. Gasdaska, S. Cochran, and G. Powis, *FEBS Lett.*, 1995, **373**, 5.
221. V. N. Gladyshev and D. L. Hatfield, unpublished results.
222. K. E. Hill, G. W. McCollum, M. E. Boeglin, and R. F. Burk, *Biochim. Biophys. Acta*, 1997, **234**, 293.
223. A. Holmgren and M. Bjornstedt, *Methods Enzymol.*, 1995, **252B**, 199.
224. T. Tamura, V. N. Gladyshev, S.-Y. Liu, and T. C. Stadtman, *BioFactors*, 1996, **5**, 99.
225. I. Y. Kim, Z. Veres, and T. C. Stadtman, *J. Biol. Chem.*, 1982, **267**, 19 650.
226. M. J. Axley, A. Böck, and T. C. Stadtman, *Proc. Natl. Acad. Sci. USA*, 1991, **88**, 8450.
227. F. Zinoni, A. Birkman, W. Leinfelder, and A. Böck, *Proc. Natl. Acad. Sci. USA*, 1987, **84**, 3156.
228. M. J. Berry, L. Banu, and P. R. Larsen, *Nature*, 1991, **349**, 438.
229. D. L. St. Germain, R. A. Schwartzman, W. Croteau, A. Kanamori, Z. Wang, D. D. Brown, and V. A. Galton, *Proc. Natl. Acad. Sci. USA*, 1994, **91**, 7767.
230. G. E. Garcia and T. C. Stadtman, *J. Bacteriol.*, 1992, **174**, 7080.
231. R. A. Arkowitz and R. H. Abeles, *J. Am. Chem. Soc.*, 1990, **112**, 870.
232. R. A. Arkowitz and R. H. Abeles, *Biochemistry*, 1991, **30**, 4090.
233. T. C. Stadtman and J. N. Davis, *J. Biol. Chem.*, 1991, **266**, 22 147.
234. F. Ursini, M. Maiorino, R. Brigelius-Flohe, K. D. Aumann, A. Roveri, D. Schomburg, and L. Flohé, *Methods Enzymol.*, 1995, **252B**, 38.
235. J. C. Boyington, V. N. Gladyshev, S. V. Khangulov, T. C. Stadtman, and P. D. Sun, *Science*, 1997, **275**, 1305.
236. V. N. Gladyshev, J. C. Boyington, S. V. Khangulov, D. A. Grahame, T. C. Stadtman, and P. D. Sun, *J. Biol. Chem.*, 1996, **271**, 8095.
237. V. N. Gladyshev, S. V. Khangulov, M. J. Axley, and T. C. Stadtman, *Proc. Natl. Acad. Sci. USA*, 1994, **91**, 7708.
238. S. V. Khangulov, V. N. Gladyshev, G. C. Dismukes, and T. C. Stadtman, *Biochemistry*, 1998, **37**, 3518.
239. M. J. Axley, D. A. Grahame, and T. C. Stadtman, *J. Biol. Chem.*, 1990, **265**, 18 213.
240. M. J. Axley and D. A. Grahame, *J. Biol. Chem.*, 1991, **266**, 13 731.
241. J. S. Holcenberg and E. R. Stadtman, *J. Biol. Chem.*, 1969, **244**, 1194.
242. G. L. Dilworth, *Arch. Biochem. Biophys.*, 1982, **219**, 30.
243. G. L. Dilworth, *Arch. Biochem. Biophys.*, 1983, **221**, 565.
244. V. N. Gladyshev, S. V. Khangulov, and T. C. Stadtman, *Biochemistry*, 1996, **35**, 212.
245. V. N. Gladyshev and P. Lecchi, *BioFactors*, 1996, **5**, 93.
246. V. N. Gladyshev, S. V. Khangulov, and T. C. Stadtman, *Proc. Natl. Acad. Sci. USA*, 1994, **91**, 232.
247. M. K. Eidsness, R. A. Scott, B. C. Prickril, D. V. DerVartanian, J. LeGall, I. Moura, J. J. G. Moura, and H. D. Peck, *Proc. Natl. Acad. Sci. USA*, 1989, **86**, 147.
248. S. H. He, M. Teixeira, J. LeGall, D. S. Patil, I. Moura, J. J. G. Moura, D. V. DerVartanian, B. H. Huynh, and H. D. Peck, *J. Biol. Chem.*, 1989, **264**, 2678.
249. M. Maiorino, C. Roche, M. Keiss, K. Koenig, D. Gawlik, M. Matthes, E. Naldini, R. Pierce, and L. Flohé, *Eur. J. Biochem.*, 1996, **238**, 838.

250. H. W. Croteau, J. C. Davey, V. A. Galton, and D. L. St. Germain, *J. Clin. Invest.*, 1996, **98**, 405.
251. S. Kreimer and J. R. Andreesen, *Eur. J. Biochem.*, 1995, **234**, 192.
252. T. G. Senkevich, J. J. Bugert, J. R. Sisler, E. V. Koonin, G. Darai, and B. Moss, *Science*, 1996, **273**, 813.
253. S. F. Altschul, W. Gish, W. Miller, E. W. Myers, and D. J. Lipman, *J. Mol. Biol.*, 1990, **215**, 403.
254. J. L. Shisler, T. G. Senkevich, M. J. Berry, and B. Moss, *Science*, 1998, **279**, 102.

Author Index

This Author Index comprises an alphabetical listing of the names of the authors cited in the text and the references listed at the end of each chapter in this volume.

Each entry consists of the author's name, followed by a list of numbers, for example

Templeton, J. L., 366, 385^{233} (350, 366), 387^{370} (363)

For each name, the page numbers for the citation in the reference list are given, followed by the reference number in superscript and the page number(s) in parentheses of where that reference is cited in the text. Where a name is referred to in text only, the page number of the citation appears with no superscript number. References cited in both the text and in the tables are included.

Although much effort has gone into eliminating inaccuracies resulting from the use of different combinations of initials by the same author, the use by some journals of only one initial, and different spellings of the same name as a result of the transliteration processes, the accuracy of some entries may have been affected by these factors.

Subject Index

PHILIP AND LESLEY ASLETT
Marlborough, Wiltshire, UK

Every effort has been made to index as comprehensively as possible, and to standardize the terms used in the index in line with the IUPAC Recommendations. In view of the diverse nature of the terminology employed by the different authors, the reader is advised to search for related entries under the appropriate headings.

The index entries are presented in letter-by-letter alphabetical sequence. Compounds are normally indexed under the parent compound name, with the substituent component separated by a comma of inversion. An entry with a prefix/locant is filed after the same entry without any attachments, and in alphanumerical sequence. For example, 'diazepines', '1,4-diazepines', and '2,3-dihydro-1,4-diazepines' will be filed as:-

    diazepines
    1,4-diazepines
    1,4-diazepines, 2,3-dihydro-

The Index is arranged in set-out style, with a maximum of three levels of heading. Location references refer to volume number (in bold) and page number (separated by a comma); major coverage of a subject is indicated by bold, elided page numbers; for example;

    triterpene cyclases, **299–320**
      amino acids, 315

See cross-references direct the user to the preferred term; for example,

    olefins *see* alkenes

See also cross-references provide the user with guideposts to terms of related interest, from the broader term to the narrower term, and appear at the end of the main heading to which they refer, for example,

    thiones
      see also thioketones

WITHDRAWAL